CENOZOIC FORAMINIFERA AND CALCAREOUS NANNOFOSSIL BIOSTRATIGRAPHY OF THE NIGER DELTA

CENOZOIC FORAMINIFERA AND CALCAREOUS NANNOFOSSIL BIOSTRATIGRAPHY OF THE NIGER DELTA

Edited by

O.S. Adegoke

A.S. Oyebamiji

J.J. Edet

P.L. Osterloff

O.K. Ulu

ELSEVIER

AMSTERDAM • BOSTON • HEIDELBERG • LONDON • NEW YORK • OXFORD • PARIS
SAN DIEGO • SAN FRANCISCO • SINGAPORE • SYDNEY • TOKYO

Elsevier
Radarweg 29, PO Box 211, 1000 AE Amsterdam, Netherlands
The Boulevard, Langford Lane, Kidlington, Oxford OX5 1GB, United Kingdom
50 Hampshire Street, 5th Floor, Cambridge, MA 02139, United States

Library of Congress Cataloging-in-Publication Data
A catalog record for this book is available from the Library of Congress

British Library Cataloguing-in-Publication Data
A catalogue record for this book is available from the British Library

ISBN: 978-0-12-812161-0

Publisher: Cathleen Sether
Acquisition Editor: Amy Shapiro
Editorial Project Manager: Tasha Frank
Production Project Manager: Anitha Sivaraj
Designer: Greg Harris

Typeset by Thomson Digital

Contents

Enclosures will be available in their entirety at http://booksite.elsevier.com/9780128121610

List of Figures

PART I

PART II

List of Tables

PART I

PART II

Enclosures

PART I

PART II

List of Contributors

Part 1: Cenozoic Foraminifera Biostratigraphy of the Niger Delta

O.S. Adegoke

S.W. Petters

E.A. Fayose

A.S. Oyebamiji

I.K. Ajisafe

A.I. Tiamiyu

O.T. Fajemila

K.M. Lawal

A.C. Oyede (Late)

J.J. Edet

Marie Stella Sonuga

O.K. Ulu

O.A. Ekun

P.L. Osterloff

A.A. Adesida

C.S. Nwajide

Wale Yussuph

Rachel O. Adebiyi

C.O. Okpo

Part 2: Cenozoic Calcareous Nannofossil Biostratigraphy of the Niger Delta

A.S. Oyebamiji

O. Ogunjobi

A. Bakare

A.E. Adebiyi

L.S. Fadiya

J.J. Edet

O.S. Adegoke

P.L. Osterloff

O.K. Ulu

M.S. Sonuga

A.C. Oyede (Late)

O. Ekun

R.O. Adebiyi

C.O. Okpo

Preface

The Niger Delta region is attractive to petroleum exploration companies primarily because of its prolific hydrocarbon wealth. Since the first commercial discovery of hydrocarbon from the Oloibiri Field in 1965 by Shell–BP Petroleum Development Company, the Niger Delta region has been an area of intensive oil and gas exploration activity. The vast amount of subsurface material and data accumulated over the past six decades, have hitherto been proprietary to the different operating companies. This book *Cenozoic Foraminifera and Calcareous Nannofossil Biostratigraphy of the Niger Delta* is the first attempt to harmonize, in a single volume, the biostratigraphic data of the individual companies that operate in the region.

The technical and research information documented in this book will appeal to both local and global professional and academic audiences, not only involved in petroleum exploration (biostratigraphers, exploration geologists, and exploration geophysicists) in the oil and gas industry, but also to other professional geoscience consultants and practitioners, reservoir geoscientists, as well as geoscientists in the academia (students, researchers, and consultants). This book, thus, serves as a practical guide to understand the rudiments of the Niger Delta geology, biostratigraphy, and sequence stratigraphy.

The book's technical content and organization present the means of tackling one of exploration's biggest challenges, which is that of consistently correlating borehole sections across major structural faults to accurately define prospectivity and plays. Furthermore, the stratigraphic framework established for the Niger Delta in this project allows for major achievements, namely, biostratigraphic age calibrations in millions of years and application of sequence stratigraphic concepts that are linked to the global cycle chart, that is updated up to year 2012, for the delineation of deltaic packages and basin analysis.

It is important to note that much of the sequence stratigraphic and chronostratigraphic framework outlined in this book have been developed primarily from onshore and conventional offshore domain data of the Niger Delta. However, some strategic deep offshore wells, with well-preserved and highly diverse foraminifera and Calcareous nannofossils species, were deliberately integrated to fill in noticeable stratigraphic gaps.

My commendations go to the Joint Venture Companies who are members of the Stratigraphic Committee of the Niger Delta (StratCom), namely, Shell Petroleum Developing Company of Nigeria Limited, Mobil Producing Nigeria Unlimited (now a subsidiary of ExxonMobil Corporation), Chevron Nigeria Limited, Elf Petroleum Nigeria Limited (now Total E&P Nigeria Limited), Nigerian Agip Oil Company Limited, Texaco Overseas Petroleum Company of Nigeria Unlimited (now defunct), Ashland Petroleum Limited (now Addax Petroleum Development Nigeria Limited), and Pan Ocean Oil Corporation (Nigeria) for their cooperation and contribution to the success of this study. Many of these companies made annual budgetary provisions for the works of the Stratigraphic Committee of the Niger Delta. We are much indebted to Dr. T.J.A. Reijers, the foundation Chairman of StratCom who proposed the idea of this study, together with Dr. Adekunle A. Adesida who served as foundation Secretary in 1995. We thank the Directors, past and present, of the Department of Petroleum Resources (DPR), for coordinating the works of StratCom in their capacities as Chairpersons.

Finally, I commend Prof. O.S. Adegoke who voluntarily offered the facilities of his company, Mosunmolu Limited, as the workplace of StratCom over the years.

J.J. Edet, PhD
Total E&P Nigeria Limited,
Lagos, Nigeria

Acknowledgments

PART I

The Editorial Board acknowledges with thanks, the cooperation and contribution of the Joint Venture Companies that were initially invited as members of the Stratigraphic Committee of the Niger Delta (StratCom), namely Shell Petroleum Development Company of Nigeria Ltd., Mobil Producing Nigeria Unlimited (now ExxonMobil), Chevron Nigeria Ltd., Elf (now Total), Nigerian Agip Oil Company Ltd., Texaco (now defunct), Ashland Petroleum Ltd. (now Addax), and Pan Ocean Oil Corporation (Nigeria). Some of these companies provided the ditch cuttings samples and the wireline logs of the wells studied for this book. Many of the companies also made annual budgetary provisions for the works of the Stratigraphic Committee of the Niger Delta. We thank the Director of Petroleum Resources, in his capacity as the Chairman, for coordinating the works of StratCom.

The cooperation of the Center for Electron Microscopy and Microanalysis (CEMMA) of the University of Southern California, Los Angeles, CA is acknowledged. The Cenozoic Foraminifera volume lead author, Prof. O.S. Adegoke was accorded unrestrained access to use the scanning electron microscope and related facilities on same terms as full-time faculty of the University. All the benthonic foraminifera and some of the planktonic foraminifera displayed in Plates 1–35 were prepared by the lead author with the assistance of the staff or the Center. We thank in particular, Alicia Thompson, Paul Webster, John Curulli, Ian McFarlane, and Yuzheng Zhang of CEM. Sopefolu Adegoke deserves special mention and appreciation for her assistance with logistics and organizing the linkages with the staff of the Center on behalf of StratCom.

The editorial team acknowledges the contributions of Shell Petroleum Development Company Limited, Port Harcourt, Nigeria and Dr. Peter Osterloff, Dr. Kelly Maguire, and Mrs. Patricia Ukeri for assisting the project team in generating and producing the Global Cycle Charts shown as Enclosures 1.18 and 1.19.

PART II

The Editorial Board acknowledges with thanks, the cooperation and contributions of the Joint Venture Companies that were initially invited as members of the Stratigraphic Committee of the Niger Delta (StratCom), namely Shell Petroleum Development Company of Nigeria Ltd., Mobil Producing Nigeria Unlimited (now ExxonMobil), Chevron Nigeria Ltd., Elf (now Total), Nigerian Agip Oil Company Ltd., Texaco (now defunct), Ashland Petroleum Ltd. (now Addax), and Pan Ocean Oil Corporation (Nigeria). Some of these companies provided the ditch cuttings samples and the wireline logs of the wells studied for this book. The Companies also made annual budgetary provisions for the works of the Stratigraphic Committee of the Niger Delta. We thank the Director of Petroleum Resources, in his capacity as the Chairman, for coordinating the works of StratCom.

Project Personnel

PART I

Project Consultants—Members of the Foraminifera Project Consortium

1. Mosunmolu Limited

O.S. Adegoke
A.S. Oyebamiji
H.A. Ogunleye
I.K. Ajisafe
A.I. Tiamiyu
O.T. Fajemila
K.M. Lawal
Rachael O. Adebiyi
C.O. Okpo

2. Others

S.W. Petters
E.A. Fayose
Y. Adewale
C.S. Nwajide
A.A. Adesida

Members of the StratCom Biostratigraphic Subcommittee

A.C. Oyede	Shell Petroleum Development Company of Nigeria Limited (SPDC)—Chairman (1997–2002)
O.K. Ulu	Chevron Nigeria Limited (CNL)—Chairman (2002–2013)
M.S. Sonuga	Mobil Producing Nigeria Unlimited (MPN)
C. Iwobi	Nigerian Agip Oil Company Limited (NAOC) (1997–2005)
A.S. Oyebamiji	Nigerian Agip Oil Company Limited (NAOC) (2005 to date)—Chairman (2013 to date)
J.J. Edet	Total E and P
O.A. Ekun	Chevron Nigeria Limited (CNL) (2013 to date)
O.O. Ogunyomi	(Late) Texaco Overseas (Nig.) Petroleum Company Unlimited (TOPCON)
O.S. Adegoke	
E.A. Fayose	
M.A. Sowunmi	
C.O.C. Awgu	
O. Lawal	
O. Ogunjobi	
Bola Akinwande	
P.L. Osterloff	(Representing SPDC 2011–2013)
V.B. Adeniran	

PART II

Project Consultants

1. Mosunmolu Limited

   A.S. Oyebamiji
   O.S. Adegoke
   A. Bakare
   Abidemi E. Adebiyi
   S.L. Fadiya

2. Global Energy Company Limited

   O. Ogunjobi

Members of the StratCom Biostratigraphic Subcommittee

A.C. Oyede	Shell Petroleum Development Company of Nigeria Limited (SPDC)—Chairman (1997–2002)
O.K. Ulu	Chevron Nigeria Limited (CNL)—Chairman (2002 to date)
M.S. Sonuga	Mobil Producing Nigeria Unlimited (MPN)
C. Iwobi	Nigerian Agip Oil Company Limited (NAOC)
A.S. Oyebamiji	Nigerian Agip Oil Company Limited (NAOC)
J.J. Edet	Elf Petroleum Nigeria Limited (EPNL)
O.O. Ogunyomi	Texaco Overseas (Nig.) Petroleum Company Unlimited (TOPCON)
O.S. Adegoke	
E.A. Fayose	
M.A. Sowunmi	
C.O.C. Awgu	
O. Lawal	
O. Ogunjobi	
Bola Akinwande	

Members of the Calcareous Nannofossil Project Consortium

A.S. Oyebamiji
O.S. Adegoke
O. Ogunjobi
S.L. Fadiya
Abidemi E. Adebiyi
A. Bakare
J.J. Edet
P.L. Osterloff

PART I

CENOZOIC FORAMINIFERA BIOSTRATIGRAPHY OF THE NIGER DELTA

O.S. Adegoke, A.S. Oyebamiji, J.J. Edet, P.L. Osterloff & O.K. Ulu

CHAPTER

1

Introduction

OUTLINE

1.1 ORIGIN AND CONCEPT OF THE STRATIGRAPHIC COMMITTEE OF THE NIGER DELTA (STRATCOM)

The need to set up the Stratigraphic Committee (formerly Commission) of the Niger Delta was mooted by The Shell Petroleum Development Company of Nigeria (SPDC) through the auspices of Dr. T.J.A. Reijers (Chairman), with Mr. Adekunle A. Adesida as Secretary in 1995. The proposal was warmly accepted and supported by other operating Joint Venture companies. The idea was then presented to the Department of Petroleum Resources (DPR) on September 5, 1995, the exploration OPTS (October 4, 1995), Nigerian Association of Petroleum Explorationists (NAPE) (17/11/95), management of OPTS (20/03/96), and NAPIMS (22/03/96). All these bodies endorsed and welcomed the concept.

The Committee was formally inaugurated on October 20, 1995. Membership was made of senior representatives of the Department of Petroleum Resources (DPR), National Petroleum Investment Services (NAPIMS), Shell Petroleum Development Company Ltd. (SPDC), Chevron Nigeria Limited, Nigerian Petroleum Development Company (NPDC), ELF (now Total), Nigerian Agip Oil Company Ltd. (NAOC), Mobil Producing Nigeria Unlimited (now ExxonMobil), TEXACO (now defunct), and ASHLAND (now ADDAX) distributed in four technical subcommittees. The DPR is Chairman with NAPIMS and Geological Survey of Nigeria taking Vice Chairmanship positions. The structure includes an Editorial Board

Cenozoic Foraminifera and Calcareous Nannofossil Biostratigraphy of the Niger Delta. http://dx.doi.org/10.1016/B978-0-12-812161-0.00001-6

that is to take care of intermediate and final publications. The Committee is to be aided by the Advisory Board in which representatives of four Nigerian Universities and at least six major service companies are represented. In addition, Nigerian Association of Petroleum Explorationists (NAPE) and Nigerian Mining and Geosciences Society (NMGS) are represented on the Committee.

The concept at inception was to pull together and make full use of the vast, but diverse, array of exploration and production data on the Niger Delta, which, until then, was exclusively in the archives of the various International Operating Companies and the Department of Petroleum Resources. In doing this, the major operating companies hoped to standardize the data and thereby develop a unified geological, structural, and chronostratigraphic scheme that can be applied throughout the Niger Delta.

The objectives of the proposed StratCom project include:

1. Formalization of a hierarchically organized computer-compatible lithostratigraphy;
2. Establishment of chronostratigraphic and sequence stratigraphic standards for the entire Niger Delta;
3. Identification, naming, and harmonization of the nomenclature of stratigraphically relevant main structural elements of the Niger Delta; and
4. Linking genetic sequences through seismo-stratigraphy.

The Scope of Work as originally defined was to:

1. Carry out detailed lithostratigraphic, biostratigraphic, and sequence stratigraphic analysis and reanalysis of about 150 selected wells; and
2. Carry out detailed structural and seismosequence stratigraphic interpretation of at least nine dip lines and four strike lines released for this purpose by the operators.

Though this work scope was modified over time, the requirements initially defined still form the lodestone for the effort of the Stratigraphic Committee of the Niger Delta.

To achieve the laudable objectives set by StratCom, the then six major Joint Ventures operators with the Nigerian National Petroleum Corporation (NNPC) entered into a working agreement on the March 22, 1996. The overall objective was to complete all defined studies and publish the results before the beginning of the new millennium (2000).

1.2 THE ORGANIZATION THAT OVERSEES STRATCOM ACTIVITIES

As set up, the activities of StratCom are to be overseen jointly by the Department of Petroleum Resources (DPR) and the National Petroleum Investments Management Services (NAPIMS) of the Federal Ministry of Petroleum Resources, with the DPR as Chairman. In its position as Chairman of the Committee, the DPR steers and coordinates the activities of StratCom. NAPIMS as Supervisor of the Joint Venture partners that include most of the participating companies, and in its capacity as the 1st Vice Chairman assures that the yearly StratCom budgetary allocations contained in the studies budget of each of the participating companies are justified and accounted for at the end of each year.

1.3 THE COMMITTEES OF STRATCOM

The various technical committees set up by StratCom for the implementation of its assignments include an Advisory Board and four multidisciplinary Committees as follows:

1. The Advisory Board;
2. Lithostratigraphic Committee;
3. Biostratigraphic Committee that includes the Foraminifera, Calcareous Nannofossil and Palynology Consortia;
4. Seismo-Sequence Stratigraphic Committee; and
5. Structural Committee.

1.4 THE FORAMINIFERA CONSORTIUM

By an Agreement dated December 20, 1998, the Stratigraphic Committee of the Niger Delta (StratCom) comprised of the following companies:

- Shell Petroleum Development Company of Nigeria Limited (SPDC);
- Chevron Nigeria Limited (CNL);
- Elf Petroleum Nigeria Limited (EPNL);
- Nigerian Agip Oil Company Limited (NAOC);
- Mobil Producing Nigeria Unlimited (MPN); and
- Texaco Overseas (Nig.) Petroleum Company Unlimited (TOPCON).

commissioned the "Foraminifera Consortium," represented by Mosunmolu Limited to carry out a Foraminifera Taxonomic study of the Cenozoic Niger Delta. The three members of the Consortium are Prof. O.S. Adegoke, Prof. E.A. Fayose, and Prof. S.W. Petters.

The Cenozoic Foraminifera study project as originally defined, was based on a study of 39 wells, selected along 4 regional seismic transects across the Niger Delta depobelts (Fig. 1.1). Five additional wells were studied and included in the final report because of the need to broaden the scope of coverage especially with respect to the planktonic foraminifera (Tables 1.1 and 1.2). The initial version of the report was completed by 2002. An unusually long delay in publication and critical review of the manuscript by the publisher necessitated the intensive revision of the manuscript in 2014 permitting the authors to capture, not only the more recent concepts and developments in global sequence stratigraphy and chronostratigraphy particularly those contained in Gradstein et al. (2012), but also the more recent geological literature on the Niger Delta.

Among the consortium members, Prof. E.A. Fayose was initially responsible for the Paleogene foraminifera zonation and taxonomy; Prof. S.W. Petters was responsible for much of the planktonic foraminifera taxonomy, zonation, and SEM photomicrographs; while Prof. O.S. Adegoke, assisted by the biostratigraphers in his Mosunmolu Ltd. office, was responsible for the Neogene taxonomy, zonation, and the scanning electron photomicrographs of the benthonic foraminifera and some of the planktonics. He eventually took responsibility for completing the unfinished aspects of the works of the other authors who withdrew from

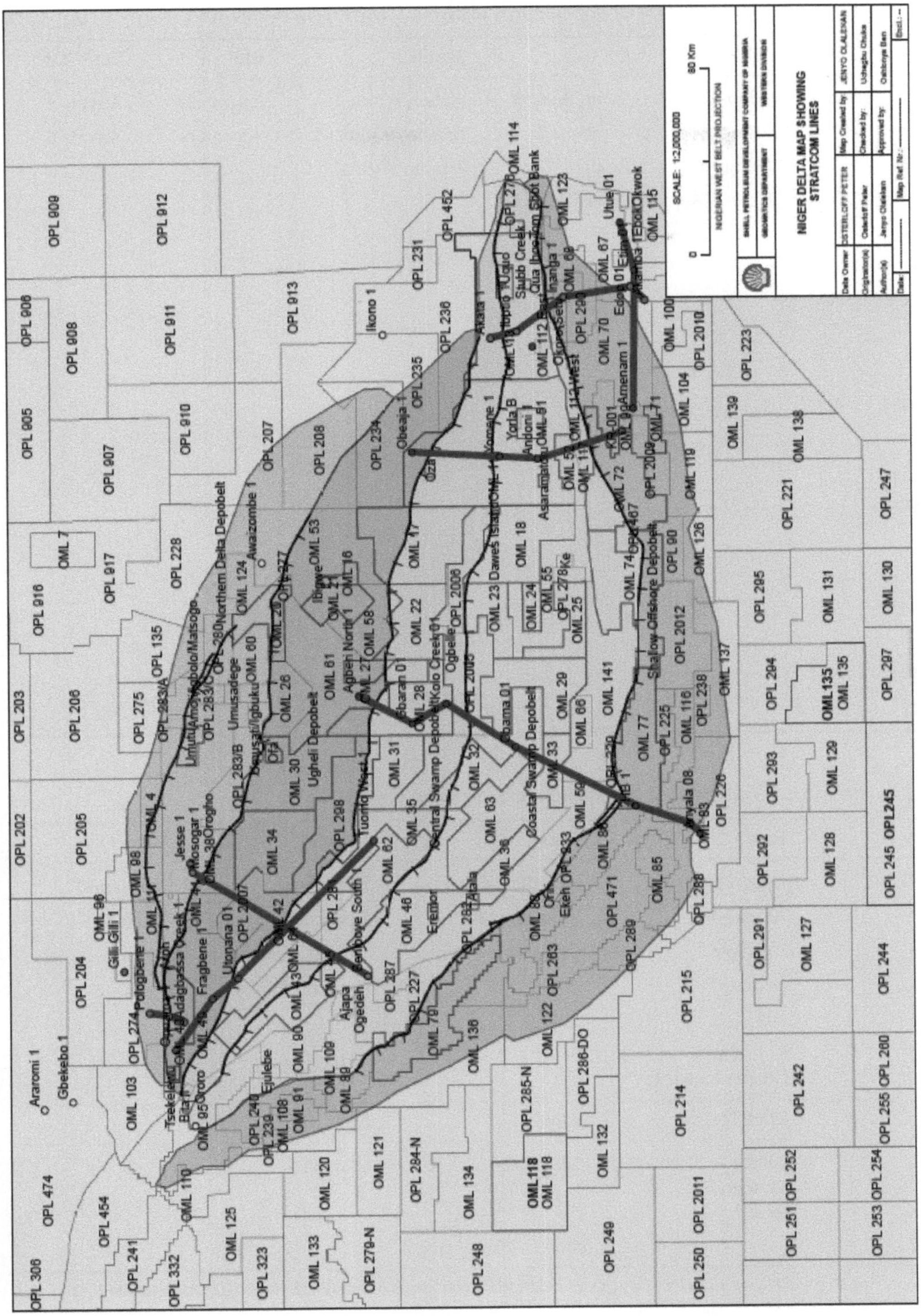

FIGURE 1.1 **Map of Niger Delta showing StratCom lines in the project area.**

TABLE 1.1 StratCom Wells Studied for Foraminifera Listed by Company Contributions

SPDC	MPN	CNL	NAOC	ELF	TOPCON
Adagbassa Creek-1	Edop-33A	Benin River-3	Agberi North-1	Akamba-1	Anyala-4
Akata-1	Etim-1	Fragbene-1	Beniboye North-1	Amenam-1	Anyala-8
Amakubu-1	Inanga-1	Ihuo-1	Tuomo West-1		
Andoni-1	Utue-17A	Utonana-1			
Batan-1					
Gilli Gilli-1					
HB-1					
Ibotio-1					
Jesse-1					
Kolo Creek-1					
KR-1					
Mosogar-1					
Obama North-1					
Obeaja-1					
Opuama-1					
Pologbene-1					
Tonjor-1					
Utapate-1					
Yomene-1					

TABLE 1.2 StratCom Wells Studied for Foraminifera Listed by Traverse Lines

Line B1	Line F	Line I	Line L	Line O
Akamba-1	Agberi North-1	Adagbassa Creek-1	Fragbene-1	Ameena-1
Akata-1	Anyala-4	Amakubu-1	Opuama-1	Etim-1
Edop-1	Anyala-8	Pologbene-1	Tuomo West-1	Utue-1
Ibotio-1	Gbaran-1		Utonana-1	
Inanga-1	HB-1			
	Kolo Creek-1			
	Obama-North 1			
LINE C	***LINE H***	***LINE J***	***LINE M***	
Andoni-1	Beniboye North-1?	Gilli Gilli-1	Utapate-1	
KR-1	Mosogar-1			
Obeaja-1	Jesse-1			
Yomene-1				

Note: The following five wells were added to the wells studied for the planktonic foraminifera content: (1) Awaizombe-1, (2) Araromi-1, (3) Bita-9, (4) Gbekebo-1, and (5) Ikono-1.

StratCom soon after the initial stage. He also was responsible for the overall rewriting and editing of the text of this publication.

1.5 PROJECT OBJECTIVES AND METHODOLOGY

1.5.1 Objectives

The objectives of the project, as defined by the Stratigraphic Committee of the Niger Delta (StratCom) include:

1. Harmonization of the taxonomic nomenclature and zonal schemes of all the StratCom member companies;
2. Identification, where possible, of the Maximum Flooding Surface (MFS) Condensed Sections in the Cenozoic Niger Delta from Paleocene to Pleistocene;
3. Recognition of new MFSs, if any;
4. Characterization of the MFSs identified using chronostratigraphically significant benthonic and planktonic foraminifera species;
5. Preparation of scanning electron photomicrographs of key marker species which are to be presented in PhotoCatalogues;
6. Preparation of a detailed foraminifera biofacies and paleoenvironmental synthesis for the Cenozoic Niger Delta;
7. Preparation of the Final Report containing the new Niger Delta Zonal Scheme and the detailed taxonomic descriptions of the chronostratigraphically and paleoenvironmentally significant species, using scientific nomenclature; and
8. Calibration of the Niger Delta Zonal Scheme, where possible, with the globally accepted time-scales of the International Commission on Stratigraphy (ISC).

To achieve the previously mentioned objectives, the Foraminifera Consortium carried out the following tasks:

- Comprehensive review of the literature and the reports in the archives of the StratCom member companies;
- Collection of ditch cutting samples of the wells selected for the project;
- Compilation of the synonyms of key foraminiferal species and the codes used by different member companies;
- Preparation of detailed checklists of old wells for which complete data was available, highlighting the First Downhole Occurrences of chronostratigraphically important foraminifera species (Enclosures 1.1–1.17);
- Analysis of over 700 selected ditch cutting samples to enhance the definition of MFS Condensed Sections;
- Preparation of plots of species diversity and population abundance for all the wells for which data was available (Figs. 4.7–4.23);
- Identification of candidate MFSs from the biofacies plots and support their candidature through delineation using wireline logs (Table 4.1; Figs. 4.7–4.23);
- Taxonomic description of identified marker species, both from the prepared samples and the company collections;

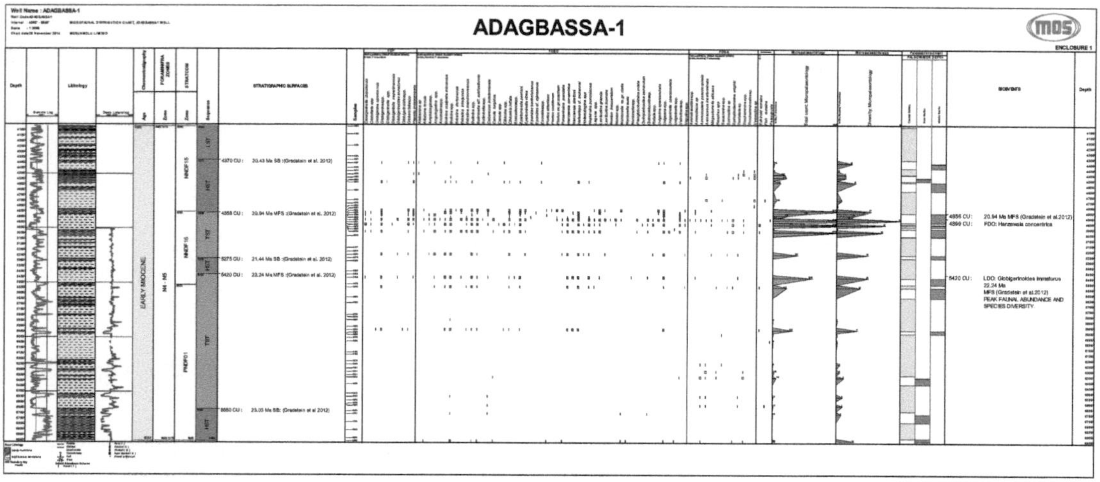

ENCLOSURE 1.1 Microfaunal checklist for the Adagbassa Creek-1 well.

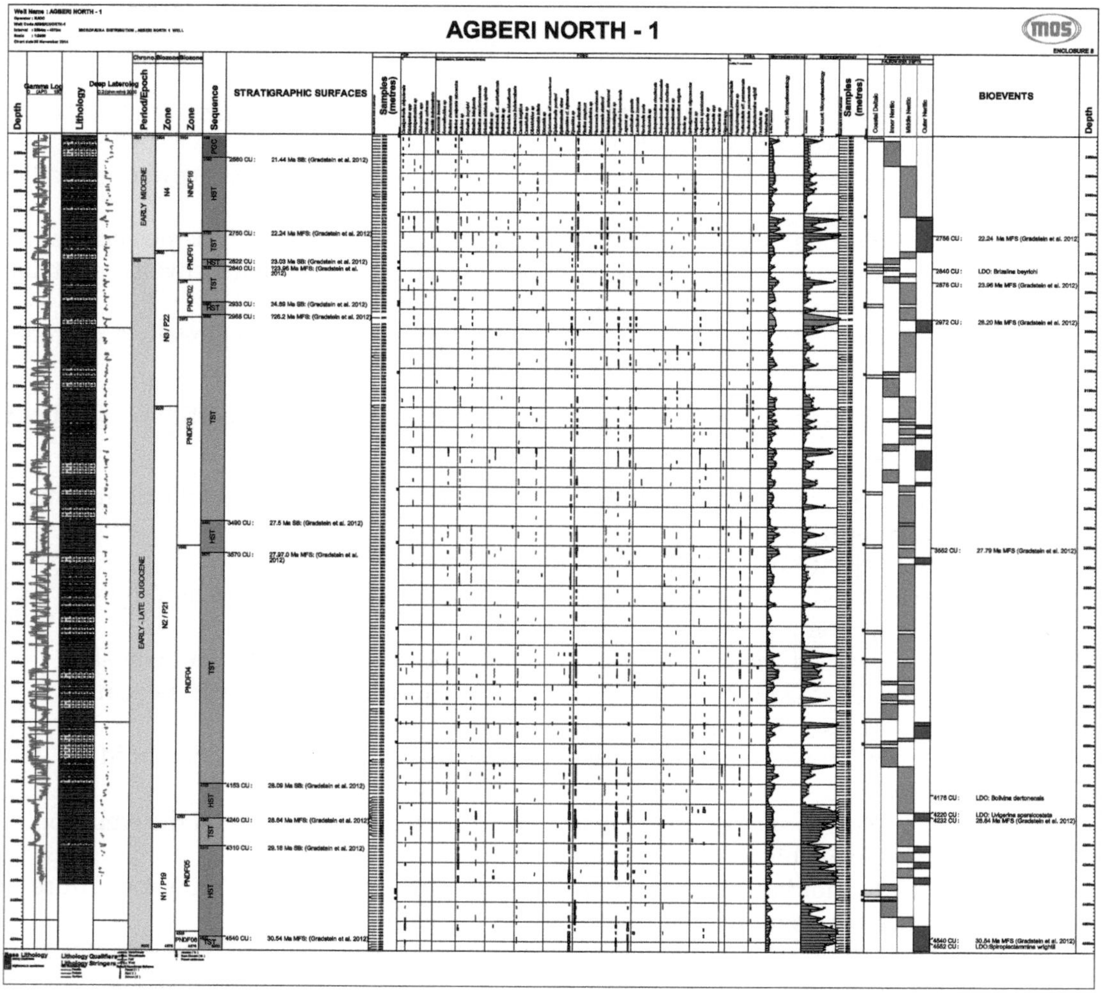

ENCLOSURE 1.2 Microfaunal checklist for the Agberi North-1 well.

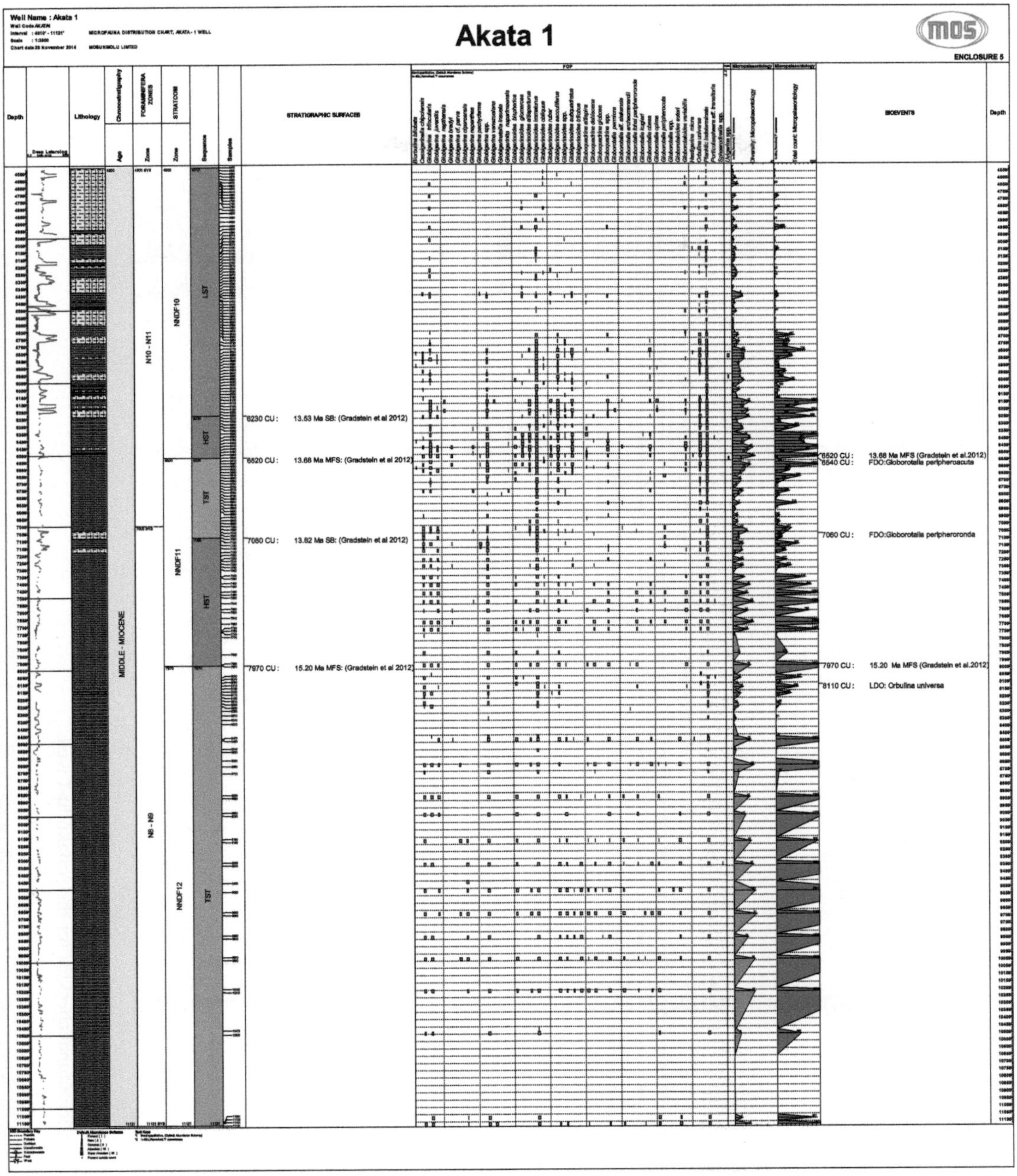

ENCLOSURE 1.3 Microfaunal checklist for the Akata-1 well.

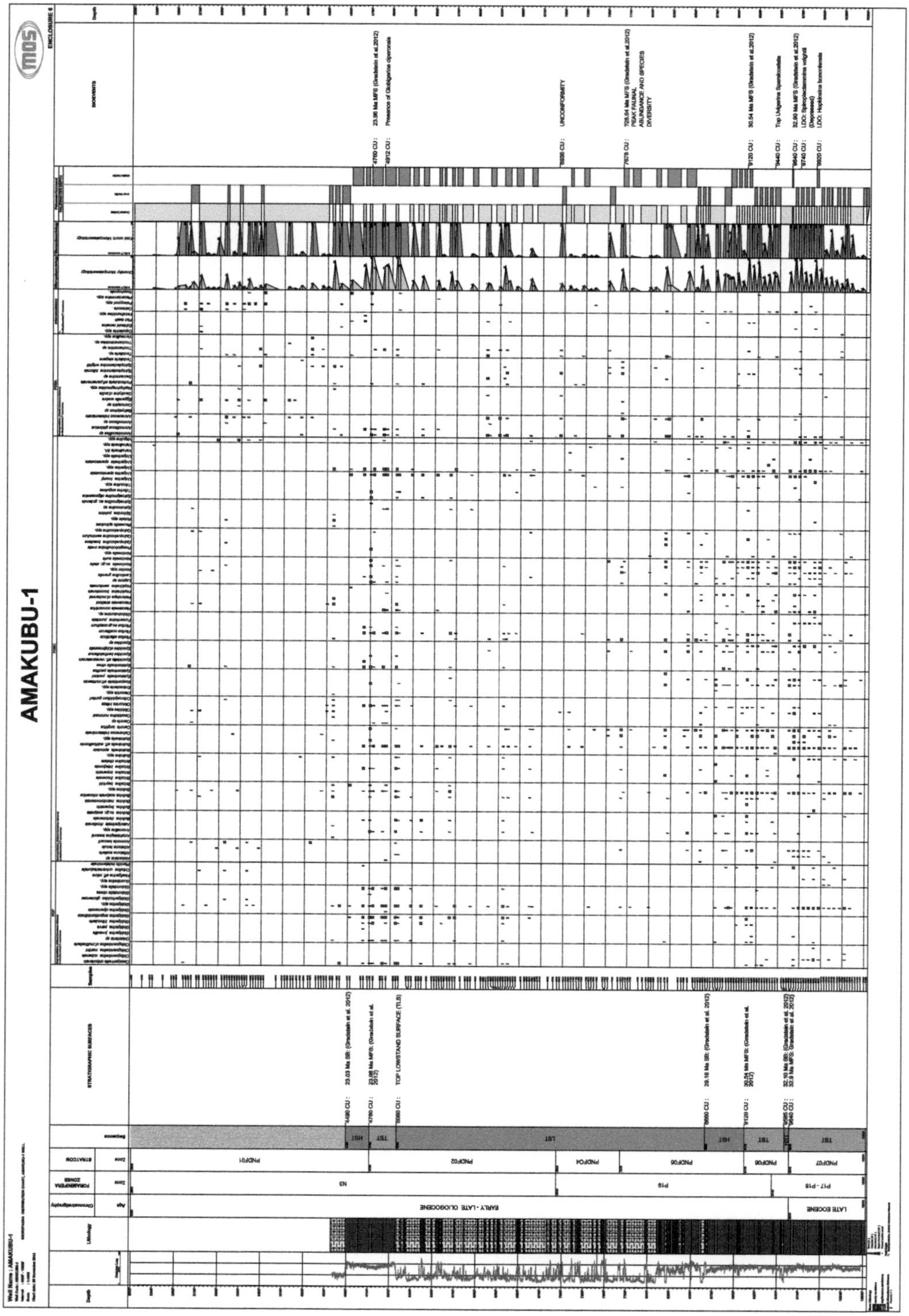

ENCLOSURE 1.4 Microfaunal checklist for the Amakubu-1 well.

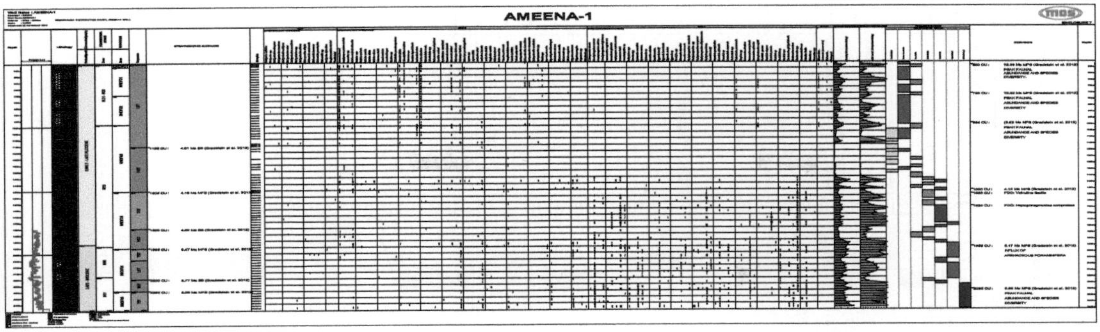

ENCLOSURE 1.5 Microfaunal checklist for the Ameena-1 well.

- Scanning electron photomicrographs of key chronostratigraphic and environmentally significant species/specimens; and
- Preparation of a PhotoCatalogue (Plates 1–35) and a Final Report incorporating the zonal schemes (Enclosures 1.18 and 1.19, available online at http://booksite.elsevier.com/9780128121610), (Enclosures 1.20 and 1.21), taxonomic descriptions, and the paleoenvironmental synthesis (Tables 5.1–5.16; Enclosure 1.22) with emphasis on species of environmental significance.

1.5.2 Methodology

The ditch cutting samples were routinely analyzed micropaleontologically at 10-m or 30-ft. intervals, or at the available sampling distances. In some rare cases sidewall samples were utilized. The processing follows standard micropaleontological preparation techniques. The unwashed ditch cutting samples are initially rinsed in distilled water to remove drilling mud and then dried over a hot plate. A standard weight (20 g) of each dried sample is soaked for 4 hours in kerosene, followed by water soaking overnight. The disaggregated sample is then washed under a shower of water over a 63-μm mesh sieve. The washed residue is then dried over a hot plate and sieved into three fractions (coarse, medium, and fine) prior to picking. During picking very rich samples are usually subjected to splitting with the split factor recorded. All the foraminifera, ostracodes, shell fragments, and other biota seen are picked and counted. The statistical data obtained is computerized using the StrataBugs software (TM from StrataData Ltd.). Barplots of the abundance and species diversity are made from which candidate MFSs are selected. Their positions are later supported by interpretation of the wireline logs. The complete micropaleontological data is plotted in color using StrataBugs. Faunal associations including benthonics, planktonics, benthonic/planktonic ratios (normalized) agglutinated/calcareous foraminiferal ratios, etc. are plotted.

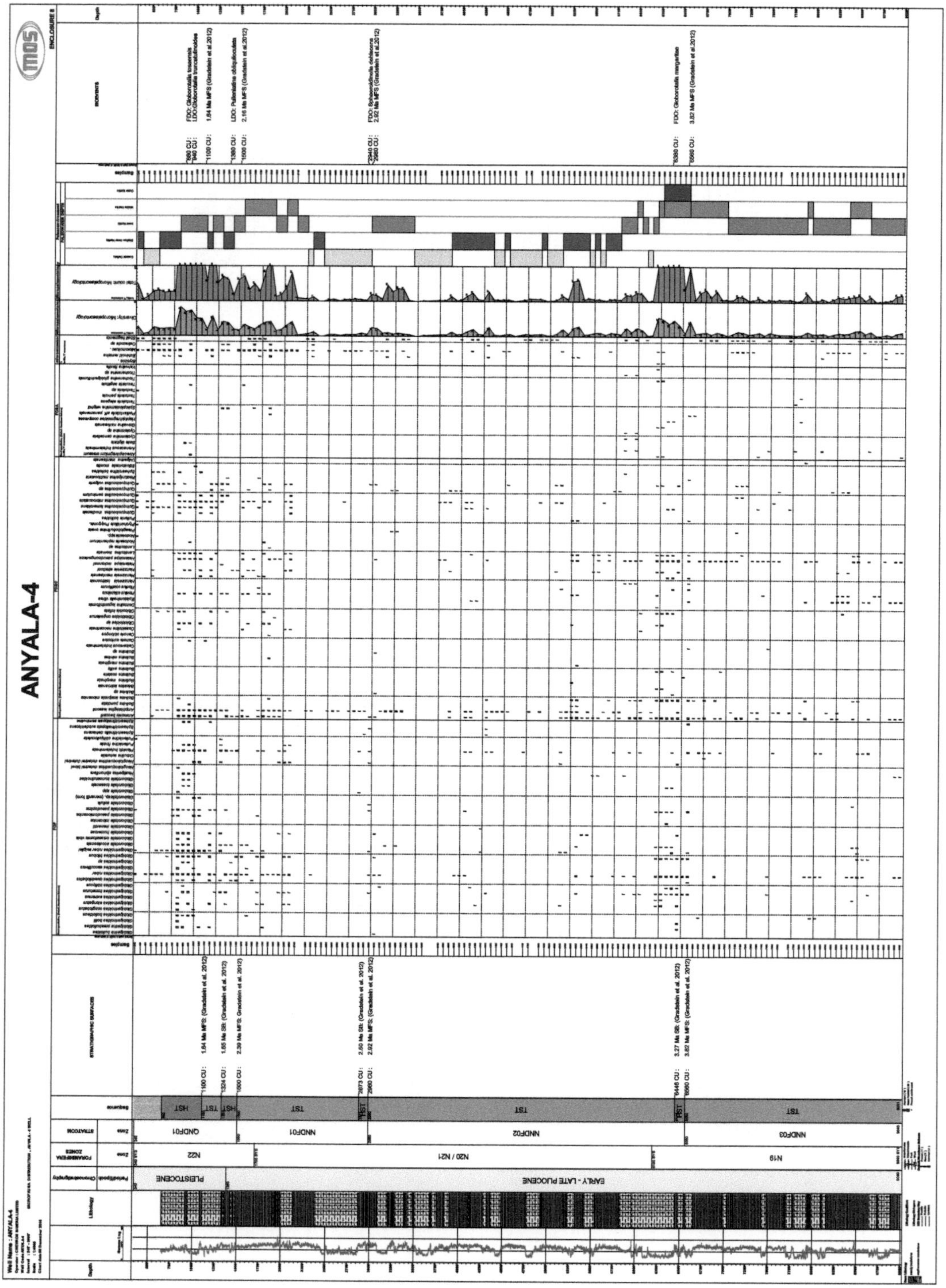

ENCLOSURE 1.6 Microfaunal checklist for the Anyala-4 well.

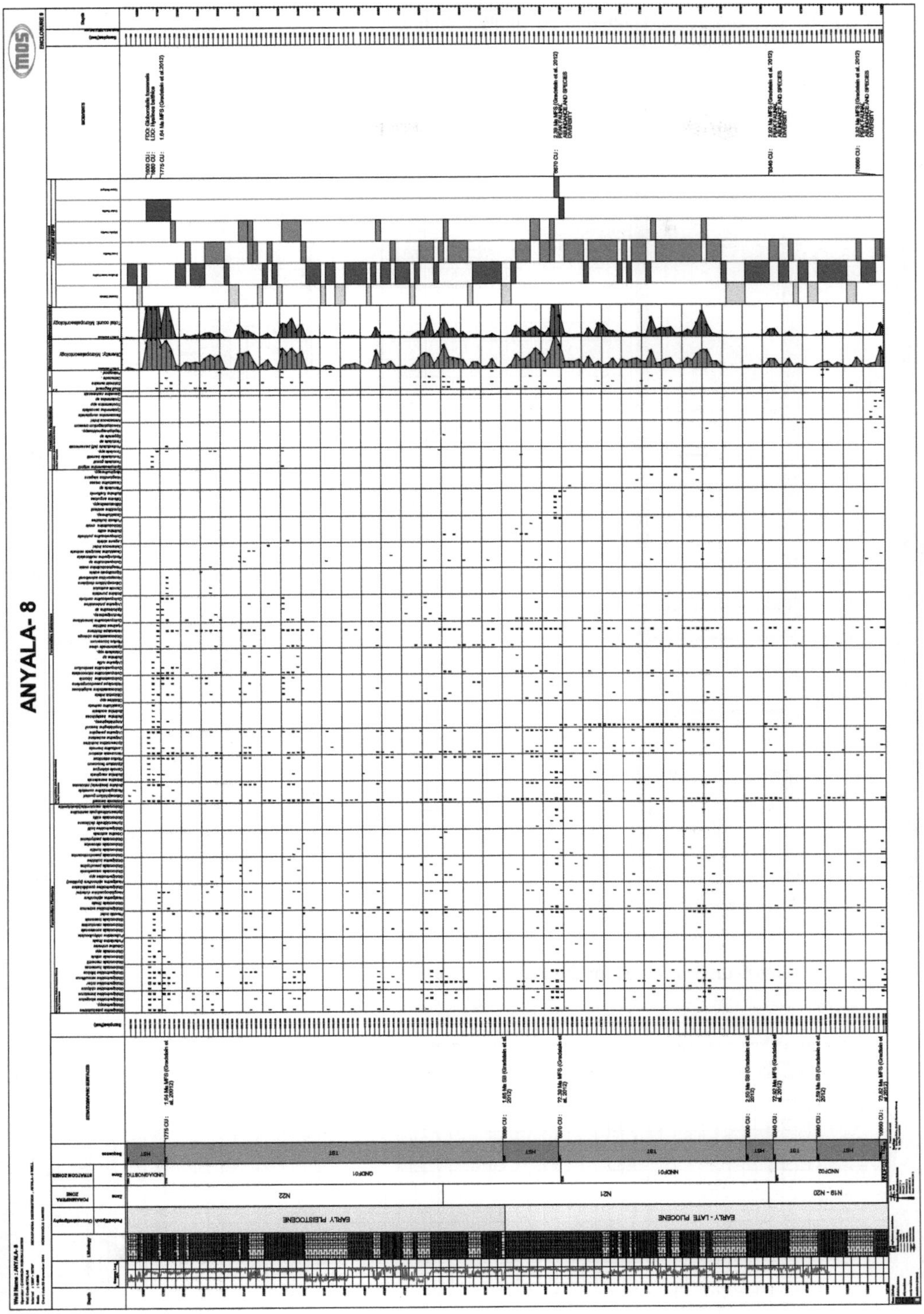

ENCLOSURE 1.7 Microfaunal checklist for the Anyala-8 well.

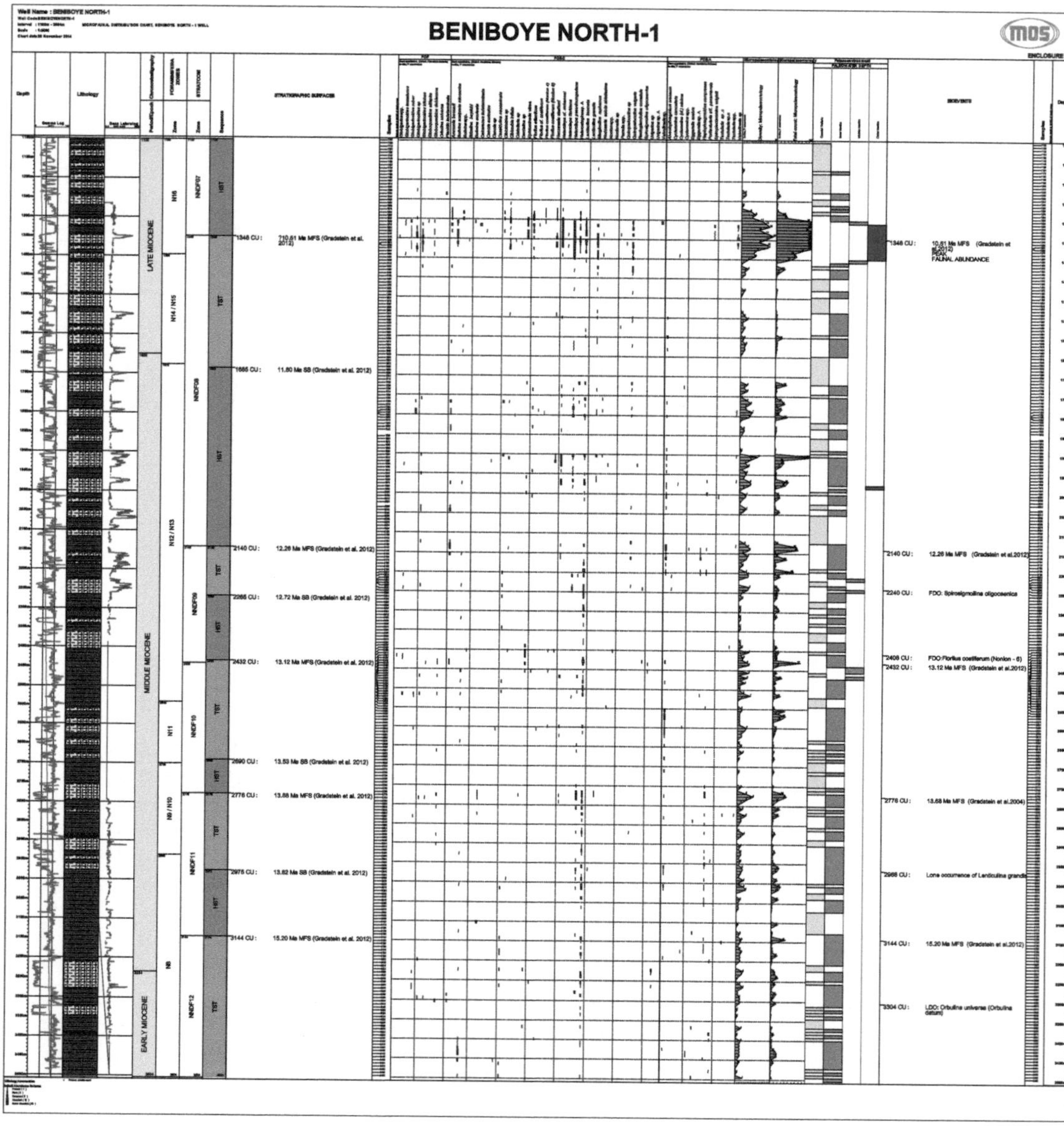

ENCLOSURE 1.8 Microfaunal checklist for the Beniboye North-1 well.

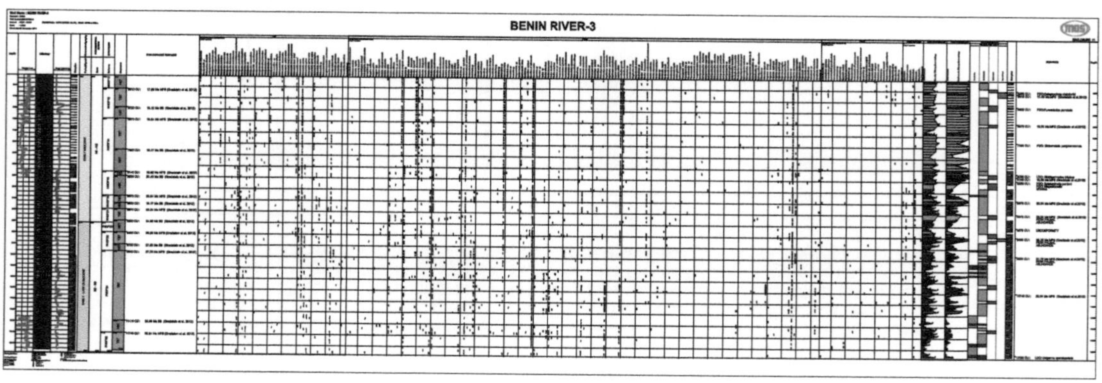

ENCLOSURE 1.9 Microfaunal checklist for the Benin River-3 well.

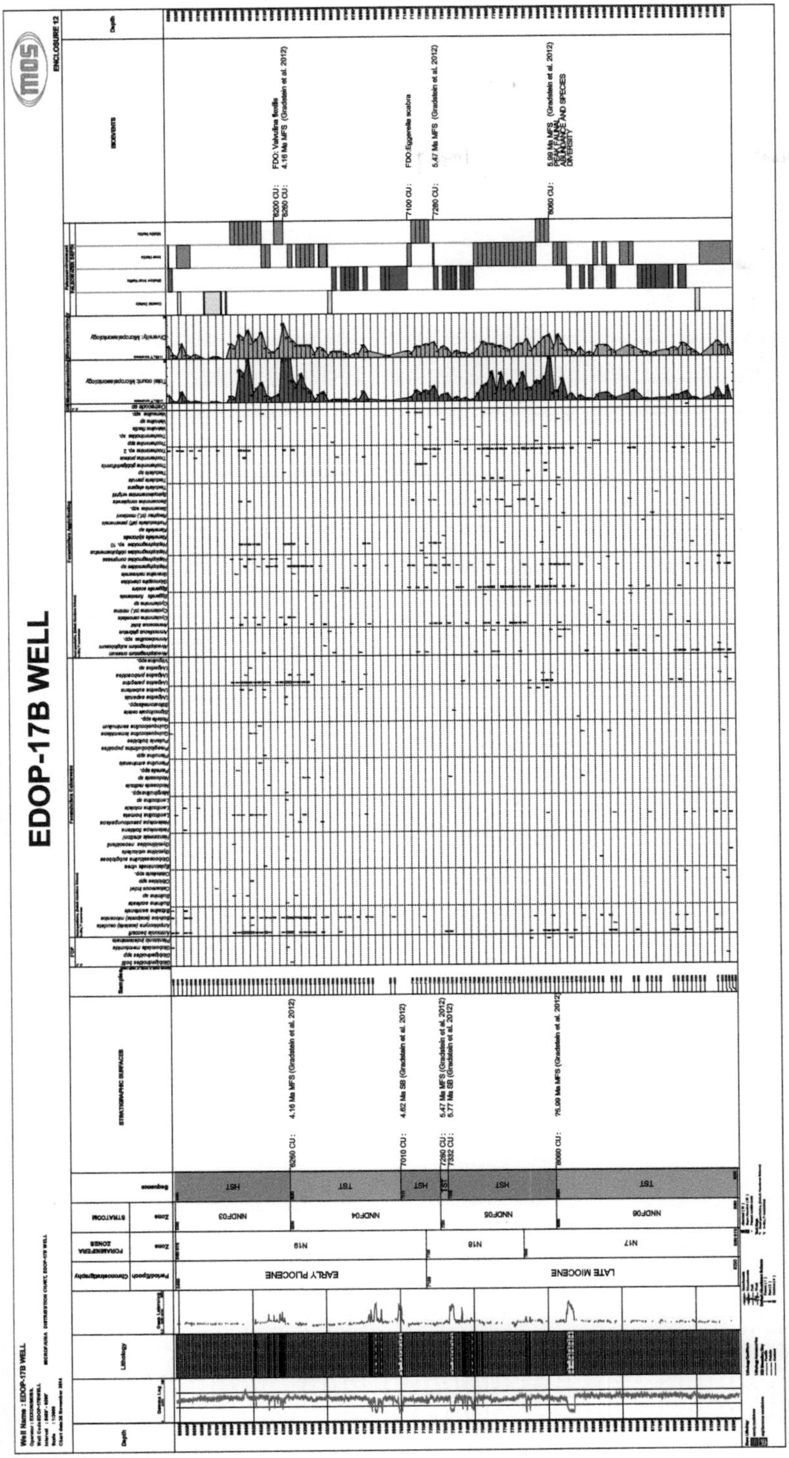

ENCLOSURE 1.10 Microfaunal checklist for the Edop-17B well.

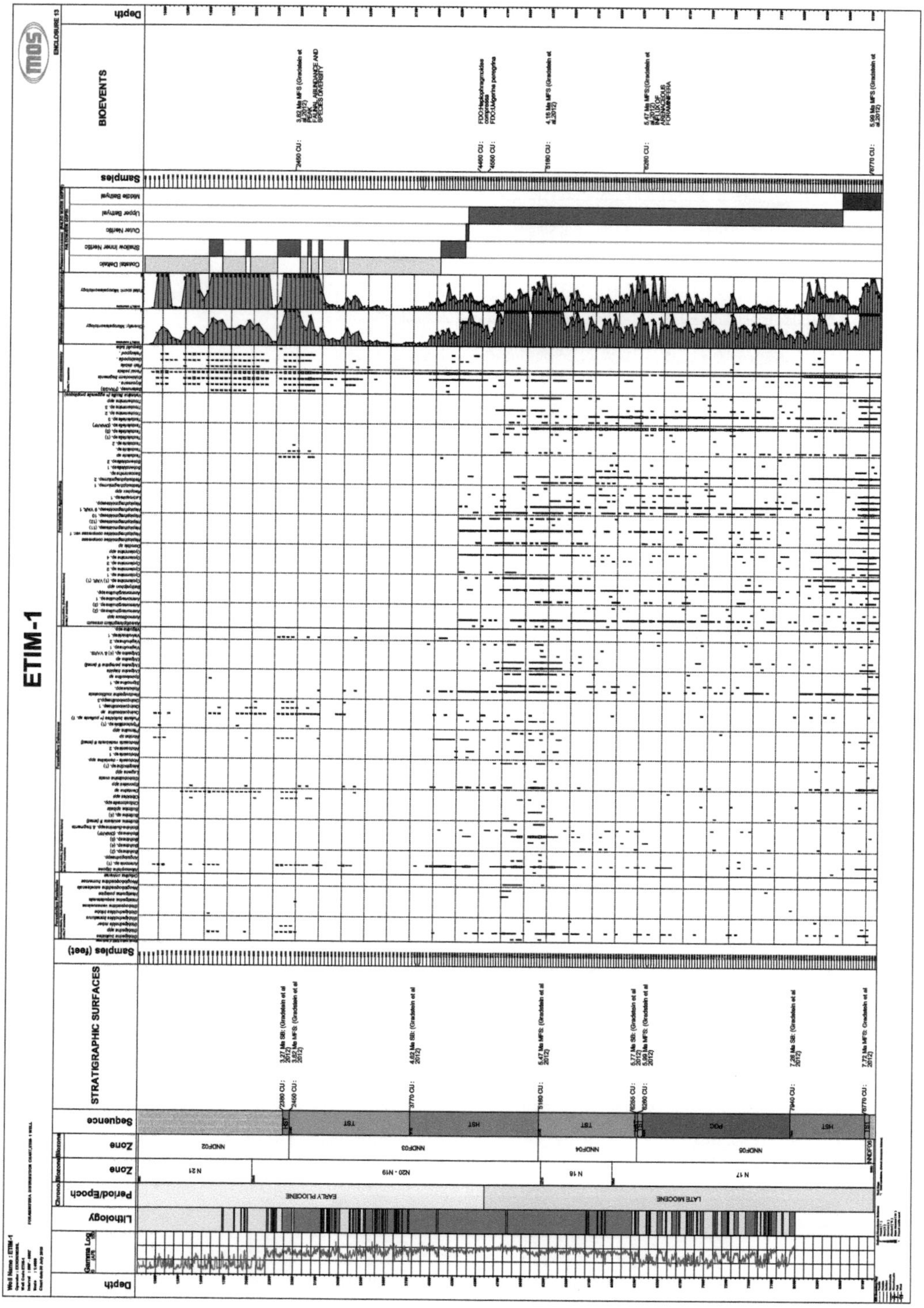

ENCLOSURE 1.11 Microfaunal checklist for the Etim-1 well.

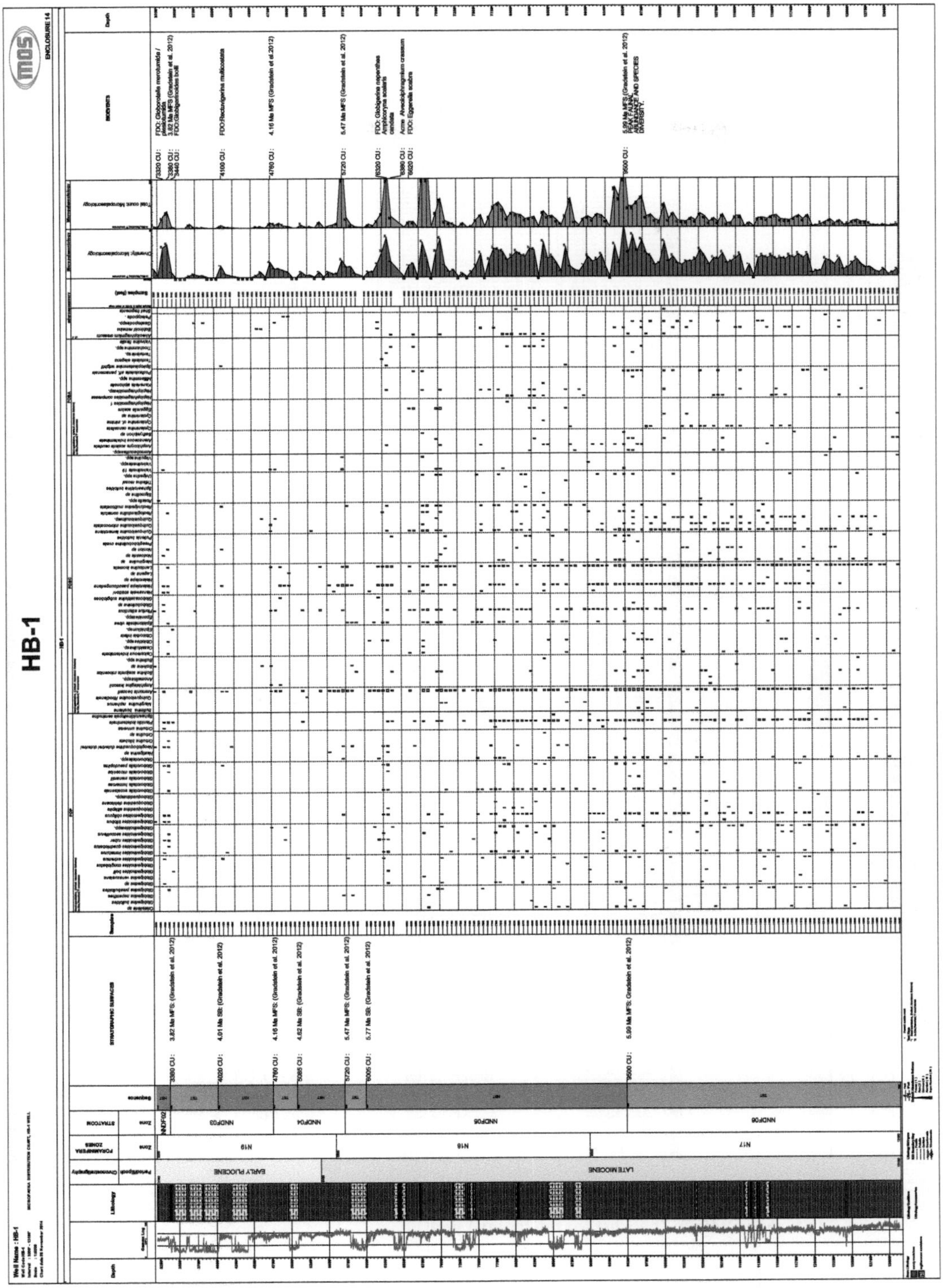

ENCLOSURE 1.12 Microfaunal checklist for the HB-1 well.

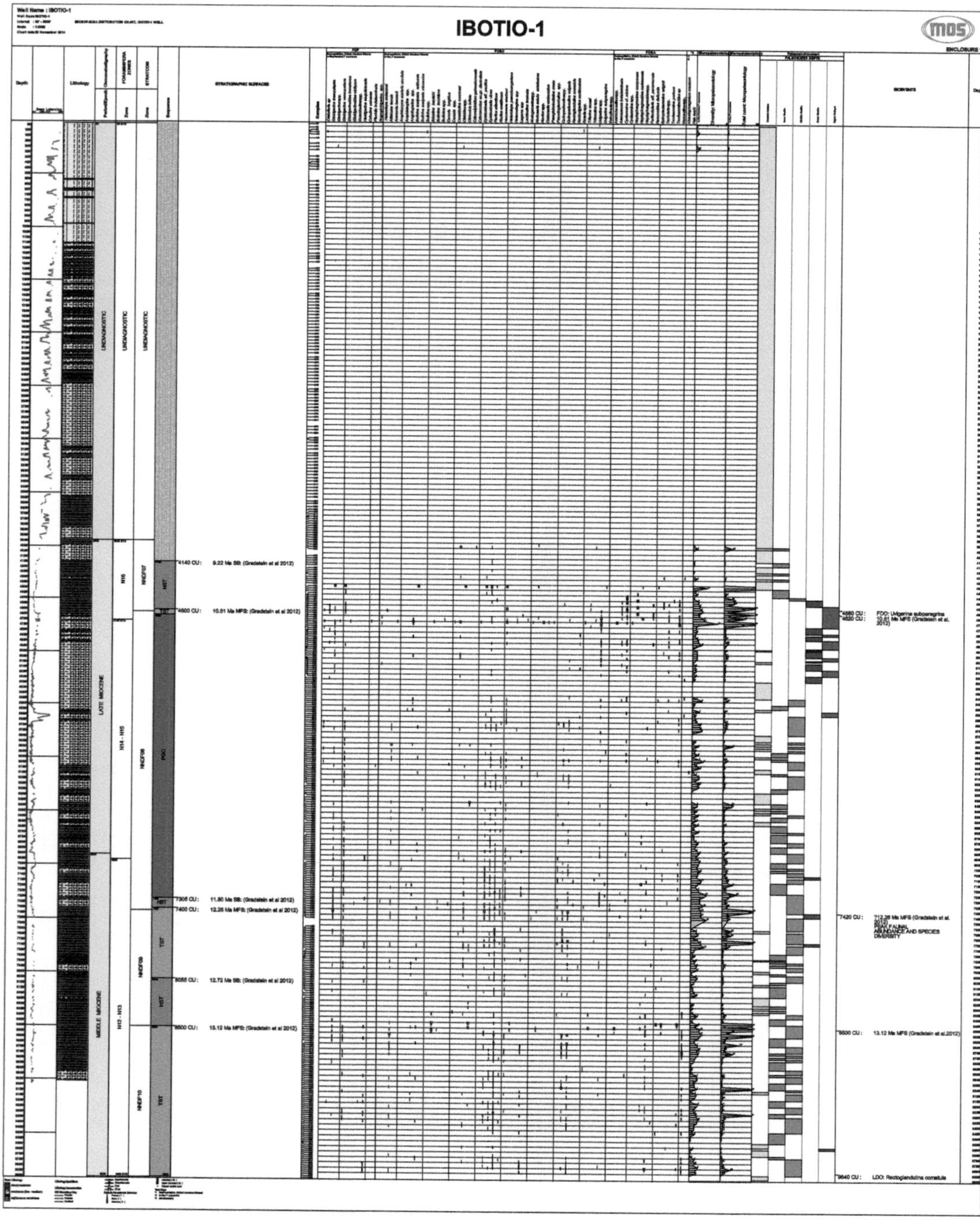

ENCLOSURE 1.13 Microfaunal checklist for the Ibotio-1 well.

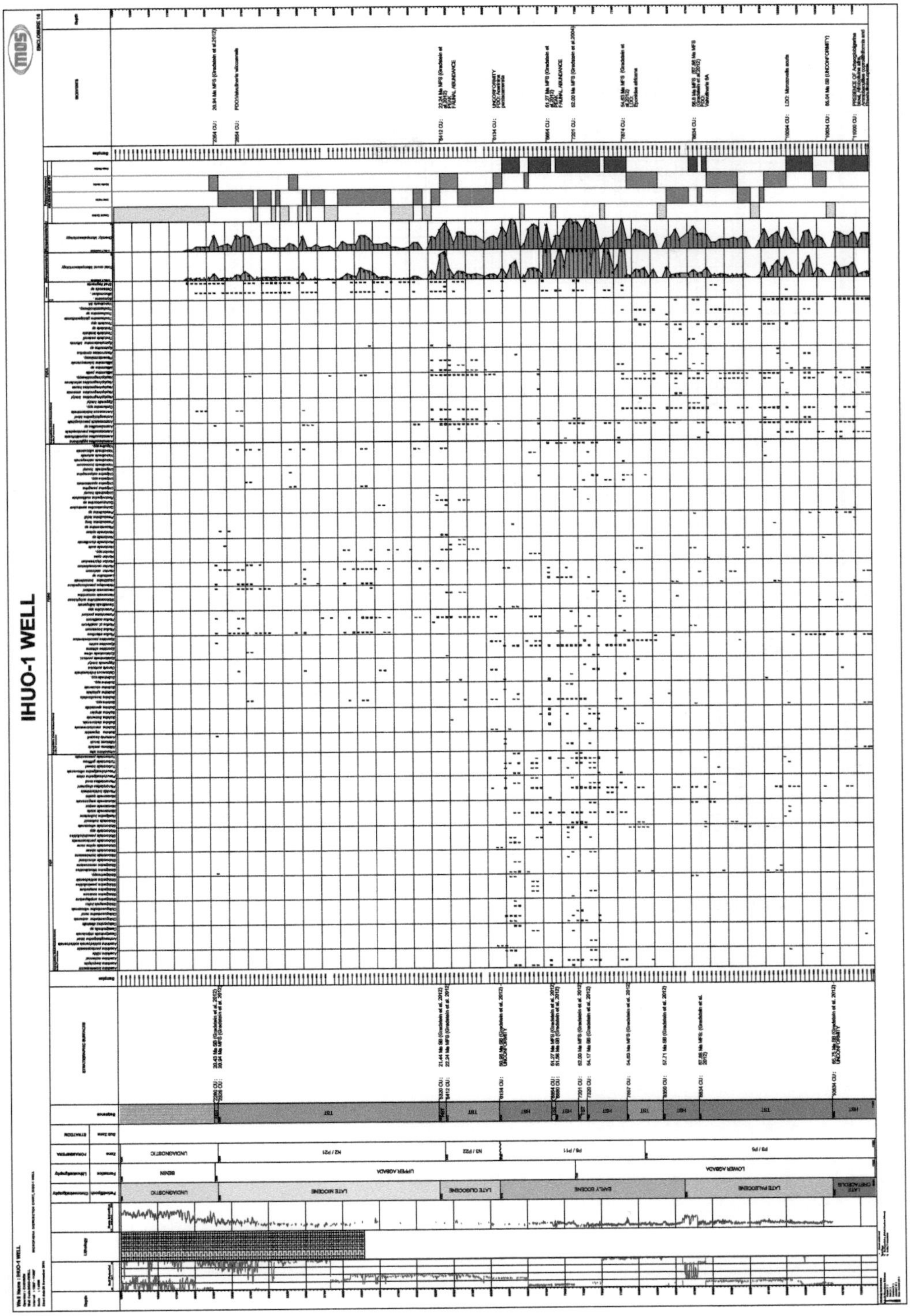

ENCLOSURE 1.14 Microfaunal checklist for the Ihuo-1 well.

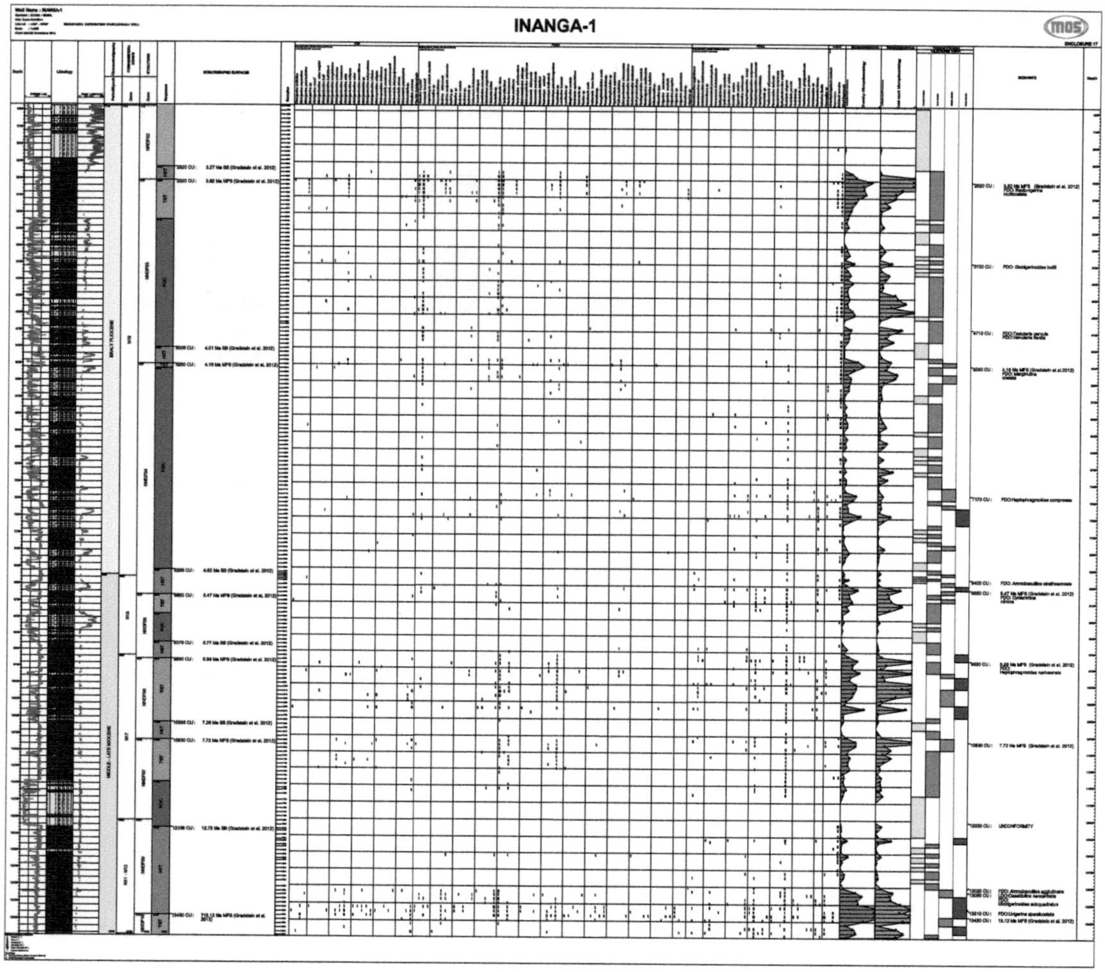

ENCLOSURE 1.15 Microfaunal checklist for the Inanga-1 well.

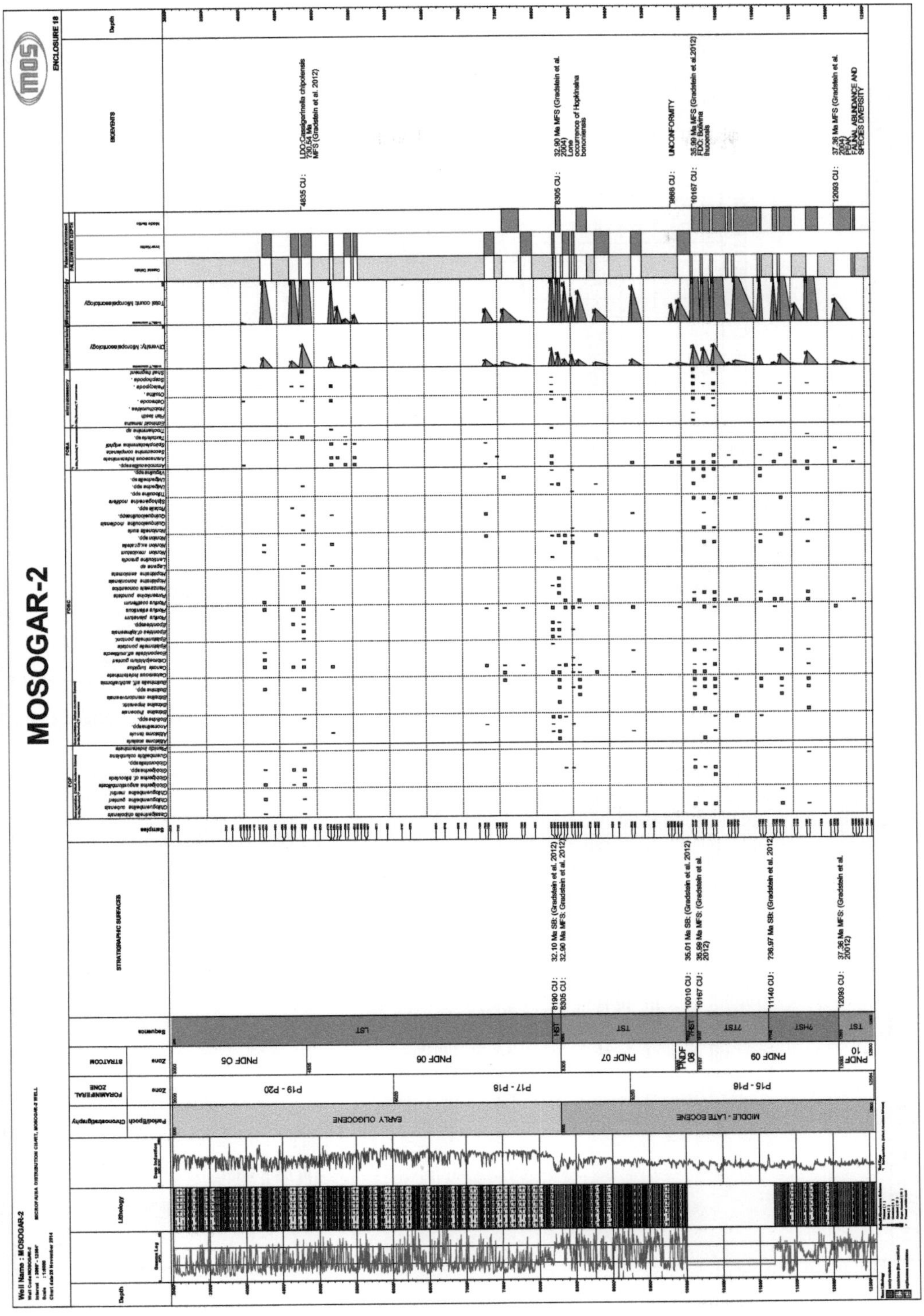

ENCLOSURE 1.16 Microfaunal checklist for the Mosogar-2 well.

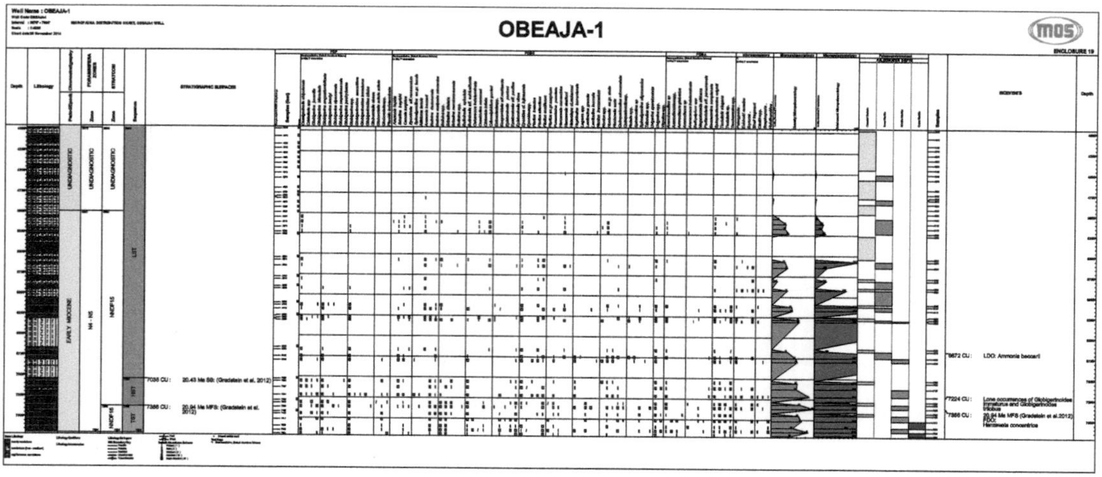

ENCLOSURE 1.17 Microfaunal checklist for the Obeaja-1 well.

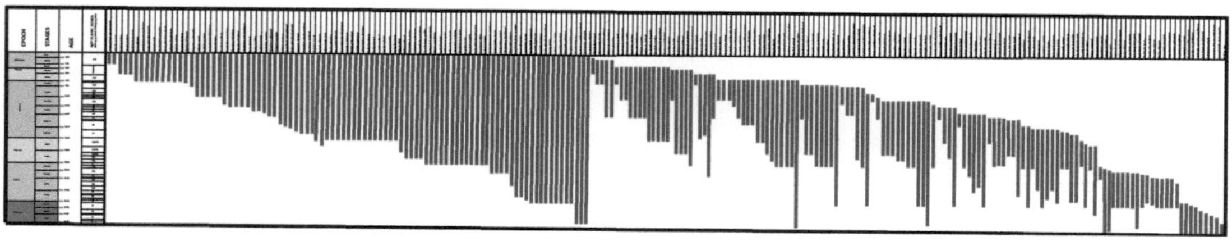

ENCLOSURE 1.20 Stratigraphic ranges of selected benthonic foraminifera species in the Cenozoic Niger Delta.

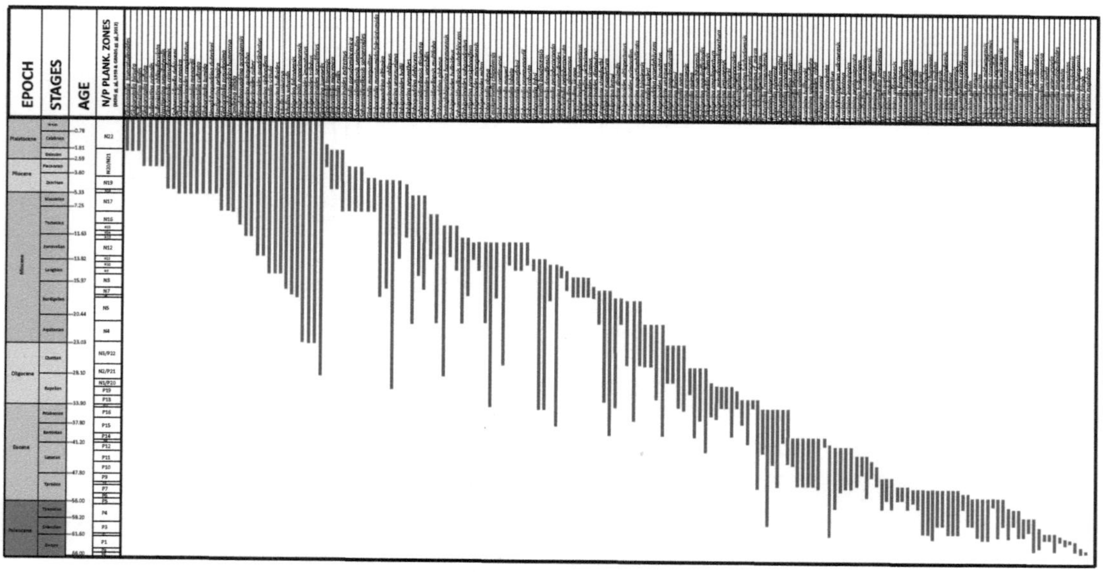

ENCLOSURE 1.21 Stratigraphic ranges of selected planktonic foraminifera species in the Cenozoic Niger Delta.

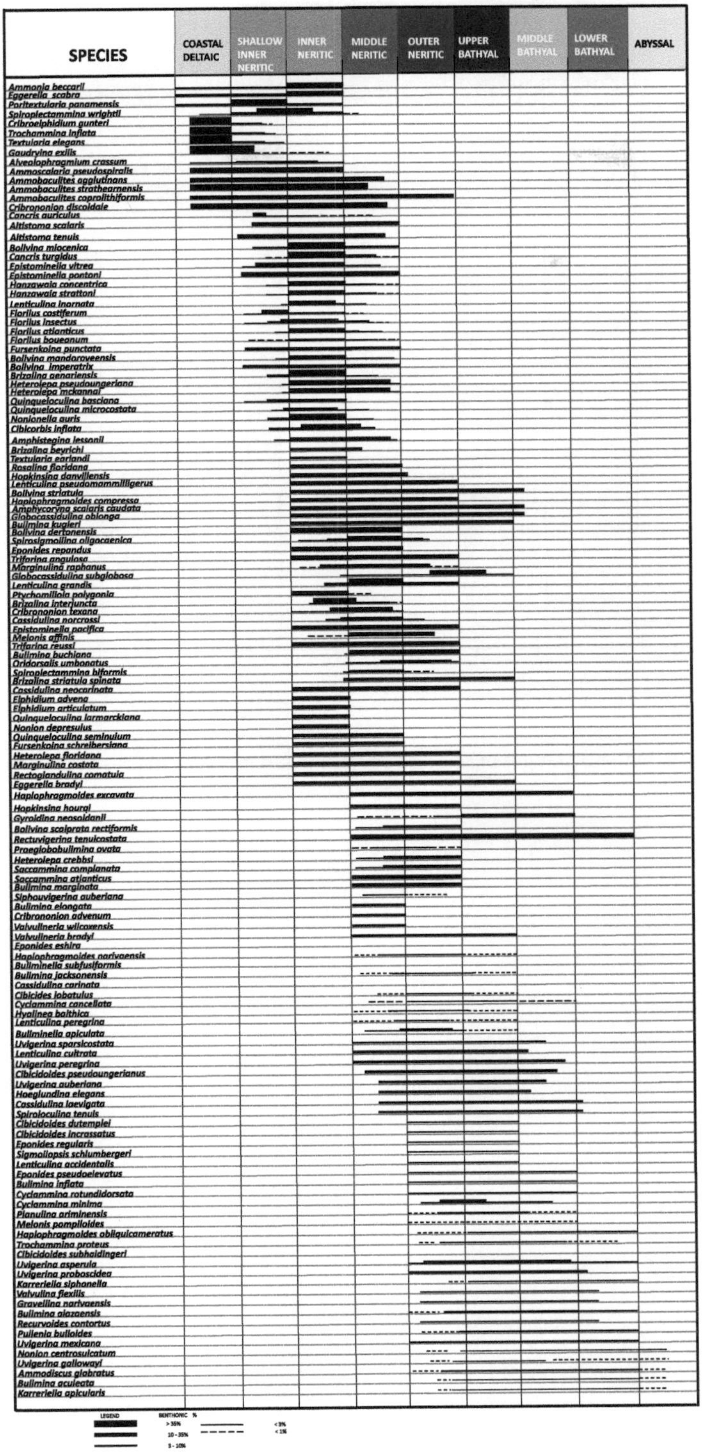
SPECIES
COASTAL DELTAIC
SHALLOW INNER NERITIC
INNER NERITIC
MIDDLE NERITIC
OUTER NERITIC
UPPER BATHYAL
MIDDLE BATHYAL
LOWER BATHYAL
ABYSSAL
Ammonia beccarii
Eggerella scabra
Paritextularia panamensis
Spiroplectammina wrightii
Cribroelphidium gunteri
Trochammina inflata
Textularia elegans
Gaudryina exilis
Alveolophragmium crassum
Ammoscalaria pseudospiralis
Ammobaculites agglutinans
Ammobaculites strathearnensis
Ammobaculites coprolithiformis
Cribrononion discoidale
Cancris auriculus
Altistoma scalaris
Altistoma tenuis
Bolivina miocenica
Cancris turgidus
Epistominella vitrea
Epistominella pontoni
Hanzawaia concentrica
Hanzawaia strattoni
Lenticulina inornata
Florilus costiferum
Florilus insectus
Florilus atlanticus
Florilus boueanum
Fursenkoina punctata
Bolivina mandoroveensis
Bolivina imperatrix
Brizalina aenariensis
Heterolepa pseudoungeriana
Heterolepa mckannai
Quinqueloculina basciana
Quinqueloculina microcostata
Nonionella auris
Cibicorbis inflata
Amphistegina lessonii
Brizalina beyrichi
Textularia earlandi
Rosalina floridana
Hopkinsina danvillensis
Lenticulina pseudomammilligerus
Bolivina striatula
Haplophragmoides compressa
Amphycoryna scalaris caudata
Globocassidulina oblonga
Bulimina kugleri
Bolivina dertonensis
Spirosigmoilina oligocaenica
Eponides repandus
Trifarina angulosa
Marginulina raphanus
Globocassidulina subglobosa
Lenticulina grandis
Ptychomiliola polygonia
Brizalina interjuncta
Cribrononion texana
Cassidulina norcrossi
Epistominella pacifica
Melonis affinis
Trifarina reussi
Bulimina buchiana
Oridorsalis umbonatus
Spiroplectammina biformis
Brizalina striatula spinata
Cassidulina neocarinata
Elphidium advena
Elphidium articulatum
Quinqueloculina larmarckiana
Nonion depresulus
Quinqueloculina seminulum
Fursenkoina schreibersiana
Heterolepa floridana
Marginulina costata
Rectoglandulina comatula
Eggerella bradyi
Haplophragmoides excavata
Hopkinsina hourqi
Gyroidina neosoldanii
Bolivina scalprata rectiformis
Rectuvigerina tenuicostata
Praeglobobulimina ovata
Heterolepa crebbsi
Saccammina complanata
Saccammina atlanticus
Bulimina marginata
Siphouvigerina auberiana
Bulimina elongata
Cribrononion advenum
Valvulineria wilcoxensis
Valvulineria bradyi
Eponides eshira
Haplophragmoides narivaensis
Buliminella subfusiformis
Bulimina jacksonensis
Cassidulina carinata
Cibicides lobatulus
Cyclammina cancellata
Hyalinea balthica
Lenticulina peregrina
Buliminella apiculata
Uvigerina sparsicostata
Lenticulina cultrata
Uvigerina peregrina
Cibicidoides pseudoungerianus
Uvigerina auberiana
Hoeglundina elegans
Cassidulina laevigata
Spiroloculina tenuis
Cibicidoides dutemplei
Cibicidoides incrassatus
Eponides regularis
Sigmoilopsis schlumbergeri
Lenticulina accidentalis
Eponides pseudoelevatus
Bulimina inflata
Cyclammina rotundidorsata
Cyclammina minima
Planulina ariminensis
Melonis pompilioides
Haplophragmoides obliquicameratus
Trochammina proteus
Cibicidoides subhaidingeri
Uvigerina asperula
Uvigerina proboscidea
Karreriella siphonella
Valvulina flexilis
Gravellina narivaensis
Bulimina alazaensis
Recurvoides contortus
Pullenia bulloides
Uvigerina mexicana
Nonion centrosulcatum
Uvigerina gallowayi
Ammodiscus glabratus
Bulimina aculeata
Karreriella apicularis
LEGEND
BENTHONIC %
> 35%
10 - 35%
5 - 10%

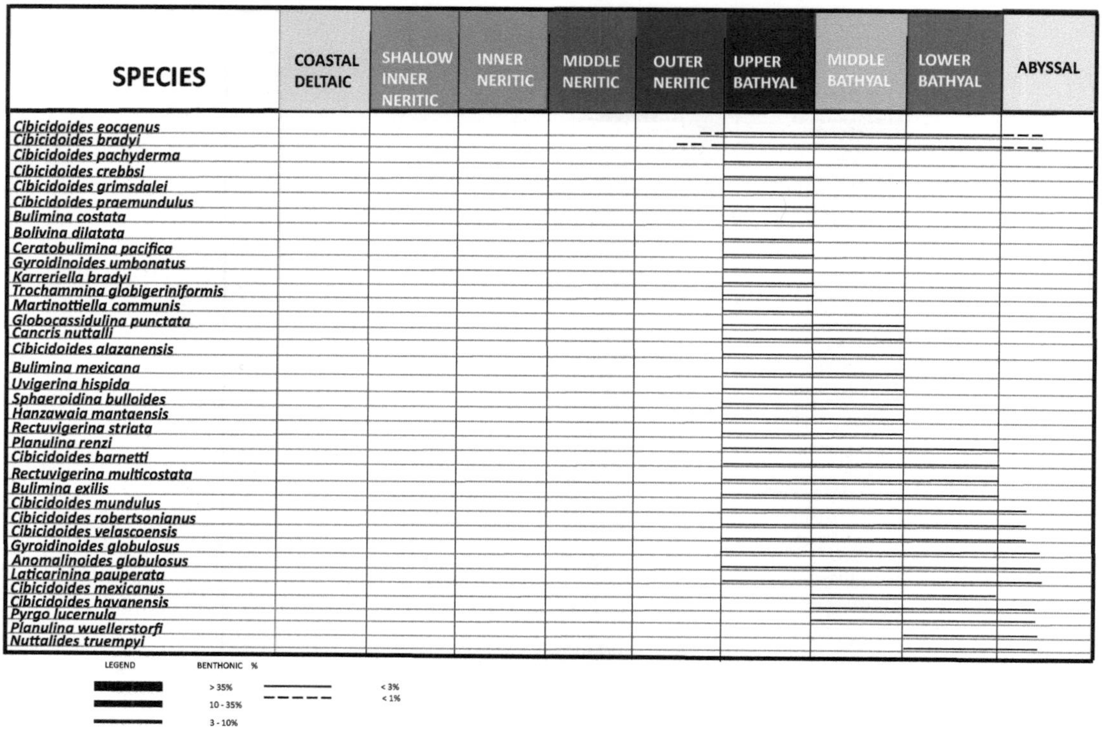

ENCLOSURE 1.22 Bathymetric ranges of selected benthonic foraminifera species in the Niger Delta.

CHAPTER

2

Geology of the Niger Delta Basin

OUTLINE

2.1 DEFINITION AND CLASSIFICATION OF THE NIGER DELTA

The Cenozoic Niger Delta overlies an area that is over 256,000 km^2. It is the youngest of the three large sediment bodies that filled the aulacogen formed after the separation of the African and South American plates. It was initially built out over a transgressive Paleocene prodelta as river-dominated lobes which later coalesced and became high-energy, wave-dominated, and tide-influenced depobelts. The progradation was facilitated by rapid sediment accumulation that substantially exceeded the rate of subsidence. Thereafter, delta construction proceeded in discrete structural units that constituted minibasins, the development of which was driven by clay diapirism and growth faulting rather than by the N–S

Cenozoic Foraminifera and Calcareous Nannofossil Biostratigraphy of the Niger Delta. http://dx.doi.org/10.1016/B978-0-12-812161-0.00002-8

and NW–SE structural trends of the underlying basement. The delta grossly consists of three subsurface lithostratigraphic units—the marine Akata Shales, the paralic Agbada Formation, and the continental Benin Formation. The outcropping correlatives are the Imo Shale, Ameki Group, and Benin Sands, respectively. The best studied of these is the petroliferous Agbada Formation with its varied lithofacies associations that range from reservoir sands through heteroliths to mudrock seal facies.

The Niger Delta is a hydrocarbon province with ultimate recovery presently estimated at nearly 40 billion barrels of oil. This is about 70% of the overall hydrocarbon reserves of sub-Saharan Africa. The gas reserves are conservatively estimated at over 40 trillion cubic feet. The hydrocarbon was sourced from marine shales with land plant materials giving rise to mainly Types III and II/III organic matter within an oil window that varied in depth from 9,000 to 14,000 ft. The reservoirs are mainly shoreface, beach, channel sands bearing low-sulfur/nickel, light waxy, nondegraded oils. Light crude-bearing deeper reservoirs have also been substantiated in some inland blocks. (Mosunmolu Limited, 2014).

The Niger Delta is the Cenozoic gross offlap clastic succession built out on top of the Anambra Basin. The delta was formed as a major continental embankment that spreads out over the cooling and subsiding oceanic crust generated as the African and the South American lithospheric plates separated. At the initial stage of development, from Eocene to Early Miocene, the delta built out as three distinct depocenters, each a lobate-elongate, river-dominated delta. From mid-Miocene onward, the three depocenters coalesced into one unit with a lobate-arcuate front that developed in a high-energy constructive, wave-dominated, and tide-influenced basin. Progradation was controlled by the rate of sediment deposition (R_d) that was generally higher than the rate of subsidence (R_s). Delta construction took place in discrete structural units referred to as depobelts that effectively behaved as minibasins. Their development was driven by the activity of large growth faults and clay diapirs (Knox and Omatsola, 1987). The diapiric mud bulges and the basinward mass movements and toe thrusting of the Akata mud/shale enabled successive deltas to prograde since the Oligocene. According to Whiteman (1982, p. 290) and Ajakaiye and Bally (2002), without the large-scale diapirism and the delta toe thrusts, the delta complex would not have developed the way it did.

In the early Paleocene, a major marine transgression across most of southern Nigeria led to the deposition of marine shales that formed the prodelta on which the Early Eocene (Ypresian) regression deposited the sediments of the "modern Niger Delta." It follows therefore, that the area underlain by Cenozoic strata in much of southern Nigeria can be regarded as integral parts of the Niger Delta. The western limit of the delta is the Okitipupa Ridge (Adegoke, 1969; Maron, 1969; Merki, 1972; Benkhelil, 1989; Gebhardt, 1997). The Anambra Basin underlies the delta basin and both are demarcated by an unconformity (Avbovbo and Ayoola, 1981). Thus the delta complex (onshore and offshore, and in the subaerial and subaqueous) is delimited onshore by the contact between the Imo and Nsukka Formations, in the offshore to the southeast by the Cameroon Volcanic Line, to the west by the Okitipupa Ridge, and to the southwest by the Guinea Abyssal Plain. The total area covered exceeds 256,000 km^2 (Whiteman, 1982, p. 163). Fig. 2.1 shows the location of the present-day Niger Delta in the triangular classification scheme by Galloway (1975). It provides a vivid explanation of the reference to the delta as being wave dominated and tide influenced.

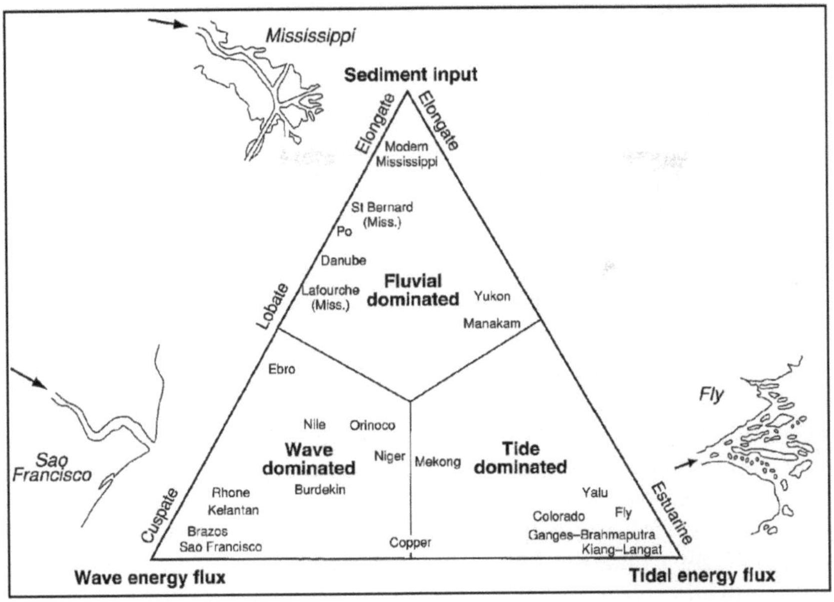

FIGURE 2.1 **Morphological classification of deltas using the relative influence of fluvial, tidal, and wave processes.** *Source: After French, P., 2007. Deltaic and estuarine environments. In: Perry, C., Taylor, K. (Eds.), Environmental Sedimentology. Blackwell Publishing, Oxford, pp. 223–262, based on Galloway, G.E., 1975. Process framework for describing the morphologic evolution of deltaic depositional systems. In: Broussard, M.L. (Ed.), Deltas: Models for Exploration. Houston Geol. Soc., Houston, p. 8798.*

2.1.1 Previous Work

Considerable geological information is available on the subsurface geology of the wetland Niger Delta from the activity and reports of the oil exploration and development companies (Frankl and Cordry, 1967; Short and Stauble, 1967; Maron, 1969; Weber, 1971; Merki, 1972; Murat, 1972; Weber and Daukoru, 1975; Weber et al., 1978; Evamy et al., 1978; Opara, 1981; Haack et al., 2000; Reijers, 2011).

The outcropping parts of the delta have also been studied mostly by the geologists of the Geological Survey of Nigeria and university researchers (Parkinson, 1907; Wilson, 1925; White, 1926; Wilson and Bain, 1928; Tattam, 1944; Du Preez, 1947; Grove, 1951; Simpson, 1955; Reyment and Barber, 1956). More recent studies include those by Nwajide (1979, 1980), Whiteman (1982), Oti and Postma (1995), Knox and Omatsola (1987), Reijers et al. (1997), and Nwajide (2013).

2.1.2 Physiography

The dry land sector of the Niger Delta is a part of the interior lowlands of southern Nigeria lying between 30 and 300 m above sea level. The Niger River valley bisects the territory longitudinally into eastern and western blocks with different topographic characteristics. The western part is dominated by a southwest sloping plain lying between 150 and 300 m above sea level referred to as the Ishan-Asaba Plateau (Udo, 1970, p. 38). The plateau is deeply incised on both the southern and eastern margins by some south-bound first-order streams. The lower terrain to the south consists of a sandy coastal plain that is generally below 120 m above sea level within which lie the towns, Agbor and Benin City. The wetland delta lies to the south of this. Sapele, Koko, and Gwato are prominent towns within this wetland. To the east, the Niger valley is coextensive with the alluvial valley of the Anambra River. The area oversteps the Imo Formation and the Ameki Group. Its depositional origin is thought to be mostly from the suspension load of both the Anambra and the Niger Rivers (Aiyuk, 1993). The dry land part of the delta lying to the north of Onitsha metropolis and to the east and southeast of the alluvial deposits shows a rugged landscape. Farther to the east, the topographic features that dominate the landscape are the Awka-Orlu Upland and the sandstone members of the Imo Formation.

In the wetland part of the delta, the topographic configuration is not easily assessed because of the rather low and virtually uniform elevation above sea level. At Onitsha, the spot height is about 20 m above mean sea level with a gradual decline of topography (i.e., zero elevation) for some 200 km to the shoreline. The rivers traversing or partly originating from the dry land part of the delta are the Niger, Ase, Benin, Anambra, Imo, Kwa Ibo, and Orashi rivers (Fig. 2.2). The area to the west of the Niger River is drained largely by one integrated system—the Benin River.

In the wetland Niger Delta, the Niger River, the world's seventh longest river (4120-km long), bifurcates into the Nun and Forcados Rivers whose distributary pattern is actually responsible for the origin of the reticulate drainage pattern that characterizes the Niger Delta proper (Fig. 2.2). The drainage density is very high and has given rise to an intricate drainage texture. The various geomorphic zones of the delta exhibit southward-increasing drainage densities, except the sandy coastal barrier strip, which exhibits a low drainage density. Its higher elevation and enhanced permeability facilitate rainwater infiltration to the extent that they constitute hydrologic centers from which northward-flowing drainage lines emerge.

2.1.3 Sediment Budget

The sediments of the Cenozoic Niger Delta were derived, and are still being derived, from the predominantly acid, coarsely crystalline Basement Complex and from the Cretaceous and Cenozoic basement-derived sedimentary units outcropping in the very extensive catchment area of the Niger-Benue River system. A major change in the composition of the sediments entering the delta took place in the Neogene when volcanically derived sediments were added after the emplacement of the Cameroon volcanic line. Prior to confluencing with the Benue at Lokoja, the Niger River transported a bed load of about $0.3 \times 12^6\ m^3$ annually. The suspended load was estimated at $4.6 \times 12^6\ m^3$ annually (NEDECO, 1961). The sands contained quartz, feldspar, and a suite of heavy minerals including hornblende, epidote, sillimanite, and zircon.

The Benue River with its smaller catchment area is a more important sediment source for the Niger Delta than the Niger River. It has such large tributaries as the Gongola River, which drains the Jos Plateau and the Katsina Ala River, which heads toward the Bamenda Highlands. The Benue itself heads, via numerous tributaries toward the Mandara Mountains and the Vokre Heights, which straddle the Nigeria–Cameroon frontier. These are all high rainfall areas underlain by crystalline Basement Complex rocks, thick sandy Cretaceous sediments, and basic volcanics in the Biu and Longuda areas, as well as the very extensive volcanic rocks of the Cameroon–Bamenda–Adamawa Mountains. The Benue sediments are feldspathic and

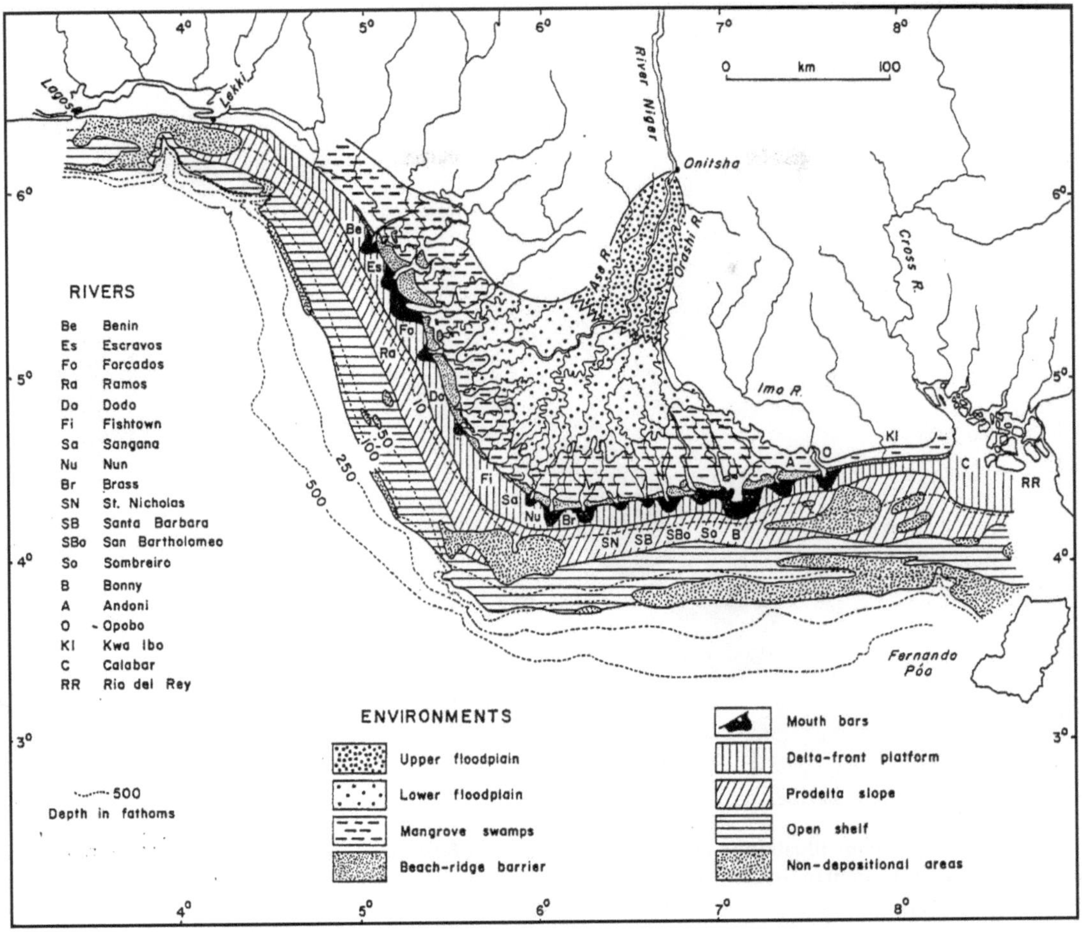

FIGURE 2.2 **Physiography and principal sedimentary environments of the modern Niger Delta.** *Source: After Allen, J.R.L., 1970. Sediments of the modern Niger Delta: summary and review. In: Morgan, J.P. (Ed.), Deltaic Sedimentation, Modern and Ancient. SEPM Special Publication No. 15, pp. 138–151.*

contain hornblende, epidote, zircon, garnet, staurolite, and sillimanite. The annual bedload discharge, which is mostly coarse sand, is about $0.6 \times 12^6\ m^3$ and the suspended load is estimated at $11 \times 12^6\ m^3$. These clearly dwarf the Niger catchment as a sediment source. Based on the provenance and sediment volumes, it would be more appropriate to call the resulting delta the "Benue Delta" (Burke, 1996). Of the combined Niger–Benue sand size load, about 65% is retained within the delta as Benin Sand facies; only about 35% reaches the sea.

2.1.4 The Geomorphic Zones and Depositional Environments

The present-day Niger Delta consists of subaerial and subaqueous belts that are subdivisible into several geomorphic units (Figs. 2.2 and 2.3) (NEDECO, 1959, 1961; Allen, 1964, 1965a,b, 1970). The subaerial part, the delta topset, consists of floodplains, mangrove swamp, and

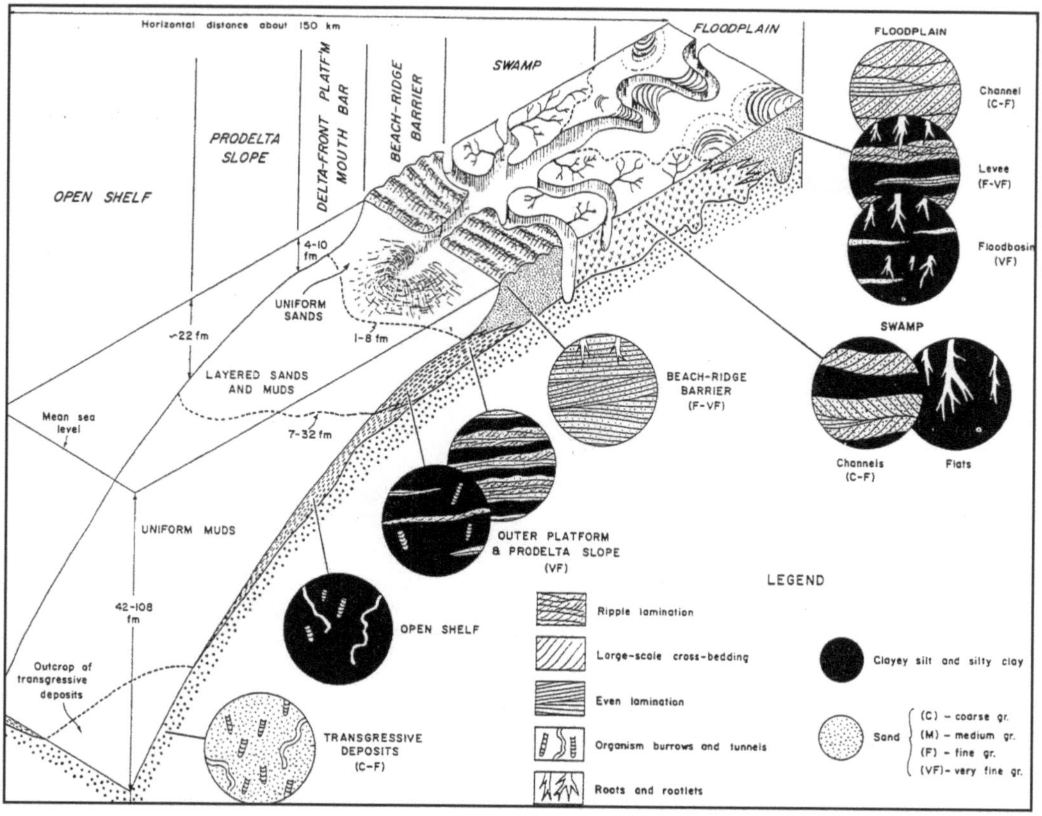

FIGURE 2.3 **Schematic illustration of the characteristics and facies relationships along an onshore–offshore longitudinal Niger Delta transect.** *Source: After Allen, J.R.L., 1970. Sediments of the modern Niger Delta: summary and review. In: Morgan, J.P. (Ed.), Deltaic Sedimentation, Modern and Ancient. SEPM Special Publication No. 15, pp. 138–151.*

beach–ridge–barrier islands. From here sediments pass down to the distributary mouth bars, delta front platform, and prodelta slope to the open shelf. Since the major regression that began in the Early Eocene, the delta has been prograding by actively accumulating sediments in its distal reaches. It thus combines the properties of a stable, partially eroding basin with those of an active sedimentary basin. The three lithostratigraphic units of the Niger Delta (the Akata, Agbada, and Benin Formations) are thus still being deposited in the present day.

2.2 ORIGIN AND LITHIC FILL OF THE NIGER DELTA BASIN

The fundamental causative event for the origin of the Niger Delta Basin was the break-up of the Gondwana supercontinent. The Benue Trough had been formed as part of the widely distributed African rift population which originated some 140 million years ago (Burke, 1996).

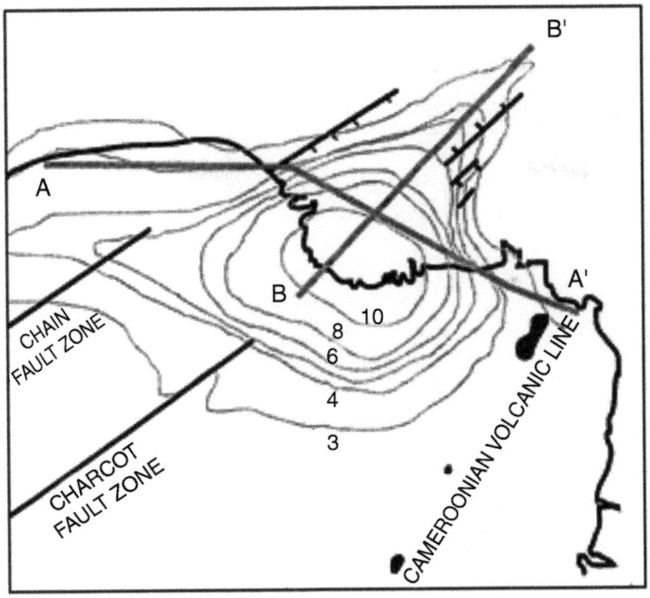

FIGURE 2.4 **Prolongation of the Chain and Charcot Fracture Zones into the African Plate.** *Source: From USGS Open File Rpt. No. 99-501.*

It is a composite of several pull-apart rifts, formed by sinistral wrench faulting. This was followed by a structural reorganization in the Santonian, thought to be a consequence of the differences in the Atlantic Mesozoic spreading rates (Burke et al., 1972; Genik, 1992, 1993). This was a thermotectonic event resulting in the folding and uplift of the southern Benue Trough and the translocation of the depositional axis.

All this involved subsidence which was not only in the area west of the Abakaliki Anticlinorium but also south and southeast of it, hence the extension of the Anambra Basin to the Calabar area. The Anambra Basin was then filled with Campanian to Early Paleocene (Danian) facies. Next was the establishment of the Niger Delta Basin consequent upon the continued subsidence of the southern area of the Benue Trough due to the thermal contraction of the lithosphere (Sleep, 1971; Turcotte, 1977; Steckler and Watts, 1978; Onuoha, 1981). As shown in Figs. 2.4–2.6, the Benue Trough, the Anambra Basin, and the Niger Delta Basin are vertically stacked in the southern reaches, but their lateral extents had been constrained and determined by the position of the fracture zones.

2.2.1 Stratigraphic Packaging

The lithic fill of the Niger Delta Basin ranges from the facies of the first major Cenozoic transgression to the depositional products of the present day, as well as materials eroded from subsurface and outcropping formations (Fig. 2.7; Tables 2.1 and 2.2).

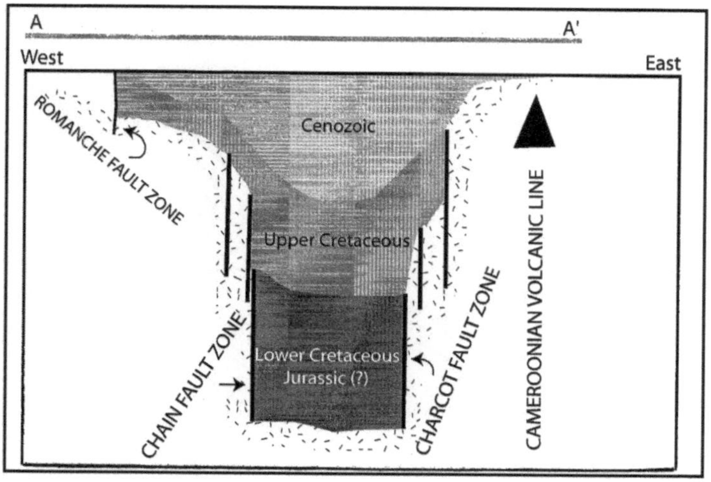

FIGURE 2.5 **Section AA′ in Fig. 2.4 showing the stacking of basins in southern Nigeria—southern Benue Trough, Anambra Basin, and the Niger Delta Basin.** *Source: From USGS Open File Rpt. No. 99-501.*

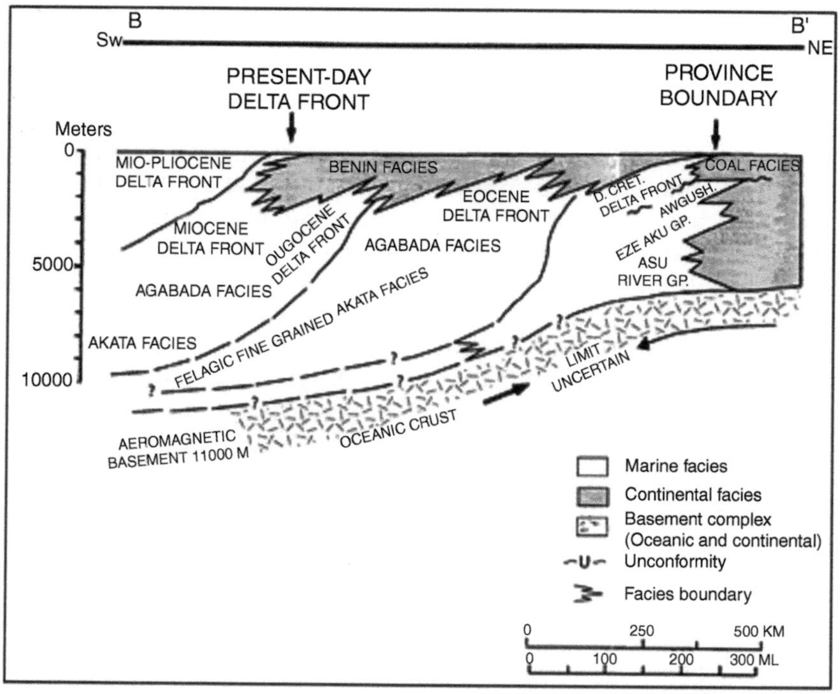

FIGURE 2.6 **NE–SW section showing the vertical and lateral successions of the component lithofacies from the southern Benue Trough, through the Anambra Basin to the Niger Delta.** *Source: From USGS Open File Rpt. No. 99-501.*

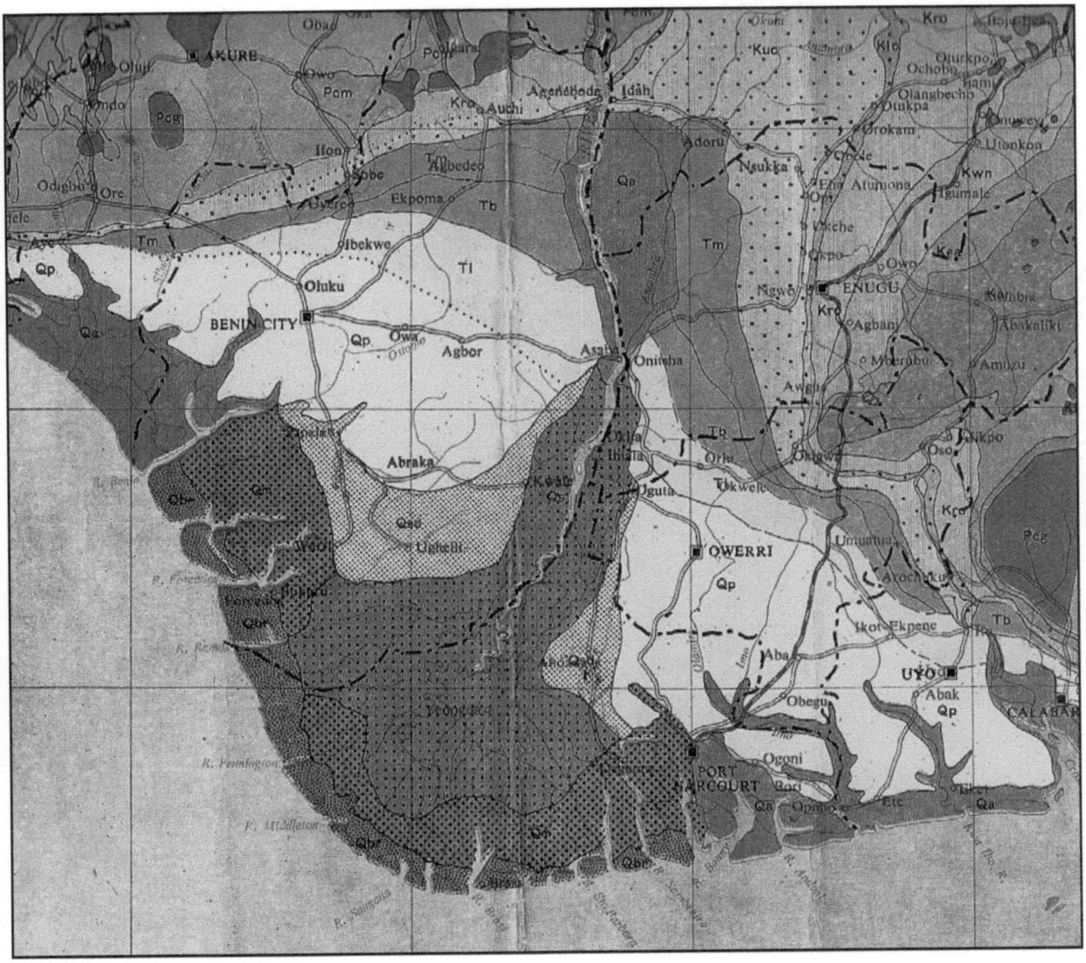

FIGURE 2.7 **Geological map of part of southern Nigeria showing the outcropping area of the Cenozoic Niger Delta.** Tm, Paleocene Imo Formation; Tb, Ameki Group; Tl, Ogwashi Asaba Formation; Qp, Benin Formation. *Source: Extracted from the 1984 Edition of the GSN Geological Map of Nigeria.*

2.2.1.1 *Outcropping Units*

2.2.1.1.1 THE IMO FORMATION

The Imo Formation is the basal unit of the Niger Delta (Figs. 2.6 and 2.7; Whiteman, 1982, p. 131). It outcrops within a concave area stretching southward from the Benin Flank in the west. From there it overlies the Nsukka Formation of the Anambra Basin and widens eastward. It swings along the Imo River near Okigwe-Umuahia Road. Its maximum thickness is estimated as 1000 m by Reyment (1965, p. 90) and 490 m by Dessauvagie (1975). Though exposures are rare, the formation consists mainly of dark gray to bluish gray shales, with occasional admixtures of clay, ironstone, thin sandstone bands, and limestone intercalations.

TABLE 2.1 Comparison of the Facies of Present-Day Delta With the Tertiary Units of the Niger Delta

Present Niger Delta	Outcrops of Tertiary strata	Subsurface formations
NEDECO Report 1954	Reyment (1965)	Short and Stauble (1967)
Continental (fluviatile) deposits, mainly sands	Benin Formation	Benin Formation with Afam Clay Member
Mixed continental brackish water and marine deposits, sands, and mudrocks	Ogwashi Asaba Formation and Ameki Formation	Agbada Formation
Marine deposits, mainly mudrocks	Imo Shales	Akata Formation

After Maron, P., 1969. Stratigraphical aspects of the Niger Delta. J. Min. Geol. 4 (1 and 2), 3–12.

TABLE 2.2 Outcropping Units of the Cenozoic Niger Delta

Age span	Lithostrat. unit		Characteristics
Oligocene–Present	Benin Fm.		Known also as the Coastal Plains Sands; cross-bedded, coarse, pebbly continental sands, with clay lenses and lignites; has marine shale breaks with foraminifera, ostracods, and molluscs
Oligocene–Miocene	Ogwashi–Asaba Fm.		Clays, silts, and sands with thin to thick lignite seams
Eocene–Early Oligocene	Ameki Gp.	Nsugbe Fm.	Mainly sands with some conglomerate bands
		Nanka Fm.	Calcareous clays and silts with thin shelly limestone, rich in foraminifera
		Ameki Fm.	Calcareous clays and silts with thin shelly limestone, rich in foraminifera, mainly sands, minor silt, and clay intercalations
Paleocene–Early Eocene	Imo Fm.		Blue–gray shales with sand lenses, marls, and fossiliferous limestones, sandstone members—Ebenebe, Umuna, and Igbaku sandstones; shales with foraminifera and ostracods

The most conspicuous lithofacies components of the formation are three sandstone bodies—the Ebenebe, the Igbaku, and the Umuna Members. They outcrop as ridges flanked by marshy low grounds that are underlain by shales. A roadside exposure of the Ebenebe Sandstone Member seen a few kilometers south of Ugwuoba is fine-grained, flat-bedded, and thickens upward, suggestive of inner shelf depositional setting. In a study of a road section exposed between Itumbuzo and Bende in Abia State, Anyanwu and Arua (1990) noted that the formation tends to become sandy upward. Here, the sandstone interval is up to 5.5-m thick, poorly consolidated, and medium to coarse grained, with very large-scale planar cross-beds with dips up to 35°, with unimodal paleocurrent to the southwest.

Limestones are not well developed in the Imo Formation. Thin carbonate beds are present with the limestone derived from fossil microfaunal tests, pelecypods, gastropods, and cephalopods. The commonest planktonic foraminifera recorded include *Globorotalia velascoensis* and *Globorotalia pseudomenardii* (Reyment, 1965 p. 88). Some of the faunal elements are age-diagnostic. The shales are composed principally of kaolinite with minor montmorillonite and illite. Based on the fossil content, especially the foraminifera, ostracod, nautiloids, and mollusca, Reyment (1965, p. 86), Adegoke (1977), and Arua (1980) assigned the formation to a neritic depositional setting within the Eocene.

2.2.1.1.2 THE AMEKI GROUP

The Ameki Group has been fairly well studied because the sediments are relatively well exposed especially along the eastern railway line and it is highly fossiliferous (Wilson, 1925; Wilson and Bain, 1928; White, 1926; Tattam, 1944; Du Preez, 1947; Jones, 1948; Grove, 1951; Simpson, 1955; Reyment and Barber, 1956; Adegoke, 1969; Adegoke and Tevesz, 1974; Adegoke et al., 1980b). Later studies (Nwajide, 1979, 1980) identified three laterally equivalent component formations of the group as the Ameki, Nanka, and Nsugbe formations.

The Ameki Formation: This formation is exposed along the railway cutting and at the Ameki railway station north of Umuahia in Abia State. Its maximum thickness is about 1500 m and consists of highly fossiliferous gray-green sandy clays with calcareous and white clayey sands and intercalations of shale and thin shelly, often coquinoid limestone. The formation displays conspicuous facies changes. Carbonaceous plant remains and lignitic horizons are present along with amber, pockets of lignite, and calcareous nodules. From the study of the abundant faunal content, the age is undoubtedly Eocene (Adegoke, 1969; Adegoke and Tevesz, 1974; Adegoke et al., 1980b). It was largely deposited in an extensive regressive shallow marine setting. It may thus be concluded that the formation consists of deltaic deposits advanced over the shallow Anambra shelf onto the continental margin.

The Nanka Formation: This formation is defined as the sand facies of the Ameki Group. It is up to 305 m thick, underlying an area of over 1400 km^2 east of the Niger River with its type area in Nanka town. Here, the formation is deeply gullied, especially on the east-facing scarp slope of the Awka-Orlu Cuesta. The formation consists of loose, flaser-bedded, fine to medium sand, with a few mudrock breaks. Elsewhere, away from the type locality, for example, along the road cut about 18 km from Onitsha along the expressway to Enugu, the depositional setting of the formation is largely subtidal (Fig. 2.8; Nwajide and Reijers, 1996; Umeji, 2003).

The Nsugbe Formation: This third lithic unit within the Ameki Group is less systematically documented than the other two formations. It is considered as a lithostratigraphic unit of formation rank in this study largely because its lithologic motif is significantly different from that of the Nanka Formation. It underlies an area over 10,000 km^2, a substantial part of which is covered by river alluvium between the Anambra River and the Niger River.

Its textural characteristics also set it apart, being composed in part of cobble size clasts within beds of conglomerate. It exhibits a degree of induration unknown in the Nanka Formation and has a prominent, fluvially dissected topographic expression. The type locality is around Nsugbe, a town about 10 km NNE of Onitsha. The area is strewn with ferruginous sandstone boulders, especially within the forest reserve north of Onitsha. Exposures can be viewed along the left bank of the Niger River about 100 m upstream of the Nkisi River mouth, as well as along the road through the prison area. Better exposures were mapped along a 1-km stretch along the right bank of the Niger River downstream from the disused ferry jetty at Asaba. The exposures form part of the 20–30 m wide terrace, which widens farther

FIGURE 2.8 **Part of the Umunya section showing clay-draped cross-bed sets, reactivation surfaces, accordion structure, and master bedding.** *Source: After Nwajide, C.S., 2013. Geology of Nigeria's Sedimentary Basins. CSS Books Ltd., p. 565.*

downstream into the floodplain at the apex of the present-day Niger Delta. Its thickness is not known, though it may not differ substantially from that of the Nanka Formation. The Asaba section is over 12-m thick and consists of two upward coarsening cycles. The lower cycle consists of flat-bedded, occasionally wave ripple-laminated and burrowed, clayey, medium-grained sandstone. It also exhibits isolated channeling and trough cross-bedding with a northeasterly azimuth. The cycle is capped off by an oligomictic, sharp-based, 1-m thick conglomerate composed of well-rounded quartz pebbles set in a matrix of coarse and poorly sorted ferruginous sandstones.

The paleogeographic setting for the deposition of the Early Coniacian to Early Eocene facies of southern Nigeria was reconstructed by Murat (1972). Particularly noteworthy is the regression that succeeded the major Paleocene inundation. The suggested depositional model is depicted in Fig. 2.9. Marginal marine conditions were prevalent and the progressive shallowing documented by the three formations of the Ameki Group appears to fit very well into the paleogeographic scheme.

The Ogwashi-Asaba Formation (The Lignite Series): This is the Oligocene–Miocene stratigraphic unit that overlies the Ameki Group. From the Okitipupa Ridge, the unit occurs in a widening outcrop pattern through Onitsha, Ozubulu, Nnewi, southeastward to Umuahia, Ikot Ekpene, Uyo, and Calabar where it is overlapped by the Benin Formation. The type locality is at Eke Mgbalimgba in Ogwashi-Asaba, Delta State. According to Okezie and Onuogu (1985, p. 5), in the Nnewi-Oba area of Anambra State, the lignites occur in the Ameki Group facies, that is, in the Nanka Formation, whereas in the Umuahia to Ikot Ekpene area, the lignite seams occur in the Benin Formation. The implication is that the uppermost strata of the Nanka Formation and the lowest beds of the Benin Formation constitute the Ogwashi-Asaba Formation. The unit consists of white, blue, and pink clays, cross-bedded sands, carbonaceous mudstones, shales, and seams of lignite. The lignites are confined to a 15–17 km wide belt stretching northwest–southeast for some 240 km from Asaba area to the east of Calabar.

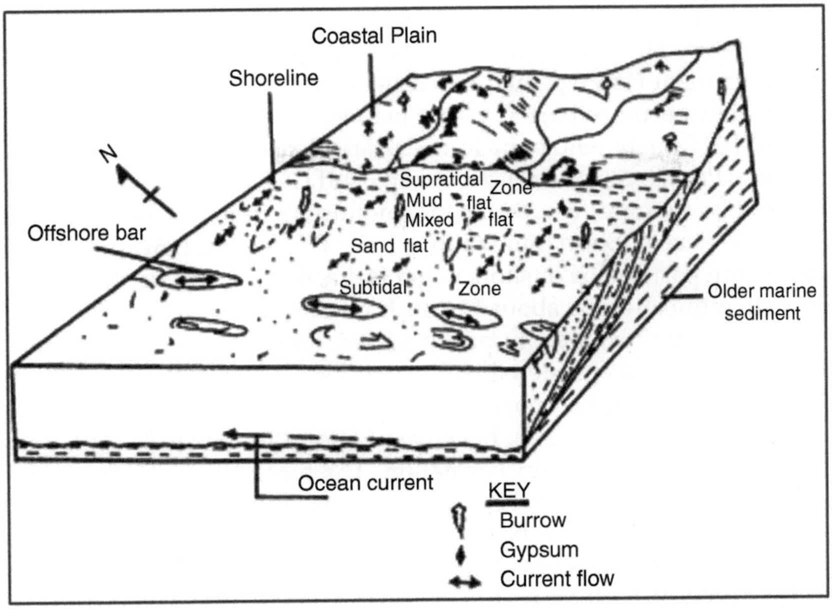

FIGURE 2.9 **A depositional model proposed for the Eocene paleogeographic setting.** *Source: After Nwajide, C.S., 1980. Eocene tidal sedimentation in the Anambra Basin, southern Nigeria. Sediment. Geol. 25, 189–207.*

Within this territory there are variably thick lignite exposures, especially at Oba and Nnewi, where the seams exceed 4 m in thickness. Seams may split and are often profusely jointed and they display conchoidal fracture. They vary from woody to earthy types and contain a considerable amount of palm debris, which accounts for their high resin and wax content. From pollen data, the lignites originated mainly from tropical and semitropical palms. An Oligo–Miocene upper flood plain depositional environment was inferred for the unit.

2.2.1.2 *Subsurface Units of the Niger Delta*

The tripartite stratigraphy of the Niger Delta Basin consists of the Akata, the Agbada, and the Benin formations.

2.2.1.2.1 THE BENIN FORMATION

This unit consists predominantly of continental fluvial sands that underlie an extensive area of southern Nigeria (Fig. 2.7) typified by the sands around Benin City where it is estimated to be 3050-m thick. Short and Stauble (1967) designated the type section as the Elele-1 well located about 40 km northeast of Port Harcourt. The unit is generally friable and consists of white, fine to coarse and pebbly, poorly sorted sands. Lignite occurs as thin streaks or as finely dispersed fragments. Within the formation are thin grayish brown shale bands containing plant fragments.

The Benin Formation is reconstructed as the upper and lower flood (delta) plain setting. Also involved are deltaic, estuarine, lagoonal, and fluviolacustrine subenvironments. Although some marine shale breaks have been identified within the formation, the bulk of the sediments were deposited in the upper delta plain as freshwater, backswamp, and meander belt facies (Allen, 1965a; Dessauvagie, 1972). The shale breaks probably are the source of the fossils on which Reyment (1965, p. 104) based the observation that the formation is "partly marine, partly deltaic, partly estuarine, partly lagoonal, and fluviolacustrine in origin." The foraminifera recorded from boreholes include *Candorbulina universa* Jedlitschka, *Globoquadrina dehiscens* Chapman, Parr, and Collins, *Cibicides* sp., and *Globigerinoides trilobus*. The formation also bears arenaceous foraminifers, ostracods, mollusc fragments, corals, and fish remains. The age is Eocene to Recent (Short and Stauble, 1967; Whiteman, 1982, p. 119).

In the southeastern parts of the offshore Niger Delta, the Benin Formation extends from the ocean bottom with a thickness of about 914 m. It retains its characteristics as a deposit consisting of fine to coarse grained, pebbly, poorly sorted, feldspathic and hematitic quartz sand, interbedded with shales, and occasional lignite streaks. The formation here is primarily nonmarine to paralic onshore, but becomes neritic in the Mobil oil block offshore (Opara, 1981). Its basal shale beds, with some sands, have been dated Pliocene to Recent, and designated informally as a member of the Benin Formation. The formation is hydrocarbon bearing in some of the Mobil fields, contrary to the situation elsewhere where it is solely water bearing.

In several parts of the subsurface Niger Delta, within the Agbada and Benin Formations, deep and relatively areally extensive erosional canyons are found filled with clays and subordinate sands. They occur in both the eastern and western re-entrants of the delta, but are more common in the eastern side (Fig. 2.10). Their location has been attributed to ocean current action at the re-entrants of the Nigerian coastline. The canyons are said to represent erosional processes during relative sea level falls, followed by canyon filling during the subsequent relative sea level rises. At those low stands of the sea, the Niger Delta built pronouncedly seaward and in fact became a shelf edge delta with sediment debouching as turbidity flows that nourished deep sea fans.

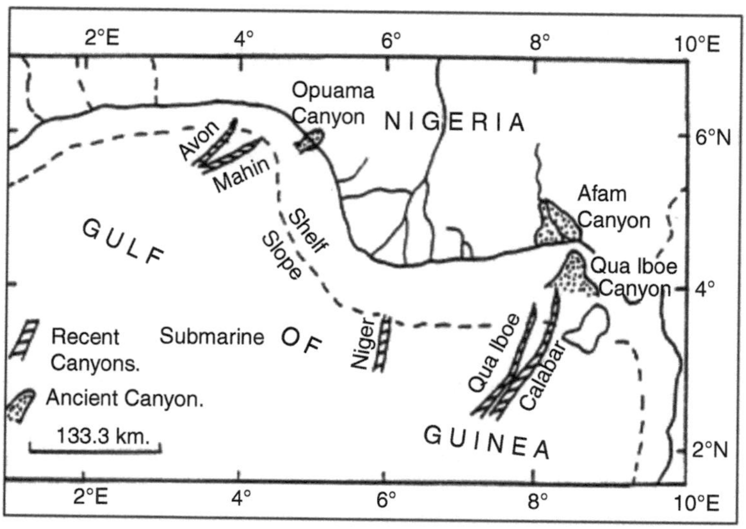

FIGURE 2.10 **Geographical distribution of the ancient and modern canyons.** *Source: After Petters, S.W., 1984. An ancient submarine canyon in the Oligocene–Miocene of the Western Niger Delta, Sedimentology 31, 805–810.*

TABLE 2.3 Lithofacies of the Niger Delta

Sediment type		Facies description	Facies code	Facies number	
Sandstone <20% clay		Pebbly sandstone	PS	1	Reservoir
		Medium- to coarse-grained cm-scale cross-bedded sandstone	Sx.m	2	
		Fine-grained mm-scale laminated sandstone	Sl.f	3	
		Fine-grained mm-scale laminated sandstone with clay laminae and drapes	Sic	4	
		Bioturbated sandstone	Sb	5	
Mixed sand and clay (heteroliths) 20–80% clay	Sand-rich heteroliths 20–50% clay	Wavy bedded sandy heterolith	SMw	6	
		Bioturbated sandy heterolith	SMb	7	
		Glauconitic or shelly muddy sandstone	SMshg	8	
	Mud-rich heteroliths 50–80% clay and silt	Lenticular bedded muddy heterolith	MSt	9	Nonreservoir
		Bioturbated muddy heterolith	MSb	10	
Mudstone >80% clay		Bedded or massive mudstone	Mb/Mm	11	
		Sideritic bedded mudstone	Msd	12	

After Reijers, T.J.A., 1995. Reservoir geological core description using standardised lithofacies and their associations in the Tertiary Niger Delta. NAPE Bull. 10, 27–39.

2.2.1.2.2 THE AGBADA FORMATION

This unit constitutes the middle part of the tripartite Niger Delta stratigraphic succession, with the type section designated in the Agbada-2 well from 5760–9500 ft., that is, 3740 ft. (1122 m) drilled about 11 km north of northwest of Port Harcourt. The formation is strongly diachronous, ranging in age from Eocene to present day. The mangrove swamp to coastal barriers and fluviomarine zones of the present-day delta (Fig. 2.6) constitute the surface exposure of Recent age.

Table 2.3 shows 12 lithofacies recognized in the Agbada Formation: 5 sandstone types, 5 heteroliths, and 2 mudstone types, all of which have been organized into 7 lithofacies associations (Reijers, 1995) as follows:

1. Coastal barrier or shoreface;
2. Back barrier (lagoonal delta or crevasse splay);
3. Distributary channel;
4. Lagoonal system;
5. Transgressive sand;
6. Suspension fall-out;
7. Allochthonous lumps.

2.2.1.2.3 THE AKATA FORMATION

The Akata Formation is the major time transgressive lithological unit of the Niger Delta. It is mainly marine mud facies with turbiditic sands and continental slope channel fills. The type section is the Akata-1 well, a wildcat drilled 80 km east of Port Harcourt. The interval of interest is 7,180–11,121 ft. (3,941 ft. or 1,182.3 m). The bottom of well did not reach the base of the formation. The formation underlies the whole of the Niger Delta complex south of the Imo Shale outcrop area and is taken as the updip equivalent of the Akata lithofacies (Whiteman, 1982, p. 131). At present the Akata facies is being deposited on the continental shelf and slope and perhaps on the lower part of the prodelta slope.

The Akata Formation consists of dark gray uniform shales, especially in its upper part. It tends to be sandy in the upper parts from where it grades into the Agbada Formation. The top of the formation is not clearly defined. It is taken arbitrarily as the deepest development of deltaic sandstones assignable to the Agbada Formation. Its base has been reached only in some wells drilled on the flanks of the delta, for example, well Awaizombe-1. Formation thickness ranges from 2,000 ft. (c. 610 m) to 20,000 ft. (6,100 m). From the abundant planktonic foraminifera, the formation is inferred to have been deposited in a relatively deep open marine shelf setting.

2.2.1.2.3.1 PETROGRAPHY OF THE AGBADA SANDSTONES. *Textures*: There is considerable variation in the textural characteristics of the Agbada sandstones, attributable to hydrodynamic processes. The coarsest and most poorly sorted sands are found within the high-energy fluvial and estuarine channels and in proximal submarine canyon channels. Tidal channel sands are generally fine to medium-grained and moderately well sorted. The beach–barrier, tidal mouth bar, and middle to upper shoreface settings are characterized by very fine to fine, well-sorted sands, with strong dominance of fine grains and low frequency of medium to coarse sand or silt and very fine grains. This is clearly due to the intense wave winnowing in the shallow depositional setting. Transgressive sands are usually poorly sorted due to the admixing of sand, coarse pellets, and bioclasts with silt and clay.

Mineralogical Composition (Table 2.4): The mineral components are mainly quartz (64–92%), with subordinate K-feldspar and minor plagioclase. Quartz occurs mainly as monocrystalline grains, with subordinate polycrystalline grains and rare chert fragments. Feldspar is mainly the potash type, plagioclase is minor. It may have been more abundant but has possibly been reduced in frequency by dissolution which probably formed oversized pores. It is more abundant in the finer sandstones, suggesting a strong grain size control. There is an increase in abundance seaward, with the general decrease in grain size associated with the transition from fluvial-dominated in the north to shallow marine dominance in the southern parts. Other components include mud chips, mica, rock fragments, and trace amounts of heavy minerals. Bioclasts and carbonaceous debris are locally abundant. The nondetrital components are the cements, all of diagenetic origin. They include quartz, K-feldspar, pyrite, hematite, siderite, calcite, dolomite, kaolinite, chlorite, illite-smectite, chert, and bitumen.

Diagenesis: The Agbada sandstones generally retain close to depositional porosity of 35%, an indication of a low degree of mechanical compaction. Following studies by Core Laboratories, the presence of locally significant amounts of kaolinite, quartz, and carbonate cements in the sandstones is attributed to the depositional setting and stratigraphic stacking pattern of the sediments. The cement types preserved within the sediment vary as a function of the depositional setting—regressive or transgressive.

TABLE 2.4 Mineral Components and Their Frequencies in the Reservoirs of the Respective Depobelts of the Agbada Formation

Mineral	Quartz (%)	Alkali Feldspar (%)	Plagioclase Feldspar (%)	Calcite (%)	Dolomite (%)	Siderite (%)	Pyrite (%)	Kaolinite (%)	Illite + Smectite + Mica (%)	Chlorite (%)
DEPOBELT										
Northern Delta	74–89	2–13	0–3	0–10	0–1	0–3	0–0.3	1–5	1–7	0–0.4
Greater Ughelli	64–80	5–9	1–5	0–2	0–1	0–8	0–0.3	3–6	5–11	0–1
Central Swamp	75–99	1–7	0–2	0–7	0–1	0–0.3	0–0.1	0–4	0–7	0
Coastal Swamp	71–92	3–10	1–8	0–6	0–1	0–1	0–0.1	0–4	1–5	0–1
Offshore	66–86	5–10	2–5	0–6	0–2	0–5	0–0.3	2–5	4–8	0
Submarine Canyon	74	7	3	0	0–50	2	0–10	4	8	0–50

Image courtesy of Core Laboratories, © 1996. Reservoir properties of Tertiary Niger Delta sediments. Nonproprietary Report, Core Laboratories Sales NV and affiliates.

In the regressive/progradational facies exposed to meteoric flushing there is reduction in ion concentrations, resulting in cementation dominated by quartz and kaolinite cements. Grains liable to dissolution (e.g., aragonite bioclasts) due to instability in meteoric pore waters disappear. This accounts for the occurrence of locally significant quartz and kaolinite cements in the fluvial sandstones and the sandstones of the upper shoreface which cap prograding (regressive) delta sequences. Transgressive/retrogradational deposits progressively become prone to flushing by marine pore waters that are characterized by high ion concentration. Thus early cementation under marine conditions is dominated by grain-coating illite-smectite, chlorite, K-feldspar overgrowths, and an increased potential for localized development of glauconite and chamosite associated with reduced cementation rates. There is also a better preservation of calcareous bioclasts since their mineralogy is more stable under marine conditions and have increased potential for calcite and dolomite cementation, with preserved bioclasts acting as nucleation sites for the carbonate cements. In land-attached deposits, such as mouth bars, mixing of marine and meteoric pore waters is likely and such cements as siderite may be produced.

2.2.1.2.3.2 PHYSICAL SEDIMENTARY STRUCTURES. The physical sedimentary structures of the Agbada Formation, as made out from slabbed cores, include parallel and ripple lamination, cross-stratification, and deformation structures. Among the stratification structures are wave-formed parallel lamination (upper regime plane beds), hummocks and swales, and low- to high-angle cross-bedding. These structures are associated with such high-energy settings as foreshore to beach/barrier subenvironments of the shallow sea. There are also those physical sedimentary structures commonly associated with tidal sedimentation—small-scale clay-draped cross-bedding and heteroliths with bifurcated double mud drapes. Any of the structures discussed earlier could undergo some degree of deformation especially before consolidation of the sediment. Such deformation is commonly in the form of slurrying of the

TABLE 2.5 Checklist of the Trace Fossils Identified in Cores of the Niger Delta

Ichnofacies	Ichnogenera	Characteristics	Paleoenvironmental significance
Skolithos	*Skolithos*[a] *Ophiomorpha*[a] *Paleophycus*[a] *Arenicolites*[a] *Monocraterion* *Diplocraterion*[a] *Thalassinoides*[a] Rooting *Conichnus*[a]	Deeply penetrating dwelling burrows (domichnia) usually more consolidated than the substrate, mostly vertical, cylindrical, or U-shaped	Relatively high levels of wave and current energy, muddy to clean, well-sorted loose substrate, such as on the foreshore and shoreface of beaches and spits, estuarine point bars, and tidal deltas
Glossifungites	*Thalassinoides*[a] *Diplocraterion*[a] *Rhizocorallium*[a] *Arenicolites*[a] *Skolithos*[a]	Vertical cylindrical tubes of suspension feeders	Environmentally wide-ranging, but here found mostly in relatively well lithified mudstone layers within friable sandstone. Generally indicative of bounding discontinuities
Cruziana	*Rhizocorallium*[a] *Teichichnus* *Ophiomorpha*[a] *Asterosoma* *Thalassinoides*[a] *Arenicolites*[a] *Planolites*[a] *Phycosiphon* *Chondrites*	Mostly horizontal, with a few vertical to steeply inclined burrows of deposit feeding and grazing tracemakers	Low energy setting

a *Member of more than one ichnoassociation.*

sediment or the distortion of its strata, commonly found in deep submarine canyons where steep slopes induce sediment remobilization.

2.2.1.2.3.3 ***BIOGENIC SEDIMENTARY STRUCTURES.*** Inspection of slabbed cores showed up various biogenic sedimentary structures in the Agbada sandstones. The trace fossils tend to occur mainly in three lithofacies: fine-grained mm-scale laminated sandstone, medium- to coarse-grained cm-scale cross-bedded sandstone, and fine-grained mm-scale laminated sandstone with clay laminae and drapes. The structures mostly belong to only 3 ichnofacies within which 15 ichnogenera were identified (Table 2.5). *Paleophycus*, *Ophiomorpha*, and *Planolites* ichnospecies are particularly abundant. The ubiquitous presence of *Ophiomorpha* in particular, and the dominance of burrows of the *Skolithos* ichnofacies in general, may have improved the connectivity of the Niger Delta reservoirs beyond the initial depositional value.

2.2.1.2.3.4 ***PETROPHYSICAL IMPLICATIONS OF THE ICHNOFABRICS.*** Bioturbation is known to dilate sediment, altering the initial porosity and determining its ultimate value and it is likely that burrows have had both negative and positive effects on reservoir connectivity. For instance, the *Glossifungites* ichnofacies occurs in such a way that the burrows tend to cut through the mudrock firmgrounds thereby connecting two separate vertically disposed sand strata. Additionally, vertically disposed burrows, especially the ichnogenera of the *Skolithos* ichnofacies are all potentially beneficial, being in a position to cut through clay drapes and flasers.

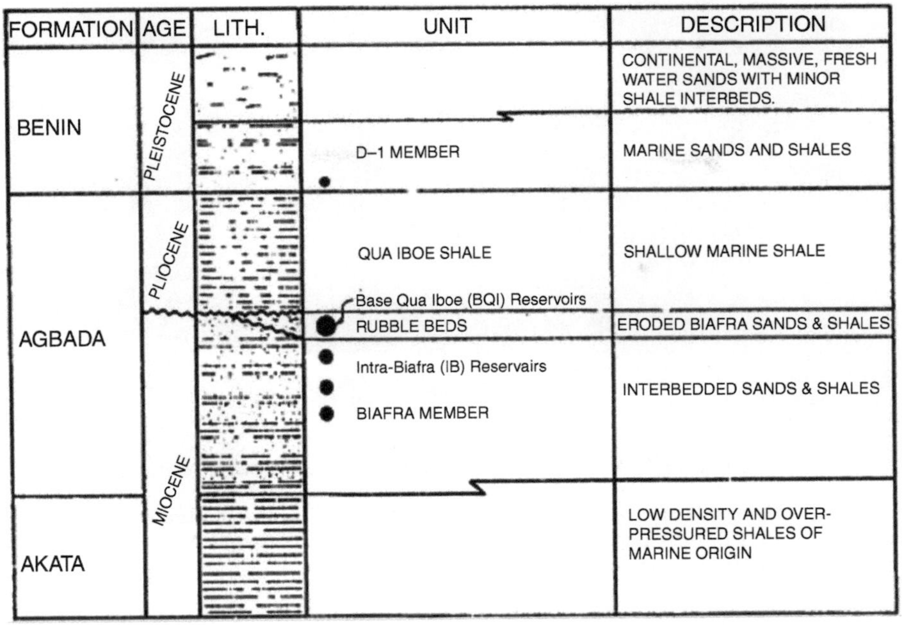

FIGURE 2.11 **Generalized stratigraphy of the Qua Iboe collapse system in southeastern Nigeria offshore.** *Source: After Opara, G.E., 1981. A review of the Tertiary Unam "B" Field in OML 67, Niger Delta. J. Min. Geol. 18, 86–95.*

2.2.1.2.3.5 MEMBERS OF THE AGBADA FORMATION. Several canyons, for example, Osare, Elelenwa, Buguma, Soku, Opuama, and the Qua Iboe collapse were incised into the Agbada Formation. The last named is remarkable given its areal extent, multiplicity of incisional events, mode of origin, and total thickness of fill. It is a stratigraphic-cum-structural feature located about 30 km in southeastern Nigeria offshore (Figs. 2.11 and 2.12) in up to over 27 m of water. The Agbada Formation here consists of three members—Biafra, Rubble Beds, and Qua Iboe. The succession originated in the overall Tertiary progradation of the Niger Delta, and the north to south diachronous stratigraphy of the delta obtains here. The Rubble Beds and the Qua Iboe members have been related to a dislocation that affected the preexisting Biafra deltaic deposits, an essentially tectonic process informally termed the "Rubble Beds Event" (Kreisa et al., 1999). The stratigraphy of the collapse system has been summarized both pictorially and synoptically as shown in Fig. 2.13.

2.2.1.2.3.6 STRATIGRAPHIC EVOLUTION OF THE AGBADA FORMATION. Beginning from the Latest Paleocene, the delta built out initially over a rifted seaward-dipping continental margin. Since the Middle Miocene, it built out over the landward-dipping oceanic lithosphere, following the stages that were meticulously worked out by Evamy et al. (1978) (Fig. 2.14).

2.2.1.2.3.7 SEDIMENTOLOGICAL EVOLUTION OF THE AGBADA. Sedimentation in the modern Niger Delta is dominantly an offlap process that produced a large spectrum of depositional facies and environments. The dominant depositional setting has been a wave–tide mix. This is in keeping with the general observation that sand preferentially accumulates in

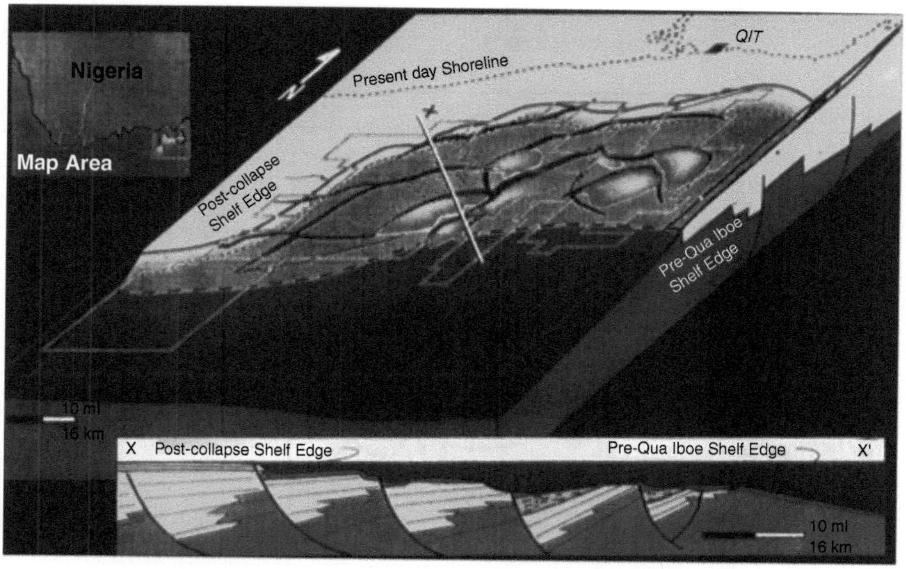

FIGURE 2.12 **Qua Iboe collapse system of SE Nigeria offshore.** QIT, Qua Iboe Terminal. *Source: After Unomah, G.I., Akinnagbe, T.A., Edet, J.T., 1999. Lithostratigraphy of the Qua Iboe collapse system (Abstr.). Niger Delta Stratigraphic Committee Workshop, Ijebu Ode, October, 1999.*

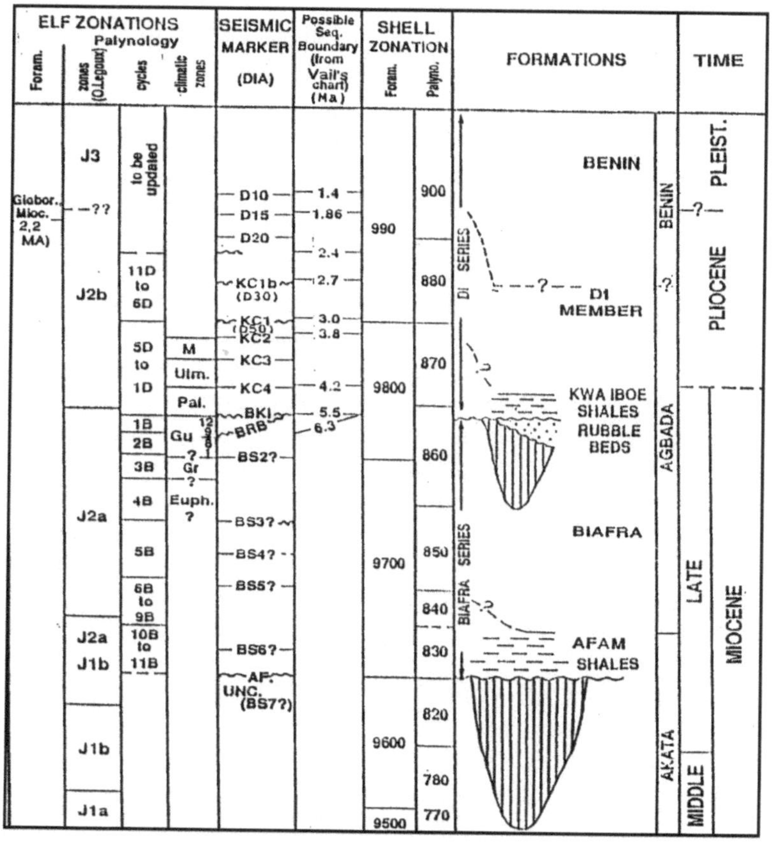

FIGURE 2.13 **Stratigraphic synopsis of the Qua Iboe collapse system.** *Source: Cited by Unomah, G.I., Akinnagbe, T.A., Edet, J.T., 1999. Lithostratigraphy of the Qua Iboe collapse system (Abstr.). Niger Delta Stratigraphic Committee Workshop, Ijebu Ode, October, 1999.*

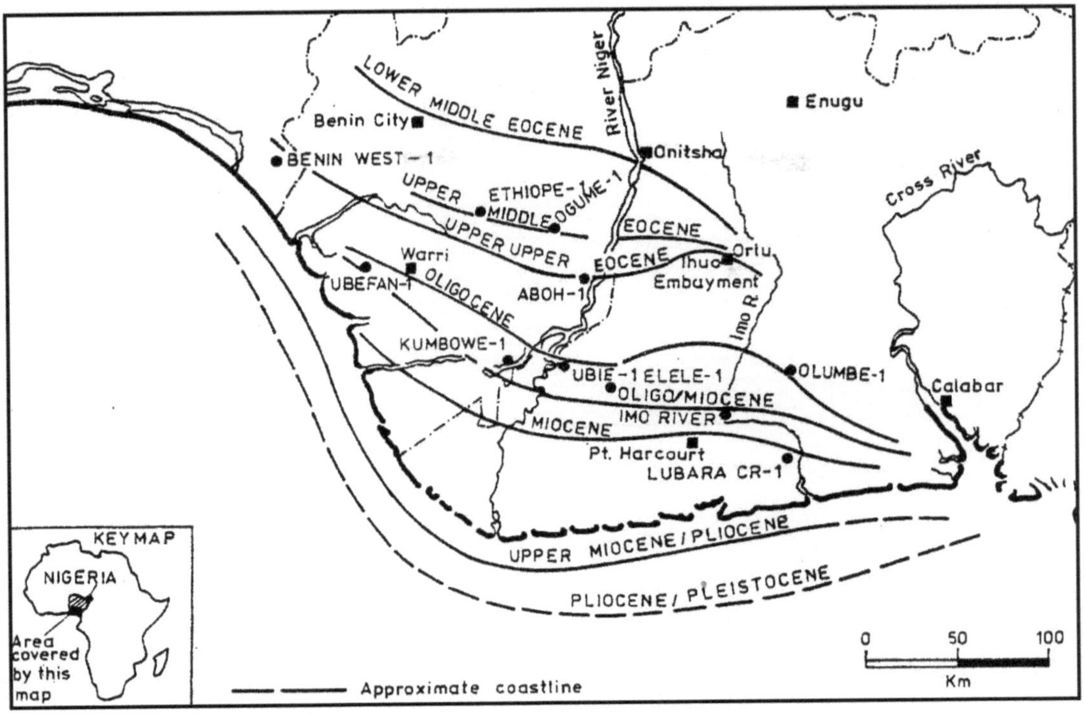

FIGURE 2.14 **Stages in the growth of the Cenozoic Niger Delta.** *Source: After Evamy, B.D., Haremboure, J., Kamerling, P., Knaap, W.A., Molloy, F.A., Rowlands, P.H., 1978. Hydrocarbon habitat of Tertiary Niger Delta. AAPG Bull. 62, 1–39.*

areas characterized by high kinetic energy, that is, in environments characterized by rapidly moving fluids, such as beaches, marine sand bars, river channels, and alluvial and submarine fans.

As established by Weber (1971) using the Uzere W-16 well and the F-1 sands of the Bomu Field (Fig. 2.15), the typical Agbada offlap rhythm typically consists of five lithologic units as presented by Whiteman (1982, p. 124). A typical offlap rhythm consists of five components separated by onlap sands as follows:

A	Transgressive marine sand (onlap)
E	Fluviatile backswamp and lagoonal sediments (offlap)
D	Barrier bar and barrier foot deposits (offlap)
C	Laminated fluviomarine sediments (offlap)
B	Marine shale (offlap)
A	Transgressive marine sand (onlap)

The construction of the rhythmic successions has been aided by the interplay of the rate of deposition (R_d) and the rate of tectonic subsidence (R_s). This interplay determines the mode of delta development as follows:

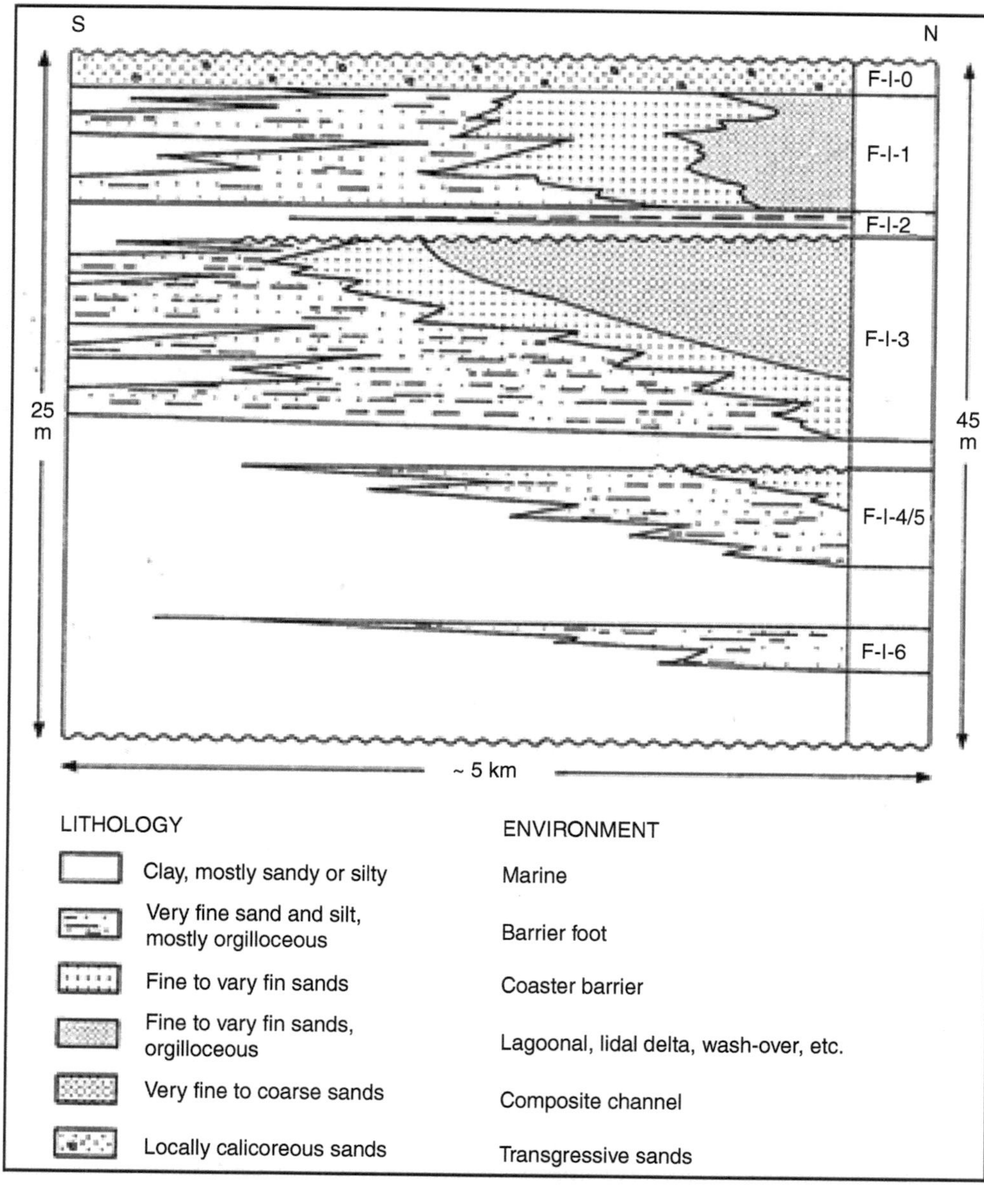

FIGURE 2.15 **A succession from Shell's Bomu field illustrating the rhythmic construction of the Agbada facies.** *Source: After Weber, K.J., 1971. Sedimentological Aspects of Oil Fields in the Niger Delta. Geologie en Mijnbouw 50, 59–67.*

1. $R_d > R_s$: delta advances, that is, builds out-progradational
2. $R_d = R_s$: delta apparently remains stationary, but actually increases in thickness, that is, builds upaggradational
3. $R_d < R_s$: delta retreats, that is, backsteps or retrogrades

These scenarios dictated whether a given depobelt would continue to receive sediments or a new depobelt would be created.

2.3 NIGER DELTA STRUCTURE AND TECTONICS

2.3.1 Introduction

The Niger Delta Basin lithic fill is considered to be self-contained in terms of its structural geology, since the N–S and NW–SE trends of the underlying basement have not controlled the arcuate, shoreline-parallel depobelt configuration dominating the delta tectonics. The penecontemporaneous deformation of the deltaic sediments has been attributed to rapid sedimentation of the Agbada facies over undercompacted and mobile shales of the Akata Formation (Short and Stauble, 1967; Evamy et al., 1978; Onyia et al., 2002). There is seismic evidence that deeper sections are not part of the ductile mass and remain undeformed. They appear to be structurally stronger older cemented clastics, carbonates, or a mix of the two. From gravity and magnetic data, it has been inferred that the underlying basement ranges from continental to oceanic crust, with abrupt relief features that are either oceanic fracture zones or due to brittle faulting associated with postrifting deformation or both (Bowen et al., 1994). Gravity and magnetic studies by Ako et al. (2004) is fairly conclusive about the structure of the underlying crystalline basement. Northeast–southwest trending lineaments running from deep offshore to onshore of the delta have been shown to be the Chain and Charcot Fracture Zones which are responsible for the horst-and-graben configuration that defines the structural highs and lows. Some of the lows have sedimentary thicknesses in excess of 14 km and have been found to correlate positively with significant hydrocarbon discoveries, especially in the deep and ultradeep offshore.

The sedimentary prism of the Niger Delta is described as a series of offlap cycles consisting of fluviomarine systems that have succeeded one another in a stepwise fashion. This reflects the basinward progradation, extending from the continental edge into the oceanic basement, in a manner likened to an escalator movement by Knox and Omatsola (1987). The delta is composed of megaunits (=depobelts), which are also characterized as self-contained entities especially with respect to stratigraphy, structure-building, and hydrocarbon distribution. In each of these entities, the succession starting from marine clays and ending with coarse and gravelly sands is observed not only vertically but also laterally basinward. Layer-cake stratigraphy is therefore ruled out across the Niger Delta. Profuse faulting precludes easy lithologic or chronocorrelation within the depobelts'.

2.3.2 Growth Faults

Growth faults are the most common structural feature observed on seismic sections of the Niger Delta. Faulting is principally syndepositional and appears to have affected mostly the paralic succession. Following Evamy et al. (1978), the synsedimentary faults of the Niger Delta (Fig. 2.16) are classified as structure-building growth faults, crestal faults, flank faults, counter-regional faults, and antithetic faults.

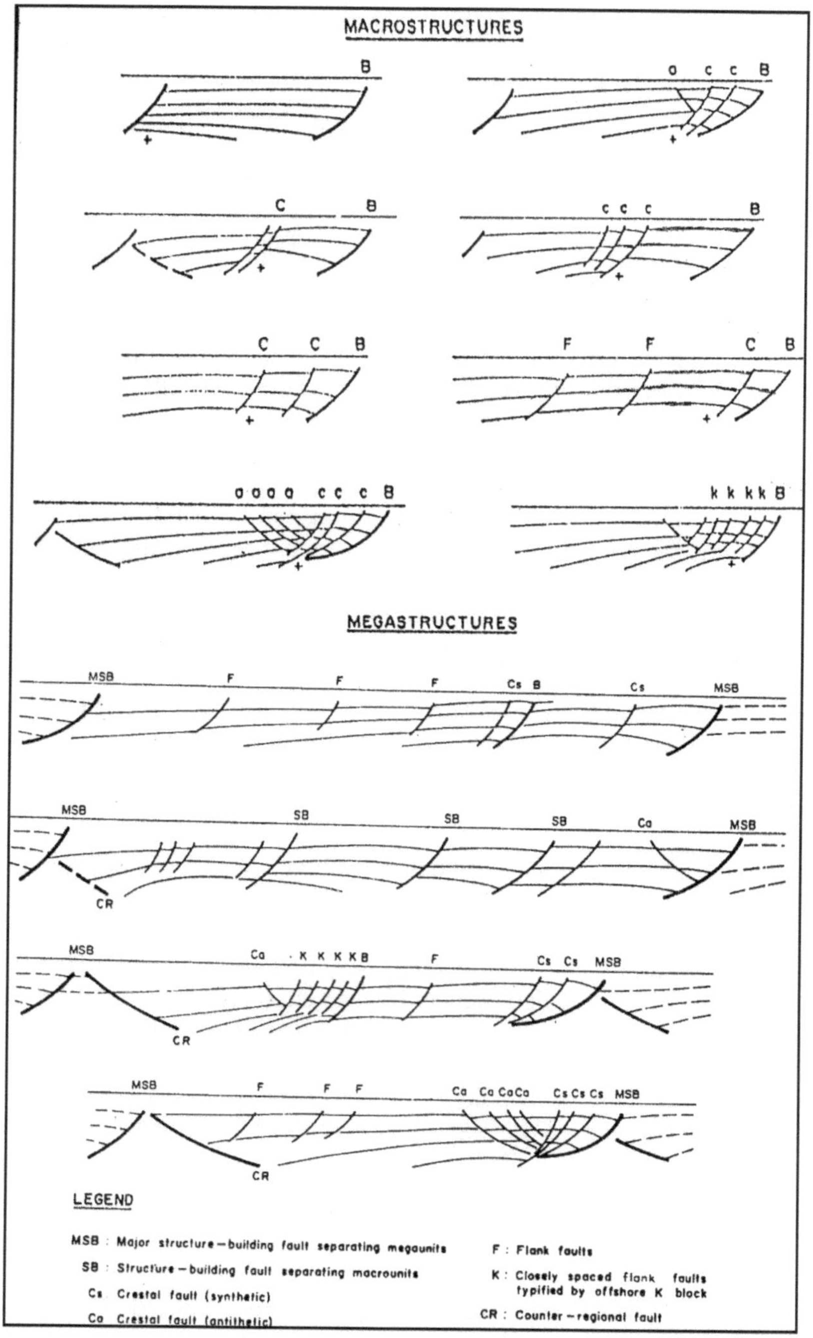

FIGURE 2.16 **Megaunits and associated sedimentary fault types in the Niger Delta.** *Source: After Evamy, B.D., Haremboure, J., Kamerling, P., Knaap, W.A., Molloy, F.A., Rowlands, P.H., 1978. Hydrocarbon habitat of Tertiary Niger Delta. AAPG Bull. 62, 1–39.*

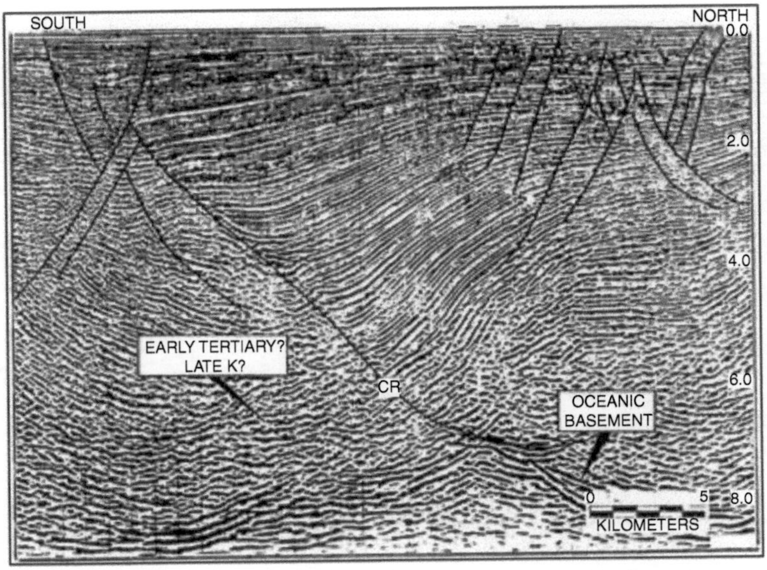

FIGURE 2.17 **A N–S section showing the general structural style in the Niger Delta.** *Source: After Bowen, B.E., Hall, D.J., Rossen, R.N., Shaffer, B.L., 1994. Sequence stratigraphic and structural framework, southeast Niger Delta shelf. NAPE Bull. 9, 51–58.*

Counter-regional growth faults (Fig. 2.17) dip landward and occur at the distal, that is, basinward parts of monoclinal flanks. They are often formed at the points of lithology change (sand to shale), and may define the basinward limit of some depobelts. They characterize the distal parts of the delta, where they separate dipping flanks from shale bulges. Antithetic faults also dip landward but are smaller in scale, of secondary structural importance, and display no growth. They are actually secondary expansion faults whose main function is some kind of compensation for overburden extension.

2.3.3 Field-Scale Structural Styles

Structural styles observable on field scale are extensional and squeeze-flow types. The former type includes rollover anticlines, collapsed crest features, and down-to-the-basin faults.

Collapsed Crest Structures (Fig. 2.18) are composed of sets of normal faults forming a graben in a rollover anticline. The faults are located in the area of maximum bed curvature, are closely spaced, and have relatively small throws. Fault spacing and throw decrease toward the center of the crest of the rollover. Crestal collapse faults do not reach basal detachment. They accommodate local strains related to bending. They are symmetrical if the rollover anticlines have resulted from bed rotations due to coeval movements on down-to-the-basin and counter-regional faults.

Squeeze Flow Structures include the counter-regionals and their derivative—the back-to-back structure (Fig. 2.19). The former type forms the seaward boundary of depobelts. The sedimentary series thickens toward it and bed rotation increases with depth. Counter-regionals tend to flatten at depth, but not enough to allow bed-parallel slip to take place as in down-to-the-basin faults.

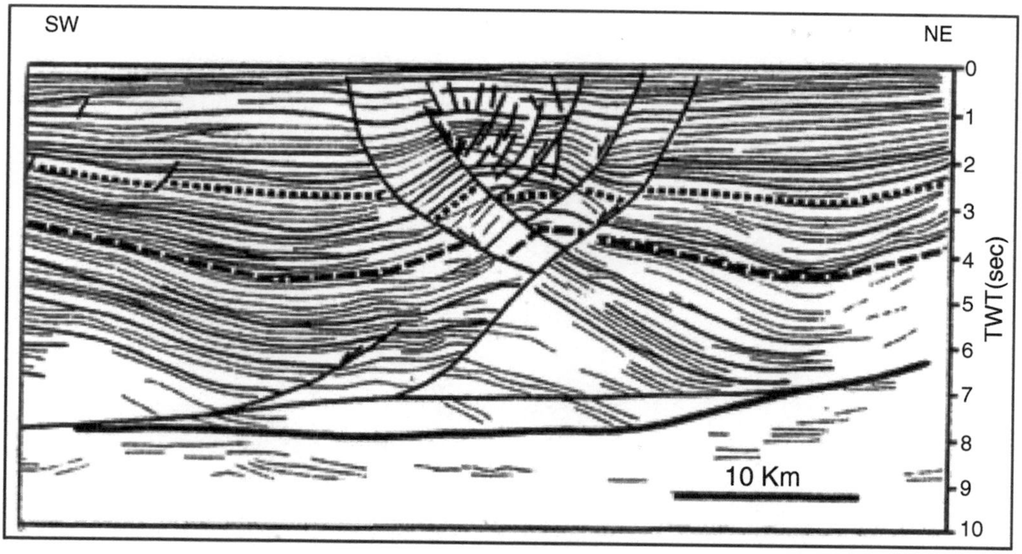

FIGURE 2.18 **A Crestal collapse system in the offshore Niger Delta.** *Source: After Ajakaiye, D.E., Bally, A.W., 2002. Course Manual and Atlas of Structural Styles on Reflection Profiles From the Niger Delta. AAPG Continuing Education Course Note Series No. 41, p. 107.*

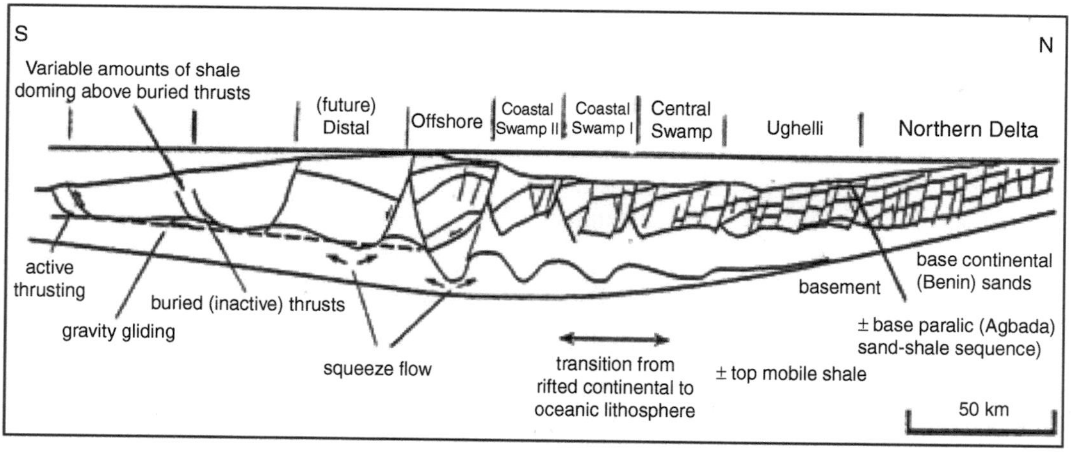

FIGURE 2.19 **A N–S section showing the basinward succession of structural styles including depobelts, squeeze flow structures, down-to-the-basin faults, and thrusts.** *Source: After Knox, G.J., Omatsola, E.M., 1987. Development of the Cenozoic Niger Delta in terms of the "Escalator Regression" model and impact on hydrocarbon distribution. Proceedings of the KNGMG Symposium: Coastal Lowlands, Geology and Geotechnology. Kluwer Academic Publishers, The Netherlands, pp. 181–202.*

Major counter-regional faults with high fault displacement rates tend to control sedimentation in shelf-edge position for periods up to 5 My. Once sediment supply surpasses structural subsidence, extension shifts basinward and a new down-to-the-basin fault forms. This results in back-to-back structures, some of which are associated with mobile shales.

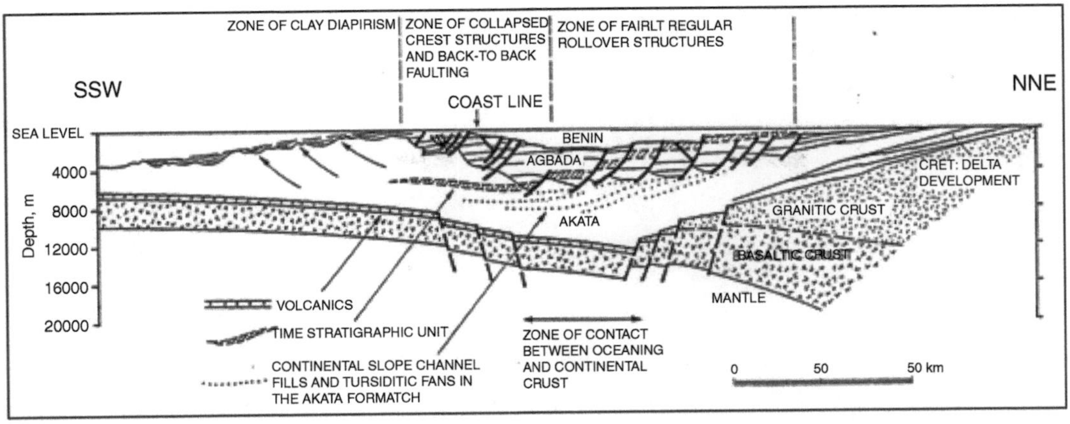

FIGURE 2.20 **Gross stratigraphy and structural provinces of the Niger Delta.** *Source: After Doust, H., Omatsola, M.E., 1989. Niger Delta. In: Edwards, J.D., Santogrossi, P.A. (Eds.), Divergent/Passive Margin Basins. AAPG Memoir, vol. 48, pp. 201–238.*

2.3.4 Regional Structural Styles

The regional structural styles of the Niger Delta may be grouped into two—depobelts (Fig. 2.19) and structural provinces (Fig. 2.20). The structural provinces are zones of regular seaward-dipping rollover structures, collapsed crest, back-to-back structures, and zone of clay diapirism.

2.3.4.1 *The Depobelts*

The subdivision of the entire Niger Delta into depobelts is based on the recognition of major regional down-to-the-basin faults. The delta is subdivided into six depobelts (Knox and Omatsola, 1987) (Fig. 2.18) which represent the successive phases of delta growth. It is noteworthy that while the proximal depobelts/structural cells exhibit extensional stress system, the distal units exhibit transitional to compressional tectonics, which is present as toe thrusts (Figs. 2.21 and 2.22).

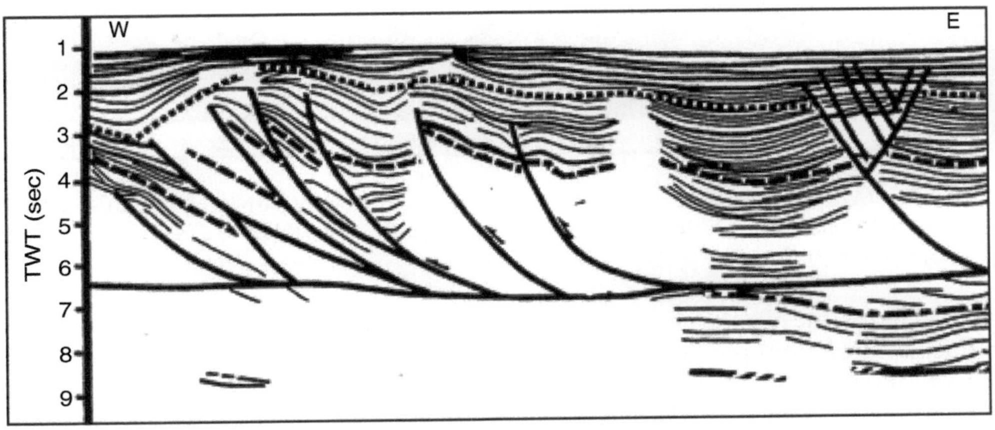

FIGURE 2.21 **East–West section in the southeastern offshore of the Niger Delta illustrating compressional lateral ramp system.** *Source: After Ajakaiye, D.E., Bally, A.W., 2002. Course Manual and Atlas of Structural Styles on Reflection Profiles From the Niger Delta. AAPG Continuing Education Course Note Series No. 41, p. 107.*

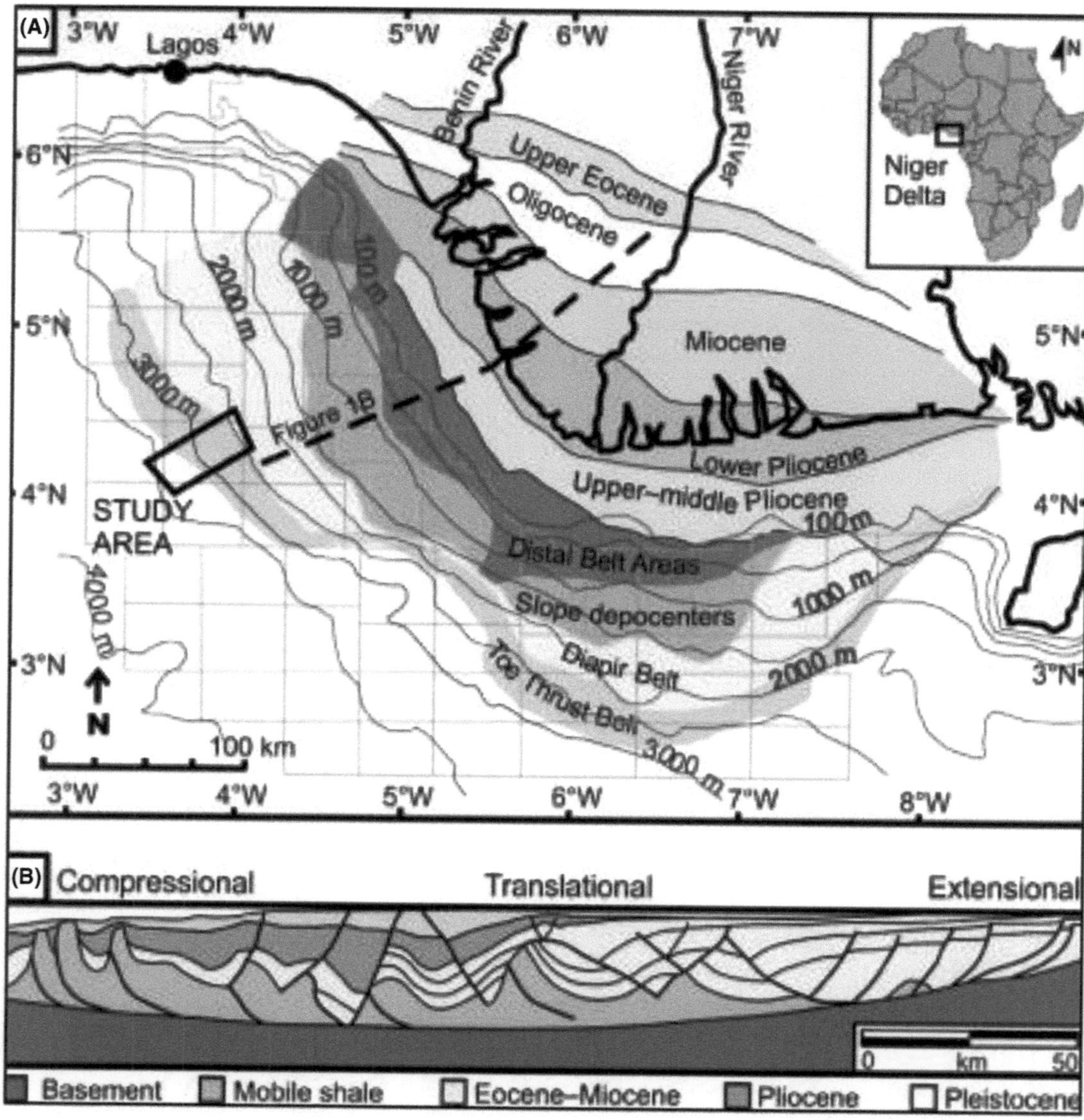

FIGURE 2.22 (A) Map showing the compressional toe-of-slope thrust belt of the Niger Delta. (B) Simplified schematic cross section from delta top to deep water. *Source: After Heinio, P., Davies, R.J., 2006. Degradation of compressional fold belts: deep-water Niger Delta. AAPG Bull. 90, 753–770.*

In broad terms, the Niger Delta shelf developed as a prograding extensional complex overlying a ductile substratum probably composed largely of overpressured marine shales. There is clearly a relationship between structure and sedimentation pattern. Rapid deposition of sands would not allow complete dewatering of the underlying Akata Shales, resulting in easy mobility of the shales and reduction in density. The consequent density inversion results in upward shale flow (shale bulge) causing an expansion (growth) fault system in the overlying strata. Such faults tend to line up parallel to the edge of the continental slope and to displace sediments in a downdip direction. Since much of the resultant fault movement is syndepositional, a considerably thicker sedimentary succession accumulates on the downthrown side of the growth fault.

A major feature of these growth faults is that the movement is not entirely translational because there is a rotational aspect to it. This means that domal structures originate in the layer cake sand and shale producing rollover anticlines themselves often internally faulted. The intensity of deformation decreases upward in the section and is usually not reflected in the Benin Formation. Back-to-back structures ("tepee" structures of Bowen et al., 1994) characterize both the onshore (from central swamp depobelt basinward) and the shelf extensional systems. In these major extensional systems, landward-dipping faults intersect seaward-dipping systems in complex interlocking fault networks. The counter-regionals may have originated as growth faults but with slip surfaces dipping away from each other. Sediment loading and consequent subsidence cause an upward propagation of the arcuate slip surface until the two intersect at some point near the sediment/water interface. Subsequent relative motion, that is, further growth, continues on only one of them while the other is cut off.

2.3.4.2 *Structural Trends in the Deep Offshore*

Omatsola and Adegoke (1982) and Ako et al. (2004) showed that the basement has a horst-and-graben configuration and that the structural lows correlate with significant hydrocarbon discoveries. The structural configuration within the sedimentary succession in the deep and ultradeep offshore is controlled largely by gravity-driven systems of linked extensional growth faults and compressional toe thrusts initiated in the Paleogene when the modern Niger Delta began to form. As illustrated by Ojo (1996), the present shelf edge is characterized by extensive counter-regional faults (Fig. 2.23). Further offshore there are back-to-back faults,

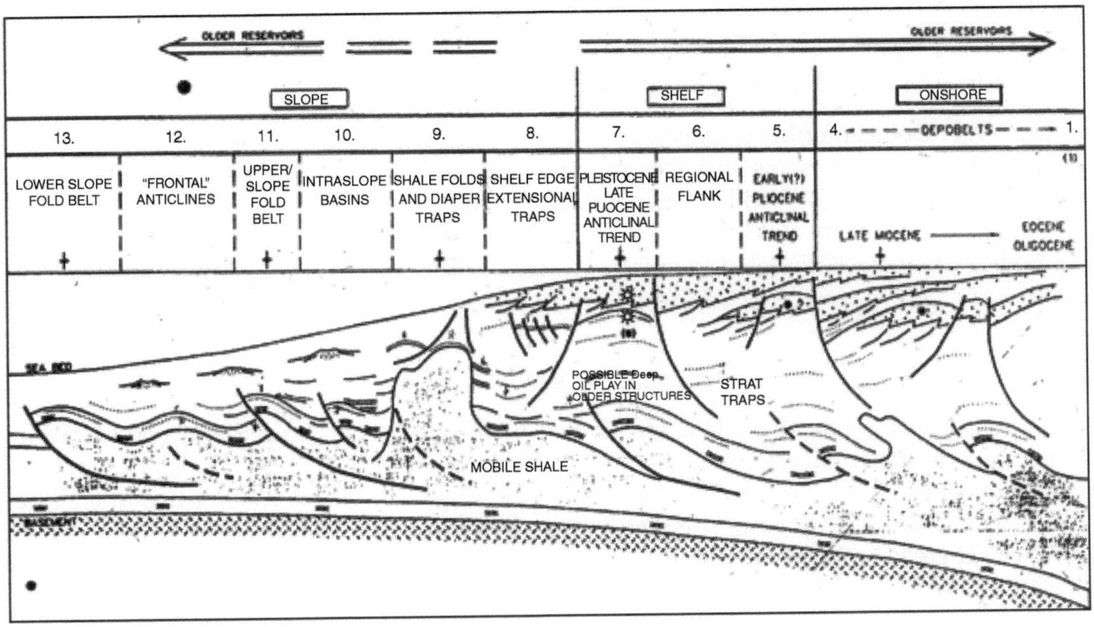

FIGURE 2.23 **Schematic longitudinal profile of the Niger Delta showing the differing structural features from onshore to deep offshore zones.** *Source: After Ojo, O.A., 1996. Pre-drill prospect evaluation in deepwater Nigeria. NAPE Bull. 11, 15–22.*

with a belt of short listric extensional faults along the shelf edge and upper slope. Further down the slope, there are gentle folds with thrusts and clay diapirs in places "outcropping" on the seabed.

2.4 THE HYDROCARBON SYSTEMS

2.4.1 Introduction

The search for oil began in Nigeria in 1908, inspired by the occurrence of tar sand seeps in the Cretaceous rocks of the Okitipupa area. Drillings around Lekki, about 90 km east of Lagos, targeting downdip Cretaceous rocks, found no hydrocarbons. Oil seeps also noted in some Cretaceous outcrops in southeastern Nigeria formed the basis for extensive reconnaissance geological and geophysical surveys by the Shell D'Arcy Exploration Company beginning in 1937. Initial discoveries were gas fields in Cretaceous rocks. The first commercial oil discovery within Tertiary rocks was made at Oloibiri, Bayelsa State.

The Niger Delta ranks as one of the major hydrocarbon provinces of the world, with ultimate recovery presently estimated at close to 40 billion barrels. This accounts for nearly 70% of the overall reserves of sub-Saharan Africa. Gas is gradually assuming greater importance and may eventually equal or even outstrip oil as a revenue source, with estimated reserves in excess of 40 trillion cubic feet. This oil province (Fig. 2.24) is just over 100,000 km^2, and constitutes a relatively small area when compared with most other petroleum provinces of the world. The individual oil fields are also relatively small in size, reservoir thickness, and volume; the larger ones being about 40 km^2 in area and 20 m thick. Some of these have 800–1000 million barrels of recoverable oil.

2.4.2 Origins of the Niger Delta Hydrocarbons

The Niger Delta completely satisfies the definition of a hydrocarbon system as an "oil and gas machine" that generates and accumulates hydrocarbons within a basin over geological time, since hydrocarbons had been generated (and are still being generated?), expelled from the source rocks (shales of Akata and Agbada Formations), transferred along faults and carrier beds, and had accumulated in mostly structural traps and subordinately in stratigraphic traps.

2.4.3 Distribution of the Hydrocarbons

On the basis of geochemical data, Haack et al. (2000) recognized three distinct hydrocarbon systems in southern Nigeria, onshore and offshore: Lower Cretaceous, Upper Cretaceous–Lower Paleocene, and Tertiary. The Tertiary system (Fig. 2.25A–B) is smaller than the Upper Cretaceous–Paleocene system. The distribution of hydrocarbons in the Niger Delta is very complex. There is however, a tendency for more gas to occur southward in a given depobelt. The following statements broadly describe the hydrocarbon distribution:

1. The central, extreme east, and extreme north locations of the delta are characterized by high to very high gas–oil ratios.

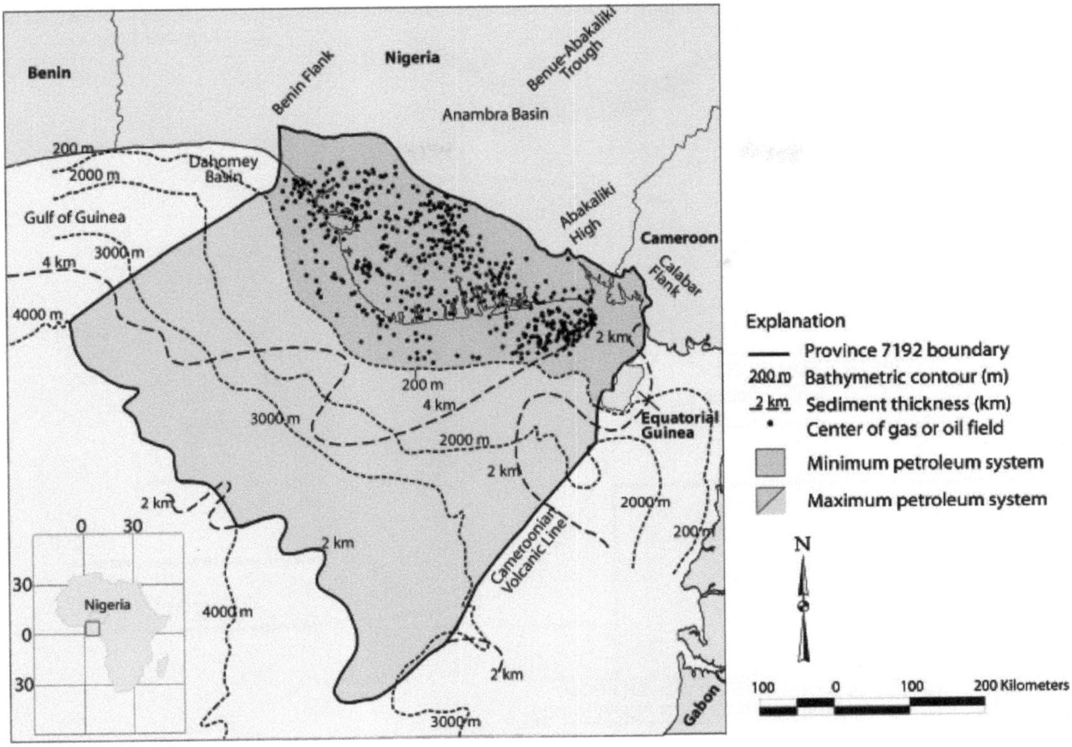

FIGURE 2.24 **General outline of the Niger Delta Petroleum Province, including shallow and deep offshore.** *Source: From USGS Open File Rpt No. 99-501.*

2. Most of the larger oil accumulations occur in either an arcuate belt extending across the delta from the northwest offshore acreage to the southwest offshore in the east, or along a number of more linear trends in the coastal swamp (Fig. 2.26).
3. Prolific north–south trends are present in the Port Harcourt and Egbema areas.
4. In individual depobelts, the gas-to-oil ratio increases seaward and along strike away from depocenters. Each depobelt thus appears to constitute a separate hydrocarbon province. This is probably due to higher shale percentages in the distal parts of depobelts, and the increased representation of marine environments that they imply.
5. Most of the commercial hydrocarbon accumulations are in the structurally highest part of any given macrostructure, despite viable hydrocarbon containment conditions down the structure.

It has been noted that gas is found mainly within the sands of transgressive systems tracts while oil tends to accumulate more in the reservoirs formed in the lowstand systems tracts, with reservoirs of the highstand systems tracts showing fairly equal volumes of oil and gas (Ozumba, 1999). Middle Miocene sequences are more gas-prone while the Middle to Late

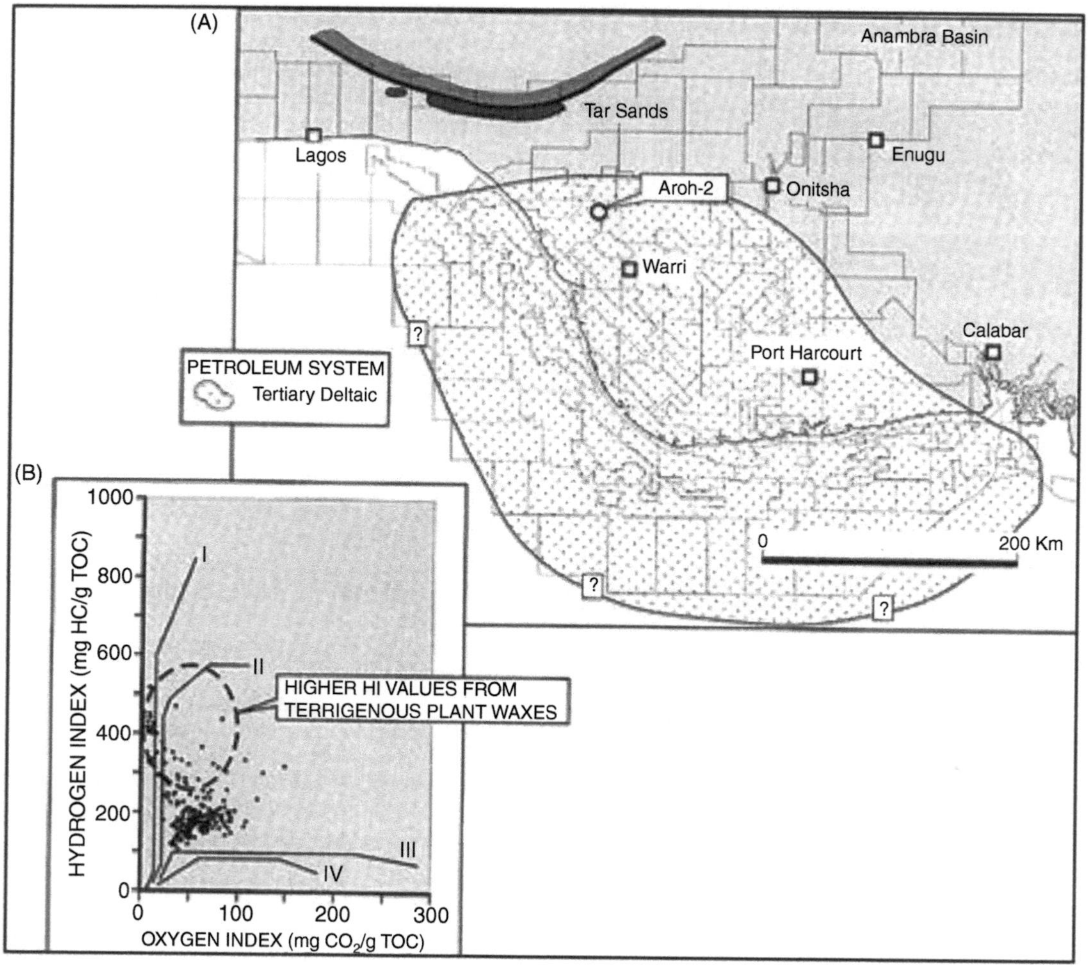

FIGURE 2.25 (A) Map of the Tertiary (deltaic) hydrocarbon system. (B) Hydrogen index–oxygen index (HI–OI) plot showing the higher HI values characterizing the richer source rock section. *Source: After Haack, R.C., Sunderarraman, P., Diedjomahor, J.O., Xiao, H., Gant, N.J., May, E.D., Kelsch, K., 2000. Niger Delta Petroleum Systems, Nigeria. In: Mello, M.R., Katz, B.J. (Eds.), Petroleum Systems of South Atlantic Margins. AAPG Memoir No. 73, pp. 213–231.*

Miocene sequences are more oil-prone probably because early generated oil was flushed updip toward the flanks by gas generated later within the deeper parts of the basin.

2.4.4 Properties of the Niger Delta Hydrocarbons

The Niger Delta hydrocarbons occur as both liquid and gas, the two often closely associated. The oils are commonly of light waxy type in deltas. Geochemical analyses suggest that the oils belong to one family, although heavier bacterially degraded oils are found at shallow depths where the formation temperature is less than 80°C.

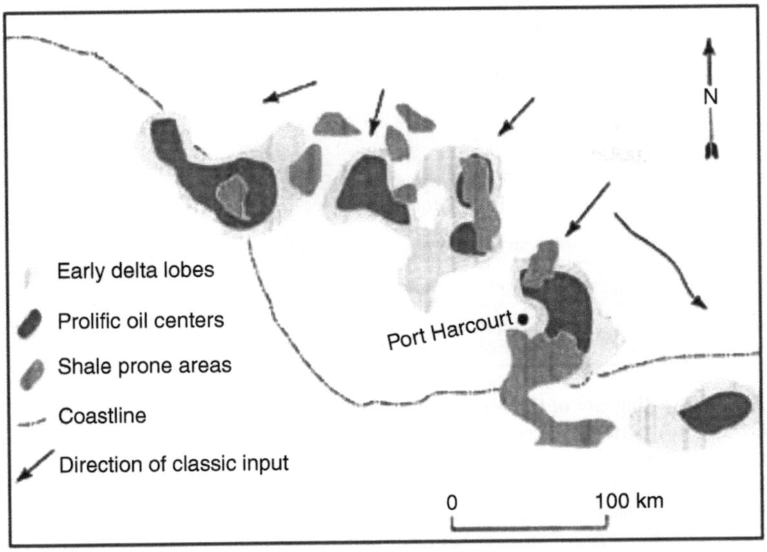

FIGURE 2.26 **Directions of clastic input and locations of early Niger Delta prolific oil centers and shale-prone areas.** *Source: From USGS Open File Rept. No. 99-501.*

The light oils (e.g., Bonny Light) are paraffinic (belonging to the homologous series of saturated aliphatic hydrocarbons with formula C_nH_{2n+2}, e.g., methane) and waxy (i.e., composed of high molecular weight hydrocarbons). Pour points range from 20 to 90°F (i.e., −6.7 to 32.2°C) and viscosities are less than 12 cP.

The degraded crudes have lower gravities, about 26° API, and are naphthenic and non-waxy, with pour points lower than −13°F (−25°C), and viscosities from 12 to 120 cP. The sulfur content is of the order of 0.1%, that is, it is quite low. The heavier crude type occurs more commonly in the cooler parts of the delta. The lighter ones, with variable pour points above 20°F, occur where subsurface temperatures are relatively higher within the same depth slice. This is attributable to the fact that bacteria would thrive only below a certain temperature, and is a confirmation that the heavier crudes originated by bacterial transformation during which paraffinic compounds were destroyed.

2.4.5 Source Rocks

Geochemical analyses indicate that the hydrocarbons of the Niger Delta belong to one family and were derived from land plants and structureless organic matter (essentially the gas-prone Type III) in rocks of Late Eocene to Pliocene age, deposited in outer neritic to bathyal setting (Fig. 2.27) (Onyia et al., 2002). Variation in depositional environment from oxic to suboxic was a major determinant of the abundance and distribution of the organic matter. Although considered poor in quality, the abundance of the source rocks has made up for this, coupled with the excellent migration routes provided by the interbedded permeable sands, and the relatively rapid rate of maturation (Bustin, 1988).

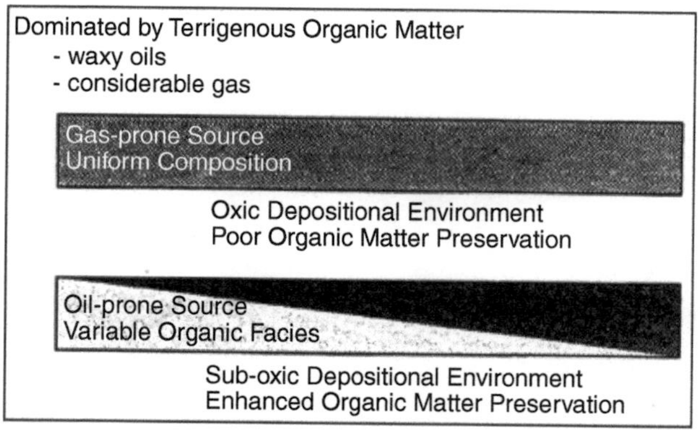

FIGURE 2.27 **Schematic diagram of the Niger Delta source rock model.** Note: Niger Delta source rocks, dominated by terrigenous organic matter, generated considerable amounts of gas and waxy oils. Oil-prone source rocks were deposited in suboxic depositional environments where hydrogenicity was preserved. Gas-prone source facies are compositionally more uniform due to degradation of organic matter in oxic environments. *Source: After Haack, R.C., Sunderarraman, P., Diedjomahor, J.O., Xiao, H., Gant, N.J., May, E.D., Kelsch, K., 2000. Niger Delta Petroleum Systems, Nigeria. In: Mello, M.R., Katz, B.J. (Eds.), Petroleum Systems of South Atlantic Margins. AAPG Memoir No. 73, pp. 213–231.*

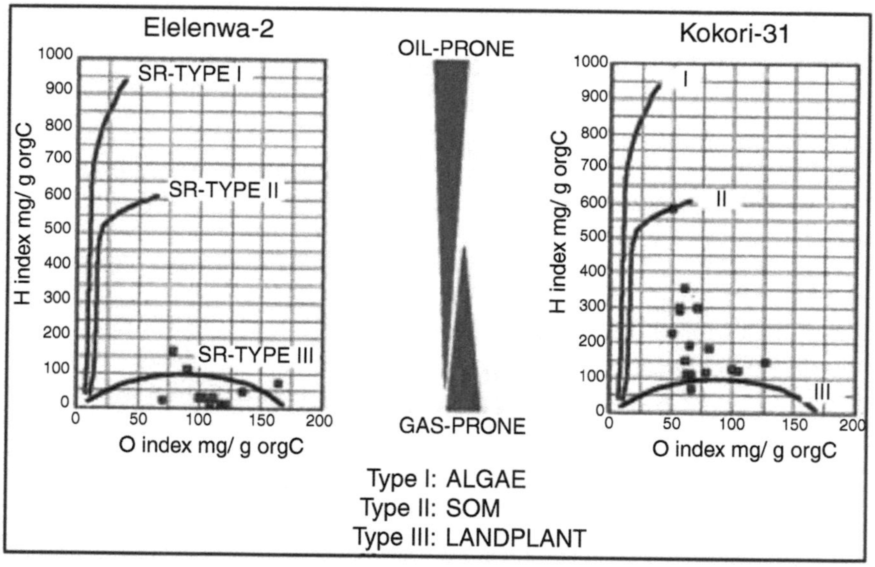

FIGURE 2.28 **Type of source rocks of the Niger Delta illustrated using samples from wells Elelenwa-2 and Kokori-31.** *Source: After Stacher, P., 1995. Present understanding of the Niger Delta hydrocarbon habitat. In: Oti, M.N., Postma, G., (Eds.), Geology of Deltas. A.A. Balkema Publishers, Rotterdam, pp. 257–267.*

Plots of the oxygen index (OI) versus hydrogen index (HI) show that most of the Niger Delta samples belong to Types II and III kerogen (Fig. 2.28). Type I character is displayed by coal and coaly shale samples from the Northern Delta.

The abundance and type of organic matter in the Niger Delta vary with age, depositional environment, and environmental depth zones. There is a decline in mean total organic carbon content from Late Eocene (2.2%) to Pliocene (0.90%) and a concurrent decrease in the HI value. These decreases are attributed to the deposition of terrestrial organic matter with generally higher sedimentation rates and increased clastic reworking in the distal parts of the delta.

The exact location and identity of the source rocks of the hydrocarbons—Akata versus Agbada shales have been controversial issues. It has been strongly suggested that the most probable source material would be located in the transition zone between the upper parts of the Akata Shales and the lower, more shaly and perhaps more mature part of the Agbada Formation. It is noteworthy that the characteristics of the Akata and those of the Agbada Shales from the transitional zone do not differ significantly. However, it is suggested that the Agbada Shales were mainly a source of gas.

2.4.6 Hydrocarbon Generation, Migration, and Containment

The burial of source rocks in the Niger Delta is modeled from the premise that oil-prone source rocks were overlain by gas-prone source rocks. Haack et al. (2000) modeled how oil-prone source rocks would generate oil as they are progressively buried and, with deeper burial, begin to generate gas (Fig. 2.28). Also, with progressive burial, the gas-prone source rocks would enter the generative window and add to the gas generated later by the oil-prone source rocks. With progressively deeper burial and consequent increasing source rock maturation there is also a concurrent change in traps, migration pathways, and connectivity of sandstone beds across faults. Thus there is a dynamic evolution of the plumbing system through which expelled hydrocarbons migrate.

Primary Migration: Weber and Daukoru (1975) postulated that the basic mode of primary migration in the Niger Delta is by oil crossing the major boundary fault at depth where the paralic succession on the downthrown side is juxtaposed against overpressured Akata Shales on the upthrown side. The geopressured shales on the upthrown side of the major structure-building and crestal faults, within which hydrocarbons have been generated, are in contact with the hydropressured section on the downthrown side. This juxtaposition generates sufficient differential pressure to enable migration of hydrocarbons into the fault zone, which then acts as a conduit. It is also thought that in zones that have favorable temperature conditions, the Agbada shales could have contributed oil that got expelled into major faults.

Secondary Migration: This most probably took place along fault zones. Some fault zones may have sealing capability caused by clay smear. Due to extensive normal faulting, long-distance lateral migration of hydrocarbons across major bounding faults is considered unlikely. The Niger Delta is thus a classic example of a vertically drained hydrocarbon system. As major faults have the deepest penetration into source rocks, they provide conduits for more mature hydrocarbons to migrate vertically and accumulate in structures formed by the major bounding faults.

There is apparently a preponderance of structural traps, specifically growth faults, in the Niger Delta. This may be attributable to the ease with which structural traps are identified or interpreted with the available exploration tools. The rollover anticlines associated with these faults usually measure several km long and a few km wide. They trend E–W in the east and NW–SE in the west, roughly parallel to the coastline. The complexity of structural deformation and containment styles increases basinward, and there is a wide variety of structural containment elements (Figs. 2.29–2.31).

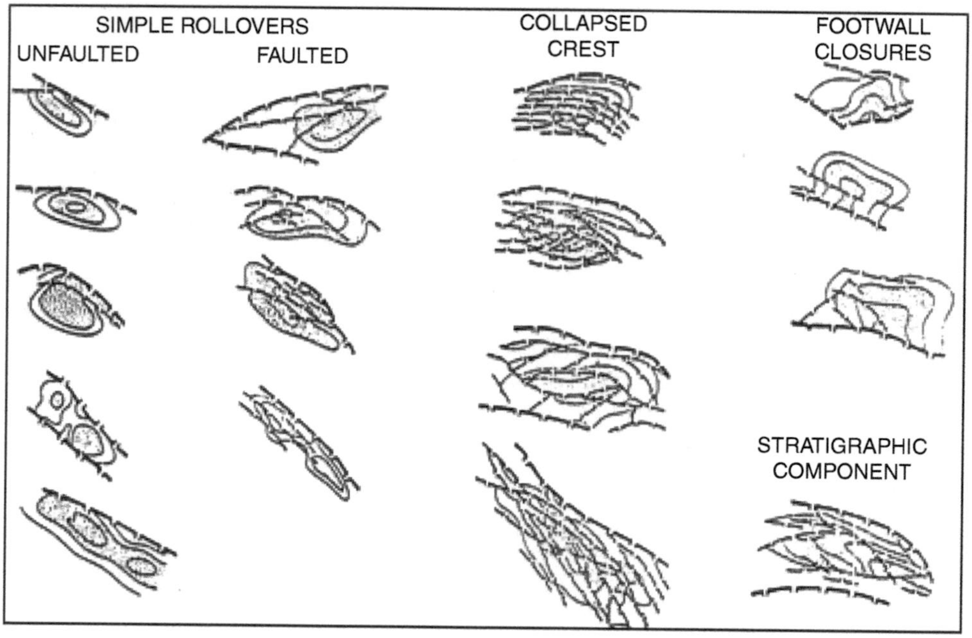

FIGURE 2.29 **Principal types of oil field structures in the Niger Delta.** *Source: After Doust and Omatsola (1989).*

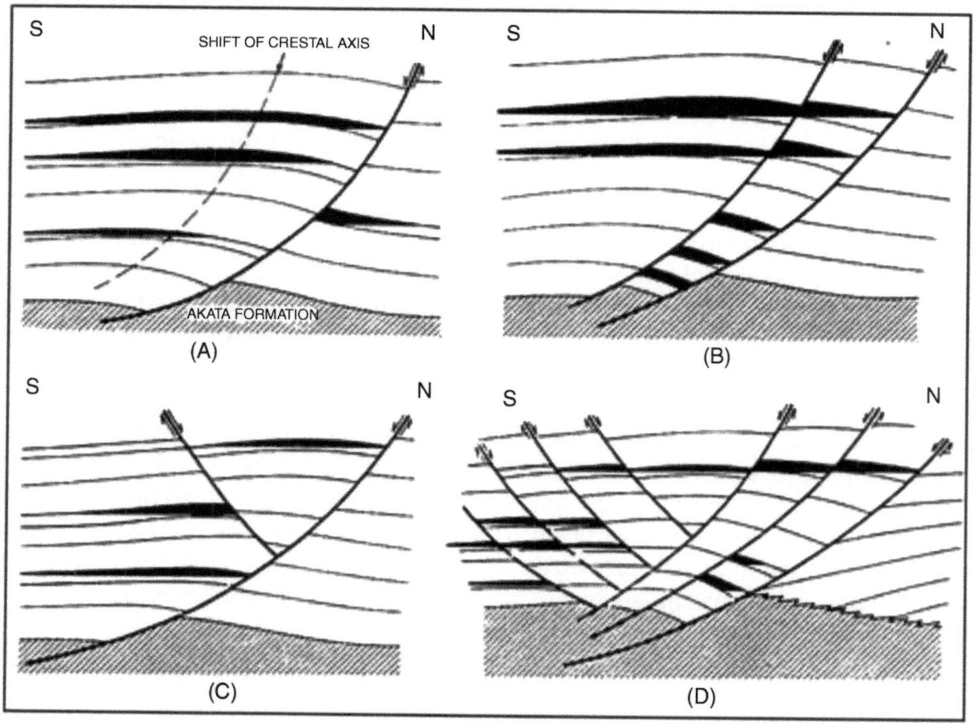

FIGURE 2.30 **Examples of trapping geometries and complexities; thick lines.** (A) Simple rollover structure; (B) Structure with multiple growth faults; (C) Structure with antithetic fault; (D) Collapsed chest structure. *Source: Okiwelu, A.A., Ude, I.A., 2012. 3D modelling and basement tectonics of the Niger delta basin from aeromagnetic data. In: Sharkov, E. (Ed.), Tectonics—Recent Advances. Published by InTech 2012. under CC BY 3.0 license. © The Author(s).*

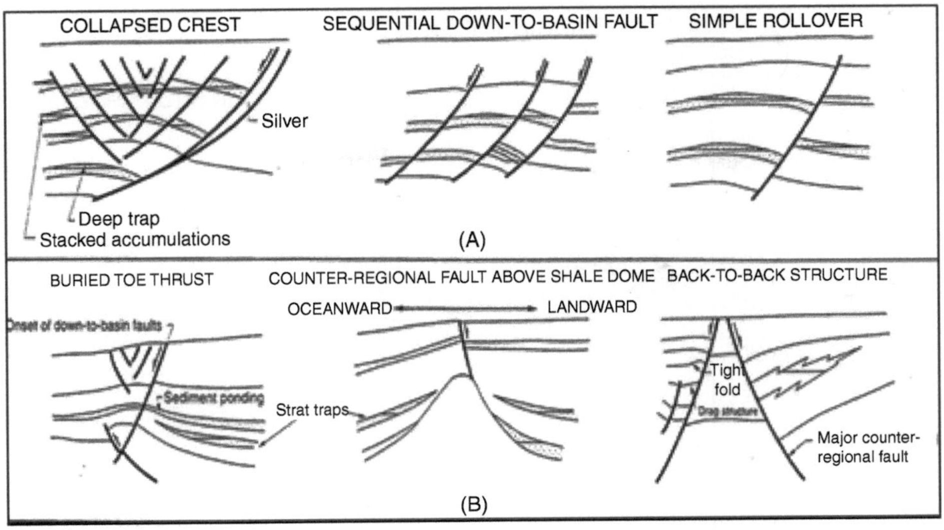

FIGURE 2.31 **Field-scale structural traps styles in the Niger Delta indicate major faults.** (A) Landward positions—extensional trap types; (B) oceanic and slope positions—diapiric and compressional traps. Hydrocarbon accumulation is stippled. Traps may have stratigraphic components where sands are truncated by clay-filled channels. *Source: After Doust, H., Omatsola, M.E., 1989. Niger Delta. In: Edwards, J.D., Santogrossi, P.A. (Eds.), Divergent/Passive Margin Basins. AAPG Memoir, vol. 48, pp. 201–238.*

The Northern Delta and Greater Ughelli depobelts have predominantly rather widely spaced simple rollover structures. The Central Swamp depobelt has faulted rollover structures. In the Coastal Swamp and Offshore depobelts, collapsed crest and K-type structures prevail. Here the influence of mobile undercompacted shales on structural deformation becomes very pronounced.

2.4.7 Reservoir Facies Characteristics

An appreciation of the general characteristics of reservoir rocks is the key to the analysis of data from a given field. Each reservoir facies is linked to its depositional environment, thus providing the basis for understanding its three-dimensional geometry and connectedness. The main factors that control reservoir properties and performance are grain size, sorting, and mud content. As a matter of general principle, reservoir facies should be identified and classified on the basis of objectively recognizable lithologies only, not on the basis of depositional environments. However, reservoir facies must be linked to depositional environments and processes. This is essential to the prediction of lateral continuity and correlation of facies required in reservoir geology and paleogeographic reconstruction.

The barrier bars have much higher permeabilities in the upper part than in the lower, resulting from the typical coarsening upward succession (Fig. 2.31). Distributary channel fills show the reverse pattern as a result of their fining upward motif (Fig. 2.32). It is thus critically important to identify the facies type prior to planning well spacing and completion intervals (Fig. 2.33).

In a study of the diagenetic attributes of the Niger Delta reservoirs, Adedokun (1981) noted that the sandstones of the Agbada Formation are mainly fine- to medium-grained subarkoses,

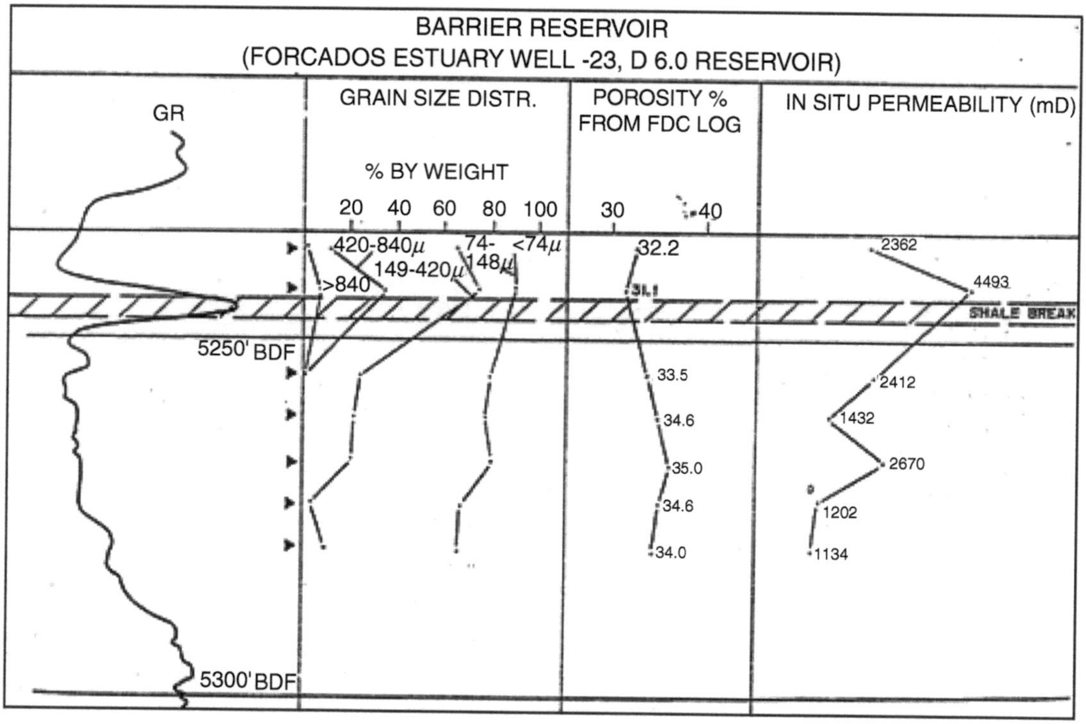

FIGURE 2.32 **Petrophysical properties of a Niger Delta barrier bar reservoir.** *Source: After Weber, K.J., 1971. Sedimentological Aspects of Oil Fields in the Niger Delta. Geologie en Mijnbouw 50, 59–67.*

sublitharenites, and quartz arenites. The framework grain contacts are usually tangential, concavo–convex, and rarely sutured or stylolitic. Quartz overgrowths are therefore, relatively rare, implying little or no pressure solution. Weber and Daukoru (1975) noted that porosity reduction with increasing burial is not simply depth dependent. Temperature and delta tectonics appear to play an important role, since each depobelt seems to have its specific porosity–depth relationships. For example, in the central swamp depobelt, the highest porosities occur at the greatest depths, whereas in the younger coastal swamp depobelt, the highest porosities are found in the shallower depths.

Nwachukwu and Odjegba (2001) noted that Niger Delta reservoir sandstones have an average compaction of 62% and an average compaction coefficient of 0.0006 m^{-1}. The implication of this is that the sandstones are undercompacted and weakly consolidated. Doust and Omatsola (1989) had similarly noted that the sands within the paralic successions are poorly consolidated and have porosities as high as 40% in some oil-bearing reservoirs.

2.4.8 Petrophysical Properties

Porosity in the Niger Delta reservoir sands has been measured directly on cores. However, log analysis using formation density logs gives reasonably accurate porosities because of the fairly constant grain density and the generally thick reservoir sands. Most sands are fairly

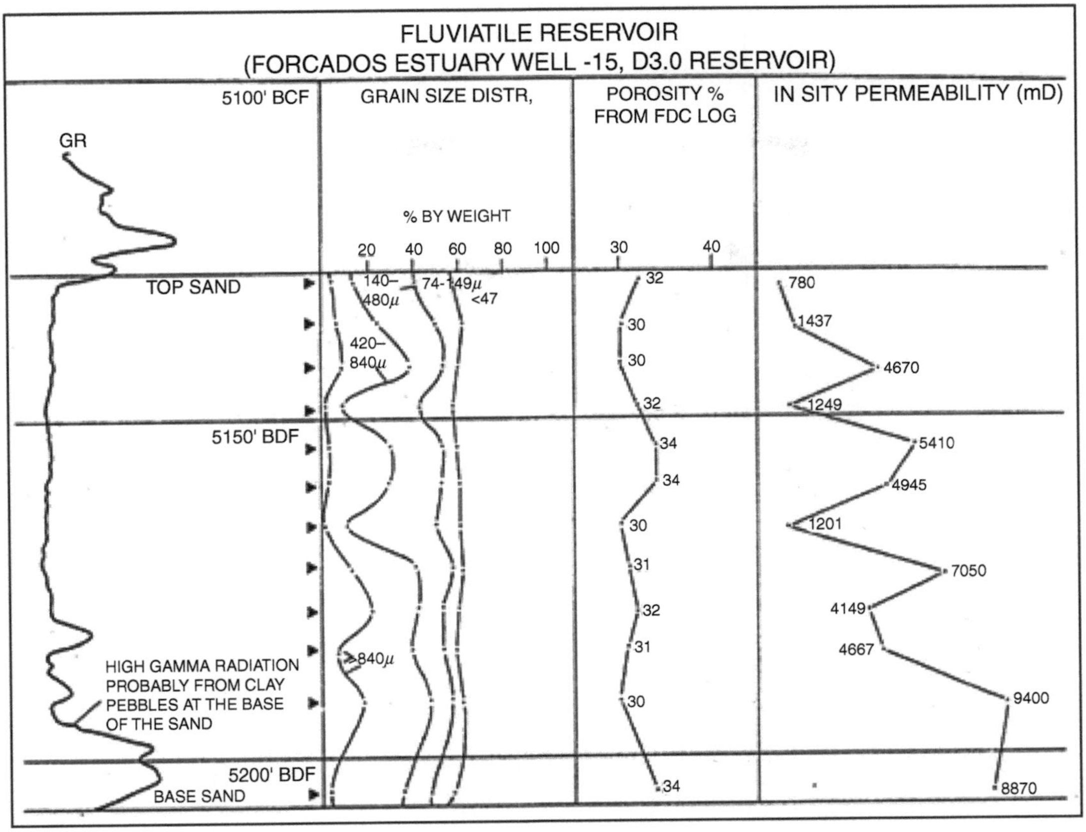

FIGURE 2.33 **Petrophysical properties of a fluviatile reservoir in the Niger Delta.** *Source: After Weber, K.J., 1971. Sedimentological Aspects of Oil Fields in the Niger Delta. Geologie en Mijnbouw 50, 59–67.*

clean and hydrocarbon saturation can be computed with the Archie formula. Permeabilities are also easily determined as part of routine core analysis. Combined core data has been used to construct porosity/permeability plots for common facies associations shown in Figs. 2.34 and 2.35. There is often a clear trend, which indicates that a useful and predictable relationship exists which may be applied both vertically and laterally over most of the delta.

According to Beard and Weyl (1973), in unconsolidated or weakly consolidated sands (which are common in the Niger Delta), reservoir properties are largely controlled by size and size sorting (which determine the dimensions of the pore throats), the area of surfaces of grains (specific surface area), and the volume of space enclosed between the grains.

2.5 SUMMARY

The Niger Delta Basin, underlying a total area of over 256,000 km^2, is a Cenozoic clastic build-out that rests on top of two older sedimentary entities (Benue Trough and Anambra Basin) within an aulacogen defined by the Chain and Charcot Fracture Zones (Fig. 2.4). Sediment

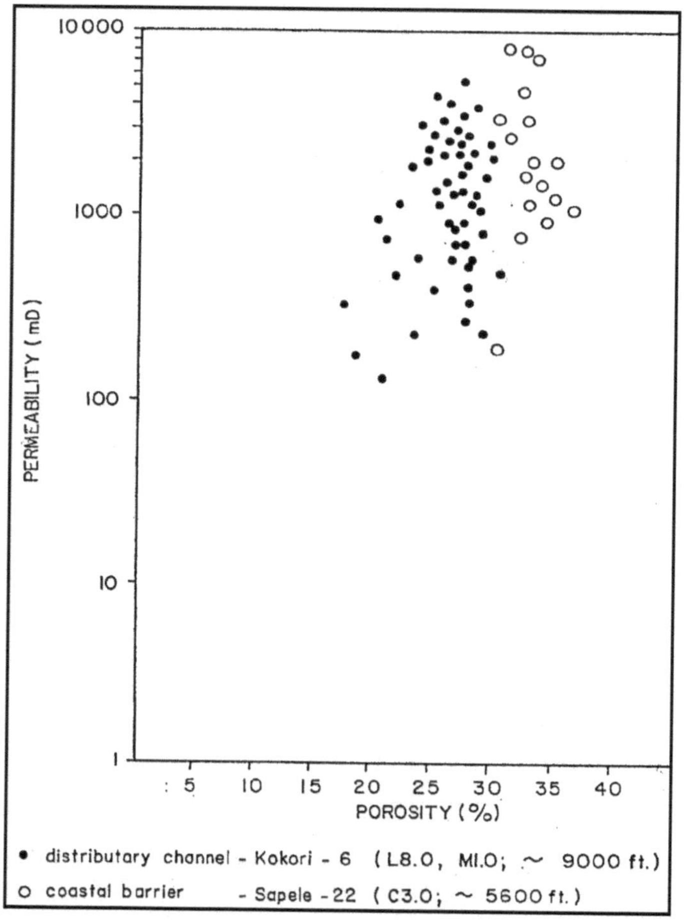

FIGURE 2.34 **Plot of porosity versus permeability for distributary channel and coastal barrier facies from wells Kokori-6 and Sapele-22 wells.** *Source: From Weber, K.J., 1971. Sedimentological Aspects of Oil Fields in the Niger Delta. Geologie en Mijnbouw 50, 59–67.*

supply had been largely by the arterial Niger–Benue system draining a very extensive area of Western Africa. Sedimentation process changed over time from fluvial to wave-dominated and tide-influenced, and had been in three zones—delta top, delta front, and bottom set, which constitute the continued deposition of the delta lithostratigraphic units, the Benin, Agbada, and Akata, respectively. This tripartite sediment packaging is considered to be the subsurface equivalent of the outcropping Benin Sands, Ameki Group, and the Imo Formation.

Considerable geological data has accumulated on the hydrocarbon-bearing Agbada Formation that has been intensely drilled, logged, and sampled. The unit is paralic—consisting mainly of marine, often richly fossiliferous clay shales (seals) and fine to coarse, cross-bedded, burrowed, feldspathic sands deposited as mouth bars, beach/barrier bars, tidal sand ridges, and distributary channel facies that constitute the reservoir facies. In both the Agbada

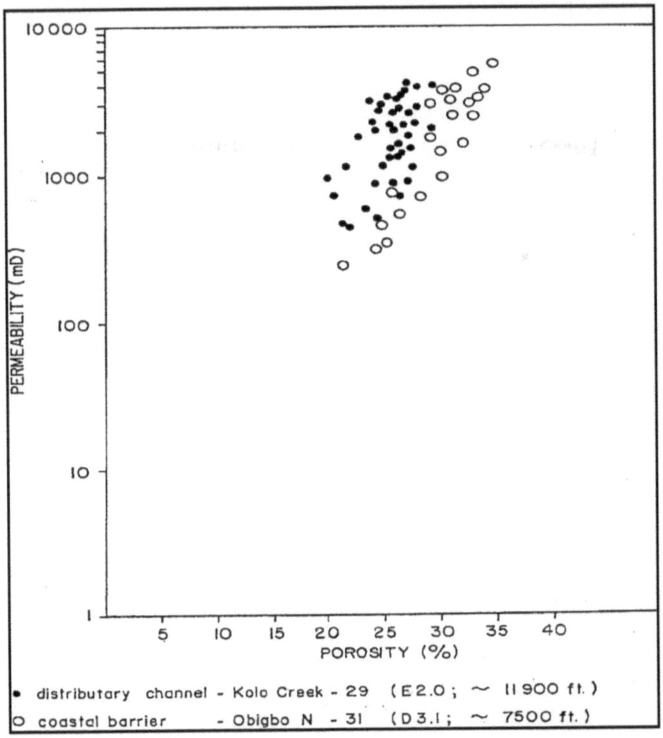

FIGURE 2.35 **Plot of porosity versus permeability for distributary channel and coastal barrier facies from wells Kolo Creek-29 and Obigbo North-31.** *Source: From Weber, K.J., 1971. Sedimentological Aspects of Oil Fields in the Niger Delta. Geologie en Mijnbouw 50, 59–67.*

and the Benin formations, deep erosional canyons are filled with clays and subordinate sands (e.g., Opuama Channel, Afam, Soku, Osare, Elelenwa, and the extensive complex called the Qua Iboe Collapse System).

The delta lithic fill is considered to be self-contained since its structural geology has not been influenced by the trends of the underlying basement. The deformational structures, mostly growth faults, formed as a result of rapid sedimentation of the Agbada facies over undercompacted and mobile Akata Shales. Growth faults are classified as structure-building faults, crestal faults, flank faults, counter-regional faults, and antithetic faults. The delta has been structurated into six depobelts demarcated by the down-to-the-basin faults in a manner likened to the movement of an escalator. Shale flowage/withdrawal is thought to be largely responsible for the movements which resulted in squeeze flow and thrusting known in the distal parts of the delta down to the deep offshore.

As a hydrocarbon province, the Niger Delta has ultimate recovery that is presently estimated at nearly 40 billion barrels, estimated to be about 70% of the overall reserves of sub-Saharan Africa, and gas reserves in excess of 40 trillion cubic feet. The Niger Delta hydrocarbons are thought to have been sourced mainly from the Akata Shales and subordinately from some of the shale intervals of the Agbada Formation, bearing land plant material that gave rise to mainly Types III and II/III organic matter (mainly gas-prone) within an oil window that varied from 9,000 to 14,000 ft. depth. The generated hydrocarbons, mainly low-sulfur/nickel, light waxy, nondegraded oils, expelled from the source rocks, migrated for relatively short distances, and vertically through the growth faults, into appropriately timed traps—largely dip closures (rollover anticlines in growth faults) and relatively rare stratigraphic and combination traps. Individual Niger Delta oil fields are small in size, but large in number. The reservoir sands are generally poorly consolidated, with porosities ranging between 10% and 35%, and permeabilities that may exceed two darcies.

CHAPTER

3

Phylogeny of the Niger Delta Cenozoic Foraminifera

OUTLINE

3.1 INTRODUCTION

The quality of preservation of foraminifera in the Niger Delta wells varies markedly. In the onshore and shallow offshore areas, microfaunal recovery and the quality of preservation vary from poor to moderately good depending on the depositional environment. In the nonmarine environments, the sediments are usually barren, while the marginal marine heterolithic sediments (sensu the Agbada Formation) contain sporadic to moderately rich foraminiferal assemblages. The near-shore shelf assemblages are almost invariably dominated by specifically nondiverse benthonics and rare planktonics.

The paucity and low diversity of the shelfal microfaunas have been attributed to the shallow environments of deposition and the dilution effect attributable to the high terrigenous clastic inputs into the depositional basins. Faunal occlusion and the occurrence of ecologically depressed tops are common, creating problems in the use of the First Downhole Occurrence and Last Downhole Occurrence datums in biochronologic correlations.

The deeper marine shales almost invariably yield richer and more diverse benthonic and planktonic microfaunas. Despite the frequent occurrence of ecologically unfavorable conditions within these ecosystems, several benthonic marker species have been shown to be useful and reliable for correlations within the Niger Delta depobelts. The chronostratigraphic scheme for both the onshore and shallow shelf areas of the Niger Delta (Enclosures 1.1, 1.2, 1.20, and 1.21) is, therefore, largely based on benthonic species.

Cenozoic Foraminifera and Calcareous Nannofossil Biostratigraphy of the Niger Delta. http://dx.doi.org/10.1016/B978-0-12-812161-0.00003-X

Microfaunal recovery improves markedly in the more distal offshore depocenters, particularly in the transgressive marine mudrocks. Many of these mudrocks reflect Maximum Flooding Surface condensed sections. Though the diversity of the benthonic forms is higher at these distal locations, their significance in chronostratigraphy assumes less importance. In contrast to the situation recorded in the shallow near-shore sediments, there is a marked increase in both the population abundance and diversity of planktonic assemblages (Figs. 4.7–4.23; Enclosures 1.3–1.19) within these ecosystems, thus permitting regional, delta-wide correlations across depobelts and major structure-building faults. It is also possible to tie the Niger Delta deeper water sequences with the global Cenozoic chronostratigraphic scheme that has been well documented in other parts of the world (Blow, 1969a, 1979; Berggren et al., 1998a,b; Wade et al., 2011; Gradstein et al., 2004a,b, 2012; Enclosures 1.20 and 1.21).

The development of the Niger Delta foraminifera assemblages bears the imprint of two water masses, namely the shallow, warm, Tropical Surface Waters and the deeper, cooler South Atlantic Central Water mass that is influenced by the Cold Benguela Current. Both paleoceanic factors influenced the paleozoogeography and biofacies of the microfaunal assemblages of the Niger Delta and have probably operated in this manner since the inception of the Niger Delta in Eocene times.

Another significant feature of the paleoceanography of the Gulf of Guinea is that several low latitude larger foraminifera (e.g., *Amphistegina* spp. and *Heterostegina* spp.), as well as, cold water indicator planktonics (e.g., *Globorotalia inflata*) are common features of the Niger Delta Neogene deeper biofacies (Adegoke et al., 1976; Murray, 1991). These forms are therefore, significant in the understanding of the Niger Delta depositional environments and chronostratigraphy.

On the whole, the quality of microfaunal preservation, particularly in the shallow, shelfal areas of the Niger Delta is generally poor. The tests of most of the foraminifera, particularly the calcareous forms, are often very badly worn (Plates 1–35). This is the result of repeated abrasion of the foraminiferal tests against the clastic sediments in the dominantly high-energy setting of the delta. In the deeper water facies, the tests of many of the agglutinated species are compressed. This is evidence of the fact that considerable pressure is exerted on foraminifera tests by the overlying sediments during diagenesis.

3.2 PHYLOGENETIC TRENDS AMONG SELECTED CENOZOIC NIGER DELTA FORAMINIFERA

3.2.1 Benthonic Foraminifera

A proper understanding of the phylogenetic relationships of some of the principal Niger Delta marker species, especially those identified in this project, is essential for a fuller appreciation of their biostratigraphic and biochronologic significance (see also Ozumba and Amajor, 1999a). A good working knowledge of the stratigraphic ranges of these forms and those of the more important cooccurring, but chronostratigraphically less significant species, are useful for the age-dating of some Niger Delta horizons in which the dominant markers are either absent or have their tops "depressed."

A few of the genera/lineages considered important are: *Lenticulina*, *Bolivina*/*Brizalina*, *Nonion*/*Florilus*, *Hanzawaia*, *Epistominella*, *Eponides berthelotianus*/*Ammonia beccarii*, and *Uvigerina*. The observed evolutionary changes in these species involve features, such as ornamentation that have probably evolved in response to changes in the environment and which, within the Niger Delta, have, not only local, but also global significance through successful correlation and or calibration with established planktonic foraminiferal and calcareous nannofossil datums.

The phylogenetic relationships of these selected benthonic marker species are shown in Fig. 3.1 and discussed succinctly as follows.

1. The *Hanzawaia* Lineage
Hanzawaia concentrica (Plate 11, Figs. 2–6) first appeared in Eocene times. It is a biconvex form with a thickened, keel-like test margin. It became extinct in the Early Miocene. The species gave rise, during the Oligocene and Early Miocene periods, to nonkeeled stocks of *Hanzawaia strattoni* (Plate 11, Figs. 7–10), designated *Valvulineria* 19 and *Valvulineria* 1A in the SPDC Catalogue. The older variant, *Valvulineria* 8B, is strongly biconvex to planispiral and became extinct during the late Early Miocene. *Valvulineria* 19 and 1A, the second variant, is readily identified by its gently concavo–convex test and is quite prominent from the Miocene to Recent.
2. The *Eponides berthelotianus*/*Ammonia beccarii* Lineage
The evolutionary datum of *E. berthelotianus* (Plate 10, Figs. 9–11) occurs close to the Eocene–Oligocene boundary. The species is strongly biconvex, large sized, and not prominently ornate. It gave rise, during the Early Miocene, to *A. beccarii* (Plate 9, Figs. 18 and 19), which is a ubiquitous rotaliid with moderately ornamented plano-convex test, which offers little or no stratigraphic values.
3. The *Epistominella* Lineage
Both *Epistominella pontoni* (Plate 8, Figs. 17–20) and its presumed descendant, *Epistominella vitrea* (Plate 9, Figs. 1 and 2) are morphologically similar. The major difference between them is in the shape of the aperture and the margin of the test. In *E. vitrea*, the dorsal lip of the aperture is more strongly arched and the margin of the test is acute. *E. pontoni* evolved close to the Paleogene–Neogene boundary with rounded margin and gave rise to *E. vitrea*, which is a long-ranging species that became quite prominent from the Early Miocene to Recent.
4. The *Lenticulina* Lineage
Species of *Lenticulina* are common throughout the Paleogene and Neogene. The majority of them are keeled. While many species are umbonate, others are nonumbonate. The Cretaceous lenticulines gave rise to two stocks in the Paleogene. *Lenticulina extruatus* evolved through *Lenticulina arkansasana*, both of which have numerous and raised sutures but lack the central umbo, to *Lenticulina pseudomamilligerus* (Plate 12, Figs. 11 and 12), which commonly has a small calcareous boss in the center. This species evolved into *Lenticulina grandis* (Plate 12, Figs. 2–6), which differs only in having a fairly well-developed central boss. One of the oldest Paleocene species of *Lenticulina* is *Lenticulina carinata*, which evolved from the Upper Cretaceous lenticulines and gave rise to *L. grandis* through *Lenticulina midwayensis*. *L. grandis* gave rise to *Lenticulina inornata* (Plate 12, Figs. 7–10) which is strongly carinate but with less prominently developed ornamentation. *L. grandis* became extinct in the Middle Miocene, while *L. inornata* ranged beyond the Middle Miocene into the Recent.

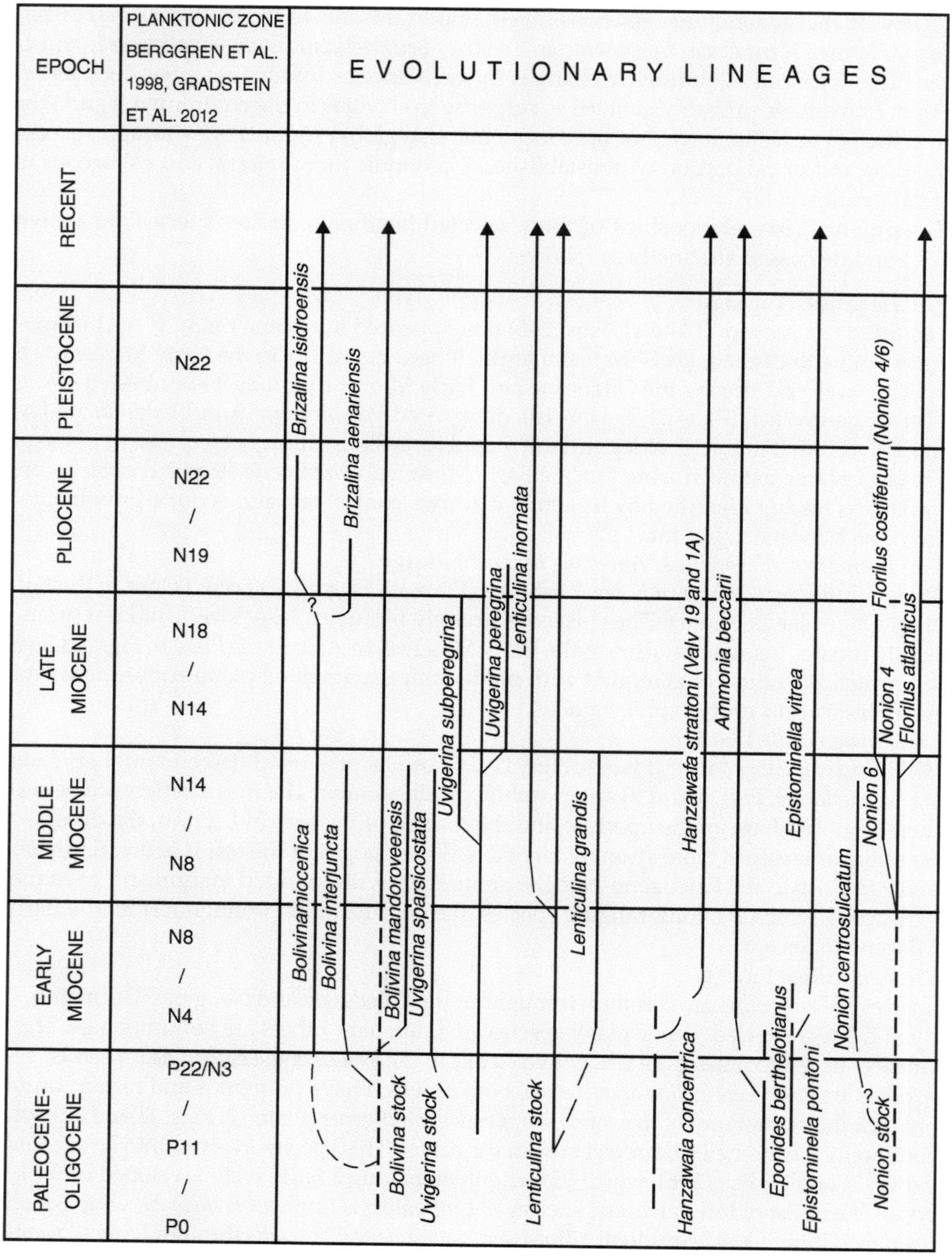

FIGURE 3.1 **Phyletic relationship of selected Niger Delta Cenozoic marker Foraminifera species.**

5. The *Bolivina/Brizalina* Lineage
Many species within the *Bolivina/Brizalina* lineage have been erroneously assigned to the genus *Brizalina*. Both *Bolivina mandoroveensis* and *Bolivina interjuncta* evolved within the Paleogene from a parent of *Bolivina* stock. While *B. interjuncta* (Plate 7, Fig. 18) is well ornamented, *B. mandoroveensis* (Plate 7, Figs. 19 and 20) is near-parallel rows of striae. Both have their Last Appearance Datums within the Middle Miocene. *Bolivina miocenica,* (Plate 8, Fig. 1) another morphotype, also evolved from the *Bolivina* parent stock sometime in the Middle Paleogene. It attained greater prominence in Early to Late Miocene times and has continued to thrive until the Recent.
Two other stratigraphically important species within the lineage have been recorded in the Niger Delta Late Miocene. These are *Brizalina aenariensis* (Plate 8, Figs. 5 and 6) and *Brizalina isidroensis* (Plate 8, Fig. 7). Both are readily distinguished morphologically. The older and longer-ranging *B. isidroensis* probably evolved from the *B. miocenica* stock near the top of the Late Miocene. The test is more cylindrical and punctate, lacking the three prominent costae that characterize *B. aenariensis.* The extinction datum of *B. isidroensis* is at the top of Late Pliocene. *B. aenariensis,* by contrast, is extremely short-ranged and stratigraphically restricted to Early Pliocene.
Its First Appearance Datum is often used to define the Miocene-Pliocene boundary in the Niger Delta.
6. The *Uvigerina* Lineage
The *Uvigerina* stock also evolved in the Paleogene. Three of the species became important markers within the Neogene of the Niger Delta. These are *Uvigerina sparsicostata, Uvigerina subperegrina,* and *Uvigerina peregrina,* all of which are distinct morphologically and with different stratigraphic ranges. *U. sparsicostata* (Plate 13, Figs. 12–14) is characterized by distinct and widely spaced costae and a densely perforate and fusiform test. It had its extinction datum within the Middle Miocene. The second species, *U. subperegrina* (Plate 13, Figs. 15 and 16) had its Last Appearance Datum in the (?)Late Miocene. It is characterized by a circular to ovate transverse section, inflated chambers which are broader than high and obscured by overlapping chamber flanges, and a finely perforate test. The third species, *U. peregrina* (Plate 13, Figs. 8 and 9) is distinguished from the others by its non-fusiform test, the faint costae, and the imperforate test. This species became prominent from the Late Miocene to Early Pliocene.
7. The *Nonion/Florilus* Lineage
The *Nonion* stock evolved in the Paleogene. Two variants of *Florilus costiferum* (Plate 6, Figs. 10 and 11), designated by SPDC as *Nonion* 6 (Plate 6, Fig. 11) and *Nonion* 4 (Plate 6, Fig. 10), attained stratigraphic significance in the Niger Delta during Middle and Late Miocene times, respectively. Though considered conspecific, both variants have subtle morphological differences. While the *Nonion* 6 variant has short umbilical sutures, these features become more elongate and pronounced in *Nonion* 4. *Nonion* 6 evolved in the early Middle Miocene and had its extinction datum in the late Middle Miocene. *Nonion* 4 evolved toward the top of the Middle Miocene and became a prominent constituent of the Late Miocene microfauna until its extinction within the middle of the Late Miocene. Their obvious successor is *Florilus atlanticus* (Plate 6, Figs. 3 and 4), which became quite prominent from the Late Miocene to the Recent. Another important Neogene marker is *Nonion centrosulcatum* (Plate 6, Figs. 5–9), a species which also evolved

during the Paleogene. It is well represented in the Early and Middle Miocene microfaunal assemblages of the Niger Delta, but had its last appearance in the early Middle Miocene.

3.2.2 Planktonic Foraminifera

Intensive worldwide planktonic foraminiferal biostratigraphic studies over the past eight decades have raised this field of study to an independent paleontological subdiscipline. Arising from the detailed taxonomic documentation which were initiated in the low latitude Caribbean region by Bolli (1957a) and Blow (1979), as well as studies on the biology of modern planktonic foraminifera (Be, 1977; Be et al., 1980) and deep-sea sediment investigations of Berggren and Amdurer (1973), and Berggren and Poore, evolutionary trends and phylogenetic relationships have been unraveled among Cenozoic planktonic foraminifera. Several regional studies in various parts of the world have yielded excellent monographs upon which the paleozoogeographic distribution of evolutionary events can be based (Ozumba, 2011).

3.2.2.1 *Evolutionary Paleoecologic Models*

Stable isotope investigations, notably oxygen-18 (O^{18}), have linked evolutionary diversification of planktonic foraminifera to global paleoclimatic events. Periods of marked adaptive radiation, for example, in the Eocene and the Middle Miocene produced new morphogroups

TABLE 3.1 Intervals of High Evolutionary Planktonic Foraminifera Speciation in the Cenozoic Niger Delta

1. Late Miocene–Pliocene	(1) *Turborotalia continuosa* → *Turborotalia acostaensis* → *Turborotalia pseudopima* → *Neogloboquadrina dutertrei* (2) *Globorotalia menardii* → *Globorotalia merotumida* → *Globorotalia plesiotumida* → *Globorotalia tumida* (3) *Sphaeroidinellopsis disjuncta* → *Sphaeroidinellopsis multiloba* → *Sphaeroidinellopsis seminulina* → *Sphaeroidinellopsis sphaeroides* → *Sphaeroidinella*
2. Middle Miocene	(4) *Turborotalia foshi* lineage, *Turborotalia peripheroronda*, *Turborotalia peripheroacuta* → *T. foshi*, *Turborotalia praefohsi* → *Turborotalia lobata* → *Turborotalia robusta*
3. Late–Early Miocene	(5) Evolution of *Orbulina* from *Globigerinoides* (*Globigerinoides bisphericus* → *Globigerinoides sicanus* → *Praeorbulina curva* → *Praeorbulina glomerosa* → *Praeorbulina circularis* → *Orbulina suturalis* → *Orbulina universa*)
4. Latest Oligocene–Early Miocene	(6) Evolution of *Globigerinoides* from *Globigerina praebulloides* (*G. praebulloides* → *Globigerinoides primordius* → *Globigerinoides triloba* → *G. sicanus*) (7) *Turborotalia opima* → *Turborotalia mayeri* → *T. peripheroronda* (8) *Turborotalia nana* → *T. continuosa*
5. Latest Eocene–Early Oligocene	(9) *Globigerina* radiation—new lineages include: *Globigerina officinalis* → *Globigerina angulioffcinalis/angulisuturalis* (10) *Globigerina occlusa* → *G. praebulloides* → *Globigerina leroyi* (11) *Globigerina ouachitaensis* → *Globigerina angustiumbilicata* → *Globigerina ciperoensis*
6. Latest Paleocene Early Eocene	(12) Diversification among *Chiloguembelina* (13) Evolution of *Pseudohastigerina wilcoxensis* from *Globanomalina chapmani*
7. Middle–Late Paleocene	(14) Diversification of *Morozovella*; appearance of new genera; and general recovery of planktonic foraminifera after the terminal Cretaceous mass extinction

among the planktonics. An example of this was the *Orbulina* bioseries. Conversely, cool global paleoclimatic trends triggered extinctions, particularly during the Late Eocene–Early Oligocene.

Paleoclimatically induced evolutionary trends are ubiquitous in the Niger Delta. The present detailed investigation on Niger Delta planktonic taxonomy has further confirmed and elucidated known globally recognized trends. Broad periods of extinction, evolutionary divergence, and repopulation of marine ecosystems through adaptive radiation are discernible from the enclosed range charts (Enclosure 1.21) of Cenozoic planktonic foraminifera of the Niger Delta.

Another factor which seems to account for the diversification of planktonic foraminifera is the expansion and reduction of ecological space (ecospace). Whereas periods of extensive marine transgressions offered greater ecospace and diverse niches, regressive cycles with lowstands of sea level caused marine habitats to shrink below the shelf edge. In the Niger Delta, the transgressive marine shales associated with the maximum flooding surfaces carry the most diverse planktonic assemblages, which suggests that speciation probably occurred during such marine transgressions.

3.2.2.2 *Periods of Pronounced Speciation*

Intervals of high evolutionary turnover in the Niger Delta, which also have biostratigraphic significance, are shown in Table 3.1.

CHAPTER

4

Chronostratigraphy of the Niger Delta Cenozoic Foraminifera

OUTLINE

4.1 HARMONIZATION OF THE FORAMINIFERA CHRONOSTRATIGRAPHIC SCHEMES OF STRATCOM MEMBER COMPANIES

A few of the companies operating in the Niger Delta have in-house biostratigraphic schemes developed within their respective areas of operation (Figs. 4.1–4.5). The Shell Petroleum Development Company Ltd. (SPDC) scheme (Fig. 4.1), for example, is based on the recognition of a succession of cyclic transgressive sequences for the Cenozoic (Paleocene to Holocene) and the identification and age dating of the Maximum Flooding Surfaces (MFS). This was superimposed on sequentially numbered Foraminiferal Zones (F-Zones) that were based mainly on numerically numbered benthonic markers, whose stratigraphic ranges have been carefully documented within the onshore and shallow-offshore areas of the Niger Delta in which most of the company's concessions were located. Until recent years, the identity of Shell's markers was kept strictly confidential from other operators and researchers.

Planktonic forms played only a complementary role in the old SPDC chronostratigraphic scheme (Table 4.1). More recently, refinements of the Shell foraminifera zones were

Cenozoic Foraminifera and Calcareous Nannofossil Biostratigraphy of the Niger Delta. http://dx.doi.org/10.1016/B978-0-12-812161-0.00004-1

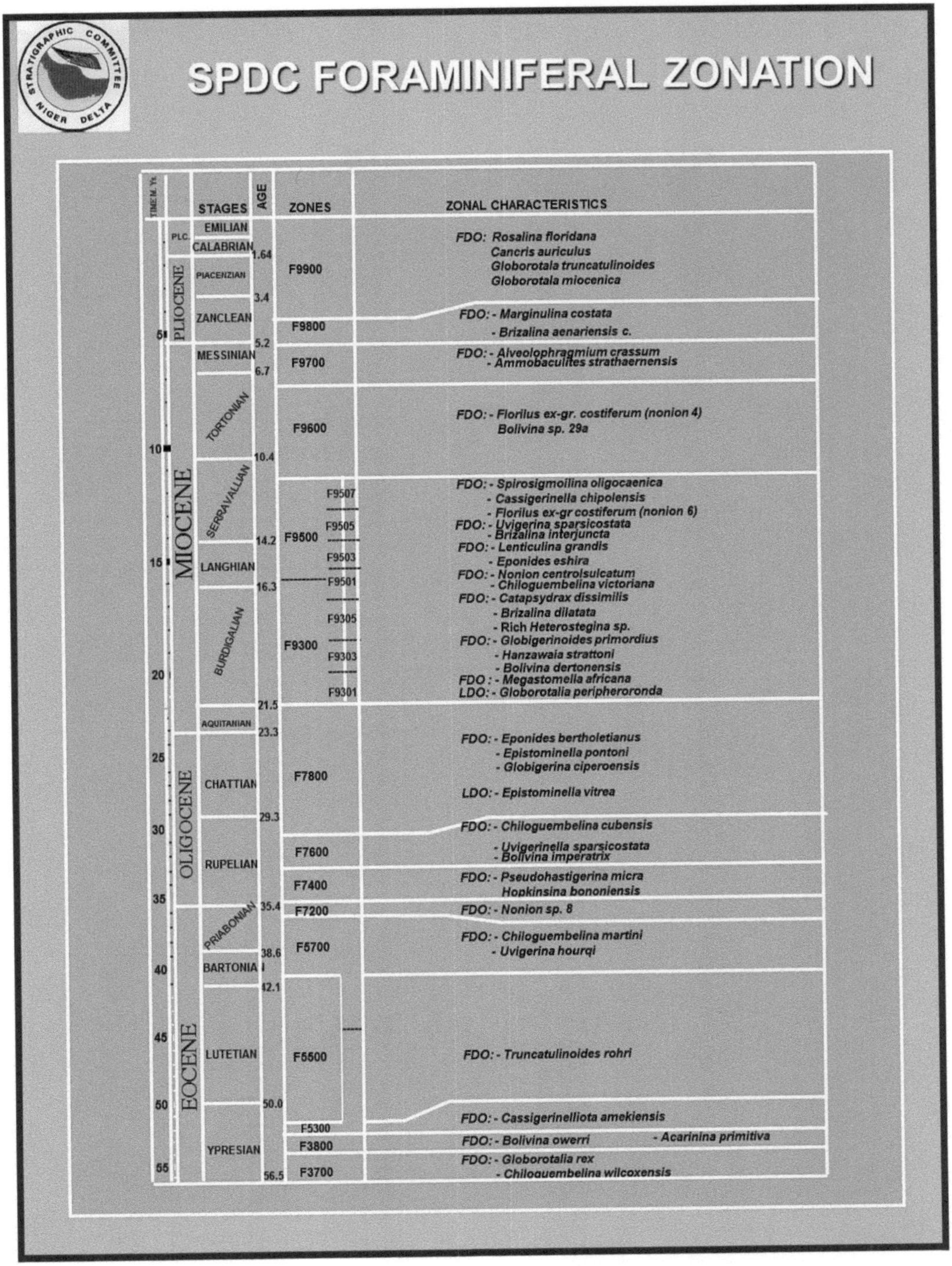

FIGURE 4.1 **Shell Petroleum Development Company Ltd. (SPDC) Foraminiferal Zonation.**

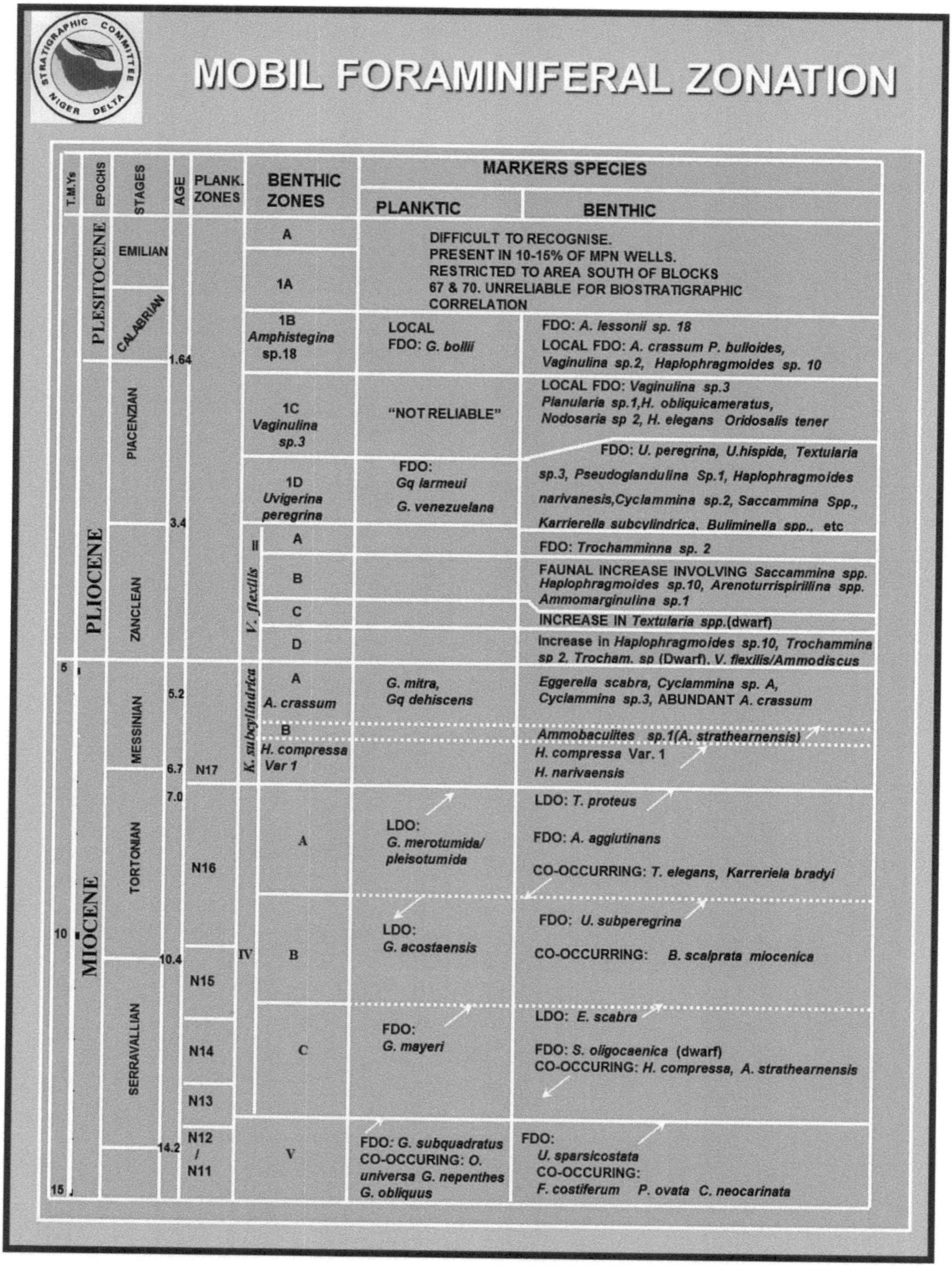

FIGURE 4.2 **Mobil Producing Nigeria (MPN) Ltd. Foraminiferal Zonation.**

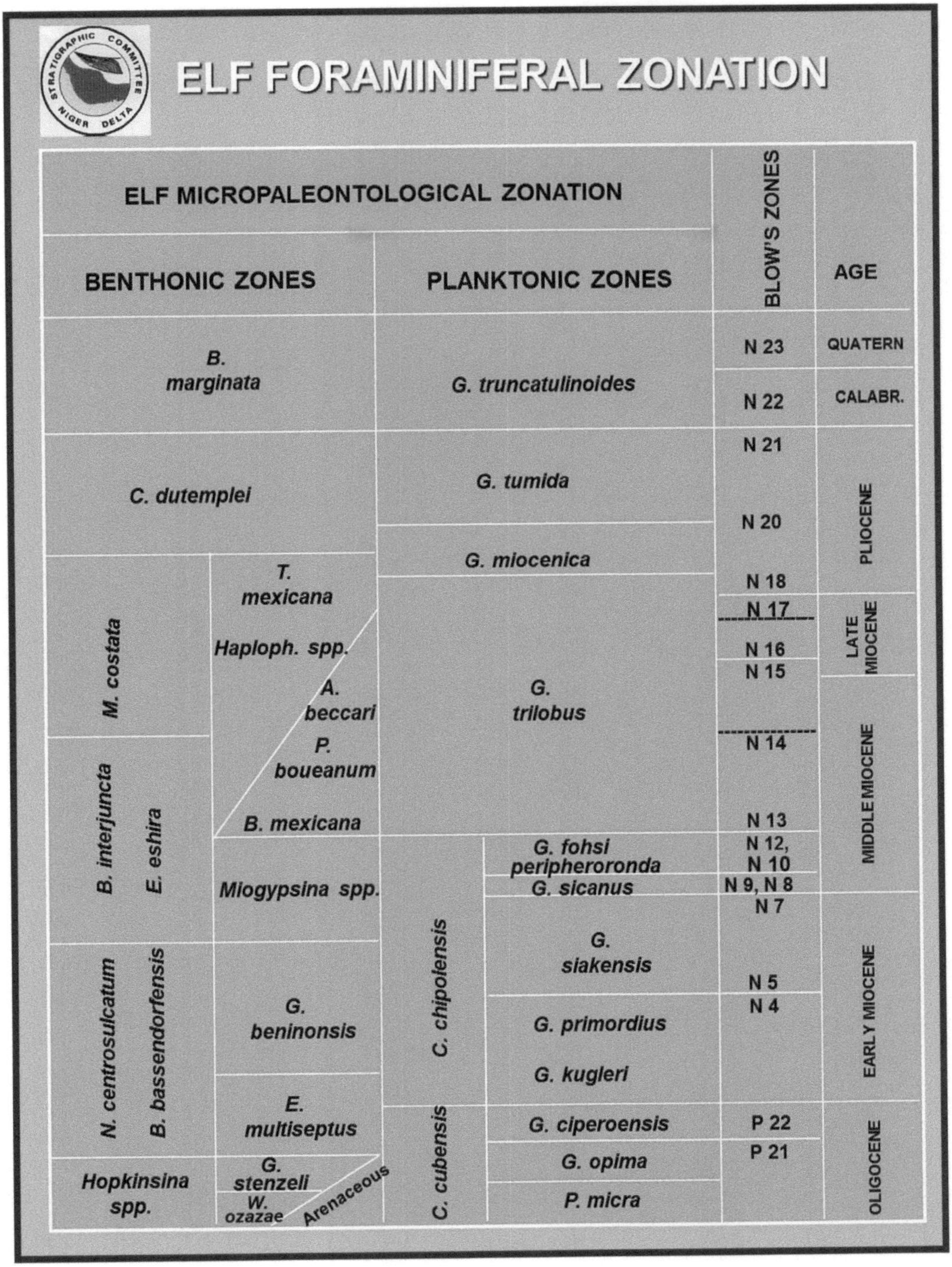

FIGURE 4.3 **ELF Nigeria Ltd. Foraminiferal Zonation.**

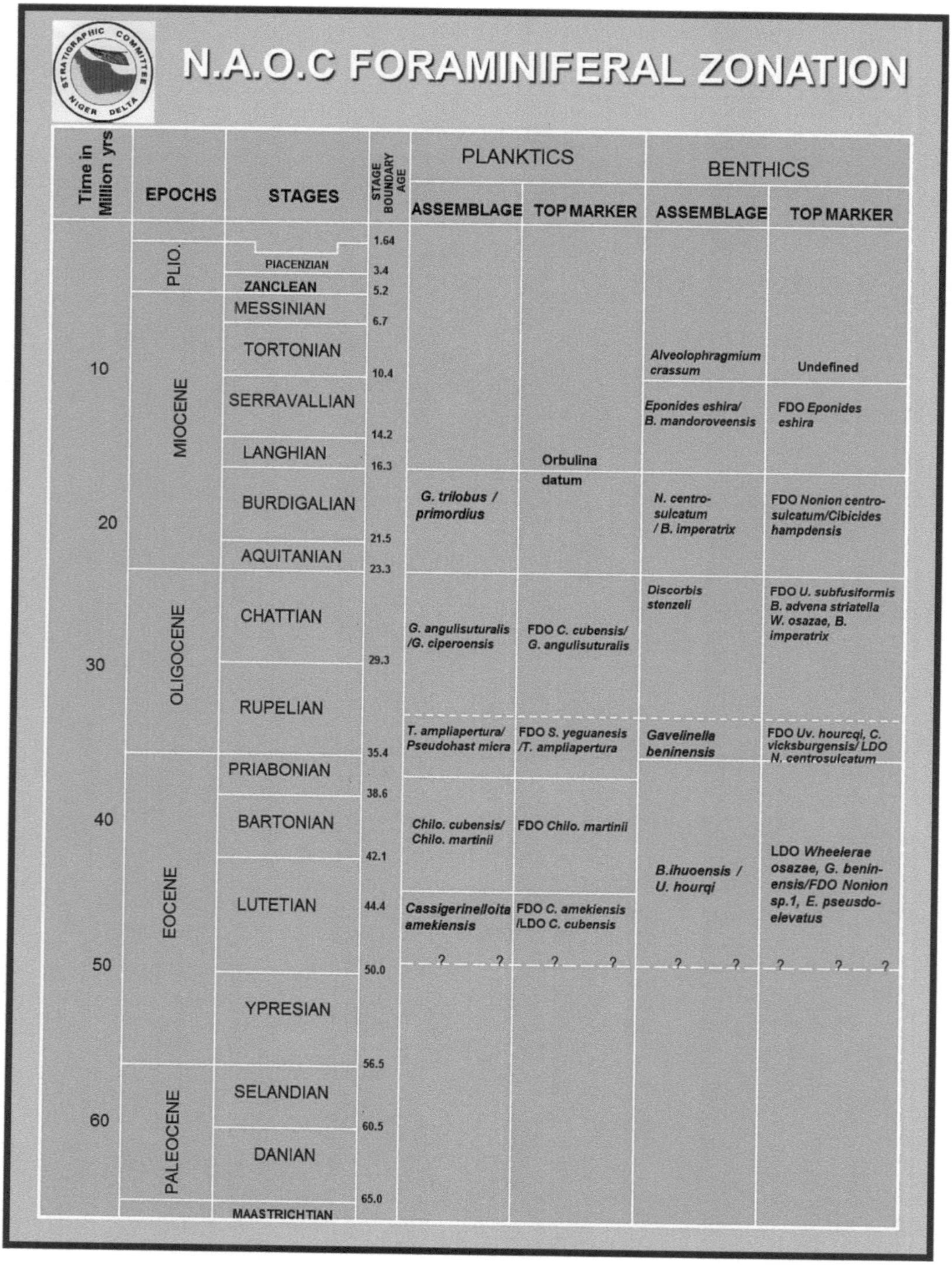

FIGURE 4.4 **Nigerian Agip Oil Company Limited (NAOC) Foraminiferal Zonation.**

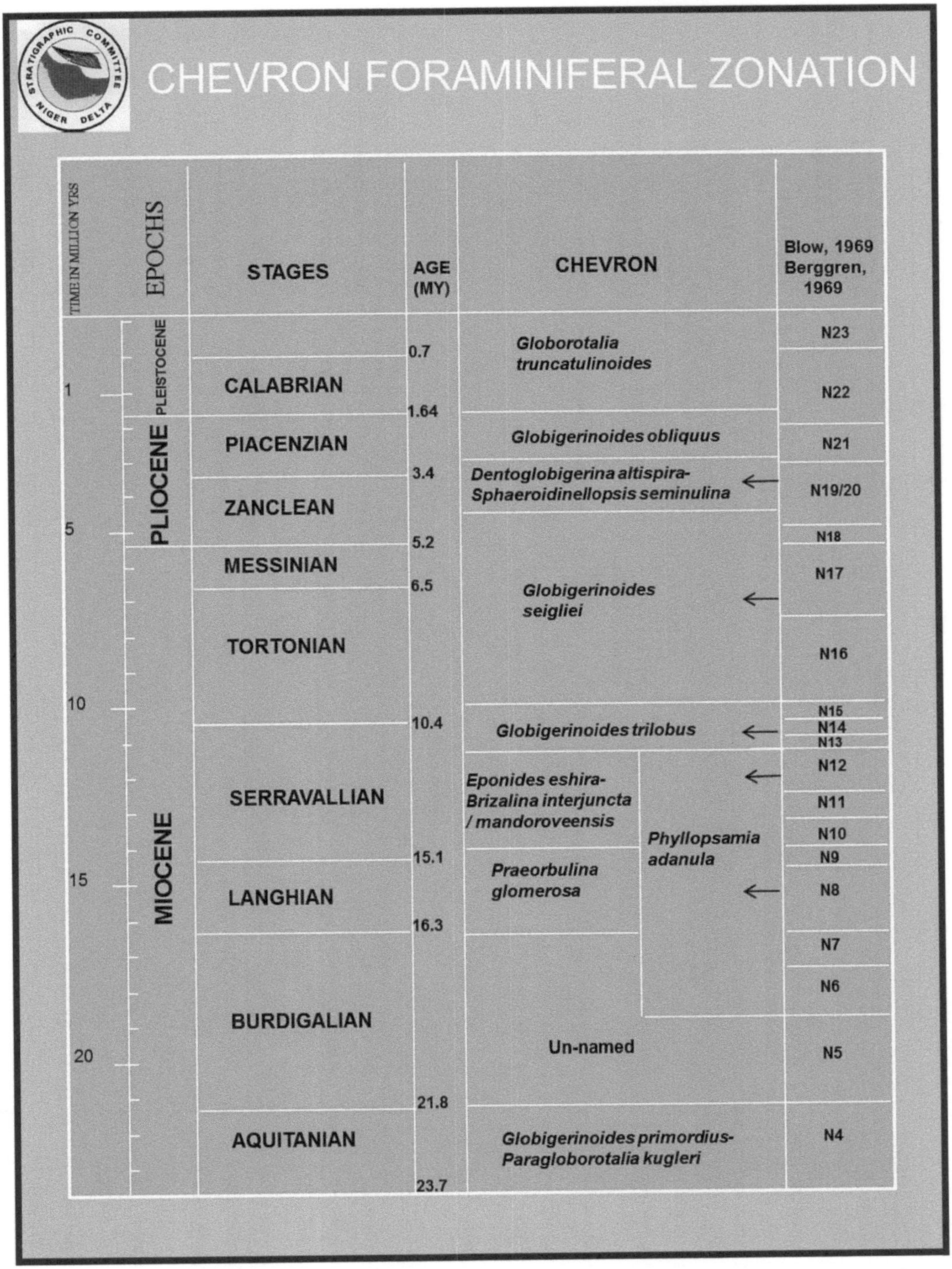

FIGURE 4.5 **Chevron Nigeria Ltd. Foraminiferal Zonation.**

TABLE 4.1 Synonyms of Selected Species From the Biostratigraphic Schemes of StratCom Members

No.	SPDC	Mobil Nigeria Ltd.	Chevron Nigeria	Total/ELF	NAOC/AGIP	Scientific names
BENTHONIC FORMS						
1.	***Alabamina 1***	No synonym	No synonym	No synonym	No synonym	***Epistominella vitrea***
2.	***Alabamina 2***	No synonym	No synonym	No synonym	No synonym	***Megastomella africana***
3.	***Ammobaculites 21, 22***	*Ammobaculites agglutinans*	No synonym	No synonym	No synonym	***A. agglutinans***
4.	***Ammobaculites 8/17***	No synonym	No synonym	No synonym	*Ammoscalaria pseudospiralis*	***A. pseudospiralis***
5.	***Ammodiscus 2***	No synonym	No synonym	No synonym	No synonym	***Ammodiscus glabratus***
6.	***Anomalina 24***	No synonym	No synonym	*Nonion centrosulcatum*	*N. centrosulcatum*	***N. centrosulcatum***
7.	***Bolivina 12***	No synonym	No synonym	No synonym	*Bolivina imperatrix*	***B. imperatrix***
8.	***Bolivina 12A***	No synonym	No synonym	No synonym	*Brizalina ihuoensis*	***Br. ihuoensis***
9.	***Bolivina 25***	No synonym	*Brizalina interjuncta*	*Br. interjuncta*	No synonym	***Br. interjuncta***
10.	***Bolivina 25A***	No synonym	*Bolivina mandoroveensis*	No synonym	*B. mandoroveensis*	***B. mandoroveensis***
11.	***Bolivina 26***	No synonym	No synonym	No synonym	*Brizalina beyrichi carinata*	***B. beyrichi***
12.	***Bolivina 27***	No synonym	No synonym	No synonym	*Brizalina paralica*	***Bolivina dertonensis***
13.	***Bolivina 46***	No synonym	No synonym	No synonym	*Brizalina aenariensis*	***B. aenariensis***
14.	***Bulimina 11***	No synonym	No synonym	No synonym	*Uvigerinella subfusiformis*	***Buliminella* aff. *subfusiformis***
15.	***Cancris 5***	No synonym	No synonym	No synonym	*Cancris auriculus*	***C. auriculus***
16.	***Cassidulina 7***	*Cassidulina neocarinata*	No synonym	No synonym	*C. neocarinata*	***C. neocarinata***
17.	***Cassigerinella 1***	No synonym	No synonym	*Cassigerinella chipolensis*	*C. chipolensis*	***C. chipolensis***
18.	***Cibicides 14***	No synonym	No synonym	No synonym	*Heterolepa pseudoungeriana*	***H. pseudoungeriana***
19.	***Cibicides 18***	No synonym	No synonym	No synonym	*Heterolepa floridana*	***H. floridana***
20.	***Cristellaria 13***	No synonym	No synonym	No synonym	No synonym	***Lenticulina grandis***
21.	***Cristellaria 14***	No synonym	No synonym	No synonym	No synonym	***Lenticulina pseudomamilligerus***
22.	***Cyclammina 3A, 3B***	*Haplophragmoides compressa*	No synonym	No synonym	No synonym	***H. compressa***

23.	***Elphidium 1***	No synonym	No synonym	No synonym	No synonym	***Cribrononion texana***
24.	***Eponides 11***	No synonym	No synonym	No synonym	*Eponides pseudoelevatus*	***E. pseudoelevatus***
25.	***Eponides 12***	No synonym	No synonym	No synonym	No synonym	***Cibicorbis inflata***
26.	***Eponides 13A***	No synonym	No synonym	No synonym	*H. pseudoungeriana*	***H. pseudoungeriana***
27.	***Globobulimina 1***	*Praeglobobulimina ovata*	No synonym	No synonym	*P. ovata*	***P. ovata***
28.	***Haplophragmoides 24***	*Haplophragmoides narivaensis*	No synonym	No synonym	No synonym	***H. narivaensis***
29.	***Karreriella 1, 2***	No synonym	No synonym	No synonym	*Karreriella siphonella*	***K. siphonella***
30.	***Nodosaria 9***	No synonym	No synonym	No synonym	No synonym	***Rectoglandulina comatula***
31.	***Nodosaria 9B***	No synonym	No synonym	No synonym	*Amphicorina scalaris*	***Am. scalaris caudata***
32.	***Nonion 4, 6.***	*Florilus ex gr. costiferum*	No synonym	No synonym	*F. costiferus*	***F. costiferum***
33.	***Nonion 16***	No synonym	No synonym	*Florilus boueanum*	*F. boueanum*	***F. boueanum***
34.	***Pullenia 3***	No synonym	No synonym	No synonym	*Pullenia bulloides*	***P. bulloides***
35.	***Rectuvigerina 1***	No synonym	No synonym	No synonym	*Rectuvigerina multicostata*	***R. multicostata***
36.	***Reophax 6***	No synonym	No synonym	No synonym	No synonym	***Reophax morrisoni***
37.	***Rotalia 1A***	No synonym	No synonym	*Ammonia beccarii*	*A. beccarii*	***A. beccarii***
38.	***Saccammina 2***	No synonym	No synonym	No synonym	No synonym	***Saccammina atlantica***
39.	***Sphaeroidina 1***	No synonym	No synonym	No synonym	*Sphaeroidina bulloides*	***S. bulloides***
40.	***Textularia 3***	No synonym	No synonym	No synonym	*Spiroplectammina wrightii*	***S. wrightii***
41.	***Uvigerina 5***	*Uvigerina sparsicostata*	No synonym	No synonym	*Uvigerina sparsicostata*	***Uvigerina sparsicostata***
42.	***Uvigerina 8***	*Uvigerina subperegrina*	No synonym	No synonym	No synonym	***U. subperegrina***
43.	***Uvigerinella 1***	No synonym	No synonym	No synonym	*Hopkinsina bononiensis*	***H. bononiensis***
44.	***Uvigerinella 3A***	No synonym	No synonym	No synonym	*Hopkinsina semiornata*	***H. semiornata***
45.	***Uvigerinella 5***	No synonym	No synonym	No synonym	No synonym	***Uvigerinella sparsicostata***
46.	***Uvigerinella 8***	No synonym	No synonym	No synonym	*Uvigerinella hourqi*	***Uvigerina hourqi***
47.	***Valvulineria 14 or 19***	No synonym	No synonym	No synonym	No synonym	***Hanzawaia strattoni***
48.	***Valvulina 1***	*Valvulina flexilis*	No synonym	No synonym	*V. flexilis*	***V. flexilis***
49.	***Virgulina 2***	No synonym	No synonym	No synonym	No synonym	***Fursenkoina punctata***
50.	***Virgulina 3***	No synonym	No synonym	No synonym	*Altistoma* cf. *A. tenuis*	***A. tenuis***
51.	***Virgulina 4, 11***	No synonym	No synonym	No synonym	*Altistoma scalaris*	***Al. scalaris***

(*Continued*)

TABLE 4.1 Synonyms of Selected Species From the Biostratigraphic Schemes of StratCom Members (*cont.*)

No.	SPDC	Mobil Nigeria Ltd.	Chevron Nigeria	Total/ELF	NAOC/AGIP	Scientific names
PLANKTONIC FORMS						
52.	*Globorotalia 20*	No synonym	No synonym	*Globorotalia opima*	No synonym	***Turborotalia opima***
53.	*Globorotalia 52*	No synonym	No synonym	No synonym	No synonym	***Turborotalia fohsi***
54.	*Globorotalia 56*	*Globorotalia mayeri*	No synonym	No synonym	*G. mayeri*	***Turborotalia mayeri***
55.	*Guembelina 4*	No synonym	No synonym	No synonym	*Chiloguembelina cubensis*	***C. cubensis***
56.	*Guembelina 5*	No synonym	No synonym	No synonym	*Chiloguembelina martini*	***C. martini***
57.	*Guembelina 12*	No synonym	No synonym	No synonym	No synonym	***Chiloguembelina wilcoxensis***
58.	*Guembelina 14*	No synonym	No synonym	No synonym	No synonym	***Chiloguembelina midwayensis***
59.	*Chiloguembelina 3*	No synonym	No synonym	No synonym	No synonym	***Chiloguembelina victoriana***
60.	*Cassigerinella 1*	No synonym	No synonym	No synonym	*C. chipolensis*	***C. chipolensis***
61.	*Globigerina 6*	No synonym	No synonym	No synonym	*Cassigerinelloita amekii*	***Cassigerinelloita amekiensis***
62.	*Globigerina 41*	No synonym	No synonym	No synonym	No synonym	***Globigerina angustiumbilicata***
63.	*Globigerina 42*	No synonym	No synonym	No synonym	No synonym	***Globigerina angulosuturalis***
64.	*Globigerina 66*	No synonym	No synonym	No synonym	No synonym	***Globigerina ouachitaensis***
65.	*Globorotalia 5*	No synonym	No synonym	No synonym	No synonym	***Morozovella angulata***
66.	*Globorotalia 16*	No synonym	No synonym	*Turborotalia peripheroronda*	No synonym	***T. peripheroronda***
67.	*Globorotalia 58*	No synonym	No synonym	No synonym	No synonym	***Turborotalia nana***
68.	*Globorotalia 20*	No synonym	No synonym	*T. opima*	No synonym	***T. opima***
69.	*Globorotalia 26*	No synonym	*Turborotalia kugleri*	*T. kugleri*	No synonym	***T. kugleri***
70.	*Globorotalia 27*	No synonym	No synonym	No synonym	No synonym	***Globorotalia archeomenardii***
71.	*Globorotalia 34*	No synonym	No synonym	No synonym	No synonym	***Globorotalia juvenelis***
72.	*Globorotalia 50*	No synonym	No synonym	No synonym	No synonym	***Globorotalia crassaformis***
73.	*Globorotalia 52*	No synonym	No synonym	No synonym	No synonym	***T. fohsi***
74.	*Globorotalia 63*	No synonym	No synonym	*Globorotalia tumida*	No synonym	***G. tumida***
75.	*Guembelitria 3*	No synonym	No synonym	No synonym	*Guembelitria columbiana*	***G. columbiana***
76.	*Pseudohastigerina micra*	No synonym	No synonym	*P. micra*	*P. micra*	***P. micra***

carried out for the Oligocene to Pliocene based largely on sequence stratigraphic principles and the stratigraphic ranges of both benthonic and planktonic forms (Mosunmolu Limited, 1991, 1992a,b, 1996). More recently, Reijers (2011) proposed a delta-wide biostratigraphic framework with alphanumeric codes developed from the biostratigraphic zones of SPDC and calibrated with data from ELF (Durand, 1995). This zonation allows correlation across all facies types from the continental zone to deeper marine. Currently, the SPDC zonation contains 18 foraminiferal biozones and 12 subzones. Hierarchically, nine partial range zones, six taxon zones, two assemblage zones, and one concurrent range zone are included. Fifteen of these zones are based on index planktonic foraminifera recognized from transgressive marine shales (Petters, 1983), while eight zones are based on benthonic foraminifera.

Another elaborate chronostratigraphic scheme was developed by Mobil Producing Nigeria (MPN) Unlimited (Fig. 4.2) applicable to the company's offshore concessions at the extreme eastern margin of the Niger Delta. The five Faunal Zones (FZ I–V) recognized were erected on the basis of peak faunal abundances and characterized by well-documented, stratigraphically constrained, benthonic and planktonic species that include a significant number of endemic forms. The MPN Faunal Zones thus anticipated the use of sequence stratigraphy in chronostratigraphy.

ELF Petroleum Nigeria Limited (EPNL) (Fig. 4.3), the NAOC (Fig. 4.4), and Chevron Nigeria Limited (CNL) (Fig. 4.5) also have well-developed chronostratigraphic schemes ranging from Oligocene to Holocene, Eocene to Miocene, and Miocene to Holocene, respectively, based on benthonic and planktonic bioevents. The TOPCON zonal scheme covers the period Early Miocene to Pleistocene and is based on correlation with Blow's (1979) N-series planktonic foraminiferal zones.

The Integrated Chronostratigraphic scheme resulting from the harmonization of the chronostratigraphic charts of the StratCom member companies and the inferred synonyms are shown in Fig. 4.6, Enclosure 1.2, and Table 4.1.

4.2 THE NIGER DELTA CHRONOSTRATIGRAPHIC SCHEME (ENCLOSURE 1.1)

The most significant objective of this project is to develop and calibrate a unified Foraminifera Biozonal Scheme that can be applied in all concessions throughout the Niger Delta. Arising from the harmonization of the Foraminifera Chronostratigraphic schemes of the member companies of StratCom (Fig. 4.6) and taking into consideration the industry practice elsewhere, it was decided to propose a new Chronostratigraphic Scheme that is Sequence Stratigraphy driven. By carefully selecting wells that penetrated successive Epochs from Paleocene to Pleistocene across the different Niger Delta onshore and shallow-offshore Depobelts (Tables 1.1 and 1.2), it was possible to document the various cyclic marine events that have characterized the Niger Delta since the beginning of the Cenozoic Era. The biofacies plots for selected representatives of these wells are shown in Figs. 4.7–4.23. It should, however, be pointed out that only a few of the wells selected for study penetrated the Paleogene.

The onshore and shallow-offshore shelf sequences expectedly contain abundant benthonic microfauna with little or no planktonic species. The zones recognized in these

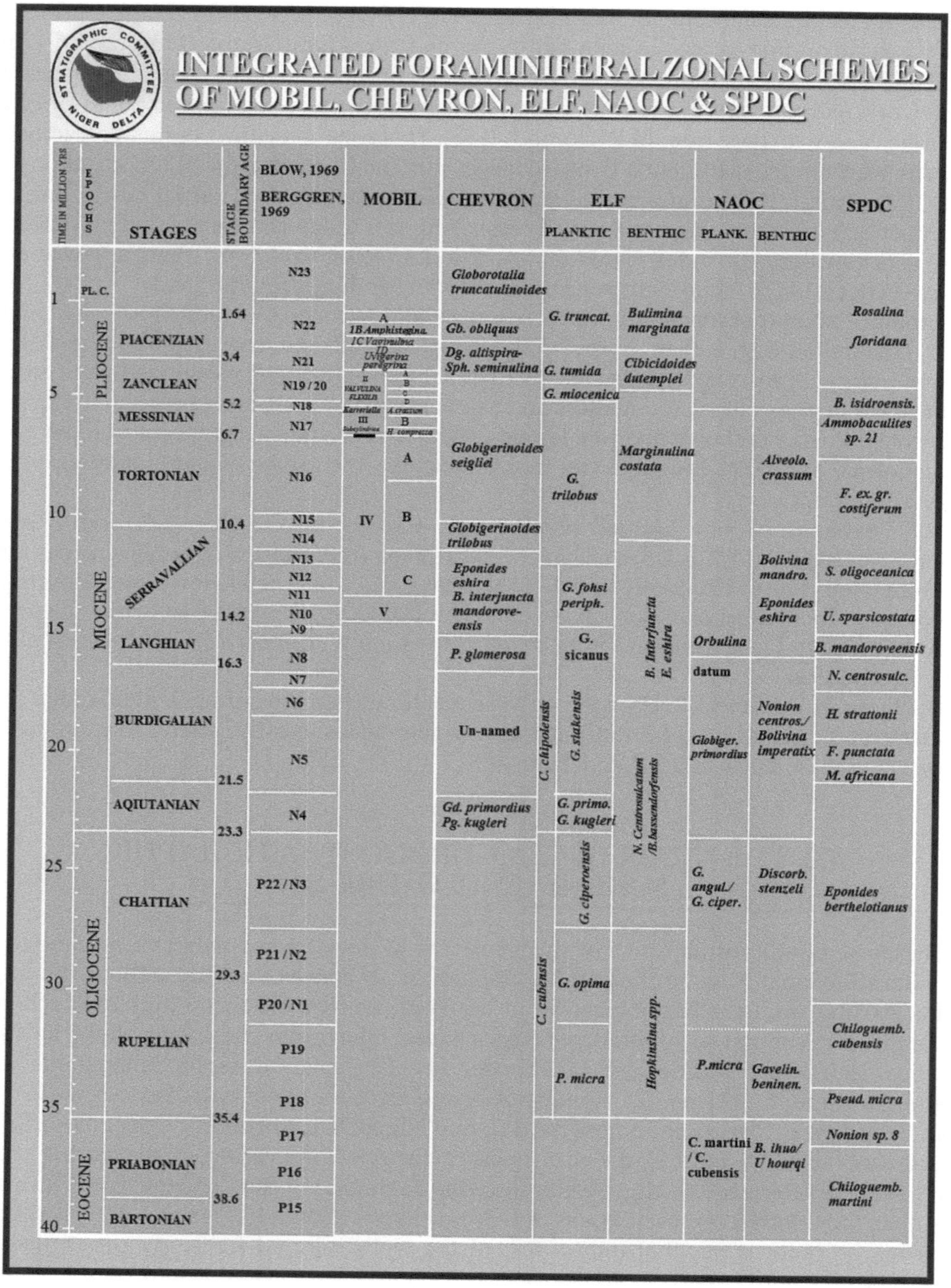

FIGURE 4.6 **Integrated Foraminiferal Zonal Schemes of Mobil, Chevron, ELF, NAOC, and SPDC.**

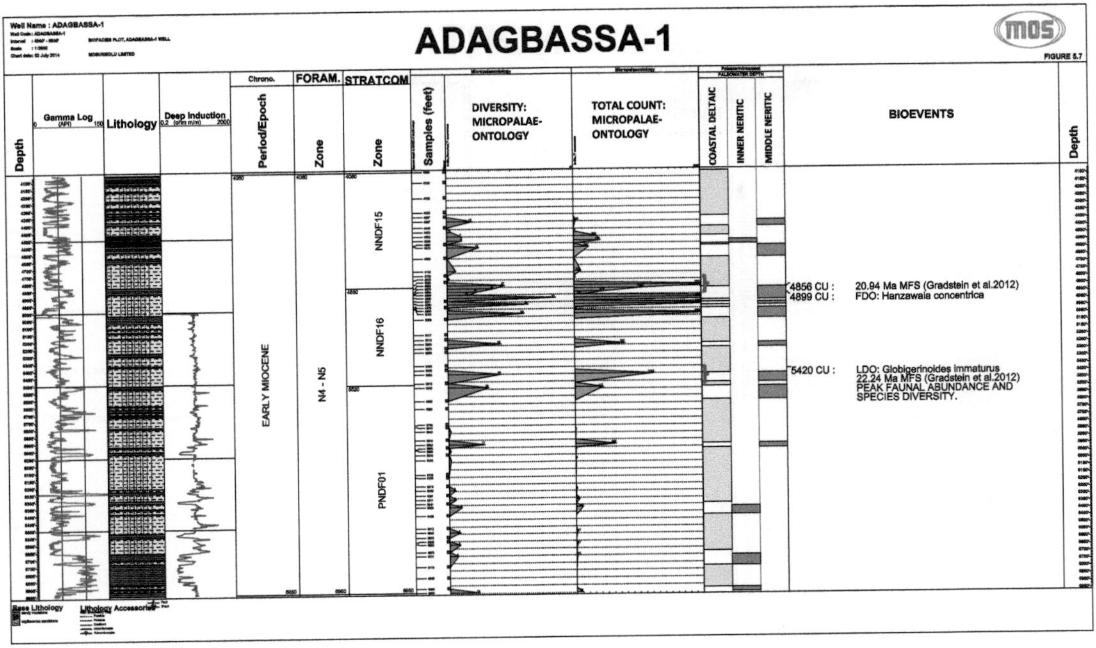

FIGURE 4.7 **Biofacies plot, Adagbassa-1 well.**

intervals were, therefore, characterized by well-documented benthonic bioevents, complemented by a few, well-tested Niger Delta planktonic biodatums and bioevents. In the distal and deeper offshore sections in which planktonic foraminifera dominates the microfauna, the abundant planktonics, which include many globally recognized marker species, formed the basis for defining the new Zones. Due to their ease of recognition and their usefulness for both local and regional correlations, MFS were used as zonal boundaries in this study.

The Niger Delta-Wide Zones recognized are code numbered as indicated:

- QNDf 01 = Quaternary Niger Delta Planktonic Foraminiferal Zone
- QNDF 01 = Quaternary Niger Delta Benthonic Foraminiferal Zone
- NNDf 01–13 = Neogene Niger Delta Planktonic Foraminiferal Zones
- NNDF 01–16 = Neogene Niger Delta Benthonic Foraminiferal Zones
- PNDf 01–19 = Paleogene Niger Delta Planktonic Foraminiferal Zones
- PNDF 01–25 = Paleogene Niger Delta Benthonic Foraminiferal Zones

The proposed new, Delta-wide Foraminifera Chronostratigraphic Scheme based on these codes is shown in Enclosure 1.1. The Foraminifera Zones proposed in this study are correlated with the Nannofosssil and Palynomorph Zones of the Niger Delta in the Integrated Biostratigraphic Zonation Chart displayed as Enclosure 1.2 in this report. The biofacies plots of selected critical Niger Delta wells are shown in Figs. 4.7–4.23.

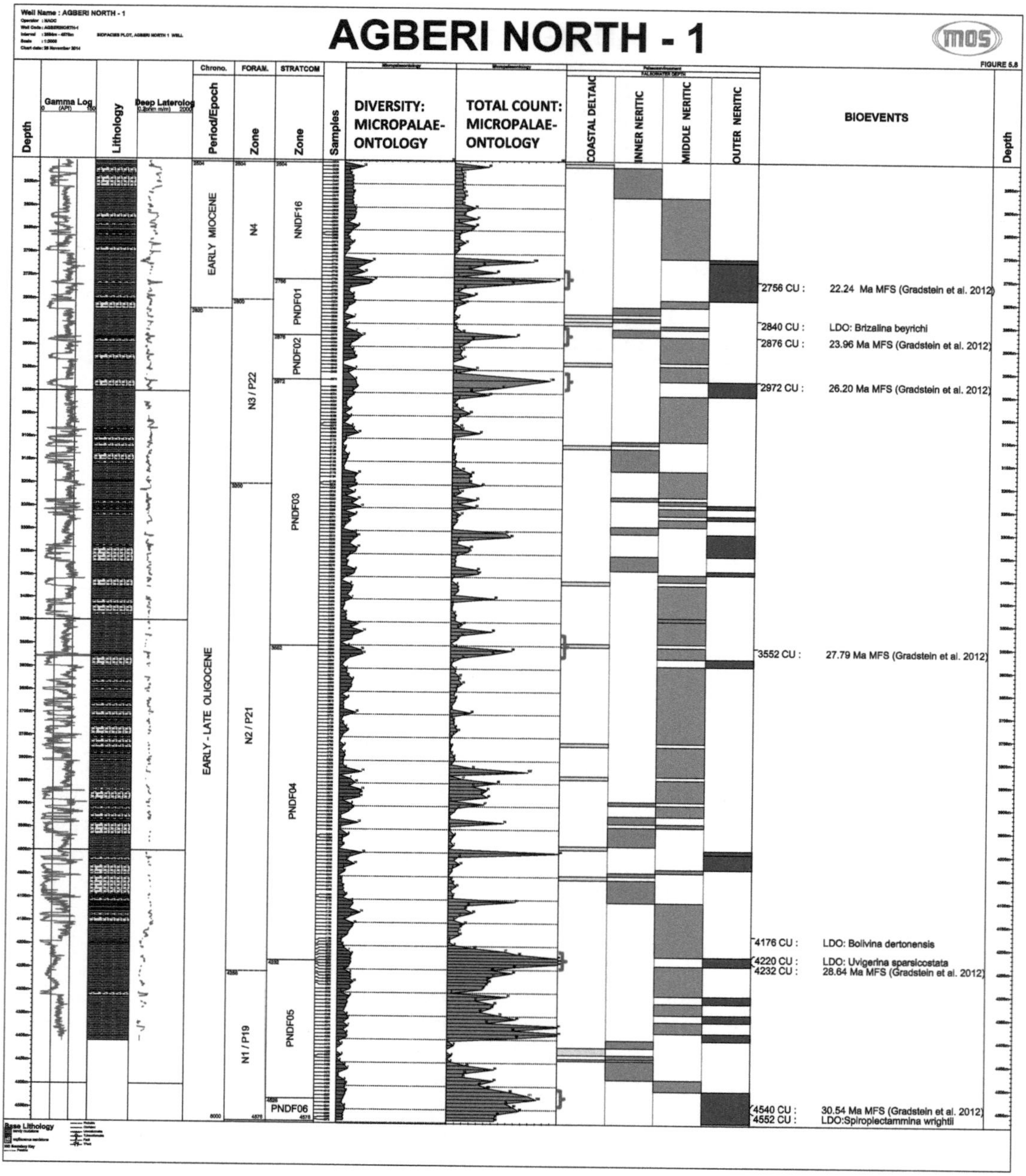

FIGURE 4.8 **Biofacies plot, Agberi North-1 well.**

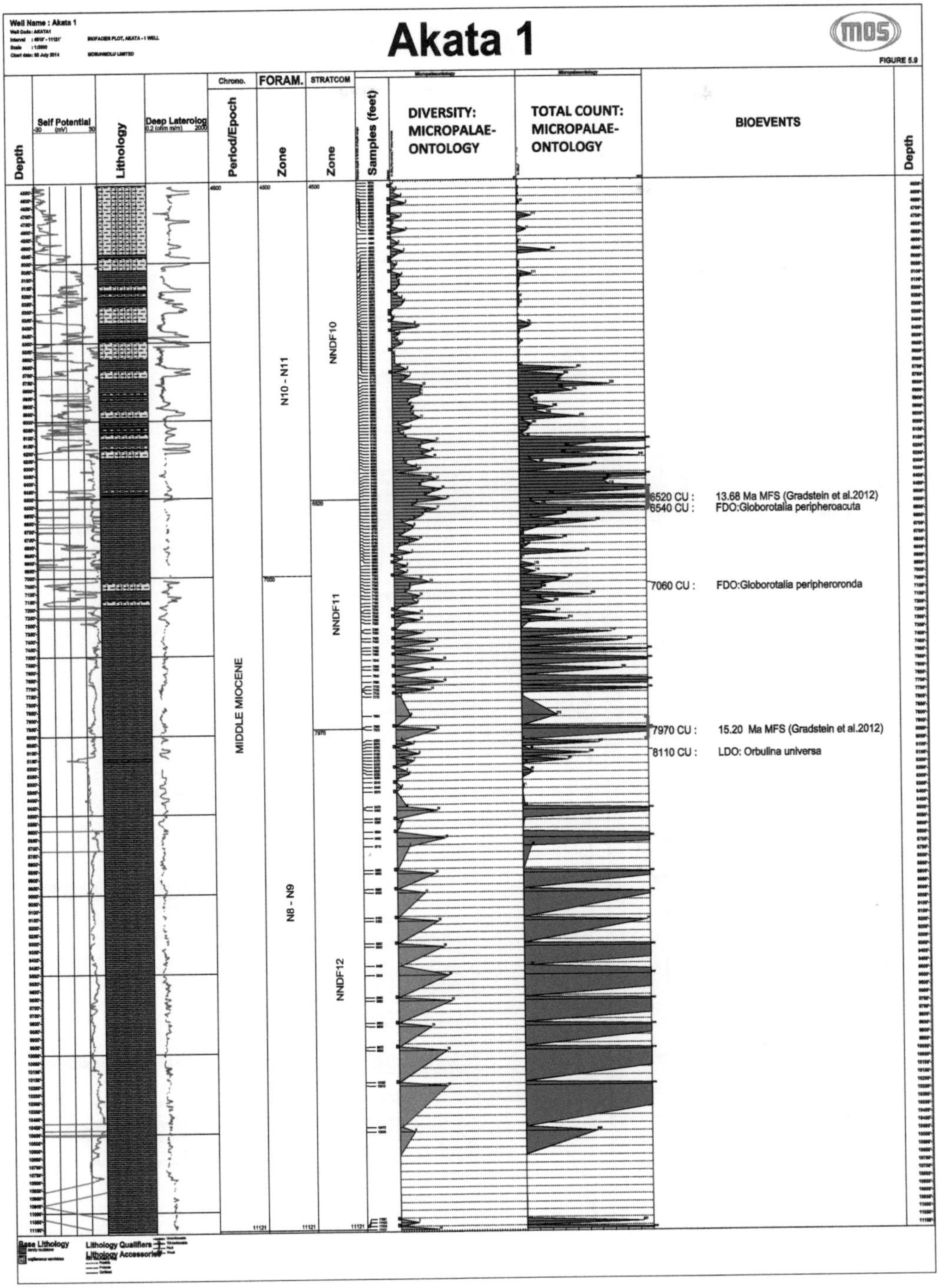

FIGURE 4.9 **Biofacies plot, Akata-1 well.**

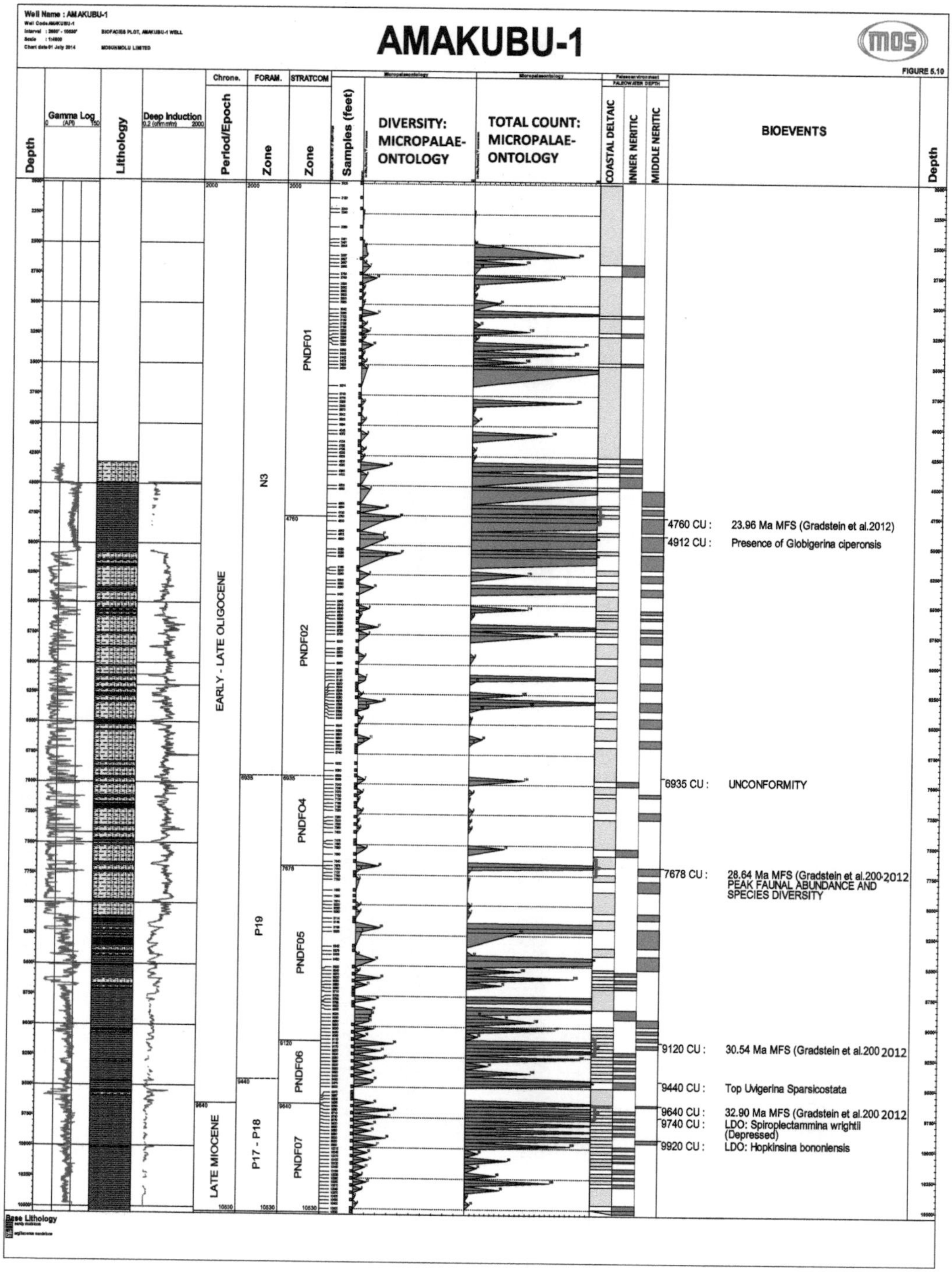

FIGURE 4.10 **Biofacies plot, Amakubu-1 well.**

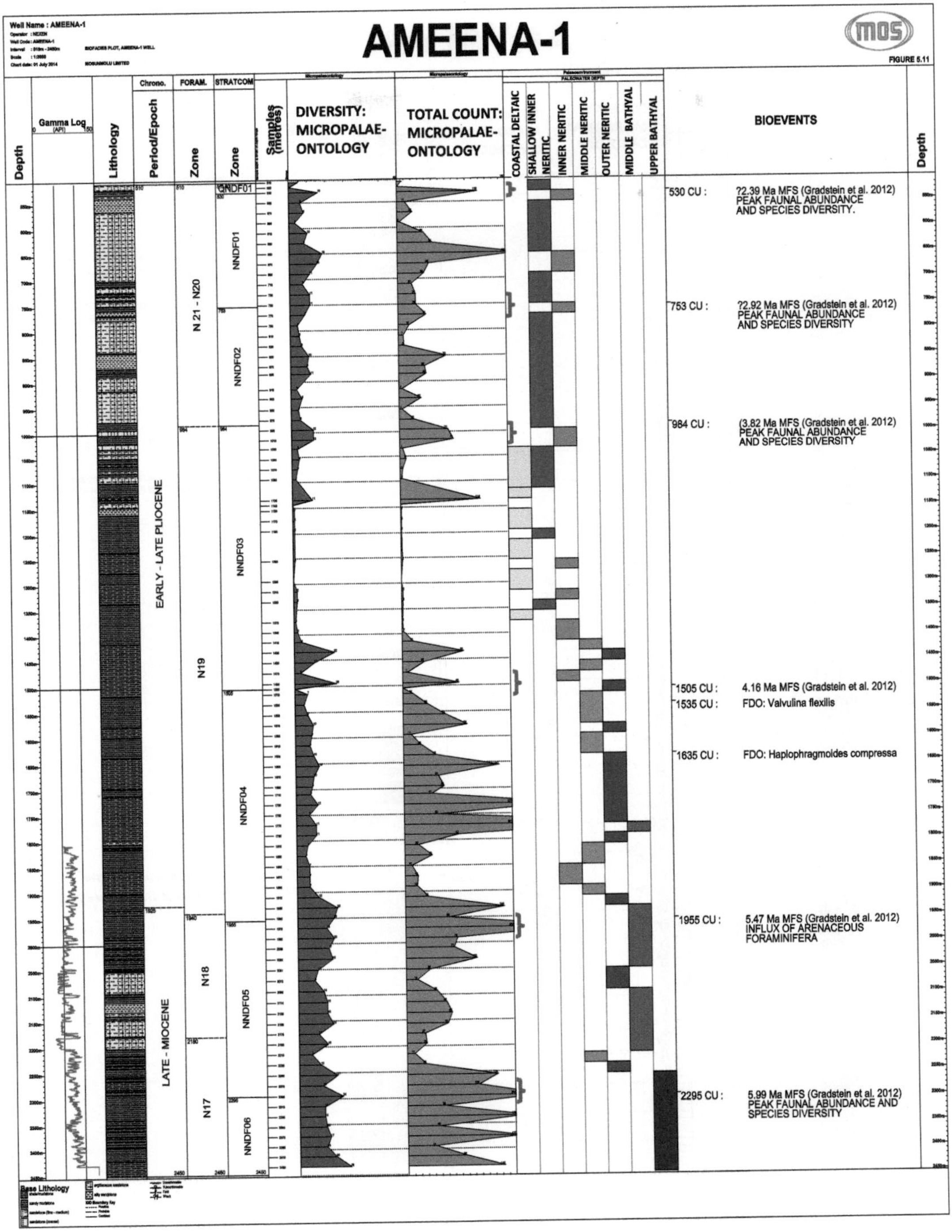

FIGURE 4.11 **Biofacies plot, Ameena-1 well.**

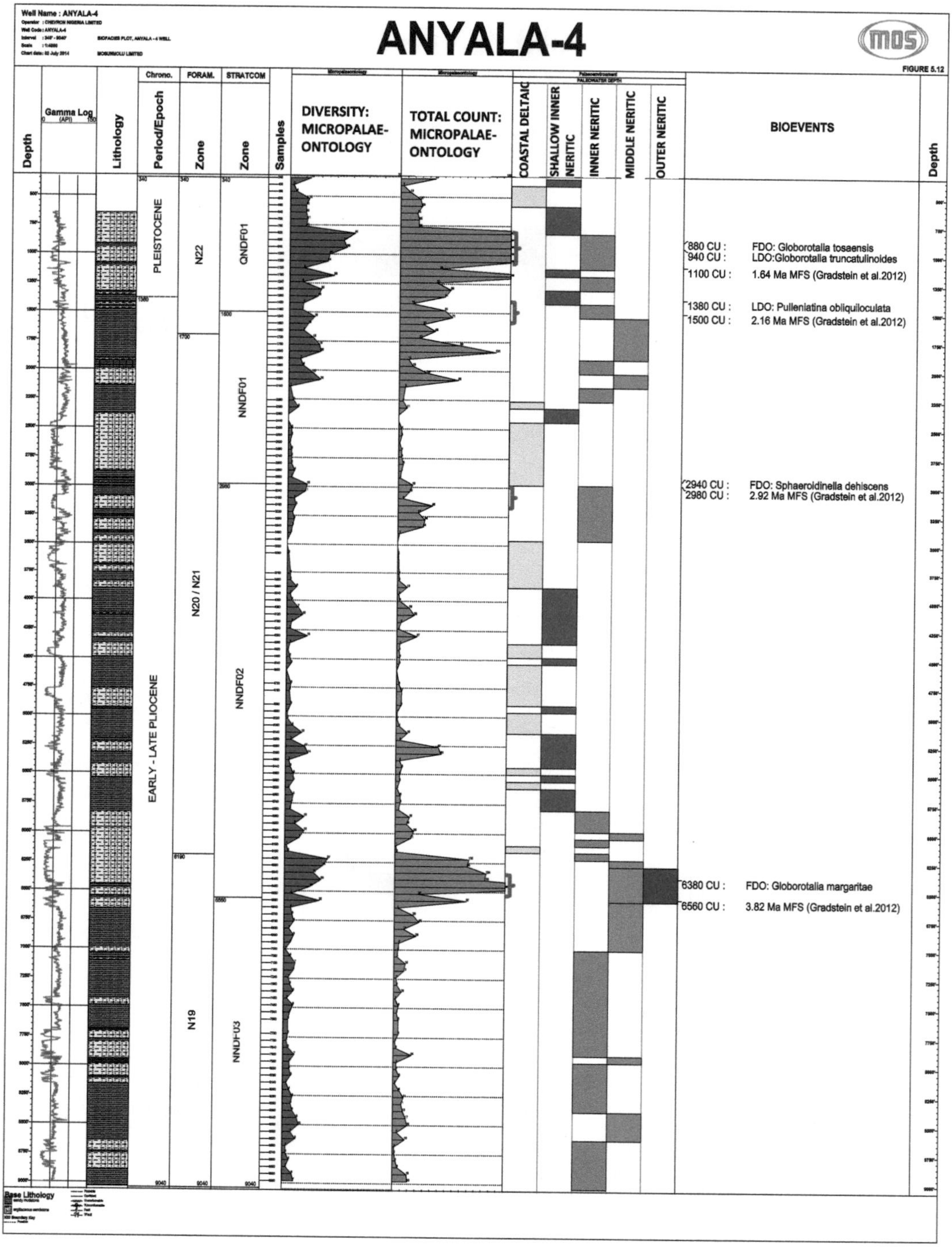

FIGURE 4.12 **Biofacies plot, Anyala-4 well.**

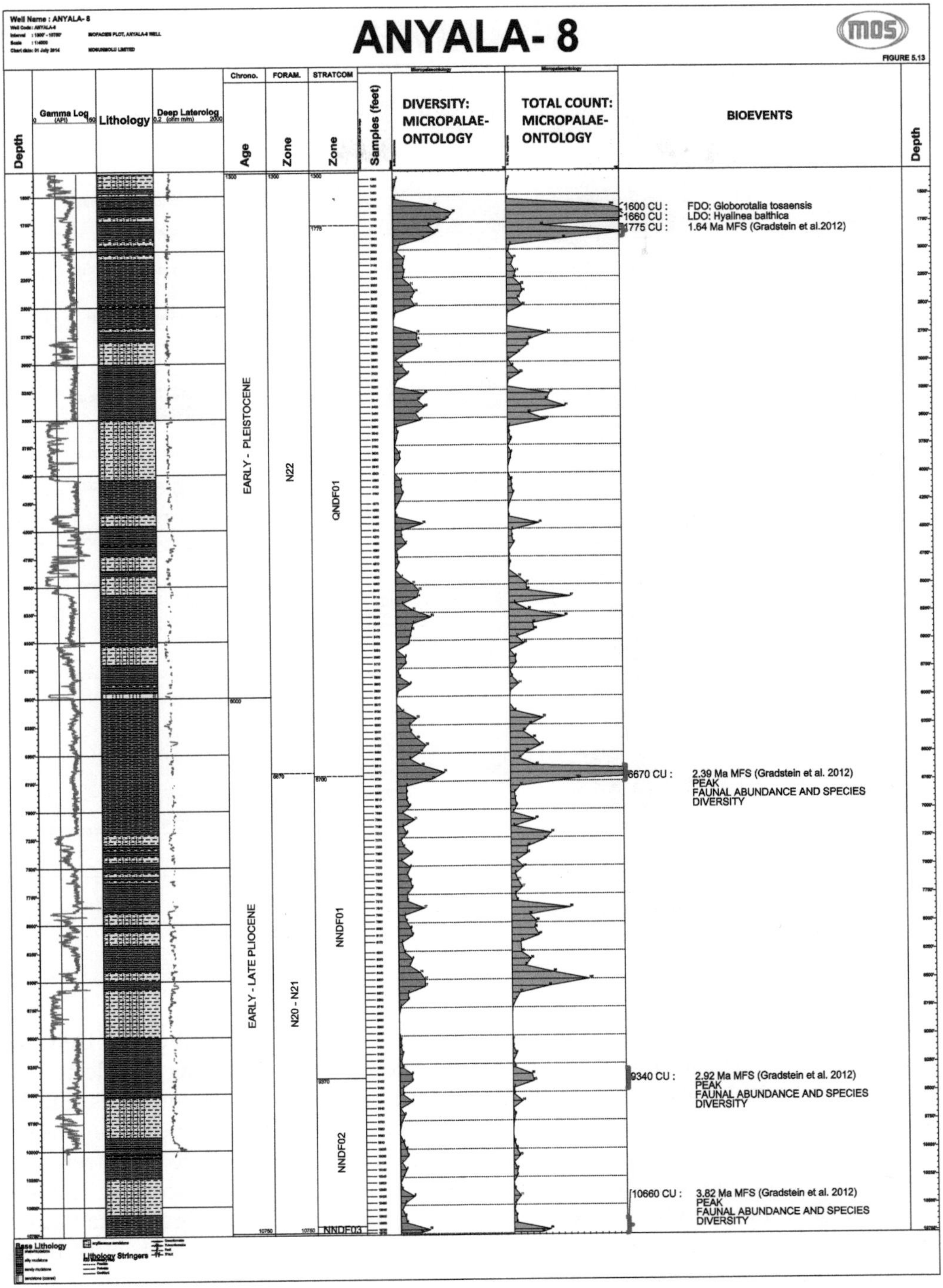

FIGURE 4.13 **Biofacies plot, Anyala-8 well.**

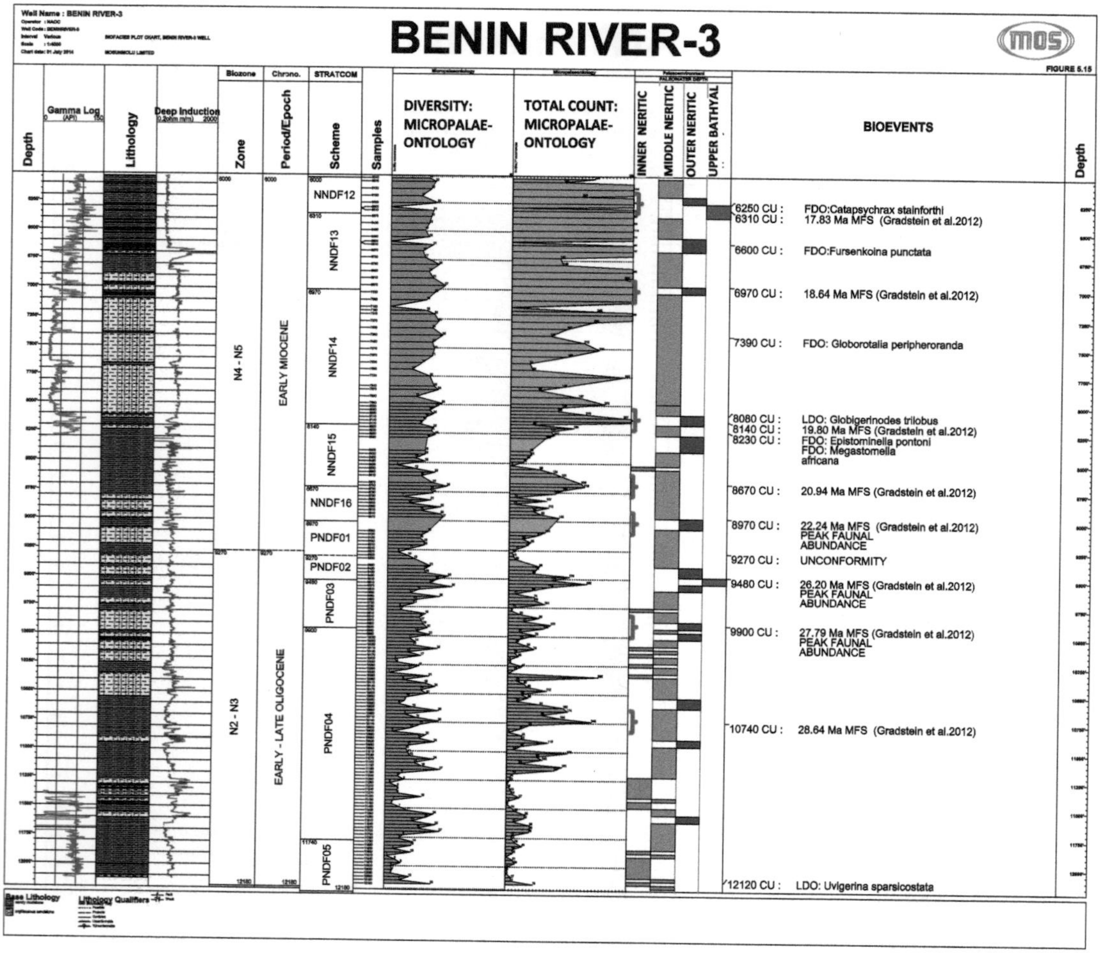

FIGURE 4.14 **Biofacies plot, Benin River-3 well.**

4.2.1 General Principles

The High-Resolution Biostratigraphic interpretation of the 36 studied Niger Delta wells was based on sequence stratigraphic principles. The detailed checklists of selected wells (Enclosures 1.3–1.19) were studied for the First and Last Downhole Occurrences (FDO and LDO, respectively) of chronostratigraphically significant foraminifera species. From the peaks of population abundance and species diversity (Figs. 4.7–4.23), candidate MFS were identified. Their positions were confirmed on wireline logs. These and the associated condensed sections represent the periods of maximum marine incursion into the Niger Delta basin that invariably coincide with periods of subsidence and deepening of the basin of sedimentation (Wornardt and Vail, 1990; Mitchum et al., 1977). These periods were accompanied by deposition of vast thicknesses of clays and shales that are seismically mappable and are very well-known in the Niger Delta (e.g., Ogara Shale and *Chiloguembelina* Shale of SPDC).

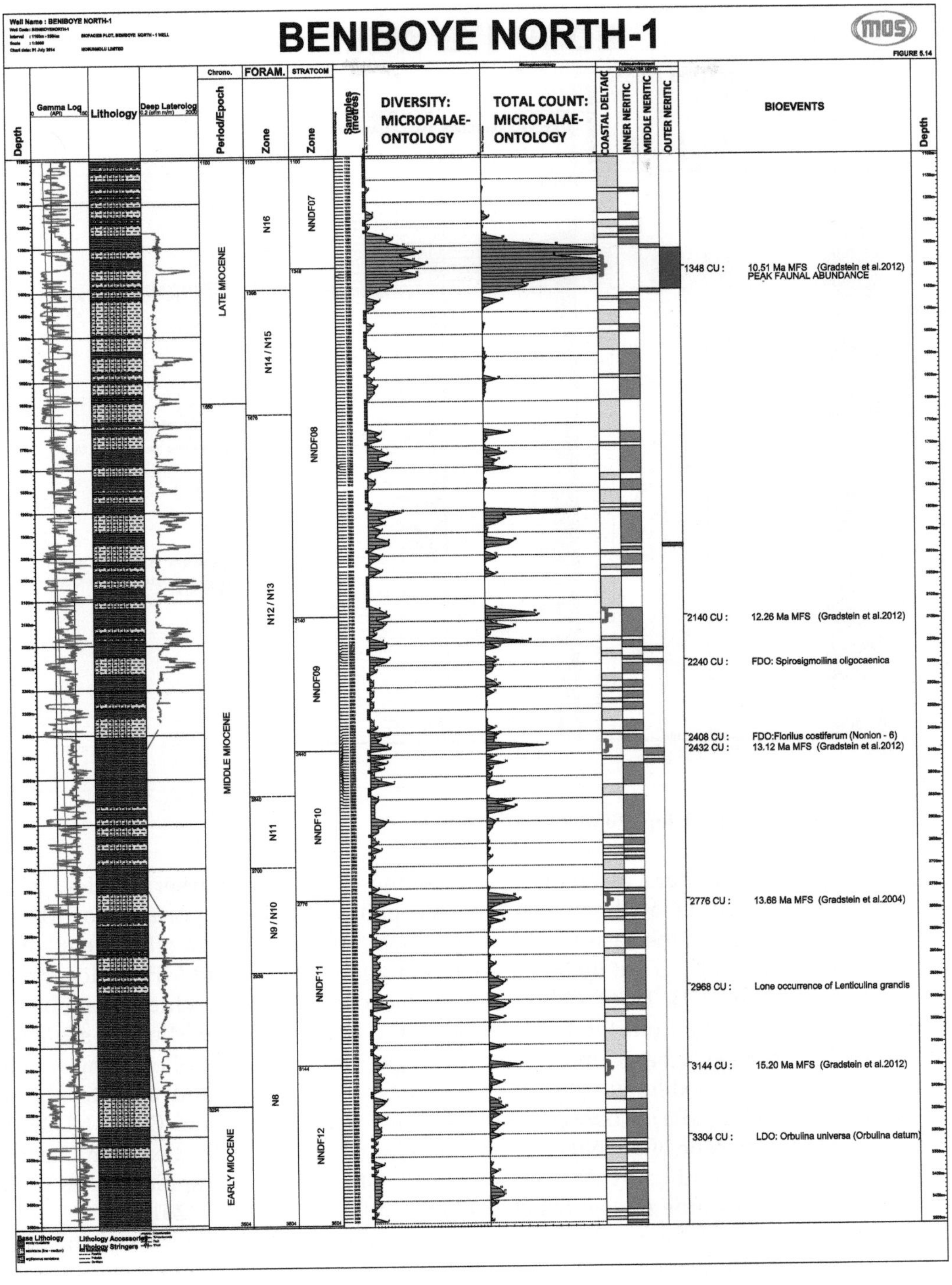

FIGURE 4.15 **Biofacies plot, Beninboye North-1 well.**

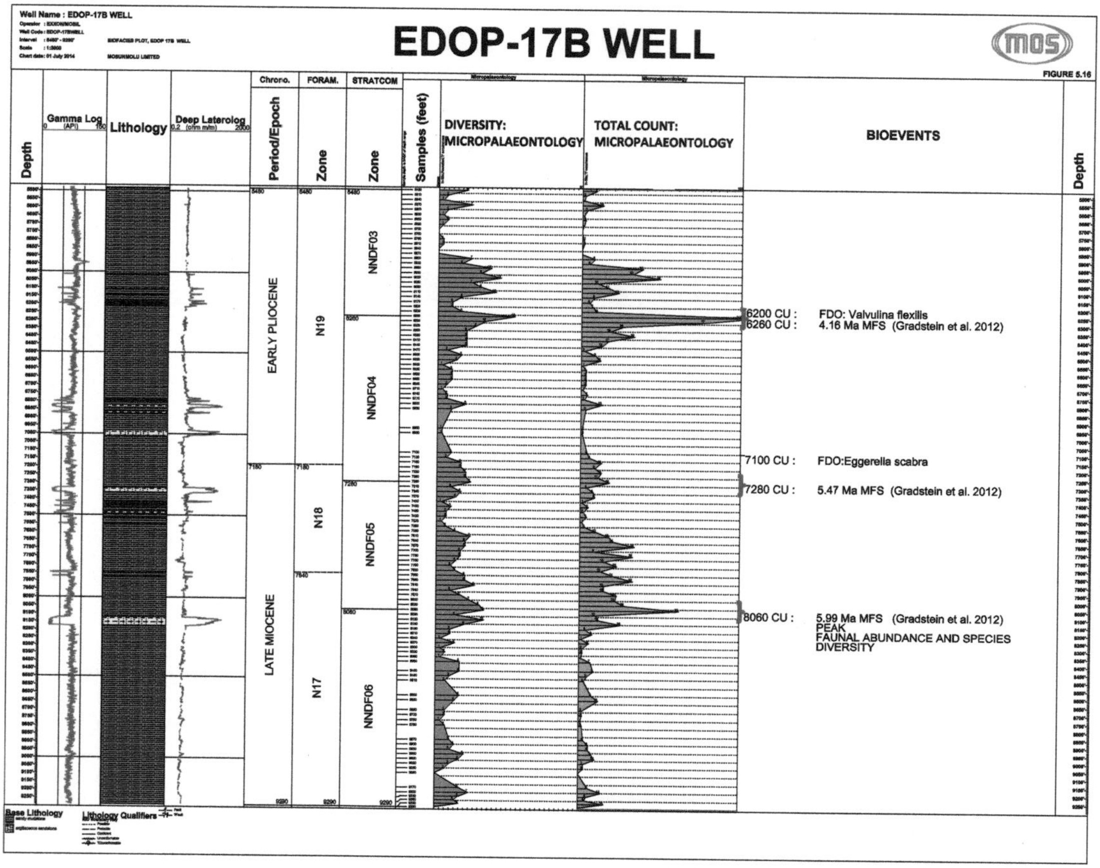

FIGURE 4.16 **Biofacies plot, Edop-17B well.**

They often have foraminifera assemblages containing diagnostic marker species, which permit the accurate dating of the shales. The recognition of the importance of MFS in chronostratigraphy formed the basis for the "high-resolution foraminifera refinement" of the Niger Delta sequences carried out by Mosunmolu Limited for SPDC and MPN (Mosunmolu Limited, 1991, 1992a,b, 1995, 1996). Other relevant biozonal schemes include Ozumba and Amajor (1999b) who carried out the biozonation of the Middle to Late Miocene sequences of the western Niger Delta based mainly on monospecific, facies-controlled, long-ranging forms with only a few moderately preserved age diagnostic benthic and planktic species. More recently, Ozumba (2011) proposed 23 planktonic foraminifera zones (ND1–ND23) for the Paleogene to Pleistocene of the Niger Delta that is largely based on the taxonomic scheme of Bolli et al., as later modified by Zachariasse (1990). Other biostratigraphic studies on the Cenozoic Niger Delta include Akaegbobi et al. (2009) and Adegoke et al. (2011).

The importance of recognizing MFS in the high-resolution biostratigraphy of the Niger Delta cannot be overemphasized because they are known to represent major cyclic bioevents of regional and stratigraphic significance. Due to their ease of recognition from biostratigraphic, seismic, as well as well log data, they have assumed tremendous importance in both

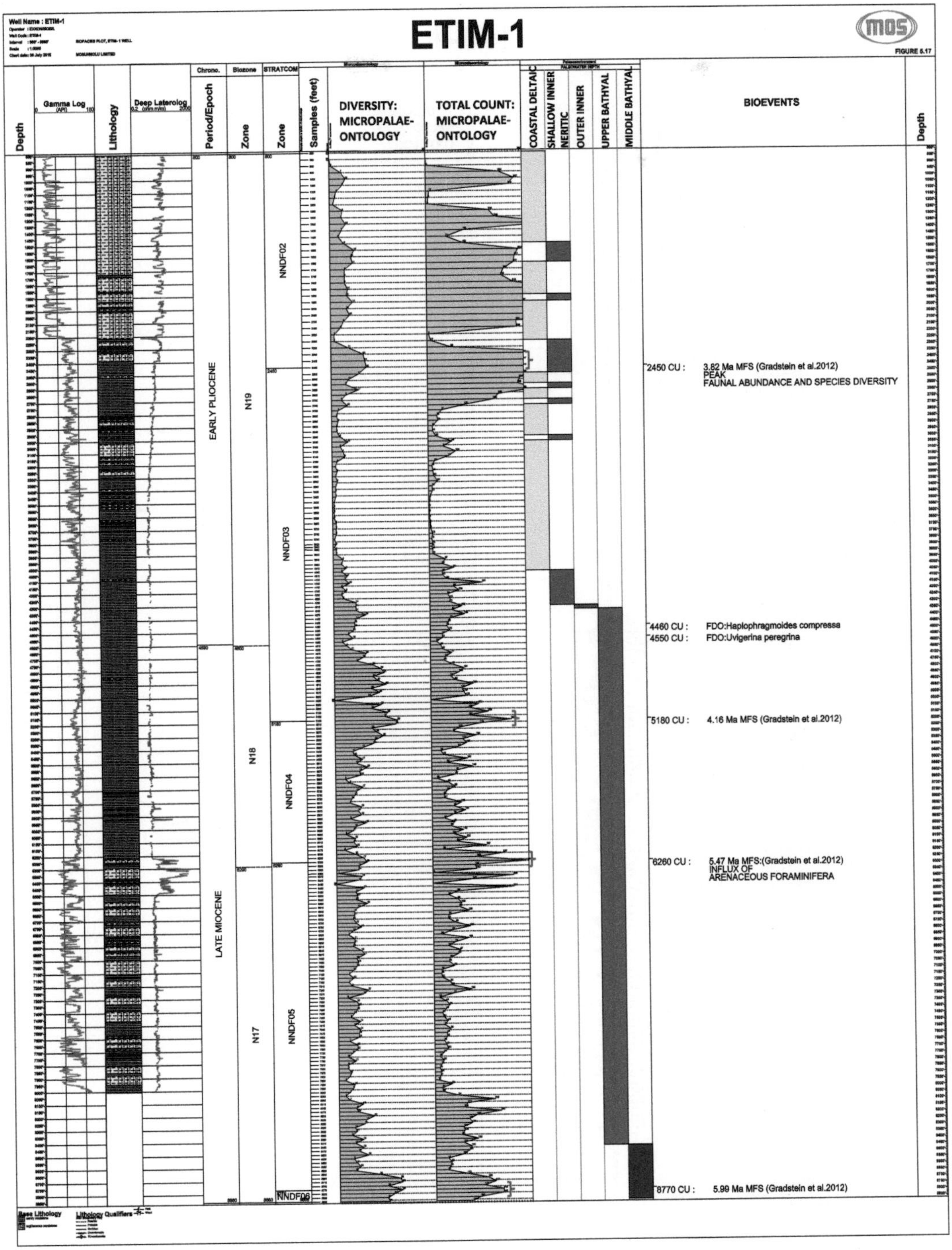

FIGURE 4.17 **Biofacies plot, Etim-1 well.**

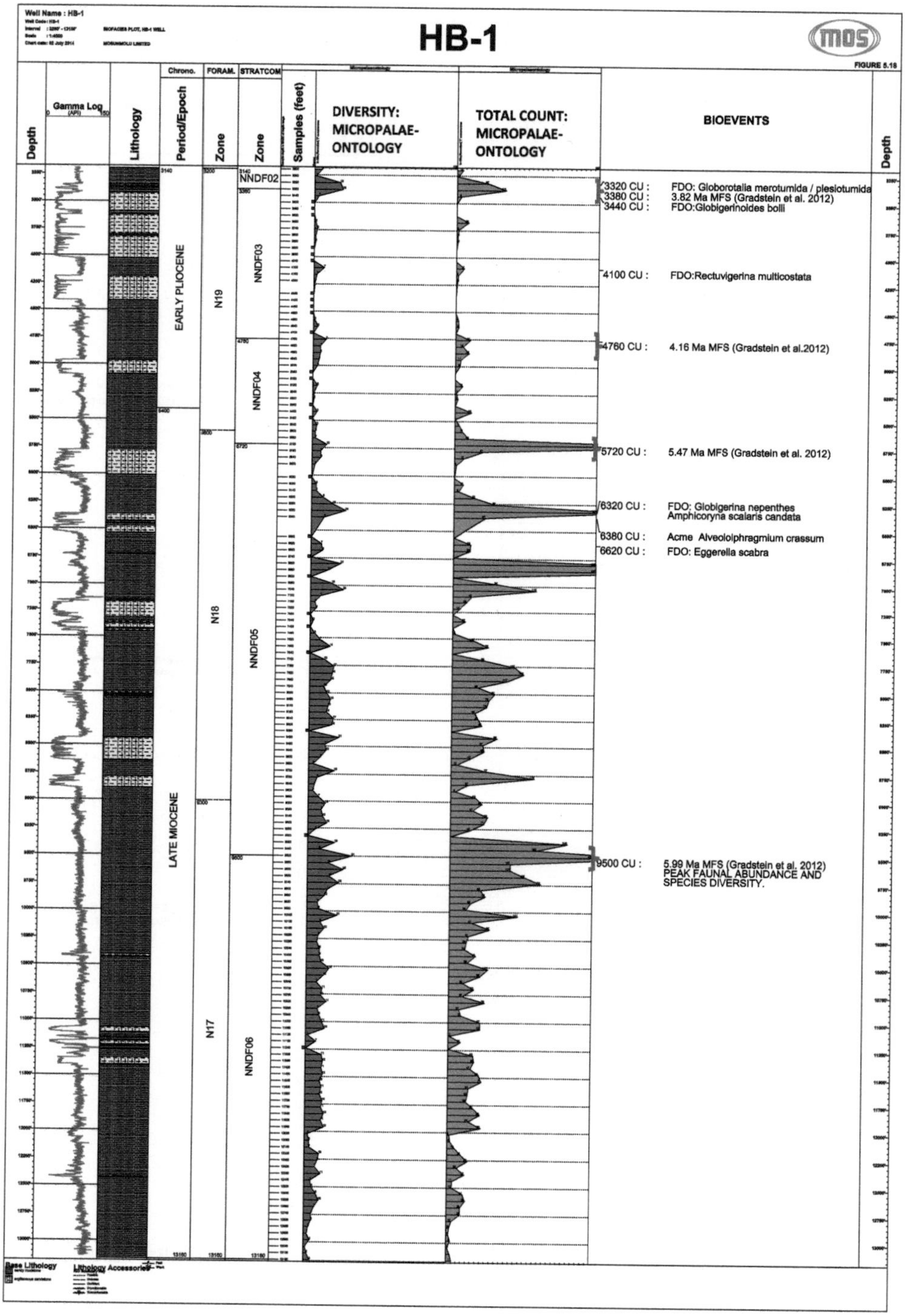

FIGURE 4.18 **Biofacies plot, HB-1 well.**

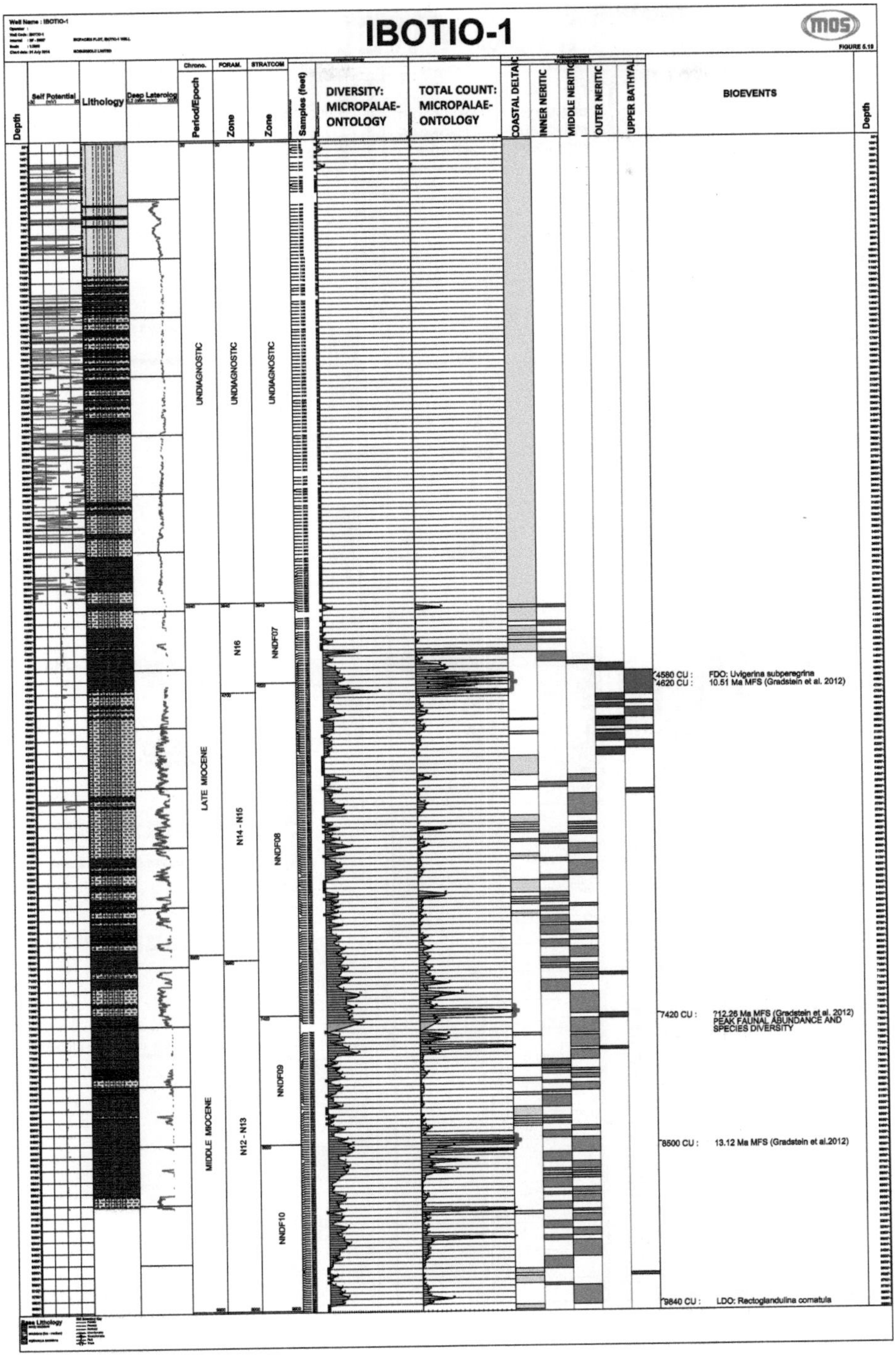

FIGURE 4.19 **Biofacies plot, Ibotio-1 well.**

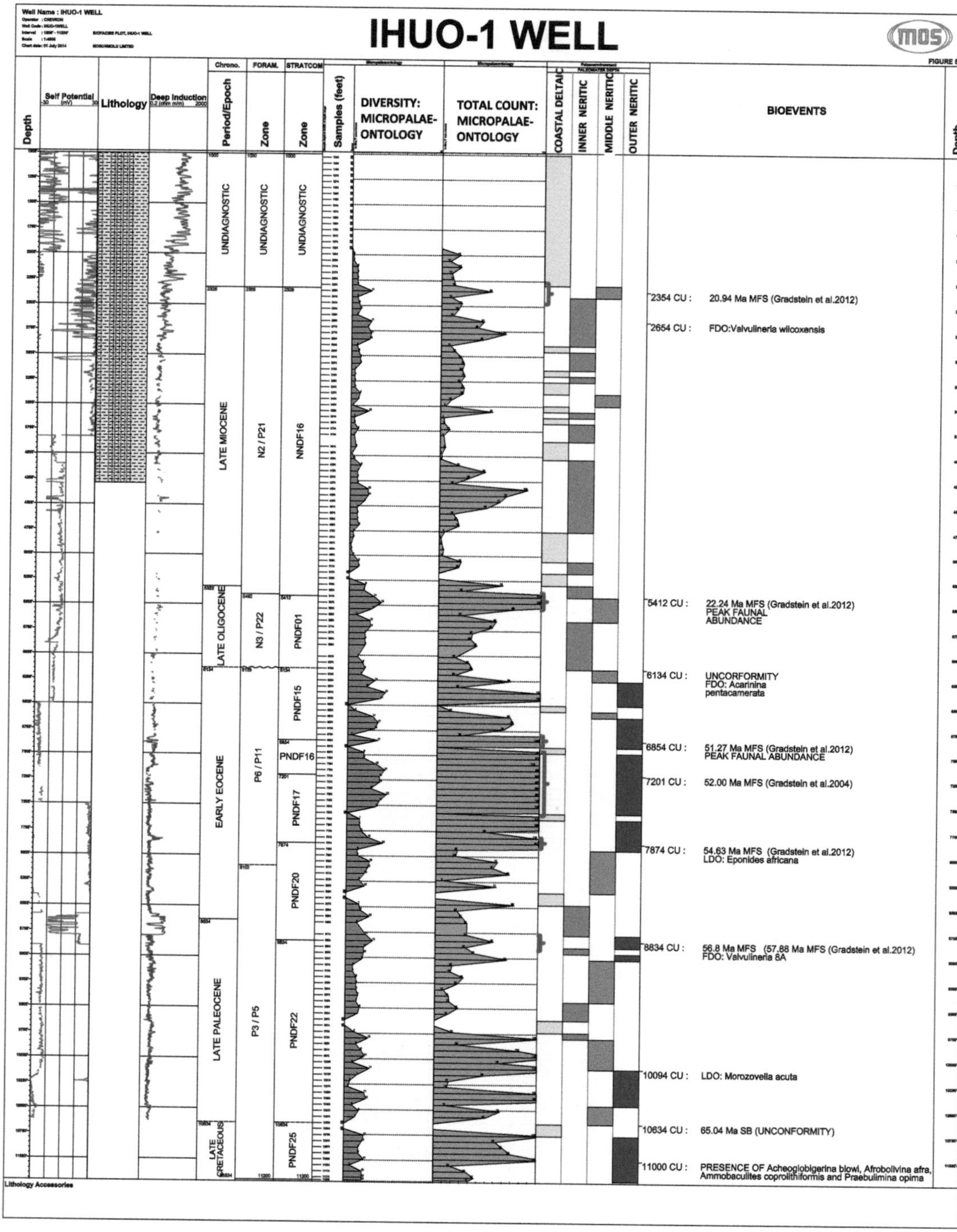

FIGURE 4.20 **Biofacies plot, Ihuo-1 well.**

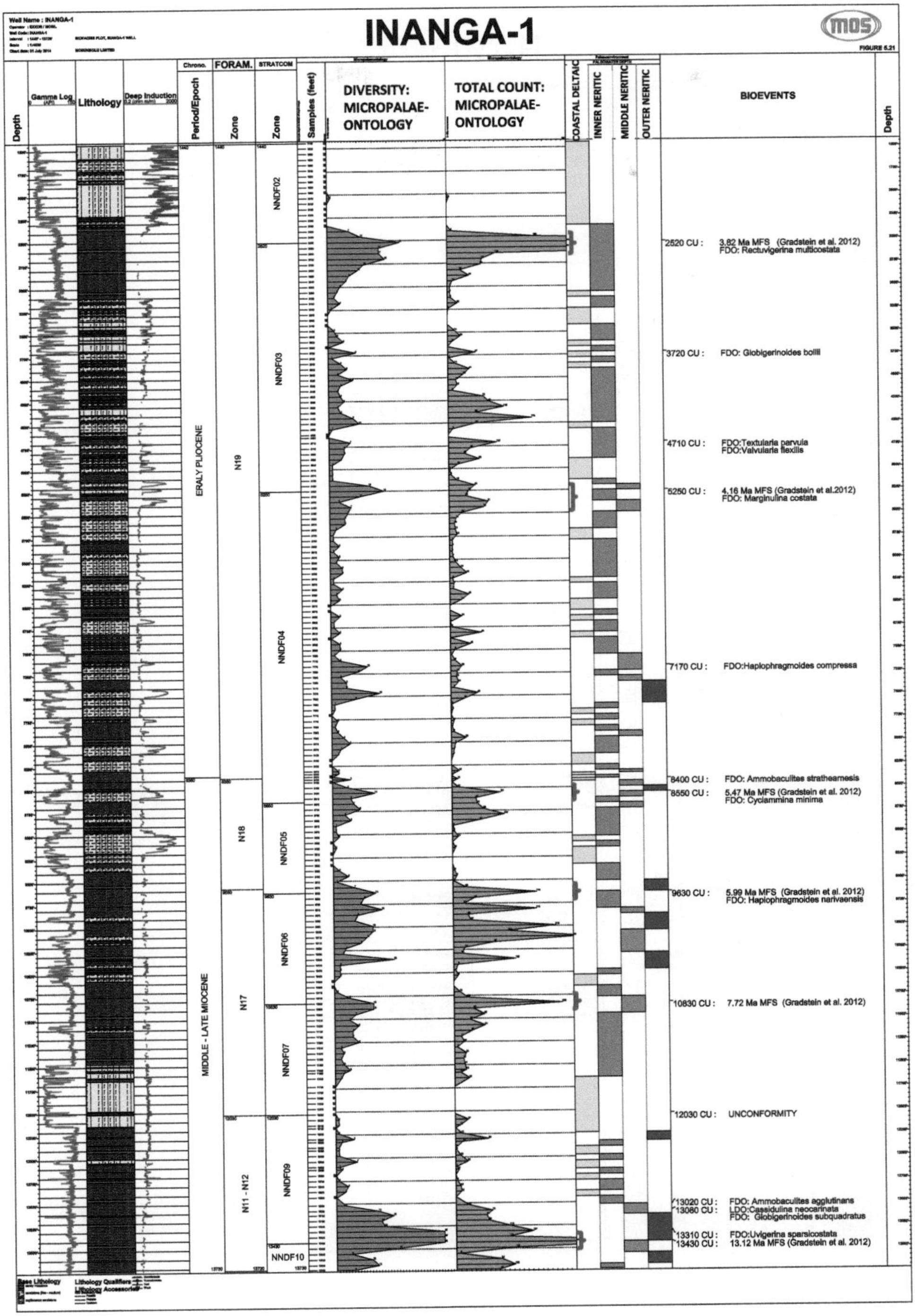

FIGURE 4.21 **Biofacies plot, Inanga-1 well.**

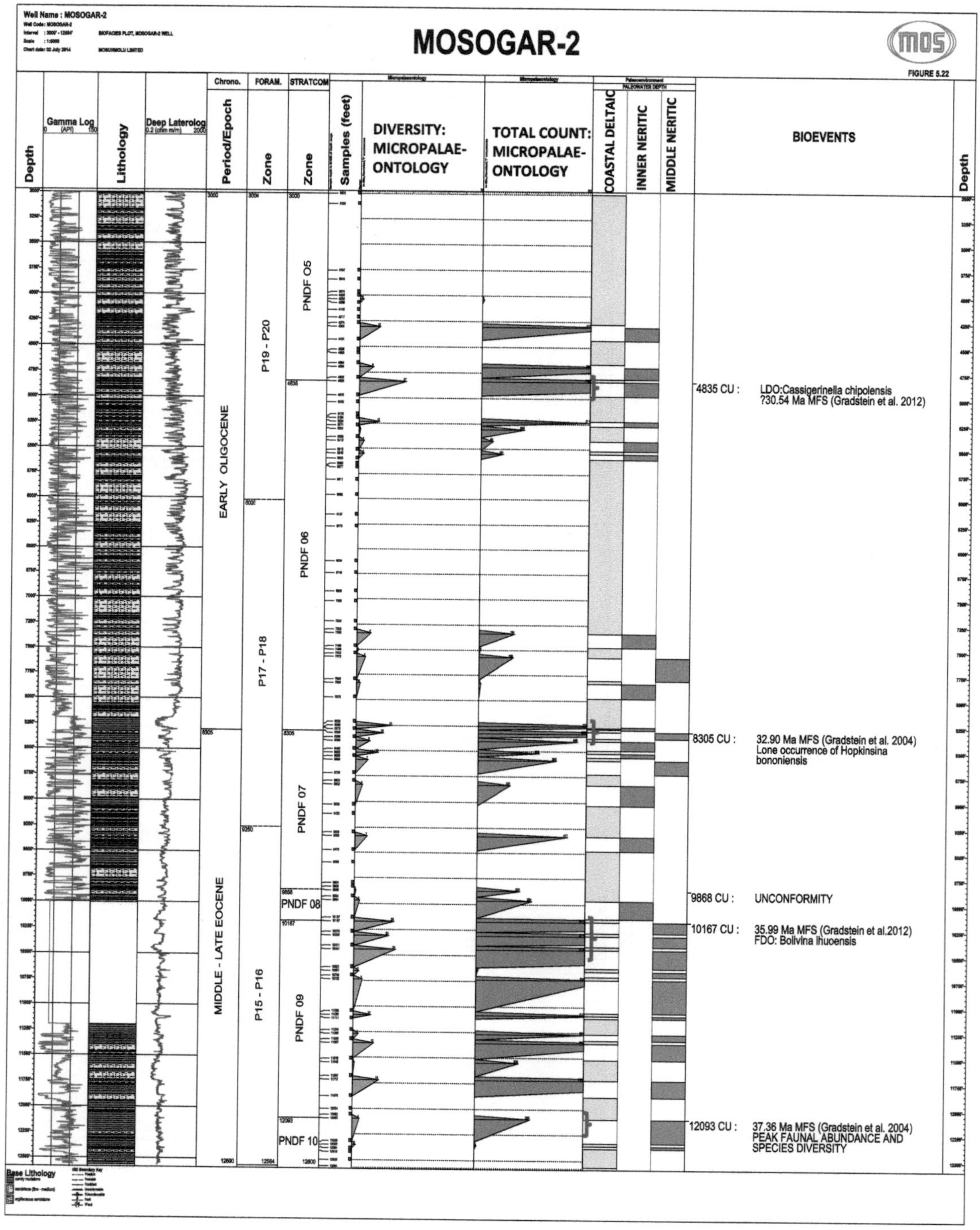

FIGURE 4.22 **Biofacies plot, Mosogar-2 well.**

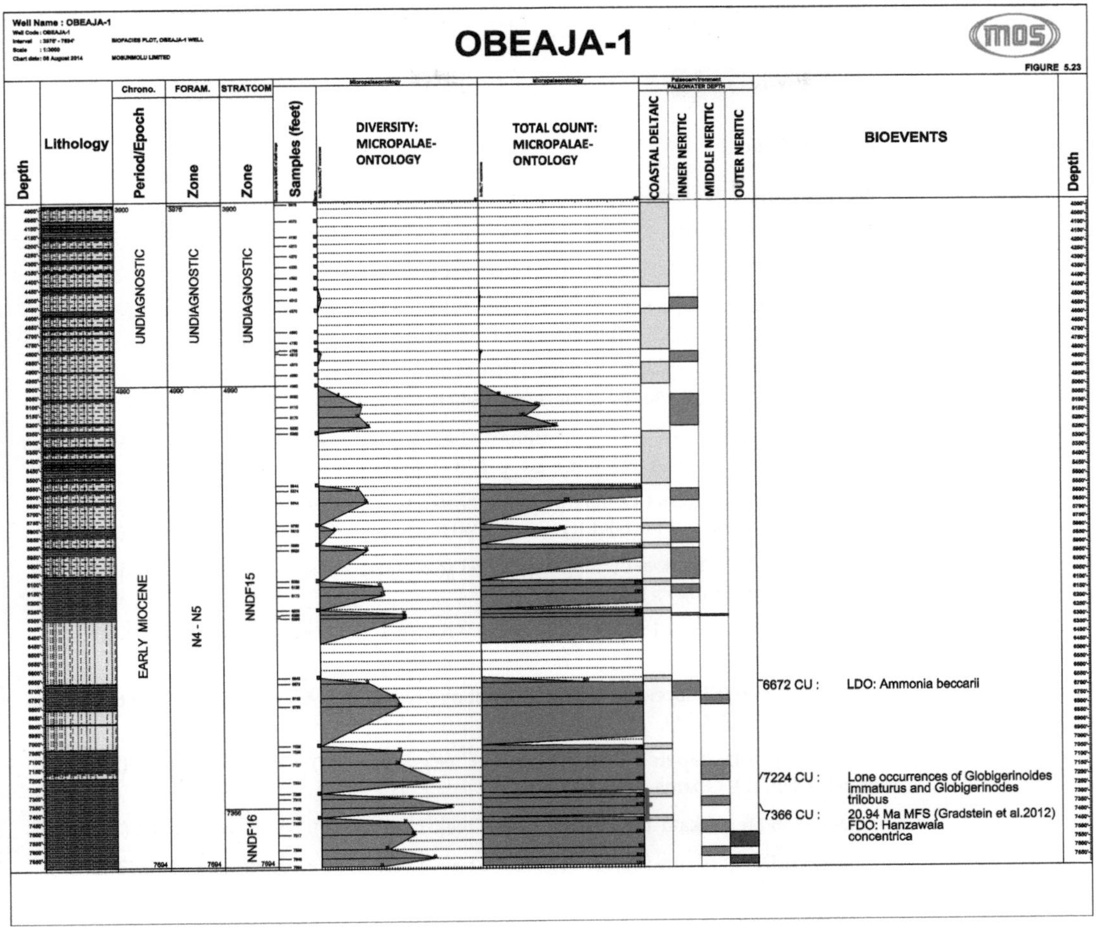

FIGURE 4.23 **Biofacies plot, Obeaja-1 well.**

regional and global correlation. The MFS are, therefore, mostly used as boundaries for the Foraminifera Zones established for the Cenozoic Niger Delta in this Report.

A total of 44 MFS were documented in the Cenozoic Niger Delta in this study. Of these, 26 are in the Paleogene, 17 in the Neogene, and 1 in the Quartenary (Holocene?). A few of the older Paleogene surfaces were not characterized adequately because only a few of the wells selected for study (Gilli Gilli-1 and Ihuo-1) penetrated the Paleocene and Eocene Epochs. The regional and stratigraphic distributions of these MFS are shown in Table 4.2.

4.2.2 Correlation Datums

The five principal biostratigraphic Correlation Datums, usually employed in high-resolution biostratigraphic refinement projects (Armentrout, 1991, 1995; Wornardt and Vail, 1990; Adegoke, 2000; Adegoke et al., 2011) were used in this report. These include the recognition of bioevents marked by:

TABLE 4.2 Stratigraphic and Regional Distribution of Maximum Flooding Surfaces (MFS) in the Cenozoic Niger Delta

No.	MFS (Ma) (Gradstein et al., 2012)	Well/Depth (ft./m)
1.	1.64	Anyala-4 (870 ft.); Anyala-8 (1,660 ft.); Akamba-1 (145 m)
2.	2.39	Anyala-4 (1,770 ft.); Anyala-8 (5,180 ft.); Akamba-1 (435 m); Ameena-1 (530 m)
3.	2.92	Anyala-4 (2,970 ft.); Anyala-8 (6,450 ft.); Ameena-1 (753 m)
4.	3.82	Anyala-4 (6,390 ft.); Anyala-8 (8,530 ft.); Akamba-1 (1,160 m) HB-1 (3,430 ft.); Inanga-1 (2,460 ft.); Edop-33A (5,325 ft.); Ameena-1 (984 m)
5.	4.16	Anyala-4 (8,310 ft.); Anyala-8 (10,620 ft.); HB-1 (4,920 ft.); Inanga-1 (5,190 ft.); Akamba-1 (1,323 m); Edop-33A (8,180 ft.); Utue-17A (5,640 ft.); Edop-17B (6,260 ft.); Ameena-1 (1,505 m)
6.	5.47	HB-1 (5,780 ft.); Inanga-1 (8,430 ft.); Akamba-1 (1,615 m); Etim-1 (5,200 ft.); Edop-33A (9,050 ft.); KR-1, (2,518 ft.); Utue-17A (8,180 ft.); Edop-17B (7,280 ft.); Ameena-1 (1,955 m)
7.	5.99	Inanga-1 (9,570 ft.); Akamba-1 (2,180 m); Utue-1 (9,460 ft.); Etim-1 (6,270 ft.); Edop-33A (?11,080 ft.); Kr-1 (8,970 ft.); Yomene-1 (4,960 ft.); HB-1 (10,040 ft.); Edop-17B (8,060 ft.); Ameena-1 (2,295 m)
8.	7.72	Inanga-1 (10,770 ft.); Akamba-1 (2,355 m); Etim-1 (8,800 ft.), Yomene-1 (8,100 ft.), Andoni-1 (6,479 ft.); Obama-1 (8,775 ft.)
9.	10.51	Utapate-1 (8,030 ft.); Ibotio-1 (4,700 ft.); Beniboye North-1 (1,300 m); Andoni-1 (8,479 ft.); Obama-1 (9,500 ft.)
10.	12.26	Kolo Creek-1 (6,860 ft.); Ibotio-1 (7,400 ft.); Beniboye North-1 (1,900 m); Utapate-1 (8,600 ft.); Obama-1 well (11,600 ft.)
11.	13.12	Inanga-1 (13,250 ft.); Utapete-1 (9,650 ft.); Kolo Creek-1 (6,860 ft.); Beniboye North-1 (2,600 m); Ibotio-1 (8,500 ft.); Obama-1 (12,800 ft.)
12.	13.68	Opuama-1 (3,380 ft.); Akata-1 (6,400 ft.); Kolo Creek-1 (10,350 ft.); Beniboye North-1 (2,776 m)
13.	15.20	Opuama-1 (4,237 ft.); Akata-1 (7,830 ft.); Kolo Creek-1 (12,683 ft.); Tuomo West-1 (2,520 m); Utonana-1 (5,200 ft.); Beniboye North-1 (3,144 m)
14.	17.83	Opuama-1 (4,557 ft.); Benin River-3 (6,310 ft.) Tuomo West-1 (3,565 m); Utonana-1 (7,850 ft.)
15.	18.64	Opuama-1 (5,722 ft.), Pologbene-1 (2,650 ft.); Benin River-3 (6,970 ft.); Tuomo West-1 (4,110 m), Fragbene-1 (5,170 ft.)
16.	19.80	Benin River-3 (8,140 ft.), Tuomo West-1 (4,595 m)
17.	20.94	Opuama-1 (6,148 ft.); Obeaja-1 (7,320 ft.); Benin River-3 (8,640 ft.); Ihuo-1 (2,350 ft.); Adagbassa Creek-1 (4,900 ft.)
18.	22.24	Adagbassa Crk-1 (5,425 ft.); Benin River-3 (8,910 ft.); Agberi North-1 (2,750 m); Ihuo-1 (5,412 ft.)
20.	23.96	Amakubu-1 (4,900 ft.); Ihuo-I (5,675 ft.); Agberi North-1 (2,870 m)
21.	26.20	Fragbene-1 (8,510 ft.); Benin River-3 (9,480 ft.); Agberi North-1 (2,965 m)
22.	27.79	Fragbene-1 (9,110 ft.); Benin River-3 (10,000 ft.); Agberi North-1 (3,570 m)
23.	28.64	Fragbene-1 (9,390 ft.), Obeaja-1 (7,360 ft.); Opuama-1 (7,690 ft.); Amakubu-1 (7,680 ft.); Benin River-3 (10,740 ft.); Agberi North-1 (4,240 m)
24.	30.54	Fragbene-1 (11,230 ft.); Mosogar-2 (5,200 ft.); Amakubu-1 (9,120 ft.); Benin River-3 (11,160 ft.); Adagbassa Creek-1 (11,390 ft.); Opuama-1 (8,562 ft.); Agberi North-1 (4,550 m)
25.	32.90	Amakubu-1 (9,640 ft.); Mosogar-2 (8,580 ft.) Obeaja-1 (7,520 ft.); Jesse-1 (7,560 ft.), Pologbene-1 (5,720 ft.)

TABLE 4.2 Stratigraphic and Regional Distribution of Maximum Flooding Surfaces (MFS) in the Cenozoic Niger Delta (*cont.*)

No.	MFS (Ma) (Gradstein et al., 2012)	Well/Depth (ft./m)
26.	34.45	Pologbene-1 (5,970 ft.); Jesse-1 (8,830 ft.); Opuama-1 (9,940 ft.)
27.	35.99	Jesse-1 (11,390 ft.); Pologbene-1 (9,445 ft.); Gilli Gilli-1 (8,040 ft.); Mosogar-2 (10,167 ft.)
28.	37.36	Amakubu-1 (9,910 ft.); Gilli Gilli-1 (8,760 ft.); Pologbene-1 (7,460 ft.); Mosogar-2 (12,093 ft.)
29.	38.46	Gilli Gilli-1 (9,200 ft.); Mosogar-2 (10,150 ft.); Pologbene-1 (9,249 ft.)
30.	40.40	Gilli Gilli-1 (9,730 ft.)
31.	42.64	Gilli Gilli-1 (10,230 ft.)
32.	44.75	Gilli Gilli-1 (10,520 ft.)
33.	46.82	Ihuo-1 (6,730 ft.)
34.	51.27	Ihuo-1 (6,850 ft.)
35.	52.00	Ihuo-1 (6,955 ft.)
36.	53.01	Ihuo-1 (7,350 ft.)
37.	53.67	Gilli Gilli-1 (10,790 ft.)
38.	54.63	Ihuo-1 (7,780 ft.)
39.	55.20	Gilli Gilli-1 (11,020 ft.)
40.	57.89	Ihuo-1 (8,800 ft.)
41.	58.45	Gilli Gilli-1 (11,290 ft.)
42.	59.26	Ihuo-1 (9,950 ft.)
43.	60.65	Gilli Gilli-1 (11,480 ft.)

1. The FDO of selected chronostratigraphically important marker species.
2. The LDO of important marker species.
3. Maximum and, in some cases, minimum faunal abundance and species diversity peaks.
4. Faunal assemblages and associations. In this regard, it is better to describe faunal assemblages and associations than try to cover absent types with terminologies, such as "suppressed/depressed tops" caused by the absence of known marker species at expected chronostratigraphic horizons largely due to observable ecological disruptions.
5. Evolutionary trends within selected chronostratigraphically significant lineages.

4.3 NIGER DELTA FORAMINIFERA BIOZONATION

4.3.1 Benthonic Foraminifera Zones

Forty-two benthonic foraminifera Zones are recognized in the Cenozoic Niger Delta, 1 in the Quarternary, 16 in the Neogene, and 25 in the Paleogene. The zones are characterized in further details and described in descending stratigraphic order.

A. *Neogene Benthonic Foraminiferal Zonation*

QUARTERNARY–PLEISTOCENE

1. *Hyalinea balthica/Rosalina floridana* Zone (QNDF 01)

Diagnosis: This is the only benthonic foraminifera zone recognized in the Pleistocene sequence of the Niger Delta. Its microfaunal characterization is weak. Though the top of the Zone cannot be readily defined on the basis of foraminifera data, the FDO of *H. balthica* is generally believed to be very close to the zonal top. The base of the Zone is defined by the FDO of the planktonic species *Globorotalia miocenica*. Very close to the zonal base in the Niger Delta is the LDO of the zonal marker, *H. balthica*. Other important bioevents that help to characterize the Zone include the FDO of *Globorotalia tosaensis*, the LDO of *Globorotalia truncatulinoides*, the acme of *Globigerinoides ruber*, and the cooccurrence of abundant *Globigerinoides trilobus*, *Globigerinoides immaturus*, *Globorotalia humerosa*, *Neogloboquadrina dutertrei*, *Neogloboquadrina acostaensis*, and *G. tumida* with modest numbers of *Quinqueloculina lamarckiana*, *Quinqueloculina microcostata*, *A. beccarii*, *R. floridana*, *C. auriculus*, *Lenticulina inornata*, *H. strattoni*, *H. floridana*, *Bolivina miocenica*, *C. inflata*, *Quinqueloculina bicornis*, *E. vitrea*, *Quinqueloculina vulgaris*, *Quinqueloculina seminulum*, *Florilus atlanticus*, and *H. pseudoungeriana*.

Age and Correlation: The foraminiferal peak abundance recognized within this Zone is dated 1.64 Ma (Calabrian 1) (Gradstein et al., 2012 chart) and correlated with the N22 Planktonic foraminifera Zone of Berggren et al. (1998a,b). The Zone is Pleistocene in age (1.64 Ma and/or younger).

Regional/Stratigraphic Distribution: This Zone is recognized only in the Anyala-4 (340–870 ft.) and Akamba-1 (145–435 m) wells. Only the basal part of the Zone is exposed at the top of the Anyala-8 well.

NEOGENE–PLIOCENE

2. *Cibicidoides dutemplei* Zone (NNDF 01)

Diagnosis: This is the youngest benthonic foraminiferal Zone defined in the Pliocene of the Niger Delta. Unlike the older Zones, especially those defined in the Miocene or Early Pliocene, benthonic species are generally unreliable for the chronostratigraphic characterization of the younger Zones in the Niger Delta. The top of the Zone is defined by the FDO of *G. miocenica*, while the base is defined by the FDOs of *Bulimina marginata* and *Cyclammina cancellata. The* LDO of *H. balthica* is recorded very close to the upper boundary of this zone. *C. dutemplei* is also a prominemt constituent of the fauna of this zone.

Other common cooccurring foraminifera include *B. miocenica*, *Uvigerina proboscidea*, *Planulina ariminensis*, *Trifarina reussi*, *Am. scalaris caudata*, *C. auriculus*, *H. floridana*, *Cibicidoides pseudoungerianus*, *L. inornata*, *Q. microcostata*, *Q. lamarckiana*, *H. pseudoungeriana*, *Hanzawaia mantaensis*, *C. inflata*, *R. floridana*, *Poritextularia panamensis*, *S. wrightii*, and species of *Pulleniatina*, as well as abundant planktonics, including *G. ruber*, *Globigerinoides extremus*, *Globigerinoides elongatus*, *N. dutertrei*, and *Orbulina universa*.

Age and Correlation: The top and the base of this Zone are correlated with the global cyclic events dated 2.39 Ma; Gelasian 1, and 2.92 Ma; Piacenzian 1, Gradstein et al. (2012), respectively. The age is Late Pliocene and the Zone is correlated with the "upper" N20/N21 planktonic foraminifera Zone of Berggren et al. (1998a,b).

Regional/Stratigraphic Distribution: The Zone was recognized in this study only in the Anyala-4 (1770–2970 ft.), Anyala-8 (5180–6450 ft.), Akamba-1 (620–1020 m), and Amenam-1 (140–440 m) wells.

3. ***Bulimina marginata/Cyclammina cancellata* Zone (NNDF 02)**

Diagnosis: The top of the *B. marginata/C. cancellata* Zone is defined by the FDO's of *B. marginata* and *C. cancellata*. The zonal base is defined by the bioevents that marked the top of the underlying *Amphistegina lessonii* NNDF 03 Zone, namely, the FDO's of *A. lessonii*, *R. multicostata*, and *Globorotalia margaritae*. The Zone is also further characterized by the occurrence of the LDO's of *Sphaeroidinellopsis dehiscens* and *G. tosaensis*. within it. Cooccurring foraminifera species include *C. auriculus*, *Am. scalaris caudata*, *B. miocenica*, *H. floridana*, *E. vitrea*, *Alveolophragmium crassum*, *C. inflata*, *H. pseudoungeriana*, *L. inornata*, *H. strattoni*, *Hanzawaia isidroensis*, *H. mantaensis*, *R. floridana*, *A. beccarii*, *F. atlanticus*, *Ptychomiliola separans*, *Uvigerina peregrina*, *Q. lamarckiana*, *P. panamensis*, and *S. wrightii*. Arenaceous species are generally scanty within the Zone.

Age and Correlation: The top of this Zone is placed at the 2.92 Ma MFS (Piacenzian 1) (Gradstein et al., 2012) downlap surface, while the base is placed at the 3.82 Ma MFS (Zanclean 2) (Gradstein et al., 2012). It is Late to Early Pliocene in age and it is correlated with the "uppermost" N19–N20/N21 Planktonic zones of Berggren et al. (1998a,b).

Regional/Stratigraphic Distribution: This Zone is recognized in the wells Anyala-4 (2970–6390 ft.), Anyala-8 (6450–8530 ft.), Amenam-1 (440–930 m), and Akamba-1 (1020–1180 m).

4. ***Amphistegina lessonii* Zone (NNDF 03)**

Diagnosis: The top of this Zone is defined by the FDO's of *A. lessonii*, *R. multicostata*, and *G. margaritae*. These bioevents have been correlated with the 3.82 Ma MFS, Zanclean 2 (Gradstein et al., 2012). The base is defined by the FDO's of *Marginulina costata* and *H. compressa*. The top of the "costate nodosarids," for example, *Amphicoryna caudata*, *R. comatula*, as well as species of *Haplophragmoides* and *Trochammina* occur within the Zone. The zonal assemblage is dominated by shallow-water calcareous species, such as *F. atlanticus*, *A. beccarii*, *C. auriculus*, *B. miocenica*, *H. floridana*, *H. pseudoungeriana*, *Q. vulgaris*, *Am. scalaris caudata*, *Nonionella auris*, *Q. lamarckiana*, *R. comatula*, *C. inflata*, *H. strattoni*, *E. vitrea*, *F. atlanticus*, *Q. microcostata*, *S. wrightii*, and *L. inornata*. Planktonic species are represented by *G. ruber*, *G. trilobus*, *Globorotalia pseudomiocenica*, *N. dutertrei*, and *Globorotalia pseudopima*.

Age and Correlation: This Zone is dated Early Pliocene (4.16–3.82 Ma) (Gradstein et al., 2012) and is correlated with the "upper" N19 Planktonic Zone of Berggren et al. (1998a,b).

Regional/Stratigraphic Distribution: The Zone is recognized in the following studied wells: Akamba-1 (1,180–1,380 m), Amenam-1 (930–1,545 m), Anyala-8 (8,530–10,620 ft.), Anyala-4 (6,390–8,310 ft.), HB-1 (3,430–4,920 ft.), Edop-33A (5,325–8,180 ft.), and Inanga-1 (2,460–5,190 ft.).

5. ***Marginulina costata* Zone (NNDF 04)**

Diagnosis: This is the oldest benthonic foraminifera zone recognized in the Pliocene of the Niger Delta. The top is defined by the FDO's of *M. costata*, *R. comatula*, *U. peregrina*, and *H. compressa*. The base of the Zone is defined by the FDO of *Cyclammina minima* and the LDO of *G. margaritae*. Other important bioevents recorded within the Zone include the FDO's of *V. flexilis* and *Arenoturrispillina* sp. The zonal assemblage is commonly dominated by *A. beccarii*, *B. miocenica*, *Marginulina raphanus*, *H. pseudoungeriana*, *Q. microcostata*, *Q. lamarckiana*, *A. crassum*, *Alveolophragmium subglobosum*, *Eggerella bradyi*, *Eggerella scabra*, *Textularia parvula*, *Bulimina inflata*, *Bulimina minima*, *F. atlanticus*, *Gyroidina soldanii*, *H. strattoni*, and *C. cancellata*.

Age and Correlation: The zonal top and base are respectively age-dated 4.16 and 5.47 Ma (Zanclean 1–Messinian 2) (Gradstein et al., 2012). The Zone correlates with the "upper" N19–N18 Planktonic zones of Berggren et al. (1998a,b).

Regional/Stratigraphic Distribution: The Zone is recognized in the HB-1 (4920–5780 ft.), Inanga-1 (5190–8430 ft.), Akamba-1 (1380–1615 m), and Amenam-1 wells. The base in the KR-1 well is at 2518 ft. and in the Edop-33A well at 9050 ft. Only the top of the zone is encountered at the base of the Inanga-1 well from 8310 ft. to TD.

6. ***Cyclammina minima* Zone (NNDF 05)**

Diagnosis: The *C. minima* Zone is the youngest benthonic Zone recognized in the Late Miocene of the Niger Delta. The top of the Zone is defined, particularly in the deep water setting, by the FDO's of *C. minima* and *Ammobaculites strathearnensis*, the top abundant occurrence of *A. crassum* and the LDO of *G. margaritae*. In the shallow-water setting, it is characterized by the FDO of *E. scabra*. Close to the top of the Zone are the LDO's of *Brizalina isidroensis* and *B. aenariensis*. Another remarkable feature of the zonal top is the major influx of arenaceous foraminifera accompanied by a concomitant reduction in both the abundance and diversity of calcareous forms, a change attributable to the effect of the terminal "Messinian Salinity Crisis event."

The base of the Zone is defined by the FDO of *H. narivaensis*, a bioevent correlated with the 5.99 Ma MFS (Tortonian 3/Messinian 1) *sensu* Gradstein et al. (2012). The Zone is further characterized by the top regular occurrence of *H. compressa*, *V. flexilis*, and *Textularia elegans* all cooccurring with rare calcareous forms, such as *A. beccarii*, *B. miocenica*, *F. atlanticus*, *E. vitrea*, *C. inflata*, *L. inornata*, *Q. microcostata*, *Q. lamarckiana*, and *H. pseudoungeriana*.

Age and Correlation: The top of this Zone has been correlated with the 5.47 Ma MFS (Messinian 2) (Gradstein et al., 2012), while the base is dated 5.99 Ma MFS, (Tortonian 3/Messinian 1) (Gradstein et al., 2012). This Zone is assigned to the Late Miocene and correlated with the "upper" N17–"lower" N18 Planktonic zones of Berggren et al. (1998a,b).

Regional/Stratigraphic Distribution: The Zone is well developed in the Inanga-1 (8,430–9,570 ft.), Etim-1 (5,200–6,270 ft.), KR-1 (2,518–8,970 ft.), Edop-33A (9,050–11,080 ft.), HB-1 (5,780–10,040 ft.), Amenam-1 (2,210–2,690 m), and Akamba-1 (1,615–2,180 m) wells. The zonal base was encountered in the Yomene-1 well at 4950 ft.

7. ***Haplophragmoides narivaensis* Zone (NNDF 06)**

Diagnosis: The top of the *H. narivaensis* Zone is defined by the FDO of *H. narivaensis*, while the zonal base is defined by the FDO's of *A. agglutinans* and *F. ex gr. costiferum* (*Nonion*-4). The top occurrence of *Saccammina complanata* and *A. glabratus* are other significant bioevents recognized within the Zone. The Zone is also further marked by the preponderance of arenaceous foraminifera, particularly *E. scabra*, *E. bradyi*, *T. elegans*, *A. crassum*, *A. subglobosum*, *A. strathearnensis*, *T. parvula*, *Trochammina globigeriniformis*, *Trochammina proteus*, *H. compressa*, *Haplophragmoides obliquicameratus*, *V. flexilis*, *C. minima*, *C. cancellata*, *K. siphonella*, *Martinottiella communis*, *Sigmoilopsis schlumbergeri*, and *Gravellina narivaensis* over the calcareous forms, which are rare within the Zone. Thus, the arenaceous/calcareous ratio for this Zone is less than 1 (<1), a fact taken to indicate that the Zone represents the peak of the "Messinian Salinity Crisis" (Haq et al., 1988; Kennett, 1960; Ingle, 1967, 1973) that seems to be well represented in the Niger Delta.

Both the lower and the upper boundaries of the Zone coincide with well-defined faunal abundance and diversity peaks dated (7.72 Ma (Tortonian 2) and 5.99 Ma (Tortonian 3/Messinian 1) sensu Gradstein et al. (2012), respectively.

Age and Correlation: This Zone is dated Late Miocene (7.72–5.99 Ma sensu Gradstein et al., 2012) and corresponds with the "middle" N17 Zone of Berggren et al. (1998a,b).

Regional/Stratigraphic Distribution: The Zone was recognized in the following studied wells: HB-1 (10,040–13,160 ft.), Yomene-1 (4,950–8,100 ft.), Etim-1 (6,270–8,800 ft.), Inanga-1 (9,570–10,770 ft.), Edop-33A (zonal top only from 11,080 ft. to TD at 11,300 ft.), Akamba-1 (2,180–2,355 m), and Amenam-1 (2,690–3,005 m). The zonal base occurs at 6479 ft. in the Andoni-1 well and 8775 ft. in the Obama-1 well.

8. *Ammobaculites agglutinans/Florilus costiferum* Zone (NNDF 07)

Diagnosis: The top of this Zone, which is correlated with the 7.72 Ma MFS (Tortonian 2) (Gradstein et al., 2012), is commonly defined in the Niger Delta by the FDO's of *A. agglutinans* and *F. costiferum* (*Nonion*-4). The base is defined by the bioevent that characterizes the top of the underlying *U. subperegrina* Zone, that is, the FDO of *U. subperegrina*. Other significant foraminifera occurring within the Zone include the regular occurrences of *H. pseudoungeriana, L. inornata, E. vitrea, U. peregrina, U. hispida, F. atlanticus, E. scabra, A. crassum, H. compressa, K. siphonella, C. minima, V. flexilis, T. parvula, Textularia barretti*, and *T. elegans*. Planktonic foraminifera are generally rare in this Zone.

Age and Correlations: This Zone is dated Late Miocene (10.51–7.72 Ma) (Gradstein et al., 2012) and is correlated with the N16 and "lower" N17 zones of Berggren et al. (1998a,b).

Regional/Stratigraphic Distribution: The Zone is recognized in many of the studied wells, particularly Andoni-1 (6,479–8,479 ft.); Yomene-1 (8,100–9,500 ft. TD); Obama-1 (8,775–9,500 ft.); Utapate-1, and Inanga-1 (10,770–12,180 ft.) wells. The zonal base is at 8030 ft. in the Utapate-1 well and 1300 m in the Beniboye North-1 well. The top could not be documented because samples were not available for study. Only the extreme top of the Zone was encountered at the base of the Akamba-1 well at 2395 m.

9. *Uvigerina subperegrina* Zone (NNDF 08)

Diagnosis: This is oldest benthonic foraminifera Zone recognized in the Late Miocene of the Niger Delta. The zonal top is a well developed marine event dated 10.51 Ma (Serravallian 3/Tortonian 1 of Gradstein et al., 2012) and is characterized by the FDO of *U. subperegrina*. The base is marked by the FDO's of *Spirosigmoilina oligocaenica* and *C. chipolensis*. Another significant bioevent recognized within the Zone is the LDO of *N. acostaensis*. The Zone is further characterized by the presence of the assemblage of deep water calcareous and arenaceous forms, such as *S. bulloides, Uvigerina auberiana, U. peregrina, Gyroidinoides neosoldanii, C. minima, Karreriella* sp., *V. flexilis*, and *G. narivaensis*. Other significant cooccurring foraminifera include *E. scabra, A. crassum, L. inornata, H. pseudoungeriana, Heterolepa mckannai, A. beccarii, Bolivina scalprata rectiformis, B. miocenica, F. atlanticus, H. strattoni, Q. vulgaris, H. compressa, P. panamensis, S. wrightii, T. proteus, Recurvoides deformis, M. raphanus, E. vitrea*, and *F. costiferum*.

Age and Correlation: The Zone is assigned to the Late Miocene (12.26–10.51 Ma) (Gradstein et al., 2012) age. It is equivalent to the "upper" N12–"lowermost" N16 Planktonic foraminiferal Zones of Berggren et al. (1998a,b).

Regional/Stratigraphic Distribution: This Zone is recognized in the wells Utapate-1 (8,030–8,600 ft.), Andoni-1 (8,479–10,030 ft. TD), Ibotio-1 (4,700–7,400 ft.), Obama-1 (9,500–11,600 ft.), and Beniboye North-1 (1,300–1,900 m).

10. *Spirosigmoilina oligocaenica* Zone (NNDF 09)

Diagnosis: The top of the *S. oligocaenica* Zone is defined by the FDO's of *S. oligocaenica* and *C. chipolensis*, while the base is defined by the foraminiferal biofacies that characterized the top of the *Uvigerina sparsicostata* Zone, that is, the FDO's of *Uvigerina sparsicostata*, *Bolivina interjuncta*, and *Globigerinoides subquadratus*. Other diagnostic bioevents defined within the Zone include the FDO's of *Turborotalia continuosa*, *T. mayeri*, *Turborotalia obesa*, and *Hastigerina praesiphonifera*. Recorded cooccurring foraminifera include *F. costiferum*, *A. beccarii*, *E. vitrea*, *B. miocenica*, *B. scalprata rectiformis*, *Bulimina buchiana*, *Cribroelphidium galvestonensis*, *F. atlanticus*, *H. strattoni*, *Q. vulgaris*, *R. comatula*, *U. subperegrina*, *A. crassum*, *P. panamensis*, *H. compressa*, *S. wrightii*, *H. mckannai*, *L. inornata*, *H. pseudoungeriana*, and *C. inflata*.

Age and Correlation: The age of the Zone is Middle Miocene (13.12–12.26 Ma) (Gradstein et al., 2012) and it is correlated with the "middle" N12 Planktonic Foraminifera Zone of Berggren et al. (1998a,b). The top is correlated with the 12.26 Ma MFS (Serravallian 3) (Gradstein et al., 2012).

Regional/Stratigraphic Distribution: Many of the wells studied, particularly the Kolo Creek-1 (upper part of the well above 6,860 ft.), Beniboye North-1 (1,900–2,600 m), Inanga-1 (?surface of unconformity between 12,180 and 13,250 ft.), Obama-1 (11,600–12,800 ft.), Utapate-1 (8,600–9,650 ft.), and Ibotio-1 (7,400–8,500 ft.) penetrated the Zone. The basal portion of the Zone was encountered in the Utonana-1 well between 4800 and 5200 ft.

11. *Uvigerina sparsicostata* Zone (NNDF 10)

Diagnosis: The top of the *Uvigerina sparsicostata* Zone is routinely defined by the FDO's of *Uvigerina sparsicostata*, *B. interjuncta*, *F. costiferum* (*Nonion*-6), and *G. subquadratus*. The zonal base is defined by the FDO's of *L. grandis*, *B. mandoroveensis*, *Eponides eshira*, and *T. peripheroronda*. The stratigraphic occurrence of *C. neocarinata* in the Niger Delta is restricted to this Zone. The LDO of *R. comatula* is also a significant bioevent recorded within the Zone. Other cooccurring important foraminifera include *H. mckannai*, *F. costiferum* (*Nonion*-4), *F. boueanum*, *F. atlanticus*, *H. strattoni*, *Q. vulgaris*, *E. vitrea*, *H. narivaensis*, *H. compressa*, *T. parvula*, *V. flexilis*, *H. floridana*, *S. wrightii*, and *P. panamensis*.

Age and Correlation: This Zone is assigned to the Middle Miocene. It is correlated with the "upper" N10–"lower" N12 Planktonic Zones of Berggren et al. (1998a,b). The lower and the upper boundaries are correlated with the global seismic cyclic events (sequence boundaries or flooding surfaces) dated 13.68 and 13.12 Ma (Langhian 2/Serravallian 1–Serravallian 2 events sensu Gradstein et al., 2012).

Regional/Stratigraphic Distribution: The Zone was penetrated by the Akata-1 (4,510–6,400 ft.), Kolo Creek-1 (6,860–10,320 ft.), Inanga-1 (13,250–13,770 ft. TD), Utapate-1 (9,650 ft.–TD), and Beniboye North-1 (from 2600 m to TD) wells. Only the upper part of the Zone was recognized in the Ibotio-1 well (8500–9950 ft. TD).

12. ***Bolivina mandoroveensis* Zone (NNDF 11)**
Diagnosis: The top of this Zone, which correlates with the 13.68 Ma MFS (Langhian 2/Serravalian 1) (Gradstein et al., 2012) is marked by the FDO's of *L. grandis, B. mandoroveensis, E. eshira,* and *Turborotalia peripheroacuta.* The FDO's of *T. peripheroronda* and *T. fohsi* are other important bioevents recognized within the Zone. The base of the Zone is the top of the *N. centrosulcatum* Zone, which is marked by the FDO's of *N. centrosulcatum, Heterostegina* sp. 1 sensu SPDC, *H. bononiensis, C. victoriana, Praeorbulina glomerosa,* and *Globigerinoides bisphericus,* as well as the LDO of *O. universa/suturalis.* The zonal base is correlated with the 15.20 Ma MFS (Burdigalian 5/Langhian 1) (Gradstein et al., 2012).
The Zone is further characterized by the cooccurring assemblage of *Uvigerina sparsicostata, L. inornata, H. floridana, Q. vulgaris, A. crassum, P. panamensis, S. wrightii, E. scabra, Bo. interjuncta, H. mckannai, C. inflata,* and *C. chipolensis.*
Age and Correlation: The Zone is Middle Miocene (15.20–13.68 Ma) (Gradstein et al., 2012) in age and is equivalent to the "upper" N8/M5b–"lower" N10 Planktonic Zones of Berggren et al. (1998a,b).
Regional/Stratigraphic Distribution: Of all the wells studied, only the wells Akata-1 (6,400–7,830 ft.), Opuama-1 (3,380–4,237 ft.), and Kolo Creek-1 (10,350–12,683 ft.) penetrated this Zone. The top of the Zone is encountered in the Utonana-1 well at 7850–8540 ft., while the base is encountered in the youngest samples (younger than 2520 feet) studied in the Tuomo West-1 well.
13. ***Nonion centrosulcatum* Zone (NNDF 12)**
Diagnosis: This is the oldest benthonic foraminiferal Zone recognized in the Middle Miocene of the Niger Delta. The top of the Zone is commonly defined by the FDOs of *N. centrosulcatum, Heterostegina* sp. 1, *H. bononiensis, C. victoriana* (?*Streptochilus dubeyi*), *P. glomerosa,* and *G. bisphericus,* as well as the LDO's of *O. universa* and *O. suturalis.* This zonal top is correlated with the global cyclic event dated 15.20 Ma MFS (Burdigalian 5/Langhian 1) (Gradstein et al., 2012). The base of the Zone marks the top of the Early Miocene, which is correlated with the 17.83 Ma MFS (Burdigalian 3 of Gradstein et al., 2012).
Other important foraminifera recorded within the Zone include *L. grandis, B. mandoroveensis, E. eshira, Bo. interjuncta, A. beccarii, C. inflata, F. costiferum, F. atlanticus, E. vitrea, L. inornata, A. crassum, E. scabra, S. wrightii,* and *P. panamensis.*
Age and Correlation: The Zone is correlated with the N6/M3–"lower" N8/M5b Planktonic Zones of Berggren et al. (1998a,b). The age is Middle Miocene (Burdigalian–Langhian; 17.83–15.20 Ma, Gradstein et al., 2012).
Regional/Stratigraphic Distribution: This Zone is penetrated in the Utonana-1 (5,200–7,850 ft.), Tuomo West-1 (2,520–3,565 ft.), Benin River-3 (above 6310 ft.), Opuama-1 (4,237–4,557 ft.), and Akata-1 (7,830–11,121 ft.) wells.
14. ***Buliminella subfusiformis* Zone (NNDF 13)**
Diagnosis: The *B. subfusiformis* Zone is the youngest benthonic foraminifera zone recognized in the Early Miocene of the Niger Delta. Its top is characterized by the FDO's of *B. subfusiformis* and *Catapsydrax dissimilis.* The zonal base is marked by the FDO's of *B. dertonensis, A. tenuis, F. punctata, Globigerina ciperoensis, Globigerinoides primordius,* as well as the LDO of *Globigerinoides altiaperturus.* Other important cooccurring

foraminifera species include *E. eshira, C. inflata, Eponidella venezuelanum, H. mckannai, B. miocenica, Bo. interjuncta, B. mandoroveensis, H. strattoni, L. inornata, N. centrosulcatum, H. semiornata, Uvigerina sparsicostata, U. subperegrina, Al. scalaris, A. glabratus, T. globigeriniformis, V. flexilis, Heterostegina* sp. 1, and *C. chipolensis*.

The top of the Zone is correlated with the global bioevent dated 17.83 Ma (Burdigalian 3 of Gradstein et al., 2012), while the base is correlated with the 18.64 Ma MFS (Burdigalian 2, Gradstein et al., 2012).

Age and Correlation: The age of the Zone is Early Miocene (18.64–17.83 Ma) (Gradstein et al., 2012) and it is correlated with the "uppermost" N5/M2 planktonic foraminiferal scheme of Berggren et al. (1998a,b). The N5/M2 zones are defined twice in literature based on the premise of uncertain position of the Neogene (N) and Paleogene (P) boundary.

Regional/Stratigraphic Distribution: This Zone is recognized in the following studied wells: Benin River-3 (6310–6970 ft.), Opuama-1 (4557–5722 ft.), and Tuomo West-1 (3565–4110 ft.).

15. ***Fursenkoina punctata* Zone (NNDF 14)**

Diagnosis: The base of the *F. punctata* Zone coincides with the global cyclic MFS event dated 19.80 Ma. It is defined by the suite of markers, which characterize the top of the *M. africana* Zone, namely, the FDO's of *M. africana* and *Epistominella pontoni* and the LDO's of *Quinqueloculina rhodiensis* and *T. peripheroronda*. The zonal top is characterized by the FDO's of *B. dertonensis, A. tenuis, F. punctata, G. ciperoensis, G. primordius*, as well as the LDO of *G. altiaperturus*. These bioevents are correlated with global cyclic events and the MFS dated 19.4 Ma on the Haq et al. (1988) scheme and equated with the 18.64 Ma MFS (Burdigalian 2 of the Gradstein et al., 2012) scheme.

Cooccurring stratigraphically significant foraminifera species include *H. strattonii, Bolivina beyrichi, Bo. interjuncta, B. mandoroveensis, Uvigerina sparsicostata, S. wrightii, H. bononiensis, Eponides iojimaensis, B. scalprata, B. miocenica, B. subfusiformis, C. inflata, E. vitrea, L. grandis, L. inornata, N. centrosulcatum, Q. rhodiensis, Nonion ecuadoranum, Uvigerina gallowayi, H. semiornata, U. subperegrina, Al. scalaris, A. glabratus*, and *V. flexilis*.

Age and Correlation: The age of the Zone is Early Miocene (19.80–18.64 Ma) (Gradstein et al., 2012). It is correlated with the "upper" N5/M2 planktonic foraminiferal Zone of Berggren et al. (1998a,b).

Regional/Stratigraphic Distribution: The *F. punctata* Zone is recognized in the Benin River-3 (6970–8140 ft.) and Tuomo West-1 (4110–4595 ft.) wells.

16. ***Megastomella africana* Zone (NNDF 15)**

Diagnosis: The top of the *M. africana* Zone in the Niger Delta is characterized by the FDO's of *M. africana* and *E. pontoni*, as well as the LDO's of *Q. rhodiensis* and *T. peripheroronda*. The FDO of *Globorotalia kugleri* also occurs within the Zone. The base of the Zone is defined by the bioevents that mark the top of the underlying *Hanzawaia concentrica* Zone, namely, the FDO of *H. concentrica* and *Valvulineria wilcoxensis*, and the LDO's of *G. trilobus, G. immaturus*, and *A. beccarii*. Other important concurrently ranging species include *F. atlanticus, L. grandis, B. scalprata, B. miocenica, C. inflata, E. vitrea, N. ecuadoranum, N. auris, P. ovata, U. subperegrina, Bo. interjuncta, B. dertonensis, Uvigerina sparsicostata, B. mandoroveensis, P. panamensis*, and *B. subfusiformis*.

The top of the Zone is coincident with the base of Magnetic Polarity Chron 6, dated 19.80 Ma MFS (Aquitanian 2/Burdigalian 1) (Gradstein et al., 2012) on the Mesozoic–Cenozoic Chart of Berggren et al. (1998a,b).

Age and Correlation: The Zone is correlated with the "middle" N5/M2 Planktonic Zones of Berggren et al. (1998a,b). It is suggested as Early Miocene (20.94–19.80 Ma) (Gradstein et al., 2012) in age.

Regional/Stratigraphic Distribution: The *M. africana* Zone was recognized in the Obeaja-1 well (above 7320 ft.), Benin River-3 well (8140–8640 ft.), and Tuomo West-1 well (below 4595 ft.).

17. ***Hanzawaia concentrica* Zone (NNDF 16)**
Diagnosis: The *H. concentrica* Zone is the oldest benthonic foraminiferal zone recognized in the Neogene section of the Niger Delta. The base of the Zone is defined by the FDO of *Eponides berthelotianus* and the LDO's of *B. beyrichi* and *Globoquadrina dehiscens*. The surface coincides with the 22.24 Ma MFS (Chattian 3/Aquitanian 1 of Gradstein et al., 2012), which is considered here as the Oligocene/Miocene boundary. The zonal top is marked by the 20.94 Ma MFS (Aquitanian 2 of Gradstein et al., 2012) and is defined by the FDO of *H. concentrica* and *V. wilcoxensis*, as well as the LDO of *A. beccarii*, *G. immaturus*, and *G. trilobus*. Other significant cooccurring foraminifera species include *E. eshira*, *B. dertonensis*, *B. scalprata*, *B. mandoroveensis*, *E. vitrea*, *N. ecuadoranum*, *N. auris*, *E. iojimaensis*, *S. complanata*, *T. elegans*, *C. inflata*, *L. grandis*, *B. subfusiformis*, *Nonionella danvillensis*, and *Uvigerina sparsicostata*.

Age and Correlation: The Zone is assigned an Early Miocene 22.24–20.94 Ma age (Gradstein et al., 2012) and correlated with the "upper" N4/M1b–"lower" N5/M2 Planktonic Zones of Berggren et al. (1998a,b).

Regional/Stratigraphic Distribution: This Zone is recognized in the Ihuo-1 (2350–4050 ft.), Adagbassa Creek-1 (4900–5425 ft.), Obeaja-1 (7320–7694 ft.), Benin River-3 (8640–8910 ft.), and Agberi North-1 (2500–2750 m) wells. In the Ihuo-1 well, the base is placed at the unconformity between the Paleogene and Neogene at (?)4050 ft.

18. ***Eponides berthelotianus* Zone (PNDF 01)**
Diagnosis: The top of the *E. berthelotianus* Zone is characterized by the datums of a suite of markers including the FDO of *E. berthelotianus* and the LDO of *B. beyrichi* coinciding with the 22.24 Ma MFS, Chattian 3/Aquitanian 1 of Gradstein et al. (2012). The base is also defined by the FDO of another suite of markers, which include *B. imperatrix* and *Bolivina tenuicostata* and dated 23.96 Ma MFS (Chattian 3) (Gradstein et al., 2012). Associated foraminifera species include *B. dertonensis*, *B. miocenica*, *B. scalprata*, *Bo. interjuncta*, *B. mandoroveensis*, *B. subfusiformis*, *C. inflata*, *H. strattoni*, *H. mckannai*, *H. pseudoungeriana*, *N. auris*, *N. danvillensis*, *U. gallowayi*, *Uvigerina sparsicostata*, *V. wilcoxensis*, *H. bononiensis*, *Ammoscalaria pseudospiralis*, *P. panamensis*, *S. oligocaenica*, *E. vitrea*, *L. grandis*, and *Al. scalaris*.

Age and Correlation: The Zone is assigned a Late Oligocene–Early Miocene (Late Chattian–Early Aquitanian, 23.96–22.24 Ma) (Gradstein et al., 2012) age and correlated with the "upper" N3/P22–"lower" N4/M1a Planktonic Zones of Berggren et al. (1998a,b).

Regional/Stratigraphic Distribution: The Zone is recognized in the Benin River-3 (8910–9120 ft.), Agberi North-1 (2750–2870 m), and Ihuo-1 (4000–5675 ft.) wells at the depths indicated.

PALEOGENE ZONES

19. ***Bolivina imperatrix* Zone (PNDF 02)**
Diagnosis: The top of the *B. imperatrix* Zone is defined by the FDO's of *B. imperatrix* and *B. tenuicostata*. This correlates with the 23.96 Ma MFS (Chattian 3 of Gradstein et al., 2012). The base is marked by the LDO of *E. vitrea*. Cooccurring with the markers are the following foraminifera species: *M. africana, Bo. interjuncta, B. miocenica, B. beyrichi, B. subfusiformis, C. inflata, F. atlanticus, H. strattoni, H. bononiensis, Q. vulgaris, Cancris turgidus, E. vitrea, E. berthelotianus, F. costiferum, Nonionella stella, P. ovata, Trifarina angulosa, S. oligocaenica, Uvigerina sparsicostata, A. glabratus, H. concentrica, P. panamensis,* and *A. tenuis*.
Age and Correlation: The age of the Zone is Late Oligocene (26.20–23.96 Ma) (Gradstein et al., 2012) and it is correlated with the "middle" N3/P22 Planktonic Zone of Berggren et al. (1998a,b).
Regional/Stratigraphic Distribution: The Zone is recognized in Jesse-1, Amakubu-1, and Gilli Gilli-1, Benin River-3 (9350–9480 ft.), and Agberi North-1 (2870–2965 m) wells. The zonal boundary was not defined in the Ihuo-1 well because many of the observed MFS could not be dated accurately. The top is at 4900 ft. in the Amakubu-1 well.
20. ***Bolivina dertonensis–Epistominella pontoni* Assemblage Zone (PNDF 03)**
Diagnosis: This Assemblage Zone is characterized by the abundance of the two nominate marker species, *B. dertonensis* and *E. pontoni*. The top is marked by the LDO of *E. vitrea*, which coincides with the 26.20 Ma MFS (Chattian 2) (Gradstein et al., 2012). The base is defined by the LDO of *B. dertonensis*. The following species occur commonly within the Zone: *B. subfusiformis, F. costiferum, L. grandis, L. inornata, B. imperatrix, Bo. interjuncta, B. mandoroveensis, C. inflata, Eponides ornatus, E. iojimaensis, Fursenkoina schreibersiana, N. centrosulcatum, Q. rhodiensis, Q. vulgaris, S. oligocaenica, A. pseudospiralis, P. panamensis, S. wrightii, E. eshira, H. concentrica, H. strattoni,* and *U. subperegrina*.
Age and Correlation: The age of the Zone is Late Oligocene (Early Chattian, 27.79–26.20 Ma) (Gradstein et al., 2012) and it is correlated with the "upper" N2/P21b–"lower" N3/P22 Planktonic Zones of Berggren et al. (1998a,b).
Regional/Stratigraphic Distribution: The Zone was encountered in the Agberi North-1 (2,695–3,570 m) and Benin River-3 (9,480–10,000 ft.) wells. The zonal boundaries were not be established in the Amakubu-1 well because markers for the 26.20 Ma MFS (Chattian 2) (Gradstein et al., 2012) were not found.
21. ***Bolivina tenuicostata/Bolivina interjuncta* Assemblage Zone (PNDF 04)**
Diagnosis: This Assemblage Zone is marked at the top by the LDO of *B. dertonensis*, which coincides with the 27.79 Ma MFS (Rupelian/Chattian 1) (Gradstein et al., 2012) and below by the LDO of *Uvigerina sparsicostata*. Associated key foraminifera include *B. mandoroveensis, B. imperatrix, B. dertonensis, Brizalina dilatata, B. subfusiformis, C. auriculus, C. inflata, E. pontoni, E. vitrea, H. bononiensis, Q. rhodiensis, Q. vulgaris, N. ecuadoranum, N. auris, Hopkinsina hourqi, Uvigerina mantaensis, U. subperegrina, S. wrightii, T. parvula, Bo. interjuncta,* and *M. africana*.
Age and Correlation: The Zone is dated Early-Late Oligocene (28.64–27.79 Ma) (Gradstein et al., 2012) and correlated with the "lower" N2/P21a Planktonic Zone of Berggren et al. (1998a,b).

Regional/Stratigraphic Distribution: The zone was recognized in the Agberi North-1 (3,570–4240 m), Benin River-3 (10,000–10,740 ft.), and Pologbene-1 (2,650–4,400 ft.) wells. The Zone was not defined in the Amakubu-1 well because the markers for dating the 27.79 Ma MFS, Rupelian/Chattian 1 (Gradstein et al., 2012) were absent.

22. *Spiroplectammina wrightii–Bulimina elongata* Assemblage Zone (PNDF 05)

Diagnosis: The top of the Zone is marked by the LDO of *Uvigerina sparsicostata* and is dated by the 28.64 Ma MFS (Rupelian 3) (Gradstein et al., 2012), while the base is defined by the LDO of *S. wrightii*, which coincides with the 30.54 Ma MFS, (Rupelian 2) (Gradstein et al., 2012). Associated with the marker species are *B. mandoroveensis, B. scalprata, B. miocenica, B. imperatrix, B. interjuncta, Bulimina alazanensis, H. bononiensis, N. stella, N. auris, N. ecuadoranum, Rectuvigerina tenuicostata, S. oligocaenica, U. mantaensis, Q. seminulum, Q. rhodiensis, A. tenuis*, Bathysiphon spp., *Spiroplectammina biformis, P. panamensis, Brizalina mexicana, B. dilatata*, and *Quinqueloculina boueana.*

Age and Correlation: The age of the Zone is Early Oligocene (30.54–28.64 Ma) (Gradstein et al., 2012) and it is correlated with the N1/P20 planktonic Zone of Berggren et al. (1998a,b).

Regional/Stratigraphic Distribution: The Zone is penetrated by the wells Amakubu-1 (7,680–9,120 ft.), Mosogar-2 (above 5,200 ft.), Agberi North-1 (4,240–4,540 m), Benin River-3 (10,740–11,160 ft.), and Pologbene-1 (4,400–5,250 ft.).

23. *Hopkinsina bononiensis–Rectuvigerina tenuicostata* Assemblage Zone (PNDF 06)

Diagnosis: The top of this Assemblage Zone is defined by the LDO of *S. wrightii*, while the base is defined by the LDO of *H. bononiensis*. The following associated species are common within the Zone: *B. imperatrix, B. miocenica, B. mandoroveensis, B. scalprata, Al. scalaris, F. punctata, C. turgidus, E. iojimaensis, H. mckannai, H. semiornata, L. grandis, L. inornata, A. tenuis, E. eshira, E. pontoni, Q. rhodiensis, S. wrightii, S. biformis, B. subfusiformis, N. danvillensis, N. auris*, and *Nonion stachei*.

Age and Correlation: The age of the Zone is Early Oligocene (32.90–30.54 Ma) (Gradstein et al., 2012) and it is correlated with the P18–P19 Planktonic Zones of Berggren et al. (1998a,b).

Regional/Stratigraphic Distribution: The Zone is recognized in the Amakubu-1 (9120–9640 ft.), Mosogar-2 (5200–8270 ft.), and Agberi North-1 (4540 m–TD) wells. In the Jesse-1 well, the basal part of the Zone is above 7,700 ft., while the top of the Zone was encountered in the Benin River-3 well at 11,160–12,180 ft. (TD).

24. *Uvigerina gallowayi–Nonionella danvillensis* Assemblage Zone (PNDF 07)

Diagnosis: This Assemblage Zone is marked at the top by the LDO's of *E. berthelotianus* and *H. bononiensis* and dated 32.90 Ma MFS (Priabonian/Rupelian 1) (Gradstein et al., 2012). The base is defined by the LDO of *B. imperatrix* and the FDO of *Nonion oyae*. Associated foraminifera include *Al. scalaris, F. punctata, B. imperatrix, B. mandoroveensis, B. subfusiformis, C. turgidus, C. inflata, E. iojimaensis, S. biformis, S. wrightii, N. auris, A. tenuis*, and *N. stachei*.

Age and Correlation: The Zone is assigned a Late Eocene–Early Oligocene age (34.45–32.90 Ma) (Gradstein et al., 2012) and is correlated with the "upper" P16–P18 Planktonic Zones of Berggren et al. (1998a,b).

Regional/Stratigraphic Distribution: The Zone is recognized in the Pologbene-1 (5,720–5,970 ft.), Jesse-1 (7,700–8,830 ft.), and Mosogar-2 (8,270–10,150 ft.) wells. The top of the Zone was also encountered towards the base of the Amakubu-1 well at 9640 ft.

25. *Nonion oyae* Zone (PNDF 08)

Diagnosis: The top of the *N. oyae* Zone is marked by the FDO of *N. oyae* and the base by the FDO of *Bolivina ihuoensis*. Associated foraminiferal species include *Eponidella nigeriana, Q. seminulum, V. wilcoxensis, F. costiferum, N. auris, C. inflata, Uvigerina havanensis, U. gallowayi, Siphogenerina nodifera, E. pontoni, H. hourqi, Globocassidulina subglobosa, S. wrightii, Anomalinoides abuillotensis*, and *Lenticulina degolyeri*.

Age and Correlation: The Zone is dated Late Eocene 35.99–34.45 Ma (Gradstein et al. 2012) and is correlated with the "middle" of the P16 planktonic Zone of Berggren et al. (1998a,b).

Regional/Stratigraphic Distribution: The Zone is recognized in the Gilli Gilli-1 (above 8,040 ft.), Mosogar-2 (10,150–10,450 ft.), and Jesse-1 (8,830–10,580 ft.) wells. The top of the Zone was recognized in the Pologbene-1 well at 5970 ft. The base was not well defined because markers were absent.

26. *Bolivina ihuoensis–Bolivina denticulocamerata* Zone (PNDF 09)

Diagnosis: The zonal top is defined by the FDO of *Bo. ihuoensis* and the base by the FDO of *H. hourqi*. Associated foraminifera species include *L. grandis, Uvigerina jacksonensis, B. mandoroveensis, B. subfusiformis, C. turgidus, F. atlanticus, F. costiferum, F. punctata, N. auris, Q. rhodiensis, S. nodifera, Al. scalaris, A. tenuis*, and *Valvulineria jacksonensis*.

Age and Correlation: The Zone is assigned a Late Eocene age (37.36–35.99 Ma) (Gradstein et al., 2012) and is correlated with the "upper" P15–"lower" P16 Planktonic Zones of Berggren et al. (1998a,b).

Regional/Stratigraphic Distribution: The Zone occurs in the Gilli Gilli-1 (8,040–8,760 ft.) and Mosogar-2 (10,450–11,357 ft.) wells. The zonal base was encountered in the Pologbene-1 well at 7,460 ft. and the top in the Jesse-1 well from below 10,580 ft. down to TD.

27. *Hopkinsina hourqi* Zone (PNDF 10)

Diagnosis: The *H. hourqi* Zone is marked at the top by the FDO of *H. hourqi* and at the base by the underlying FDO of *Eponides africana*. Cooccurring with the markers are *H. concentrica, B. imperatrix, B. beyrichi, U. havanensis, U. gallowayi, Uvigerina asperula, G. subglobosa, B. subfusiformis, S. complanata, Q. seminulum, V. wilcoxensis, N. auris, E. iojimaensis, Bo. ihuoensis, C. turgidus, N. oyae, H. hourqi, H. strattoni*, and *E. eshira*.

Age and Correlation: The Zone is dated Middle Eocene (38.46–37.36 Ma) (Gradstein et al., 2012) and is correlated with the "middle" P15 Planktonic Zone of Berggren et al. (1998a,b).

Regional /Stratigraphic Distribution: The Zone is identified in the wells: Gilli Gilli-1 (8,760–9,200 ft.), Pologbene-1 (7,460–9,250 ft.), and Mosogar-2 (11,357 ft.–TD).

28. *Eponides africana* Zone (PNDF 11)

Diagnosis: The top of the *E. africana* Zone is marked by the FDO of *E. africana*, which is dated 38.46 Ma MFS (Bartonian 1) (Gradstein et al., 2012), while the base is defined by the FDO of *Hopkinsina danvillensis* dated with the 40.40 Ma MFS (Lutetian 4 of Gradstein et al., 2012). Associated foraminifera include *U. havanensis, C. turgidus, F. costiferum,*

N. oyae, H. hourqi, H. strattoni, A. pseudospiralis, Milliammina petila, N. auris, Bo. ihuoensis, U. gallowayi, H. danvillensis, F. boueanum, Bulimina kugleri, and *E. eshira.*

Age and Correlation: The Zone is assigned a Middle Eocene age (40.40–38.46 Ma) (Gradstein et al., 2012) and is correlated with the P13–"lowermost" P15 Planktonic Zone of Berggren et al. (1998a,b).

Regional/Stratigraphic Distribution: The zone is recognized in the Gilli Gilli-1 well (9200–9730 ft.). The top of the Zone is also encountered at the base of the Pologbene-1 well between 9250 ft. and TD.

29. ***Hopkinsina danvillensis* Zone (PNDF 12)**

Diagnosis: The top of the *H. danvillensis* Zone is marked by the FDO of *H. danvillensis*, which coincides with the 40.40 Ma MFS (Lutetian 4 of Gradstein et al., 2012). The base is defined by the LDO of *U. gallowayi*. Associated foraminifera include *R. tenuicostata, H. hourqi, F. costiferum, F. atlanticus, H. pseudoungeriana, V. wilcoxensis, S. wrightii, Nonion obducum, Planulina renzi, N. ecuadoranum, N. oyae, Bo. ihuoensis, N. auris, N. danvillensis, Valvulineria trochoidea, Q. seminulum, U. havanensis,* Bathysiphon spp., *Haplophragmoides canariensis, Valvulineria rugosa, Valvulineria bradyi, A. agglutinans, F. boueanum, Al. scalaris,* and *A. tenuis.*

Age and Correlation: The Zone is assigned a Middle Eocene (42.64–40.40 Ma) (Gradstein et al., 2012) age and is correlated with the "upper" P12 Planktonic Zone of Berggren et al. (1998a,b).

Regional/Stratigraphic Distribution: The Zone is recognized in the Gilli Gilli-1 well at 9,730–10,230 ft.

30. ***Altistoma tenuis–Uvigerina havanensis* Assemblage Zone (PNDF 13)**

Diagnosis: The zonal top is defined by the LDO of *U. gallowayi*, which correlates with the 42.64 Ma MFS (Lutetian 3) (Gradstein et al., 2012). The base is marked by the LDO of *A. tenuis* age dated 44.75 Ma (Lutetian 2) (Gradstein et al., 2012). Cooccurring with the markers are *Fursenkoina wilcoxensis, N. obducum, V. wilcoxensis, B. subfusiformis, Al. scalaris, E. africana, H. canariensis, Q. seminulum, S. wrightii, A. agglutinans, Bo. ihuoensis,* and *B. kugleri* in the Middle Eocene.

Age and Correlation: The age of the Zone is Middle Eocene (44.75–42.64 Ma) (Gradstein et al., 2012) and is correlated with the "upper" P11–"lower" P12 Planktonic Zones of Berggren et al. (1998a,b).

Regional/Stratigraphic Distribution: The Gilli Gilli-1 well penetrated this Zone between 10,230 and 10,520 ft.

31. ***Eponides eshira–Fursenkoina wilcoxensis* Assemblage Zone (PNDF 14)**

Diagnosis: This Assemblage Zone is marked at the top by the LDO of *A. tenuis*, which coincides with the 44.75 Ma MFS (Lutetian 2) (Gradstein et al., 2012) while the base coincides with the 48.9 Ma MFS (46.82 Ma MFS, Lutetian 1) (Gradstein et al., 2012) and is marked by the LDO of *E. eshira*. Associated foraminifera species include *E. africana, H. danvillensis, Cancris oblongus, V. wilcoxensis, A. agglutinans, A. strathearnensis,* and *B. kugleri.*

Age and Correlation: The Zone is assigned a Middle Eocene age (46.82–44.75 Ma) (Gradstein et al., 2012) and is correlated with the "upper" P10–"lower" P11 Planktonic Zones of Berggren et al. (1998a,b).

Regional/Stratigraphic Distribution: Though the Zone is most likely present in the Gilli Gilli-1 well, its boundaries could not be defined with any degree of certainty.

32. ***Bolivina simplex–Bulimina kugleri* Assemblage Zone (PNDF 15)**
Diagnosis: This Assemblage Zone is marked at the top by the LDO of *E. eshira*, which coincides with the 46.82 Ma MFS, Lutetian 1 (Gradstein et al., 2012). The base is marked by the LDO of *B. simplex*, which coincides with the 51.27 Ma MFS, Ypresian 7 (Gradstein et al., 2012). Cooccurring with the nominate marker species are *Sagrina circumspinosa, N. danvillensis, F. atlanticus, N. oyae, Valvulineria midwayensis*, and *E. africana*.
Age and Correlation: The Zone is assigned a Middle Eocene, 51.27–46.82 Ma age (Gradstein et al., 2012) and is correlated with the "upper" P7–"lower" P10 Planktonic Zones of Berggren et al. (1998a,b).
Regional/Stratigraphic Distribution: The Zone is recognized in the Ihuo-1 well (basal portion above 6850 ft.) and Gilli Gilli-1 well. The boundaries were not clearly established because of missing MFS.

33. ***Sagrina circumspinosa–Uvigerina jacksonensis* Assemblage Zone (PNDF 16)**
Diagnosis: This Assemblage Zone is defined at the top by the LDO of *B. simplex*, which coincides with the 51.27 Ma MFS, Ypresian 7 (Gradstein et al., 2012). The base is defined by the LDO of *S. circumspinosa*, which coincides with the 52.00 Ma MFS, Ypresian 6 (Gradstein et al., 2012). Associated foraminifera include *E. africana, N. oyae, Al. scalaris, Bo. ihuoensis, F. boueanum, F. atlanticus, N. obducum, A. pseudospiralis*, and *H. concentrica*.
Age and Correlation: The Zone is dated Early Eocene (52.00–51.27 Ma) (Gradstein et al., 2012) and correlated with the upper portion of the "middle" P7 Planktonic Zone of Berggren et al. (1998a,b).
Regional Stratigraphic Distribution: The Zone is recognized in the Ihuo-1 well (6850–7710 ft.). The boundaries could not be established in the Gilli Gilli-1 well.

34. ***Altistoma scalaris–Valvulineria jacksonensis* Assemblage Zone (PNDF 17)**
Diagnosis: The top of this Assemblage Zone is defined by the LDO of *S. circumspinosa*, which coincides with the 52.00 Ma MFS (Ypresian 6) (Gradstein et al., 2012). The base is marked by the FDO of *Fursenkoina pontoni*, which coincides with the 53.01 Ma MFS (Ypresian 5) (Gradstein et al., 2012). Associated foraminifera species include *V. jacksonensis, Bo. ihuoensis, E. pontoni, E. eshira, E. africana, F. costiferum, H. concentrica, Haplophragmoides sahariense*, and *A. abuillotensis*.
Age and Correlation: The Zone is dated Early Eocene 53.01–52.00 Ma (Gradstein et al., 2012) and is correlated with the "upper" P6b to "lower" P7 Planktonic Zones of Berggren et al. (1998a,b).
Regional/Stratigraphic Distribution: The Zone is recognized in the Gilli Gilli-1 well and Ihuo-1 (top is below 7210 ft.) well. The boundaries could not be defined because of presumed unconformities.

35. ***Fursenkoina pontoni* Zone (PNDF 18)**
Diagnosis: The top of this Zone is defined by the FDO of *F. pontoni*, which coincides with the 53.01 Ma MFS (Ypresian 5) (Gradstein et al., 2012). The base is defined by the FDO of *Haplophragmoides excavata* and the LDO of *Al. scalaris*, both of which coincide with the 53.67 Ma MFS, Ypresian 4 (Gradstein et al., 2012). Foraminiferal species that cooccur

with the marker species include *Uvigerina peruviana, U. havanensis, U. jacksonensis,* and *Nonion boueanum.*

Age and Correlation: The Zone is dated Early Eocene, 53.67–53.01 Ma (Gradstein et al., 2012) and is correlated with the "Middle" of the P6b Planktonic Zone of Berggren et al. (1998a,b).

Regional/Stratigraphic Distribution: The Zone was recognized in both the Ihuo-1 and Gilli Gilli-1 wells (below the 53.67 Ma MFS) (Gradstein et al., 2012). Their limits could not be clearly established because of presumed unconformities.

36. ***Haplophragmoides excavata* Zone (PNDF 19)**

Diagnosis: The top of the *H. excavata* Zone is marked by the FDO of *H. excavata* and the LDO of *Al. scalaris* while the base is marked by the FDO of *E. pseudoelevatus* and the LDO of *E. africana.* The Zone thus ranges in age from the 54.63 to the 53.67 Ma MFS (Gradstein et al., 2012). The following foraminiferal species cooccur with the marker species, *E. africana, A. agglutinans,* and *Valvulineria suturalis.*

Age and Correlation: The Zone is dated Early Eocene 54.63–53.67 Ma age (Gradstein et al., 2012) and correlated with the P6a–b Planktonic Zones of Berggren et al. (1998a,b).

Regional/Stratigraphic Distribution: The top of the Zone is recognized in the Gilli Gilli-1 well at 10,790 ft. The base was also encountered in the Ihuo-1 well above 7,780 ft.

37. ***Eponides pseudoelevatus* Zone (PNDF 20)**

Diagnosis: The top of the *E. pseudoelevatus* Zone coincides with the global cyclic event dated 54.63 Ma MFS (Ypresian 3) (Gradstein et al., 2012). It is defined by the FDO of *E. pseudoelevatus* and the LDO of *E. africana,* while the base is defined by the FDO of *L. pseudomamilligerus* and the LDO of *L. grandis.* Associated foraminifera include *H. concentrica, V. trochoidea, E. eshira, F. atlanticus, G. subglobosa, H. pseudoungeriana, Haplophragmoides bradyi, H. sahariense, H. excavata, Milliammina telemaquensis, Textularia laminata, T. globigeriniformis,* and *V. jacksonensis.*

Age and Correlation: The Zone is dated Early Eocene age (55.20–54.63 Ma) (Gradstein et al., 2012) and is correlated with the P6a Planktonic Zone of Berggren et al. (1998a,b).

Regional/Stratigraphic Distribution: The top of the Zone is penetrated by the Ihuo-1 well below 7780 ft. The base was not defined because the 55.20 Ma MFS (Gradstein et al., 2012) was missing. The basal part of the Zone was present in the Gilli Gilli-1 well above the 55.20 Ma MFS (Gradstein et al., 2012).

38. ***Lenticulina pseudomamilligerus* Zone (PNDF 21)**

Diagnosis: The *L. pseudomamilligerus* Zone straddles the Paleocene–Eocene boundary and ranges in age from 55.20 to 57.88 Ma (Ypresian 2–Thanetian 3) (Gradstein et al., 2012). The top of the Zone is marked by the FDO of *L. pseudomamilligerus* and the LDO of *L. grandis,* while the base is marked by the FDO of *Valvulineria martinezensis.* Important associated forms include *N. danvillensis, H. concentrica, F. punctata, F. wilcoxensis, Lagena striata,* and *A. abuillotensis.*

Age and Correlation: The Zone is dated Late Paleocene–Early Eocene (57.88–55.20 Ma of Gradstein et al., 2012) and is correlated with the P4c–P5 Planktonic Zones of Berggren et al. (1998a,b).

Regional/Stratigraphic Distribution: The base of the Zone was penetrated by the Ihuo-1 well at 8820 ft. The zonal top was also encountered in the Gilli Gilli-1 well at 11020 ft. (55.20 Ma) (Gradstein et al., 2012).

39. ***Valvulineria martinezensis* Zone (PNDF 22)**
Diagnosis: The top of the *V. martinezensis* Zone is marked by the 57.88 Ma MFS, Thanetian 3 (Gradstein et al., 2012) defined by the FDO of *V. martinezensis*. The base is marked by the 58.45 Ma MFS (Gradstein et al., 2012) defined by the FDO of *C. texana*. The following foraminifera species cooccur with the marker species: *E. pseudoelevatus*, *H. pseudoungeriana*, *N. danvillensis*, *Praebulimina fang*, *Praebulimina laddi*, *V. midwayensis*, *A. pseudospiralis*, *H. bradyi*, *H. sahariense*, *M. petila*, *Recurvoides contortus*, *T. globigeriniformis*, *Spiroplectammina earlandi*, and *H. excavata*.
Age and Correlation: The Zone is dated Late Paleocene age (58.45–57.88 Ma) (Gradstein et al., 2012) and correlated with the P4b Planktonic Zone of Berggren et al. (1998a,b).
Regional/Stratigraphic Distribution: The top of the Zone is penetrated by the Ihuo-1 well at 8820 ft. The base is probably defined by the unconformity at 9650 ft. The limits of the Zone in the Gilli Gilli-1 well could not be defined because the 57.88 Ma MFS was not identified.
40. ***Cribrononion texana* Zone (PNDF 23)**
Diagnosis: The top of the *C. texana* Zone is marked by the FDO of *C. texana*. This is dated 58.45 Ma MFS (Thanetian 2 sensu Gradstein et al., 2012). The base of the Zone is defined by the bioevents that delineate the top of the underlying *Ammobaculites coprolithiformis* Zone, namely the LDO of *B. tenuicostata* and the FDO of *A. coprolithiformis*. Associated foraminifera species include *B. tenuicostata*, *H. sahariense*, *Lenticulina midwayensis*, and *V. midwayensis*.
Age and Correlation: The Zone is dated Late Paleocene (59.26–58.45 Ma of Gradstein et al., 2012) and correlated with the P4a–b Planktonic Zones of Berggren et al. (1998a,b).
Regional/Stratigraphic Distribution: The Zone is penetrated by the Ihuo-1 well (9650–9950 ft.). The topmost part of the Zone is encountered at the base of the Gilli Gilli-1 well from 11,290 to 11,540 ft. (TD).
41. ***Ammobaculites coprolithiformis* Zone (PNDF 24)**
Diagnosis: The top of the *A. coprolithiformis* Zone is defined by the FDO of *A. coprolithiformis* and the LDO of *B. tenuicostata* both of which coincide with the 59.26 Ma MFS, Selandian 2/Thanetian 1 (Gradstein et al., 2012). The base is marked by the FDO of *Anomalinoides madrugaensis*. Other significant associated foraminifera species include *L. midwayensis*, *H. sahariense*, and *Valvulinera midwayensis*.
Age and Correlation: The Zone is assigned a Late Paleocene age (60.65–59.26 Ma) (Gradstein et al., 2012) and correlated with the P3a–P4a Planktonic Zones of Berggren et al. (1998a,b).
Regional/Stratigraphic distribution: The zone was penetrated by the Ihuo-1 well between 9,950 and 10,870 ft.
42. ***Anomalinoides madrugaensis* Zone (PNDF 25)**
Diagnosis: This is the oldest benthonic foraminiferal zone recognized in the Paleogene section of the Niger Delta. The Zone extends from the 60.65 Ma MFS (Selandian 1) (Gradstein et al., 2012) to the 64.40 Ma MFS (Danian 2) (Gradstein et al., 2012). The base is marked by the FDO of the heavily costate *Bolivina afra*, while the top is defined by the FDO of *A. madrugaensis*. Cooccurring with the species are *V. martinezensis*,

Lenticulina extruatus, *Eponides lysi*, and *Bolivina africana*. No marker species was found to characterize the 63.30 Ma MFS (Danian 3) (Gradstein et al., 2012).

Age and Correlation: The Zone is dated Early Paleocene (64.40–60.65 Ma) (Gradstein et al., 2012) and is correlated with the P1b–c to P3a Planktonic Zones of Berggren et al. (1998a,b).

Regional/Stratigraphic Distribution: The top of the Zone is recognized in the Ihuo-1 well below 10,870 ft.

4.3.2 Planktonic Foraminifera Zones

Of the 38 characterized worldwide planktonic foraminifera Zones that have been established for the Cenozoic (Stainforth et al., 1975), only 10 were previously recognized in the Niger Delta. However, in the course of the present comprehensive regional foraminiferal investigation, 33 planktonic foraminiferal zones were identified in the Niger Delta, 19 for the Paleogene, 13 for the Neogene, and 1 in the Quaternary. Several planktonic foraminiferal zonal schemes for the Cenozoic have been published by the petroleum industry, notably, Blow (1969a,b) from British Petroleum, Postuma (1971) from Shell, and Stainforth et al. (1975) from the Exxon Group. Although the Bolli Zonation also grew out of oil exploration in the Caribbean region, his preference for deeper water planktonic assemblages in zonal definitions greatly limits the application of his zonation within a largely shallow-water basin, such as the Niger Delta. The aforementioned zonations are by far more relevant to the Niger Delta than that of Berggren et al., which was based on oceanic planktonic foraminifera from the global Ocean Drilling Project.

Using a quantitative biostratigraphic technique known as Constrained Optimization, Daniel et al. (2007), analyzed planktonic foraminifera data from the continental margin of the southwest Caribbean and proposed 13 planktonic foraminifera zones and 8 subzones for the Eocene–Pliocene interval. The zonal scheme relies exclusively on last occurrences. The zones recognized are readily applied to petroleum exploration.

Ozumba (2011), on the basis of the frequency of well-known index planktonic foraminifera recorded at several areas in the Niger Delta proposed 23 planktonic foraminifera zones (ND1–ND23) for the Niger Delta. These range in age from Paleocene to Pleistocene. Many of these zones are readily correlated with the zones proposed in this study

Planktonic foraminifera zones from the Exxon group (Stainforth et al., 1975) were defined using, as much as possible, the total assemblages, a greater percentage of which have faunal elements that are represented in the Niger Delta, in contrast to the zonal criteria of Postuma (1971) and Blow (1969a,b). Therefore, just as the global sea-level curve of Exxon (Vail et al., 1977) has been successfully adapted to the Niger Delta sequence, the Exxon planktonic foraminiferal zonation has been mostly found to be useful in the present study. This is because both the sea-level curve and the planktonic zones were based on continental margin sequences similar to the Niger Delta. However, for the Paleocene, Postuma's (1971) zonation is more appropriate. International zonal names have been retained even for the zones where the nominal taxa were not encountered in the course of this investigation to enhance communication and correlation with the standard Cenozoic Chronostratigraphic Chart, thus making room for the fact that the nominal taxa could have been encountered by the oil companies operating in Nigeria and used for their in-house zonations (Figs. 4.1–4.6).

As the international zonal names have acquired the status of formal chronostratigraphic and geochronologic units, they can be applied worldwide, like the standard European Stage nomenclature, which is used in other parts of the world outside the European stratotypes, regardless of local lithologic differences. Similarly, the standard planktonic foraminiferal biochronological names can be retained in the Niger Delta even if the nominal taxon has not yet been reported.

The Tropical Surface Water Mass produced warm, shallow-water benthonic assemblages where most of the endemic foraminiferal species in the Niger Delta and the Gulf of Guinea are found. This also contains the Late Neogene "West Indian" fauna characterized by *Amphistegina* and *Heterostegina* (Harris, 1981). Whereas the Tropical Surface Water Mass in the Recent Niger Delta occurs at depths shallower than 50 m, as an unstratified water mass in which foraminiferal biofacies changes are small and sometimes hardly noticeable, the deeper parts of the Niger Delta that were influenced by the South Atlantic Central Water and the Benguela Current, contain predominantly cosmopolitan benthonic species, especially in the outer shelf and slope environments. Benthonic faunal similarity is particularly strong between the Niger Delta and the Caribbean region thereby enhancing correlations (Petters, 1982).

Although Stolk (1965) identified over 55 planktonic foraminiferal species in the Paleocene–Eocene of the Niger Delta basin and other workers (Harris, 1981; Petters, 1983) uncovered rich planktonic assemblages in the Oligocene–Recent, the planktonic foraminifera of the Niger Delta remained underutilized in biostratigraphy. As noted by Harris (1981), planktonic foraminifera are common even under shelfal conditions in the Niger Delta to the extent that they distort the planktonic/benthonic ratio as a paleobathymetric indicator. Consequently, attention was paid to the planktonics in this study, thereby uncovering in this study the richest planktonic assemblages ever known or described in the region.

Planktonic foraminifera paleozoogeographic affinities have also been demonstrated, with a preponderance of Southern Ocean High-Latitude elements during the major sea level highstands in the Paleogene when the influence of the cold Benguela Current was strong. This accounts for the dominance of *Acarinina*, *Subbotina*, and *Cassigerinelloita* in the Eocene of the Niger Delta. With the waning influence of the Benguela Current during the Neogene, low-latitude planktonic assemblages predominate with *Orbulina*, *T. fohsi*, and *Sphaeroidinellopsis* bioseries and *G. tumida* and *G. menardii* among the evolutionary lineages that are well represented in the Niger Delta.

The zones proposed here are Range Zones based on index planktonic foraminifera or associations of species that are quite abundant and diverse in the Niger Delta, especially in the transgressive marine shales. These shales have been correlated with major global flooding events, hence they coincide with the MFS in the Niger Delta chronostratigraphic chart. Due to high-microfaunal diversity in the transgressive marine shales, integration of two separate planktonic and benthonic biozonations has been accomplished (Enclosure 1.1) for the Paleogene and Neogene of the Niger Delta. This has allowed the extension of the Maximum Flooding Surface ages into shallow-marine facies, even when the index planktonic foraminiferal species are absent on account of facies change. Also the numerical planktonic foraminiferal zonation of Blow (1969a,b, 1979) has been correlated with the benthonic zones in the Niger Delta (Enclosure 1.1).

A major constraint to the application of planktonic foraminifera in the Niger Delta, apart from their previously poor taxonomic documentation, has been deltaic facies changes, which often falsify bioevents. The global geochronological and chronostratigraphic scheme (Berggren et al., 1998a,b, Gradstein et al., 2004a,b, 2012; Wade et al., 2011) from which that of the Niger Delta is derived is based on worldwide synchronous evolutionary first and last appearance datums in continuous marine sections. Evolutionary appearance and extinctions have been calibrated in numerical, radiometric ages where known. But as continuous, uninterrupted marine sedimentation did not prevail in the delta, first appearance and extinction datums may not be perfectly valid stratigraphic parameters in the Niger Delta. The absolute ages that have been assigned to both the benthonic and planktonic foraminiferal zones are, therefore, approximate numerical notations. Reference should be made to the enclosed planktonic foraminiferal range chart for other zonal taxa that are not specified in the zonal definitions.

A. *Neogene Planktonic Foraminiferal Zonation*

PLIOCENE–PLEISTOCENE ZONES

1. ***Globorotalia truncatulinoides* Zone (QNDf 01)**

   Diagnosis: In the absence of *G. truncatulinoides*, this Zone can be recognized on negative evidence, notably above the extinction of forms, such as *Globigerinoides obliquus* and *G. extremus*.

   Age and Correlation: This is the youngest planktonic foraminiferal zone commonly encountered in the Niger Delta. It spans the Pliocene–Pleistocene interval and corresponds with the N22–N23 Zones of Blow (1979), with the 2.39 Ma MFS (Gradstein et al., 2012) near the base and the 1.64 Ma MFS (Gradstein et al., 2012) occurring within the Zone.

   Regional/Stratigraphic Distribution: This Zone occurs in the Anyala-4 well, 340–2140 ft.

PLIOCENE ZONES

2. ***Pulleniatina obliquiloculata* Zone (NNDf 13)**

   Diagnosis: The zonal top is defined by the extinction of *Globoquadrina altispira* and the genus *Sphaeroidinellopsis*. Species that appear in this Zone for the first time include *G. miocenica* and *G. inflata*.

   Age and Correlation: This Zone is of Middle Pliocene age and it correlates with Zone N21 (Blow, 1979). The base is defined by the 3.82 Ma MFS (Gradstein et al., 2012) and the top approximates the 2.39 Ma MFS (Gradstein et al., 2012).

   Regional/Stratigraphic Distribution: This Zone occurs in the Inanga-1 well, 2380–2980 ft., Anyala-8 well, 8300–8600 ft., Anyala-4 well, 6140–6740 ft., and HB-1 well, 3140–3340 ft.

3. ***Globorotalia margaritae* Zone (NNDf 12)**

   Diagnosis: The base of this Zone is characterized by the appearance of *G. magaritae* and *G. crassaformis*. *G. tumida* also appears for the first time in this Zone. *Neogloboquadrina humerosa* is abundant.

   Age and Correlation: This Zone is of Early Pliocene age (Zones N19–N21 sensu Blow, 1979). As defined in this report, the zonal base and top are respectively correlated with the 5.99 and 3.82 Ma MFS (Gradstein et al., 2012), with the 5.47 and 4.16 Ma MFS (Gradstein et al., 2012) falling within the Zone.

   Regional/Stratigraphic Distribution: This Zone is found in the Etim-1 well, 4800–5000 ft.

LATE MIOCENE ZONES

4. ***Turborotalia acostaensis* Zone (NNDf 11)**

Diagnosis: This Zone is characterized by the range of *T. acostaensis* and the association of *Globigerinoides seiglei*, *Sphaeroidinellopsis seminulina*, *Globorotalia merotumida*, *G. plesiotumida*, and *N. dutertrei*.

Age and Correlation: This Late Miocene Zone encompasses Blow (1979) Zones N16–N18, and is defined by the 10.51 and 5.99 Ma MFS (Gradstein et al., 2012).

Regional/Stratigraphic Distribution: This Zone is found at the Inanga-1 well, 9,580–11,180 ft.

5. ***Globorotalia menardii* Zone (NNDf 10)**

Diagnosis: This Zone embraces the interval between the extinction of *T. siakensis* and the evolutionary appearance of *N. acostaensis*. The planktonic assemblages include abundant *G. trilobus*, *G. bulloides*, *G. sacculifer*, and *Globigerina falconensis*.

Age and Correlation: This is an early Late Miocene Zone, which spans Zones N15 and N16, with the 10.51 Ma MFS (Gradstein et al., 2012) at the top.

Regional/Stratigraphic Distribution: This Zone occurs in the Utapate-1 well, 7800–9500 ft.

MIDDLE MIOCENE ZONES

6. ***Turborotalia mayeri/Turborotalia siakensis* Zone (NNDf 09)**

Diagnosis: This is an Assemblage Zone in which *T. siakensis*, *Hastigerina siphonifera*, *Globigerinita incrusta*, and *T. obesa* cooccur.

Age and Correlation: The Zone correlates with the N15 Zone and lies across the Middle–Late Miocene boundary, with the 12.26 Ma MFS at the top. The zonal top is probably correlatable with the 12.26 Ma MFS (Gradstein et al., 2012).

Regional/Stratigraphic Distribution: This Zone is found in the Kolo Creek-1 well between 6700 ft and 7300 ft.

7. ***Turborotalia robusta–lobata* Zone (NNDf 08)**

Diagnosis: The *T. robusta–lobata* Zone is defined on the presence of the nominal taxa, *T. robusta* and *T. lobata*, which are the most highly evolved members of the *T. fohsi* evolutionary lineage.

Age and Correlation: This is a late Middle Miocene Zone that correlates with Zone N12 at the base and N13–N14 at the top. The 13.12 Ma MFS (Gradstein et al., 2012) is within the Zone.

Regional/Stratigraphic Distribution: This Zone occurs in the Bita-9 well at 9170 ft. This is the only occurrence recorded in this study.

8. ***Turborotalia fohsi* Zone (NNDf 07)**

Diagnosis: This Zone is characterized by the presence of the marker species *T. fohsi*.

Age and Correlation: This Zone correlates with Blow's (1979) N10–N11 zones within the Middle Miocene. The base, as defined in this report correlates with the 13.68 Ma MFS (Gradstein et al., 2012), while the top is older than the 13.12 Ma MFS (Gradstein et al., 2012).

Regional/Stratigraphic Distribution: This Zone was encountered in the Manuso-1 well, 6920 ft.; Ewan-1 well, 6120 ft.; and Etelebou-1 well, 6930 ft.

9. ***Turborotalia peripheroronda* Zone (NNDf 06)**

Diagnosis: This Zone is defined by the overlap in the ranges of *T. peripheroronda* and *T. peripheroacuta*.

Age and Correlation: This Zone correlates with Blow's (1979) N9 Zone in the Middle Miocene and is marked by the 13.68 Ma MFS (Gradstein et al., 2012) at the top. The base lies between the 13.68 and 15.20 Ma MFS (Gradstein et al., 2012).

Regional/Stratigraphic Distribution: This Zone occurs in the Benin River-3 well, 6,440–6,640 ft.; Kolo Creek-1 well, 9,900–10,300 ft.; and in the Inanga-1 well, 13,380–13,580 ft.

10. ***Praeorbulina glomerosa* Zone (NNDf 05)**

Diagnosis: The base of this Zone is defined by the evolutionary appearance of *O. suturalis* and *O. universa*, while the top is marked by the disappearance of *P. glomerosa* and *P. circularis*. *T. peripheroronda* first appears in this Zone.

Age and Correlation: This Zone defines the base of the Middle Miocene (N8 Zone in Blow's, 1979 zonation) and occurs between the 13.68 and 15.20 Ma MFS (Gradstein et al., 2012).

Regional/Stratigraphic Distribution: This Zone occurs in the Akata-1 well, 6310–6510 ft.

EARLY MIOCENE ZONES

11. ***Globigerinatella insueta* Zone (NNDf 04)**

Diagnosis: This Zone can be easily recognized in the Niger Delta based on the presence of the short-ranging marker species, *S. dubeyi*. Other cooccuring taxa include *G. bisphericus*, *Praeorbulina curva*, and *Globigerinoides sicanus*.

Age and Correlation: This Zone falls at the top of the Early Miocene (N7 Zone; Blow, 1979). The top is well defined by the 15.20 Ma MFS (Gradstein et al., 2012) and the base by the 17.83 Ma MFS (Gradstein et al., 2012).

Regional/Stratigraphic Distribution: This Zone is widespread across the Niger Delta, with the distinctive planktonic foraminiferal assemblage recorded in the Batan-1 well at 15,100 ft.

12. ***Catapsydrax stainforthi* Zone (NNDf 03)**

Diagnosis: The top of this Zone is marked by the extinction of *C. dissimilis* and *C. stainforthi* and the asssociation of *Globigerina euapertura* and *G. ampliapertura*, which also disappear at the top.

Age and Correlation: This Zone falls within Blow's (1979) Zone N6, in the Early Miocene. It is found in the upper part of the Ogara Shale, with its base just above the 18.64 Ma MFS (Gradstein et al., 2012) and below the appearance of the marker *S. dubeyi*. The zonal top correlates with the 17.83 Ma MFS (Gradstein et al., 2012).

Regional/Stratigraphic Distribution: This Zone occurs in the Benin River-3 well at 7040–7240 ft.

13. ***Catapsydrax dissimilis* Zone (NNDf 02)**

Diagnosis: This Zone is marked by the first explosive development and occurrence of species of *Globigerinoides* and *Globoquadrina*. These include *G. trilobus*, *Globigerinoides sacculifer*, *G. subquadratus*, *G. obliquus*, and *Globigerinoides inusitatus*. Others include *G. dehiscens*, *G. altispira*, and *G. globosa*. The base of the Zone is placed at the appearance of *Globigerina bulloides*, *G. falconensis*, and *Globigerinita juvenelis*. *C. stainforthi*, *T. peripheroronda*, and *G. archaeomenardii* appear in this Zone. Taken together, the aforementioned planktonic assemblage defines the *C. dissimilis* Zone.

Age and Correlation: The *C. dissimilis* Zone is of mid-Early Miocene age and correlates with Zone N5 of Blow (1979). The base is further defined by the 20.94 Ma MFS (Gradstein et al., 2012) with the 18.64 Ma MFS (Gradstein et al., 2012) marking the upper part of this Zone.

Regional/Stratigraphic Distribution: This Zone is well represented across the Niger Delta in the Akata-1 well, 8510–9110 ft.; Opuama-1 well, 4890–6210 ft.; and Benin River-3 well, 7240–8040 ft.

B. *Paleogene Planktonic Foraminiferal Zonation*

OLIGOCENE ZONES

A threefold subdivision of the Oligocene into Early, Middle, and Late Oligocene is possible.

LATEST OLIGOCENE–EARLY MIOCENE ZONES

14. *Turborotalia kugleri* Zone (NNDf 01)

Diagnosis: The top of the *T. kugleri* Zone is marked by the disappearance of *G. primordius* and *G. angustiumbilicata* with the top of *T. kugleri* slightly below the upper zonal boundary. The rich planktonic foraminiferal assemblage includes *G. euapertura*, *Globigerinoides praebulloides*, *Globigerinoides occlusus*, *T. obesa*, *T. mayeri*, and *Catapsydrax unicavus*.

Age and Correlation: This Zone marks the earliest Miocene and is equivalent to the N4 Zone (of Blow, 1979), with the 22.24 Ma MFS at the base and the 20.94 Ma MFS (Gradstein et al., 2012) at the top.

Regional/Stratigraphic Distribution: This Zone is well developed in the Obeaja-1 well, 7200–7300 ft. and in the Opuama-1 well at 6600–7100 ft.

MIDDLE OLIGOCENE ZONES

15. *Globigerina ciperoensis* Zone (PNDf 19)

Diagnosis: The zonal top is marked by the *Globigerinoides* datum. The extinction of *G. angulosuturalis*, *G. ciperoensis*, and *Globigerina officinalis* occur in this Zone.

Age and Correlation: The Zone correlates with the top part of the Middle Oligocene, upper N2 Zone, and the lower N3 Zone sensu Blow (1979). The base and top are respectively dated 27.79 and 22.24 Ma MFS (Gradstein et al., 2012).

Regional/Stratigraphic Distribution: This Zone occurs in the Amakubu-1 well at 4880 ft.; Obeaja-1 well between 7300 and 7690 ft.; and Opuama-1 well between 6660 and 7100 ft.

16. *Turborotalia opima* Zone (PNDf 18)

Diagnosis: The Zone is characterized by the presence of *T. opima*, *G. angulisuturalis*, *G. ouachitaensis*, and *G. officinalis*.

Age and Correlation: Middle Oligocene, N2 Zone (Blow, 1979) and ranges between the 30.54 and 27.79 Ma MFS (Gradstein et al., 2012).

Regional/ Stratigraphic Distribution: The *T. opima* Zone occurs in the Amakubu-1 well, 9080–9680 ft.; Obeaja-1 well, 7220–7690 ft.; Opuama-1 well, 7150–9900 ft.; Meta-1 well, 9440–9500 ft.; Pologbene-1 well, 4440–5720 ft.; and Benin River-3 well, 9440–10040 ft.

EARLY OLIGOCENE ZONES

17. *Globigerina ampliapertura* Zone (PNDf 17)

Diagnosis: This Zone is marked by the first appearance of *G. opima*, abundance of *G. ampliapertura* and the earliest appearance of *G. angulisuturalis*.

Age and Correlation: This Zone falls within the middle Early Oligocene P17/18 Zones with the top and base dated 30.54 and 32.90 Ma MFS (Gradstein et al., 2012), respectively.

Regional/Stratigraphic Distribution: The *G. ampliapertura* Zone is documented in the Gilli Gilli-1 well at 8460–8640 ft.

18. ***Cassigerinella chipolensis–Pseudohastigerina micra* Zone (PNDf 16)**
Diagnosis: The top is marked by the extinction of *P. micra*, *Globigerina yeguaensis*, *Globigerina gnaucki*, *C. victoriana*, *S. cubensis*, *Subbotina corpulenta*, and the association of *C. chipolensis* and *G. ciperoensis*. The base is marked by the disappearance of *Pseudohastigerina danvillensis*, *Turborotalia cerroazulensis*, *S. linaperta*, *Streptochilus martini*, and *Streptochilus* cf. *multicellaris*.
Age and Correlation: This Zone correlates with P17 Zone. The base is dated 34.45 Ma, (Gradstein et al., 2012) in the Early Oligocene while the top correlates with the 32.90 Ma MFS (Gradstein et al., 2012).
Regional/Stratigraphic Distribution: This Zone is well represented in the Niger Delta particularly in the Pologbene-1 well at 5800–7400 ft.

LATE EOCENE ZONES

19. ***Turborotalia cerroazulensis* Zone (PNDf 15)**
Diagnosis: The FDOs of *T. cerroazulensis*, *T. pomeroli*, *Subbotina linaperta*, and *S.* cf. *multicellaris* define this Zone.
Age and Correlation: The zonal age is Late Eocene (P15–P16 Zones). The Zone correlates with the interval between the 37.36 and 34.45 Ma MFS (sensu Gradstein et al., 2012).
Regional/Stratigraphic Distribution: This Zone is well developed in the Tonjor-1 well at 8450 ft.

MIDDLE EOCENE ZONES

20. ***Truncorotaloides rohri* Zone (PNDf 14)**
Diagnosis: This Zone is marked by the disappearance (LAD) of *T. rohri* at the top and the association of *G. columbiana* and *Acarinina bullbrooki*.
Age and Correlation: This Zone correlates with the late Middle Eocene, P14 Zone. The base of the Zone, as defined in this project, is younger than the 40.40 Ma MFS, while the top approximates the 37.36 Ma MFS (Gradstein et al., 2012).
Regional/Stratigraphic Distribution: This Zone is well developed in the Tonjor-1 well, 7,340–8,450 ft. and Mosogar-2 well at 11,700 ft.
21. ***Turborotalia pomeroli* Zone (PNDf 13)**
Diagnosis: The LDO of *T. pomeroli* defines the Zone.
Age and Correlation: The zonal age equates with the Middle Eocene P13 Zone of Blow (1979). The Zone is bounded by the 42.64 and 40.40 Ma MFS (Gradstein et al., 2012).
Regional/Stratigraphic Distribution: The Zone occurs in the Gilli Gilli-1 well (9,730-10,420 ft.) and the Oziengbe South-1 well (2,970- 3,080 m).
22. ***Morozovella lehneri/Chiloguembelina cubensis* Zone (PNDf 12)**
Diagnosis: The Zone is characterized by the FDO of *M. lehneri* and the LDO of *C. cubensis* and *Orbulinoides beckmanni*.
Age and Correlation: The Zone falls within the Middle Eocene P12 Zone. Its base and top are respectively correlated with the 44.75 and 42.64 Ma MFS sensu Gradstein et al. (2012).
Regional/Stratigraphic Distribution: The Zone is well developed in the Gilli Gilli-1 well.
23. ***Hantkenina aragonensis* Zone (PNDf 11)**
Diagnosis: Although the index taxon was not found, this zone is identified based on the association of *A. bullbrooki*, *Acarinina inaequispira*, "*Hastigerina*" *bolivariana*, and *C. amekiensis*.

Age and Correlation: Early Middle Eocene, P10 Zone (lower part). Its base falls at the 46.82 Ma MFS (Gradstein et al., 2012), while the top is at the 44.75 Ma MFS (Gradstein et al., 2012).

Regional/Stratigraphic Distribution: It occurs in the Gilli Gilli-1 well between 9300 and 9480 ft.

EARLY EOCENE ZONES

24. *Acarinina pentacamerata* Zone (PNDf 10)

Diagnosis: The association of *A. pentacamerata* and *A. frontosa* characterizes this Zone. Other species that appear in this Zone include *S. martini*, *S.* cf. *multicellaris*, and *T. rohri*.

Age and Correlation: Late Early Eocene, P9 Zone of Blow (1979). The Zone is defined by the 51.27 Ma MFS, at the base and the 46.82 Ma MFS (Gradstein et al., 2012) at the top.

Regional/Stratigraphic Distribution: This Zone is well developed in the Ihuo-1 well between 6550 and 6750 ft.

25. *Morozovella aragonensis* Zone (PNDf 09)

Diagnosis: This Zone is characterized by the association of *A. pentacamerata*, *Morozovella subbotinae*, and *M. lensiformis*.

Age and Correlation: This zonal base falls at the 52.00 Ma MFS (Gradstein et al., 2012) and the Zone correlates with the P8 Zone *sensu* Blow (1979). The top is at the 51.27 Ma MFS (Gradstein et al., 2012).

Regional/ Stratigraphic Distribution: This Zone is well developed on the Western Delta flank in the Gbekebo-1 well at 2420 ft. and in the Ihuo-1 well at 6950 ft.

26. *Morozovella formosa* Zone (PNDf 08)

Diagnosis: The *M. formosa* Zone is recognizable in the Niger Delta based on an assemblage comprising *Morozovella quetra*, *A. pentacamerata*, *A. primitiva*, and *A. bulbrooki*, with *Morozovella gracilis* ranging into the lower part.

Age and Correlation: This Zone is of mid-Early Eocene age and correlates with the upper P7 Zone of Blow (1979), and falls between the 53.01 and 52.00 Ma MFS (Gradstein et al., 2012).

Regional/Stratigraphic Distribution: The Zone is well-developed on both flanks of the delta and was recognized on the eastern flank in the Ikono-1 well at 3060 ft. and in the Ihuo-1 well between 7350 and 7550 ft.

27. *Morozovella subbotinae* Zone (PNDf 07)

Diagnosis: As defined by Stainforth et al. (1975), this Zone is characterized by the association of *Morozovella aequa*, *M. subbotinae*, *M. marginodentata*, *M. gracilis*, and *Acarinina wilcoxensis*. Species of *Chiloguembelina* are diverse.

Age and Correlation: This Zone correlates with the Early Eocene P6 Zone (Blow, 1979) and falls between the 55.20 and 53.01 Ma MFS sensu Gradstein et al. (2012). More detailed stratigraphic work would lead to further refinement of this relatively thick zone.

Regional/Stratigraphic Distribution: The Zone was recorded in the Gilli Gilli-1 well at 10,500–10,560 ft. and in the Ihuo-1 well between 7,750 and 7,950 ft.

PALEOCENE ZONES

28. *Morozovella velascoensis* Zone (PNDf 06)

Diagnosis: This represents the interval from the last occurrence of *Globoanomalina pseudomenardii* to the top of *M. velascoensis*.

Age and Correlation: This Zone correlates with the Late Paleocene and with Zone P5 sensu Blow (1979). The Zone as defined ranges between the 57.88 and 55.20 Ma MFS sensu Gradstein et al. (2012).

Regional/Stratigraphic Distribution: The *M. velascoensis* Zone occurs in the Gilli Gilli-1 well at 10,980–11,100 ft.

29. ***Globoanomalina pseudomenardii* Zone (PNDf 05)**
Diagnosis: This Zone is best identified on the basis of the presence of *G. pseudomenardii.*
Age and Correlation: The *G. pseudomenardii* Zone correlates with the Late Paleocene P4 Zone of Blow (1979). The lower boundary is dated 58.45 Ma MFS, while the upper boundary is dated 58.45 Ma MFS (Gradstein et al., 2012).
Regional/Stratigraphic Distribution: This Zone is found in the Gilli Gilli-1 well between 11,160 and 11,280 ft. and in the Ikono-1 well at 2,860 ft.

30. ***Igorina pusilla* Zone (PNDf 04)**
Diagnosis: This Zone is characterized by the association of *I. pusilla, Morozovella conicotruncata*, and *Globanomalina chapmani.*
Age and Correlation: This Zone falls in the late Middle Paleocene and correlates with the upper part of the P3 Zone of Blow (1979) between the MFS age dated 59.26 and 58.45 Ma (Gradstein et al., 2012).
Regional/Stratigraphic Distribution: This Zone occurs in the Ituk-2 well at 3,710 ft. on the eastern flank of the delta and in Gilli Gilli-1 well between 11,360 and 11,420 ft.

31. ***Morozovella angulata* Zone (PNDf 03)**
Diagnosis: This Zone is recognized by the presence of *M. angulata* and the cooccurring species, *M. conicotruncata.*
Age and Correlation: The *M. angulata* Zone correlates with the early Middle Paleocene, that is, the early part of the P3 Zone of Blow (1979). The base and top are respectively correlated with the 60.65 and 59.26 Ma MFS sensu Gradstein et al. (2012).
Regional/Stratigraphic Distribution: The top of this Zone was identified in the Araromi-1 well at 9850 ft.

32. ***Praemurica uncinata* Zone (PNDf 02)**
Diagnosis: This Zone is identified in the Niger Delta based on the association of *Globanomalina compressa, Subbotina inconstans*, and *P. uncinata.* This planktonic foraminiferal assemblage was also used by Stolk (1965) to characterize the Early Paleocene of the Niger Delta.
Age and Correlation: This Zone correlates with the late Early Paleocene (Postuma, 1971) and with the Blow (1979) P2 Zone. Its base is close to the 63.30 Ma MFS (Danian 3) (Gradstein et al., 2012), while the top is equated with the 61.61 Ma MFS (Selandian 1) (Gradstein et al., 2012).
Regional/Stratigraphic Distribution: This Zone occurs in the Awaizombe-1 well from 9580 to 9600 ft.

33. ***Globoconusa daubjergensis* Zone (PNDf 01)**
Diagnosis: Postuma (1971) defined this zone based solely on the range of *G. daubjergensis.* This species has also been used to recognize this zone in the Niger Delta. This definition and usage were upheld by Stainforth et al. (1975).
Age and Correlation: This Zone correlates with the Lower Paleocene and is equivalent to Berggren et al. (1998a,b) Zone P1. In the informal SPDC foraminiferal

zonation, *G. daubjergensis* was used to define the F3100 Zone. The base of this Zone correlates with the 64.40 Ma MFS (Danian 2) (Gradstein et al., 2012), while the top is correlated with the 63.30 Ma MFS (Danian 3) (Gradstein et al., 2012).

Regional/Stratigraphic Distribution: This Zone is best developed on the western flank of the Niger Delta. It was recognized in the Gilli Gilli-1 well from 11,480 to 11,600 ft. *G. daubjergensis* was also found in the Araromi-1 well at 780 ft. It can thus be inferred that the zone occurs there. The Araromi-1 well is located further west of Gilli Gilli-1 on a basement high, the Okitipupa Ridge, separating the Niger Delta from the eastern Dahomey Basin.

4.4 BIOFACIES ANALYSIS OF THE STUDIED WELLS

4.4.1 Generation of Biofacies Plots

The Biofacies plots of selected analyzed Niger Delta wells are highlighted in this study (Figs. 4.7–4.23). These biofacies plots vividly provide the detailed summary of the critical geological, stratigraphic, and chronostratigraphic information derived from the high-resolution biostratigraphic, log, and sequence stratigraphic analyses of the wells. When plotted along with wireline logs and details of unconformities, they enhance the visualization and selection of major and minor flooding events, the sequence boundaries and also help in discerning and reconstructing the depositional environment.

In preparing the biofacies plots, the following important steps were followed:

Step 1: A plot is generated containing the details of the population abundance (total count of specimens per sample) in the well, as well as the species diversity (total number of species identified in each sample). The data is plotted using the StrataBugs software or other appropriate biostratigraphic software applications. The plot of the population abundance and species diversity peaks are then imported into Powerpoint software. Chronostratigraphic and biozonation data, the wireline log (Gamma Ray and Resistivity) data, and lithostratigraphic data are added.

Step 2: Important or major Condensed Sections and the minor Condensed Sections from the population abundance and species diversity peaks are identified (Adegoke et al., 2011). Data are calibrated with the electric logs.

Step 3: Sequence Boundaries are identified from the wireline logs.

Step 4: Age date the MFS are identified from the abundance and diversity peaks.

Step 5: The environment of deposition of the sediments are interpreted using the bathymetric ranges of selected environmentally significant benthonic foraminiferal species supplemented with palynomorph data. The data generated are entered into the biofacies plots as detailed further.

4.4.2 Biofacies Plots of Selected Niger Delta Wells

The Biofacies Plots of the studied Niger Delta wells are shown in Figs. 4.7–4.23.

CHAPTER

5

Paleoenvironmental Synthesis

OUTLINE

5.1 INTRODUCTION

Much of our knowledge of the sedimentary processes and environments of deposition of the "geological" Niger Delta is derived from detailed studies of the modern Niger Delta. The more important works include NEDECO (1954, 1961); Allen (1965a,b); Short and Stauble (1967); Weber (1971); Weber and Daukoru (1975); Adegoke and Stanley (1972); Adegoke (1975, 1977), Adegoke et al. (1971, 1980a,b,c, 1987); Asseez et al. (1974); Whiteman (1982), and Doust and Omatsola (1989). A lithogenetic model has been established that relates the facies variations observed in the Niger Delta to a high-energy, wave-dominated, constructional, arcuate/lobate delta (Figs. 5.1 and 5.2) (see also Chapter 2 and Reijers, 1996).

Allen (1965a,b) recognized six superenvironments in the modern Niger Delta. These include:

1. *The Subaerial Delta*, which is made up of the Upper Floodplain, the Lower Floodplain, the Mangrove Swamp, and the Beach environments.
2. *The Marginal Barrier Island/Lagoon Complex*, which includes the Lagoon, Lagoon Delta, Channel and Creek, Marginal Swamp, and Beach.
3. *The Marginal Estuary Complex*, consisting of the Marginal Swamp and Open Water.
4. *The Continental Shelf under Delta influence*, made up of the River Mouth Bar, Delta Front Platform, Prodelta Slope, and Open Shelf.
5. *The Continental Slope*.
6. *The Nondepositional Superenvironment*.

The latter two were not subdivided into environments.

Cenozoic Foraminifera and Calcareous Nannofossil Biostratigraphy of the Niger Delta. http://dx.doi.org/10.1016/B978-0-12-812161-0.00005-3

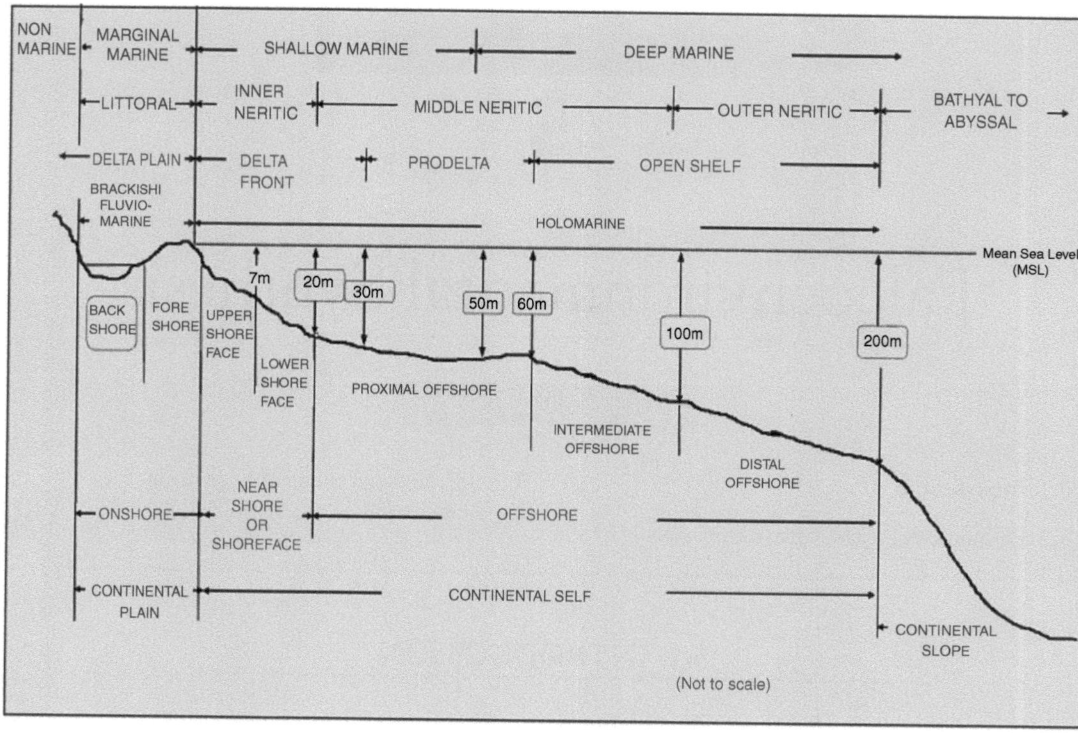

FIGURE 5.1 **Sketch showing the main sedimentary environments of the Niger Delta.** *Source: Modified from Allen, 1965a,b, 1970.*

Short and Stauble (1967) regrouped Allen's (1965a,b) subdivisions and related them to the three major lithofacies frequently encountered in the Niger Delta, thus:

1. The Continental Benin Formation facies, consisting of the:
   a. Upper Deltaic Plain and Alluvial Valley sequences made up essentially of the Upper and Lower Flood Plains of Allen (1965a,b), and
   b. Lower Deltaic Plain made up of the marginal barrier islands of Allen (1965a,b).
2. The Paralic Agbada Formation facies, recognized in the:
   a. Coastal Belt consisting of the lagoonal complex and marginal estuaries, and the
   b. Fluviomarine environment consisting of the River Mouth Bars, Delta Front Platform, and Prodelta Slope.
3. The Holomarine depositional environment, which is essentially recognized in the marine Akata Formation facies and is made up of the Open Shelf and Continental Slope of Allen (1965a,b).

Adegoke (1975), Adegoke et al. (1971), Adegoke et al. (1976), Adegoke et al. (1987), and Asseez et al. (1974) studied the foraminifera faunas of the Nigerian coastlines and characterized the various environments from the Coastal and Near-shore to the Lower Bathyal environments. The open shelf environments were subdivided into three major subunits, namely:

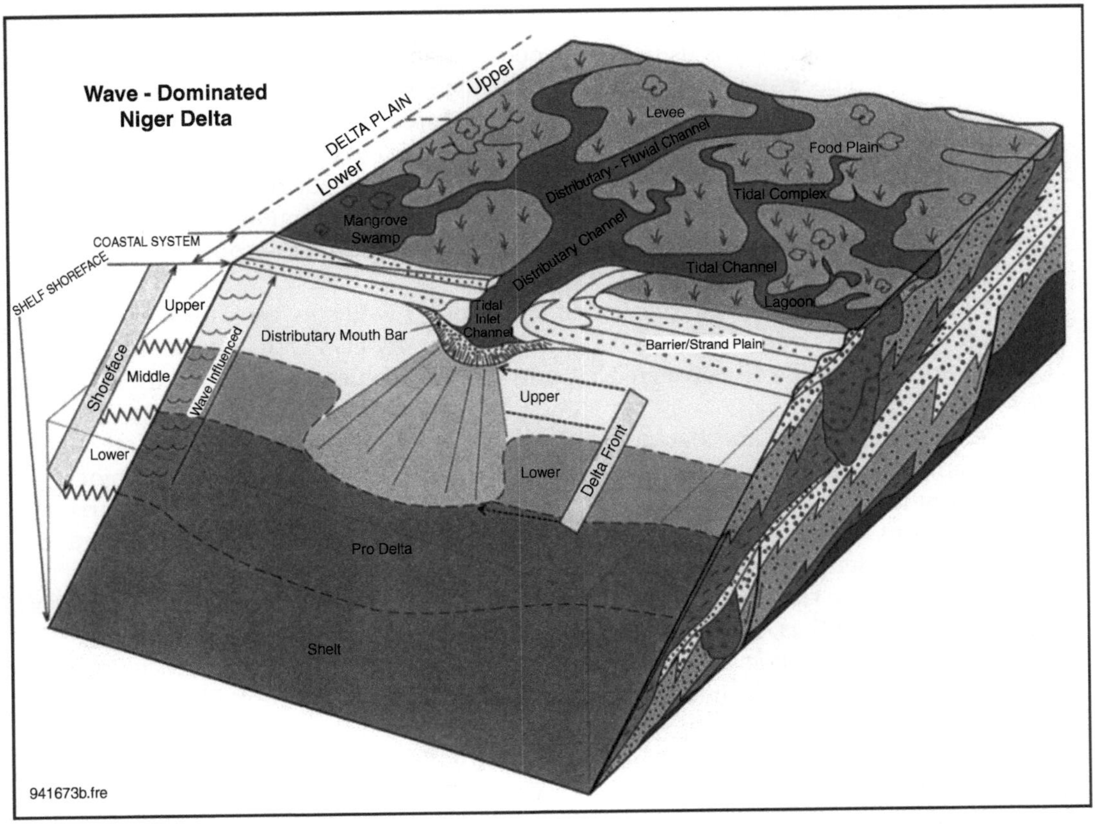

FIGURE 5.2 The main subenvironments of the Modern Niger Delta. *Source: Modified from Oomkins, 1974. Sedimentology, v. 21, p. 115–222.*

1. Upper Continental Shelf	Inner Neritic
2. Mid-Continental Shelf	Middle Neritic
3. Lower Continental Shelf	Outer Neritic

while the Bathyal environment was recognized on the Continental Slope.

The various works highlighted earlier and the recent review of the Geology of the Niger Delta by Nwajide (2013) synthesized in Chapter 2 of this volume provide a convenient and viable basis for the palaeoenvironmental reconstruction of the ancient Niger Delta using the principles enunciated in the classical studies of the bathymetric ranges of recent foraminifera by Bandy et al. (1967), Bandy and Arnal (1957), Phleger (1960), Boltovskoy et al. (1980), and Murray (1991). The palaeoenvironments that are recognized in the Niger Delta would, thus, be characterized microfaunally as detailed further:

1. *Delta Top Biofacies*
   This facies includes the various geomorphological features or units, such as the meandering river, tidal channel, distributary channel, coastal lagoon, tidal plain, shoreface

barrier, and barrier bars. The sequences consist mainly of sand bodies with subordinate shale. Microfaunal recovery is usually very poor and is commonly represented by shallow water, nearshore species, such as *Amphistegina lessonii, Florilus atlanticus, Ammonia beccarii, Lenticulina inornata*, and species of *Quinqueloculina*. Asseez et al. (1974) and Dublin-Green (1996) showed that *Miliammina fusca, Ammonia tepida, Arenoparrella mexicana*, and *Ammotium salsum* are common constituents of the component subenvironments.

The tidal creeks and estuaries in the Recent Niger Delta are similarly dominated by simple arenaceous genera, such as *Ammobaculites* spp., *Textularia* spp., and *Haplophragmoides* spp. (Adegoke et al., 1976).

The delta top biofacies is exemplified in the Inanga-1 well between 11,600 and 12,700 ft., Edop–33A well between 5,000 and 6,200 ft., Anyala-4 well between 340 and 1,100 ft., and Anyala-8 well between 3,520 and 4,980 ft. (Tables 5.1–5.4).

2. *Delta Front Biofacies*

The Delta Front deposits are characterized by sand/shale alternations, typical of the Agbada Formation facies, ranging across different stratigraphic ages. The major subenvironments recognized here include the river mouth bar, barrier bar, and barrier foot. Abundant microfauna is commonly recorded in the shales but most of the sandy sediments are often poorly fossiliferous or totally barren of foraminifera.

The characteristic faunal assemblage includes *Textularia parvula, Eggerella scabra, Ammonia beccarii, Florilus atlanticus, Quinqueloculina microcostata, Epistominella vitrea, Heterolepa pseudoungeriana, R. comatula*, as well as species of *Haplophragmoides* and *Ammobaculites*. This biofacies is recognized in the Inanga-1 well between 3,095 and 5,005 ft. and 7,220–8,270 ft., and in the Anyala-8 well between 9,010 and 10,243 ft. (Tables 5.2 and 5.4).

TABLE 5.1 Paleoenvironmental Synthesis, Anyala-4 Well

Interval (ft.)	Depositional facies	Dominant fauna	Paleoenvironment
1100–1340	(?) Distributary/tidal channel sands	Abundant *A. beccarii, Hanzawaia strattoni, F. atlanticus, Heterolepa pseudoungeriana, A. lessonii*, and *Uvigerina peregrina*	Shallow Inner Neritic/Inner Neritic
1340–2470	Hemipelagic shales interbedded with lower shoreface sands and silts	Abundant *Hanzawaia mantaensis, H. pseudoungeriana, F. atlanticus*, and species of *Cibicorbis, Quinqueloculina*, and *Bulimina*	Inner to Middle Neritic at the base, shoaling to Inner Neritic at the top of the interval
2470–2860	Fluviatile/estuarine sands	Sporadic *Quinqueloculina* spp. and *A. lessonii*	Predominantly Coastal Deltaic
2860–3140	Hemipelagic shales interbedded with shoreface sands	Rare *A. lessonii, A. beccarii*, and *Quinqueloculina vulgaris*	Shallow Inner Neritic
3140–6420	Prograding complex	Rare *F. atlanticus, Cibicorbis inflata, H. strattoni, A. lessonii, A. beccarii*, and *Quinqueloculina* spp.	Coastal Deltaic/Inner Neritic
6420–9000	Hemipelagic shales with interbedded, lower shoreface sands	Abundant *H. pseudoungeriana, A. lessonii, Quinqueloculina* spp., *Bulimina minima, Hanzawaia isidroensis*, and *U. peregrina*	Inner Neritic at the base, deepening to Middle Neritic and Outer Neritic toward the top of the interval

TABLE 5.2 Paleoenvironmental Synthesis, Anyala-8 Well

Interval (ft.)	Depositional facies	Dominant fauna	Paleoenvironment
1,300–1,560	Distribudary/tidal channel deposits	Barren	Coastal Deltaic
1,560–3,520	Transgressive marine shales/mudstones with distributary/tidal channel deposits	Abundant *F. atlanticus*, *A. beccarii*, *A. lessonii*, *H. pseudoungeriana*, *L. inornata*, *Quinqueloculina inornata*, *Quinqueloculina lamarckiana*, *Quinqueloculina microcostata*, and *Sphaeroidina bulloides*	Predominantly Inner to Middle Neritic settings with Upper Bathyal incursions for the more shaly horizons
3,520–4,980	Distributary/tidal channel deposits	Very rare *A. beccarii*, *F. atlanticus*, *H. pseudoungeriana*, *Quinqueloculina oblonga*, and species of *Bulimina* and *Vaginulopsis*	Coastal Deltaic/Shallow and Inner Neritic
4,980–7,190	Marine shale/mudstone with distributary/river mouth bar deposits	Abundant *A. beccarii*, *Bulimina miocenica*, *H. strattoni*, *S. bulloides*, *L. inornata*, and *Bolivina punctata*	Inner to Outer Neritic
7,190–8,560	Marine shales with tidal channel, and barrier bar/barrier foot deposits	Common *L. inornata*, *H. strattoni*, *Q. oblonga*, *Heterolepa crebbsi*, *S. bulloides*, and species of *Bulimina* and *Pseudovalvulineria*	Inner Neritic, with occasional Middle to Outer Neritic influence
8,560–9,010	Distributary channel deposits	Barren	Coastal Deltaic
9,010–10,243	Barrier bar/foot, tidal channel, and Delta fringe deposits	Common *Bolivina scalprata miocenica*, *B. punctata*, *A. beccarii*, *L. inornata*, *A. lessonii*, and *Brizalina isidroensis*	Inner Neritic
10,243–10,776	Marine, mudstones, and shales	Abundant *Gravellina narivaensis* and *Textularia parvula*	Outer Neritic/Upper Bathyal (Adegoke et al., 1976)

TABLE 5.3 Paleoenvironmental Synthesis, Edop-17B Well

Interval (ft.)	Depositional facies	Dominant fauna	Paleoenvironment
5500–6800	Shallow marine shales interbed with regressive sandstones	*Trochammina proteus*, *Haplophragmoides compressa*, *Cyclammina cancellata*, *U. peregrina*, *L. inornata*, *B. scalprata miocenica*, and *A. beccarii*	Environments that ranges from Coastal Deltaic to Inner Neritic deepening to Middle Neritic at some horizons (Adegoke et al., 1976; Murray, 1991)
6800–8125	Marine shales with distributary channel sandstones	*Trochammina globigeriniformis*, *Saccammina complanata*, *H. compressa*, *Eggerella scabra*, *Ammodiscus glabratus*, *Alveolophragmium crassum*, *U. peregrina*, *B. scalprata miocenica*, and *A. beccarii*	Environments of deposition that fluctuates between Shallow Inner Neritic and Inner Neritic, deepening to Middle Neritic at some horizons (Adegoke et al., 1976; Murray, 1991; Bandy, 1967)
8125–9290	Marine shales overlain by channel sandstones	*A. beccarii*, *B. scalprata miocenica*, *A. crassum*, *E. scabra*, *L. inornata*, as well as species of *Haplophragmoides* and *Trochammina*	Environment of deposition that fluctuates between Shallow Inner Neritic and Inner Neritic

TABLE 5.4 Paleoenvironmental Synthesis, Inanga-1 Well

Interval (ft.)	Depositional facies	Dominant fauna	Paleoenvironment
9,40–2,330	Distributary/tidal channel and back-swamp deposits	Shell fragments	Coastal deltaic, Continental back-swamp environment, close to a mangrove-infested coastline
2,330–3,095	Hemipelagic shales	Abundant *A. beccarii*, *Epistominella vitrea*, *F. atlanticus*, *Brizalina beyrichi*, *Fursenkoina schreibersiana*, *Rectoglandulina comatula*, and *B. scalprata miocenica*	Inner Neritic
3,095–5,005	Distributary/tidal channel and barrier bar deposits	Scanty *A. beccarii*, *E. vitrea*, *F. atlanticus*, *Br. beyrichi*, and *R. comatula*	Shallow Inner Neritic
5,005–5,210	Hemipelagic shales interbedded with sands and silts	Common *A. beccarii*, *F. atlanticus*, *L. inornata*, and *Q. microcostata*	Depositional environment fluctuating between Coastal Deltaic and Shallow Inner Neritic (Adegoke et al., 1976)
5,210–7,220	Prograding complex, upper shoreface sands alternating with shales	Abundant *A. beccarii*, *E. vitrea*, *Q. microcostata*, *F. atlanticus*, and *Cibicorbis*	Shallow Inner Neritic (Adegoke et al., 1976; Murray, 1991)
7,220–8,270	Prograding complex alternating marine shales, barrier bar, and distributary channel deposits	Common to abundant *T. parvula*, *Valvulina flexilis*, *A. crussum*, and *Ammobaculites* spp.	Inner Neritic (Adegoke et al., 1976; Murray, 1991)
8,270–8,620	Hemipelagic shales/mudstone with shoreface sand interbeds	*C.* cf. *minima*, *H. compressa*, *Haplophragmoides* spp., and *E. vitrea*	Middle Neritic
8,620–9,160	Delta front deposits	*T. parvula*, *Haplophragmoides* spp., *E. scabra* with rare calcareous forms including *F. atlanticus*, *Q. microcostata*, and *E. vitrea*	Inner Neritic
9,160–9,905	Hemipelagic shales and silts	*T. parvula*, *Ammobaculites strathearnensis*, *T. proteus*, *Haplophragmoides* spp., and *Ammodiscus* spp.	Inner to Outer Neritic environment of deposition (Adegoke and Stanley, 1972; Murray, 1991)
9,905–10,410	Prograding complex	*Uvigerina mantaensis*, *Pullenia bulloides*, *V. flexilis*, *Globocassidulina oblonga*, *H. compressa*, and *Haplophragmoides narivaensis*	Depositional environment flunctuating between shallow marine and Middle to Outer Neritic
10,410–11,600	Marine shales with interbedded channel sands	Common occurrence of: *F. atlanticus*, *Florilus costiferum*, *L. inornata*, *T. parvula*, *H. pseudoungeriana*, and *Haplophragmoides* spp.	Inner to Middle Neritic (Adegoke and Stanley, 1972; Murray, 1991)

Interval (ft.)	Depositional facies	Dominant fauna	Paleoenvironment
11,600–12,700	Deltaic plain sands	Shell fragments and Echinoid remains	Shallow, nearshore Coastal Deltaic to marginal marine depositional environment (Adegoke and Stanley, 1972; Murray, 1991)
12,700–13,720	Marine shales and mudstone	*H. compressa*, *H. narivaensis*, *Uvigerina aculeata*, *Uvigerina sparsicostata*, *E. vitrea*, and *Bolivina* spp.	Marine depositional environment fluctuating between Inner Neritic and Outer Neritic (Adegoke et al., 1976; Bandy, 1967; Murray, 1991)

3. *Prodelta and Open Shelf Biofacies*
These units contain predominantly marine shales with rare interbedded silty and sandy levels. The main feature is the relatively high-benthonic foraminiferal diversity and the abundant planktonic species. Adegoke et al. (1976) subdivided the unit into four major subunits, that are characterized faunally as discussed:

a. Upper Continental Shelf (Inner Neritic) Biofacies
Sediments of the Inner Neritic biofacies are generally silty clay with occasional sandy intervals. The biofacies is characterized by the common occurrence of *Ammonia beccarii*, *Florilus atlanticus*, *Floriluscostiferum*, *Lenticulina inornata*, *Heterolepa pseudoungeriana*, *Heterolepa floridana*, *Cribroelphidium gunteri*, *Cancris auriculus*, *Quinqueloculina* spp., *Poritextularia panamensis*, *Spiroplectamina wrightii*, *Bolivina scalprata miocenica*, *Marginulina costata*, *Epistominella vitrea*, *Hanzawaia strattonii*, and *Cibicorbis inflata*. This is usually a high-energy environment, with clastic sediments that are normally rich in shell fragments but devoid of mica flakes (Adegoke and Stanley, 1972). Planktonic species are well represented in the biofacies.

The biofacies is recognized in the Anyala-8 well between 9,010 and 10,243 ft., Inanga-1 well between 10,410 and 11,600 ft., Beniboye North-1 well between 1,545 and 1,720 m; Tuomo West-1 well between 2,385 and 2,516 m, 2,797 and 2,904 m, and 3,029 and 3,106 m (Tables 5.2, 5.4, and 5.5).

b. Mid Continental Shelf (Middle Neritic) Biofacies
Middle Neritic sediments are often composed of fine to very fine-grained sand, silt, and clay. The biofacies is characterized by assemblages dominated mostly by deep water middle to outer shelfal foraminiferal assemblages, such as *Cyclammina minima*, *Ammodiscus glabratus*, *Haplophragmoides compressa*, *Vavulina flexilis*, *Trochammina globigeriniformis*, *Uvigerina peregrina*, *Uvigerina sparsicostata*, *Globobulimina ovata*, *Buliminella subfusiformis*, *Eponides eshira*, *Cibicorbis inflata*, *Bolivina mandoroveensis*, *Bolivina interjuncta*, *Bulimina marginata*, *Bulimina fusiformis*, and *Rosalina floridana*.

The biofacies is recognized in the Inanga-1 well (9,905–10,115 ft.), Edop-33A well (7,910–9,850 ft.), Utue-17A well (10,220–10,460 ft.), Etim-1 well (4,090–62,00 ft.), Benin River-3 well (6,900–8,800 ft.), and Tuomo West-1 well (3,222–3,560 m; 3,725–3,922 m) (Tables 5.4, 5.6, and 5.7).

c. Lower Continental Shelf (Outer Neritic) Biofacies
This deeper water outer shelf biofacies is characterized by the association of *Globocassidulina subglobosa*, *Planulina ariminensis*, *Praeglobobulimina ovata*, *Pullenia bulloides*, *Gyroidina neosoldanii*, *Sphaeroides bulloides*, *Uvigerina proboscidea*, *Uvigerina*

TABLE 5.5 Paleoenvironmental Synthesis, Beniboye North-1 Well

Interval (m)	Depositional facies	Dominant fauna	Paleoenvironment
1100–1420	Nearshore marine shales interbedded with tidal channel sands	*S. wrightii, Q. vulgaris, Marginulina raphanus, L. inornata, H. pseudoungeriana, Heterolepa* cf. *mckannai, F. atlanticus, C. inflata,* and *B. scalprata miocenica*	Environment that varied from Coastal Deltaic to Middle Neritic, deepening to Outer Neritic
1420–2175	Upper shoreface sands interbedded with shallow marine shales	*P. panamensis, H. compressa, A. crassum, Q. vulgaris, M. raphanus, L. inornata, H. pseudoungeriana, H.* cf. *mckannai, A. beccarii,* and *H. strattoni*	Environment of deposition that fluctuated predominantly between Coastal Deltaic and Inner Neritic (Adegoke and Stanley, 1972)
2175–3200	Nearshore marine shales with interbedded sands	*P. panamensis, H. compressa, A. crassum, L. inornata, F. costiferum, E. vitrea,* and *H.* cf. *mckannai*	Environment that fluctuated between Coastal Deltaic and Inner Neritic, deepening to Middle Neritic at some horizons (Adegoke et al., 1976; Murray, 1991)
3200–3500	Marine shales with barrier bar and upper shoreface sands	*S. wrightii, C. cancellata, A. crassum, L. inornata, E. vitrea, Cassidulina neocarinata, F. atlanticus,* and *F. costiferum*	Environment of deposition that fluctuated between Costal Deltaic and Inner Neritic (Adegoke and Stanley, 1972)

TABLE 5.6 Paleoenvironmental Synthesis, Benin River-3 Well

Interval (ft.)	Depositional facies	Dominant fauna	Paleoenvironment
6,040–6,900	Marine shales and silts	*E. eshira, Lenticulina grandis, U. sparsicostata, A. glabratus,* and *T. globigeriniformis*	Middle Neritic with considerable Outer Neritic influence (Adegoke et al., 1976; Murray, 1991)
6,900–8,100	Massive channel sands with shale breaks	*E. eshira, Lenticulina* sp., *U. sparsicostata, C. inflata, Brizalina dertonensis, Brizalina interjuncta,* and *B.* aff. *subfusiformis*	Middle Neritic (Bandy, 1967)
8,100–8,800	Marine shales and silts	*E. eshira, U. sparsicostata, Nonion centrosulcatum, B.* aff. *subfusiformis,* and *Brizalina mandoroveensis*	Middle Neritic (Phleger, 1960; Bandy, 1967)
8,800–10,500	Prograding complex	*E. eshira, C. inflata, L. grandis, Hanzawaia concentrica,* and *Brizalina interjuncta*	Middle Neritic (Adegoke et al., 1976; Murray, 1991)
10,500–12,180	Slope fan complex (offshore shale and silt)	*E. eshira, Hopkinsina bononiensis, U. sparsicostata,* and *B.* aff. *subfusiformis*	Inner to Outer Neritic (Phleger, 1960; Adegoke et al., 1976; Bandy, 1967)

auberiana, Bulimina aculeata, Globocassidulina oblongata, Karreriella subcylindrica, and *Cyclammina minima* (Adegoke et al., 1976; Murray, 1991). This biofacies was identified in the Utue-17A well between 8380 and 8860 ft. and between 5400 and 7070 ft.

d. Continental Slope (Bathyal) Biofacies

The Continental Slope sediments are mostly fine grained and are dominantly composed of oozes made up of calcareous and siliceous tests of foraminifera and

TABLE 5.7 Paleoenvironmental Synthesis, Etim-1 Well

Interval (ft.)	Depositional facies	Dominant fauna	Paleoenvironment
800–1430	Fluviatile/Deltaic deposit	Barren	Coastal Deltaic
1430–2380	Upper shoreface sands	Rare *M. raphanus*, *A. beccarii*, *Cibicidoides pseudoungerianus*, and *Lenticulina* spp.	Shallow Inner Neritic
2380–4090	Shallow marine shales with distributary channel/mouth bar deposits	Rare *C. inflata*, *F. costiferum*, *F. atlanticus*, *C. pseudoungerianus*, and species of *Epistominella*, *Textularia*, *Nonion*, and *Nodosaria*	Coastal Deltaic/ Shallow Inner Neritic
4090–6200	Marine shale interbedded with offshore sands and silts	*U. peregrina*, *G. ovata*, and *Cyclammina* sp., *Trochammina* spp., *Haplophragmoides* sp., and *Gravellina* spp.	Middle Neritic
6200–8380	Hemipelagic deposits	Abundant *T. parvula*, *Trochammina* spp. and *Haplophragmoides* spp.	Upper to Middle Bathyal
8380–8890	Shallow marine sequence	*M. raphanus*, *Lenticulina rotulata*, *C. pseudoungerianus*, and *Textularia elegans* and species of *Quinqueloculina* and *Lenticulina*	Inner to Middle Neritic

other microorganisms. The biofacies is faunally characterized by the cooccurrence of *Hoeglundina elegans*, *Oridorsalis umbonatus*, *Globocassidulina subglobosa*, *Sphaeroides bulloides*, *Globobulimina ovata*, *Bulimina aculeata*, *Gyroidina soldanii*, *Gyroidinoides neosoldanii*, *Bulimina inflata*, *Uvigerina hispida*, *Uvigerina proboscidea*, *Uvigerina auberiana*, *Uvigerina canariensis*, *Haplophragmoides obliquicameratus*, *Cyclammina cancellata*, *Recurvoides contortus*, and *Glomospira charoides*. The biofacies is recognized in the Utue-17A well between 8020 and 8380 ft. and between 9960 and 9152 ft.

The most important of the environmentally significant foraminifera species from the Niger Delta wells studied are illustrated in Plates 1–41 and succinctly described along with other species in the taxonomic section of this monograph.

5.2 FORAMINIFERAL PALEOBATHYMETRIC MODEL

Implicit in the foraminiferal biofacies characterization of the Niger Delta sediments discussed earlier and in the classical studies of Bandy (1967) and Phleger (1960) is the fact that water depth exercises important controls on the distribution of benthonic foraminifera. The biofacies studies of the modern estuarine, deltaic, and marine sediments off the Nigerian coast (1987; Dublin-Green, 1996; Harris, 1981; Larsen, 1982) and the neighboring Gulf of Guinea marine realm (Asseez et al., 1974; Fayose, 1976; Debenay, 1990; Debenay and Basov, 1993; Debenay and Redois, 1997; Debenay et al., 1996) have provided a model for the reconstruction of the paleoenvironments of the ancient Niger Delta. As most of the living foraminiferal genera recorded in the Niger Delta evolved in the Late Tertiary and Neogene, the same model can validly be applied throughout the Cenozoic Niger Delta. The various environments recognized in these studies and the bathymetric ranges of selected environmentally significant benthonic foraminiferal species (Enclosure 1.22) used in the paleobathymetric interpretations are summarized succinctly in the following paragraphs.

1. *Lagoonal Environment biofacies*
The lagoonal environment is characterized by variable salinities and lithofacies. The faunal assemblage is dominated by agglutinated foraminifera, particularly species of *Ammobaculites*, *Haplophragmoides*, and *Textularia*. These cooccur with large numbers of *Ammonia beccarii*, *Cribroelphidium gunteri*, *Quinqueloculina lamarckiana*, as well as species of *Bulimina*, *Eponides*, *Hanzawaia*, and *Nonionella* (Adegoke, 1975; Adegoke et al., 1976).
2. *Nearshore (Fluviomarine) Environment biofacies (Marsh, Estuaries, and Shoreline) 0–8 m*
This environment is characterized by shallow, turbulent waters. The bottom sediments are generally sandy, though pockets of high-organic clays are frequently encountered. The characteristic microfaunal assemblage includes *Miliammina fusca*, *Ammonia beccarii*, *Poritextularia panamensis*, *Cribroelphidium gunteri*, *Florilus atlanticus*, *Hanzawaia concentrica*, *A. mexicana*, *Textularia* spp., *Haplophragmoides* spp., and *Ammobaculites* spp.
3. *Inner Neritic (Upper Continental Shelf) Environment biofacies (8–55 m)*
The Inner Neritic Environment ranges in depth from 8–55 m. The sediments are silty clays with occasional sandy intervals. The environment is dominated by *Bolivina miocenica*, *Florilus atlanticus*, *Ammonia beccarii*, *Lenticulina inornata*, *Florilus costiferum*, *Spiroplectammina wrightii*, *Heterolepa pseudoungeriana*, *Hanzawaia concentrica*, *Hanzawaia strattoni*, *lessonii*, *Rosalina floridana*, *Poritextularia panamensis*, *Textularia earlandi*, *Nonion depressulus*, *Ptychomiliola separans*, *Quinqueloculina microcostata*, *Quinqueloculina lamarckiana*, *Epistominella vitrea*, *Fursenkoina schreibersiana*, *Rectuvigerina tenuicostata*, *Rectuvigerina multicostata*, *Cribroelphidium gunteri*, *Eponides repandus*, *Amphicoryna caudata*, *Gavellinopsis lobatulus*, *Reusella aequa*, *Cancris auriculus*, *Rosalina globularis*, *Hanzawaia solei*, *Bolivina striatula*, and *Lagena costareticulata* (Asseez et al., 1974; Fayose, 1976; Adegoke et al., 1976).
4. *Middle Neritic (Middle Continental Shelf) Environment biofacies (40–100 m)*
The Middle Neritic environment characteristically contains fine to very fine-grained sands, silt, and clay. The rate of deposition is generally slow. The dominant taxa include: *Cibicidoides pseudoungerianus*, *Cibicidoides crebbsi*, *A. crassum*, *Haplophragmoides narivaensis*, *Haplophragmoides compressa*, *Cibicides lobatulus*, *Brizalina aenariensis*, *Eponides eshira*, *Eponides pseudoelevatus*, *Marginulina raphanus*, *Bolivina mandoroveensis*, *Bulimina subfusiformis*, *Bulimina elongata*, *Bulimina marginata*, *Uvigerina sparsicostata*, *Uvigerina auberiana*, *Uvigerina peregrina*, *Hopkinsina hourqi*, *Cassidulina laevigata*, *Globocassidulina subglobosa*, *Spiroloculina tenuis*, *Hoeglundina elegans*, *Lenticulina grandis*, *Bolivina dertonensis*, *Cancris turgidus*, *Ammodiscus planorbis*, *Bolivina spinescens*, *Bolivina elongata*, *Bolivina ordinaria*, *Bolivina aenariensis*, *Heterolepa pseudoungeriana*, *Cassidulina laevigata*, *Cribostomoides orbicularis*, *Dentalina albatrossi*, *Globobulimina australiensis*, *Globulina myristiformis*, *Guttulina communis*, *Lagena williamsoni*, *Miliolina sublineata*, and *Quinqueloculina* cf. *pulchella*.
5. *Outer Neritic (Lower Continental Shelf) Environment biofacies (100–250 m)*
This biofacies occupies a narrow bathymetric range (see Table 5.6; Fig. 5.1) and is characterized by a distinctive benthonic foraminiferal assemblage that is generally absent from other environments. The dominant species include: *Bulimina marginata*, *Buliminella apiculata*, *Cassidulina carinata*, *Epistominella exigua*, *Trifarina angulosa*, *Nonion centrosulcatum*, *Cyclammina minima*, *Cyclammina cancellata*, *Saccammina complanata*, *Ammodiscus glabratus*, *Uvigerina subperegrina*, *Uvigerina gallowayi*, *Bolivina retiformis*, *Oridorsalis umbonatus*, *Praeglobobulimina ovata*, *Haplophragmoides narivaensis*, *Gravellina*

narivaensis, Vavulina flexilis, Sphaeroides bulloides, Pullenia bulloides, Recurvoides contortus, Globocassidulina oblonga, Karreriella siphonella, Karreriella subcylindrica, Buliminellita bassendorfensis, Elphidium discoidale, Fissurina globosa, Gyroidina umbonata, Marginulina glabra, Miliolinella circularis, Oolina melo, Rosalina bradyi, and *Triloculina trigonula.*

6. *Canyon (?)Environment biofacies*
   Canyons and channels feature prominently in the geological history of the Cenozoic Niger Delta and have been shown to hold appreciable reserves of hydrocarbon. Though they may not conventionally be regarded as a discrete environment, studies of the Nigerian Continental Shelf by Adegoke et al. (1976) show that a distinctive benthonic foraminiferal assemblage was found in both the Avon and Mahin Canyons on the Nigerian shelf. The Canyon biofacies is characterized by *Bolivina goessi, Uvigerina peregrina, Bulimina aculeata, Cassidulina carinata, Brizalina aenariensis, Heterolepa pseudoungeriana,* and *Sphaeroides bulloides.* The cooccurrence of relatively high numbers of shallower water species, particularly *Hanzawaia concentrica* and *Poritextularia panamensis* were also noted. Their occurrence at these depths is as a result of passive downslope transport of sediments into the canyons by turbidity currents.
7. *Upper Slope (Upper Bathyal) Environment biofacies (250–500 m)*
   Benthonic species are relatively few in the Upper Bathyal environment, the fauna being dominantly composed of planktonic species. *Gyroidina neosoldanii, Bulimina aculeata, Uvigerina hispida, Uvigerina proboscidea, Uvigerina mediterrenea, Uvigerina asperula, Poritextularia ariminensis, Trochammina proteus, Haplophragmoides obliquicameratus, Haplophragmoides compressa, Haplophragmoides narivaensis, Textularia parvula, Textularia angulosa, Brizalina aenariensis, Bolivina retiformis, Oridorsalis umbonatus, Praeglobobulimina ovata, Hyalinea balthica, Cibicidoides crebbsi, Hanzawaia mantaensis, N. centrosulcatum, Lenticulina cultrata, Bolivinopsis floridanus, Brizalina subaeneriensis, Elphidium advenum, Lagena gracilis, Lenticulina peregrina, Lenticulina calcar, Lenticulina iota, Nonionella atlantica, Nonionella minuta, Quinqueloculina auberiana, Quinqueloculina enoplostoma,* and *T. tricarinata* are the commonest benthonic species recorded within this environment.
8. *Lower Slope (Lower Bathyal) Environment biofacies (500–2000 m)*
   Sediments of the Lower Bathyal Environment are mostly foraminiferal oozes enmeshed in fine-grained clays and silts. The dominant foraminifera include *Ceratobulimina pacifica, Uvigerina asperula, Uvigerina auberiana, Epistominella exigua, Eggerella bradyi, Karreriella siphonella, Melonis pompilioides, Nonion centrosulcatum, Planulina wuellerstorfi, Rectuvigerina multicostata, Gyroidina neosoldanii, Globocassidulina subglobosa, Pullenia bulloides, Cassidulina laevigata, Haplophragmoides obliquicameratus, Trochammina proteus, Ammodiscus glabratus, Sigmoilina tenuis, Lenticulina cultrata, Globocassidulina inflata, Sphaeroidina bulloides, Recurvoides contortus, Cyclammina cancellata, Bulimina aculeata,* and *Oridorsalis umbonatus.* Identical microfaunal assemblages were recovered at deep offshore sites within OPL 246 in the Niger Delta at depths between 1800 and 2000 m.
9. *Abyssal Biofacies (>2000 m)*
   At abyssal depths, the rate of sediment deposition is extremely slow, the sediments being dominantly composed of the tests of planktonic foraminifera (oozes). The rare benthonic foraminifera found include: *Pullenia bulloides, Melonis pompilioides, Planulina wuellerstorfi, Epistominella exigua, Sigmoilopsis schlumbergeri, Globocassidulina subglobosa, Eponides tenera, Globocassidulina neosoldanii, Nonion centrosulcatum, Bulimina aculeata, Uvigerina gallowayi, Eponides bradyi, Alabamina* spp., and *Rhabdammina abyssorum.*

5.3 PALEOENVIRONMENTAL SYNTHESIS OF SELECTED NIGER DELTA WELLS

The paleoenvironmental synthesis of selected Niger Delta wells presented further is based on the integration of the textural characteristics of the ditch-cutting samples of the studied wells with wireline log analytical data and the qualitative evaluation of the bathymetric ranges of selected environmentally significant benthonic foraminifera (Adegoke et al., 1976; Murray, 1991; Mosunmolu Ltd.'s several unpublished reports). Details of the paleoenvironmental synthesis are presented on interval basis for the selected wells (Tables 5.1–5.16). Key foraminifera species of environmental significance are illustrated and specifically highlighted in the Plates.

TABLE 5.8 Paleoenvironmental Synthesis, Adagbassa Creek-1 Well

Interval (ft.)	Depositional facies	Dominant fauna	Paleoenvironment
4060–5635	Distributary channel sandstone with marginal marine mudstone	*S. wrightii*, *Megastomella africana*, *Altistoma tenuis*, *U. sparsicostata*, *P. ovata*, *L. grandis*, *H. bononiensis*, *H. mckannai*, *Epistominella pontoni*, *B. subfusiformis*, and *Br. mandoroveensis*	Environment that ranges from Coastal Deltaic to Middle Neritic
5635–6655	Channel sandstone interbedded with thin sandstone	*Ammoscalaris pseudospiralis*, *H. mckannai*, *H. strattoni*, *H. concentrica*, *E. pontoni*, *Br. mandoroveensis*, *B. subfusiformis*, and *B. scalprata miocenica*	Environment that fluctuates predominantly between Coastal Deltaic and Inner Neritic, deepening to Middle Neritic (Adegoke et al., 1976)
6655–6950	Marginal marine/shelf mudstone with interbedded thin sandstone	*S. wrightii*, *A. pseudospiralis*, *A. tenuis*, *Nonionella auris*, *C. turgidus*, *F. atlanticus*, and *Fursenkoina punctata*	Environment fluctuates between Coastal Deltaic and Inner Neritic

TABLE 5.9 Paleoenvironmental Synthesis, Agberi North-1 Well

Interval (m)	Depositional facies	Dominant fauna	Paleoenvironment
2504–3080	Distributary channel sandstone with marine shale	*U. sparsicostata*, *L. grandis*, *L. inornata*, *F. atlanticus*, *H. bononiensis*, *H.* cf. *mckannai*, *H. strattoni*, *C. turgidus*, *C. inflata*, and *A. beccarii*	Environment that fluctuates between Inner Neritic and Middle Neritic, deepening to Outer Neritic at some horizons (Adegoke and Stanley, 1972)
3080–3600	Marine shales with interdistributary channel sandstone	*C. turgidus*, *B. interjuncta*, *C. inflata*, *F. atlanticus*, *H. bononiensis*, *L. inornata*, *H.* cf. *mckannai*, *L. grandis*, and *U. sparsicostata*	Environment ranges from Inner Neritic to Outer Neritic
3600–4095	Channel sandstone interbedded with marine shales	*F. atlanticus*, *U. sparsicostata*, *P.* aff. *panamensis*, *H. bononiensis*, *C. turgidus*, *Bo. dertonensis*, *Br. interjuncta*, and *L. grandis*	Environment fluctuates between Middle Neritic and Inner Neritic, deepening to Outer Neritic at some horizons
4095–4576	Marine shales overlain by distributary channel sandstone	*Br. mandoroveensis*, *B.* aff. *subfusiformis*, *C. turgidus*, *F. atlanticus*, *H. bononiensis*, *L. grandis*, *Quinqueloculina rhodiensis*, and *S. wrightii*	Environment fluctuates between Middle Neritic and Outer Neritic, shallowing to Coastal Deltaic at some horizons

TABLE 5.10 Paleoenvironmental Synthesis, Amakubu-1 Well

Interval (ft.)	Depositional facies	Dominant fauna	Paleoenvironment
2,000–4,500	Massive channel sandstone with shale interbeds	*A. tenuis, E. scabra, S. wrightii, H. mckannai, H. strattoni, E. vitrea, A. beccarii,* and *B. scalprata miocenica*	Environment that fluctuated between Coastal Deltaic and Inner Neritic
4,500–5,080	Shallow marine shales	*Bo. dertonensis, B. scalprata miocenica, Br. beyrichi, Br. interjuncta, B. subfusiformis, C. inflata, E. vitrea, F. costiferum,* and *U. sparsicostata*	Environment that is predominantly Middle Neritic (Adegoke et al., 1976; Murray, 1991)
5,080–8,110	Channel sandstone interbedded with thin shales	*Spiroplectamina biformis, U. sparsicostata, F. costiferum, L. grandis, N. auris, C. turgidus, B. subfusiformis,* and *Bo. dertonensis*	Environment that ranged from Coastal Deltaic to Middle Neritic (Adegoke and Stanley, 1972)
8,110–10,500	Marginal marine shales with thin sandstone interbeds	*S. biformis, A. tenuis, Altistoma scalaris, U. sparsicostata, N. auris, F. costiferum, H. bononiensis,* and *B. subfusiformis*	Environment that ranged from Coastal Deltaic to Middle Neritic (Adegoke and Stanley, 1972)

TABLE 5.11 Paleoenvironmental Synthesis, Ameena-1 Well

Interval (m)	Depositional facies	Dominant fauna	Paleoenvironment
510–628	Upper shoreface sands alternating with shallow marine shales	*A. beccarii, F. atlanticus, Cribroelphidium decipiens, Haplophragmoides pseudoungeriana, E. vitrea,* and *H. strattoni*	Predominantly Shallow Inner Neritic/Inner Neritic
628–778	Offshore sand bars with interbedded marine shales	*A. beccarii, F. atlanticus, C. decipiens, C. inflata, H. strattoni, E. vitrea, H. floridana, Q. microcostata,* and *P. panamensis*	Shallow Inner Neritic to Inner Neritic
778–973	Upper shoreface sands alternating with shallow marine shales and mudstones	*A. beccarii, F. atlanticus, C. inflata, E. vitrea, H. pseudoungeriana,* and *S. wrightii*	Shallow Inner Neritic
973–1022	Marine shales interbedded with offshore sand bars	*F. atlanticus, A. beccarii, C. inflata, B. scalprata miocenica, E. vitrea, H. floridana,* and *S. wrightii*	Shallow Inner Neritic to Inner Neritic
1022–1405	Lowstand wedge deposits	Rare *F. atlanticus, A. beccarii, H. strattoni, H. floridana,* and *Lenticulata inornata*	Coastal Deltaic deepening to Shallow Inner Neritic to Inner Neritic
1405–1810	Shelfal shales/mudstones interbedded with (?) transgressive sands	*S. complanata, T. proteus, T. parvula, A. glabratus, Eg. bradyi, T. globigeriniformis, V. flexilis, K. siphonella, U. auberiana, B. scalprata miocenica, L. inornata, A. beccarii,* species of *Trochammina,* and *Haplophragmoides*	Outer Neritic at the base shoalint to Inner Neritic (Adegoke et al., 1976)
1810–2047	Alternating marine shales/mudstones and silts overlain by offshore sands	Common to abundant *S. complanata, K. siphonella, C. minima, G. narivaensis, Eg. bradyi, A. crassum, U. auberiana, H. floridana,* and species of *Trochammina* and *Haplophragmoides*	Upper Bathyal shoaling to Outer Neritic up dip
2047–2200	Lowstand channel sands	*S. complanata, G. charoides, Martinottiella communis, Eg. bradyi, C. minima, V. flexilis,* and *K. siphonella*	Predominantly Upper Bathyal shoaling to Outer Neritic updip
2200–2450	Hemipelagic shales/mudstones alternating	*C. minima, Eg. bradyi, T. globigeriniformis, V. flexilis, T. proteus, S. complanata, C. cancellata, G. charoides,* and *Bathysiphon* spp.	Predominantly Middle Bathyal shoaling through Outer Neritic to Middle Neritic

TABLE 5.12 Paleoenvironmental Synthesis, Hb-1 Well

Interval (m)	Depositional facies	Dominant fauna	Paleoenvironment
3,200–7,580	Tidal channel sands with interbedded shales	*H. pseudoungeriana, L. inornata, A. beccarii, F. atlanticus, Q. lamarckiana, E. scabra, A. crassum, S. wrightii,* and *P. panamensis*	Shallow Inner Neritic to Inner Neritic
7,580–8,540	Marine shales/silts	*S. bulloides, M. raphanus, P. bulloides, V. flexilis, C. minima, H. compressa, L. inornata, F. atlanticus,* and *A. beccarii*	Inner Neritic to Middle Neritic with some Outer Neritic influence
8,540–9,380	Marine shales/ mudstones	*H. pseudoungeriana, A. beccarii, L. inornata, F. atlanticus, E. vitrea, Q. lamarckiana,* and *C. cancellata*	Shallow Inner Neritic to Inner Neritic
9,380–9,920	Marine shale	*P. ovata, P. bulloides, Bulimina buchiana, C. cancellata, Trochammina* spp., *H. pseudoungeriana, A. beccarii,* and *F. atlanticus*	Inner Neritic
9,920–13,160	Marine shale/mudstone interbedded with shoreface sands	*F. atlanticus, A. beccarii, H. pseudoungeriana, L. inornata, Q. lamarckiana, P. panamensis,* and *Q. microcostata*	Shallow Inner Neritic to Inner Neritic

TABLE 5.13 Paleoenvironmental Synthesis, Ibotio-1 Well

Interval (ft.)	Depositional facies	Dominant fauna	Paleoenvironment
30–1145	Stacked channel sandstones	Barren	Coastal Deltaic
1145–2475	Lagoonal mud with distributary channel sandstone	Barren	Coastal Deltaic
2475–3820	Channel sandstones with thin shale inter beds	Barren	Coastal Deltaic
3820–4650	Marine mudstone with distributary channel sandstone	*T. proteus, Recurvoides deformis, H. narivaensis, H. compressa, U. subperegrina, H. pseudoungeriana, F. atlanticus,* and *E. vitrea*	Environment that deepens from Coastal Deltaic through Outer Neritic to Upper Bathyal (Adegoke and Stanley, 1972; Bandy, 1967)
4650–6080	Stacked channel sandstones with thin shales	*T. proteus, R. deformis, C. minima, U. subperegrina, Q. vulgaris, H. pseudoungeriana, F. atlanticus, B. scalprata miocenica,* and *A. beccarii*	Environment of deposition that fluctuated predominantly between Outer Neritic and Upper Bathyal at the upper part while it ranges from Coastal Deltaic to Middle Neritic at the lower part
6080–7400	Marine shales with upper shoreface sands	*E. vitrea, F. atlanticus, F. costiferum, H. strattoni, H. pseudoungeriana, Q. vulgaris, R. comatula, U. subperegrina,* and *P. panamensis*	Environment that fluctuated predominantly between Inner Neritic and Middle Neritic
7400–9900	Marine shales with channel sandstones	*T. proteus, P. panamensis, R. comatula, H. strattoni, Q. vulgaris, H. pseudoungeriana, F. costiferum, F. atlanticus, E. vitrea, B. scalprata miocenica,* and *A. beccarii*	Environment that ranges from Coastal Deltaic to Middle Neritic deepening to Upper Bathyal at some horizons

TABLE 5.14 Paleoenvironmental Synthesis, Ihuo-1 Well

Interval (ft.)	Depositional facies	Dominant fauna	Paleoenvironment
5,54–2,280	Massive channel sands with thin shale breaks	Barren of foraminifera	Continental (Upper Deltaic Plain) environment
2,280–2,865	Alternating distributary channel sands and marine shales	Common *F. atlanticus*, *H. pseudoungeriana*, *Nonion obducum*, and *H. strattoni*	Inner Neritic environment (Adegoke et al., 1976; Murray, 1991)
2,865–3,030	Marine shale/mudstone	Common *F. atlanticus*, *H. pseudoungeriana*, and *H. strattoni*	Shallow Inner Neritic Environment
3,030–4,035	Distributary/tidal channel sands	*F. atlanticus*, *F. costiferum*, *H. strattoni*, *Ammobaculites* spp., and abundant shell fragments	Depositional environment fluctuating between Coastal Deltaic and Inner Neritic
4,035–6,135	Prograding complex (shale/mudstone with delta front deposits)	*F. costiferum*, *H. pseudoungeriana*, *N. obducum*, *H. concentrica*, *E. pontoni*, *Quinqueloculina seminulum*, and *Ammobaculites* spp.	Shallow Inner Neritic with Coastal Deltaic environment at the top 4405–4092 ft.
6,135–6,865	Marine shale/mudstone with (?)turbidites	*Bo. mandoroveensis*, *Bolivina ihuoensis*, *E. eshira*, *Eponides africana*, and *Bulimina* spp.	Outer Neritic environment (Adegoke et al., 1976; Murray, 1991
6,865–7,150	Slope fan complex I	*Bo. mandoroveensis*, *H. bononiensis*, *Bo. ihuoensis*, *E. africana*, and *E. eshira*	Outer Neritic environment (Adegoke et al., 1976; Murray, 1991)
7,150–8,600	Marine shale/mudstone with turbidite sands	*Valvulineria wilcoxensis*, *E. eshira*, *E. africana*, *Uvigerina* spp., *Trochammina* spp., and *Miliammina telemaquensis*	Outer Neritic environment (Adegoke et al., 1976; Murray, 1991)
8,600–8,800	Slope fan/turbidite complex II	Marked reduction in the diversity and population of arenaceous benthics with *Textularia* spp. and *Haplophragmoides* spp. predominating	Inner Neritic environment with some deepening toward the top of the interval
8,800–9,250	Slope fan/turbidite complex I	*Ammobaculites* spp., *Haplophragmoides excavata*, *Trochammina* spp., and *Haplophragmoides sahariense*	Inner Neritic environment with some deepening toward the top of the interval
9,250–11,200	Hemipelagic shales and mudstones	*Valvulineria 8A*, *Ammobaculites* spp., *A. pseudospiralis*, *Miliammina petila*, *Trochammina* spp., *Valvulina palmerae*, and *R. contortus*	Outer Neritic environment at the base, shoaling to Middle and Inner Neritic toward the top of the interval Inner–Outer Neritic

TABLE 5.15 Paleoenvironmental Synthesis, Mosogar-2 Well

Interval (ft.)	Depositional facies	Dominant fauna	Paleoenvironment
3,000–8,190	Stacked channel sand stone with thin shale interbeds	The upper and lower parts of the interval are barren; middle part contains *S. wrightii, C. turgidus,C. gunteri, F. atlanticus, F. costiferum, H. concentrica*, and *Nonion stella*	Upper and lower parts are Coastal Deltaic while the middle and the lowermost parts fluctuated between Coastal Deltaic and Inner Neritic, deepening to Middle Neritic environment
8,190–10,000	Tidal, distributary channel sandstones with marginal marine shales	*A. tenuis, A. scalaris, N. stella, F. atlanticus, F. costiferum, E. pontoni, Eponides* cf. *iojimaensis, Br. mandoroveensis*, and *Brizalina imperatrix*	Environments that fluctuate between Coastal Deltaic and Inner Neritic, deepening to Middle Neritic
10,000–12,504	Shallow marine shales with regressive sandstones	*N. auris, N. stella, F. punctata, F. costiferum, F. atlanticus, B. subfusiformis, C. turgidus, Br. mandoroveensis*, and *Brizalina ihuoensis*	Environment that varied from Coastal Deltaic to Middle Neritic

TABLE 5.16 Paleoenvironmental Synthesis, Obeaja-1 Well

Interval (ft.)	Depositional facies	Dominant fauna	Paleoenvironment
4000–6070	Channel, tidal sandstone interbedded with thin mudstones	*A. beccarii, Bolivina beyrichi, C. gunteri, Eponides gunteri, Eponides* cf. *iojimaensis, F. costiferum, H. strattoni*, and *S. wrightii*	Depositional environment is predominantly Coastal Deltaic at the upper part while the lower part fluctuates between Coastal Deltaic and Inner Neritic
6070–7035	Intercalation of shallow marine mudstones with channel sandstones	*Bo. beyrichi, C. turgidus, A. beccarii, Bo. dertonensis, E. vitrea, F. costiferum, F. punctata, Q. microcostata*, and *S. wrightii*	Environment of deposition that fluctuates between Coastal Deltaic and Inner Neritic (Adegoke et al., 1976; Murray, 1991)
7035–7694	Marine mudstones with thin interbedded sandstones	*C. turgidus, Bo. beyrichi, B., subfusiformis, B. scalprata miocenica, Br. mandoroveensis, Eponides iojimaensis, H. bononiensis, F. punctata, F. atlanticus*, and *S. wrightii*	Depositional environment that fluctuated between Middle Neritic and Outer Neritic (Bandy, 1967; Adegoke and Stanley, 1972)

CHAPTER

6

Systematic Paleontology: Niger Delta Foraminifera

OUTLINE

6.1 INTRODUCTION

The taxonomic descriptions of the significant benthonic and planktonic foraminiferal species recovered from the Niger Delta wells studied are presented in following section in a way that should enhance routine identification of species by micropalaeontologists. To achieve this, the genus was used as basis for the arrangement of taxa in the PhotoCatalogue. All the genera within the same family are arranged alphabetically for ease of reference. The families are arranged according to the suborders to which they belong.

A complete listing of synonymies was not carried out for all the species. Instead, the original reference to the species is given in each case wherever possible, followed by selected references to describe forms that either illustrate the concept of the species as adopted in this report or which have specific bearing to the Niger Delta. Also, in the synonymies, coded species names contained in the PhotoCatalogues prepared by some of the oil companies are harmonized where necessary. The generic diagnoses of Cifelli and Scott (1986), Bolli (1957a–c), and Loeblich and Tappan (1987) as modified by Olsson et al. (1999) and Blow (1979) are followed especially for the planktonic foraminifera.

Cenozoic Foraminifera and Calcareous Nannofossil Biostratigraphy of the Niger Delta. http://dx.doi.org/10.1016/B978-0-12-812161-0.00006-5

6.2 BENTHONIC FORAMINIFERA

Family: SACCAMMINIDAE Brady, 1884
Genus: SACCAMMINA Sars, 1869
Saccammina complanata Franke
Plate 1, Figs. 1 and 2

Saccammina 2: (*Saccammina complanata* Franke), SPDC Catalogue, 1980, Stereoscan Photos, Paleontological Type Collection, Nigeria, pl. 4, figs. D1, E2.

Diagnosis: Test free, composed of a single globular chamber with subcircular outline, often compressed as a result of diagenesis; wall with pseudochitinous base and an outer agglutinated layer, firmly cemented; aperture variable, rounded, or may be produced on a short neck.

Range: Oligocene–Late Miocene.

Saccammina sphaerica Brady, 1871
Plate 1, Fig. 3

Saccammina sphaerica Brady, 1871. On the rhizopodal fauna of the deep sea. Proceedings of the Royal Society of London, vol. 18, 1868, No. 114, pp. 59–62.

Saccammina sphaerica Brady, 1871. Ann. Mag. Nat. Hist., ser. 4, 7, 177–184, pl. 12.

Saccammina sphaerica Sars, 1869. Cushman, 1918. The Foraminifera of the Atlantic Ocean, 1918, p. 44, pl. 16, figs. 4, 5; pl. 19, figs. 2–5.

Diagnosis: Test free, single globular/spherical chamber; wall of coarse sand grains, firmly cemented, interior smooth, exterior usually smooth in specimens with very large sand grains and small test, somewhat roughened. Aperture single, circular with a very slight neck protruding from the surface of the test.

Range: Paleocene–Pleistocene.

Saccammina atlantica
Plate 1, Fig. 4

Proteonina atlantica Cushman, 1944, Cushman Lab. Foramin. Res., Special Publication No. 12, p. 5, pl. 1, fig. 4.

Saccammina atlantica Cushman, 1944, Todd and Bronnimann, 1957, p. 22, pl. 1, fig. 14. In: Paleoecology, Biostratigraphy, Paleoceanography and Taxonomy of Agglutinated foraminifera, p. 335. Edited by Christoph Hemleben, Michael A. Kaminski, Wolfgang Kuhnt, D.B. Scott.

Saccammina 2: (*Saccammina atlantica* Cushman, 1944) SPDC Catalogue, 1980, Stereoscan Photos, Paleontological Type Collection, Nigeria, pl. 4, fig E1.

Diagnosis: Test free, single globular chamber; wall with pseudochitinous base and agglutinated layer, firmly cemented; aperture rounded, may be produced on short neck.

Range: Paleocene–Pleistocene.

Genus: AMMODISCUS Reuss, 1862
Ammodiscus glabratus Cushman and Jarvis, 1928
Plate 1, Fig. 5

Ammodiscus glabratus Cushman and Jarvis, 1928, Contrib. Cushman Lab. Foramin. Res. 4, 86, pl. 12, fig. 6.

Ammodiscus glabratus Bubik, 1995, In: Kaminski, M.A., Geroch, S., Gasinski, M.A. (Eds.), Proceedings of the Fourth International Workshop on Agglutinated Foraminifera, p. 111, pl. 10, figs. 6a,b.

Ammodiscus 2: (*Ammodiscus glabratus*, Cushman and Jarvis), SPDC Catalogue, 1980, Stereoscan Photos, Paleontological Type Collection, Nigeria, pl. 4, fig. B2.

Diagnosis: Test free, discoidal, proloculus followed by undivided planispirally enrolled tubular chamber; wall agglutinated; aperture at open end of tubular chamber.

Range: Oligocene–Late Miocene.

Family: AMMODISCIDAE Reuss, 1862
Subfamily: AMMODISCINAE Reuss, 1862
Genus: GLOMOSPIRA Rzehak, 1885
Glomospira charoides Jones and Parker, 1860
Plate 1, Fig. 6

Glomospira charoides Jones and Parker, 1860, Q. J. Geol. Soc. London 16, 292–307.

Glomospira charoides Jones and Parker, AGIP SPA, 1982, Foraminifera Padani, pl. 1, figs. 3, 3p, 4, 4p.

Glomospira charoides, Jones and Parker, 1860, Jenkins and Murray, 1989, Stratigraphical Atlas of Fossil Foraminifera, p. 455, pl. 9.1, fig. 17.

Diagnosis: Test free, fine grained, approximately spherical in shape, coiling of early whorls is fairly regular, but the last whorl may be much larger, encircling test equatorially or in a plane at 90° to previous whorls; aperture at the end of tubular chamber.

Range: Paleocene–Early Pliocene.

Glomospira gordialis Jones and Parker, 1860
Plate 1, Fig. 7

Glomospira gordialis Jones and Parker, 1860, Q. J. Geol. Soc. London 16, 292–307.

Glomospira gordialis Jones and Parker, AGIP SPA, 1982, Foraminifera Padani, pl. 1, figs. 5, 5 p.

Diagnosis: Test free, fine grained, coiling in one plane, stretospiral, or irregular aperture at the end of tubular chamber.

Range: Paleocene–Middle Miocene.

Genus: REOPHAX Montfort, 1808
Reophax sp. 1
Plate 1, Fig. 8

Diagnosis: Test free, elongate, nearly straight, chambers few, increasing in size as added; sutures nearly horizontal; wall agglutinated, grains held in place with comparatively little cement; surface rough, aperture terminal, rounded, without a neck.

Remarks: This species is larger and with more globular chambers than *Reophax morrisoni*. The species is frequently encountered in shallow water deposits in the Cenozoic Niger Delta.

Reophax morrisoni Cushman and Ellisor, 1945
Plate 1, Fig. 9

Reophax morrisoni Cushman and Ellisor, 1945, The foraminifera fauna of the Anahuac Formation. J. Paleontol. 19 (6), 545, pl. 71, figs. 2–4.
Reophax 6: (*Reophax* cf. *R. morrisoni* Cushman and Ellisor), SPDC Catalogue, 1980, Stereoscan Photos, Paleontological Type Collection, Nigeria, pl. 4, fig. C3.

Diagnosis: Test free, uniserial, elongate, slightly tapering, cylindrical, circular in transverse section; chambers five to six in the adult test, increasing gradually in size as added; sutures distinct, somewhat depressed; wall coarsely agglutinated, formed by large, well-cemented sand grains, surface rough; aperture terminal, rounded, without a neck.
Range: Middle Oligocene–Miocene.

Genus: ARENOTURRISPIRILLINA Tairov, 1956
Arenoturrispirillina sp. 1
Plate 1, Figs. 10 and 11

Diagnosis: Test conical, with regular trochospiral coil, evolute and not close coiled or high-spired; aperture at open end of tube.
Range: Paleocene–Eocene.

Family: LITUOLIDAE de Blainville, 1825
Genus: AMMOBACULITES Cushman, 1910
Ammobaculites sp. 1
Plate 1, Fig. 12

Diagnosis: Test free, streptospiral, early portion coiled, later whorls slightly uncoiled; whorls rounded in section; sutures straight, slightly distinct; wall coarsely agglutinated, aperture terminal, rounded.
Remarks: This species differs from *Ammobaculites agglutinans* and *Ammobaculites coprolithiformis* because it lacks the well developed, straight uncoiled later portion of the test of the latter. It is frequently found in shallow waters in the Neogene of the Niger Delta.

Ammobaculites strathearnensis Cushman and LeRoy, 1938
Plate 1, Figs. 13–16

Ammobaculites strathearnensis Cushman and LeRoy, 1938, A microfauna from the Vagueros Formation Lower Miocene, Simi Valley, Ventura County, California. J. Paleontol. 2, 122, pl. 22, figs. 1, 2.

Diagnosis: Test free, streptospiral, compressed, with seven rapidly enlarging chambers in the last whorl; the last chamber is slightly uncoiled; sutures flush with the surface, wall coarsely agglutinated.

Range: Middle Miocene–Late Miocene.

Ammobaculites agglutinans d'Orbigny, 1846
Plate 1, Figs. 17 and 18

Spirulina agglutinans d'Orbigny, 1846, Foraminiferes fossiles du Bassin Tertiare de Vienne (Autriche). Gide et Comp. Paris, p. 137, pl. 7, figs. 10–12.

Ammobaculites agglutinans (d'Orbigny, 1846), Bender, 1989, Gehauseaufbau, Gehausegenese und Biologie agglutinierter Foraminiferen (Sarcodina:Textulariina). Jahrbuch Geologische Bundesanstalt, 132 (2), 297, pl. 5, fig. 6; pl. 16, fig. 18.

Ammobaculites 21 and 22: (*Ammobaculites agglutinans*, d'Orbigny, 1846), SPDC Catalogue, 1980, Stereoscan Photos, Paleontological Type Collection, Nigeria, pl. 2, figs. A10, A11.

Diagnosis: Test free, streptospiral, early portion close coiled, later uncoiled and rectilinear, rounded in section; sutures straight; wall agglutinated, interior simple; aperture terminal, rounded.

Range: Eocene–Late Miocene.

Ammobaculites coprolithiformis Schwager, 1868
Plate 1, Fig. 19

Haplophragmium coprolithiformis Schwager, 1868, *Haplophragmium coprolithiforme*. Beneck's Georg. Paleont. Beitrage, 1, 654, pl. 34, fig. 1.

Ammobaculites 14, SPDC Catalogue, 1980 Stereoscan Photos, Paleontological Type Collection, Nigeria, pl. 2, fig. A7.

Ammobaculites coprolithiformis Petters, 1982, Central West Africa Cretaceous-Tertiary benthic foraminifera and stratigraphy. Palaeontographica, Bd. 179, Abs. A., v. 179, no. 1–3, pl. 2, figs. 13, 14.

Diagnosis: Test arenaceous, elongate, initial coil distinct, composed of coarse grains, later uniserial, and consists of five chambers; sutures horizontal, depressed, aperture round, terminal.

Range: Maastrichtian–Paleocene.

Genus: AMMOSCALARIA Hoglund, 1947
Ammoscalaria pseudospiralis Williamson, 1953
Plate 1, Fig. 20

Ammoscalaria pseudospiralis Williamson, 1953. Taxonomic Notes in Society of Economic Paleontologists and Mineralogists 1960, p. 68, pl. XXIII, figs. 1–4.

Ammobaculites 8/17: (*Ammoscalaria pseudospiralis* Williamson, 1953) SPDC Catalogue, 1980, Stereoscan Photos, Paleontological Type Collection, Nigeria, pl. 2, figure A5.

Diagnosis: Test free, elongate, early portion planispiral, later uncoiling and rectilinear, original development as tubular test with secondarily formed septa and resultant chamber developing; sutures indistinct at surface, internal septa extremely thin, straight, and pseudochitinuous; exterior wall coarsely agglutinated and thick. Aperture rounded to slightly elongate and may be produced on distinct neck. *Ammoscalaria* differs from *Ammobaculites* in its secondary formed thin pseudochitinuous sector.

Range: Pleistocene.

Genus: CYCLAMMINA Brady, 1879
Cyclammina cancellata Brady, 1879
Plate 2, Figs. 1–5

Cyclammina cancellata Brady, 1879, p. 62–63, Type figure in Brady, 1884, Report on the foraminifera dredged by HMS Challenger during the years 1873–1876. Report of Scientific Results of the Exploration Voyage of HMS Challenger, Zoology, vol. 9, pp. 1–814, pl. 37, figs. 8–16.

Cyclammina cancellata Loeblich and Tappan, 1964, Treatise on Invertebrate Palaeontology, Part C. Protista 2. Sarcodina-Chiefly "Thecamoebians" and Foraminiferida. vols. 1 and 2. Geological Society of America; University of Kansas Press, Lawrence, KS, p. C229, pl. 142, figs. 1–4.

Cyclammina cancellata LeRoy and Hodgkinson, 1975, Benthonic foraminifera and some Pteropoda from a deep-water dredge sample, northern Gulf of Mexico. Micropaleontology, 21 (4), 432, pl. 4, figs. 15–17.

Cyclammina cancellata Kohl, 1985, Early Pliocene benthic forminifers from the Salina Basin, Southeastern Mexico. Bull. Am. Paleontol., 88 (322), 116, pl. 1, fig. 4.

Cyclammina 4: (*Cyclammina cancellata* Brady), SPDC Catalogue, 1980, Stereoscan Photos, Paleontological Type Collection, Nigeria, pl. 3, fig. B5.

Diagnosis: Test free, large, planispiral, biconvex, involute, with labyrinthic interior; periphery rounded, subcircular in side view, with 9–10 chambers in the last whorl, slightly inflated, gradually increasing in size as added; sutures radial, slightly depressed; wall finely agglutinated; aperture multiple, consisting of an equatorial interiomarginal slit and a series of rounded pores scattered over the apertural face.

Range: Middle Oligocene–Late Pliocene.

Cyclammina minima LeRoy, 1939
Plate 2, Figs. 6 and 7

Cyclammina minima LeRoy, 1939, Some small foraminifera, ostracoda and otoliths from the Neogene ("Miocene") of the Rokan-Tapanoeli area, Central Sumatra. Natuurkwetenschappelijk Tijdschrift voor Nederlandsch Indie, 99 (6), 228, pl. 9, figs. 15, 16.

Cyclammina 1, 5: (*Cyclammina* cf. *C. minima*, LeRoy), SPDC Catalogue, 1980, Stereoscan Photos, Paleontological Type Collection, Nigeria, pl. 3, figs. B1, B6.

Diagnosis: Test free, circular in transverse section, compressed, slightly umbilicate, periphery acute; chambers distinct, 9–10 in last whorl; sutures distinct, somewhat limbate; wall finely agglutinated, occasionally showing interior labyrinthic structures; aperture near the base of last chamber on the periphery.

Range: Middle Miocene–Late Miocene.

Genus: HAPLOPHRAGMOIDES Cushman, 1910
Haplophragmoides sp. 1
Plate 2, Fig. 8

Diagnosis: Test free, planispiral, involute, compressed, periphery acute; chambers distinct, eight in the final whorl, increasing gradually in size, each chamber slightly depressed near each suture; sutures radial to slightly curved; wall finely agglutinated; aperture and equatorial interiomarginal slit at the base of last chamber.

Remarks: This species differs from *Haplophragmoides compressa* because it lacks a distinct keel. The species is a deepwater arenaceous form.

Haplophragmoides compressa LeRoy, 1939
Plate 2, Fig. 9

Haplophragmoides compressa LeRoy, 1939, Some small foraminifera, ostracoda and otoliths from the Neogene ("Miocene") of the Rokan-Tapanoeli area, Central Sumatra. Natuurkwetenschappelijk Tijdschrift voor Nederlandsch Indie, 99 (6), 227, pl. 5, figs. 24, 25.

Cyclammina 3A, 3B: (*Haplophragmoides compressa* LeRoy), SPDC Catalogue, 1980, Stereoscan Photos, Paleontological Type Collection, Nigeria, pl. 3, figs. B3, B4.

Diagnosis: Test free, planispiral, involute, compressed, periphery acute, chambers distinct, five to six in the final whorl, increasing gradually in size as added, chamber slightly depressed near each suture; sutures radial to slightly curved, flush with surface; wall smooth, finely agglutinated, with distinct keel, aperture, and equatorial interiomarginal slit.

Range: Miocene–Early Pliocene.

Haplophragmoides excavata Cushman and Waters, 1927
Plate 2, Figs. 10–12

Haplophragmoides excavata Cushman and Waters, 1927, Some arenaceous foraminifera from the Upper Cretaceous of Texas, Acta Palaeontol. Pol. 2 (4), 82, pl. 10, fig. 3.

Haplophragmoides excavata Fayose, 1970, Stratigraphical paleontology of Afowo-1 well, Southwestern Nigeria. J. Min. Geol. 5 (1 and 2), 10, pl. 1, fig. 1.

Haplophragmoides 9, SPDC Catalogue, 1980, Stereoscan Photos, Paleontological Type Collection, Nigeria, pl. 3, fig. A4.

Diagnosis: Test planispiral, compressed; periphery angular; eight chambers visible in the last-formed whorl, increasing gradually in size as added; sutures limbate, raised, radial, slightly curved, forming a thickened ridge around a depressed central portion; aperture interiomarginal; wall agglutinated.

Range: Paleocene–Eocene.

Haplophragmoides narivaensis Bronnimann, 1953
Plate 2, Figs. 13 and 14

Haplophragmoides narivaensis Bronnimann, 1953, Arenaceous foraminifera from the Oligocene–Miocene of Trinidad, Contrib. Cushman Found. Foramin. Res. 4, 96, pl. 15, fig. 4.

Haplophragmoides 24: (*Haplophragmoides narivaensis* Bronnimann), SPDC Catalogue, 1980, Stereoscan Photos, Paleontological Type Collection, Nigeria, pl. 3, fig. A10.

Diagnosis: Test free, planispiral, and involute, test composed of about two volutions the last of which is five to seven chambered; umbilicus deep and small; the chambers tend to become elongate in the axial direction and they increase rapidly in size so that the end chamber is about twice the size of the first chamber of the last volution; sutures indistinct, periphery weakly lobulate; aperture not visible but has been observed in an equatorial section where it appears to be interiomarginal; wall finely to coarsely arenaceous, surface smoothly finished.

Range: Oligocene–Late Miocene.

Haplophragmoides obliquicameratus Marks, 1951
Plate 2, Figs. 15–17

Haplophragmoides obliquicameratus Marks, 1951, A revision of the smaller foraminifera from the Miocene of the Vienna Basin. Contrib. Cushman Found. Foramin. Res., 35, pl. 5, fig. 1.

Trochamminoides 1A: (*Haplophragmoides obliquicameratus* Marks), SPDC Catalogue, 1980, Stereoscan Photos, Paleontological Type Collection, Nigeria, pl. 2, fig. C2.

Diagnosis: Test free, planispiral, vertically and tangentially compressed, giving a very irregular appearance, almost involute, umbilical region strongly depressed, periphery subacutely rounded; chambers very slightly inflated, 8–10 chambers visible in the last whorl, increasing gradually in size as added; sutures straight, radiate, slightly depressed; wall arenaceous, consisting of fine sandy materials that are firmly cemented; aperture a very narrow elongate slit at the base of the last-formed chamber on the ventral side of the test.

Range: Late Miocene–Early Pliocene.

Haplophragmoides bradyi Robertson, 1891
Plate 2, Figs. 18 and 19

Trochammina bradyi Robertson, 1891. Ann. Mag. Nat. Hist. 6, 388.

Haplophragmoides bradyi Petters, 1982. Central West African Cretaceous-Tertiary benthic foraminifera and stratigraphy, Palaeogeographica, 179, 97, pl. 1, figs. 4–5.

Diagnosis: Test free, wedge-shaped chambers in the last whorl; lobulate pheriphery; sutures depressed; wall coarsely agglutinated and strongly distorted; aperture, a very narrow elongated slit at the base of the last-formed chamber.

Range: Late Miocene–Pliocene.

Haplophragmoides canariensis d'Orbigny, 1839
Plate 2, Fig. 20; Plate 3, Fig. 1

Noninina canariensis d'Orbigny, Barker-Webb, and Berthelot, Hist. Nat. lles Canaries, vol. 2, part 2, figs. 33, 34.

Haplophragmoides canariensis Cushman, 1918, US Natl. Museum Bull. 71, pt. 1, 1910, p. 101, fig. 149.

Haplophragmoides canariensis d'Orbigny, 1839, Cushman, 1918, The Foraminifera of the Atlantic Ocean, p. 38, pl. 8, fig. 1.

Diagnosis: Test free, planospiral, composed of a few coils partially involute, umbilicate; chambers subglobular, six or seven chambers in the final coil, the last chamber larger than the preceding ones; sutures distinct, periphery lobulated, wall arenaceous, made up of sand grains, smoothly finished, thin; aperture at the base of the last-formed chamber narrow, the overhanging portion of the wall slightly extended, forming a thin lip.

Range: Paleocene–Pleistocene.

Genus: ALVEOLOPHRAGMIUM Shchedrina, 1936
Alveolophragmium crassum
Plate 3, Figs. 2 and 3

Alveolophragmium crassum AGIP SPA, 1988, Foraminifera of the Niger Delta, pl. 17, figs. 3a–d.

Diagnosis: Test free, large, planispiral, involute with complex interior, wall with inner alveolar structure below imperforate outer layer; aperture equatorial and areal, with bordering lips.

Range: Late Oligocene–Late Miocene.

Alveolophragmium scitulum Brady, 1954
Plate 3, Figs. 4 and 5

Alveolophragmium scitulum Brady, 1954. Bull. Mus. Comp. Zool., Harvard, vol. 111, No. 10, p. 487.

Alveolophragmium scitulum Brady, 1954. Taxonomic notes in Society of Economic Paleontologists and Mineralogists 1960, p. 70, pl. XXXIV, figs. 11–13.

Diagnosis: Test free, planispiral involute with complex interior, 9–10 chambers in the final whorl, coarsely agglutinated, with inner alveolar structure; sutures distinct, slightly depressed; aperture equatorial and areal with bordering lip.

Range: Pliocene–Pleistocene.

Alveolophragmium subglobosum Sars, 1910
Plate 3, Figs. 6 and 7

Alveolophragmium subglobosum Sars, 1910. US Natl. Museum Bull. 71, 105.

Alveolophragmium subglobosum Sars, 1910. Taxonomic notes. In: Society of Economic Paleontologists and Mineralogists 1960, p. 70, pl. XXXIV figs. 7, 8, 10, and 14.

Diagnosis: Test free, planispiral involute with complex interior, wall agglutinated with alveolar structure below imperforate outer layer; seven chambers in the final whorls; chambers increasing rapidly in size as added; sutures distinct, slightly depressed; aperture equatorial and areal with bordering lip.

Range: Middle Miocene–Late Miocene.

Family: ATAXOPHRAGMIIDAE Schwager, 1877
Genus: EGGERELLA Cushman, 1933
Eggerella sp. 1
Plate 3, Fig. 8

Diagnosis: Test free, globular, trochoid, spire in the early stage with five chambers in a whorl, later reduced to four, and in the adult to three; wall finely arenaceous with fine pores; aperture a low arched slit at the base of the inner margin.

Remarks: The species is frequently found in the Miocene–Pliocene sequence of the Niger Delta. It differs from other species of *Eggerella* in its broader, shorter form and the rough, nonpolished surface.

Eggerella scabra Williamson, 1922
Plate 3, Fig. 9

Verneuilina scabra (Williamson) Cushman, 1922, US Natl. Museum Bull. 104 (pt. 3), 55.

Eggerella scabra (Williamson) Cushman, 1937 Cushman Laboratory for Foraminiferal Research, Special Publication No. 8, p. 48, pl. 5, figs. 10, 11.

Verneuilina 4, 4a: (*Eggerella scabra* Williamson), SPDC Catalogue, 1980, Stereoscan Photos Paleontological Type Collection, Nigeria, pl. 5, figs. H4, H5.

Diagnosis: Test free, conical, at an early stage a trochospiral coil reducing to triserial in the adult; chambers increasing rapidly in size as added; wall finely agglutinated with fine pores; sutures distinct, slightly depressed; aperture a low interiomarginal thin slit.

Range: Middle Miocene–Late Miocene.

Eggerella bradyi Cushman, 1933
Plate 3, Figs. 10–12

Verneuilina bradyi Cushman, 1911, 1910–17. A monograph of the foraminifera of the North Pacific Ocean. Bull. 71 (pt.2). Textulariidae (1911), 108 p., 156 text-figs.

Eggerella bradyi Cushman, 1933. Some new foraminiferal genera. Contrib. Cushman Lab. Foramin. Res. 9, (pt. 2), 32–38, pls. 3, 4.

Eggerella bradyi Cushman, 1933, Treatise on Invertebrate Paleontology, Loeblich and Tappan, 1964, p. C276, figs. 2–5

Eggerella bradyi, Cushman, 1933. AGIP SPA, 1982, Foraminifera Padani, pl. IV, figs. 2, 2a.

Diagnosis: Test free, trochospirally coiled, with three to five chambers to whorl; wall finely agglutinated on pseudochitinous base, may be of calcareous particles in calcareous cement; aperture a low interiomarginal slit.

Range: Oligocene–Pleistocene.

Eggerella forestensis
Plate 3, Figs. 13 and 14

Eggerella forestensis Bronnimann, 1953. Contrib. Cushman Found. Foramin. Res. 4, (pt. 3), 93, pl. 15, fig. 8.

Verneuillina 2: (*Eggerella* ex. gr. *forestensis* Bronnimann) SPDC Catalogue, 1980, Stereoscan Photos, Paleontological Type Collection, Nigeria, pl. V, figure H 2.

Diagnosis: Test free, subconical, early stage trochospiral then triserial at a later stage. The subglobular chambers increase in size rapidly as added; wall, finely agglutinated, sutures distinct, depressed; aperture, a narrow slit near the base of the apertural face of the last chamber.

Range: Eocene–Pleistocene.

Eggerella propinqua Brady, 1937
Plate 3, Figs. 15

Eggerella propinqua Brady, 1937. Cushman Laboratory Foraminifera Research, Special Publication No. 8, p. 53.

Eggerella propinqua Brady, 1937. Taxonomic notes. In: Society of Economic Paleontologists and Mineralogists Bull. 1960, p. 96, pl. XLVII, figs. 8–12.

Diagnosis: Test in trochospiral coil with five chambers to whorl in early stage of microspheric form, gradually reduced to three whorls in adult, wall coarsely agglutinated on pseudochitinuous base, maybe of calcareous particles in calcareous cement; sutures distinct, slightly depressed; aperture a low interiomarginal slit.

Range: Eocene–Pleistocene.

Gravellina narivaensis Bronnimann, 1953
Plate 3, Fig. 16

Gravellina narivaensis Bronnimann, 1953. Arenaceous foraminifera from the Oligo–Miocene of Trinidad. Contrib. Cushman Found. Foramin. Res. 4, 87–100, pl. 15, text-fig. 1–15.

Gravellina narivanensis Bronnimann, 1953. Treatise on Inverbrate paleontology (Leoblich and Tappan, 1964), p. C276, figures 7, 8.

Valvulina 1A: (*Gravellina narivaensis* Bronnimann, 1953) SPDC Catalogue, 1980, Stereoscan Photos, Paleontological Type Collection, Nigeria, pl. V, figure A2.

Gravellina narivaensis Bronnimann, 1953. Benthic foraminifera biostratigraphy of the South Caribbean Region, Bolli, Beckmann and Saunders, Cambridge University Press, 1994, p. 258, fig. 64, 4–7.

Diagnosis: Test free, quadriserial throughout; wall finely agglutinated; suture distinct and slightly depressed; aperture an intermarginal arch.

Range: Oligocene–Miocene.

Genus: KARRERIELLA Cushman, 1933
Karreriella siphonella Reuss, 1937
Plate 3, Fig. 17

Karreriella siphonella Reuss, 1937, pl. 9.2, fig. 3; In: Paleoecology, Biostratigraphy, Paleoceanography and Taxonomy of Agglutinated foraminifera, p. 194. Edited by Christoph Hemleben, Michael A. Kaminski, Wolfgang Kuhnt, D.B. Scott.

Karreriella 1 and 2: *(Karreriella siphonella* Reuss), SPDC Catalogue, 1980, Stereoscan Photos, Paleontological Type Collection, Nigeria, pl. 5, figs. B1, B2.

Diagnosis: Test free, elongate, early chambers form a short and tapering trochoid spire followed by a well-developed longer biserial stage; chambers inflated and subglobular; wall finely agglutinated, smooth with calcareous cement; sutures distinct in biserial portion and depressed, less distinct in trochoid portion; aperture circular or elliptical, with a short neck placed above the base of the apertural face.

Range: Early Oligocene–Early Miocene.

Karreriella subcylindrica Nuttall, 1928
Plate 3, Fig. 18

Gaudryina subcylindrica Nuttall, 1928, Q. J. Geol. Soc., 84, 76, pl. 3, figs. 17, 18.

Karreriella subcylindrica (Nuttall) Acosta, 1940, Algunos foraminifeos nuevos de la costas Cubanas, Torreia, Havana, Cuba, No. 5, p. 3, pl. 1, figs. 1–3.

Karreriella subcylindrica Cushman, 1937, Special Publication No. 8, p. 128, pl. 15, fig. 25.

Karreriella subcylindrica Cushman, 1947a, The foraminiferal fauna of the Oligocene Ste. Croix formation of Trinidad, B.W.I., Cushman Laboratory for Foraminiferal Research, Special Publication No. 22, p. 44, pl. 7, figs. 24–26.

Karreriella subcylindrica Kohl, 1985, Early Pliocene benthic forminifers from the Salina Basin, Southeastern Mexico. Bull. Am. Paleontol. 88 (322) 118, pl. 3, fig. 6.

Karreriella subcylindrica AGIP SPA, 1982, Foraminiferi Padani, pl. V, figs. 5, 5a, 5p.

Diagnosis: Test free, elongate, cylindrical, circular in transverse section, early chambers form a trochoid spire one-fifth the length of test, followed by a well-developed biserial portion twisted about the axis; sutures indistinct in trochoid portion, distinct and depressed in biserial portion; wall finely agglutinated, smooth, finely perforate; aperture rounded, located on the terminal face of the ultimate chamber.

Range: Middle Eocene–Early Pliocene.

Karreriella bradyi Cushman, 1937
Plate 3, Fig. 19

Karreriella bradyi Cushman, 1937. Cushman Laboratory for foraminifera Research. Special Publication No. 8, p. 98.

Karreriella bradyi Cushman, 1937. Taxonomic notes in Society of Economic Paleontologists and Mineralogists 1960, p. 94, pl. XLVI, figs. 1–4.

Karreriella bradyi Cushman, 1937. AGIP SPA, 1982 Foraminifera Padani, pl. V, fig. 1, 1a, 1 p.

Diagnosis: Test free, elongate, early chambers in trochoid spire of one or more whorls followed by well-developed biserial stage. Wall finely agglutinated smoothly finished, aperture rounded in terminal face of final chamber bordered by lip; sutures distinct in biserial portion and depressed less distinct in the trochoid portion.

Range: Middle Oligocene–Early Miocene.

Karreriella gaudryinoides Fornasini, 1885
Plate 3, Fig. 20; Plate 4, Figs. 1 and 2

Karreriella gaudryinoides Fornasini, 1885. AGIP SPA, 1982, Foraminifera Padani, pl. V, fig 3, 3a.

Diagnosis: Test free, elongate, early chambers form a trochoid spire followed by a well-developed biserial portion, twisted about the axis; sutures indistinct in the trochoid portion, distinct and depressed in the biserial portion; wall finely agglutinate, smooth, finely perforate, aperture rounded, located on the terminal face of the ultimate chamber.

Range: Miocene–Pliocene.

Genus: VALVULINA d'Orbigny, 1826
Valvulina flexilis Cushman and Renz, 1941
Plate 4, Figs. 3–5

Valvulina flexilis Cushman and Renz, 1941, p. 7, pl. 1, figs. 16, 17.

Valvulina flexilis Petters, 1982, p. 96, pl. 1, fig. 27.

Valvulina flexilis AGIP SPA, 1988, Foraminifera of the Niger Delta, pl. 16, figs. 1a–e.

Valvulina 1: (*Valvulina flexilis* Cushman and Renz), SPDC Catalogue, 1980, Stereoscan Photos, Paleontological Type Collection, Nigeria, pl. 5, fig. A1.

Diagnosis: Test free, triserial, moderately flared with a strongly lobulate periphery; chambers expand fairly rapidly and are distorted and pushed toward the proloculus so that they form overlapping lobes; sutures horizontal and depressed; wall agglutinated, fairly smooth with angular quartz grains of uniform size which are set in abundant matrix; aperture wide and high on an excavated apertural face.

Range: Early Oligocene–Early Pliocene.

Genus: VULVULINA d'Orbigny, 1826
Vulvulina pennatula Batsch, 1791
Plate 4, Fig. 6

Vulvulina pennatula Batsch, 1791. Taxonomic notes. In: Society of Economic Paleontologists and Mineralogists, 1960, p. 92, pl. XLV, figs. 1–8.

Vulvulina pennatula Batsch, 1791. AGIP SPA, 1982, Foraminifera Padani, pl. III, figs.4, 4a, 5, 5bis.

Diagnosis: Test free, flaring to elongate, lozenge-shaped, rhomboidal in section, lateral margins acutely angled; chambers increasing rapidly in size as added, biserially arranged, broad and low, somewhat arched over highly coiled primary chambers, recurved laterally, final chambers uniserial; sutures distinct, commonly thickened and elevated in early portion later moderately depressed; wall agglutinated but very finely grained and smoothly finished; aperture in early stage broad, low interiomarginal arch; in uniserial stage, becoming elongate, narrow, terminal slit.

Range: Early Miocene–Early Pliocene.

Martinottiella communis d'Orbigny, 1826
Plate 4, Fig. 7

Martinottiella communis d'Orbigny, 1826. Ann. Sci. Nat. Paris, ser. 1, 7, 245–314; Atlas, pl. 10–17, crochard.

Martinottiella communis d'Orbigny, 1826, Jenkins and Murray, 1989, Stratigraphical Atlas of Fossil Foraminifera, p. 456, pl. 9.2, figs. 4, 5.

Martinottiella communis d'Orbigny, 1826. AGIP SPA, 1982, Forminifera Padani, pl. v, figs. 6a, 6, 6sl.

Diagnosis: Test free, elongate, cylindrical, early chambers trochospiral with four or five to a whorl, progressively reduced to triserial, biserial, and uniserial, adult with relatively elongate uniserial development; wall finely agglutinated; aperture terminal, elongate slit, commonly arcuate, with bordering lip.

Range: Early Eocene–Pleistocene.

Family: TEXTULARIIDAE Ehrenberg, 1838
Genus: PORITEXTULARIA Loeblich and Tappan, 1952
Poritextularia panamensis Cushman, 1918
Plate 4, Figs. 8 and 9

Textularia panamensis Cushman, 1918a, Some Pliocene and Miocene foraminifera of the Coastal Plain of the U.S. US Geol. Surv. Bull. 676, 53, pl. 10, fig. 1.

Textularia panamensis Lalicker and McCulloch, 1940, Some Textulariidae of the Pacific Ocean. Allan Hancock Pacific Exped., 6 (2), University of Southern California Press, p. 136, pl. 15, figs. 18a–e.

Poritextularia aff. *P. panamensis* (Cushman) Loeblich and Tappan, 1952, v. 42, no. 8, p. 264.

Textularia 11: (*Poritextularia* aff. *P. panamensis*, Cushman), SPDC Catalogue, 1980, Stereoscan Photos, Paleontological Type Collection, Nigeria, pl. 1, fig. A7.

Diagnosis: Test free, biserial throughout, rhomboid in outline, severely compressed, side flat; chambers numerous, low and broad, increasing gradually in size from the rounded proloculus; very oblique, highest at midportion of the test and hanging downward to the margins; sutures distinct, depressed, strongly oblique, slightly curved upward; wall finely arenaceous, with occasional large black mineral grains included; aperture in the early stages consisting of an elongate slit at the base of the last chamber, parallel with the flat sides of the test which in the adult may be bridged by pillars so that the former slit is broken up into a linear series of small pore-like openings extending from the base across the terminal portion of the final chamber.

Range: Middle Oligocene–Early Miocene.

Genus: SPIROPLECTAMMINA Cushman, 1927
Spiroplectammina wrightii Silvestri, 1903
Plate 4, Figs. 10 and 11

Spiroplecta wrightii Silvestri, 1903, Dimorfis oe nomenclatra d'una Spiroplecta. Altre notizie sulla struttura della Siphogenerina columcllaris. Atti della Pontificia Acaddemia Romana dei Nuovi Lincei 59, 59–66, text figs. 1–6.

Spiroplectammina wrightii, Cushman, 1949, vol. 6, figs. 2–4. In: Palaeontologia Electronica 2012, Article number: 15.2.16A: A taxonomic guide to modern benthic shelf foraminifera of the western Mediterranean Sea.

Textularia 3: (*Spiroplectammina wrightii* Silvestri) SPDC Catalogue, 1980, Stereoscan Photos, Paleontological Type Collection, Nigeria, pl. 1, fig. A2.

Diagnosis: Test biserial, elongate, expanding gradually from the bluntly pointed initial end, compressed, thickest along the axis; periphery slightly lobate; 16 chambers visible, compressed, increasing gradually in size as added; sutures distinct, slightly curved and depressed; test arenaceous; aperture interiomarginal, forming a crescentic opening at the inner margin of the last chamber.

Range: Early Oligocene–Miocene.

Genus: TEXTULARIA Defrance, 1824
Textularia elegans Lacroix, 1931
Plate 4, Fig. 12

Textularia elegans Lacroix, 1931, Microtexture du test des Textularidae. Bulletin de I'Institut Oceanographique Monaco 588, p. 15, text-fig. 11.

Textularia elegans Lacroix, 1932, Textularidae du plateau continental Mediterraneen entre Saint-Raphael et Monaco. Bulletin de I'Institut Oceanographique, Monaco 591, p. 8, text-figs. 4–6.

Texturalia 7: (*Textularia elegans* Lacroix), SPDC Catalogue, 1980, Stereoscan Photos, Paleontological Type Collection, Nigeria, pl. 1, fig. A.

Diagnosis: Test free, biserial, elongate, peripheral margin rounded, lobate, chambers inflated; wall coarsely agglutinated, composed of poorly sorted grains; sutures depressed; aperture lunate, on convex apertural face.

Range: Late Miocene–Early Pliocene.

Textularia soldanii Fornasini, 1883
Plate 4, Fig. 13

Textularia soldanii Fornasini, 1883, Note preliminare sui foraminiferi della marna Pliocenica del pouticallo di savana nel Holoehasa. Soc. Geol. Ital. 2, 183, pl. 2, figs. 2a–c.

Textularia soldanii AGIP SPA, 1982, Foraminiferi Padani, pl. 11, figs. 10, 10a, 10p, 10sl.

Diagnosis: Test free, elongate, conical, early portion narrow and compressed, later portion step-like, peripheral margin acute, serrate, final pair of chambers rounded; wall agglutinated, rough, composed of diverse materials; sutures slightly depressed and slightly distinct; aperture lunate, on convex apertural face.

Range: Late Miocene–Pleistocene.

Textularia parvula
Plate 4, Fig. 14

Textularia parvula Cushman, 1922, The Foraminifera of the Atlantic Ocean, World Record of Marine Species, p. 11 pl. 6, figs. 1, 2.

Textularia parvula Cushman, 1922, Petters, 1982, Central West African Cretaceous-Tertitary benthic foraminifera and stratigraphy, Palaeontographica, 179, 45, pl. 2, figs 41–42.

Textularia 8: (*Textularia parvula* Cushman) SPDC Catalogue, 1980, Stereoscan Photos, Paleontological Type Collection, Nigeria, pl. 1, fig. A5.

Diagnosis: Test free, elongate, biserial gradually tapering from the bluntly pointed apical end to the broadly rounded apertural end. Chambers numerous, generally closely appressed; sutures distinct, wall agglutinated, simple aperture slightly rounded at the central part of the inner margin of the last-formed chamber.

Range: Miocene–Pliocene.

Textularia gramen d'Orbigny, 1846
Plate 4, Fig. 15

Textularia gramen d'Orbigny, 1846. Foraminiferes fossils du Bassin Tertiaire de Vienne (Autriche)-Gide et Compe (Paris), p. 1–312.

Textularia pseudogramen Chapman and Parr, 1937. Australasian Antarctic Exped. 1911–1914, Sci Repts. vol. 1, p. 153.

Textularia gramen d'Orbigny, 1846. Taxonomic notes. In: Society Economic Paleontologists and Mineralogists 1960, p. 88, pl. XLIII, figs. 9, 10.

Textularia gramen d'Orbigny, 1846. Central West African Cretaceous-Tertiary benthic foraminifera and stratigraphy, Petters, 1982, Palaeontographica, 179, p. 44, pl. 2, figs. 39, 40.

Diagnosis: Test free, elongate, biserial, peripheral margin acute lobate, chambers inflated; wall coarsely agglutinated; sutures depressed, distinct; aperture lunate, on convex apertureal face.

Range: Miocene–Pliocene.

Textularia sagittula Defrance, 1824
Plate 4, Figs. 16–18

Textularia sagittula Defrance, 1824. Dictionnaire des sciences naturelles: 1824, v. 32, p. 1–567.

Textularia sagittula Defrance, 1824. Taxonomic notes. In: Society of Economic Paleontologists and Mineralogists 1960, p. 86, pl. XLII, figs. 17, 18.

Textularia 18: (*Textularia sagittula* Defrance, 1824) SPDC Catalogue, 1980, Stereoscan Photos, Paleontological Type Collection, Nigeria, pl. 1, figure A 11.

Textularia sagittula Defrance, 1824. AGIP SPA, 1982, Foraminifera Padani, pl. 11, figs 5, 5a, 5p.

Diagnosis: Test free, elongate, biserial, generally more or less compressed in plane of biseriahty or rarely oval to circular in cross section; chambers numerous, generally closely aprised; wall agglutinated, simple; aperture single low arch at the base of last chamber. Sutures distinct and depressed.

Range: Oligocene–Pleistocene.

Subfamily: PAVONITININAE Loeblich and Tappan, 1961
Genus: TEXTULARIELLA Cushman, 1927
Textulariella barrettii Jones and Parker, 1876
Plate 4, Figs. 19 and 20; Plate 5, Fig. 1

Textulariella barrettii Jones and Parker, 1876. Notice sur les foraminiferes vivants et fossils de Jamaique: Suivie de la description dune espece nouvelle [Tinoporus pillaris] des Couches Miocenes dela jamaique, par H.B. Brady: Soc Malacol. Belg., Ann., vol. II, Mein., p. 91–103.

Textulariella barrettii Jones and Parker, 1876. Taxonomic notes. In: Society of Economic Paleontologists and Mineralogists 1960, p. 88, pl. XLIII fig. 17.

Textulariella barrettii Jones and Parker, 1876. Treatise on Invertebrate Paleontology (Leoblich and Tappan, 1964), p. C295, Figs. 202, 1, 2.

Diagnosis: Test subconical, early stage trochospiral with three or more chambers in whorl, later biserial with interior labyrinthic, aperture low interiomarginal arch.

Range: Middle Oligocene–Early Miocene.

Family: TROCHAMMINIDAE Schwager, 1877
Genus: TROCHAMMINA Parker and Jones, 1859
Trochammina sp. 1
Plate 5, Figs. 2 and 3

Diagnosis: Test free, trochospiral, globular to ovate; chambers increasing gradually in size as added; wall agglutinated; aperture a low, interiomarginal extraumbilical–umbilical arch which may have narrow bordering lip.

Remarks: This arenaceous species is frequently found in deep waters in the Neogene of the Niger Delta. The species differs from other *Trochammina* species in its small body size, the finely agglutinated texture of the test and the compressed to globular test.

Trochammina proteus Karrer, 1865
Plate 5, Fig. 4

Trochammina proteus Karrer, Sitz. Akad. Wiss. Wien, v. 52 (Abth. 1), 1856 (1866), p. 494, pl. 1, fig. 8.

Ammodiscus proteus Rhumbler, Arch. Prot., v. 3, 1903, p. 281, fig. 131.

Trochamminoides proteus Cushman, Bull. 71, US Natl. Museum Bull. 71, pt. 1, 1910, p. 98, figs. 142–144.

Trochamminoides 2: (*Trochammina proteus*, Karrer) SPDC Catalogue, 1980, Stereoscan Photos, Paleontological Type Collection, Nigeria, pl. 2, figs. C3V, C3D.

Diagnosis: Test free, typically planospiral, composed of several coils, the earlier ones usually unconstricted and undivided, the later ones divided into several chambers with corresponding constrictions of the wall; wall of fine sand and a yellowish-brown cement; aperture simple at the end of the last-formed chamber.

Range: Paleocene–Pleistocene.

Trochammina globigeriniformis Parker and Jones, 1876
Plate 5, Figs. 5 and 6

Trochammina globigeriniformis Parker and Jones, 1865. On some foraminifera from the North Atlantic and Arctic Oceans including Davis Straits and Baffin's Bay, Philosophical Transactions 155, 325–441, pl. 12–19.

Trochammina globigeriniformis Parker and Jones, 1865. Taxonomic notes. In: Society of Economic Paleontologists and Mineralogists 1960, p. 72, pl. xxxv, figs. 10, 11.

Trochammina globigeriniformis Parker and Jones, 1865. Treatise on Invertebrate Paleontology. Loeblich and Tappan (1964), p. C260, figure 1 and 2.

Diagnosis: Test free, trochospiral, globular to ovate chambers increasing gradually in size as added; wall agglutinated; aperture low interiomarginal, extraumbilical–umbilical arch which may have narrow bordering lip.

Range: Middle Oligocene–Early Miocene.

Recurvoides deformis Andreae, 1934
Plate 5, Figs. 7 and 8

Trochamminoides 1A: (*Recurvoides deformis* Andrea, 1934) SPDC Catalogue, 1980, Stereoscan Photos, Paleontological Type Collection, Nigeria, Pl. 2, fig. C1

Diagnosis: Test free, subglobular, coiling streptospiral with few chambers in each whorl, wall agglutinated, thin; aperture small, areal with distinct bordering lip.

Range: Miocene–Pleistocene.

Superfamily: RHIZAMMININAE Rhumbler, 1895
Genus: BATHYSIPHON Sars, 1872
Bathysiphon sp. Sars, 1872
Plate 5, Fig. 9

Bathysiphon sp. Sars, 1872. Treatise on Invertebrate Paleontology, Loeblich and Tappan, 1964, p. C188, figs. 4–10.

Bathysiphon sp. Stratigraphical Atlas of fossil foraminifera. Jenkins and Murray, 1989, p. 24, plate 2.1, fig. 4.

Diagnosis: Test free, cylindrical, large, up to 50 mm in length, elongate, narrow, more or less flexible tube, which may have annular constrictions; wall agglutinated, commonly of siliceous sponge spicules and fine sand or other mineral matter in calcareous cement; aperture at both ends of the tube.

Range: Paleocene–Recent.

Family: MILIOLIDAE Ehrenberg, 1839
Genus: PTYCHOMILIOLA Eimer and Fickert, 1899
Ptychomiliola polygonia Schlumberger, 1890
Plate 5, Fig. 10

Adelosina polygonia Schlumberger, 1890, Note aur *l'Adelosina polygonia*. Bulletin de la Societe Zoologique de France, 15, 139–144, tfs 1–11.
Ptychomiliola 2: (*Ptychomiliola polygonia* Schlumberger, 1890), SPDC Catalogue, 1980, Stereoscan Photos, Paleontological Type Collection, Nigeria, pl. 23, fig. D1.

Diagnosis: Test free, early chambers milioline in arrangement, later ones planispiral and evolute, about three chambers to a whorl; wall calcareous porcellaneous, imperforate, with longitudinal costae; aperture terminal, rounded, produced at the end of a short neck, with lip and distinct bifid tooth.
Range: Pliocene–Recent.

Subfamily: QUINQUELOCULININAE Cushman, 1917
Genus: QUINQUELOCULINA d'Orbigny, 1826
Quinqueloculina lamarckiana d'Orbigny, 1839
Plate 5, Figs. 11 and 12

Quinqueloculina lamarckiana d'Orbigny, 1839, Foraminiferes des Iles Canaries, In: Barker-Webb, P., Berthelot, S., Historie Naturelle des Canaries. Paris, vol. 2, Part 2, pp. 119–146, pl. 11, figs. 14, 15.
Quinqueloculina lamarckiana Boltovskoy and Wright, 1980, Atlas of Benthic Foraminifera of the Southwest Atlantic. Dr. W. Junk b.v., The Hague, p. 114, pl. 28, figs. 10–12.

Diagnosis: Test free, rounded at the prolocular part, truncated at apertural end, triangular in transverse section; chambers angulate and carinate; three chambers visible from the exterior on one side of test, four visible on the opposite side; wall calcareous porcellaneous, imperforate; sutures almost flush with surface; aperture terminal, ovate, with long simple tooth.
Range: Eocene–Pleistocene.

Quinqueloculina microcostata Natland, 1938
Plate 5, Figs. 13–15

Quinqueloculina microcostata Natland, 1938, New Species of Foraminifera from off the West Coast of North America and from the Later Tertiary of the Los Angeles Basin, California, Technical Series, vol. 4, p. 142, pl. 4, figs. 6a–c.
Quinqueloculina 11: (*Quinqueloculina microcostata* Natland, 1938), SPDC Catalogue, 1980, Stereoscan Photos, Paleontological Type Collection, Nigeria, Pl. 22, fig. A8.

Diagnosis: Test free, broadly fusiform, periphery acute but not carinate; chambers distinct, with sides slightly convex; sutures depressed; wall calcareous porcellaneous, ornamented with very fine, irregular, longitudinal costae; aperture formed by an extension of the final chamber to form an oval opening with a narrow elongated tooth.

Range: Miocene–Pleistocene.

Quinqueloculina seminulum Linne, 1767
Plate 5, Figs. 16–20; Plate 6, Fig. 1

Serpula seminulum Linne, 1767, Syst. Nat., 12 ed., p. 1264, no. 791.
Quinqueloculina seminulum (Linne). Boltovsky et al., 1980, Atlas of Benthic Foraminifera of the Southwest Atlantic. Dr. W. Junk b.v., The Hague, pl. 29, figs. 7–13.
Quinqueloculina seminulum AGIP SPA, 1982. Foraminiferi Padani, pl. VI, fig. 6.

Diagnosis: Test free, with rounded peripheral margin, chambers of uniform diameter throughout length of test; wall smooth, shiny; sutures depressed; aperture semicircular with rim and thin bifid tooth.
Range: Oligocene–Pleistocene.

Quinqueloculina rhodiensis Parker, 1953
Plate 6, Fig. 2

Quinqueloculina rhodiensis 1953, J. Paleontol. Soc. India, 2002, 47, 55, pl. 1, fig. 14.
Quinqueloculina rhodiensis Parker. Matoba. 1970, Science Reports of the Tohoku University, Sendai, (Ser 2), 42, p. 1–85, pl. 2, figs. 11a, 11b.
Quinqueloculina 5: (*Quinqueloculina rhodiensis* Parker) SPDC Catalogue, 1980, Stereoscan Photos, Paleontological Type Collection, Nigeria, pl. 22, fig. A4.

Diagnosis: Test free, broadly fusiform, periphery rounded, chamber distinct; sutures depressed, wall calcareous, porcellaneous, ornamented with fine longitudinal costae; aperture semicircular with a narrow elongated tooth.
Range: Miocene–Recent.

Quinqueloculina bicornis Walker and Jacob, 1798
Plate 6, Fig. 3

Quinqueloculina bicornis (Walker and Jacob, 1798), Taxonomic notes. In: Society of Economic Paleontologists and Mineralogists 1960, p. 13, pl. 6, figures 9a, 9b, 9c.
Quinqueloculina bicornis Walker and Jacob, 1798, AGIP SPA, 1982, Forminifera Padani, pl. V, figure 10, 10a, 10p 10bis, bis-a.

Diagnosis: Test free, broadly fusiform, periphery acute, chambers distinct, suture depressed, wall calcareous; porcellaneous, ornamented with a narrow elongated tooth.
Range: Pliocene–Pleistocene.

Quinqueloculina vulgaris d'Orbigny, 1826
Plate 6, Fig. 4

Quinqueloculina 11A: (*Quinqueloculina vulgaris* d'Orbigny, 1826) SPDC Catalogue, 1980, Stereoscan Photos, Paleontological Type Collection, Nigeria, pl. 22, fig. A9.

Quinqueloculina vulgaris d'Orbigny, 1826. AGIP SPA, 1982, Foraminifera Padani, pl. VI, figs. 8, p. 8.

Diagnosis: Test free, broadly fusiform, periphery rounded; three chambers visible from the exterior on both sides, slightly convex; sutures depressed; wall calcareous, porcellaneous, surface smooth; aperture terminal, semicircular with simple tooth.

Range: Pliocene–Pleistocene.

Genus: SIGMOILOPSIS Finlay, 1947
Sigmoilopsis schlumbergeri Silvestri, 1904
Plate 6, Figs. 5–7

Sigmoilopsis schlumbergeri Silvestri, 1904, Ricerche strutturali su alcune forme dei trubi di Bonfornello (Palermo), Accad. Pont. Romana Nuovi Lincei, Mem., Roma, Italia, 22, 267, 269. Type figure from Schlumberger, 1887, Boll. Soc. Zool. France, v. XII, pl.7, figs. 12–14.

Sigmoilopsis schlumbergeri Silvestri, Kohl, 1985, Bull. Am. Paleontol., 88 (322), 120, pl. 5, fig 6.

Sigmoilopsis schlumbergeri Silvestri, Cenozoic Cosmoplitan Deep water Benthic foraminifera (Elf Aquitaine) 1986, p. 58, pl. 18, figs. la–1e.

Sigmoilopsis schlumbergeri Silvestri, AGIP SPA, 1982, Foraminifera Padani, pl. VII, figs. 6, 6a.

Diagnosis: Test free, distinctly granular in texture, ovate, with chambers one half coil in length, at first quinqueloculine in arrangement. Wall agglutinated with calcareous cement enclosing sand, calcareous shell fragments, and sponge spicules; aperture terminal, rounded with small tooth.

Range: Middle Miocene–Pleistocene.

Sigmoilopsis celata Costa, 1855
Plate 6, Figs. 8–11

Sigmoilopsis celata Costa, 1855, AGIP SPA, 1982, Foraminifera Padani, pl. vii, figures 4, 4p.

Diagnosis: Test free, ovate, chambers one and a half coil in length at first, quinqueloculine in arrangement, then with increasing angle between plane of coiling of successive chambers, wall agglutinated with calcareous cement, aperture terminal, rounded with small tooth.

Range: Oligocene–Pleistocene.

Genus: SPIROSIGMOILINA Parr, 1942
Spirosigmoilina tenuis
Plate 6, Figs. 12 and 13

Spiroloculina 6: (*Spirosigmoilina* ex. gr. *tenuis* CZJZEK) SPDC Catalogue, 1980, Stereoscan Photos, Paleontological Type Collection, Nigeria, pl. 23, fig. B4.

Diagnosis: Test free, ovate in outline, with sigmoid curve, chambers with broad lateral extensions which obscure all preceding chambers except penultimate one; wall calcareous, thick, imperforate, porcellaneous; aperture terminal rounded with tooth.

Range: Late Eocene–Pleistocene.

Spirosigmoilina oligocaenica Cushman
Plate 6, Figs. 14–16

Spiroloculina 4A (*Spirosigmoilina oligocaenica* Cushman): SPDC Catalogue, 1980, Stereoscan Photos, Paleontological Type Collection, Nigeria, pl. 23, fig. B3.

Diagnosis: Test of moderate size, early chambers arranged in a sigmoid arrangement, quinqueloculine, later chambers flattened and added in single plane on alternate sides; about six chambers visible; wall porcellaneous, imperforate, terminal aperture at open end of final chamber, with bifid tooth.

Range: Oligocene–Middle Miocene.

Family: NUBECULARIIDAE Jones, 1875
Genus: SPIROLOCULINA d'Orbigny, 1826
Spiroloculina depressa d'Orbigny, 1826
Plate 6, Fig. 17

Spiroloculina depressa d'Orbigny, 1826, Tableau methodique de la classe des cephalopodes. Ann. Sci. Nat. Paris, ser. 1, 7, 298, text-fig. 1.

Spiroloculina depressa (d'Orbigny), Parker, Jones, and Brady, 1871, On the nomenclature of the foraminifera. Part XIV—The species enumerated by d'Orbigny in the 'Annales des Sciences, Naturelles' 1826, v. 7. Annals and Magazine of Natural History, London, Ser. 4, v. 8, 145, 179, 283–266, pl. 8, fig. 23.

Spiroloculina depressa (d'Orbigny), Boltovskoy et al., 1980, Atlas of Benthic Foraminifera of the Southwest Atlantic. Dr. W. Junk b.v., The Hague, p. 120, pl. 31, figs. 19, 20.

Spiroloculina depressa (d'Orbigny), AGIP SPA, 1982, Foraminiferi Padani, pl. VII, fig. 1.

Diagnosis: Test free, elliptical in outline, slightly longer than wide, peripheral margin angulate; chambers markedly quadrate in section, ending in straight line where they abut against earlier chambers thus forming a zigzag line along longitudinal axis, last chamber with neck on outside margin; wall calcareous, porcellaneous, rough; sutures depressed; aperture elongate, large, almost rectangular with lip and short narrow simple tooth.

Range: Pliocene–Pleistocene.

Spiroloculina excavata d'Orbigny, 1846
Plate 6, Fig. 18

Spiroloculina excavata d'Orbigny, 1846, Foraminiferes fossiles du Bassin Tertiare de Vienne (Autriche). Gide et Comp. Paris, p. 271, pl. 16, figs. 19–21.

Spiroloculina excavata (d'Orbigny), AGIP SPA, 1982, Foraminiferi Padani, pl. VII, figs. 2, 2p.

Diagnosis: Test free, elongate to circular in side view, rectangular in transverse section; chambers distinct, six to seven in adult, quadrangular in section, gradually increasing in size, the outer carinate margins distinctly raised above the surface of the previous chambers producing an excavated or channeled appearance; sutures distinct, depressed; wall calcareous, porcellaneous; aperture terminal, slightly produced, an elongate opening with a short simple tooth at the inner margin.

Range: Late Eocene–Pleistocene.

Family: NODOSARIIDAE Ehrenberg, 1838
Genus: NODOSARIA Lamarck, 1812
Nodosaria radicula Linne, 1758
Plate 6, Figs. 19 and 20

Nodosaria radicula Linne, 1758. Taxonomic notes. In: Society of Economic Paleontologists and Mineralogists 1960, pl. LXI, p. 128, figs. 28–31.

Nodosaria radicula Linne, 1758. Treatise on Invertebrate paleontology (Loeblich and Tappan, 1964), p. C512, figure 400, 1a, 1b.

Nodosaria radicula Linne, 1758. AGIP SPA, 1982, Foraminifera Padani, pl. XIV figs. 2, 2a.

Diagnosis: Test free, uniserial, three chambers arranged in straight format; wall calcareous, finely perforate; aperture terminal, rounded or radiate.

Range: Middle Miocene–Pliocene.

Nodosaria raphanistrum Linne, 1758
Plate 7, Figs. 1 and 2

Nodosaria raphanistrum Linne, 1758. AGIP SPA, 1982, Foraminifera Padani, pl. XIV, figs. 3, 3bis.

Diagnosis: Test free, multilocular, rectilinear, rounded in section, sutures distinct and commonly perpendicular to the axis of the test, surface with longitudinal costae that terminates at the sutures; wall calcareous; aperture terminal, centrally radiate.

Range: Middle Miocene–Early Pleistocene.

Genus: DENTALINA Risso, 1826
Dentalina leguminiformis Batsch, 1791
Plate 7, Figs. 3 and 4

Dentalina leguminiformis Batsch, 1791. AGIP SPA, 1982, Foraminifera Padani, pl. LII, figs. 5, 5a, 5p, 5bis, 5bis-p.

Diagnosis: Test free, elongate, acute, uniserial, sutures commonly oblique, some straight, some curved, aperture radiate terminal, may be concentric or nearly central. Wall calcareous, hyaline, finely perforate.

Range: Miocene–Pleistocene.

Family: EOUVIGERINIDAE Cushman, 1927
Genus: STILOSTOMELLA Guppy, 1927
Stilostomella advena Cushman and Laiming, 1931
Plate 7, Fig. 5

Stilostomella advena Cushman and Laiming, 1931. AGIP SPA, 1982, Foraminifera Padani, pl. XXXV, figs. 5, 5a.

Diagnosis: Test free, elongate, uniserial, and rectilinear, with gradually enlarging subglobular chambers; wall calcareous, hyaline, finely perforate, surface smooth; aperture terminal and rounded.
Range: Middle Miocene–Middle Pliocene.

Stilostomella monilis Silvestri, 1872
Plate 7, Figs. 6 and 7

Stilostomella monilis Silvestri, 1872. AGIP SPA, 1982, Foraminifera Padani, pl. XXXVI, figs. 3, 3a, 4.

Diagnosis: Test free, elongate, uniserial and rectilinear with gradually enlarging subglobular chambers; wall calcareous hyaline, finely perforate, surface smooth; aperture terminal, produced on a short neck with phialine lip and slight indentation at one side owing to surface reflection of internal spatulate tooth.
Range: Pliocene–Early Pleistocene.

Genus: LAGENA Walker and Jacob, 1798
Lagena striata d'Orbigny, 1839
Plate 7, Fig. 8

Oolina striata d'Orbigny, 1839, Voy. Amer. Merid. v. 5, pt. 5, p. 21, pl. 5, fig. 12.
Lagena striata (d'Orbigny), Cushman and Stainforth, 1945, The Foraminifera of the Cipero Marl Formation of Trinidad, British West Indies. Cushman Laboratory for Foraminiferal Research, Special Publication No. 14, p. 29, pl. 4, fig. 15.
Lagena striata Boltovskoy et al., 1980, Atlas of Benthic Foraminifera of the Southwest Atlantic. Dr. W. Junk b.v., The Hague, p. 98, pl. 20, figs 11–14.
Lagena striata AGIP SPA, 1982, Foraminiferi Padani, pl. XIX, figs. 5, 5a.

Diagnosis: Test free, unilocular, spherical, circular in transverse section; wall calcareous, hyaline, finely perforate, ornamented with longitudinal costae which extend from the base of the test to the periphery of the smooth neck, aperture terminal, a simple circular opening at the tip of the neck.
Range: Eocene–Pleistocene.

Subfamily: OOLININAE Loeblich and Tappan, 1961
Genus: OOLINA d'Orbigny, 1839
Oolina faveolata Seguenza, 1862
Plate 7, Fig. 9

Oolina faveolata Seguenza, 1862. AGIP SPA, 1982, Foraminifera Padani, pl. XVII, figs. 6, 6a.

Diagnosis: Test single globular to ovate chamber, surface ornamented with reticulations; wall calcareous, coarsely perforate; aperture rounded and terminal.

Range: Middle Miocene–Early Pleistocene.

Family: NONIONIDAE Schultze, 1854
Genus: ELPHIDIUM Montfort, 1808
Elphidium incertum Williamson,1858
Plate 7, Figs. 10 and 11.

Elphidium incertum (Williamson). Costello, 2001, European Register of Marine Species: Collection patrimonies Naturels, 50, pp. 60–75.

Elphidium incertum (Williamson). Kohl, 1985 Bull. Am. Paleontol., 88 (322), 143, pl. 28, fig. 6.

Diagnosis: Test free, planispiral, bilaterally symmetrical, involute, periphery rounded, chambers distinct about 10–12 in the final whorl, slightly inflated increasing in size gradually as added; umbilical area depressed with a central boss, sutures depressed, slightly curved with tubular perforations; wall calcareous, hyaline, smooth, finely perforate; aperture a row of pores at the base of septa face.

Range: Miocene–Pleistocene.

Elphidium flexuosum d'Orbigny, 1846
Plate 7, Figs. 12 and 13

Elphidium flexuosum d'Orbigny. AGIP SPA, 1982, Foraminiferi Padani, pl. xxii, figs. 3, 3a.

Diagnosis: Test free, planispiral, bilaterally symmetrical, involute, slightly compressed, periphery acute to subacute, narrowly keeled, slightly lobulate, umbilical area depressed with a central boss; chambers distinct, about 15–16 in the final whorl, increasing in size gradually as added; sutures depressed, slightly curved; wall calcareous, hyaline, smooth, finely perforate; aperture a series of small pores at the base of septal face, earlier chambers with areal pores.

Range: Miocene–Pleistocene.

Elphidium advenum Cushman, 1922
Plate 7, Fig. 14

Polystoma advena Cushman, 1922d, p. 56, figs. 11–12.

Elphidium advenum (Cushman) Cushman and Cahill, 1933, p. 22, pl. 7, figs. 10a,b.

Elphidium advenum Cushman, 1939, A Monograph of the foraminiferal family Nonionidae. United States Geological Survey, Professional Paper, No. 191, p. 60, pl. 16, figs. 31–35.

Diagnosis: Test free, planispiral, bilaterally symmetrical, involute, slightly compressed; periphery acute to subacute, narrowly keeled, slightly lobulate; umbilical area depressed with a central boss, with numerous retral processes along septal border; chambers distinct, about 10 in the final whorl; slightly inflated, increasing in size gradually as added; sutures depressed, slightly curved to radial, obscured by retral processes; wall calcareous, hyaline, smooth, finely perforate; aperture a series of small pores at base of septal face, earlier chambers with areal pores.

Range: Miocene–Pleistocene.

Genus: CRIBROELPHIDIUM Cushman and Bronnimann, 1948
Cribroelphidium decipiens Costa
Plate 7, Figs. 15 and 16

Cribroelphidium decipiens Costa, AGIP SPA, 1982, Foraminiferi Padani, pl. xxii, figs. 2, 2a.

Diagnosis: Test free, panispiral, bilaterally symmetrical, involute; chamber distinct, about 10–12 in the final whorl, slightly inflated, increasing in size gradually as added; sutures depressed slightly curved to radial, obscured by retral processes; wall calcareous, hyaline, smooth, finely perforate; apertures one or more at the base of the apertural face with a series of supplementary apertures consisting of small rounded openings.
Range: Miocene–Pleistocene.

Florilus atlanticus Cushman, 1947
Plate 7, Figs. 17 and 18

Nonionella atlanticus Cushman, 1947, New species and varieties of Foraminifera from the southeastern Coast of the United States. Contrib. Cushman Lab. Foramin. Res., 23, 90, pl. 20, figs. 4–5.

Nonionella atlanticus Cushman, 1947, Hansen and Lykke-Andersen, 1976, Wall structure and Classification of fossil and recent elphidiid and nonionid formainifera. Fossil and Strata, Oslo, No. 10, p. 23, pl. 21, figs. 9–12.

Pseudononion atlantica (Cushman, 1947) Andersen, 1961, Genesis and paleontology of the Mississippi River Mudlumps, Lower Mississippi River Delta, Louisiana. Louisiana Geological Society, Bulletin 35, p. 84, pl. 18, figs. 1a,b, 2a–c.

Florilus atlanticus (Cushman, 1947) Akers and Dorman, 1964, Pleistocene foraminifera of the Gulf Coast. Tulane Stud. Geol., 3 (1), 34, pl. 6, figs. 26, 27.

Florilus atlanticus (Cushman, 1947) Kohl, 1985, Early Pliocene benthic forminifers from the Salina Basin, Southeastern Mexico. Bull. Am. Paleontol. 88 (322), 147, pl. 32, figs. 2a–c.

Nonion 3: (*Florilus atlanticus* Cushman) SPDC Catalogue, 1980, Stereoscan Photos, Paleontological Type Collection, Nigeria, pl. 6, fig. A3.

Diagnosis: Test free, compressed, asymmetrical, dorsal side showing earlier whorls which are covered on the ventral side, periphery rounded, umbilical region on ventral side slightly depressed, filled with granular skeletal material, which extends along the sutures on the ventral and dorsal sides; chambers distinct, inflated, 9–10 in final whorl, increasing in size as added; sutures curved, depressed, incised near umbilical area, filled with granular skeletal material; wall calcareous, hyaline, finely perforate, smooth except for areas near the sutures, umbilicus, and base of apertural face are papillate; aperture obscured by skeletal deposits, but is probably an interiomarginal opening.
Range: Early Miocene–Recent.

Florilus costiferum Cushman, 1926
Plate 7, Figs. 19 and 20

Nonionina costifera Cushman, 1926, Cushman Lab. Foram. Res. Contrib.1, (4), 90, pl. 13, fig. 2.

Nonion costiferum Cushman, 1939, A Monograph of the foraminiferal family Nonionidae. United States Geological Survey, Professional Paper, No. 191, p. 15, pl. 4, fig. 5.

Nonion costiferum Cushman, 1926, Kleinpell, 1938, Miocene Stratigraphy of California. American Association of Petroleum Geologists, Tulsa, OK, p. 229, pl. 15, figs. 13a, b.

Nonion costiferum Cushman, 1926, Fayose, 1970, Stratigraphical paleontology of Afowo-1 well, Southwestern Nigeria. J. Min. Geol. 5 (1 and 2), 32, pl. 10, fig. 3.

Nonion 4 and 6: (*Florilus* ex. gr. *costiferum* Cushman) SPDC Catalogue, 1980, Stereoscan Photos, Paleontological Type Collection, Nigeria, pl. 6, figs. A4 and A6.

Florilus costiferum (Cushman), AGIP SPA, 1988, Foraminifera of the Niger Delta, pl. 7, figs. 6a–c.

Florilus costiferum (Cushman), Ozumba and Amajor, 1999a, Evolutionary relationships in some benthonic foraminifera of the Middle to the Late Miocene, Niger Delta. NAPE Bull., 14 (2), 161, fig. 4.

Diagnosis: Test free, planispiral, asymmetrical, involute, subcircular in outline, compressed, periphery rounded, umbilical region slightly depressed, filled with big canonical boss at the center, surrounded by granular skeletal material (*Nonion*-4); while *Nonion*-6 has a center filled with granular skeletal material/retral process, chambers distinct, broad, 10–16 in final whorl, increasing gradually in size as added; sutures curved, raised, and incised near umbilical areas; wall calcareous, hyaline, finely perforate, smooth except for areas near the sutures, umbilicus, and base of apertural face papillate; aperture, a low arched slit at the base of the last chamber.

Range: Middle Miocene–Late Miocene.

Genus: FLORILUS de Montfort, 1808

Florilus boueanum d'Orbigny, 1846

Plate 8, Fig. 1

Nonionina boueana d'Orbigny, 1846, Ann. Sci. Nat. Paris, ser. 1, 7, 245–314; Atlas, pl. 10–17 Crochard (Paris).

Florilus boueanum d'Orbigny, 1846, Jenkins and Murray, 1989, Stratigraphical Atlas of Fossil Foraminifera, p. 476, pl. 9.7, figs. 7, 8.

Nonion 16: (*Florilus* ex. gr. *boueanum* d'Orbigny, 1846) SPDC Catalogue, 1980, Stereoscan Photos, Paleontological Type Collection, Nigeria, pl. 6, fig A12.

Florilus boueanum d'Orbigny, 1846, AGIP SPA, 1982, Foraminifera Padani, pl. XX, figs. 9a, 9.

Diagnosis: Test free, planispiral, involute on both sides; periphery rounded to angular. Chambers increase rapidly as added; 10–15 chambers in the last whorl, umbilical region slightly depressed, filled with granular skeletal materials which may extend slightly along the sutures; sutures radial to curved; wall calcareous, finely perforate, granular in structure. Aperture, a narrow, interiomarginal, equatorial opening.

Range: Early Miocene–Pleistocene.

Florilus insectus Schwager

Plate 8, Fig. 2

Nonion 5: (*Florilus insectus* Schwager) SPDC Catalogue, 1980, Stereoscan Photos, paleontological Type Collection, Nigeria, pl. 6, figures A5.

Diagnosis: Test free, small planispiral involute, slightly compressed, periphery rounded, peripheral outline lobulate, chambers inflated increasing gradually in size as added with a highly inflated ultimate chamber; sutures distinct, depressed slightly curved; wall calcareous, finely perforate, surface smooth; aperture an interiomarginal equatorial slit, umbilical–extraumbilical.

Range: Early Miocene–Pleistocene.

Genus: NONION de Montfort, 1808
Nonion centrosulcatum de Klasz, Le Calvez, and Rerat, 1964
Plate 8, Figs. 3–7

Nonion centrosulcatum de Klasz, Le Calvez, and Rerat, 1964, Deux Nouveaux genres de Foraminiferes du Gabon (Afrique Equatoriale). C.R. Somm. Soc. Geol. France, Fasc. 6, 194, figs. 2a–b, 3.

Nonion centrosulcatum AGIP SPA, 1988, Foraminifera of the Niger Delta, pl. 7, fig. 1.

Anomalina 24: (*Nonion centrosulcatum* de Klasz, Le Calvez, and Rerat, 1964), SPDC Catalogue, 1980, Stereoscan Photos, Paleontological Type Collection, Nigeria, pl. 9, figs. A5V, A5D.

Diagnosis: Test free, planispiral and involute, slightly compressed, biumbonate, periphery rounded, peripheral outline lobulate; chambers strongly inflated increasing gradually in size as added, with a highly inflated ultimate chamber; sutures distinct, depressed, radial, slightly curved; wall calcareous, finely perforate, granular in structure, surface smooth; aperture an interiomarginal, equatorial slit, umbilical–extraumbilical, extending under the lobe of the ultimate chamber.

Range: Middle Oligocene–Middle Miocene.

Nonion oyae Reyment, 1966
Plate 8, Fig. 8

Planulina oyae Reyment, 1966, p. 287, pl. 1, figs. 45–48. In: Okosun et al., 2015, Taxonomic notes on selected Late Paleocene to Early Eocene foraminifera from Eastern Benin Basin, Southwestern Nigeria, J. Basic Appl. Res. Int., 8 (1) p. 41–55, 2015.

Nonion 1:, SPDC Catalogue, 1980, Stereoscan Photos, Paleontological Type Collection, Nigeria, pl. 6, fig. A1.

Nonion oyae (Reyment), Petters, 1982, Central West Africa Cretaceous-Tertiary benthic foraminifera and stratigraphy. Palaeontographica Part A, 179 (1–3), pl. 2, figs. 6, 7, pp. 1–105.

Diagnosis: Test planispiral, slightly elongate, compressed; periphery rounded, slightly lobulate; chambers numerous, 11 in number, increasing gradually in size as added; sutures raised, curved, emanating from a big central calcareous boss; aperture interiomarginal.

Range: Paleocene–Late Eocene.

Nonion ecuadoranum Cushman and Stainforth, 1951
Plate 8, Fig. 9

Nonion ecuadoranum de Klasz, I., Le Calvez, Y., and Rerat, D. 1964b. Deux importantes especes de foraminiferes du Miocene inferieur de l'Afrique occidentale, C.R. Somm. Soc. Geol. France, 5, 194, figs. 2, 3.

Nonion ecuadoranum Cushman and Stainforth, 1951. Central West Africa Cretaceous-Tertiary benthic foraminifera and stratigraphy, p. 79, pl. 12, figs. 1 and 2.

Diagnosis: Test free, planispiral involute, biumbonate, periphery rounded, peripheral outline lobulate, chambers strongly inflated, increasing gradually in size as added, seven to eight chambers in the final whorl, with a highly inflated ultimate chamber; sutures distinct, depressed, radial, slightly curved; wall calcareous, finely perforate, granular in structure, surface smooth; aperture an interiomarginal, equatorial slit, umbilical–extraumbilical, extending under the lobe of the ultimate chamber.

Range: Middle Oligocene–Middle Miocene.

Family: ELPHIDIIDAE Galloway, 1933
Genus: CRIBRONONION Thalmann, 1947
Cribrononion texana Cushman
Plate 8, Fig. 10

Elphidium 1: (*Cribrononion* aff. *texana* Cushman) SPDC Catalogue, 1980, Stereoscan Photos, Paleontological Type Collection, Nigeria, pl. 7, fig. B1.

Diagnosis: Test planispiral, bilaterally symmetrical, involute; eight chambers, increasing gradually in size as added; sutures excavated; aperture at the base of the apertural face.

Range: Paleocene.

Genus: NONIONELLA Cushman, 1926
Nonionella sp. 1
Plate 8, Fig. 11

Diagnosis: Test free, trochospiral, slightly compressed, periphery rounded, spiral side partially evolute with umbonal boss, opposite side involute with final chamber overhanging umbilical region; wall calcareous, perforate, granular in structure; aperture interiomarginal, a low arch near periphery extending onto umbilical side.

Remarks: This species differs from *Nonionella danvillensis* in the absence of a distinct umbilical flap. It differs from species of *Nonion* in being asymmetrical and trochospiral and in possessing an asymmetrically developed final chamber.

Nonionella auris d'Orbigny, 1839
Plate 8, Figs. 12 and 13

Valvulina auris d'Orbigny, 1839. Voy. Amer. Merid., v. 5, pt. 5, p. 47, pl. 2, figs. 15–17.

Nonionella auris (d'Orbigny) 1839. Atlas of Benthic Shelf Foraminifera of the Southwest Atlantic 1980, pl. 22, figs. 10–12.

Anomalina 8: (*Nonionella auris* d'Orbigny) SPDC Catalogue, 1980, Stereoscan Photos, Paleontological Type Collection, Nigeria, pl. 8, figs. A5V, A5D, A6V, A6D.

Diagnosis: Test free, dorsal side trochospiral, periphery rounded, dorsal side partially evolute, ventral side planispiral, involute with final chamber overhanging umbilical region.

Chambers inflated, slightly curved umbilical region; 9–11 in final whorl, sutures very depressed. Wall calcareous, perforate, granular in structure; aperture interiomarginal, a low arch near periphery extending on to umbilical side.

Range: Oligocene–Miocene.

Nonionella danvillensis Howe and Wallace, 1932
Plate 8, Figs. 14–16

Nonionella danvillensis Howe and Wallace, 1932, A Monograph of the foraminiferal family Nonionidae, No. 191, Louisiana Department of Conservation, Geological Bulletin 2, p. 52, pl. 9, fig. 4.

Nonionella danvillensis (Howe and Wallace), Petters, 1982, Central West Africa Cretaceous-Tertiary benthic foraminifera and stratigraphy. Palaeontographica, Bd. 179, Abt. A, p. 103, pl. 12, fig. 4.

Nonionella cf. *danvillensis* Howe and Wallace, AGIP SPA, 1988, Foraminifera of the Niger Delta, pl. 8, figs. 1a–b.

Diagnosis: Test free, trochospiral, periphery rounded, spiral side partially evolute, ventral side involute with final chamber overhanging umbilical region; chambers inflated, slightly curved, six in final whorl, increasing in size rapidly as added; sutures distinct, slightly curved, depressed; wall calcareous, hyaline, smooth, finely perforate; aperture an interiomarginal opening, umbilical to extraumbilical, extending under the lobe of the ultimate chamber.

Range: Middle Eocene–Early Miocene.

Genus: PULLENIA Parker and Jones, 1862
Pullenia bulloides d'Orbigny, 1846
Plate 8, Figs. 17–19

Nonionina bulloides d'Orbigny, 1846, Foraminiferes fossiles du Bassin Tertiare de Vienne (Autriche). Gide et Comp. Paris, p. 107, pl. 5, figs. 9, 10.

Pullenia bulloides (d'Orbigny, 1846), Galloway and Morrey, 1929, A lower Tertiary foraminiferal fauna from Manta, Ecuador. Bull. Am. Paleontol. 15 (55), 43, pl. 6, fig. 16.

Pullenia bulloides (d'Orbigny, 1846), Galloway and Hemingway, 1941, Tertiary foraminifera of Porto Rico New York Academy Sciences. Sci. Surv. Porto Rico Virgin-Islands, 3 (Pt. 4), 360, pl. 15, fig. 4.

Pullenia bulloides (d'Orbigny, 1846), Kohl, 1985, Early Pliocene benthic forminifers from the Salina Basin, Southeastern Mexico. Bull. Am. Paleontol. 88 (322), 147, pl. 32, fig. 5.

Pullenia bulloides AGIP SPA, 1982, Foraminiferi Padani, pl. XLIV, figs. 8, 8a.

Pullenia 3: (*Pullenia bulloides* d'Orbigny, 1846) SPDC Catalogue, 1980, Stereoscan Photos, Paleontological Type Collection, Nigeria, pl. 12, fig. C1.

Diagnosis: Test free, planispiral, involute, spheroidal, circular in side view, ovate in apertural view; chambers distinct, inflated, four to five in the final whorl increasing gradually in size as added; sutures distinct, flush or slightly depressed between the last few chambers; wall calcareous, hyaline, very finely perforate; aperture a narrow crescentic interiomarginal slit, extending to the umbilical region.

Range: Early Oligocene–Pleistocene.

Pullenia quinqueloba Reuss, 1851
Plate 8, Fig. 20

Pullenia quinqueloba Reuss, 1851. Taxonomic notes. In: Society of Economic Paleontologists and Mineralogists, 1960, p. 174, pl. LXXXIV, figs. 14a, 14b, 15.
Pullenia quinqueloba Reuss, 1851. AGIP SPA, 1982, Foraminifera Padani, pl. XLIV, figs. 10, 10a.
Pullenia quinqueloba Reuss, 1851. Bull. Am. Paleontol., 1985, 88 (322), 147, pl. 32, figs. 6a, 6b.
Pullenia quinqueloba Reuss, 1851. Benthic foraminiferal biostratigraphy of the South Caribbean region, Bolli, Beckmann, and Saunders, Cambridge University Press, 1994, p. 214, figs. 31–32.

Diagnosis: Test free, sphaeroidal to compressed, planispiral involute, ovate in apertural view; chambers distinct, four to five in the final whorl, increasing gradually in size as added; sutures distinct, flushed or slightly depressed between the last few chambers; wall calcareous, finely perforate, granular in structure; aperture a narrow, crescentic interiomarginal slit extending nearly from umbilicus on one side to that opposite.
Range: Eocene–Pleistocene.

Pullenia salisburyi Stewart, 1930
Plate 9, Fig. 1

Pullenia salisburyi Stewart, 1930. AGIP SPA, 1982, Foraminifera Padani, pl. XLV, figs. 1, 1a.
Pullenia salisburyi Stewart, 1930. Bull. Am. Paleontol., 1985, 88 (322), 147, pl. 32, figs. 7a, 7b.

Diagnosis: Test free, sphaeroidal to highly compressed, planispiral involute, ovate in apertural view; chambers distinct, six to seven in the final whorl, increasing gradually in size as added; sutures distinct, flushed or slightly depressed between the last few chambers; wall calcareous, finely perforate, granular in structure; aperture a narrow, crescentic interiomarginal slit extending nearly from umbilicus on one side to that opposite.
Range: Late Oligocene–Late Pliocene.

Family: ALABAMINIDAE Hofker, 1951
Genus: GYROIDINA d'Orbigny, 1826
Gyroidina soldanii d'Orbigny, 1826
Plate 9, Figs. 2 and 3

Gyroidina soldanii d'Orbigny, 1826, Modeles, no. 36, p. 278, no. 5.
Gyroidina soldanii d'Orbigny, 1826, Petters, 1982, Central West African Cretaceous-Tertiary benthic foraminifera and stratigraphy, 81, pl. 8, figs. 14, 20.
Gyroidina soldani d'Orbigny, AGIP SPA. 1982, Foraminifera Padani, pl. XL, figs. 1v, 1p, 1d.

Diagnosis: Test free, trochospiral, periphery rounded, to subtruncate, spiral side flattened with all chambers visible, opposite side elevated and umbilicate with only chambers of final whorl visible; chambers rhomboidal in section, suture radial to oblique, flush to depressed; wall calcareous, perforate, granular in structure; primary aperture a low interiomarginal slit restricted to midportion of apertural face, bordered by narrow lip, with small secondary apertures umbilical in position.
Range: Miocene–Pleistocene.

Genus: GYROIDINA d'Orbigny, 1826
Gyroidina orbicularis d'Orbigny, 1826
Plate 9, Figs. 4–8

Gyroidina orbicularis d'Orbigny, 1826. Treatise of Invertebrate Paleontology (Loeblich and Tappan, 1964), p. C749, figs. 5a–5c, 6.

Gyroidina orbicularis d'Orbigny, 1826. Bull. Am. Paleontol., 1985, 88 (322), 148, pl. 33, figs. 1a–1c, 2a–2b.

Diagnosis: Test free, trochospiral, periphery rounded to subtruncate, spiral side flattened with all chambers visible, opposite side elevated and umbilicate with only chambers of the final whorl visible, chambers rhomboidal in section with angled umbilical shoulder; sutures radial to oblique, flush to depressed; wall calcareous, perforate, granular in structure; primary aperture a low interiomarginal slit, restricted to midportion of apertural face, bordered by narrow lip, previous chamber whorl with projecting umbilical flap extending backward over it.

Range: Eocene–Pleistocene.

Family: *Osangulariidae* Brotzen, 1940
Genus: GYROIDINOIDES Brotzen, 1942
Gyroidinoides neosoldanii Brotzen, 1936
Plate 9, Figs. 9 and 10

Gyroidinoides neosoldanii (Brotzen, 1936) Gross, 2001. (Accepted name *Gyroidina neosoldanii* Brotzen, 1936), foraminifera. In: Costello, M.J., et al. (Ed. 2001). Collection Partrimoines Naturels, 50, p. 60–75

Gyroidinoides neosoldanii Brotzen, 1936, AGIP SPA, 1982, Foraminifera Padani, pl. xxxix, figs. 10v, 10p, 10d.

Diagnosis: Test free, trochospiral, spiral side flattened, umbilical side elevated, periphery rounded; chambers rhomboidal in section, sutures radial to curved, flush to depressed; wall calcareous, perforate, bilamellar, granular in structure; aperture a continuous, low interiomarginal slit extending from periphery to umbilicus, umbilical portion partially obscured by umbilical flap from each chamber.

Range: Early Miocene–Pliocene.

Family: ALABAMINIDAE Hofker, 1951
Genus: ORIDORSALIS Andersen, 1961
Oridorsalis umbonatus Reuss, 1851
Plate 9, Figs. 11–13

Oridorsalis umbonatus Reuss, 1851. AGIP SPA, 1982, Foraminifera Padani, pl. XL, figs. 10v, 10d, 10p.

Oridorsalis umbonatus Reuss, 1851. Bull. Am. Paleontol., 1985, 88 (322), 148, pl. 33, figs. 6a–6d.

Oridorsalis umbonatus Reuss, 1851. Benthic foraminiferal biostratigraphy of the South Caribbean region, Bolli, Beckmann, and Saunders, Cambridge University Press, 1994, p. 246, fig. 58, 10a–10c, 11a, 11b, 12, 13a, 13b.

Diagnosis: Test free, lenticular, periphery carinate, chambers arranged in low trochospiral coil, chambers broad, low, all visible in spiral side but only those of final whorl visible on the ventral side; sutures radial, slightly curved on spiral side, strongly sinuate on the umbilical side; wall calcareous, finely perforate, granular in structure; primary aperture interiomarginal, extending from periphery nearly to close umbilicus of umbilical side, small secondary sutural openings on spiral side.

Range: Eocene–Pleistocene.

Family: SPHAEROIDINIDAE Cushman, 1927
Genus: SPHAEROIDINA d'Orbigny, 1826
Sphaeroidina bulloides d'Orbigny, 1826
Plate 9, Fig. 14

Sphaeroidina bulloides d'Orbigny, 1826, Tableau methodique de la classe des cephalopodes. Ann. Sci. Nat. Paris, ser. 1, 7, 267, models no. 65.

Sphaeroidina bulloides Cushman, 1924, The foraminifera of the Atlantic Ocean, Part 5—Chilostomellidae and Globigerinidae. US Natl. Museum Bull. 104 (Pt. 5), 36, pl. 7, figs. 1–6.

Sphaeroidina bulloides Cushman and Todd, 1945, Miocene foraminifera from Buff Bay, Jamaica. Cushman Laboratory for Foraminiferal Research, Special Publication No. 15, p. 65, pl. 11, fig. 9.

Sphaeroidina bulloides Cushman and Todd, Kohl, 1985, Early Pliocene benthic forminifers from the Salina Basin, Southeastern Mexico. Bull. Am. Paleontol. 88 (322), 129, pl. 14, fig. 6.

Sphaeroidina bulloides Cushman and Todd, AGIP SPA, 1982, Foraminiferi Padani, pl. XLV, figs. 2, 2a, 2p.

Sphaeroidina 1: (*Sphaeroidina bulloides* d'Orbigny) SPDC Catalogue, 1980, Stereoscan Photos, Paleontological Type Collection, Nigeria, pl. 12, figs. AIV, A1D.

Diagnosis: Test free, subglobular, early portion trochospiral, later whorls streptospiral; chambers inflated, distinct, hemispherical, rapidly increasing in size as added, embracing most of the preceding ones, four to six chambers visible, final chamber comprising almost one-half the size of the test; sutures distinct, depressed; wall calcareous, hyaline, smooth, finely perforate; aperture interiomarginal, a crescentic slit at the basal margin of the final chamber, with a lip and a flat tooth.

Range: Early Oligocene–Pleistocene.

Genus: SARACENARIA Defrance, 1824
Saracenaria nuttalli Sansores and Flores-Covarrubias, 1972
Plate 9, Fig. 15

Saracenaria nuttalli Sansores and Flores—Covarrubias 1972. Bull. Am. Paleontol. 1985, 88 (no. 322), 126, pl. 11, figs. 7a–7c.

Diagnosis: Test free, planispiral in early stage, later with tendency to uncoil; triangular in section, with broad, flat apertural face, the outer margin and two angles of face may be acute or rounded; aperture at peripheral angle radiate.

Range: Pliocene.

Genus: LATICARININA Galloway and Wissler, 1927
Laticarinina pauperata Parker and Jones, 1865
Plate 9, Figs. 16 and 17

Laticarinina pauperata Parker and Jones, 1865. Taxonomic notes. In: Society of Economic Paleontologists and Mineralogists, 1960, p. 214, pl. CIV, figs. 3–11.

Laticarinina pauperata Parker and Jones, 1865. Treatise in Invertebrate Paleontology (Loeblich and Tappan, 1964), p. C581, figs. 2, 3a, 3b, 4a, 4b, and 4c.

Laticarinina pauperata Parker and Jones, 1865. AGIP SPA,1982, Foraminifera Padani, pl. L, figs. 8d, 8p, 8bis-v.

Laticarinina pauperata Parker and Jones, 1865. Bull. Am. Paleontol., 1985, 88 (322), 141, pl. 26, figs. 1a–1d.

Diagnosis: Test free, planispiral, broad peripheral keel; chambers saddle-shaped, anterior margin of keel forming separation between two lobes of next developed chamber, lobes larger on one side, and closely appressed, final chamber with broad attachment flange somewhat loosely attached at posterior umbilical margin; wall calcareous, finely perforate, keel imperforate; peripheral aperture at one side of keel, low slit perpendicular to periphery with supplementary openings beneath posterior umbilical margin of smaller lobes of later chambers beneath the umbilical chamber flaps.

Range: Middle Oligocene–Early Pliocene.

Genus: RECTOGLANDULINA Loeblich and Tappan, 1955
Rectoglandulina comatula Cushman, 1923
Plate 9, Figs. 18 and 19

Nodosaria comatula Cushman, 1923, The foraminifera of the Atlantic Ocean. US Natl. Museum Bull. 104 (pt. 4), 83, pl. 14, fig. 5.

Pseudoglandulina comatula (Cushman, 1923), Hedberg, 1937, Foraminifera of the Middle Tertiary Carapita Formation of northeastern Venezuela, J. Paleontol., 11 (8), 673, pl. 91, figs. 9–10.

Pseudoglandulina comatula (Cushman, 1923), Cushman and Todd, 1945, Miocene foraminifera from Buff Bay, Jamaica. Cushman Laboratory for Foraminiferal Research, Special Publication No. 15, p. 29, pl. 4, fig. 25.

Pseudoglandulina comatula (Cushman, 1923), Todd and Low, 1976, Smaller foraminifera from deep wells on Puerto Rico and St. Croix. United States Geological Survey, Professional Paper No. 863, p. 24, pl. 3, fig. 18.

Rectoglandulina comatula (Cushman, 1923), Barker, 1960, Taxonomic notes on the species figured by H. B. Brady in his report on the foraminifera dredged by H.M.S Challenger during the years 1873–1876. Society of Economic Paleontologist and Mineralogist, Special Publication No. 9, p. 134, pl. 64, figs. 1–5.

Nodosaria 9: (*Rectoglandulina comatula* Cushman) SPDC Catalogue, 1980, Stereoscan Photos, Paleontological Type Collection, Nigeria, pl. 33, fig. A2.

Diagnosis: Test free, elongate, uniserial, rectilinear, periphery slightly lobate, circular in transverse section; chambers inflated, strongly embracing, three to five chambers in the adult; sutures distinct, somewhat depressed, horizontal; wall calcareous, hyaline, finely perforate, ornamented with longitudinal costae, which are continuous from one chamber to another; aperture terminal, radiate.

Range: Middle Miocene–Early Pliocene.

Genus: MARGINULINA d'Orbigny, 1826
Marginulina costata Batsch, 1791
Plate 9, Fig. 20

Marginulina costata Batsch, 1791, Testaceorum arenulae marinae tabulae sex—(Sechs Kupfertafeln mit Conchylien des Seesandes, gezeichnet undgestochen). Jena, 6, p. 1, 4, pl. 1, figs. 1a–g.

Marginulina costata Schroeter, 1803, Archiv. Zool. Zoot., Braunschweig, Part 2, p. 63, 66.

Marginulina costata AGIP SPA, 1982, Foraminiferi Padani, pl. XII, figs. 2, 2a, 2p, 3, 3p.

Diagnosis: Test free, elongate, initial portion slightly coiled, later portion rectilinear, circular in transverse section, periphery broadly rounded; chambers inflated, strongly embracing, 8–10 in number; sutures depressed, somewhat oblique; wall calcareous, finely perforate, ornamented with longitudinal costae which run continuously along the entire length of the test; aperture terminal, radiate, at the end of a short neck.

Range: Middle Miocene–Early Pliocene.

Family: NODOSARIIDAE Ehrenberg, 1838
Genus: AMPHICORYNA Schlumberger, 1881
Amphicoryna scalaris caudata Silvestri, 1791; Boltovsky, 1980
Plate 10, Fig. 1

Nautilus (*Orthoceras*) *scalaris* Batsch, 1791, Testaceorum arenulae marinae tabulae sex—(Sechs Kupfertafeln mit Conchylien des Seesandes, gezeichnet undgestochen). Conchyl. Seesand, p. 1, pl. 2, fig. 4.

Amphicoryna scalaris (Batsch) Boltovskoy et al., 1980, Atlas of Benthic Foraminifera of the Southwest Atlantic. Dr. W. Junk b.v., The Hague, p. 60, pl. 1, figs. 10–12.

Nodosaria 9B: (*Amphicoryna scalaris caudata* Silvestri), SPDC Catalogue, 1980, Stereoscan Photos, Paleontological Type Collection, Nigeria, pl. 33, fig. A3.

Diagnosis: Test free, uniserial, initial chamber with a short spine, three to four globose chambers, increasing in size as added; sutures depressed and very visible; wall calcareous, hyaline, finely perforate, ornamented by long longitudinal costae running from the base to the neck; aperture terminal, a circular opening at the end of the neck with a lip.

Range: Oligocene–Early Pliocene.

Family: BOLIVINITIDAE Cushman, 1927
Genus: ALTISTOMA de Klasz and Rerat, 1962
Altistoma tenuis de Klasz and Rerat, 1962
Plate 10, Figs. 2 and 3

Altistoma tenuis de Klasz and Rerat, 1962, Quelques nouveaux foraminiferes du Cretace et du Tertiare du Gabon (Afrique Equatoriale). Rev. Micropaleontol. 4, 180, pl. 1, figs. 6a,b.

Virgulina 3: (*Altistoma tenuis* de Klasz and Rerat, 1962), SPDC Catalogue, 1980, Stereoscan Photos, Paleontological Type Collection, Nigeria, pl. 28, fig. A4.

Altistoma cf. *A. tenuis* de Klasz and Rerat, AGIP SPA, 1988, Foraminifera of the Niger Delta, pl. 10, fig. 2.

Diagnosis: Test free, biserial, strongly overlapping, with slightly twisted lower margin; sutures strongly oblique; wall calcareous, finely peforate, surface smooth; aperture arched, bordered by thickened lip.

Range: Middle Eocene–Early Miocene.

Altistoma scalaris de Klasz and Rerat, 1962
Plate 10, Fig. 4

Altistoma scalaris de Klasz and Rerat, 1962, Quelques nouveaux foraminiferes du Cretace et du Tertiare du Gabon (Afrique Equatoriale). Rev. Micropaleontol. 4, 180, pl. 1, fig. 5.

Altistoma scalaris (de Klasz and Rerat) Petters, 1982, Central West Africa Cretaceous-Tertiary benthic foraminifera and stratigraphy. Palaeontographica Part A, 179 (1–3), pl. 10, figs. 26, 27.

Altistoma scalaris AGIP SPA, 1988, Foraminifera of the Niger Delta, pl. 10, fig. 3.

Virgulina 4 and 11: (*Altistoma scalaris* de Klasz and Rerat), SPDC Catalogue, 1980, Stereoscan Photos, Paleontological Type Collection, Nigeria, pl. 28, figs. A5, A10.

Diagnosis: Test free, biserial with strongly overlapping chambers with lobulate lower margin; suture depressed; wall calcareous, finely perforate, surface smooth; aperture large, forming a highly symmetrical arch bordered by thickened lip.

Range: Middle Eocene–Early Miocene.

Genus: BOLIVINA d'Orbigny, 1839
Bolivina afra Reyment, 1959
Plate 10, Fig. 7

Afrobolivina afra Reyment, 1959, The foraminiferal genus *Afrobolivina* gen. nov. and *Bolivina* in the Upper Cretaceous and Lower Tertiary of West Africa. Stockh. Contr. Geol., 3 (1), pls. I, II, III, IV, figs. 1a,b, pl. V, figs. 1–3, 5, 6, 7.

Bolivina afra (Reyment) Fayose, 1970, Stratigraphical paleontology of Afowo-1 well, Southwestern Nigeria. J. Min. Geol. 5, pl. 5, fig. 9.

Bolivina afra (Reyment) Petters, 1982, Central West Africa Cretaceous-Tertiary benthic foraminifera and stratigraphy. Palaeontographica Part A, 179 (1–3), pl. 11, fig. 33, 34.

Diagnosis: Test elongate, with almost parallel sides, greatest width formed by the last pair of chambers; chambers inflated, 14 in number, increasing gradually in size as added; chamber overlaps, most prominent near the aperture; ornament, consisting of strong longitudinal costae which are continuous over the suture and anastomosing along the midline; aperture large, nearly oval, on the final chamber.

Range: Paleocene.

Bolivina africana Graham, de Klasz, and Rerat, 1965
Plate 10, Fig. 8

Bolivina explicata Cushman and Hedberg *africana* Graham, de Klasz, and Rerat, 1965, Quelques importants Foraminiferes du Tertiaire du Gabon (Afrique Equatoriale). Rev. Micropaleontol. 4, 72, pl. 1, fig. 6 a,b.

Bolivina africana Petters, 1982, Central West Africa Cretaceous-Tertiary benthic foraminifera and stratigraphy. Palaeontographica, Bd. 179, Abs. A, nos. 1–3, p. 59, pl. 11, figs. 27–28.

Bolivina explicata africana Brun et al., 1984, Evolution and morphological variations of the principal species of Bolivinitidae in the Tertiary of the Gulf of Guinea. Geologie méditerranée Tome XI, no. 1, 1984, p. 24, pl. 1, figs. 2 a–l.

Diagnosis: Test elongate subcylindrical in outline; gradually tapering toward the umbilicus, slightly compressed; chambers low, vaguely distinct, sutures, not visible in the early part of the test, slightly depressed in later portions, crenulate with globular extensions over previous chambers; wall coarsely perforate; aperture a broad slit, terminal.

Range: Paleocene.

Bolivina beyrichi Reuss, 1851
Plate 10, Fig. 9

Bolivina beyrichi Reuss, 1851, Ueber die fossilen Foraminiferen un Entomostraceen der Septarienthone der Umgegend von Berlin. Deutsche geologische Gessellschaft, Zeitschrift 3, 83, pl. 6, fig. 51.

Bolivina 26: (*Brizalina beyrichi* Reuss), SPDC Catalogue, 1980, Stereoscan Photos, Paleontological Type Collection, Nigeria, pl. 26, fig. A19.

Brizalina beyrichi Petters, 1982, Central West Africa Cretaceous-Tertiary benthic foraminifera and stratigraphy. Palaeontographica, Bd. 179, Abs. A, nos. 1–3, p. 59, pl. 11, fig. 13.

Diagnosis: Test free, biserial, compressed, periphery carinate with downward-curving projections at the basal portion of each chamber; 10 pairs of chambers in the adult, increasing gradually in size as added, somewhat inflated; sutures distinct, curved, slightly depressed; wall calcareous, hyaline, smooth, finely perforate; aperture loop-shaped, extending up from the base of the final chamber with a small tooth.

Range: Early Miocene–Pliocene.

Bolivina denticulocamerata Graham, de Klasz, and Rerat, 1965
Plate 10, Fig. 10

Eouvigerina denticulocamerata Graham, de Klasz, and Rerat, 1965, Quelques importants Foraminiferes du Tertiaire du Gabon (Afrique Equatoriale). Rev. Micropaleontol. 4, 74, pl. 1, figs. 12, 13.

Uvigerinella 6: SPDC Catalogue, 1980, Stereoscan Photos, Paleontological Type Collection, Nigeria, pl. 30, fig. A6.

Bolivina denticulocamerata Petters, 1982, Central West Africa Cretaceous-Tertiary benthic foraminifera and stratigraphy. Palaeontographica, Bd. 179, Abs. A, nos. 1–3, p. 60, pl. 7, fig. 21.

Diagnosis: Test elongate, biserial, tapering gradually to the proloculus, twisted in the early part; chambers greater in width than height in the early portion of the test, later becoming narrow and high, overlapping with distinct posterior chamber extensions; sutures horizontal, slightly depressed, axially zigzag; wall smooth; aperture, a wide loop along apertural face.

Range: Eocene.

Bolivina dertonensis Gianotti, 1953
Plate 10, Figs. 11–13

Bolivina dertonensis Gianotti, 1953, Microfaune della serie Tortoniana del Rio Mazzapiedi Castellania (Tortona—Alessandria). Riv. Ital. Paleontol. Stratigr. 6, 40 (1), p. 42, figs. 1–7.

Bolivina 27: (*Bolivina dertonensis* Gianotti, 1953), SPDC Catalogue, 1980, Stereoscan Photos, Paleontological Type Collection, Nigeria, Pl. 27, fig. A2.

Brizalina paralica (Perconig), AGIP SPA, 1988, Foraminifera of the Niger Delta, pl. 14, fig. 2.

Diagnosis: Test biserial, tapering; chambers 15 in number, increasing gradually in size as added, overlapping and ending in some cases in blunt spines; sutures depressed, gently curved in the early portion of the test, becoming strongly curved toward the apertural end; aperture an opening at the base of the last chamber; test ornamented with longitudinal striae.

Range: Late Oligocene–Early Miocene.

Bolivina ihuoensis Reyment, 1959
Plate 10, Figs. 14 and 15

Bolivina ihuoensis Reyment, 1959, The foraminiferal genus *Afrobolivina* gen. nov. and *Bolivina* in the Upper Cretaceous and Lower Tertiary of West Africa. Stockh. Contr. Geol. 3 (1), 43, pl. 7, figs. 7–10.

Bolivina 12A, SPDC Catalogue, 1980, Stereoscan Photos, Paleontological Type Collection, Nigeria, pl. 26, fig. A9.

Brizalina ihuoensis Brun et al., 1984, Evolution and morphological variations of the principal species of Bolivinitidae in the Tertiary of the Gulf of Guinea. Geologie Méditerranée Tome XI, No. 1, p. 28, pl. IV, figs. 10 a,b, pl. 5, fig. 10.

Brizalina ihuoensis Brun et al., 1984, AGIP SPA,1988, Foraminifera of the Niger Delta. (S. Donato Milanese). pl. 13, figs. 2 a–k.

Diagnosis: Test compressed, with almost parallel sides, periphery keeled; ornamented with sharp bifurcating ribs that run from the proloculus to the penultimate chamber; biserial throughout; sutures distinct, arched; aperture a loop at the base of the last chamber.

Range: Paleocene–Late Eocene.

Bolivina interjuncta Cushman and Renz, 1941
Plate 10, Fig. 16

Bolivina interjuncta Cushman var. *simplex* Cushman and Renz, 1941, New Oligocene–Miocene foraminifera from Venezuela. Contrib. Cushman Lab. Foramin. Res. 17 (pt. 1), 20, pl. 3, fig. 15.

Bolivina interjuncta Cushman var. *simplex* Cushman and Renz Okosun and Liebau, 1999, Foraminiferal biostratigraphy of eastern Niger Delta, Nigeria. NAPE Bull. 14, 148, pl. 1, figs. 1–6, 8–9.

Bolivina 25: (*Brizalina interjuncta* Cushman and Renz, 1980) SPDC Catalogue, 1980, Strereoscan photos, Paleontological Type Collection, Nigeria, pl. 26, fig. 17.

Bolivina simplex Petters, 1982, Central West Africa Cretaceous-Tertiary benthic foraminifera and stratigraphy. Palaeontographica, Bd. 179, Abs. A, no. 1–3, p. 62, pl. 10, figs. 6, 7.

Bolivina interjuncta Ozumba and Amajor, 1999a, Evolutionary relationships in some benthonic foraminifera of the Middle to the Late Miocene, Niger Delta. NAPE Bull. 14 (2), 162, fig. 5.

Diagnosis: Test compressed, kite-shaped, tapering toward the proloculus, giving a subacute initial end, slightly ovate at apertural end; periphery acute in last-formed chambers where it is somewhat rounded; 20 chambers increasing gradually in size as added; sutures, broad, limbate slightly curved; wall thin except at sutures where it is thickened and opaque so that they appear to be very white; aperture narrow, at the base of the last chamber.

Range: Early Miocene–Middle Miocene.

Bolivina mandoroveensis Graham, de Klasz, and Rerat, 1965
Plate 10, Figs. 17 and 18

Bolivina interjuncta (Cushman) var. *mandoroveensis* Graham, de Klasz, and Rerat, 1965, Quelques importants Foraminiferes du Tertiaire du Gabon (Afrique Equatoriale). Rev. Micropaleontol. 4 (2), 82, pl. 2, fig. 4.

Bolivina 25A: (*Brizalina mandoroveensis* Graham, de Klasz, and Rerat), SPDC Catalogue, 1980, Stereoscan Photos, Paleontological Type Collection, Nigeria, pl. 26, fig. A16.

Brizalina mandoroveensis (Graham, de Klasz and Rerat) Ozumba and Amajor, 1999a, Evolutionary relationships in some benthonic foraminifera of the Middle to the Late Miocene, Niger Delta. NAPE Bull. 14 (2), 162, fig. 5.

Diagnosis: Test biserial, elongate, tapering gradually from the aperture to the proloculus, slightly compressed with a subrounded periphery; sutures depressed, slightly curved creating an axial zigzag feature sandwiched between two widely spaced longitudinal costae, the rest of the test covered by longitudinal striae except for the final pair of chambers; seven pairs of chambers visible; aperture a loop on the apertural face.

Range: Early Miocene–Middle Miocene.

Bolivina miocenica Macfayden, 1930
Plate 10, Fig. 19

Bolivina scalprata Schwager var. *miocenica* Macfayden, 1930, Miocene Foraminifera from the Clysmic Area of Egypt and Sinai. Egyptian Geological Survey, p. 61, pl. 4, figs. 22a,b.

Bolivina scalprata Kohl, 1985, Early Pliocene benthic forminifers from the Salina Basin, Southeastern Mexico. Bull. Am. Paleontol. 88 (322), 132, pl. 17, figs. 2–3.

Bolivina 11 and 29: (*Bolivina scalprata miocenica* Macfayden, 1930), SPDC Catalogue, 1980, Stereoscan Photos, Paleontological Type Collection, Nigeria, pl. 26, fig. A11, pl. 27, fig. A4.

Bolivina scalprata miocenica Ozumba and Amajor, 1999a, Evolutionary relationships in some benthonic foraminifera of the Middle to the Late Miocene, Niger Delta. NAPE Bull. 14 (2), 162, fig. 5.

Diagnosis: Test free, compressed, biserial, twice as long as broad, initial end rounded, periphery subacute; chambers with nodes at the basal margins, distinct, with 9–10 pairs in the adult, chambers increase gradually in size as added; sutures limbate, raised, and strongly curved, giving the chambers a depressed appearance; wall calcareous, hyaline, coarsely perforate, ornamented by a raised median ridge that extends from the initial chambers to the penultimate chamber; aperture an elongate narrow slit in the plane of compression, extending from the base of the ultimate chamber to a terminal position with a toothplate.

Range: Early Miocene–Pleistocene.

Bolivina punctata d'Orbigny, 1839
Plate 10, Fig. 20

Bolivina punctata d'Orbigny, 1839, Foraminiferes des Iles Canaries. In: Barker-Webb, P., Berthelot, S., Historie Naturelle des Canaries. Paris, vol. 2, Part 2, p. 63, pl. 8, figs. 10–12.

Bolivina punctata Cushman, 1937b, A monograph of the foraminiferal family Valvulinidae Cushman Laboratory for Foraminiferal Research, Special Publication No. 8, p. 124, pl. 13, fig. 12.

Bolivina punctata AGIP SPA, 1982, Foraminiferi Padani, pl. XXX, figs. 9, 9a, 9p.

Diagnosis: Test free, biserial, elongate, 4 times as long as broad, rounded in transverse section, broadest near apertural end, initial end rounded; 10–12 pairs of chambers in the adult, slightly compressed, increasing gradually in size as added; sutures distinct, subcarinate, strongly oblique; wall calcareous, coarsely perforate; aperture a simple slit, extending from the base of the ultimate chamber to a terminal position, with a narrow tooth.

Range: Middle Miocene–Pliocene.

Bolivina scalprata Schwager, 1883
Plate 11, Figs. 1 and 2

Bolivina scalprata Schwager, 1883. Paleontographica, 30, 114, pl. 25 (6), fig. 12.

Bolivina scalprata Liebus, 1911, Sitz. Akad. Wiss. Wien, 120, 60, pl. 1, fig. 13.

Diagnosis: Test about twice as long as broad, much compressed, periphery sharply keeled, thickest along the median line, broadly fusiform in front view, greatest breadth at or slightly above the middle; chambers low and broad, increasing slightly in size as added; sutures fairly distinct, slightly limbate, strongly oblique and somewhat curved; wall finely perforate, strongly costate in the median axis on each face, becoming less costate toward apertural end; aperture narrow, elongate.

Range: Early Miocene–Pleistocene.

Bolivina simplex Cushman and Renz, 1941
Plate 11, Fig. 3

Bolivina interjuncta Cushman var. *simplex* Cushman and Renz, 1941, New Oligocene–Miocene foraminifera from Venezuela. Contrib. Cushman Lab. Foramin. Res., 17 (pt. 1), 20, pl. 3, fig. 15.

Bolivina 25, SPDC Catalogue, 1980, Stereoscan Photos, Paleontological Type Collection, Nigeria, pl. 26, fig. A17.

Bolivina simplex Petters, 1982, Central West Africa Cretaceous-Tertiary benthic foraminifera and stratigraphy. Palaeontographica, Bd. 179, Abs. A, no. 62, pl. 10, figs. 6, 7.

Diagnosis: Test compressed, kite-shaped, tapering toward the proloculus giving a subacute initial end, slightly ovate at apertural end; periphery acute in last-formed chambers where it is somewhat rounded; 20 chambers, increasing gradually in size as added; sutures, broad, limbate, slightly curved; wall thin except at sutures where it is thickened and opaque so that they appear very white; aperture narrow, at the base of the last chamber.

Range: Oligocene–Middle Miocene.

Bolivina tenuicostata de Klasz, Le Calvez, and Rerat, 1964
Plate 11, Figs. 4 and 5

Bolivina tenuicostata de Klasz, Le Calvez, and Rerat, 1964b, Deux importantes especes de foraminifers du Miocene inferieur de l'Afrique occidentale. C.R. Somm. Soc. Geol. France, Fasc. 5, 194, fig. 1.

Bolivina 38. SPDC Catalogue, 1980, Stereoscan Photos, Paleontological Type collection, Nigeria.

Bolivina tenuicostata Petters, 1982, Central West Africa Cretaceous-Tertiary benthic foraminifera and stratigraphy. Palaeontographica, Bd. 179, Abs. A, p. 62, pl. 6, figs. 12, 13.

Diagnosis: Test short, triserial, widest toward the proximal end; chambers inflated, 10 in number; sutures slightly curved, depressed; test covered with widely spaced longitudinal costae; aperture distinct, a loop in the apertural face.

Range: Early Eocene–Late Oligocene.

Genus: BRIZALINA Costa, 1856
Brizalina imperatrix Graham, de Klasz, and Rerat, 1965
Plate 11, Fig. 6

Bolivina imperatrix Graham, de Klasz, and Rerat, 1965, Quelques importants Foraminiferes du Tertiaire du Gabon (Afrique Equatoriale). Rev. Micropaleontol. 4, 78, pl. 2, figs. 3a,b.

Bolivina 12: (*Brizalina imperatrix* Graham, de Klasz, and Rerat), SPDC Catalogue, 1980, Stereoscan Photos, Paleontological Type Collection, Nigeria, pl. 26, fig. A8.

Brizalina imperatrix Graham, de Klasz, and Rerat, AGIP SPA, 1988, Foraminifera of the Niger Delta, pl. 13, figs. 3a–f.

Diagnosis: Test free, biserial, compressed, tapering, widest near apertural end, periphery subacute, carinate, initial chamber rounded; seven to nine pairs of chambers in the adult; chambers slightly inflated, increasing gradually in size as added; sutures distinct, slightly limbate and strongly oblique; wall calcareous, coarsely perforate, ornamented by many longitudinal narrow costae extending from the proloculus to the ultimate chamber; aperture elongate, extending from the base of the ultimate chamber to a terminal position, with a narrow toothplate.

Range: Early Oligocene–Late Oligocene.

Brizalina spinata Cushman, 1936
Plate 11, Fig. 7

Bolivina striatula Cushman var. *spinata* Cushman, 1936a, Some new species of *Nonion*. Contrib. Cushman Lab. Foramin. Res. 12, 59, pl. 8, figs. 9a,b.

Brizalina striatula spinata (Cushman) Phleger and Parker, 1951, Ecology of foraminifera of northwest Gulf of Mexico. Geological Society of America, Memoir 46, Part 2, p. 14, pl. 7, fig. 7.

Bolivina spinata (Cushman) Anderson, 1961, Genesis and paleontology of the Mississippi River Mudlumps, Lower Mississippi River Delta, Louisiana. Louisiana Geological Society, Bulletin 35, 1–208, 29 pls. p. 93, pl. 20, figs. 11a and b.

Brizalina spinata Cushman Akers and Dorman, 1964, Pleistocene foraminifera of the Gulf Coast. Tulane Stud. Geol., 3 (1), 26, pl. 8, fig.10.

Brizalina striatula spinata (Cushman) Kohl, 1985, Early Pliocene benthic forminifers from the Salina Basin, Southeastern Mexico. Bull. Am. Paleontol. 88 (322), 132, pl. 17, fig. 8.

Diagnosis: Test free, biserial, elongate, rounded in transverse section, broadest near apertural end, initial end with short spine; chambers distinct, usually 12–13 pairs, very slightly inflated, increasing gradually in size as added; sutures slightly limbate, curved, slightly depressed; wall calcareous hyaline, coarsely perforate, ornamented by numerous, longitudinal narrow costae extending from proloculus to almost two-thirds the length of the test; aperture elongate, extending from the base of the ultimate chamber to a terminal position.

Range: Pleistocene.

Brizalina aenariensis Costa, 1856
Plate 11, Figs. 9 and 10

Brizalina aenariensis Costa, 1856, Paleontologia del regno di Napoli, Parte II—Academia Pontaniana, Atti, Napoli, 7, 2, p. 297, pl. 15, figs. 1A, B.

Bolivina 46: (*Brizalina aenariensis* Costa) SPDC Catalogue, 1980, Stereoscan Photos, Palaeontological Type Collection, Nigeria, Pl. 27, fig. A15.

Brizalina aenariensis Costa, AGIP SPA, 1982, Foraminifera Padani, Pl. XXVIII, fig. 9.

Diagnosis: Test free, biserial, compressed, length twice the width; widest near apertural end, tapering toward the spiny initial end; periphery acute, carinate; seven to nine pairs of chambers in the adult; chambers narrow, slightly curved, increasing gradually in size as added; suture distinct, slightly depressed, strongly curved; wall calcareous, hyaline, finely perforate; test ornamented with two or more longitudinal costae, the two costae along the central axes extend from the initial whorls to two-third the length of the test, all other costae shorter, on either side of the two prominent costae; aperture loop shaped, extending up from the base of the final chamber, with a narrow toothplate.

Range: Middle Miocene–Late Miocene.

Brizalina isidroensis Cushman and Renz, 1941
Plate 11, Fig. 11

Brizalina isidroensis Cushman and Renz, 1941, New Oligocene–Miocene foraminifera from Venezuela. Contrib. Cushman Lab. Foramin. Res. 17 (pt. 1), 17, pl. 3, figs. 8a,b.

Bolivina 47: (*Brizalina isidroensis* Cushman and Renz) SPDC Catalogue, 1980, Stereoscan Photos, Paleontological Type Collection, Nigeria, pl. 27, fig. A16.

Diagnosis: Test free, biserial, elongate, tapering evenly, broadest near apertural end, initial end rounded, periphery broadly rounded, ovate in transverse section; eight to nine pairs of chambers in the adult; chambers slightly inflated, increasing gradually in size as added; sutures distinct, very strongly oblique, slightly limbate, not depressed; wall calcareous, hyaline, coarsely perforate, with coarse surface ornamentation consisting of an irregular, rough, pitted surface for the initial half of test but almost smooth near the apertural end; aperture elongate, extending from the base of the final chamber to a terminal position, surrounded by an elevated lip and an imperforate area with a narrow toothplate.

Range: Late Miocene–Pleistocene.

Brizalina dilatata Reuss, 1850
Plate 11, Figs. 12 and 13

Bolivina 13A: (*Brizalina dilatata* Reuss, 1850) SPDC Catalogue, 1980, Stereoscan Photos, Paleontological Type Collection, Nigeria, pl. 26, figure A10.

Brizalina dilatata Reuss, 1850. AGIP SPA, 1982, Foraminifera Padani, pl. XXIX, fig 9, 9p.

Diagnosis: Test free, elongate, tapering, commonly compressed and laterally carinate, rounded in transverse section, broadest near apertural end; chambers distinct usually seven to eight pairs, increasing gradually in size as added, sutures slightly limbate, curved, slightly depressed; wall calcareous hyaline, finely perforate; aperture elongate, loop shaped, extending up from base to final chamber with toothplate.

Range: Oligocene–Early Miocene.

Family: BULIMINIDAE Jones, 1875
Genus: BULIMINA d'Orbigny, 1826
Bulimina sp. 1
Plate 11, Fig. 14

Diagnosis: Test fusiform, wide near the proximal end, tapering on both sides; chambers inflated, 11 in number; sutures horizontal, obscured by longitudinal striae which cover the entire test; aperture, a depression on the apertural face.

Range: Oligocene–Miocene.

Bulimina aculeata d'Orbigny, 1826
Plate 11, Figs. 15–17

Bulimina aculeata d'Orbigny, 1826, Ann. Sci. Nat. Paris, ser. 1, 7, 269, no. 7 Fornasini, 1902, Acc. Sci. Bologna, Mem., ser. 5, v. 9, p. 153, fig. 4.

Bulimina aculeata d'Orbigny, Kohl, 1985, Bull. Am. Paleontol. 88 (322), 135, pl. 20, fig. 1.

Bulimina aculeata d'Orbigny, Atlas of Benthic Shelf foraminifera of the South West Atlantic, 1980, p. 69, pl. 5, figs. 1–3.

Bulimina aculeata d'Orbigny Cenozoic Cosmopolitan deep water benthic foraminifera (Elf Aquitaine) 1986, p. 33, pl. 7, figures 1–3

Bulimina aculeata d'Orbigny, AGIP SPA, 1982, Foraminifera Padani, pl. XXV, figures 8, 8a.

Diagnosis: Test free, tapering from the widest portion in the last whorl, initial end subacute, consisting of four to six whorls; chambers distinct, regularly triserial, increasing rapidly in size as added, rounded, inflated, especially those of the last whorl. The test has a distinctive elongate, tapering shape, the lower third of which may be moderately to heavily spinose. The sutures are distinct, depressed; wall of initial portion ornamented by heavy spines, sometimes only at the base but often extending as far as half way up the test, otherwise smooth, very finely perforate, polished; aperture broadly loop shaped, in a depression of the apertural face, with a raised lip.

Range: Late Miocene–Pleistocene.

Bulimina costata d'Orbigny, 1852
Plate 11, Fig. 18

Bulimina costata d'Orbigny, 1852, Prodrome de Paleontologie stratigraphique universelle des animaux mollusques et rayonnes Paris, France, V. Masson, Ann. Sci. Nat., 3, 103.

Bulimina costata d'Orbigny, Fornasini, 1901, Le Bulimine e le Cassiduline fossili d'Italia. Soc. Geol. Ital., 20, 174, text-fig. 1.

Bulimina costata d'Orbigny, Cushman and Parker, 1938, Contrib. Cushman Lab. Foramin. Res. 14, 54, pl. 9, figs. 1–2.

Bulimina costata d'Orbigny, AGIP SPA, 1982, Foraminiferi Padani, pl. XXVII, figs. 6, 6a.

Diagnosis: Test free, triserial, triangular in profile, rounded in cross section, expanding rapidly in size from the proloculus; chambers slightly inflated, increasing gradually in size as added, ornamented by longitudinal costae that run the entire length of the test with short blunt basal spines; sutures depressed; wall calcareous hyaline, finely to coarsely perforate, radial in structure; aperture an elongate slit with toothplate, extending up from the base of ultimate chamber, test with free border that has a slightly elevated rim.

Range: Middle Miocene–Pleistocene.

Bulimina marginata d'Orbigny, 1826
Plate 11, Figs. 19 and 20

Bulimina marginata d'Orbigny, 1826, Ann. Sci. Nat., Ser. 1., 7, 269, no. 4, pl. 12, figs. 10–12.

Bulimina marginata d'Orbigny, Murray, 1971, An Atlas of British Recent Foraminiferids. Elsevier Publishing Company, Amsterdam, p. 119, pl. 49, figs. 1–2.

Bulimina marginata d'Orbigny, Kohl, 1985, Early Pliocene benthic forminifers from the Salina Basin, Southeastern Mexico. Bull. Am. Paleontol. 88 (322), 20, fig. 3.

Bulimina marginata d'Orbigny, Morkhoven, Berggren, and Edwards, 1986, Cenozoic cosmopolitan deep-water benthic foraminifera. Bull. Centres Rech. Explor. Prod. Elf Aquitaine, Mem. II, p. 20, pl. 2, fig. 1.

Bulimina marginata d'Orbigny, AGIP SPA, 1982, Foraminiferi Padani, pl. XXVII, fig. 8.

Diagnosis: Test free, triserial, expanding rapidly from the apical to apertural end, step-like in profile, the last few chambers occupying the major portion of test; spire with four to five volutions; chambers inflated, subtrigonal with angulate lower border and crenulated chamber margins; wall calcareous, hyaline, finely perforate, smooth, margins of chambers ornamented with short downward projecting spines which decrease in number on later chambers; aperture a loop-shaped slit with toothplate in the apertural face at the inner margin of the final chamber, surrounded by a raised lip.

Range: Late Pliocene–Pleistocene.

Bulimina minima Tedeschie and Zanmatti, 1957
Plate 12, Figs. 1 and 2

Bulimina minima Tedeschi and Zanmatti, AGIP SPA, 1982 Foraminiferi Padani, pl. xxv, figs. 10, 10a.

Diagnosis: Test free, triserial, slightly longer than broad, widest near middle of test, ovate in transverse section, chambers inflated, rounded, four to five whorls, with chambers increasing rapidly in size; sutures distinct, depressed, generally obscured by ornamentation on early portion of test; wall calcareous hyaline, finely perforate, lower half of test ornamented with fine spines; chambers of last whorl smooth, with a few spines along the lower portion of each chamber; aperture a loop-shaped slit with toothplate in the apertural face at the inner margin of the ultimate chamber.

Range: Late Miocene–Late Pliocene.

Bulimina alazanensis Cushman, 1927
Plate 12, Fig. 3

Bulimina alazanensis Cushman, 1927. AGIP SPA,1982, Foraminifera Padani, pl. XXVI, fig 2, 2a.

Bulimina alazanensis Cushman, 1927. Bull. Am. Paleontol. 1985, 88 (322), 135, pl. 20, fig 2a–2c.

Diagnosis: Test free, triserial, rounded in cross section, expanding gradually in size from proloculus, chambers slightly inflated, increasing gradually in size as added. Ornamented by longitudinal costae that run ¾ of the entire length of the test with short blunt basal spine. Wall calcareous hyaline, finely perforate, aperture an elongated slit with toothplate extending up from the base of the ultimate chamber.

Range: Middle Eocene–Middle Miocene.

Bulimina basispinosa Tedeschi and Zanmatti, 1957
Plate 12, Fig. 4

Bulimina basispinosa Tedeschi and Zanmatti, 1957. AGIP SPA, 1982, Foraminifera Padani, pl. XXV, figs. 9, 9a.

Diagnosis: Test free, large tapering from the widest portion in the last whorl, initial end subacute, consisting of five to six whorls; chambers distinct, regularly triserial, increasing rapidly in size as added, rounded, inflated, especially those of the last whorl. The test has a distinctive elongate tapering shape, the lower one-third moderately spinose; the sutures are distinct, depressed; aperture broadly loop shaped in a depression of the apertual face with a raised lip.

Range: Middle Pliocene–Pleistocene.

Bulimina buchiana d'Orbigny, 1846
Plate 12, Fig. 5

Bulimina buchiana d'Orbigny, 1846. Taxonomic notes. In: Society of Economic Paleontologists and Mineralogists 1960, p. 104, pl. L1, figs. 18, 19.

Diagnosis: Test free, tapering from the widest portion in the last whorl, consisting of five to six whorls, chambers distinct, triserial, increasing rapidly in size as added, rounded inflated, especially those of the last whorl, ornamented by fewer longitudinal costae which are continuous over the sutures and run three quarter of the entire length of the test with short blunt basal spine; wall calcareous hyaline, finely perforate, aperture an elongated slit with toothplate extending up from the base of the ultimate chamber.

Range: Late Oligocene–Late Pliocene.

Bulimina exilis Brady, 1884
Plate 12, Figs. 6–8

Bulimina exilis Brady, 1884. Report on the foraminifera dredged by H.M.S. Challenger, during the years 1873–1876. Challenger Exped. 1873–1876, Rept., London, Zool., 9 (22), 399, pl. 50, figs 5, 6.

Bulimina exilis Brady, 1884. Taxonomic notes. In: Society of Economic Paleontologists and Mineralogists 1960, p. 102, pl. L, fig. 5, 6.

Bulimina exilis Brady, 1884. AGIP SPA, 1982, Foraminifera Padani, pl. xxvii, fig. 2.

Bulimina exilis Brady, 1884. Cenozoic cosmopolitan deep-water benthic foraminifera (Elf Aquitaine) 1986, p. 24, pl. 4, figs 1, 2.

Diagnosis: Test free, elongated, slender, tapering, the length sometimes as much as 5 times the diameter with a basal spine; chambers elongated, sharply angled, oblique, very slightly inflated, slightly depressed; wall calcareous, smooth, polished, and finely perforate; aperture, broad, loop shaped, placed at the apex of the test, pointing directly downward to meet the junction of the second and third chambers.

Range: Pliocene–Middle Pleistocene.

Bulimina inflata Seguenza, 1862
Plate 12, Figs. 9 and 10

Bulimina inflata Seguenza, 1862. Taxonomic notes. In: Society of Economic Paleontologists and Mineralogists 1960, p. 104, pl. LI, figs. 10–12.

Bulimina inflata Seguenza, 1862. AGIP SPA, 1982, Foraminifera Padani, pl. xxvi, fig. 5.

Diagnosis: Test free, triserial, expanding rapidly from the apical to the apertural end, step-like in profile, the last chamber occupies the major portion of the test, spire with four to five volutions; chambers inflated, subtrigonal, and angulate lower border with crenulated chamber margins; wall calcareous, hyaline, finely perforate, ornamented with short longitudinal costae that end with each volution of the chamber; aperture a loop-shaped slit in the apertural face at the inner margin of the final chamber rounded by the raised lip.

Range: Middle Pliocene–Middle Pleistocene.

Bulimina spicata Phleger and Parker, 1951
Plate 12, Figs. 11 and 12

Bulimina spicata Phleger and Parker, 1951. Bull. Am. Paleontol. 1985, 88 (322), 135, pl. 20, fig. 5a–5d.

Diagnosis: Test free, triserial, medium to large, tapering from the widest point in the last whorl, step-like in profile, last few chambers occupy the major portion of the test, spire with five to six volutions; chambers inflated, with rounded border, smooth, finely perforate ornamented with short, downward projecting spines; wall calcareous hyaline coarsely perforate; aperture a loop-shaped slit with toothplate in the apertural face at the inner margin of the ultimate chamber.

Range: Pliocene–Pleistocene.

Genus: PRAEGLOBOBULIMINA Hofker, 1951
Praeglobobulimina ovata d'Orbigny, 1846
Plate 12, Fig. 13

Bulimina ovata d'Orbigny, 1846, Foraminiferes fossiles du Bassin Tertiare de Vienne (Autriche). Gide et Comp. Paris, p. 185, pl. 11, figs. 13, 14.

Bulimina ovata d'Orbigny, 1846, LeRoy, 1941, Small Foraminifera from Late Tertiary of the Netherlands East Idies. Colo. Sch. Mines Q. Golden, 36 (1), 13–62.

Bulimina ovata d'Orbigny, 1846, Cushman and Parker, 1947, *Bulimina* and related foraminiferal genera. United States Geological Survey Professional Paper, 189-D, p. 106, pl. 25, figs. 8–9.

Praeglobobulimina ovata (d'Orbigny) Haynes, 1954, Taxonomic position of some British Paleocene Buliminidae. Contrib. Cushman Found. Foramin. Res. 5 (4), 190, Text figs. 9–12, 17–19.

Protoglobobulimina ovata (d'Orbigny) Belford, 1966, Bull. Am. Paleontol., 68, pl. 6, figs. 6–9.

Praeglobobulimina ovata (d'Orbigny) Kohl, 1985, Early Pliocene benthic forminifers from the Salina Basin, Southeastern Mexico. Bull. Am. Paleontol. 88 (322), 136, pl. 21, fig. 3.

Praeglobobulimina ovata (d'Orbigny) AGIP SPA, 1983, Foraminiferi Padani, pl. XXVII, fig. 9; *Globobulimina* 1. (*Praeglobobulimina ovata* d'Orbigny), SPDC Catalogue, 1980, Stereoscan Photos, Paleontological Type Collection, Nigeria, pl. 24, fig. C1.

Diagnosis: Test free, triserial, consisting of two to three whorls, elongate, last whorl forming one-half or more of the test, circular in transverse section; chambers somewhat inflated, strongly overlapping on previous ones; sutures distinct, slightly depressed, oblique; wall calcareous, hyaline, smooth, finely perforate; aperture loop shaped, with a lip and distinct toothplate.

Range: Eocene–Pleistocene.

Praeglobobulimina pupoides d'Orbigny, 1846
Plate 12, Figs. 14–16

Praeglobobulimina pupoides d'Orbigny, 1846. AGIP SPA, 1982, Foraminifera Padani, pl. XXVIII, figs. 2, 2a.

Diagnosis: Test free, elongate, triserial with strongly overlapping chambers at the upper part, wall calcareous, perforate, radial in structure; sutures distinct, slightly depressed; aperture elongate, loop-shaped with a lip.

Range: Middle Eocene–Pleistocene.

Genus: GLOBOBULIMINA Cushman, 1927
Globobulimina sp. 1
Plate 12, Fig. 17

Diagnosis: Test globular to ovate, triserial; chambers inflated, later chambers strongly overlap earlier ones; wall calcareous, thin, finely perforate; aperture loop shaped, with a tooth or plate and internal tube.

Remarks: This species is frequently found in the Neogene sequences of the Niger Delta as well as in the deeper sediments of the Niger Delta. The lower part of the toothplate touches the projected tip of the toothplate of the preceding chamber.

Genus: GLOBOBULIMINA Cushman, 1927
Globobulimina ovula d'Orbigny, 1839
Plate 12, Fig. 18

Gbobobulimina ovula d'Orbigny, 1839. AGIP SPA, 1982, Foraminifera Padani, pl. XXVII, figs. 10, 10a.

Diagnosis: Test free, globular to ovate, chamber triserially arranged, strongly overlapping earlier whorls; wall calcareous, thin, finely perforate, radial in structure; surface, smooth; sutures distinct, slightly depressed; aperture loop shaped, terminal, toothplate doubly-folded pillar-like trough, joined to apertural border at one side.

Range: Middle Pliocene–Middle Pleistocene.

Family: CERATOBULIMINIDAE Cushman, 1927
Genus: CERATOBULIMINA Toula, 1915
Ceratobulimina pacifica Cushman and Harris, 1927
Plate 12, Figs. 19 and 20; Plate 13, Fig. 1

Ceratobulimina pacifica Cushman and Harris, 1927. Taxonomic notes. In: Society of Economic Paleontologists and Mineralogists, 1960, p. 112, pl. LIV, figs. 18a and 18b.

Diagnosis: Test free, trochospiral, deeply umbilicate, chambers enlarging rapidly as added, whorls few, coiling dextrally; wall calcareous, finely perforate, laminated; surface smooth, polished; aperture umbilical, consisting of elongate slit extending in groove up face of final chamber on umbilical side; sutures distinct, depressed, and slightly curved.

Range: Middle Oligocene–Pleistocene.

Family: TURRILINIDAE Cushman, 1927
Genus: BULIMINELLA Cushman, 1911
Buliminella subfusiformis Cushman, 1925
Plate 13, Figs. 2–4

Buliminella subfusiformis Cushman, 1925a, Some new foraminifera from the Velasco Shale of Mexico. Contrib. Cushman Lab. Foramin. Res. 1, 33, pl. 5, fig. 12.

Bulimina 11: (*Buliminella* aff. *subfusiformis* Cushman, 1925) SPDC Catalogue, 1980, Stereoscan Photos, Paleontological Type Collection, Nigeria, pl. 25, fig. A5, pl. 29, fig. A4.

Bulimina 11: (*Buliminella* aff. *subfusiformis* Cushman, 1925) Ozumba and Amajor, 1999a, Evolutionary relationships in some benthonic foraminifera of the Middle to the Late Miocene, Niger Delta. NAPE Bull. 14 (2), 163, fig. 6.

Diagnosis: Test free, elongate, tapering rapidly, many chambers to a whorl, three whorls in all; sutures distinct, depressed; wall calcareous, smooth, finely perforate; aperture loop shaped with internal toothplate.

Range: Eocene–Early Miocene.

Family: DISCORBIDAE Ehrenberg, 1838
Genus: DISCORBIS Lamarck, 1804
Discorbis sp. 1
Plate 13, Fig. 5

Diagnosis: Test typically plano-convex, the ventral side flattened, sometimes showing a long *Spirillina*-like, second chamber of several coils before division; chambers often produced to partially cover the umbilical area; wall calcareous perforate; aperture at the base of the umbilical margin on the ventral side of the chamber.

Remarks: This species is frequently found in the Eocene–Pliocene sequences of the Niger Delta. It differs from other species of *Discorbis* in the rapidly increasing size of the chambers which extend, on the ventral side, to almost half of the periphery. There is an angular projection over the umbilical region.

Genus: EPISTOMINELLA Husezima and Maruhasi, 1944
Epistominella pontoni Cushman, 1930
Plate 13, Figs. 6–9

Pulvinulinella pontoni Cushman, 1930a, The foraminifera of the Choctawhatchee Formation of Florida. Florida State Geol. Surv. Bull. 4, 57, pl. 11, fig. 2.

Epistominella pontoni (Cushman) Petters, 1982, Central West Africa Cretaceous-Tertiary benthic foraminifera and stratigraphy. Palaeontographica, Bd. 179, Abs. A, p. 103, pl. 13, figs. 19, 20.

Alabamina 1A: (*Epistominella pontoni* Cushman) SPDC Catalogue, 1980, Stereoscan Photos, Paleontological Type Collection, Nigeria, pl. 11, fig. A2.

Diagnosis: Test free, trochospiral, six to eight chambers in final whorl, plano-convex, periphery rounded; sutures on spiral side slightly swept back, those on umbilical side radial; wall calcareous, perforate; aperture arched on the face of the final chamber.

Range: Oligocene–Early Miocene.

Epistominella vitrea Parker, 1953
Plate 13, Figs. 10 and 11

Epistominella vitrea Parker, 1953, Ecology of foraminifera from San Antonio Bay and environs, Southwest Texas. Cushman Foundation for Foraminiferal Research, Special Publication No. 2, p. 9, pl. 4, figs. 34–36.

Alabamina 1, SPDC Catalogue, 1980, Stereoscan Photos, Paleontological Type Collection, Nigeria, Pl, 11, fig. A1.

Epistominella vitrea Petters, 1982, Central West Africa Cretaceous-Tertiary benthic foraminifera and stratigraphy. Palaeontographica, Bd. 179, Abs. A, nos. 1–3, p. 70, pl. 7, figs. 3, 4.

Diagnosis: Test small, plano-convex, convex on the dorsal side, depressed toward the umbilicus on the ventral side; periphery acute; chambers slightly inflated, six in the last whorl; sutures slightly depressed and curved on the dorsal side, more depressed on the ventral side, earlier ones radial, later ones slightly curved; aperture long, narrow, and curved on apertural face.

Range: Late Oligocene–Pliocene.

Epistominella pacifica Cushman, 1927
Plate 13, Fig. 12

Parrella 3: (*Epistominella pacifica* Cushman) SPDC Catalogue, 1980, Stereoscan Photos, paleontological type collection, Nigeria, pl. 20, figures A2V, A2D.

Diagnosis: Test free, trochospiral, biconvex in edge view, periphery acute with six to seven chambers in the final whorl, sutures distinct limbate, curved; wall calcareous hyaline, smooth, finely perforate; aperture an interiomarginal arch.

Range: Late Miocene–Pliocene.

Genus: CANCRIS de Montfort, 1808
Cancris auriculus Fichtel and Moll, 1803
Plates 13, Figs. 13–15

Nautilus auriculus Fichtel and Moll, 1803, Testacea microscopica aliaque minuta ex generibus, Argonauta et Nautilus Adnaturam delineata et descripta; Milroskopische und Andere Kleine Schalthiere aus den Geschlectern Argonaute and Schiffer, nach der Natur gezeichnet und Beschrieben, Wien, Camesina, p. 108, pl. 20, figs. a–c, d–f.

Cancris auriculus (Fichtel and Moll, 1803) Jenkins and Murray, 1989, Stratigraphical Atlas of fossil Foraminifera. Ellis Horwood Publishers; John Wiley and Sons, New York, NY, p. 470, pl. 9.5, figs. 12, 13.

Cancris auriculus (Fichtel and Moll, 1803) AGIP SPA, 1982, Foraminifera Padani, pl. XLII, figs. 4v, 4d, 4p.

Cancris 5: (*Cancris auriculus* Fichtel and Moll, 1803) SPDC Catalogue, 1980, Stereoscan Photos, Paleontological Type Collection, Nigeria, pl. 21, figs. A3V, A3D.

Diagnosis: Test free, trochospiral, unequally biconvex, slightly convex on spiral side, highly convex on the umbilical side, periphery acutely lobate, with a distinct narrow keel; seven to eight chambers in the final whorl, all visible on the spiral side, increasing rapidly in size as added, inflated ventrally, lobate, the final chamber making up over one-half the size of the test in side view; umbilical area closed by an umbilical flap projecting downward from the final chamber; sutures, distinct, strongly to slightly curved on umbilical side; wall calcareous, hyaline, coarsely perforate; aperture interiomarginal, umbilical to extraumbilical, extending from a point near the periphery to the umbilicus.

Range: Early Miocene–Pleistocene.

Cancris turgidus Cushman and Todd, 1942
Plate 13, Figs. 16 and 17

Cancris turgidus Cushman and Todd, 1942. Contrib. Cushman Lab. Foramin. Res. 18, 92, pl. 11, fig. 2.

Cancris turgidus Cushman and Todd, 1942. Institute du desert d' Egypte 1954 v. 4–6, pl. 3, fig. 23 *Cancris turgidus* Cushman and Todd, 1942. Institute of Geological Sciences, issues 7–11, issues 82, 1983, p. 29.

Diagnosis: Test free, trochospiral, biconvex, periphery rounded, six chambers on the last whorl, all visible on the spiral and umbilical side, increasing rapidly in size as added, inflated lobate, the final chamber enlarged making up one-half the size of the wall calcareous hyaline, coarsely perforate; aperture interiomarginal, umbilical, extending from a point near the periphery to the umbilicus.

Range: Lower Oligocene–Late Miocene.

Genus: ROSALINA d'Orbigny, 1826
Rosalina floridana Cushman, 1922
Plate 13, Figs. 18 and 19

Discorbis floridana Cushman, 1922, The foraminifera of the Atlantic Ocean, Part 3. Textulariidae. US Natl. Museum Bull. 104, 39, pl. 5, figs. 11, 12.

Discorbis floridana Cushman, 1931, New late Tertiary foraminifera from Vitilevu, Fiji. Contrib. Cushman Lab. Foramin. Res. 7 (pt. 2), 21, pl. 4, figs. 7, 8.

Discorbis floridana Cushman and Parker, 1931, Recent foraminifera from the Atlantic Coast of South America. Proc. US Natl. Museum, 80, 18, pl. 4, fig. 5.

Rosalina floridana (Cushman) Parker, 1954, Bull. Mus. Comp. Zool. III (10), 524, pl. 8, figs. 19, 20.

Rosalina floridana (Cushman) Todd, 1965, US Natl. Museum Bull. 161 (part 4), pl. 3, figs. 1, 3; pl. 4, fig. 5.

Valvulineria 23: (*Rosalina floridana* Cushman) SPDC Catalogue, 1980, Stereoscan Photos, Paleontological Type Collection, Nigeria, pl. 14, figs. A8V, A8D.

Diagnosis: Test free, trochospiral, plano-convex, wall calcareous, hyaline, with medium size perforations on the spiral side, finely perforate to imperforate on the umbilical side, area above the umbilicus is imperforate; aperture an interiomarginal, extraumbilical–umbilical arch with lip; periphery broadly rounded, lobate; six to seven chambers in the final whorl, increasing rapidly in size as added; suture, limbate, flush, becoming simple and somewhat depressed on the spiral side, all sutures are deeply incised on the umbilical side; umbilicus, large and partially covered by very large triangular chamber flaps.

Range: Early Pliocene–Pleistocene.

Genus: VALVULINERIA Cushman, 1926

Valvulineria sp. 1

Plate 13, Fig. 20; Plate 14, Figs. 1 and 2

Diagnosis: Test trochoid, close coiled, ventrally umbilicate; chambers numerous, wall calcareous, finely perforate, aperture ventral, large, extending from the umbilicus toward the periphery, often with a thin plate that fills the umbilical area, sometimes extending into the face of the chamber.

Remarks: This species is commonly found in the Eocene–Early Miocene sequences of the Niger Delta. It differs from other species in the less inflated, nondistinct chambers and sutures, the coarsely perforate, apertural flap projecting slightly toward the umbilical side.

Valvulineria jacksonensis Cushman, 1933

Plate 14, Figs. 3–5

Valvulineria jacksonensis Cushman, 1933, New foraminifera from the Upper Jackson Eocene of the Southeastern Coastal Plain region of the United States. Contrib. Cushman Lab. Foramin. Res. 9 (pt. 1), 18, pl. 2, fig. 9.

Valvulineria jacksonensis (Cushman, 1933) Fayose, 1970, Stratigraphical paleontology of Afowo-1 well, Southwestern Nigeria. J. Min. Geol., 5 (1 and 2), 22, pl. 7, fig. 5.

Diagnosis: Test biconvex to concavo-convex; equatorial periphery ovate, weakly lobulate; ventrally convex toward the periphery but depressed at the umbilicus which is somewhat papillate; periphery rounded, chambers distinct, eight in the final whorl, increasing gradually in size as added; on the sutures distinct on spiral side, gently curved, limbate; nearly straight on the umbilical side, oblique, very slightly depressed; wall smooth; aperture, ventral, beneath the umbilical lobe of the last chamber.

Range: Paleocene–Early Eocene.

Valvulineria martinezensis Cushman and Bermudez, 1948

Plate 14, Figs. 6 and 7

Anomalina martinezensis Cushman and Bermudez, 1948, Some Paleocene foraminifera from the Madruga Formation of Cuba. Contrib. Cushman Lab. Foramin. Res. Sharon Mass 24, 74, pl. 12, figs. 14–16.

Valvulineria 8A: SPDC Catalogue, 1980, Stereoscan Photos, Paleontological Type Collection, Nigeria, pl. 13, fig. A6V.

Diagnosis: Test trochoid, biconvex; both spiral and umbilical sides are depressed in the median portion, periphery broadly rounded, tending to become slightly angular; chambers distinct, inflated, eight chambers in the last whorl, increasing very gradually in size as added; sutures distinct, limbate, and raised, dorsally oblique, curved, ventrally nearly radiate; wall coarsely perforate; aperture interiomarginal, extending to the umbilical side with a lip.

Remarks: The species which is known to occur in the Niger Delta is very common in the Paleocene of Cuba and the Gulf Coastal Region of the United States.

Range: Early Paleocene–Late Paleocene.

Valvulineria trochoidea Plummer, 1927
Plate 14, Fig. 8

Truncatulina midwayensis Plummer var. *trochoidea* Plummer, 1927, Foraminifera of the Midway Formation in Texas. Bulletin Univ. Texas Bureau Econ. Geol. Technol. 2644, 142, pl. 9, fig. 8.

Valvulineria 5: (*Valvulineria* cf. *midwayensis trochoidea* Plummer, 1927) SPDC Catalogue, 1980, Stereoscan Photos, Paleontological Type Collection, Nigeria, Pl. 13, figs. A3V, A3D.

Diagnosis: Test trochoid, concavo-convex; dorsal side evolute, strongly convex; 10 chambers visible in the last whorl with strongly raised and curved sutures; periphery lobulate; ventral side depressed in the central portion; suture raised and curved, leaving a round and deep umbilicus in the center; aperture, at the base of the last chamber, extending to the umbilical side.

Range: Eocene.

Valvulineria wilcoxensis Cushman and Ponton, 1932
Plate 14, Fig. 9

Valvulineria wilcoxensis Cushman and Ponton, 1932, The Foraminifera of the Upper, Middle and Lower part of the Lower Miocene of Florida. Bull. Florida State Geol. Surv. 9, 70, pl. 9, fig. 6.

Valvulineria wilcoxensis (Cushman and Ponton, 1932) Petters, 1982, Central West Africa Cretaceous-Tertiary benthic foraminifera and stratigraphy. Palaeontographica, Bd. 179, Abs. A, p. 104, pl. 14, figs. 5, 6, 10.

Diagnosis: Test free, trochospiral, umbilicate, periphery acute, chambers increasing gradually in size as added; sutures radial; wall calcareous, finely perforate, radial in structure, monolamellid, surface smooth, aperture interiomarginal, extraumbilical to umbilical, with broad, thin apertural flap projecting over the umbilicus.

Range: Eocene–Early Miocene.

Family: ROTALIIDAE Ehrenberg, 1839
Genus: AMMONIA Brunnich, 1772
Ammonia beccarii Linnaeus, 1758
Plate 14, Figs. 10–14

Nautilus beccarii Linnaeus, 1758, Systema naturae per regna tria naturae, secundum classes, ordines, genera, species, cum characteribus, differentiis, synonymis, locis. In: Engleman, G., tenth ed., vol. 1, Lipsiae, p. 710, pl. 1, figs 1a–c.

Rotalia beccarii (Linnaeus, 1758), Parker, Phleger, and Pierson, 1953, Ecology of foraminifera from San Antonio Bay and environs, Southwest Texas. Cushman Foundation for Foraminiferal Research, Special Publication No. 2, p. 13, pl. 4, figs. 29, 30.

Ammonia beccarii (Linnaeus, 1758) Matoba, 1970, Distribution of Recent shallow-water foraminifera of Matsushima Bay. Miyagi Prefecture, northwest Japan. Tohoku Univ. Sci. Rep., 2nd series (Geology), 42 (1), 48, pl. 5, figs. 11a–c.

Ammonia beccarii AGIP SPA, 1982, Foraminiferi Padani, pl. XLI, figs. 3V, 3d, 3p.

Ammonia beccarii Okosun and Liebau, 1999, Foraminiferal biostratigraphy of eastern Niger Delta, Nigeria. NAPE Bull. 14, 136–156, pl. 2, figs. 17, 30.

Rotalia 1A: (*Ammonia beccarii* Linnaeus, 1758), SPDC Catalogue, 1980, Stereoscans Photos, Paleontological Type Collection, Nigeria, pl. 18, figs. A2V, A2D.

Diagnosis: Test free, trochospiral, periphery rounded, spiral side evolute, umbilical side involute with a broad open umbilicus and a small umbilical plug; chambers inflated, eight to nine in final whorl, increasing gradually in size as added, last few chambers with an extension into the open umbilicus around the umbilical plug; sutures distinct, radial to slightly oblique and flush on the spiral side, radial on the umbilical side and incised as suture approaches the umbilicus; wall calcareous, hyaline, smooth, finely perforate; aperture an interiomarginal extraumbilical–umbilical slit.

Range: Early Miocene–Pleistocene.

Ammonia tepida Cushman, 1926
Plate 14, Figs. 15 and 16

Rotalia 21: (*Ammonia tepida* Cushman, 1926) SPDC Catalogue, 1980, Stereoscan Photos, Paleontological Type Collection, Nigeria, pl. figure A6.

Ammonia tepida Cushman, 1926. AGIP SPA, 1982, Foraminifera Padani, pl. XLI, fig 6, 6 p.

Ammonia tepida Cushman, 1926. Bull. Am. Paleontol. 1985, 88, 143, pl. 28, fig 31–33c.

Diagnosis: Test free, periphery rounded, spiral side evolute, umbilical side involute with a broad open umbilicus; chambers inflated, six chambers in the final whorl, increasing gradually in size as added; sutures distinct, radial to slightly oblique and flush on the spiral side, radial on the umbilical side and incised as sutures; wall calcareous, finely perforate, aperture an interiomarginal extraumbilical–umbilical slit.

Range: Miocene–Pleistocene.

Family: AMPHISTEGINIDAE Cushman, 1927
Genus: AMPHISTEGINA d'Orbigny, 1826
Amphistegina lessonii d'Orbigny, 1826
Plate 14, Figs. 17 and 18

Amphistegina lessonii d'Orbigny, 1826, Ann. Sci. Nat., 7, 304, pl. 17, figs. 1–4.

Amphistegina lessonii Phleger and Parker, 1951, Geological Society of America, Memoir 46, Part 2, pl. 13, figs. 13,14; pl. 14, fig. 1.

Amphistegina lessonii AGIP SPA, 1982, Foraminifera Padani, pl. XLII, figs. 10d, 10p, 10v, 10se, 10st.

Diagnosis: Test free, lenticular, trochoid, generally unequally biconvex, multichambered; chambers arranged in complex spiral which splits up into chamberlets on the ventral side and extends into peripheral flange; surface smooth, granulate; aperture a narrow slit at inner margin of last chamber.

Range: Late Oligocene–Early Pliocene.

Family: CASSIDULINIDAE d'Orbigny, 1839
Genus: CASSIDULINA d'Orbigny
Cassidulina crassa d'Orbigny, 1839
Plate 14, Fig. 19

Cassidulina crassa d'Orbigny, 1839, Foraminiferes des Iles Canaries. In: Barker-Webb, P., Berthelot, S., Historie Naturelle des Canaries. Paris, vol. 2, Part 2, p. 56, pl. 7, figs. 18–20.

Cassidulina crassa Boltovskoy et al., 1980, Atlas of Benthic Foraminifera of the Southwest Atlantic. Dr. W. Junk b.v., The Hague, p. 72, pl. 6, figs. 19–23; pl. 7, figs. 1–3.

Cassidulina crassa AGIP SPA, 1982, Foraminiferi Padani, pl. XLII, figs. 4, 4a.

Diagnosis: Test free, oval to circular, peripheral margin broadly rounded; chambers slightly inflated, the last chamber rectangular to trapezoidal; sutures slightly depressed; wall thick, calcareous, hyaline, smooth to finely perforate; aperture ovate, almost perpendicular to basal suture, flanked by a broad rim.

Range: Early Miocene–Pleistocene.

Cassidulina laevigata d'Orbgny, 1826
Plate 14, Fig. 20; Plate 15, Figs. 1 and 2

Cassidulina laevigata d'Orbigny, 1826. Taxonomic notes. In: Society of Economic Paleontologists and Mineralogists 1960, p. 110, pl. LIV, figs. 2, 3.

Cassidulina laevigata d'Orbigny, 1826. Treatise on Invertebrate Paleontology (Loeblich and Tappan, 1964) p. C737, figures 1, 2.

Cassidulina laevigata d'Orbigny, 1826. Petters, 1982 Central West African Cretaceous-Tertiary benthic foraminifera and stratigraphy, Palaeontographica, p. 77, pl. 9, figs. 1, 2

Cassidulina laevigata d'Orbigny, 1826. AGIP SPA, 1982 Foraminifera Padani, pl. XLIII, figs 5, 5a.

Diagnosis: Test free, lenticular, commonly biumbonate, with clear central bosses; chamber biserially arranged in coil, chambers alternating on each side of periphery, each reaching boss on one side and only extending part way to boss of opposite side, succeeding chamber extending to center on alternate sides; wall calcareous, hyaline, perforate, granular in structure, surface generally smooth aperture an elongate slit, extending from base of final chamber upward in curve paralleling anterior margin of chamber with narrow bordering lip on lower margin.

Range: Early Miocene–Early Pliocene.

Cassidulina neocarinata Thalmann, 1950
Plate 15, Figs. 3 and 4

Cassidulina neocarinata Thalmann, 1950. Taxonomic notes. In: Society of Economic Paleontologists and Mineralogists 1960, p. 110, pl. LIII, figs. 2, 3

Cassidulina neocarinata Thalmann, 1950. AGIP SPA, 1982, Foraminifera Padani, pl. XLIII, figs. 6, 6a.

Cassidulina neocarinata Thalmann, 1950. Bull. Am. Paleontol. 1985, 88 (322), 145, pl. 30, fig 1a–1c.

Diagnosis: Test free, oval to circular in shape, peripheral margin acute chambers compressed, biserially arranged in coil, chambers alternating on each side of periphery with slight keel; sutures slightly depressed; wall, thin, calcareous hyaline, smooth to finely perforate; aperture ovate, almost perpendicular to basal suture.

Range: Middle Miocene–Pleistocene.

Genus: GLOBOCASSIDULINA Voloshinova, 1960
Globocassidulina subglobosa Bradyi, 1884
Plate 15, Figs. 5–7

Globocassidulina subglobosa Bradyi, 1884. AGIP SPA, 1982, Foraminifera Padani, pl. XLIII, figs. 8, 8a.

Globocassidulina subglobosa Bradyi, 1884. Petters, 1982, Central West African Cretaceous-Tertiary benthic foraminifera and Stratigraphy, Palaeontographica, p. 78, pl. 9, figs. 5, 6, and 8.

Globocassidulina subglobosa Bradyi, 1884. Bull. Am. Paleontol. 1985, 88 (322), 145, pl. 30, figs. 3a–3d, 4a–4b.

Diagnosis: Test free, subglobular, periphery margin rounded, umbilicus closed; chambers biserially arranged and enrolled; wall calcareous, finely perforate, granular in structure, surface commonly smooth; sutures distinct, slightly depressed; aperture a narrow slit extending up the face of final chamber, may have narrow in-folded rim but no apertural toothplate.

Range: Late Eocene–Pleistocene.

Globocassidulina oblonga Reuss, 1850
Plate 15, Figs. 8 and 9

Globocassidulina oblonga Reuss, 1850. Neuss Foraminiferenaus den schichten des Wiener Tertiarbeckens: K. Akad. Wiss. Wien, math-naturwiss. Cl., Denkschr., v. 1, p. 365–390, pl. 46–51.

Globocassidulina oblonga Reuss, 1850. Treatise on Invertebrate Paleontology (Loeblich and Tappan, 1964), p. C737, fig. 604; 6,7.

Globocassidulina oblonga Reuss, 1850. AGIP SPA, 1982, Foraminifera Padani, pl. XLIII, figs. 7, 7a.

Diagnosis: Test free, subglobular, oblong in shape, peripheral margin rounded, umbilicus closed; chambers biserially arranged and enrolled; wall calcareous, finely perforate, granular in nature, surface smooth; aperture a narrow slit extending up the face of final chamber.

Range: Pliocene–Early Pleistocene.

Family: CIBICIDIDAE Cushman, 1927
Genus: CIBICIDES de Montfort, 1808
Cibicides lobatulus Walker and Jacob, 1798
Plate 15, Figs. 10 and 11

Nautilus lobatulus Walker and Jacob, 1798, In: Kanmacher, F. (Ed.), Adams—Essays on the microscope. second ed., with considerable additions and improvements. Dillon and Keating, London, p. 642, pl. 14, fig. 36.

Cibicides lobatulus (Walker and Jacob, 1798) Boltovskoy et al., 1980, Atlas of Benthic Foraminifera of the Southwest Atlantic. Dr. W. Junk b.v., The Hague, p. 76, pl. 9, figs. 1–4.

Cibicides lobatulus Jenkins and Murray, 1989, Stratigraphical Atlas of fossil Foraminifera. Ellis Horwood Publishers; John Wiley and Sons, New York, p. 517, pl. 10.4, figs. 14–16.

Cibicides lobatulus AGIP SPA, 1982, Foraminiferi Padani, pl. L1, figs. 6d, 6p, 6v.

Diagnosis: Test trochospiral, spiral side flat to concave, evolute, umbilical side convex, involute, peripheral margin acute to carinate; chambers inflated, six to seven in final whorl; umbilicus depressed; wall calcareous, coarsely perforate, sutures limbate on spiral side, depressed on umbilical side; aperture with distinct lip and extending rearward to the penultimate chamber along the spiral suture, also extending along the base of the apertural face on the umbilical side.

Range: Eocene–Pleistocene.

Genus: HYALINEA Hofker, 1951
Hyalinea balthica Schroeter, 1783
Plate 15, Figs. 12 and 13

Nautilus balthicus Schroeter, 1783, Einleitung in die Couchylien Kenntniss nach Linne, Bd. 1.-J. J. Gebauer, Halle, p. 20, pl. 1, fig. 2.

Hyalinea balthica (Schroeter), Morkhoven, Berggren, and Edwards, 1986, Cenozoic cosmopolitan deep-water benthic foraminifera. Bull. Centres Rech. Explor. Prod. Elf Aquitaine, Mem. II, 1–421, p. 21–23, pl. 3, figs. 1, 2a–b, 3.

Hyalinea balthica AGIP SPA, 1982, Foraminiferi Padani, pl. XLIX, figs. 5V, 5P, 5d.

Diagnosis: Test free, planispirally coiled and strongly compressed; 9–12 chambers in the outer whorl; sutures curved, limbate, and merges with the peripheral imperforate keel; wall calcareous finely perforate, with septa and marginal nonperforate keel; aperture an interiomarginal slit bordered by a thickened lip, secondary apertures open beneath umbilical flaps on both sides of the test.

Range: Late Pliocene–Pleistocene.

Family: EPONIDIDAE Hofker, 1951
Genus: EPONIDES de Montfort, 1808
Eponides africana de Klasz and Rerat, 1963
Plate 15, Figs. 14 and 15

Eponides africana de Klasz and Rerat, 1963, *Mandjina* Nov. Gen. Nouveau Foraminifer de l'Eocene superieur du Gabon (Afrique Equatoriale), C.R. Som. Séance Soc. Geol. France, Fasc. 4, 184, pl. 3, figs. 1, 2.

Eponides africana (de Klasz and Rerat, 1963) Petters, 1982, Central West Africa Cretaceous-Tertiary benthic foraminifera and stratigraphy. Palaeontographica, Bd. 179, Abs. A, nos. 1–3, pl. 14, figs. 15–17, 21.

Diagnosis: Test trochoid, unequally biconvex; periphery lobulate, dorsal side more convex than ventral side; ventral side with a wide umbilicus filled with shell material; dorsal side evolute, having two and half whorls; six chambers in the last whorl; periphery lobulate; aperture a low arch at the base of the last septal face.

Range: Late Paleocene–Middle Eocene.

Eponides berthelotianus d'Orbigny, 1932
Plate 15, Figs. 16–18

Rotalia 2, SPDC Catalogue, 1980, Stereoscan Photos, Paleontological Type Collection, Nigeria, pl. 18, fig. A3.

Diagnosis: Test trochoid biconvex, periphery acute; seven chambers in the final whorl increasing slowly in size as added; ventral side involute; sutures limbate, slightly curved, emanating from a central calcareous boss; dorsal side convex, evolute; sutures raised, gently curved; test coarsely perforate; aperture an interiomarginal slit.

Range: Early Oligocene–Early Miocene.

Eponides eshira de Klasz and Rerat, 1962
Plate 15, Figs. 19 and 20; Plate 16, Fig. 1

Eponides eshira de Klasz and Rerat, 1962, Quelques nouveaux foraminiferes du Cretace et du Tertiare du Gabon (Afrique Equatoriale). Rev. Micropaleontol. 4, 112, fig. 1.

Eponides eshira (de Klasz and Rerat, 1962) Petters, 1982, Central West Africa Cretaceous-Tertiary benthic foraminifera and stratigraphy. Palaeontographica, Bd. 179, Abs. A, nos. 1–3, p. 104, pl. 14, figs. 19, 20.

Eponides 1: (*Eponides eshira* de Klasz and Rerat, 1962) SPDC Catalogue, 1980, Stereoscan Photos, Paleontological Type Collection, Nigeria, pl. 16, figs. A1V, A1D.

Diagnosis: Test free, trochospiral, eight to nine chambers in the final whorl, biconvex, periphery angled and carinate, with broad depression in umbilical region, septa double with interseptal passages on the spiral side; sutures curved on the spiral side, nearly radial to curved on the umbilical side; wall calcareous, finely perforate, radial in structure; aperture an interiomarginal arch without internal toothplate.

Range: Late Eocene–Middle Miocene.

Eponides iojimaensis
Plate 16, Figs. 2 and 3

Anomalina 9A (*Eponides* cf. *iojimaensis* Asano) SPDC Catalogue, 1980, Stereoscan Photos, Paleontological Type Collection, Nigeria, pl. 8, figs. A7V, A7D.

Diagnosis: Test trochoid biconvex, periphery rounded; nine chambers in the final whorl, increasing slowly in size as added, ventral side planispiral, involute, dorsal side convex, trochospiral evolute, sutures depressed, gently curved; test calcareous, finely perforate; aperture interiomarginal slit.

Range: Eocene–Miocene.

Eponides pseudoelevatus Graham, de Klasz, and Rerat, 1965
Plate 16, Figs. 3–5

Eponides pseudoelevatus Graham, de Klasz, and Rerat, 1965, Quelques importants Foraminiferes du Tertiaire du Gabon (Afrique Equatoriale). Rev. Micropaleontol. 4, 76, pl. 1, figs. 14 a–c.

Eponides pseudoelevatus Graham, de Klasz, and Rerat, 1965, Fayose, 1970, Stratigraphical paleontology of Afowo-1 well, Southwestern Nigeria. J. Min. Geol., 5 (1 and 2), 30, pl. 6, fig. 2.

Eponides pseudoelevatus Graham, de Klasz, and Rerat, 1965, Petters, 1982, Central West Africa Cretaceous-Tertiary benthic foraminifera and stratigraphy. Palaeontographica, Bd. 179, Abs. A, 1–105, nos. 1–3, pl. 7, figs. 7, 8.

Eponides 11 (*Eponides pseudoelevatus* Graham, de Klasz, and Rerat, 1965), SPDC Catalogue, 1980, Stereoscan Photos, Paleontological Type Collection, Nigeria, pl. 16, fig. A8.

Diagnosis: Test biconvex, circular in outline; periphery weakly lobulate with a narrow keel; ventral side strongly convex, nine chambers forming the last whorl and increasing gradually in size as added; sutures curved, limbate, raised, and thickened to form a high ridge round the umbilicus; dorsal side, moderately convex; sutures curved, strongly raised; aperture interiomarginal, forming a low arch.

Range: Paleocene–Eocene.

Eponides ornatus d'Orbigny
Plate 16, Fig. 6

Eponides 19: (*Eponides ornatus* d'Orbigny) SPDC Catalogue, 1980, Stereoscan Photos, Paleontological Type Collection, Nigeria, pl. 17, figures A3V, A3D.

Diagnosis: Test free, trochospiral, eight to nine chambers in the final whorl biconvex in edge view, periphery acute, with broad depression in the umbilical side, sutures, curved on the spiral side, nearly radial to curved on the umbilical side; wall calcareous, finely perforate; aperture an interiomarginal arch.

Range: Miocene–Pliocene.

Genus: CIBICORBIS Hadley, 1934
Cibicorbis inflata d'Orbigny, 1846
Plate 16, Figs. 7–10

Eponides 12: (*Cibicorbis inflata* d'Orbigny, 1846), SPDC Catalogue, 1980, Stereoscan Photos Paleontological Type Collection, Nigeria.

Diagnosis: Test small, subcircular in outline, concavo-convex, ventral side convex, dorsal side concave; seven chambers visible in the last whorl, slightly inflated; sutures raised, strongly curved, with a calcareous boss on the ventral side; aperture obscured at the base of the last chamber.

Range: Eocene–Oligocene.

Family: CAUCASINIDAE Bykova, 1959

Genus: FURSENKOINA Loeblich and Tappan, 1961

Fursenkoina howei Cushman, 1936

Plate 16, Fig. 11

Virgulina recta Cushman var. *howei* Cushman, 1936, New genera and species of the families Verneuilinidae and Valvulinidae and of the subfamily Virgulininae. Cushman Laboratory for foramininfera Research, Special Publication No. 6, p. 47, pl. 7, fig. 4.

Fursenkoina howei (Cushman) Petters, 1982, Central West Africa Cretaceous-Tertiary benthic foraminifera and stratigraphy. Palaeontographica, Bd. 179, Abs. A, no. 1–3, p. 76, pl. 5, fig. 22.

Diagnosis: Test elongate, fusiform; chambers distinctly angulated, with overhanging margins, 11 chambers present, early chambers twisted and biserial; sutures horizontal, strongly depressed; surface smooth; aperture narrow, elongate, extending up the face of the final chamber.

Range: Middle Oligocene–Early Miocene.

Fursenkoina punctata d'Orbigny, 1839

Plate 16, Fig. 12

Virgulina punctata d'Orbigny, 1839, Foraminiferes des Iles Canaries. In: Barker-Webb, P., Berthelot, S., Historie Naturelle des Canaries. Paris, vol. 2, Part 2, p. 139, pl. 1, figs. 35, 36.

Virgulina 2: (*Fursenkoina punctata* d'Orbigny) SPDC Catalogue, 1980, Stereoscan Photos, Paleontological Type Collection, Nigeria, pl. 28, fig. 3A; pl. 31, fig. D1.

Diagnosis: Test free, narrow, elongate, early portion in highly twisted biserial arrangement, later becoming less sigmoidal and more typically biserial; sutures distinct, depressed to oblique; wall calcareous, very finely perforate, granular in texture, surface smooth; aperture narrow, elongate, extending up to the face of the final chamber, comma shaped with tooth.

Range: Eocene–Early Miocene.

Fursenkoina schreibersiana Czjzek, 1848

Plate 16, Fig. 13

Virgulina schreibersiana Czjzek, 1848 Bettrag zur Kenntniss der fossili foraminiferen des Wiener. Natur Abh Wien Osteireich Bd. 2, Abth. 1, p. 147, pl. 13, figs. 18–21.

Fursenkoina schreibersiana Czjzek, 1848. AGIP SPA, 1982, Foraminifera Padani, pl. XXVIII, figs. 6, 6a, 6p.

Diagnosis: Test elongate and fusiform; periphery acute; chambers gently compressed, eight in all; early portion of the test twisted and biserial, later becoming less sigmoidal; sutures distinct, depressed, and oblique; aperture narrow, elongate, extending up the face of the last chamber.

Range: Middle Eocene–Pleistocene.

Fursenkoina wilcoxensis Cushman and Ponton, 1932
Plate 16, Fig. 14

Virgulina wilcoxensis Cushman and Ponton, 1932, The Foraminifera of the Upper, Middle and part of the Lower Miocene of Florida. Bull. Florida State Geol. Surv. 8 (122, pt. 3), 67, pl. 81, fig. 22.

Diagnosis: Text elongate, fusiform, somewhat compressed, about two and half times as long as broad; early portion irregularly spiral, later portion irregularly biserial; periphery rounded; chambers distinct, slightly inflated; sutures distinct, slightly depressed; wall smooth, with a distinctly perforate aperture, a loop on the apertural face.

Range: Paleocene–Eocene.

Genus: HOPKINSINA Howe and Wallace, 1932
Hopkinsina bononiensis Fornasini, 1888; Macfayden, 1930
Plate 16, Figs. 15–17

Uvigerina bononiensis Fornasini, 1888, Tavola Paleo-protis tografica. Soc. Geol. Ital. 7, Fasa. 1, 1–48, p. 48, pl. 3, figs. 12, 12a.

Hopkinsina bononiensis Fornasini, 1888, Macfayden, 1930, Miocene foraminifera from the Clysmic area of Egypt and Sinai. Egyptian Geological Survey, Egypt, p. 93, pl. 3, figs. 28a,b.

Hopkinsina bononiensis Fornasini, 1888, Fayose, 1970, Stratigraphical paleontology of Afowo-1 well, Southwestern Nigeria. J. Min. Geol. 5 (1 and 2), 17, pl. 4, fig. 5.

Hopkinsina bononiensis Fornasini, 1888, Petters, 1982, Central West Africa Cretaceous-Tertiary benthic foraminifera and stratigraphy. Palaeontographica, Bd. 179, Abs. A, p. 66, pl. 5, fig. 18.

Uvigerinella 1: (*Hopkinsina bononiensis* Fornasini) SPDC Catalogue, 1980, Stereoscan Photos, Paleontological Type Collection, Nigeria, pl. 26, fig. A8, pl. 30, figs. A1, A7.

Diagnosis: Test fusiform, about twice as long as wide; periphery lobulate, greatest width near the aperture; test tapering toward the base, initial triserial portion small, the rest of the test biserial; chambers six in number, closely arranged and separated by slightly depressed and curved sutures; final chamber terminating in a short apertural neck, test covered with numerous longitudinal costae, interrupted at the sutures; aperture terminal, round, with a lip.

Range: Early Oligocene–Middle Miocene.

Hopkinsina semiornata d'Orbigny, 1846
Plate 16, Figs. 18 and 19

Uvigerina semiornata d'Orbigny, 1846, Foraminiferes fossiles du Bassin Tertiare de Vienne (Autriche). Gide et Comp. Paris, p. 189, pl. 11, figs. 23, 24.

Uvigerina semiornata Cushman and Edwards, 1938, Notes on the Oligocene species of *Uvigerina* and *Angulogerina*. Contrib. Cushman Lab. Foramin. Res. 14 (pt. 4), 89, pl. 14, figs. 11, 12.
Uvigerinella 3A: (*Hopkinsina semiornata* d'Orbigny) SPDC Catalogue, 1980, Stereoscan Photos, Paleontological Type Collection, Nigeria, pl. 30, fig. A4.
Hopkinsina semiornata d'Orbigny, AGIP SPA, 1988, Foraminifera of the Niger Delta, pl. 11, fig. 5.

Diagnosis: Test free, elongate, early-stage triserial and compressed, later biserial; chambers inflated; wall calcareous, perforate, surface ornamented by thick longitudinal costae; sutures depressed; aperture terminal, at the end of the ultimate chamber or may be slightly produced on a short neck.
Range: Oligocene–Middle Miocene.

Hopkinsina danvillensis Howe and Wallace, 1932
Plate 16, Fig. 20

Hopkinsina danvillensis Howe and Wallace, 1932, A Monograph of the foraminiferal family Nonionidae, No. 191, Louisiana. Louisiana Department of Conservation, Geological Bulletin 2, p. 52, pl. 9, figs. 5a–c.
Hopkinsina danvillensis Howe and Wallace, 1932, Fayose, 1970, Stratigraphical paleontology of Afowo-1 well, Southwestern Nigeria. J. Min. Geol. 5 (1 and 2), 17, pl. 4, fig. 4.
Hopkinsina danvillensis Howe and Wallace, 1932, Petters, 1982, Central West Africa Cretaceous-Tertiary benthic foraminifera and stratigraphy. Palaeontographica, Bd. 179, Abs. A, nos. 1–3, pl. 6, fig. 6.

Diagnosis: Test elongate, round in transverse section, gradually tapering toward the two ends and widest about the middle of the test; chambers closely arranged; sutures slightly curved, depressed, biserial portion long, consisting of six slightly inflated chambers which overlap, all except the last chamber covered with longitudinal striae; aperture terminal, elliptical with a short neck.
Range: Early Eocene–Middle Eocene.

Hopkinsina hourqi Graham, de Klasz, and Rerat, 1965
Plate 17, Figs. 1 and 2

Uvigerina hourqi Graham, de Klasz, and Rerat, 1965, Quelques importants Foraminiferes du Tertiaire du Gabon (Afrique Equatoriale). Rev. Micropaleontol. 4, 75, pl. 1, fig. 9.
Uvigerina hourqi Graham, de Klasz, and Rerat, 1965, Petters, 1982, Central West Africa Cretaceous-Tertiary benthic foraminifera and stratigraphy. Palaeontographica, Bd. 179, Abs. A, nos. 1–3, p. 67, pl. 7, figs. 19, 20.
Uvigerina hourqi Graham, de Klasz, and Rerat, 1965, AGIP, 1988 SPA Foraminifera, pl. 11, figs. 1 and 2.
Uvigerinella 8, SPDC Catalogue, 1980, Stereoscan Photos, Paleontological Type Collection, Nigeria, pl. 30, figs. A18.

Diagnosis: Test elongate, cylindrical in shape, initial portion of the test triserial, later portion biserial; test tapering toward the proloculus; chambers increasing gradually in size as added;

sutures inconspicuous, depressed, and horizontal; wall ornamented with widely spaced costae; aperture terminal, round, at the end of a short neck.

Range: Paleocene–Late Eocene.

Family: ANOMALINIDAE Cushman, 1927
Genus: HANZAWAIA Asano, 1944
Hanzawaia mantaensis Galloway and Morrey, 1929
Plate 17, Fig. 3

Anomalina mantaensis Galloway and Morrey, 1929. A lower Tertiary foraminiferal fauna from Manta, Equador. Bull. Am. Paleontol, 28, pl. 4, fig. 5.

Hanzawaia mantaensis (Galloway and Morrey, 1929) Petters, 1982, Central West African Cretaceous-Tertiary benthic foraminifera and stratigraphy, Palaeontographica, Bd. 179, Abs. A, p. 84, pl. 13, figs. 9, 10.

Planulina 6: (*Hanzawaia mantaensis*, Galloway) SPDC Catalogue, 1980, Stereoscan Photos, Paleontological Type Collection, Nigeria, pl. 9, figs. B5V, B5D.

Diagnosis: Test free, trochoid, plano-convex, periphery moderately angled with keel, ventral side partially involute with elevated flaps on lower margin of chamber. Spiral side involute but without open umbilicus, central area with clear boss; sutures strongly curved, thickened; wall calcareous, granular in microstructure coarsely perforate; aperture an arch on periphery, extending onto convex involute side.

Range: Middle Oligocene–Early Miocene.

Hanzawaia concentrica Cushman, 1918
Plate 17, Figs. 4–8

Truncatulina concentrica Cushman, 1918a, Some Pliocene and Miocene foraminifera of the Coastal Plain of the U.S. US Geol. Surv. Bull. 676, 64, pl. 21, figs. 3a–c.

Cibicides concentricus (Cushman), Renz, 1948, Bull. Am. Paleontol., 127, pl. 10, figs. 8a–c.

Hanzawaia concentrica (Cushman), Puri, 1953, Contribution to the study of the Miocene of the Florida Panhandle. Florida Geol. Surv. Bull. 36, Part II, Foraminifera, p. 140, pl. 12, figs. 7–9.

Hanzawaia concentrica (Cushman), Poag, 1981, Ecologic Atlas of Benthic Foraminifera of the Gulf of Mexico, Marine Science International, Woods Hole, p. 68, pl. 39, fig. 1, pl. 40, fig. 1.

Valvulineria 8B: (*Hanzawaia concentrica* Cushman) SPDC Catalogue, 1980, Stereoscan Photos, Paleontological Type Collection, Nigeria, pl. 13, figs. A7V, A7D.

Diagnosis: Test free, trochospiral, plano-convex to unequally biconvex, spiral side flattened and partially evolute, umbilical side convex and involute; periphery angular, chambers inflated, seven to nine in the final whorl, increasing rapidly in size as added; sutures distinct, slightly curved on spiral and ventral sides, slightly depressed between later chambers on the ventral side; wall calcareous hyaline, smooth, finely perforate except for apertural face and umbilical extension on the spiral side; aperture an arch on the periphery with a distinct lip.

Range: Eocene–Early Miocene.

Hanzawaia strattoni Applin, 1925
Plate 17, Figs. 9–12

Truncatulina americana Cushman var. *strattoni* (Applin, 1925) Applin, Ellisor, and Kniker, 1925, In: Applin, E.S., Ellisor, A.C., Kniker, H.T., Sub-surface stratigraphy of the coastal plain of Texas and Louisiana. AAPG Bull. 9 (1), 99, pl. 3, fig. 3.

Cibicidina strattoni (Applin, 1925), Parker, 1954, Distribution of the foraminifera in the northeast Gulf of Mexico. Harvard University, Museum Comp. Zool. Bull. 111 (10), 544, pl. 13, figs. 8, 11.

Hanzawaia strattoni (Applin, 1925), Bandy, 1954, Distribution of some shallow-water foraminifera in the Gulf of Mexico. United States Geological Survey, Professional Paper No. 254-F, p. 136, pl. 31, fig. 4.

Hanzawaia strattoni (Applin, 1925), Akers and Dorman, 1964, Pleistocene foraminifera of the Gulf Coast. Tulane Stud. Geol., 3 (1), 37, pl. 15, figs. 13, 14.

Hanzawaia strattoni (Applin, 1925), Poag, 1981, Ecologic Atlas of Benthic Foraminifera of the Gulf of Mexico, Marine Science International, Woods Hole, p. 67, pl. 39, fig. 2, pl. 40, fig. 2.

Valvulineria 14: (*Hanzawaia strattoni* Applin, 1925) SPDC Catalogue, 1980, Stereoscan Photos, Paleontological Type Collection, Nigeria, pl. 13, figs. A9V, A9D.

Diagnosis: Test free, trochoid, ventral side slightly involute, periphery rounded, lobate in side view; chambers inflated, six chambers in final whorl, increasing in size rapidly as added, chambers on spiral side have imperforate flaps on the proximal margin of each chamber which merge to cover the entire umbilicus; sutures distinct, depressed on both the spiral and umbilical sides of the last few chambers, slightly curved to radial; wall calcareous hyaline, smooth, coarsely perforate, except for the flaps on spiral side which are imperforate; aperture an interiomarginal arch on the periphery of the whorl with a distinct lip, extending to spiral side under the apertural flaps, supplementary openings on spiral side along the sutures and outer margin of the central umbilical ring.

Range: Miocene–Pleistocene.

Hanzawaia isidroensis Cushman and Renz
Plate 17, Figs. 13 and 14

Cibicides 21: (*Hanzawaia isidroensis* Cushman and Renz). SPDC Catalogue, 1980, Stereoscan Photos, Paleontological Type Collection, Nigeria, pl.10, figures A9V, A9D.

Hanzawaia isidroensis Cushman and Renz. Bull. Am. Paleontol., 1985, 88 (322), 151, pl. 36, figure 2.

Diagnosis: Test free, trochospiral, plano-convex to unequally biconvex, spiral side flattened and partially evolute, umbilical side convex and involute; periphery acute; chambers, 9–12 in the final whorl, increasing rapidly in size as added; sutures distinct, slightly curved on both spiral and ventral sides; wall, calcareous hyaline, smooth, finely perforate, aperture an arch on the periphery.

Range: Miocene–Pliocene.

Genus: HETEROLEPA Franzenau, 1884
Heterolepa floridana Cushman, 1918
Plate 17, Figs. 15–17

Truncatulina floridana Cushman, 1918a, Some Pliocene and Miocene foraminifera of the Coastal Plain of the U.S. US Geol. Surv. Bull. 676, 62, pl. 19, figs. 2a–c.

Truncatulina floridana (Cushman) Nuttall, 1928, Tertiary foraminifera from the Naparima region of Trinidad (British West Indies). Q. J. Geol. Soc. London 84 (33), 98, pl. 7, figs. 14, 16.

Cibicides floridanus (Cushman), Cushman, 1930, The foraminifera of the Choctawhatchee Formation of Florida. Florida State Geol. Surv. Bull. 4, 61, pl. 12, figs. 3a–c.

Cibicides floridanus (Cushman), Galloway and Heminway, 1941, Tertiary foraminifera of Porto Rico New York Academy Sciences. Sci. Surv. Porto Rico Virgin-Islands, 3 (Pt. 4), 392, pl. 23, figs. 2a–c.

Cibicidoides floridanus (Cushman) LeRoy and Levinson, 1974, A deep-water Pleistocene microfossil assemblage from a well in the northern Gulf of Mexico. Micropaleontology 20 (1), 16, pl. 8, figs. 7–9.

Cibicides 14: (*Heterolepa floridana* Cushman) SPDC Catalogue, 1980, Stereoscan Photos, Paleontological Type Collection, Nigeria, pl. 26, fig. A8., pl. 10, figs. A5V, A5D.

Diagnosis: Test free, trochospiral, biconvex in edge view, subcircular in side view, peripheral margin subacute; 10–11 chambers in the final whorl, chambers of earlier whorls obscured by secondary calcite on the spiral side; sutures on umbilical side raised, limbate, and curved; those on the spiral side curved, limbate, and raised except on the last few chambers on which they are flush or depressed; wall calcareous, hyaline, smooth, and coarsely perforate; aperture a low interiomarginal arch with a slight projecting lip extending across the periphery to the spiral side and continuing along the spiral suture the length of two or three chambers.

Range: Middle Miocene–Pleistocene.

Heterolepa pseudoungeriana Cushman, 1922
Plate 17, Figs. 18–20

Truncatulina pseudoungeriana Cushman, 1922, The foraminifera of the Atlantic Ocean, Part 3. Textulariidae. US Natl. Museum Bull. 104, 97, pl. 20, fig. 9.

Truncatulina pseudoungeriana (Cushman) Heron and Earland, 1932, Foraminifera. Part. 1. The ice free area of the Falkland Islands and adjacent seas. Discov. Rep. 4., 424, pl. 14, fig. 35.

Eponides 13A: (*Heterolepa pseudoungeriana* Cushman, 1922) SPDC Catalogue, 1980, Stereoscan Photos, Paleontological Type Collection, Nigeria, pl. 26, fig. A8; pl. 17, figs. A IV, A1D; pl. 10, figs. A5V, A5D.

Heterolepa pseudoungeriana (Cushman) AGIP SPA, 1988, Foraminifera of the Niger Delta, pl. 8, figs. 4a–b.

Diagnosis: Test free, trochospiral, biconvex in edge view, subcircular in side view, periphery subacute, with 9–11 chambers in the final whorl, chambers of earlier whorls obscured by secondary calcite on the spiral side; sutures distinct, limbate and curved on the umbilical side, those on the spiral side curved, limbate, flush with the test; wall calcareous, hyaline, smooth, coarsely perforate; aperture an interiomarginal arch with distinct lip, continuing along spiral suture the length of 2–3 chambers.

Range: Early Miocene–Late Miocene.

Heterolepa mckannai Galloway and Wissler, 1927
Plate 18, Figs. 1–3

Cibicides mckannai Galloway and Wissler, 1927, Pleistocene Foraminifera from the Lomita Quarry, Palos Verdes Hills, California. J. Paleontol. 1, 65, pl. 10, figs 5, 6.

Cibicides 22, (*Heterolepa mckannai,* Galloway and Wissler). SPDC Catalogue, 1980, Stereoscan Photos, Paleontological Type Collection, Nigeria, Plate 10, figs. A10V, A10D.

Diagnosis: Test low, trochoid, free, with subcircular outline, plano-convex to unequally biconvex, umbilical (ventral) convexity accentuated by the well-developed umbo, peripheral margin subacute; chambers low, curved, with 9–10 chambers in the final whorl, increasing gradually in size as added; spiral side covered by thick calcareous material often forming a central knob; wall calcareous, coarsely perforate; sutures raised, arcuate, thick, somewhat depressed on umbilical side, covered with clear material, aperture extending along the spiral suture over about 4 chambers.

Range: Oligocene–Miocene.

Heterolepa bellincionii Giannine and Tauani, 1960
Plate 18, Figs. 4–6

Heterolepa bellincionii Giannine and Tavani, 1960. AGIP SPA, 1982, Foraminifera Padani, pl. XL, Figs. 6d, 6p, 6v.

Diagnosis: Test free, trochospiral, biconvex in edge view, subcircular in side view, peripheral margin rounded; 10–11 chambers in the final whorl; sutures distinct, limbate, and curved on the umbilical side; those on the spiral side are curved, limbate, flush with the test; wall calcareous, hyaline, smooth, coarsely perforate; aperture an interiomarginal arch with distinct lip.

Range: Middle Miocene–Pliocene.

Genus: CIBICIDOIDES Thalmann, 1939
Cibicidoides pseudoungerianus Cushman, 1922
Plate 18, Figs. 7 and 8

Cibicidoides psendoungerianus Cushman, 1922. AGIP SPA, 1982, Foraminifera Padani, pl. LII, figs. 1v, 1p, 1d.

Cibicidoides psendoungerianus Cushman, 1922. Benthic foraminifera biostratigraphy of the south Caribbean region, Bolli, Beckmann, and Saunders, Cambridge University Press, 1944, p. 244, fig. 57, 1

Diagnosis: Test free, trochospiral, biconvex, biumbonate; all chambers visible on spiral side, only those of final whorl visible on umbilical side; wall calcareous hyaline, with series of coarse perforations on the spiral side; sutures slightly curved; aperture a low interiomarginal equatorial arch with slight projecting lip.

Range: Oligocene–Pleistocene.

Cibicidoides crebbsi Hedberg, 1937
Plate 18, Figs. 9–11

Cibicidoides crebbsi Hedberg, 1937. Foraminifera of the Middle Tertiary Carapita Formation of northeastern Venezuela. J. Paleontol. 11 (98), 679, pl. 92, fig. 1.

Cibicidoides crebbsi Hedberg, 1937. Cenozoic Cosmopolitan deep-water benthic foraminifera (Elf Aquitaine) 1986, p. 141, pl. 45A, fig 1a–2c.

Diagnosis: Test free, ventral side convex, dorsal side slightly convex, about 3 whorls visible on dorsal side, 10–13 chambers in the last whorl, periphery subacute with a narrow border of glassy shell material, chambers gradually increase in size as added, sutures clear, very sinuous on ventral side, and fused to a glassy boss at the umbo, straight on dorsal side, and tangential to the periphery of the proceeding whorl on the spiral side; wall, calcareous coarsely perforate; aperture a low opening at the base of the last chamber between the periphery and the umbo.

Range: Late Oligocene–Late Miocene.

Cibicidoides pachyderma Rzehak, 1886
Plate 18, Figs. 12–14

Cibicidoides pachyderma Rzehak, 1886. Die Foraminiferenfauna der Neogenformation der Umgebung von Mahr. Ostrau. Naturf. Ver. Brunn, verh. Brunn [Brno, Czechoslovakia], Bd. 24 (1885), p. 87, pl. 1, fig. 5a–c.

Cibicidoides pachyderma Rzehak, 1886. Cenozoic cosmopolitan deep-water benthic foraminifera (Elf Aquitaine) 1986, p. 68, pl. 22, figs. 1a–1c.

Cibicidoides pachyderma Rzehak, 1886. Stratigraphic Atlas of fossil foraminifera, Jenkins and Murray, 1989, p. 472, pl. 9.5, figs. 22–24.

Diagnosis: Test free, subcircular, biconvex, with 10–12 chambers in the final whorl, periphery moderately to strongly keeled; spiral side coarsely perforate, umbilical side very finely perforate with slightly swollen umbonal area; umbilical side with flush or slightly depressed, gently curved sutures; sutures on the spiral side flushed, weakly limbate; in edge view, the sloping angle of the last chamber, the relatively flat umbonal region and the margin, gently sloping toward the periphery give this form a characteristic profile.

Range: Early Oligocene–Pleistocene.

Genus: ANOMALINOIDES Brotzen, 1942
Anomalinoides madrugaensis Cushman and Bermudez, 1948
Plate 18, Figs. 15–17

Anomalina madrugaensis Cushman and Bermudez, 1948, Some Paleocene foraminifera from the Madruga Formation of Cuba. Contrib. Cushman Lab. Foramin. Res. Sharon Mass 24, 86, pl. 15, figs. 4–6.

Anomalina madrugaensis Cushman, 1951, US Geol. Surv. Prof. Paper 232, 64, pl. 22, figs. 1–3.
Valvulineria 20, SPDC Catalogue, 1980, Stereoscan Photos, Paleontological Type Collection, Nigeria, pl. 14, figs. A5V, A5D.

Diagnosis: Test subcircular in outline, depressed in the central area at both spiral and umbilical sides, periphery rounded; chambers distinct, somewhat inflated; six chambers in the last whorl increasing gradually in size as added; sutures distinct, depressed, slightly curved; wall smooth, coarsely perforate; aperture interiomarginal, extending slightly to the central side.

Remarks: This typical Nigeer Delta species is also common in the Paleocene sections in the Gulf Coastal Region of the United States and adjacent areas.

Range: Early Paleocene–Middle Paleocene.

Genus: MELONIS De Montfort, 1808
Melonis padanum Perconig, 1952
Plate 18, Fig. 18

Melonis padanum Perconig, 1952. AGIP SPA, 1982, Foraminifera Padani, pl. XXI, figs 3, 3a.

Diagnosis: Test free, planispiral, symmetrical, and involute, deeply bi-umbilicate with umbilicus bordered by rim of nonperforate skeletal material, periphery rounded; 11–12 chambers per whorl; sutures flush to slightly depressed, radiate, slightly curved septa double bilamellar; wall calcareous coarsely perforate, granular in structure surface smooth; aperture an elongate interiomarginal; equatorial slit, extending laterally to umbilicus on both sides of the test.

Range: Late Oligocene–Pleistocene.

Melonis affinis Reuss
Plate 18, Figs. 19 and 20

Melonis affinis Reuss. Bull. Am. Paleontol. 1985, 88 (322), p. 151, pl. 36, figures 4a, 4b, 4c.

Diagnosis: Test free, biumbilicate, planispiral involute, outline smooth, periphery rounded, 11–12 chambers in the final whorl, low and slightly compressed, increasing slowly in size as added; sutures slightly curved, slightly limbate, radiating, smooth with a glassy appearance; aperture a low interiomarginal equatorial slit with a distinct protruding lip that continues into the umbilicus; umbilicus deep and open; wall calcareous, coarsely perforate.

Range: Miocene–Pleistocene.

Subfamily: PLANULININAE Bermudez, 1952
Genus: PLANULINA d'Orbigny, 1826
Planulina wuellerstorfi Schwager, 1866
Plate 19, Figs. 1 and 2

Planulina wuellerstorfi Schwager, 1866. Fossile Foraminiferen von kar Nikobar. NOVARA Exped. 1857–1859, Wien, Geol. Theil, 2 (2), p. 258, pl. 7, figs. 105, 107.

Planulina wuellerstorfi Schwager, 1866, Cenozoic cosmopolitan deep-water benthic foraminifera (Elf Aquitaine) 1986, p. 49, pl. 14, figs. 1a–1c, 2a–2c.
Planulina wuellerstorfi Schwager, 1866. AGIP SPA, 1982, Foraminifera Padani, pl. L, fig. 7v, 7p, 7d.

Diagnosis: Test large and flattened, coarsely perforate spiral surface, dorsal side generally flat, ventral side gently convex, periphery with a distinct keel, test thick walled, coarsely and densely perforate; consists of about 3 whorls with up to 10 chambers in the final whorl; sutures strongly recurved on both sides, distinctly limbate on dorsal side; aperture a low arch at the base of the final chamber.
Range: Middle Miocene–Pleistocene.

Planulina ariminensis d'Orbigny, 1826
Plate 19, Figs. 3 and 4

Planulina ariminensis d'Orbigny, 1826. Tableau methodique de la classe des Cephalopodes. Ann. Sci. Nat. Paris, ser. 1, 7, 245–314; Atlas, pl. 10–17, Crochard (Paris).
Planulina ariminensis d'Orbigny, 1826. Treatise in Invertebrate Paleontology (Loeblich and Tappan, 1964), p. C686, figs. 552,1.
Planulina ariminensis d'Orbigny, 1826. AGIP SPA, 1982, Foraminifera Padani, pl. L, figs. 4v, 4d, 4p, 4bis-v, 4bis-d, 4bis-p.
Planulina ariminensis d'Orbigny, 1826. Cenozoic cosmopolitan deep-water benthic foraminifera (Elf Aquitaine) 1986, p. 38, pl. 10, figs. 1–4.

Diagnosis: Test free, dischoidal, compressed, low trochospiral, spiral side evolute, umbilical side partially evolute, with 10–11 chambers in the final whorl, periphery truncate, with thick, marginal imperforate keel; wall calcareous, radial in structure, coarsely perforate; sutures limbate, slightly depressed between final chambers, dorsal side strongly punctuate, ventral side less punctuate; aperture medial, extending along the spiral line under slightly crenulated edges of final chambers.
Range: Middle Miocene–Early Pleistocene.

Family: NUMMULITIDAE de Blainville, 1825
Genus: HETEROSTEGINA d'Orbigny, 1826
Heterostegina sp. 1 d'Orbigny, 1826
Plate 19, Fig. 5

Heterostegina sp. 1 d'Orbigny, 1826, Tableau methodique de la classe des cephalopodes. Ann. Sci. Nat. Paris, ser. 1, 7, 305, pl. 17, figs. 5–7.

Diagnosis: Test free, compressed, early whorls embracing and fairly thick, lenticular in side view; chambers numerous, subdivided into chamberlets by transverse partitions which are visible from the exterior, chamberlets not directly connected with one another in the same chamber but are connected by the previously formed apertures with the adjacent chamberlets of the previous chamber; the central portion of the test is decidedly thickened; while the last-formed whorl is extremely thin; aperture consists of rows of pores on the outer face of the last chamber, one pore for each of the chamberlets.

Range: Oligocene–Middle Miocene.

Family: CERATOBULIMINIDAE Cushman, 1927
Genus: HOEGLUNDINA Brotzen, 1948
Hoeglundina elegans (d'Orbigny) Phleger and Parker, 1951
Plate 19, Figs. 6–8

Rotalia (Turbinulina) elegans d'Orbigny, 1826, Ann. Sci. Nat. ser. 1, 7, 276, No. 54.

Epistomina elegans (d'Orbigny) Cushman and Jarvis, 1930, Miocene foraminifera from Buff Bay, Jamaica. J. Paleontol. 4, 365, pl. 34, figs. 1a–c.

Hoeglundina elegans (d'Orbigny) Phleger and Parker, 1951, Ecology of foraminifera of northwest Gulf of Mexico. Geological Society of America, Memoir 46, Part 2, p. 22, pl. 12, figs. 1a,b.

Hoeglundina elegans (d'Orbigny) Kohl, 1985, Early Pliocene benthic forminifers from the Salina Basin, Southeastern Mexico. Bull. Am. Paleontol. 88 (322), 129, pl. 14, figs. 4, 5.

Hoeglundina elegans (d'Orbigny) AGIP SPA, 1982, pl. XLI, figs. 10v, 10p, 10d.

Diagnosis: Test free, trochospiral, spiral side evolute, umbilical side involute, biconvex, periphery acute, six to seven chambers in final whorl, increasing in size gradually; sutures distinct, flush with surface, strongly oblique on the spiral side, oblique to radial on the umbilical side; wall calcareous, hyaline, finely perforate, smooth, often with a complex pattern of thickenings made up of blotches of white shell materials on the umbilical side; aperture an elongate slit, extending across the breadth of the final chamber.

Range: Late Eocene–Early Pleistocene.

Genus: LENTICULINA Lamarck, 1804
Lenticulina calcar Linnaeus, 1767
Plate 19, Fig. 9

Nautilus calcar Linnaeus, 1767, Syst. Nat. ed. 12, 1767, p. 1162, no. 272; Gmelin's ed. 13, 1788, p. 3770, no. 2.

Cristellaria calcar (Linnaeus) Cushman, 1918, The Foraminifera of the Atlantic Ocean, p. 115, pl. 30, fig. 7; pl. 31, figs. 4, 5.

Lenticulina calcar (Linnaeus), AGIP SPA, 1982, Foraminifera Padani, pl. IX, figs. 3, 3a, 3bis, 3bis-a.

Diagnosis: Test free, biconvex, umbonate; chambers five or six in the last-formed coil, the earlier ones often seen through the transparent umbilical region; sutures distinct, nearly straight; wall smooth, periphery with a narrow carina, with a typically long acicular spine, one from each chamber; aperture at the angle of the chamber radiate, slightly projecting.

Range: Oligocene–Pleistocene.

Lenticulina grandis Cushman, 1920
Plate 19, Figs. 10–14

Cristellaria americana Cushman var. *grandis* Cushman, 1920, Lower Miocene foraminifera of Florida. US Geol. Surv. Prof. Paper, Washington, United States No. 128B, p. 68, pl. 11, fig. 2.

Cristellaria 13: (*Lenticulina grandis* Cushman) SPDC Catalogue, 1980, Stereoscan Photos, Paleontological Type Collection, Nigeria, pl. 32, fig. A9.

Lenticulina grandis (Cushman) Ozumba and Amajor, 1999a, Evolutionary relationships in some benthonic foraminifera of the Middle to the Late Miocene, Niger Delta. NAPE Bull. 14 (2), 160, fig. 3.

Diagnosis: Test planispiral, involute, subcircular in outline, slightly keeled, made up of 12 chambers in the last whorl, increasing gradually in size as added; sutures curved, raised, ending in a raised umbonal area; surface otherwise smooth; aperture radiate, without a definite projection.

Range: Eocene–Middle Miocene.

Lenticulina inornata d'Orbigny, 1846
Plate 19, Figs. 15–18

Robulina inornata (d'Orbigny) 1846, Foraminiferes fossiles du Bassin Tertiare de Vienne (Autriche). Gide et Comp. Paris, p. 102, pl. 4, figs. 25, 26.

Robulus inornatus (d'Orbigny) Cushman, Stewart, and Stewart, 1947, Upper Coaledo (Upper Eocene) Foraminifera from Yokan Point, Coos County, Oregon. Oregon Dep. Geol. Min. Ind. Bull. 36, 60, pl. 7, fig. 3.

Robulus inornatus Sullivan, 1962, Foraminifera from the type section of the San Lorenzo Formation, Santa Cruz County, California, University of California Publications in the Geological Sciences 37, 258, pl. 5, figs. 5a,b.

Cristellaria inornata (d'Orbigny) Subbotina, 1953, Iskopaemye Foraminifery SSR. Globigerinidy, Khantkeninidy i Globorotaliidy (Fossil Foraminifera of the USSR. Globigerinidae, Hantkeninidae and Globorotaliidae). Trudy Vsesoyuznogo Neftyanogo Nauchnoissledovatel skogo Geologorazvedochnogo Instituta (VNIGRI), Ser. 76, p. 156, pl. 1, fig. 1.

Lenticulina inornata (d'Orbigny), Fairchild, Wesendunk, and Weaver, 1969, Eocene and Oligocene foraminifera from the Santa Cruz Mountains, California. Geol. Sci. 81, 106, pl. 6, fig. 1.

Lenticulina inornata (d' Orbigny), AGIP SPA, 1982, Foraminifera Padani, pl. X, figs. 4, 4a.

Lenticulina inornata Ozumba and Amajor, 1999a, Evolutionary relationships in some benthonic foraminifera of the Middle to the Late Miocene, Niger Delta. NAPE Bull. 14 (2), 160, fig. 3.

Lenticulina inornata Okosun and Liebau, 1999, Foraminiferal biostratigraphy of eastern Niger Delta, Nigeria. NAPE Bull. 14 (2), 149, pl. 2, fig. 27, p. 150, pl. 3, figs. 14–15.

Cristellaria 16: (*Lenticulina inornata* d'Orbigny) SPDC Catalogue, 1980, Stereoscan Photos, Paleontological Type Collection, Nigeria, pl. 32, fig. A11.

Diagnosis: Test free, planispiral, involute, biconvex in edge view, circular in side view; five to seven chambers in the final whorl, increasing in size as added, has a transparent outer sheet on the test with fairly visible ornamentation; sutures distinct, flush with surface; wall calcareous, hyaline, smooth, finely perforate; aperture radiate, at peripheral angle, with a vertical robuline slit extending downward in the apertural face with a fin-like projection on either side.

Range: Oligocene–Pleistocene.

Lenticulina rotulata Lamarck
Plate 19, Fig. 19

Lenticulina rotulata Lamarck, 1804, AGIP SPA, 1982, Foraminifera Padani, pl. x, figures 6, 6a.
Lenticulina rotulata Lamarck, 1804, Frenzel, 2000, The benthic foraminifera of the Rugen Chalk (Lower Maastrichtian, Northeastern Germany) Neve Palaontolische Abhandlungen, 3: Dresden 361 p., plate 14, fig. 1.

Diagnosis: Test free, planispiral, lenticular, biconvex in edge view; circular in side view; periphery angular; many chambers in the final whorl increasing in size as added, has a transparent outer sheet on the test with fairly visible ornamentation; sutures distinct, curved, flush with the surface; wall calcareous hyaline, smooth, finely perforate; aperture radiate, terminate at the end of the final chamber.
Range: Middle Miocene–Pliocene.

Lenticulina pseudomamilligerus (Plummer) Cushman, 1951
Plate 19, Fig. 20; Plate 20, Fig. 1

Cristellaria pseudomamilligerus Plummer, 1927, Foraminifera of the Midway Formation in Texas. Bull. Univ. Texas Bureau Econ. Geol. Technol. 2644 (pt. 7), 98, fig. 11.
Robulus pseudomamilligerus Cushman, 1951, Adegoke et al., 2011, Sequence Stratigraphic Correlation of the Uppermost Miocene to Pleistocene interval of the Wells A1, A2 and A3, Niger Delta, Nigeria. NAPE Bull. 23 (2), 13, pl. 4, figs. 1–5.
Cristellaria 14: SPDC Catalogue, 1980, Stereoscan Photos, Paleontological Type Collection, Nigeria, pl. 32, fig. A10.

Diagnosis: Test slightly elongate, strongly compressed; periphery with a thick keel; chambers 9–11 in final whorl distinct; sutures curved, marked by conspicuous tapering elevations emanating from an irregularly developed central boss or from a group of protuberances; aperture, radiate and protruding.
Range: Paleocene–Early Eocene.

Genus: MEGASTOMELLA Faulkner, de Klasz and Rerat, 1963
Megastomella africana Faulkner, de Klasz and Rerat, 1963
Plate 20, Figs. 2–6

Megastomella africana Faulkner, de Klasz, and Rerat, 1963, *Mandjina* Nov. Gen. Nouveau Foraminifer de l'Eocene superieur du Gabon (Afrique Equatoriale) C.R. Som. Séance Soc. Geol. France, Fasc. 4, 20, pl. 1, figs. 1–2.
Alabamina 2: (*Megastomella africana* Faulkner, de Klasz, and Rerat), SPDC Catalogue, 1980, Stereoscan Photos, Paleontological Type Collection, Nigeria, pl. 11, figs. A3V, A3D.

Diagnosis: Test free, trochospiral, periphery subangular, seven to eight chambers in the final whorl, elongate, increasing rapidly in size as added, with two prominently elongated ridges on the final chamber bordering the primary aperture, all chambers visible on spiral and umbilical sides; sutures oblique; wall calcareous, finely perforate, granular in structure;

aperture an elongate vertical slit on the apertural face, extending from near the periphery almost to the umbilicus.

Range: Oligocene–Early Miocene.

Genus: PLECTOFRONDICULARIA Liebus, 1902
Plectofrondicularia sp. 1 Cushman and Laiming, 1931
Plate 20, Fig. 7

Plectofrondicularia sp. 1 Cushman and Laiming, 1931, Miocene foraminifera from Los Sauses Creek, Ventura County. California, J. Paleontol. 5, 105, pl. 11, fig. 13.

Diagnosis: Test free, compressed, elongate, chambers distinct, only slightly inflated, earliest chambers biserial, later becoming uniserial; sutures slightly depressed, distinct; wall calcareous, hyaline, smooth, very finely perforate, ornamented by prominent longitudinal costae which extend from the prolocular end to the base of the final chamber or to the aperture; aperture elongate, at end of the last-formed chamber.

Range: Oligocene–Miocene.

Family: UVIGERINIDAE Haeckel, 1894
Genus: RECTUVIGERINA Matthews, 1945
Rectuvigerina multicostata Cushman and Jarvis, 1929
Plate 20, Fig. 8

Siphogenerina multicostata Cushman and Jarvis, 1929, p. 14, pl. 3, fig. 6. In: The University of Kansas Paleontological Contributions Article 66: Stratigraphic significance of Uvigerinid Foraminifers in the Western Hemisphere, p. 8, pl. 14: 1–3; fig. 1.

Rectuvigerina multicostata Elf Aquitaine Catalogue, p. 117, pl. 36, fig. 1, 2.

Rectuvigerina multicostata AGIP SPA, 1982, Foraminiferi Padani, pl. XXXV, fig. 2.

Rectuvigerina 1: (*Rectuvigerina multicostata* Cushman and Jarvis, 1929) SPDC Catalogue, 1980, Stereoscan Photos, Paleontological Type Collection, Nigeria, pl. 30, fig. B1.

Diagnosis: Test free, elongate, length about 3 times the width, early portion triserial, slightly tapering, later portion uniserial, entire test somewhat fusiform, greatest width at about the middle; chambers distinct, slightly inflated; sutures distinct, slightly depressed; wall ornamented with numerous highly elevated longitudinal costae; costae of each chamber independent of those of adjacent chambers, early chambers with distinct short spines that are not aligned with the later costae; aperture circular, with a cylindrical neck and wide, thin lip.

Range: Late Oligocene–Early Pliocene.

Rectuvigerina tenuicostata Becker and Dusenbury, 1985
Plate 20, Fig. 9

Rectuvigerina tenuicostata Becker and Dusenbury, 1958, Mio–Oligocene Aquitanian foraminifera from the Goajira Peninsula Columbia. Contrib. Cushman Foramin. Res., Special Publication No. 4, p. 35, pl. 4.

Rectuvigerina 1: SPDC Catalogue, 1980, Stereoscan Photos, Paleontological Type Collection, Nigeria, pl. 30, fig. B1.

Rectuvigerina tenuicostata Petters, 1982, Central West Africa Cretaceous-Tertiary benthic foraminifera and stratigraphy. Palaeontographica, Bd. 179, Abs. A, p. 1–105, nos. 1–3, pl. 5, fig. 15.

Diagnosis: Test elongate, early stage biserial, later uniserial in two-thirds to half of the test, wider in early part but narrower in later part; test subcircular in outline; chambers inflated initially, later broad and less inflated; sutures horizontal, moderately depressed; wall densely perforate and costate; aperture small, terminal, on a narrow neck.

Range: Late Miocene–Pliocene.

Genus: SAGRINA d'Orbigny, 1837
Sagrina circumspinosa de Klasz and Rerat, 1961
Plate 20, Fig. 10

Bitubologenerina circumspinosa de Klasz and Rerat, 1961, p. 186, pl. 3, figs. 3–4.

Virgulina 4A: SPDC Catalogue, 1980, Stereoscan Photos, Paleontological Type Collection, Nigeria, pl. 28, fig. A6.

Sagrina circumspinosa (de Klasz and Rerat), Petters, 1982, Central West Africa Cretaceous-Tertiary benthic foraminifera and stratigraphy. Palaeontographica, Bd. 179, Abs. A, 1–105, nos. 1–3, p. 66, pl. 6, figs. 10, 11.

Diagnosis: Test free, elongate, ovate in outline, early portion triserial, later biserial; chambers over 12 in number commonly with angular margin; wall calcareous hyaline, coarsely perforate with prominent lobes; aperture elongate with a lip extending from base up the apertural face.

Remarks: The species is common in the Eocene of the Gilli Gilli-1 and Ihuo-1 wells.

Range: Eocene–Pliocene.

Genus: TRIFARINA Cushman, 1923
Trifarina reussi Cushman, 1913 (assigned in 1913, but genus only erected in 1923)
Plate 20, Fig. 11

Trifarina reussi Cushman, 1913, A monograph of foraminifera of the North Pacific Ocean: Part 3, Lagenidae, US Natl. Museum Bull. 71 (3), 412, pl. 39, figs. 3a,b.

Reussella 1: (*Trifarina reussi* Cushman) SPDC Catalogue, 1980, Stereoscan Photos, Paleontological Type Collection, Nigeria, pl. 28, fig. B1.

Diagnosis: Test free, triserial, thick and short, triangular in cross section, initial end broadly rounded; chambers few, elongate, overlapping, with weakly developed costae, three main costae developing at the angles of the test, traversing the entire length of the chambers, terminating at the base of the short neck; sutures indistinct; wall calcareous, finely perforate; aperture terminal, circular, at the end of a short neck with thickened rim and a toothplate.

Range: Late Miocene–Pleistocene.

Siphouvigerina auberiana d'Orbigny, 1839
Plate 20, Fig. 12

Uvigerina 12: (*Siphouvigerina auberiana* attenuate Cushman) d'Orbigny, 1839. SPDC Catalogue, 1980, Stereoscan Photos, Paleontological Type Collection, Nigeria, pl. 29, fig. A8.

Siphouvigerina auberiana d'Orbigny, 1839. Bull. Am. Paleontol., 1985, 88 (322), 138, pl. 23, figs. 1a–1e.

Diagnosis: Test free, elongate, fusiform, initial portion compact with a basal spine, chambers triserial on the early part, later biserial to uniserial, inflated, last-formed uniserial chamber distinctly elongated upward with a long neck; sutures distinct, depressed; wall calcareous, coarsely hispid; aperture terminal, rounded at the end of a long, hispid neck.

Range: Miocene–Pleistocene.

Genus: UVIGERINA d'Orbigny, 1826

Diagnosis: Test elongate, triserial, rounded in section; chambers inflated, wall calcareous, perforate, surface costate; aperture terminal, rounded, with nonperforate neck and may have phialine lip; internal toothplate with distinct wing at one side.

Uvigerina auberiana d'Orbigny, 1839
Plate 20, Fig. 13

Uvigerina auberiana d'Orbigny, 1839, Taxonomic note. In: Society of Economic Paleontologists and Mineralogists, 1960, p. 157, pl. LXXV, figs. 6, 7, 9.

Uvigerina 12: (*Siphouvigerina auberiana* attenuate Cushman) SPDC Catalogue, 1980, Stereoscan Photos, Paleontological Type Collection, Nigeria, pl. 29, fig. A8.

Diagnosis: Test free, elongate, fusiform, initial portion compact with a basal spine; chambers triserial in the early part, later biserial to uniserial; inflated. Last-formed chamber distinctly elongated upward; suture distinct, depressed; wall calcareous, coarsely hispid, aperture terminal at the end of a long hispid neck with a phialine lip and toothplate.

Range: Miocene–Pleistocene.

Uvigerina pygmaea d'Orbigny, 1826
Plate 20, Figs. 14 and 15

Uvigerina pygmaea d'Orbigny, 1826, Ann. Mag. Nat. Hist. 1865, 15–41, Ser. 3, v. 16, pl. 1–3.

Uvigerina pygmaea d'Orbigny, 1826, Treatise on Invertebrate Paleontology, 1964, p. C565, figs. 1a, 1b.

Uvigerina pygmaea d'Orbigny, AGIP SPA, 1982, Foraminifera Padani, pl. XXXIV, fig. 3, 3bis, 3bis-a.

Diagnosis: Test free, elongate, triserial, rounded in section, chambers inflated; wall calcareous, perforate, surface with costae and striae; aperture terminal, rounded at the end of a short neck with phialine lip and toothplate.

Range: Middle Miocene–Pleistocene.

Uvigerina gallowayi Cushman, 1929
Plate 20, Fig. 16

Uvigerina gallowayi Cushman, 1929, A Late Tertiary fauna of Venezuela and other related regions. Contrib. Cushman Lab. Foramin. Res. 5 (pt. 4), p. 94, pl. 13, figs. 33–34.

Uvigerina gallowayi Cushman, 1929, Petters, 1982, Central West Africa Cretaceous-Tertiary benthic foraminifera and stratigraphy. Palaeontographica, Bd. 179, Abs. A, nos. 1–3, p. 67, pl. 5, fig. 9.

Uvigerina alata Galloway and Morrey, 1929, A lower Tertiary foraminiferal fauna from Manta, Ecuador. Bull. Am. Paleontol. 15, (55), 38, pl. 6, fig. 1.

Uvigerina 13: (*Uvigerina gallowayi* Cushman) SPDC Catalogue, 1980, Stereoscan Photos, Paleontological Type Collection, Nigeria, pl. 29, fig. A9.

Uvigerina gallowayi Cushman, 1929, AGIP SPA, 1982, Foraminifera Padani, pl. XXXIII, fig. 5, 5a, 5bis.

Diagnosis: Test rather short especially in the megalospheric form, more elongate and fusiform in the microspheric form; apertural end truncated; chambers inflated; sutures largely obscured by the ornamentation which consists of longitudinal costae independent of chamber lines, costae almost spinose at the base becoming smoother toward the apertural end; aperture with a short neck.

Range: Middle Eocene–Middle Miocene.

Uvigerina havanensis Cushman and Bermudez, 1936
Plate 20, Fig. 17

Uvigerina havanensis Cushman and Bermudez, 1936, Additional new species of foraminifera and a new genus from the Eocene of Cuba. Contrib. Cushman Lab. Foramin. Res. 12 (3), 59, pl. 10, figs. 19–21.

Uvigerina havanensis Cushman and Bermudez, 1936, Petters, 1982, Central West Africa Cretaceous-Tertiary benthic foraminifera and stratigraphy. Palaeontographica, Bd. 179, Abs. A, nos. 1–3, p. 67, pl. 5, fig. 14.

Uvigerina 4: SPDC Catalogue, 1980, Stereoscan Photos, Paleontological Type Collection, Nigeria, pl. 29, fig. A1.

Diagnosis: Test elongate, about 2.5 times as long as broad, tapering; periphery lobulate; chambers distinct, inflated, increasing gradually in size as added; sutures distinct depressed; wall ornamented with numerous raised costae, with those of each chamber independent of adjacent ones; aperture terminal, with a short neck.

Range: Middle Eocene–Early Miocene.

Uvigerina peregrina Cushman, 1923
Plate 20, Figs. 18 and 19

Uvigerina peregrina Cushman, 1923, The foraminifera of the Atlantic Ocean. US Natl. Museum Bull. 104 (pt. 4), 166, pl. 42, figs. 7–10.

Uvigerina peregrina Cushman, 1923, Murray, 1971, An Atlas of British Recent Foraminiferids. Elsevier Publishing Company, Amsterdam, p. 121, pl. 50, figs. 1–7.
Uvigerina peregrina Cushman, 1923, LeRoy and Levinson, 1974, A deep-water Pleistocene microfossil assemblage from a well in the northern Gulf of Mexico. Micropaleontology, 20 (1), 10, pl. 5, fig. 18.
Uvigerina peregrina Cushman, 1923, Kohl, 1985, Early Pliocene benthic forminifers from the Salinas Basin, Southeastern Mexico. Bull. Am. Paleontol. 88 (322), 139, pl. 24, fig. 7.
Uvigerina peregrina Cushman, 1923, AGIP SPA, 1982, Foraminiferi Padani, pl. XXXIV, figs. 1, 1bis, 1 bis-a.

Diagnosis: Test free, triserial; chambers inflated; sutures depressed, indistinct; wall calcareous, perforate, ornamented by numerous high, blade-like segmented or fimbriate costae that are prominent on each chamber, costae not continuous from one chamber to another; aperture terminal, circular, with a toothplate and distinct, thickened lip at the end of a short neck.
Range: Middle Miocene–Early Pliocene.

Uvigerina subperegrina Cushman and Kleinpell, 1934
Plate 20, Fig. 20; Plate 21, Fig. 1

Uvigerina subperegrina Cushman and Kleinpell, 1934, New and unrecorded foraminifera from the California Miocene. Contrib. Cushman Lab. Foramin. Res. 10 (1), 12, pl. 2, figs. 9–11.
Uvigerina 8: (*Uvigerina subperegrina* Cushman and Kleinpell, 1934) SPDC Catalogue, 1980 Stereoscan Photos, Paleontological Type Collection, Nigeria, pl. 29, fig. A5.
Uvigerina subperegrina Ozumba and Amajor, 1999a, Evolutionary relationships in some benthonic foraminifera of the Middle to the Late Miocene, Niger Delta. NAPE Bull. 14 (2), 163, fig. 6.

Diagnosis: Test free, triserial, circular to ovate in transverse section; chambers inflated, broader than high, strongly embracing; sutures depressed, indistinct, obscured by overlapping chambers; wall calcareous, hyaline, finely perforate, ornamented by low, narrow longitudinal costae which project below the basal margin without crossing the suture, giving a scalloped appearance to the base of each chamber; aperture terminal, circular, with a toothplate and distinct thickened lip at the end of a short neck.
Range: Middle Miocene–Late Miocene.

Uvigerina proboscidea Schwager, 1866
Plate 21, Figs. 2 and 3

Uvigerina proboscidea Schwager, 1866, Fossile Foraminiferen von Kar-Nicobar. Novara Exped., Geologisch Theil, v. 2, p. 250, pl. 7, fig. 96.
Uvigerina proboscidea Galloway and Morrey, 1929, A lower Tertiary foraminiferal fauna from Manta, Ecuador. Bull. Am. Paleontol. 15 (55), 39, pl. 6, fig. 4.
Uvigerina proboscidea Morkhoven, Berggren, and Edwards, 1986, Cenozoic cosmopolitan deep-water benthic foraminifera. Bull. Centres Rech. Explor. Prod. Elf Aquitaine, Mem. II, p. 29, pl. 6, figs. 1–3.

Uvigerina proboscidea AGIP SPA, 1982, Foraminiferi Padani, pl. XXXIV, fig. 2.

Diagnosis: Test free, elongate, fusiform, initial portion compact with a basal spine; chambers triserial in the early part, later biserial to uniserial, inflated, last-formed chamber distinctly elongated upward; sutures distinct, depressed; wall calcareous, finely hispid; aperture terminal, at the end of a long, hispid neck with a phialine lip and toothplate.

Range: Late Miocene–Pleistocene.

Uvigerina sparsicostata Cushman and Laiming, 1931; LeRoy, 1944
Plate 21, Figs. 4–6

Uvigerinella sparsicostata Cushman and Laiming, 1931, Miocene foraminifera from Los Sauses Creek, Ventura County. J. Paleontol. 5 (2), 112, pl. 12, fig. 12.

Uvigerina sparsicostata LeRoy, 1944, Miocene foraminifera from Sumatra and Java, Netherlands East Indies, Part. 1. Miocene foraminifera of Central Sumatra, Netherlands East Indies. Colorado Sch. Mines Q. 39, 31, pl. 1, figs. 45–46, pl. 5, fig. 25.

Uvigerina sparsicostata Cushman and Laiming, 1931, Petters, 1982, Central West Africa Cretaceous-Tertiary benthic foraminifera and stratigraphy. Palaeontographica, Bd. 179, Abs. A, no. 1–3, p. 68, pl. 5, figs. 2, 3; pl. 10, fig. 1.

Uvigerina sparsicostata Cushman and Laiming, 1931, Ozumba and Amajor, 1999a, Evolutionary relationships in some benthonic foraminifera of the Middle to the Late Miocene, Niger Delta. NAPE Bull. 14 (2), 163, fig. 6.

Uvigerina sparsicostata Cushman and Laiming, 1931, Okosun and Liebau, 1999, Foraminiferal biostratigraphy of eastern Niger Delta, Nigeria. NAPE Bull. 14, (2), 148, pl. 1, figs. 28, 30, 35, pl. 3, fig. 16.

Uvigerina 5: (*Uvigerina sparsicostata* Cushman and Laiming) SPDC Catalogue, 1980, Stereoscan Photos, Paleontological Type Collection, Nigeria, pl. 29, fig. A2; pl. 30, figs. A5.

Diagnosis: Test triserial, fusiform, slightly lobulate at the periphery, wide in the middle and tapering at both ends and almost becoming uniserial at later stages; chambers distinct, slightly inflated in the middle of the test; sutures distinct, curved, only slightly depressed; wall ornamented with costae, those of adjacent chambers being independent of one another; aperture with a distinct neck and lip.

Range: Early Oligocene–Middle Miocene.

Uvigerina asperula Czjzek, 1848
Plate 21, Figs. 7 and 8

Uvigerina asperula Czjzek, 1848. Taxonomic notes. In: Society of Economic Paleontologists and Mineralogists, 1960, p. 156, pl. LXXV, figs. 6–9.

Uvigerina 11: (*Uvigerina asperula* Czjzek, 1848) SPDC Catalogue, 1980, Stereoscan Photos, Paleontological Type Collection, Nigeria, pl. 29, fig. A7.

Uvigerina asperula Czjzek, 1848. AGIP SPA, 1982, Foraminifera Padani, pl. XXXII, figs.7, 7a.

Diagnosis: Test free, elongate, fusiform, initial portion compact with a basal spine; chambers triserial on the early part, later biserial to uniserial, inflated; sutures distinct, depressed;

wall calcareous, coarsely hispid; aperture terminal at the end of a long, hispid neck with a phialine lip and toothplate.

Range: Oligocene–Middle Miocene.

Uvigerina mantaensis Cushman and Edwards, 1938
Plate 21, Fig. 9

Uvigerina mantaensis Cushman and Edwards, 1938. Notes on the Oligocene species of *Uvigerina* and *Angulogerina*. Contrib. Cushman Lab. Foramin. Res. 14 (pt. 4), 84, pl. 14, figs. 8a,b.

Uvigerina mantaensis Cushman and Edwards, 1938. Bull. Am. Paleontol. 1985, 88 (322), 139, pl. 24, figs. 4a–4c.

Uvigerina mantaensis Cushman and Edwards, 1938. Benthic foraminiferal biostratigraphy of the South Caribbean region, Bolli, Beckmann, and Saunders, Cambridge University Press, 1994, p. 238, figs. 54, 58.

Diagnosis: Test free, elongate, fusiform, initial portion compact with a basal spine, chambers triserial on the early part, later biserial to uniserial, inflated; sutures distinct, depressed; wall calcareous hyaline, finely perforate; surface smooth; aperture terminal at the end of a long neck with a phialine lip and toothplate.

Range: Middle Oligocene–Late Miocene.

Uvigerina hispida Schwager, 1866
Plate 21, Figs. 10 and 11

Uvigerina hispida Schwager, 1866. Fossile Foraminiferen von Kar Nikobar. NOVARA Exped. 1857–1859, Wien, Geol. Theil, 2 (2), p. 249, pl. 7, fig. 95.

Uvigerina hispida Schwager, 1866. Cenozoic cosmopolitan deep-water benthic foraminifera (Elf Aquitaine) 1986, p. 62, pl. 20, figs. 1–4.

Diagnosis: Test free, elongate, fusiform, initial portion compact with a basal blunt spine to coarse papillae, chambers triserial on the early part, later biserial to uniserial, inflated; ornamented with short, coarse spines which are not aligned; sutures distinct, depressed; wall calcareous; aperture terminal at the end of a short neck with a phialine lip and toothplate.

Range: Early Miocene–Pleistocene.

6.3 PLANKTONIC FORAMINIFERA

The 156 planktonic foraminiferal species presented in this section fall under the suborder GLOBIGERINA of the Loeblich and Tappan (1987) classification. These species are described alphabetically under the various genera to which they belong. Since some planktonic foraminifera genera have restricted stratigraphic ranges, in which case Paleogene genera (e.g., *Morozovella, Subbotina, Globanomalina, Acarinina*) are distinct from Neogene genera (e.g., *Globoquadrina, Globigerinoides, Globorotalia, Sphaeroidinellopsis, Catapsydrax*), it was deemed

TABLE 6.1 Suprageneric Classification of the Niger Delta Planktonic Foraminifera

Superfamily	Family	Genus
GLOBOROTALIACEA	TRUNCOROTALOIDIDAE	*Morozovella* *Acarinina* *Truncorotaloides* *Praemurica*
	CATAPSYDRACIDAE	*Subbotina* *Cassigerinelloita* *Globorotaloides* *Globoquadrina* *Dentoglobigerina* *Catapsydrax*
	GLOBOROTALIIDAE	*Igorina* *Turborotalia* *Neogloboquadrina* *Globorotalia*
	EOGLOBIGERINIDAE	*Globoconusa*
HANTKENINACEA	GLOBANOMALINIDAE	*Globanomalina* *Pseudohastigerina*
HETEROHELICACEA	GUEMBELITRIIDAE	*Guembelitria*
	CHILOGUEMBELINIDAE	*Chiloguembelina* *Streptochilus*
HANTKENINACEA	CASSIGERINELLIDAE	*Cassigerinella*
GLOBIGERINACEA	GLOBIGERINIDAE	*Globigerina* *Globigerinoides* *Praeorbulina* *Orbulina* *Sphaeroidinellopsis*
	CANDEINIDAE	*Globigerinita* *Tenuitella*
	HASTIGERINIDAE	*Hastigerina*

preferable to present, as much as possible, the various genera in their sequence of stratigraphic incoming rather than follow a strictly taxonomic arrangement at the suprageneric level. The suprageneric classification of the 30 planktonic foraminifera genera encountered in this study is presented separately in the Table 6.1.

Genus: MOROZOVELLA McGowran, 1968
Morozovella acuta Toulmin, 1941
Plate 22, Fig. 1

Globorotalia wilcoxensis Cushman and Ponton var. *acuta* Toulmin, 1941, Eocene smaller Foraminifera from the Salt Mountain Limestone of Alabama. J. Paleontol. 15, 608, pl. 82, figs. 6–8.

Morozovella acuta Toumarkine and Luterbacher, 1985, Paleocene and Eocene planktic foraminifera. In: Bolli, H.M., Saunders, J.B., Perch-Nielsen, K. (Eds.), 1985, Plankton Stratigraphy. Cambridge University Press, Cambridge, p. 111, text-fig. 14 (7–8; 5, 7, 10).

Morozovella acuta Olsson et al., 1999, Atlas of Paleocene planktonic foraminifera. Smithsonian Contrib. Paleobiol. 85, 55, pl. 45, figs. 1–14.

Original Diagnosis: Test trochiform, plano-convex dorsal side flat, ventral side strongly convex, deeply umbilicate, periphery strongly lobate, acute, and bounded by a thick flange; chambers distinct, about 4½ in the last whorl, increasing regularly in size as added; sutures distinct, on the dorsal side slightly curved, limbate, slightly if at all depressed, on the ventral side radiate, depressed; wall roughened with minute, low spinose processes, especially along the peripheral border; aperture an arched opening on the ventral side of the final chamber extending from the periphery flange to the umbilicus.

Range: Late Paleocene–Earliest Eocene.

Morozovella acutispira Bolli and Cita, 1960
Plate 22, Figs. 2 and 3

Globorotalia acutispira Bolli and Cita, 1960, Globigerine e Globorotalie del Paleocene di Paderno d'Adda (Italia). Rivista Italiana di Paleontologiae Stratigrafia, 66, 375–377, pl. 35, figs. 3a–c.

Morozovella acutispira Petters, 1983, Gulf of Guinea planktonic foraminiferal biochronology and geological history of the South Atlantic. J. Foramin. Res., 13, 52, pl. 5, fig. 4.

Morozovella acutispira Olsson et al., 1999, Atlas of Paleocene planktonic foraminifera. Smithsonian Contrib. Paleobiol. 85, 56, pl. 46, figs. 1–15.

Diagnosis: Test lenticular to nearly circular, plano-convex to biconvex, gently lobate; axial periphery sharp, keeled; chambers petaloid, four to six in the last whorl; umbilical sutures radial and gently curved, depressed, sutures raised, beaded and curved on spiral side as extensions of beaded equatorial keel; aperture low arch, interiomarginal umbilical–extraumbilical; umbilicus deep, narrow; wall distinctly pustulose.

Range: Late Paleocene.

Morozovella aequa Cushman and Renz, 1942; Jenkins, 1971
Plate 22, Figs. 4–6

Globorotalia crassata (Cushman) var. *aequa* Cushman and Renz, 1942, Eocene Midway Foraminifera from Soldado Rock, Trinidad. Contrib. Cushman Lab. Foramin. Res. 18, 12, pl. 3, fig. 3.

Globorotalia (*Morozovella*) *aequa aequa* Jenkins, 1971, New Zealand Cenozoic planktonic foraminifera. Paleontol. Bull. Wellington 42, 100, pl. 7, figs. 167–169.

Morozovella aequa Petters, 1983, Gulf of Guinea planktonic foraminiferal biochronology and geological history of the South Atlantic. J. Foramin. Res. 13, 52, pl. 5, fig. 3.

Morozovella aequa Olsson et al., 1999, Atlas of Paleocene planktonic foraminifera. Smithsonian Contrib. Paleobiol. 85, 57, pl. 47, figs. 1–16.

Diagnosis: Test moderate, umbilico-convex trochospire, flat spiral side, strongly convex umbilically, lobate outline, with keel; three to four chambers in ultimate whorl, last chamber occupies one-third to half the test, rapidly increasing angulo-conical chambers; sutures widely depressed, radial on umbilical side; curved, depressed, or flush; deep, narrow umbilicus; aperture a low extraumbilical–umbilical arch with thin lip; wall surface pustulose, smooth in later part.

Range: Late Paleocene–Early Eocene.

Morozovella angulata White, 1928
Plate 22, Figs. 7–9

Globorotalia angulata White, 1928, Some index foraminifera of the Tampico Embayment area of Mexico. J. Paleontol. 2, p. 191–192, pl. 27, figs. 13a–c.

Globorotalia 5: SPDC Foraminiferal Type Collection.

Globorotalia angulata Fayose, 1970, Stratigraphical paleontology of Afowo-1 well, Southwestern Nigeria. J. Min. Geol. 5 (1 and 2), 52–53, pl. 13, figs. 5, 7.

Morozovella angulata Petters, 1983, Gulf of Guinea planktonic foraminiferal biochronology and geological history of the South Atlantic. J. Foramin. Res. 13, 52, pl. 5, figs. 11, 15, 19.

Diagnosis: Test umbilico-convex, flat to gently convex spiral side, convex and conical umbilical side, lobate periphery, axial periphery sharp with prominent keel; chambers angular-conical, four to five in final whorl, increased rapidly in size; sutures umbilically depressed, radial; strongly curved and depressed on spiral side; umbilicus narrow, deep, with strong umbilical shoulders above umbilicus and higher on ultimate chamber; aperture low umbilical–extraumbilical arch, with thin lip; wall roughly pustulose especially on umbilical side.

Remarks: This is a distinctive taxon characterized by oblique interlocking alignment of chambers especially on spiral side.

Range: Middle–Late Paleocene.

Morozovella conicotruncata Subbotina, 1947
Plate 22, Figs. 10–12

Globorotalia conicotruncata Subbotina, 1947, Foraminifery Datskikh i Paleogenovykh-otlozheniy severnogo Kavkaza (Foraminifera of Danian and Paleogene deposits of the northern Caucasus) in Mikrofauna Neftyanykh Mastorozhdenig Kavkaza, Embyi Sredny Azii. Leningrad: VNGRI, p. 115, pl. 4, figs. 11–13.

Morozovella conicotruncata Olsson et al., 1999, Atlas of Paleocene planktonic foraminifera. Smithsonian Contrib. Paleobiol. 85, 60–61, pl. 11, figs. 10–15; pl. 50, figs. 1–15.

Diagnosis: Test conical trochospire, flat spiral side, strongly convex umbilically; five chambers in ultimate whorl; periphery broadly lobate, acute, not keeled; sutures depressed, curved, radial on umbilical side; narrow, deep umbilicus; aperture slit-like, curved, extending from umbilicus to about midway between umbilicus and periphery; wall rough, pustulose.

Range: Middle–Late Paleocene.

Morozovella edgari Premoli Silva and Bolli, 1973
Plate 22, Fig. 13

Morozovella edgari Premoli Silva and Bolli, 1973, Late Cretaceous to Eocene Planktonic Foraminifera and Stratigraphy of Leg 15 Sites in the Caribbean Sea. Initial Reports Deep Sea Drilling Project, vol. 15, p. 526, pl. 7, figs. 10–12.

Morozovella edgari Toumarkine and Luterbacher, 1985, Paleocene and Eocene planktic foraminifera. In: Bolli, H.M., Saunders, J.B., Perch-Nielsen, K. (Eds.), 1985, Plankton Stratigraphy. Cambridge University Press, Cambridge, p. 112, figs. 15, 16.

Diagnosis: Test characteristically small, low angulo-conical, trochospire; lobate periphery; weakly developed keel; chambers wedge shaped, raised umbilically; inflated initially in early part of test; 4½ to 6 in the last whorl, exhibiting gradual size increase, with a reduction in the size of the ultimate chamber; sutures depressed and straight umbilically and on spiral side strongly bent; aperture low, umbilical–extraumbilical slit, bordered by lip; wall is pustulose and densely ornamented in early part of test; last chamber smooth.

Range: Latest Paleocene–Early Eocene.

Morozovella gracilis Bolli, 1957
Plate 22, Figs. 14 and 15

Morozovella formosa gracilis Bolli, 1957a–c, The genera *Globigerina* and *Globorotalia* in the Paleocene-Lower Eocene Lizard Springs Formation of Trinidad, B.W.I. US Natl. Museum Bull. 215, 75, pl. 18, figs. 4–6.

Morozovella gracilis Olsson et al., 1999, Atlas of Paleocene planktonic foraminifera. Smithsonian Contrib. Paleobiol. 85, 61–62, pl. 54, figs. 13–15.

Original Diagnosis: Shape of test very low trochospiral, spiral side almost flat or slightly convex, umbilical side distinctly convex, equatorial periphery lobate, axial periphery angular with a faint keel ornamented with spines; wall calcareous, perforate, surface distinctly spinose; chambers angular, inflated, about 12, arranged in 2½ to 3 whorls, the 5 or 6 chambers of the last whorl increasing rapidly in size; sutures on the dorsal side slightly curved to oblique, slightly depressed; on umbilical side radial, distinctly depressed; umbilicus fairly narrow, deep, open; aperture a low arch, interiomarginal, extraumbilical–umbilical; coiling between 90% and 100% dextral.

Range: Late Paleocene–Early Eocene.

Morozovella lensiformis Subbotina, 1953
Plate 22, Figs. 16–18

Globorotalia lensiformis Subbotina, 1953, Iskopaemye Foraminifery SSR. Globigerinidy, Khantkeninidy i Globorotaliidy (Fossil Foraminifera of the USSR. Globigerinidae, Hantkeninidae and Globorotaliidae). Trudy Vsesoyuznogo Neftyanogo Nauchnoissledovatel skogo Geologorazvedochnogo Instituta (VNIGRI), Ser. 76, p. 214, pl. 18, figs. 4, 5.

Globorotalia lensiformis Stainforth et al., 1975, Cenozoic planktonic foraminiferal zonation and characteristics of index forms. Paleontological Contribution, University of Kansas, Article 62, p. 200, fig. 61.

Diagnosis: Test moderate, umbilico-convex trochospire, flat spiral side, strongly convex umbilically; lobate equatorial periphery; sharp, keeled axial periphery; chambers in last whorl tightly arranged, four to five angulo-conical, increased rapidly in size; ultimate chamber constituting over one-third of entire test; sutures spirally curved, depressed or flush; narrow, depressed and radiate on umbilical side; small, deep umbilicus; aperture low umbilical–extraumbilical slit. Wall surface strongly rugose.

Range: Early Eocene.

Morozovella marginodentata Subbotina, 1953
Plate 22, Fig. 19

Globorotalia marginodentata Subbotina, 1953, Iskopaemye Foraminifery SSR. Globigerinidy, Khantkeninidy i Globorotaliidy (Fossil Foraminifera of the USSR. Globigerinidae, Hantkeninidae and Globorotaliidae). Trudy Vsesoyuznogo Neftyanogo Nauchnoissledovatel skogo Geologorazvedochnogo Instituta (VNIGRI), Ser. 76, p. 212–213, pl. 17, figs. 14–16, pl. 18, figs. 1–3.

Globorotalia marginodentata Stainforth et al., 1975, Cenozoic planktonic foraminiferal zonation and characteristics of index forms. Paleontological Contribution, University of Kansas, Article 62, p. 203, fig. 65.

Morozovella marginodentata Petters, 1983, Gulf of Guinea planktonic foraminiferal biochronology and geological history of the South Atlantic. J. Foramin. Res. 13, 54, pl. 5, fig. 12.

Diagnosis: Test umbilico-convex, lenticular, inner whorl higher than last whorl on spiral side, periphery strongly lobate; sharp axial periphery with broad, frilled keel; chamber compressed, angulo-conical, increased slowly in size; four to five in the last whorl; sutures on umbilical side broadly depressed, radial; curved, raised to flush on spiral side; umbilicus closed or narrow; sharp ornamented umbilical shoulders; aperture low umbilical–extraumbilical interiomarginal arch; wall rugose especially on umbilical side.

Range: Late Paleocene–Early Eocene.

Morozovella quetra Bolli, 1957
Plate 23, Figs. 1–3

Globorotalia quetra Bolli, 1957a, The genera *Globigerina* and *Globorotalia* in the Paleocene—Lower Eocene Lizard Springs Formation of Trinidad, B.W.I. US Natl. Museum Bull. 215, 79–80, pl. 19, figs. 1–6.

Globorotalia quetra Stainforth et al., 1975, Cenozoic planktonic foraminiferal zonation and characteristics of index forms. Paleontological Contribution, University of Kansas, Article 62, p. 221, fig. 80.

Original Diagnosis: Shape of test very low trochospire, spiral side flat or slightly concave, umbilical side strongly convex, angular; equatorial periphery lobate; axial periphery subacute to acute; a spiny peripheral keel is often present in the early chambers of the last whorl; ultimate and penultimate chambers acute or rounded; wall calcareous, perforate, distinctly spinose; chambers angular to subangular, inflated; about 12, arranged in 2 whorls, the 4–5 chambers of the last whorl increasing rapidly in size; sutures on spiral side oblique or curved, depressed; on umbilical side radial, depressed; umbilicus fairly narrow, deep, open; aperture a low arch; interiomarginal, extraumbilical–umbilical.

Remarks: Distinctive keeled forms with depressed spiral side and highly perforate wall occur in the Niger Delta. The prominent keel, ornamentation, spiral shape, and highly convex umbilical side distinguish *M. quetra* from *M. lensiformis.*

Range: Early Eocene.

Morozovella subbotinae Morozova, 1939
Plate 23, Fig. 4

Globorotalia subbotinae Morozova, 1939, K stratigrafii verkhnego mela i paleogena Embenskoi oblasti po faune foraminifer. Byulletin Moskovskogo Obshchestva Ispytateley Prirody, Otdel Geologicheskii 23, 80–81, pl. 2, fig. 16.

Globorotalia subbotinae Stainforth et al., 1975, Cenozoic planktonic foraminiferal zonation and characteristics of index forms. Paleontological Contribution, University of Kansas, Article 62, p. 230, fig. 89.

Morozovella subbotinae Petters, 1983, Gulf of Guinea planktonic foraminiferal biochronology and geological history of the South Atlantic. J. Foramin. Res. 13, 54, pl. 5, figs. 8, 9.

Diagnosis: Test rounded low trochospire, flat spirally and umbilically inflated with raised chamber shoulders around umbilicus; equatorial periphery even and weakly lobate; axial periphery acute, spinose keel present; chambers four to five in final whorl, crescentic to rhomboidal; increased rapidly in size, last chamber occupying one-third to one-fourth of test; sutures depressed, radial on umbilical side; recurved on spiral side varying from depressed to slightly raised and beaded; deep, narrow umbilicus; aperture a low arch, umbilical–extraumbilical; wall cancellate; spinose along peripheral margin.

Remarks: *Morozovella subbotinae* is distinguishable from *M. aequa* by its prominent keel and tighter coil; and from *M. marginodentata* by its more lenticular shape and a broader keel in the latter.

Range: Latest Paleocene–Early Eocene.

Morozovella velascoensis Cushman, 1925
Plate 23, Figs. 5–7

Pulvinulina velascoensis Cushman, 1925, Some new foraminifera from the Velasco Shale of Mexico. Contrib. Cushman Lab. Foramin. Res. 1, 19, pl. 3, figs. 5a–c.

Globorotalia velascoensis Fayose, 1970, Stratigraphical paleontology of Afowo-1 well, Southwestern Nigeria. J. Min. Geol. 5 (1 and 2), 70, pl. 15, fig. 10.

Morozovella velascoensis Petters, 1983, Gulf of Guinea planktonic foraminiferal biochronology and geological history of the South Atlantic. J. Foramin. Res. 13, 54, pl. 5, figs. 1, 2.

Diagnosis: Test circular plano-convex; spirally convex to concave; conical umbilical side, strongly convex with distinctive blunt chamber umbilical terminations; equatorial periphery gently lobate; axial periphery acute with prominent keel; chambers five to eight in final whorl, angular-conical which increases gradually and regularly in size, with ultimate chamber sometimes reduced in size; sutures depressed, radial on umbilical side; prominent, raised, beaded on spiral side; umbilicus wide deep; well-developed chamber tips comprised of fused blunt, thick pustules around umbilicus; aperture interiomarginal masked by fused calcite masses; wall coarsely perforate to pustulose.

Remarks: Unlike *M. acuta* and *M. aequa*, *M. velascoensis* has more chambers in the last whorl, more prominent umbilical shoulders, and better developed keel. Stainforth et al. (1975) suggested a morphologic link between *M. velascoensis* and *M. acuta*; the latter being the possible ancestor (Tourmarkine and Luterbacher, 1985).

Range: Late Paleocene–Early Eocene.

Genus: SUBBOTINA Brotzen and Pozaryska, 1961
Subbotina corpulenta Subbotina, 1953
Plate 23, Figs. 8–10

Globigerina corpulenta Subbotina, 1953, Iskopaemye Foraminifery SSR. Globigerinidy, Khantkeninidy i Globorotaliidy (Fossil Foraminifera of the USSR. Globigerinidae, Hantkeninidae and Globorotaliidae). Trudy Vsesoyuznogo Neftyanogo Nauchnoissledovatel skogo Geologorazvedochnogo Instituta (VNIGRI), Ser. 76, p. 76, pl. 9, figs 5–7; pl. 10, figs. 1–4.

Globigerina corpulenta Stainforth et al., 1975, Cenozoic planktonic foraminiferal zonation and characteristics of index forms. Paleontological Contribution, University of Kansas, Article 62, p. 263, fig. 112.

Subbotina corpulenta Petters, 1983, Gulf of Guinea planktonic foraminiferal biochronology and geological history of the South Atlantic. J. Foramin. Res. 13, 59, pl. 4, figs. 19, 23.

Diagnosis: Test a stout trochospire, low to moderately high spiral side; outline quadrate, moderately to strongly lobate, axial periphery rounded; chambers usually four in the last whorl, slightly inflated, final chambers small; sutures wide, depressed; wider and radial on umbilical side, slightly slanting on spiral side; umbilicus wide; aperture arched, umbilical to extraumbilical; wall cancellate.

Remarks: This species has a fairly stable morphology in the Niger Delta, with a characteristic four-chambered, low to moderately spired test, the ultimate chamber being reduced in size; and the wall cancellate. It differs from *Acarinina wilcoxensis* in the wider umbilicus and sutures; less convex umbilical side; and the smaller ultimate chamber.

Range: Late Middle–Late Eocene.

Subbotina inconstans Subbotina, 1953
Plate 23, Figs. 11 and 12

Globigerina inconstans Subbotina, 1953, Iskopaemye Foraminifery SSR. Globigerinidy, Khantkeninidy i Globorotaliidy (Fossil Foraminifera of the USSR. Globigerinidae, Hantkeninidae and Globorotaliidae). Trudy Vsesoyuznogo Neftyanogo Nauchnoissledovatel skogo Geologorazvedochnogo Instituta (VNIGRI), Ser. 76, p. 58, pl. 3, figs. 1–2.

Globorotalia inconstans Stainforth et al., 1975, Cenozoic planktonic foraminiferal zonation and characteristics of index forms. Paleontological Contribution, University of Kansas, Article 62, 163–425, p. 193, fig. 55.

Subbotina inconstans Petters, 1983, Gulf of Guinea planktonic foraminiferal biochronology and geological history of the South Atlantic. J. Foramin. Res. 13, 59, pl. 4, figs. 9, 13.

Diagnosis: Test low trochospiral; flat spiral side, sometimes depressed; ovate profile, gently lobate equatorial periphery, axially rounded shape; chambers usually five to six in the last whorl, increased slowly in size; sutures depressed, radial; umbilicus wide and shallow; aperture low arch, umbilical–extraumbilical; wall smooth, cancellate.

Range: Paleocene.

Subbotina triangularis White, 1928
Plate 23, Figs. 13–15

Globigerina triangularis White, 1928, J. Paleontol. 2, 195, pl. 28, figs. 1a–c.

Subbotina triangularis triangularis Blow, 1979, Danian to Oligocene planktonic foraminiferal biostratigraphy. The Cainozoic Globigerinida. A Study of the Morphology, Taxonomy, Evolutionary Relationships and the Stratigraphical Distribution of Some Globigerinida (Mainly Globigerinacea). E.J. Brill, Leiden, p. 1281, pl. 91, figs. 7, 9.

Subbotina triangularis Olsson et al., 1999, Atlas of Paleocene planktonic foraminifera. Smithsonian Contrib. Paleobiol. 85, 30, pl. 26, figs. 1–13.

Original Diagnosis: Test triangular, composed of two coils arranged in a low trochoid spire; chambers inflated, somewhat appressed, three subequal chambers comprising the last whorl; sutures deep; aperture must be in the small umbilicus.

Remarks: Niger Delta specimens are typical of *S. triangularis* and are distinct from *Subbotina triloculinoides* in wall structure and in the extraumbilical extension of the primary aperture.

Range: Middle Paleocene–Early Eocene.

Genus: PRAEMURICA
Praemurica uncinata Bolli, 1957
Plate 23, Figs. 16 and 17

Globorotalia uncinata Bolli, 1957a, The genera *Globigerina* and *Globorotalia* in the Paleocene—Lower Eocene Lizard Springs Formation of Trinidad, B.W.I. US Natl. Museum Bull. 215, 74, pl. 17.

Acarinina praecursoria Petters, 1983, Gulf of Guinea planktonic foraminiferal biochronology and geological history of the South Atlantic. J. Foramin. Res. 13, 59, pl. 3, figs. 14–15.

Original Diagnosis: Shape of test low trochospiral, spiral side almost flat or slightly convex, umbilical side distinctly convex; equatorial periphery distinctly lobate; axial periphery rounded to subangular; wall calcareous, inflated, laterally compressed; 12–15 chambers, arranged in about 2 whorls, the 5–6 chambers of last whorl increasing moderately in size; sutures on spiral side strongly curved, depressed; on umbilical side radial, depressed; umbilicus fairly, narrow, deep, open; aperture a low arch; interiomarginal, extraumbilical–umbilical; coiling random.

Range: Early Paleocene (Zone P2 to lower Zone P3).

Genus: IGORINA Davidzon, 1976

Igorina albeari Cushman and Bermudez, 1949
Plate 23, Figs. 18–20

Globorotalia albeari Cushman and Bermudez, 1949, Some Cuban species of *Globorotalia.* Contrib. Cushman Lab. Foramin. Res. 3 (pt. 1), no. 39, 33, pl. 6, figs. 13–15.

Igorina albeari Olsson et al., 1999, Atlas of Paleocene planktonic foraminifera. Smithsonian Contrib. Paleobiol. 85, 69–70, pl. 16, figs. 1–6, pl. 56, figs. 1–16.

Original Diagnosis: Test small for the genus, strongly biconvex, spiral side showing all the coils and umbilical side only the last-formed whorl, periphery somewhat rounded; chambers not very distinct, 9–10 in the last-formed whorl, only slightly inflated ventrally, increasing very gradually in size as added; sutures fairly distinct but only slightly depressed except in the last whorl on the ventral side, rather strongly recurved on the dorsal side; wall slightly spinose, coarsely perforate; aperture an elongated opening on the ventral side of the last-formed chamber extending from nearly the inner end to the periphery and with a distinct thin lip.

Remarks: Olsson et al. (1999) added that the test is moderately to strongly biconvex, essentially circular, cancellate, pustulose test with six to eight chambers in final whorl; intercameral sutures on umbilical side radial to weakly recurved yielding triangular-shaped chambers; strongly recurved and exhibiting distinct limbation on the spiral side, particularly between the last three to four chambers, yielding trapezoidal-shaped chambers; peripheral margin distinctly carinate, particularly on last chambers of final whorl; aperture a low interiomarginal, umbilical–extraumbilical arch extending toward, but not to the peripheral margin.

Range: Late Paleocene.

Igorina pusilla Bolli, 1957
Plate 24, Fig. 1

Globorotalia pusilla pusilla Bolli, 1957a, The genera *Globigerina* and *Globorotalia* in the Paleocene—Lower Eocene Lizard Springs Formation of Trinidad, B.W.I. US Natl. Museum Bull. 215, 78, pl. 20, figs. 8–10.

Globorotalia pusilla Fayose, 1970, Stratigraphical paleontology of Afowo-1 well, Southwestern Nigeria. J. Min. Geol. 5 (1 and 2), 65, pl. 17, fig. 7.

Igorina pusilla Olsson et al., 1999, Atlas of Paleocene planktonic foraminifera. Smithsonian Contrib. Paleobiol. 85, 70–71, pl. 57, figs. 1–16.

Original Diagnosis: Shape of test low trochospiral biconvex, compressed; equatorial periphery nearly circular, slightly lobate; axial periphery acute to subacute. Wall calcareous, perforate, surface smooth. Chambers compressed; 12–16, arranged in 2–3 whorls, the 5–6 chambers of the last whorl increasing moderately in size. Sutures on spiral side strongly curved, slightly depressed; on umbilical side radial, depressed. Umbilicus narrow, open. Aperture a low arch, with narrow lip; interiomarginal, extraumbilical–umbilical.

Range: Late Paleocene.

Genus: GLOBANOMALINA Haque, 1956, emended
Globanomalina chapmani Parr, 1938
Plate 24, Figs. 2–4

Globorotalia chapmani Parr, 1938, Upper Eocene foraminifera from the deep borings in King's Park, Perth, Western Australia. J. R. Soc. West. Aust. 24, 87, pl. 9, figs. 8, 9.

Globorotalia chapmani Fayose, 1970, Stratigraphical paleontology of Afowo-1 well, Southwestern Nigeria. J. Min. Geol. 5 (1 and 2), 54, pl. 14, fig. 2.

Globanomalina chapmani Olsson et al., 1999, Atlas of Paleocene planktonic foraminifera. Smithsonian Contrib. Paleobiol. 85, 39, pl. 34, figs. 1–7.

Diagnosis: Test small, very low trochospire, biconvex, more convex on umbilical side, strongly lobular; rounded peripheral margin; chambers angular-rhomboid, four to six in the last whorl, increasing fairly rapidly in size, last chamber forming one-third of whorl; sutures depressed, radial to slightly curved on umbilical side; curved on spiral side; umbilicus wide, shallow, with corresponding depression of inner whorls on spiral side; aperture low interiomarginal arch, umbilical to extraumbilical, with thin lip; wall smooth, densely perforate.

Remarks: Niger Delta forms show no imperforate marginal band, hence no traces of keel, but are completely perforate in wall structure. *Globanomalina chapmani* is intermediate morphologically between *G. compressa* and *G. pseudomenardii* which is keeled. Berggren et al. (1967) demonstrated that *G. chapmani* was the direct ancestor of the common shallow water planktonic species, *Pseudohastigerina wilcoxensis*.

Range: Middle Paleocene–Early Eocene.

Globanomalina compressa Plummer, 1927
Plate 24, Figs. 5–8

Globigerina compressa Plummer, 1927, Foraminifera of the Midway Formation in Texas. Bull. Univ. Texas Bureau Econ. Geol. Technol. 2644, 135, pl. 8, fig. 11.

Globorotalia compressa Fayose, 1970, Stratigraphical paleontology of Afowo-1 well, Southwestern Nigeria. J. Min. Geol. 5 (1 and 2), 55, pl. 17, fig. 3.

Globanomalina compressa Olsson et al., 1999, Atlas of Paleocene planktonic foraminifera. Smithsonian Contrib. Paleobiol. 85, 40, pl. 14, figs. 1–3; pl. 32, figs. 11–16; pl. 35, figs. 1–13, 17.

Diagnosis: Test small, low, biconvex, trochospiral, more convex on umbilical side than on spiral side; broadly circular profile; lobate periphery, axially rounded to nearly acute; chambers five in last whorl, slightly inflated, almost overlapping umbilically; sutures strongly depressed and curved on the umbilical side; aperture very low arch, wall smooth, densely perforate.

Range: Early–Middle Paleocene.

Globanomalina imitata Subbotina, 1953
Plate 24, Figs. 9 and 10

Globorotalia imitata Subbotina, 1953, Iskopaemye Foraminifery SSR. Globigerinidy, Khantkeninidy i Globorotaliidy (Fossil Foraminifera of the USSR. Globigerinidae, Hantkeninidae and Globorotaliidae). Trudy Vsesoyuznogo Neftyanogo Nauchnoissledovatel skogo Geologorazvedochnogo Instituta (VNIGRI), Ser. 76, p. 206, pl. 16, figs. 14a–c.

Globorotalia imitata Fayose, 1970, Stratigraphical paleontology of Afowo-1 well, Southwestern Nigeria. J. Min. Geol. 5 (1 and 2), 58, pl. 14, fig. 6.

Globorotalia imitata Olsson et al., 1999, Atlas of Paleocene planktonic foraminifera. Smithsonian Contrib. Paleobiol. 85, 42, pl. 10, figs. 12–14; pl. 12, figs. 10–12; pl. 36, figs. 7–12, 16.

Diagnosis: Test small, strongly lobate, quadrate, circular outline; flattened spiral side, but convex umbilically; rounded peripheral margin; chambers ovate on spiral side, umbilically triangular; four chambers in final whorl showing rapid increase in size; sutures short depressed, slightly curved on spiral side, almost straight and umbilically radiate; aperture low, umbilical to extraumbilical, bordered by thick rim; wall polished with sparse, rounded pores on spiral side; densely perforate in later part of test.

Range: Paleocene–Early Eocene.

Globanomalina planoconica Subbotina, 1953
Plate 24, Figs. 11 and 12

Globorotalia planoconica Subbotina, 1953, Iskopaemye Foraminifery SSR. Globigerinidy, Khantkeninidy i Globorotaliidy (Fossil Foraminifera of the USSR. Globigerinidae, Hantkeninidae and Globorotaliidae). Trudy Vsesoyuznogo Neftyanogo Nauchnoissledovatel skogo Geologorazvedochnogo Instituta (VNIGRI), Ser. 76, p. 210, pl. 17, figs. 4 a–c.

Globanomalina planoconica Olsson et al., 1999, Atlas of Paleocene planktonic foraminifera. Smithsonian Contrib. Paleobiol. 85, 44, pl. 10, figs. 15–17, pl. 34, figs. 8–17.

Diagnosis: Test small, ovate, plano-convex, spirally flattened, moderately convex umbilically; lobate outline, compressed in later part of test; chambers five to six in final whorl; chambers increase fairly rapidly in size in the last whorl; sutures wide, depressed, curved; aperture low, interiomarginal umbilical–extraumbilical arch bordered with short rim; wall perforate.

Remarks: *Globanomalina planoconica* in the Niger Delta developed a thin peripheral keel in the later part of test, a feature also observed by Olsson et al. (1999) who stated that the peripheral margin is scalloped, acute, and sometimes possesses a narrow keel.

Range: Upper Paleocene–Lower Eocene.

Globanomalina pseudomenardii Bolli, 1957
Plate 24, Figs. 13–15

Globorotalia pseudomenardii Bolli, 1957a–c, The genera *Globigerina* and *Globorotalia* in the Paleocene—Lower Eocene Lizard Springs Formation of Trinidad, B.W.I. US Natl. Museum Bull. 215, 77, pl. 20, figs. 14–16, pl. 20, fig. 17.

Globorotalia pseudomenardii Fayose, 1970, Stratigraphical paleontology of Afowo-1 well, Southwestern Nigeria. J. Min. Geol. 5 (1 and 2), 63, pl. 14, fig. 11.

Planorotalites pseudomenardii Petters, 1983, Gulf of Guinea planktonic foraminiferal biochronology and geological history of the South Atlantic. J. Foramin. Res. 13, 54, pl. 3, figs. 1, 2.

Diagnosis: Test very low trochospiral, biconvex to plano-convex; acute periphery with prominent keel; lobate outline especially in adult forms; chambers usually five in the last whorl, tending to be elongate and curved spirally and wedge shaped umbilically; slow rate of chamber size increase; last two chambers on spiral side less compact and wide apart; on umbilical side, chambers are even and acute angled; umbilicus may be wide and shallow in loosely coiled specimens or closed if chambers are tightly arranged; sutures depressed, recurved spirally, radiate umbilically; aperture low arch, umbilical to extraumbilical; wall smooth, densely perforate.

Range: Late Paleocene.

Globanomalina renzi Bolli, 1957
Plate 24, Figs. 16–19

Globorotalia renzi Bolli, 1957a–c, The genera *Globigerina* and *Globorotalia* in the Paleocene—Lower Eocene Lizard Springs Formation of Trinidad, B.W.I. US Natl. Museum Bull. 215, 168, pl. 38, figs. 3a–c.

Globorotalia (Planorotalites) renzi Jenkins, 1971, New Zealand Cenozoic planktonic foraminifera. Paleontol. Bull. Wellington 42, 110, pl. 9, figs. 224–226.

Planorotalites renzi Petters, 1983, Gulf of Guinea planktonic foraminiferal biochronology and geological history of the South Atlantic. J. Foramin. Res. 13, 54, pl. 3, figs. 4, 16.

Original Diagnosis: Bolli (1957a–c) originally described this species as follows: "Shape of test very low trochospiral; equatorial periphery almost circular, only very slightly lobate; axial periphery angular with a thin keel. Wall calcareous, finely perforate, surface smooth or very finally pitted. Chambers strongly compressed; 15-18 arranged in about 2½ whorls; the chambers of the last whorl, usually 6 in number, increase fairly rapidly in size. Sutures on spiral side, curved; on umbilical side radial or very slightly curved, slightly depressed between last chambers. Umbilicus very narrow, shallow. Aperture a low arch, often with a distinct lip, interiomarginal, umbilical–extraumbilical."

Remarks: *Globanomalina renzi* is slightly more inflated at least in the early part of the test than *G. planoconica.* In wall structure, *G. renzi* is more coarsely perforate than *G. planoconica,* otherwise both species are quite similar. It is likely that *G. renzi* could have evolved from *G. planoconica* in the Early Eocene.

Range: Early–Middle Eocene.

Genus: GLOBOCONUSA
Globoconusa daubjergensis Bronnimann, 1953
Plate 25, Figs. 1 and 2

Globigerina daubjergensis Bronnimann, 1953, p. 340, text-fig. 1.

Original Diagnosis: The specimens are very small for the genus (*Globigerina*). The outline of the trochoid test is distinctly lobulate. The spiral side is pointed in the initial portion. The umbilicus is small and shallow. In the final whorl, the dominant portion of the test consists of three to four gradually increasing subglobular chambers. The sutures of the final whorl are strongly incised, those of the early test are not clearly visible. The extremely small, subcircular aperture opens into the shallow umbilical depression. The rough surface is covered by minute irregularly distributed spines. The thin walls are finely perforate. The direction of coiling is undetermined; 11 out of 19 specimens are coiling to the right hand side.

Remarks: The Niger Delta specimens illustrated here are the megalospheric generation, with fewer chambers and a subdued spiral side. This species has the outer wall covered by small pustules. Olsson et al. (1999) noted that there is a wide variation in spire height; some specimens marked by relatively low trochospire, others display relatively high spire (similar to short-spired *Guembelitria cretacea*). They added that *Globoconusa daubjergensis* was abundant in high-latitude and nearshore planktonic foraminiferal assemblages. This species is also considered to be a direct descendant of *Guembelitria cretacea*, a Late Cretaceous planktonic species.

Range: Lower (Basal) Paleocene.

Genus: PSEUDOHASTIGERINA Blow and Banner, 1959
Pseudohastigerina barbadoensis Blow, 1969
Plate 25, Figs. 3 and 4

Pseudohastigerina barbadoensis Blow, 1969, Late Middle Eocene to Recent planktonic foraminiferal biostratigraphy. In: Bronnimann, P., Renz, H.H. (Eds.), Proceedings of the First International Conference on Planktonic Microfossils, vol. 1. E.J. Brill, Geneva, p. 409, pl. 53, figs. 11.

Pseudohastigerina barbadoensis Petters, 1983, Gulf of Guinea planktonic foraminiferal biochronology and geological history of the South Atlantic. J. Foramin. Res. 13, 48, pl. 3, fig. 11.

Diagnosis: Test small, moderately compressed, evolute test with wide umbilicus, ovate outline; gently rounded periphery; chambers in last whorl slightly inflated; ultimate chamber raised and somewhat projecting above general test outline; fairly rapid size increase; sutures broad, slightly curved, depressed; aperture low, bordered by faint lip; wall smooth, densely pitted.

Range: Late Eocene–Early Oligocene.

Pseudohastigerina danvillensis Howe and Wallace, 1932
Plate 25, Figs. 5 and 6

Nonion danvillensis Howe and Wallace, 1932

Pseudohastigerina danvillensis (Howe and Wallace) In: Tourmarkine and Luterbacher, 1985, Paleocene and Eocene planktic foraminifera. In: Bolli, H.M., Saunders, J.B., Perch-Nielsen, K. (Eds.) Plankton Stratigraphy. Cambridge University Press, Cambrigde, p. 119, figs. 21, 9.

Diagnosis: Test small, subovate, slightly compressed, biumbilicate; narrow, rounded periphery, lobate; chambers wedge shaped, increased moderately in size; sutures slightly curved, depressed; moderately deep umbilicus; aperture low, equatorial; wall smooth.

Remarks: Comparisons were in a previous study (1981) made in the Smithsonian collection, between paratypes of *Pseudohastigerina danvillensis* (Howe and Wallace, 1933 -1932) and *P. micra* (Cole, 1927) Banner and Blow, 1959. The former (Cushman Collection 1831) showed a distinct pinching of the chambers into a narrow margin around the test, thus conferring an inverted V-shape and an angled periphery on *P. danvillensis*. Figured holotype in Tourmakine and Luterbacher (1985) show similar test compression, whereas *P. micra* is more inflated.

Niger Delta forms that display a more rapid chamber size increase than *P. micra*, thereby being more like the paratypes of *P. danvillensis*. The latter seems to be quite distinct in the Niger Delta.

Range: Eocene.

Pseudohastigerina micra Cole, 1927; Banner and Blow, 1959
Plate 25, Figs. 7 and 8

Nonion micros Cole, 1927, A foraminiferal fauna from the Guayabal Formation in Mexico. Bull. Am. Paleontol. 14, 22, pl. 5, fig. 12.

Pseudohastigerina micra Banner and Blow, 1959, The classification and stratigraphical distribution of the Globigerinacea. Palaeontology, 21, 19–20, fig. 4 (g–i).

Pseudohastigerina micra Petters, 1983, Gulf of Guinea planktonic foraminiferal biochronology and geological history of the South Atlantic. J. Foramin. Res. 13, 48, pl. 3, fig. 3.

Pseudohastigerina micra AGIP, 1988, Foraminifera of the Niger Delta, pl. 2, fig. 9.

Diagnosis: Test small, tightly coiled planispiral; weakly lobate outline, subovate profile; chambers partially overlapping, six to seven in last whorl, slow increase in size, periphery rounded, subacute in later part of test; sutures depressed, nearly radial in early portion of test; more curved in later part; aperture low interiomarginal-equatorial arch continuing along base of apertural face to earlier whorl; wall polished, perforate.

Range: Middle Eocene.

Pseudohastigerina cf. *micra*
Plate 25, Figs. 9 and 10

Pseudohastigerina micra Petters, 1983, Gulf of Guinea planktonic foraminiferal biochronology and geological history of the South Atlantic. J. Foramin. Res. 13, 48, pl. 3, fig. 5.

Diagnosis: Test small, moderately inflated in axial view; triangular profile; periphery strongly lobate; six chambers in final whorl, wedge shaped, expanding rapidly in size especially in the last two chambers; raised ultimate chamber, slightly inflated in equatorial view; sutures, wide, depressed, straight, radiating from wide umbilicus; aperture indistinct, apparently an infilled low arch; wall very smooth, imperforate.

Remarks: This species is related to *P. micra*, with which it is transitional. It is distinct from *P. micra* by the triangular test shape, raised ultimate chamber, wider and shallower umbilicus, and more depressed sutures. *Pseudohastigerina* cf. *micra* is more strongly lobate and has a smoother wall than *P. micra*.

Range: Middle Eocene–Early Oligocene.

Pseudohastigerina wilcoxensis Cushman and Ponton, 1932
Plate 25, Figs. 11 and 12

Nonion wilcoxensis Cushman and Ponton, 1932, The Foraminifera of the Upper, Middle and part of the Lower Miocene of Florida. Bull. Florida State Geol. Surv. 8, 64, pl. 8, fig. 11.

Pseudohastigerina wilcoxensis (Cushman and Ponton). Berggen et al., 1967, Origin and evolutionary development of the foraminiferal genus *Pseudohastigerina* Banner and Blow, 1959. Micropaleontology 13, 278–280, figs. 2–6.

Pseudohastigerina wilcoxensis AGIP, 1988 Foraminifera of the Niger Delta, pl. 2, figs. 10, 11.

Diagnosis: Test small, planispiral volute; periphery moderately lobate, rounded; sutures depressed, radial to slightly curved; aperture low arch at base of ultimate chamber across periphery, with short but distinct lip; wall very smooth, with tiny pores.

Remarks: Unlike *P. micra*, this species is generally more inflated, larger, and has many intermediate forms linking it with *P. micra*.

Range: Late Paleocene–Middle Eocene.

Pseudohastigerina cf. *wilcoxensis*
Plate 25, Figs. 13 and 14

Diagnosis: Test ovate, widely biumbilicate due to evolute planispiral coil; moderately lobate; periphery pinched; chambers slightly inflated, increasing rapidly in size; sutures wide, curved; aperture at base of ultimate chamber, with faint lip; wall smooth, perforate.

Remarks: Generally similar to *P. wilcoxensis*, but with pinched margin of ultimate chamber and chambers that increase more rapidly in size. This form is intermediate between *P. wilcoxensis* and *P. micra* and was illustrated by Cordey et al. (1970, text fig. 5; 16–20).

Range: Early Eocene.

Genus: HASTIGERINA
Hastigerina bolivariana V. Petters, 1954
Plate 25, Fig. 15

Globigerina wilsoni bolivariana V. Petters, 1954, Tertiary and Upper Cretaceous foraminifera from Colombia, S.A. Contrib. Cushman Lab. Foramin. Res. 5, 39, pl. 8, figs. 9a–c.

Hastigerina bolivariana Toumarkine and Luterbacher, 1985, Paleocene and Eocene planktic foraminifera. In: Bolli, H.M., Saunders, J.B., Perch-Nielsen, K. (Eds.), 1985, Plankton Stratigraphy. Cambridge University Press, Cambridge, p. 126, fig. 27.

Subbotina bolivariana AGIP, 1988, Foraminifera of the Niger Delta, pl. 1, fig. 8.

Diagnosis: Test small, rounded, planispiral; equatorial periphery strongly lobate; axial periphery rounded; four chambers in the last whorl, globular; increased rapidly in size; last chamber constitutes about half the size of the test; sutures narrow, pronouncedly depressed, straight, radial; aperture a low arch extending across axial periphery, bordered by a short stout lip; wall smooth, cancellate.

Remarks: In wall structure, this taxon is allied to *Subbotina*, which explains why in the AGIP PhotoCatalogue, it is referred to as *Subbotina bolivariana.* The generic status is, however, uncertain; as the only species in the group is not related to the true *Hastigerina* which is found in the Neogene. Petters (1954) reported the *Globigerina wilsoni bolivariana* V. planispiral form from the Middle Eocene of Colombia, South America. Similar forms are known from Peru, Trinidad, and Southern Spain, as well as the Niger Delta, and the South Atlantic (Jenkins, 1971).

Range: Middle Eocene.

Genus: ACARININA Subbotina, 1953

Acarinina bullbrooki Bolli, 1957
Plate 25, Figs. 16 and 17

Globorotalia bullbrooki Bolli, 1957c, Planktonic foraminifera from the Eocene Navet and San Fernando formations of Trinidad, B.W.I., US Natl. Museum Bull. 215, 167–168, pl. 38, figs. 4a–5c.

Globorotalia bullbrooki Stainforth et al., 1975, Cenozoic planktonic foraminiferal zonation and characteristics of index forms. Paleontological Contribution, University of Kansas, Article 62, p. 174, fig. 40.

Acarinina bullbrooki Petters, 1983, Gulf of Guinea planktonic foraminiferal biochronology and geological history of the South Atlantic. J. Foramin. Res. 13, 50, pl. 7, figs. 13, 18, 21, 24.

Diagnosis: Test inflated, plano-convex; inflated on umbilical side; spiral side flat, to slightly convex; outline circular; periphery strongly lobate, rounded to slightly acute, not keeled; chambers inflated; four in the final whorl, increased rapidly in size, subangular; spirally imbricate; sutures depressed, radial on umbilical side, oblique on spiral side giving impression of imbrication and overlap; aperture wide, low to moderately high arch, with a distinct lip; wall spinose and cancellate; umbilically somewhat pustulose.

Remarks: Distinguished by its compact, inflated, quadrate test, wide primary aperture which is umbilical to extraumbilical, whereas similar forms, such as *Acarinina pseudotopilensis* and *A. frontosa* have more umbilical apertures and more trilobate profiles.

Range: Middle Eocene.

Acarinina esnaensis LeRoy, 1953
Plate 25, Figs. 18 and 19

Globigerina esnaensis LeRoy, In: Loeblich and Tappan, 1957, Planktonic foraminifera of Paleocene and Early Eocene age from the Gulf and Atlantic Coastal Plains. US Natl. Museum Bull. 215, 89, pl. 57, figs. 7a–c.

Globorotalia esnaensis Fayose, 1970, Stratigraphical paleontology of Afowo-1 well, Southwestern Nigeria. J. Min. Geol. 5 (1 and 2), 57, pl. 14, fig. 1.

Diagnosis: Test, small, somewhat ovate, flattened spiral side; umbilically inflated; strongly lobate, axial periphery rounded; chambers inflated, four in the final whorl; increased fairly rapidly and regularly in size; final chamber constitutes roughly one-third of test; sutures deep, wide, radial on umbilical side, slightly oblique spirally; umbilicus deep, quadrate, open; aperture low arch, interiomarginal; only slightly extraumbilical, bordered by thin lip, wall cancellate to pustulose around umbilicus.

Remarks: Fayose (1970) observed that *Globorotalia esnaensis* is a common form in Nigerian Late Paleocene–Early Eocene strata. The specimens found vary in the size of the test and the degree of flattening of the dorsal side. The species is both dextrally and sinistrally coiled across the "Paleocene–Eocene boundary," the former being predominant.

Range: Late Paleocene–Early Eocene.

Acarinina frontosa Subbotina, 1952
Plate 26, Figs. 1 and 2

Globigerina frontosa Subbotina 1953, Iskopaemye Foraminifery SSR. Globigerinidy, Khantkeninidy i Globorotaliidy (Fossil Foraminifera of the USSR. Globigerinidae, Hantkeninidae and Globorotaliidae). Trudy Vsesoyuznogo Neftyanogo Nauchnoissledovatel skogo Geologorazvedochnogo Instituta (VNIGRI), Ser. 76, p. 84, pl. 12, fig. 3–7.

Globigerina frontosa Stainforth et al., 1975, Cenozoic planktonic foraminiferal zonation and characteristics of index forms. Paleontological Contribution, University of Kansas, Article 62, p. 187, fig. 51.

Diagnosis: Test small, trochospiral, subrounded, umbilically inflated; periphery strongly trilobate; chambers inflated, subcircular, rapidly expanding, 3 to 3½ in last whorl; sutures wide, depressed, radial; umbilicus small nearly covered by enveloping the ultimate chamber; aperture low, slightly extraumbilical; wall strongly pustulose to finely cancellate.

Range: Early–Middle Eocene.

Acarinina inaequispira Subbotina, 1953
Plate 26, Figs. 3 and 4

Globigerina inaequispira Subbotina, 1953, Iskopaemye Foraminifery SSR. Globigerinidy, Khantkeninidy i Globorotaliidy (Fossil Foraminifera of the USSR. Globigerinidae, Hantkeninidae and Globorotaliidae). Trudy Vsesoyuznogo Neftyanogo Nauchnoissledovatel skogo Geologorazvedochnogo Instituta (VNIGRI), Ser. 76, p. 69, pl. 6, figs. 1–4.

Globigerina inaequispira Fayose, 1970, Stratigraphical paleontology of Afowo-1 well, Southwestern Nigeria. J. Min. Geol. 5 (1 and 2), 43, pl. 12, fig. 5.

Acarinina inaequispira Petters, 1983, Gulf of Guinea planktonic foraminiferal biochronology and geological history of the South Atlantic. J. Foramin. Res. 13, 50, pl. 4, figs. 7, 15.

Subbotina inaequispira AGIP, 1988, Foraminifera of the Niger Delta, pl. 1, figs. 6a–7b.

Diagnosis: Test low trochospire, umbilically inflated; subtriangular profile; equatorial periphery strongly lobate; rounded axial periphery; often four globular chambers in final whorl; chambers increased very rapidly in size, with final chamber constituting ½ to 1/3 of test; sutures wide, depressed axially and umbilically radiate, umbilicus shallow, narrow; aperture small, low arch, bordered by lip; wall finely cancellate to hispid especially on spiral surface which is rough.

Remarks: This species is similar to *A. linaperta*; but *A. inaequispira* is more loosely coiled, hence it is less compact than the former.

Range: Early–Middle Eocene.

Acarinina linaperta Finlay, 1939
Plate 26, Fig. 5

Globigerina linaperta Finlay, 1939, New Zealand Foraminifera: key species in stratigraphy—No. 2. Trans. R. Soc. NZ 69, 125, pl. 13, figs. 54, 56.

Subbotina cf. *linaperta* Finlay, AGIP, 1988, Foraminifera of the Niger Delta, pl. 1, fig. 9.
Globigerina 46: SPDC Foraminiferal Type Collection.
Globigerina linaperta Stainforth et al., 1975, Cenozoic planktonic foraminiferal zonation and characteristics of index forms. Paleontological Contribution, University of Kansas, Article 62, p. 201, fig. 63.

Diagnosis: Test compact, inflated, similar on both spiral and umbilical sides in degree of inflation; rounded, trilobate outline, slightly lobate; chambers inflated; 3 to 3½ chambers in final whorl; increased rapidly in size; sutures strongly depressed, radial; umbilicus deep; aperture large, low, umbilical, with a prominent lip; wall rough, spinose to partially cancellate.
Range: Late Paleocene–Late Eocene.

Acarinina mckannai White, 1928
Plate 26, Figs. 6 and 7

Globigerina mckannai White, 1928, Some index foraminifera of the Tampico Embayment area of Mexico. J. Paleontol. 2, 194, pl. 27, figs. 16a–c.
Acarinina mckannai Olsson et al., 1999, Atlas of Paleocene planktonic foraminifera. Smithsonian Contrib. Paleobiol. 85, 48, pl. 40, figs. 1–16.

Original Diagnosis: Test rotaliform, dorsal side strongly convex, ventral side very convex, umbilicate periphery rounded; chambers usually six in the last whorl, gradually increasing in size, closely appressed; sutures distinct, deep, not limbate; wall gradual or subspinose, very finely perforate; aperture an elongate opening extending from the umbilicus about half way to the peripheral margin.
Range: Late Paleocene.

Acarinina pentacamerata Subbotina, 1947
Plate 26, Figs. 8–10

Globorotalia pentacamerata Subbotina, 1947, Foraminifery Datskikh i Paleogenovykh -otlozheniy severnogo Kavkaza (Foraminifera of Danian and Paleogene deposits of the northern Caucasus) in Mikrofauna Neftyanykh Mastorozhdenig Kavkaza, Embyi Sredny Azii. Leningrad: VNGRI, p. 128–129, pl. 7, fig. 12–17, pl. 9, figs. 24–26.
Acarinina pentacamerata Petters, 1983, Gulf of Guinea planktonic foraminiferal biochronology and geological history of the South Atlantic. J. Foramin. Res. 13, 50, pl. 4, fig. 20.
Acarinina pentacamerata AGIP—Foraminifera of the Niger Delta, 1988, pl. 1, figs. 3–5b.

Diagnosis: Test small, low trochospire, with two to three whorls, flattened spiral side, inflated umbilically; periphery lobulate and broad equatorially; distinctive tendency for ultimate chamber to overhang spiral side; chambers axially elongate, moderately inflated; last two chambers subequal in the last whorl; sutures wide depressions, curved on spiral side, umbilically radial; wide, deep umbilicus; aperture umbilical to extraumbilical, low arch; wall cancellate, peripherally and umbilically hispid.
Range: Early–Middle Eocene.

Acarinina primitiva Finlay, 1947
Plate 26, Figs. 11 and 12

Globoquadrina primitiva Finlay, 1947, New Zealand Foraminifera: key species in stratigraphy—No. 5. NZ J. Sci. Technol. 28, 291, pl. 8, figs. 129–134.

Globigerina primitiva Stainforth et al., 1975, Cenozoic planktonic foraminiferal zonation and characteristics of index forms. Paleontological Contribution, University of Kansas, Article 62, p. 215, fig. 75.

Acarinina primitiva Petters, 1983, Gulf of Guinea planktonic foraminiferal biochronology and geological history of the South Atlantic. J. Foramin. Res. 13, 50, pl. 7, figs. 14, 17.

Diagnosis: Test compact, triangular profile, tightly coiled trochospire, flattened spiral side, inflated, strongly lobate periphery, quadrate, chambers inflated, three to four in final whorl, increased regularly and rapidly in size; sutures depressed; umbilically radial, slightly curved on spiral side; aperture low interiomarginal, umbilical–extraumbilical, with lip; wall rugose, perforate, with pustules, umbilicus deep.

Range: Late Paleocene–Middle Eocene.

Acarinina pseudotopilensis Subbotina, 1953
Plate 26, Fig. 13

Globorotalia pseudotopilensis Subbotina, 1953, Iskopaemye Foraminifery SSR. Globigerinidy, Khantkeninidy i Globorotaliidy (Fossil Foraminifera of the USSR. Globigerinidae, Hantkeninidae and Globorotaliidae). Trudy Vsesoyuznogo Neftyanogo Nauchnoissledovatel skogo Geologorazvedochnogo Instituta (VNIGRI), Ser. 76, p. 227–228, pl. 21, figs. 8, 9; pl. 22, figs. 1–3.

Globorotalia pseudotopilensis Stainforth et al., 1975, Cenozoic planktonic foraminiferal zonation and characteristics of index forms. Paleontological Contribution, University of Kansas, Article 62, p. 217, fig. 78.

Acarinina pseudotopilensis Petters, 1983, Gulf of Guinea planktonic foraminiferal biochronology and geological history of the South Atlantic. J. Foramin. Res. 13, 50, p. l4, figs. 18, 22, 27.

Acarinina pseudotopilensis AGIP, Foraminifera of the Niger Delta,1988, pl. 1, figs. 2a, b.

Diagnosis: Test small, distinctly quadrate profile; flattened spiral side; umbilical side strongly inflated; equatorial periphery strongly lobate; axial periphery rounded; chambers four to five in the last whorl; increased rapidly in size; sutures wide, depressed, radial on umbilical side; axial sutures slightly oblique; deep umbilicus; aperture, low, umbilical–extraumbilical, bordered by thin lip; wall spinose, especially on umbilical side; cancellate on spiral side.

Remarks: This is a highly variable taxon, the range of morphologic variation being evident in illustrations in Petters (1983), where some forms tend toward a reduction in the size of the ultimate chamber.

Range: Late Paleocene–Early Eocene.

Acarinina strabocella Loeblich and Tappan, 1957
Plate 26, Fig. 14

Globorotalia strabocella Loeblich and Tappan, 1957, Planktonic foraminifera of Paleocene and Early Eocene age from the Gulf and Atlantic Coastal Plains. US Natl. Museum Bull. 215, 195, pl. 61, figs. 6a–c.

Globorotalia (Acarinina) strabocella Jenkins, 1971, New Zealand Cenozoic planktonic foraminifera. Paleontol. Bull. Wellington 42, 84, pl. 4, figs. 102–104.

Original Diagnosis: Test free, of medium size, trochospiral, sides moderate convex, umbilical shoulder rounded, umbilicus broad and open, periphery broadly rounded, peripheral outline lobulate; chambers increasing gradually in size as added, of greater breadth than height, five to six per whorl in the adult, early whorls somewhat elevated above the level of the final whorl, each successive chamber on the spiral side added somewhat below the level of that preceding, resulting, in an imbricate appearance; sutures distinct, depressed, curved, and oblique on the spiral; wall calcareous, finely perforate, surface spinose, especially on the umbilical side; aperture an interiomarginal, extraumbilical–umbilical opening extending to the periphery.

Remarks: Niger Delta forms are typically with five to six inflated subangular chambers, and biconvex test with pustulose wall surface quite like the forms figured by Olsson et al. (1999, pl. 43, figs. 1–6).

Range: Late Paleocene.

Acarinina tribulosa Loeblich and Tappan, 1957
Plate 26, Figs. 15–17

Globorotalia tribulosa Loeblich and Tappan, 1957, Planktonic foraminifera of Paleocene and Early Eocene age from the Gulf and Atlantic Coastal Plains. US Natl. Museum Bull. 215, 195, pl. 56, figs. 3a–c; pl. 61, figs. 7a–c.

Acarinina tribulosa Fayose, 1970, Stratigraphical paleontology of Afowo-1 well, Southwestern Nigeria. J. Min. Geol. 5 (1 and 2), 68.

Acarinina tribulosa Petters, 1983, Gulf of Guinea planktonic foraminiferal biochronology and geological history of the South Atlantic. J. Foramin. Res. 13, 50, pl. 4, figs. 21, 24.

Original Diagnosis: Test free, trochospiral, biconvex, spire nearly flat; umbilicus small and deep, periphery rounded, peripheral outline lobulate; chambers globular, increasing rapidly in size, forming about two whorls, commonly with four to five chambers in the final whorl; sutures distinct; constricted or contracted, gently curved to radial; wall calcareous, distinctly perforate throughout; surface finely but prominently hispid; aperture an interiomarginal, extraumbilical–umbilical, high, broad arch.

Remarks: Most Niger Delta specimens have 4 to 4½ globular chambers in the final whorl (Fayose, 1971, p. 68; Petters, 1983, pl. 4, figs. 21, 24); otherwise they are quite typical and comparable with the holotype and paratype in the Smithsonian Type Collection.

Range: Late Paleocene–Early Eocene.

Acarinina wilcoxensis Cushman and Ponton, 1932
Plate 26, Figs. 18–20

Globorotalia wilcoxensis Cushman and Ponton, 1932, The Foraminifera of the Upper, Middle and part of the Lower Miocene of Florida. Bull. Florida State Geol. Surv. 8, 71, pl. 9, figs. Na–c.

Acarinina wilcoxensis Petters, 1983, Gulf of Guinea planktonic foraminiferal biochronology and geological history of the South Atlantic. J. Foramin. Res. 13, 50, pl. 3, figs. 17, 18.

Diagnosis: Test stout, low trochospire, flat to concave spiral side, strongly convex umbilically; strongly lobate outline; periphery rounded to subacute; chambers often four in the final whorl, increases very rapidly in size; conical to subangular chamber profile; sutures depressed, wide, radiate on umbilical side; deep, narrow umbilicus; aperture low arch, umbilical–extraumbilical; wall coarsely perforate, pustulose especially on umbilical side.

Range: Late Paleocene–Earliest Eocene.

Genus: TRUNCOROTALOIDES Bronnimann and Bermudez, 1953
Truncorotaloides rohri Bronnimann and Bermudez, 1953
Plate 27, Figs. 1 and 2

Truncorotaloides rohri Bronnimann and Bermudez, 1953, *Truncorotaloides*, a new foraminiferal genus from the Eocene of Trinidad, B.W.I., J. Paleontol. 27, 818–819, pl. 87, figs. 79.

Truncorotaloides rohri Fayose, 1970, Stratigraphical paleontology of Afowo-1 well, Southwestern Nigeria. J. Min. Geol. 5 (1 and 2), 72, pl. 17, fig. 6.

Truncorotaloides rohri Petters, 1983, Gulf of Guinea planktonic foraminiferal biochronology and geological history of the South Atlantic. J. Foramin. Res. 13, 54, pl. 7, fig. 12.

Diagnosis: Test low trochospire, flattened spiral side, moderately inflated umbilical view; equatorial periphery moderately lobate, subcircular; axial periphery broadly rounded to fairly acute; chambers subspherical to angular; four to six in last whorl; increasing gradually in size; sutures incised, depressed, curved on spiral side, radial on umbilical side; deep, wide umbilicus; secondary sutures on spiral side; primary aperture basal, umbilical to extraumbilical, with rim; wall perforate, strongly spongy.

Remarks: The distinctive features are the sutural supplementary apertures with which it is differentiated from *Acarinina.*

Range: Middle Eocene.

Genus: GUEMBELITRIA Cushman, 1933
Guembelitria columbiana Howe, 1939
Plate 27, Figs. 3–5

Guembelitria columbiana Howe, 1939, Louisiana Cook Mountain Eocene Foraminifera. Bull. Geol. Surv. Louisiana 14, 62, pl. 8, figs. 12, 13.

Guembelitria columbiana AGIP, 1988, Foraminifera of the Niger Delta, p. 2, fig. 4.

Guembelitria 3: SPDC Catalogue, 1980, Stereoscan Photos, Paleontological Type Collection, Nigeria.

Guembelitria columbiana Petters, 1983, Gulf of Guinea planktonic foraminiferal biochronology and geological history of the South Atlantic. J. Foramin. Res. 13, 43, pl. 1, figs. 19, 24.

Diagnosis: Test small, subconical, triserial, rapidly flaring; outline strongly lobate; chambers globular, inflated; sutures very distinct, strongly depressed; aperture a low interiomarginal arch at the base of the last chamber, bordered with an imperforate lip; wall polished.

Remarks: Although phylogenetically unrelated, *G. columbiana* is somewhat similar to the cooccurring species *Cassigerinelloita amekiensis* in their strongly inflated chambers and lobulate periphery, particularly when the conical, streptospiral test of *C. amekiensis* is considered. *Cassigerinelloita amekiensis* has three strongly inflated chambers in the ultimate whorl, which, from the apertural view may appear triserial as in *G. columbiana.*

Range: Middle Eocene.

Genus: CASSIGERINELLOITA Stolk, 1965

Diagnosis: This genus, originally described by Stolk (1965) from the Niger Delta, was validated by Loeblich and Tappan (1987). The test is small, trochospiral to streptospiral, with few rapidly expanding globular chambers that appear to be arranged triserially in a low trochospire. The walls of *Cassigerinelloita* are pustulose and microperforate whereas a similar genus, *Cassigerinella* is smooth walled.

Cassigerinelloita amekiensis Stolk, 1965
Plate 27, Figs. 6 and 7

Cassigerinelloita amekiensis Stolk, 1965, Contribution a l'etude des correlations microfauniques du Tertiaire inferieur de la Nigeria meridionale. Memoires du Bureau de Recherches Geologiques et Minieres 32, 264–266, pl. 1, pl. 2, pl. 3.

Globigerina 6: SPDC, Foraminiferal Type Collection.

Cassigerinelloita amekii AGIP, 1988, Foraminifera of the Niger Delta, pl. 2, figs. 7, 3.

Cassigerinelloita amekiensis Li and Radford, 1994, Notes on the topotype and deep-sea specimens of the planktonic foraminifer *Cassigerinelloita amekiensis* Stolk. J. Micropaleontol. 13, 157–159, pl. 1.

Cassigerinelloita amekiensis Petters, 1983, Gulf of Guinea planktonic foraminiferal biochronology and geological history of the South Atlantic. J. Foramin. Res. 13, 48, pl. 3, figs. 24, 25.

Diagnosis: Test small, subspherical to polygonal in outline; trochospiral or slightly streptospiral. Chambers few, rapidly expanding, globular; those in final whorl obscuring those in earlier whorl; sutures deep, narrow; aperture small, arched, may have a narrow lip and sometimes a bulla over the umbilicus, with secondary apertures on some adult forms; wall calcareous, finely perforate, smooth to rugose.

Remarks: Li and Radford (1994) observed that the most prominent feature of *C. amekiensis* is its apparently double coiling axes, in which chambers are triserial in a low trochospire, forming a polygonal to subspherical test. Seen from the apertural end, there are three prominent chambers constituting a trigonal outline.

Stolk (1965) believed that *Cassigerinelloita* was closely related to *Cassigerinella*, hence the generic name. Li and Radford (1994) linked *C. amekiensis* with the cooccurring planktonic species *Guembelitria triserata*, and suggested that the former probably evolved from the latter. Both forms occur in flood abundance in deep-sea cores from Antarctica, where they flourished in cold and upwelling waters (Li and Radford, 1994) from the uppermost Early Eocene to lower Middle Eocene. Northward transport of *C. amekiensis* by the cold Benguela current probably brought this species to the Niger Delta.

Range: Late Early–Early Middle Eocene.

Genus: CASSIGERINELLA Pokorny, 1955
Cassigerinella boudecensis Pokorny, 1955
Plate 27, Figs. 8 and 9

Cassigerinella boudecensis Pokorny, 1955, *Cassigerinella boudecensis* n. gen., n. sp. (Foraminifera, Protozoa) zolgocenu Zdanickeho flyse. Vestnik. Ustredniho Ustara Geologickeho, 30, 136–140.

Cassigerinella boudecensis Carbonnel et al., 1986, Microfaunes et milieux de depots des niveaux oligocenes surmontant la 'discordance oligocene' sur la partite terrestre du bassin sedimentaire cotier du Benin. Actes Colloques Angers 1994, p. 240, pl. 1, fig. 3.

Diagnosis: Test small, biserial enrolled, ovate outline and strongly lobulate periphery; chambers inflated subspherical, the ultimate chamber enlarged and protruding; sutures markedly depressed, narrow; aperture high, broad, and circular; wall smooth and finely perforate.

Remarks: Carbonnel et al. (1986) referred to this species "Oligocene cassigerinellids" with high and broad primary apertures. Similar forms occur in the Niger Delta, which are distinct from the cooccurring form, *C. chipolensis.*

Range: Late Eocene–Middle Miocene.

Cassigerinella chipolensis Cushman and Ponton, 1932
Plate 27, Figs. 10–12

Cassidulina chipolensis Cushman and Ponton, 1932, The Foraminifera of the Upper, Middle and part of the Lower Miocene of Florida. Bull. Florida State Geol. Surv. 8, 98, pl. 15, figs. 2a–c.

Cassigerinella 1: SPDC, Foraminiferal Type Collection.

Cassigerinella chipolensis Petters, 1983, Gulf of Guinea planktonic foraminiferal biochronology and geological history of the South Atlantic. J. Foramin. Res. 13, 48, pl. 3, figs. 19, 23.

Cassigerinella chipolensis AGIP, 1988, Foraminifera of the Niger Delta, pl. 2, figs. 6a,b.

Diagnosis: Test small, biserial enrolled, circular outline, strongly lobulate periphery; eight or more chambers visible, strongly inflated, expanding rapidly; sutures incised, distinct; aperture a wide but U-shaped arch, central on apertural face of last chamber.

Range: Oligocene–Middle Miocene.

Genus: CHILOGUEMBELINA Loeblich and Tappan, 1956
Chiloguembelina crinita Glaessner, 1937; Beckmann, 1957
Plate 27, Figs. 13 and 14

Guembelina crinita Glaessner, 1937, Studien uber Foraminiferen aus der Kreide und dem Tertidr. Des Kaukasus. I. Die Foraminiferen der altesten Tertiarschichten des Nordwest-Kaukasus. Problemy Paleontologii, Paleontologicheskaya Laboratoriya Moskovskogo Gosudarstvennogo Universiteta, 2–3, 383, pl. 14, fig. 34.

Chiloguembelina crinita Beckmann, 1957, *Chiloguembelina* Loeblich and Tappan and related foraminifera from the Lower Tertiary of Trinidad, B.W.I. US Natl. Museum Bull. 215, 89, pl. 21, fig. 4, text-fig. 14 (1–4).

Guembelina 15: SPDC Foraminiferal Type Collection.
Chiloguembelina crinita Olsson et al., 1999, Atlas of Paleocene planktonic foraminifera. Smithsonian Contrib. Paleobiol. 85, 90, pl. 69, figs. 1–8.

Diagnosis: Test small, rapidly flaring, biserial; slightly lobulate periphery; chambers slightly inflated, ovate, rapidly increasing in size; last chamber slightly overhangs the previous one and is considerably enlarged; sutures distinct, moderately depressed; aperture moderate to high.
Remarks: *Chiloguembelina crinita* can be distinguished from *C. midwayensis* by the higher more inflated chambers. It is believed to have evolved from *C. midwayensis*.
Range: Late Paleocene–Early Eocene.

Chiloguembelina midwayensis Cushman, 1940; Beckmann, 1957
Plate 27, Figs. 15–17

Guembelina midwayensis Cushman, 1940, Midway foraminifera from Alabama. Contrib. Cushman Lab. Foramin. Res. 16, 65, pl. 11, fig. 15.
Chiloguembelina midwayensis midwayensis Beckmann, 1957, *Chiloguembelina* Loeblich and Tappan and related foraminifera from the Lower Tertiary of Trinidad, B.W.I., US Natl. Museum Bull. 215, 90, pl. 21, fig. 1, text-fig. 14 (24–27).
Guembelina 14: SPDC Foraminiferal Type collection.
Chiloguembelina midwayensis Olsson et al., 1999, Atlas of Paleocene planktonic foraminifera. Smithsonian Contrib. Paleobiol. 85, 90, pl. 13, figs. 9, 10, 12, 13.

Diagnosis: Test small, compressed, slender to rapidly tapering, twice as long as broad, maximum width at the last pair of chambers; periphery rounded, lobulate; chambers inflated, interlocking; sutures depressed, slightly curved; aperture high arch, bordered by distinct lateral flanges; wall heavily pustulose.
Remarks: Beckmann (1957) proposed subspecies for *Chiloguembelina midwayensis*, comprising *C. midwayensis midwayensis*, *C. midwayensis strombiformis*, and *C. midwayensis subcylindrica*. These are treated here as distinct species.
Range: Paleocene–Early Eocene.

Chiloguembelina parallela Beckmann, 1957
Plate 27, Figs. 18–20

Chiloguembelina parallela Beckman, 1957, *Chiloguembelina* Leoblich and Tappan and related foraminifera from the Lower Tertiary of Trinidad, B.W.I. US Natl. Museum Bull. 215, 91, pl. 21, fig. 8, text-fig. 15 (36–38).

Diagnosis: Test thick, short, rapidly flaring, almost as long as broad, slightly compressed; rectangular outline, not strongly indented except toward apertural end; chambers inflated toward aperture; projecting into penultimate chamber; subglobular chambers, 8–12 in number rapidly increasing in size; sutures distinct, depressed, and oblique; aperture narrow, high, symmetrical, rimmed by two parallel lateral flanges from which the species derives its name; wall polished and microperforate.

Remarks: This distinctive species was reported from the Niger Delta by Stolk (1965) on his range chart. It is readily distinguished from other species of *Chiloguembelina* by its wide, rapidly flaring test and protruding penultimate chamber.

Range: Early Eocene.

Chiloguembelina strombiformis Beckmann, 1957
Plate 28, Fig. 1

Chiloguembelina midwayensis strombiformis Beckmann, 1957, *Chiloguembelina* Loeblich and Tappan and related foraminifera from the Lower Tertiary of Trinidad, B.W.I. US Natl. Museum Bull. 215, 90, pl. 21, fig. 6, text fig. 14 (28–31).

Diagnosis: Test rapidly flaring, twisted, almost as wide as long, with rounded periphery and smooth outline; chambers narrow, inflated slightly in the early part, the last pair of chambers strongly inflated with an abrupt size increase, constituting one-third of test; sutures depressed, regular, slightly oblique; aperture wider than high with collar; apertural face not at right angles to the plane of maximum test width, hence the twisted appearance of test; wall smooth, finely perforate.

Remarks: The chambers of *C. strombiformis* expand more rapidly than those of *C. midwayensis*, but toward the apertural end of the latter, chamber width tends to exceed chamber height. Sutures of *C. strombiformis* are more oblique than in *C. midwayensis.*

Range: Paleocene–Early Eocene.

Chiloguembelina subtriangularis Beckmann, 1957
Plate 28, Figs. 2–4

Chiloguembelina subtriangularis Beckmann, 1957, *Chiloguembelina* Loeblich and Tappan and related foraminifera from the Lower Tertiary of Trinidad, B.W.I. US Natl. Museum Bull. 215, 91, pl. 21, fig. 5, text-fig. 15 (39–42).

Original Diagnosis: Test small, subtriangular, pointed at the base, compressed, with a subangular periphery; chambers biserial, very slightly inflated; sutures nearly horizontal, slightly depressed, at least in the later stages; wall smooth; aperture commonly slightly eccentric, semicircular to subquadrangular, may have a slight collar.

Remarks: The species is easily distinguished from others in the genus by its compressed, subtriangular test.

Range: Early–Late Paleocene (*Praemurica uncinata* Zone to base of *Morozovella velascoensis* Zone).

Chiloguembelina wilcoxensis Cushman and Ponton, 1932
Plate 28, Figs. 5–9

Guembelina wilcoxensis Cushman and Ponton, 1932, The Foraminifera of the Upper, Middle and part of the Lower Miocene of Florida. Bull. Florida State Geol. Surv. 8, 66, pl. 8, figs. 16, 17.

Guembelina 12: SPDC Foraminiferal Type Collection.

Chiloguembelina wilcoxensis Petters, 1983, Gulf of Guinea planktonic foraminiferal biochronology and geological history of the South Atlantic. J. Foramin. Res. 13, 45, pl. 1, fig. 30.

Chiloguembelina wilcoxensis Olsson et al., 1999, Atlas of Paleocene planktonic foraminifera. Smithsonian Contrib. Paleobiol. 85, 92, pl. 13, figs. 19, 20; pl. 70, figs. 11–13, 15–18.

Diagnosis: Test gradually tapering, elongated, wider at apertural end, periphery broad, rounded, strongly lobulate; chambers much inflated, with rapid size increase, the last two pairs of chambers constituting over half the test; sutures distinct, broadly depressed, horizontal, prominent straight median suture; aperture low, bordered by short thick rim, sometimes reaching to about half the apertural face; wall distinctly papillose and perforate.

Remarks: This species can be easily distinguished from other chiloguembelinids by its globular, somewhat loosely arranged chambers, strongly lobulate periphery, and wide apertural end.

Range: Late Paleocene–Early Eocene.

Genus: STREPTOCHILUS Bronnimann and Resig, 1971

Diagnosis: Bronnimann and Resig (1971) originally described *Streptochilus* as follows: "Test biserial, sometimes twisted; wall calcareous perforate; aperture high arch, exocentric in position, extending from the base of the last chamber onto the aperture. Near the base of the inside margin, the collar and apertural edge are turned inward, producing a plate-like connection with the proximal margin of the collar of the previous aperture."

Remarks: Bronnimann and Resig (1971) suggested that this genus probably evolved from *Chiloguembelina* which it resembles, through infolding of the inner margin of the aperture. In view of its apparent close affinity to *Chiloguembelina*, the rather simple modification of the interior does not seem to warrant exclusion of this genus from classification within the Heterohelicidae. However, Loeblich and Tappan (1987) in their authoritative encyclopedia of foraminifera did not consider *Streptochilus* to be a valid genus, hence they placed it under *Laterostomella* which de Klasz and Rerat (1962) proposed as a West African benthonic genus. In their subsequent review of this genus, de Klasz et al. (1989) disagreed and confirmed that *Streptochilus* was planktonic and distinct from the benthic genus *Laterostomella* and is still living in the Indian Ocean. These authors went on to add: "We agree with Poore and Gosnell (1985) to include Upper Paleogene species in *Streptochilus*. These forms may have derived from Paleogene *Chiloguembelina*."

Since Paleogene biserial planktonic foraminifera (*Chiloguembelina*, *Streptochilus*) are so diverse in the Niger Delta, which apparently provided a more conducive environment (Petters, 1983) than the oceans, it has been deemed necessary to review the classification of these taxa in considerable detail.

Stolk (1965) demonstrated the radiation of biserial planktonic foraminifera in the Niger Delta during the Late Paleogene, with common species, such as *Chiloguembelina cubensis*, *C.* cf. *multicellaria*, and *C. martini*, to which can be added *C. victoriana*; and the excellent Middle Miocene index species, *Streptochilus dubeyi* (Petters, 1983). The Late Paleogene forms are here placed under *Streptochilus* as suggested by Poore and Gosnell (1985). The species under *Streptochilus* in the study, in addition to their possessing apertural collars, all exhibit smooth perforate tests, at least in the Niger Delta.

Streptochilus sp. 1
Plate 28, Figs. 19 and 20

Textularia globigera Schwager, 1866, Fossile Foraminiferen von Kar-Nicobar. Novara Exped., Geologisch Theil, 2, 252, pl. 7, fig. 100.

Streptochilus globigerum (Schwager, 1866) Kennett and Srinivasan, 1983, Neogene Planktonic Foraminifera: A Phylogenetic Atlas. Hutchinson Ross, Stroudsburg, PA, p. 21, pl. 1, figs. 2, 6–8.

Diagnosis: Test small, biserial with six to seven pairs of inflated chambers, tending to become globular in the latter stages; wall calcareous, perforate, ornamented with rugae; aperture a high arch extending from the base of the last chamber onto the apertural face and bordered by a distinct lip.

Range: Middle Miocene–Early Pliocene.

Streptochilus cubensis Palmer, 1934
Plate 28, Figs. 10–13

Guembelina cubensis Palmer, 1934, Some large fossil foraminifera from Cuba. Memorias de la Sociedad Cubana de Historia Natural, 8, 74, text figs. 1–6.

Guembelina 4: SPDC 1967, Preliminary Range Chart of Selected Foraminifera, Niger Delta.

Chiloguembelina cubensis Petters, 1983, Gulf of Guinea planktonic foraminiferal biochronology and geological history of the South Atlantic. J. Foramin. Res. 13, 32–59, pl. 1, fig. 31.

Chiloguembelina cubensis AGIP, 1988, Foraminifera of the Niger Delta, pl. 2, figs. 1–3.

Diagnosis: Test small, thick, somewhat inflated, moderately flaring; rounded periphery, lobulate outline; chambers moderately inflated, expanding rapidly in size after initial few pairs of chambers, last pair of chambers strongly inflated, raised ultimate chamber; sutures wide, depressed, subhorizontal; prominent sutural depression along median line; aperture low asymmetrical, rimmed by short thick collar; wall smooth, perforate.

Remarks: Forms illustrated in the AGIP PhotoCatalogue of the Foraminifera of the Niger Delta (pl. 2, figs. 1–3) as *Chiloguembelina cubensis* are too rugose and pustulose to be included in the taxon. They are more like *Streptochilus victoriana.*

Range: Eocene–Middle Oligocene.

Streptochilus dubeyi Petters, 1983
Plate 28, Figs. 14–18

Streptochilus dubeyi Petters, 1983, Gulf of Guinea planktonic foraminiferal biochronology and geological history of the South Atlantic. J. Foramin. Res. 13, 45, pl. 1, figs. 28, 29.

Bolivina 43: SPDC 1967, Preliminary Range Chart of Selected Foraminifera, Niger Delta Sediments.

Diagnosis: Original description of this species (Petters, 1983) was as follows: "Test small, biserial, narrow, gradually flaring, moderately compressed, and sometimes slightly twisted. Periphery narrow; mildly lobulate outline. 7 to 8 interlocking pairs of chambers, initial chambers indistinct and strongly compressed but later chambers are slightly inflated. Sutures horizontal, narrow, and incised. Aperture narrow and with high arch, bordered by an inward twisted collar. Wall densely perforate. Length of holotype 0.23 mm, maximum; width 0.11 mm."

Remarks: A distinctive marker species, resembling the bolivinids superficially, but with simpler sutures, without the proximally directed chamber margin digitations of *Bolivina*. The test is usually slender with the chambers regularly disposed and increasing slowly in size thereby resembling *S.* cf. *multicellaris* which differs in size.

Range: Late Early Miocene *Praeorbulina glomerosa* Zone.

Streptochilus heterostoma Bermudez, 1937
Plate 29, Figs. 1–3

Gumbelina cubensis Palmer var. *heterostoma* Bermudez, 1937, Nuevas especies de foraminiferos del Eocene de Cuba. Memorias de la Sociedad Cubana de Historia Natural "Felipe Poey", 11 (2), 143, pl. 17, figs. 5–7.

Chiloguembelina heterostoma Petters, 1983, Gulf of Guinea planktonic foraminiferal biochronology and geological history of the South Atlantic. J. Foramin. Res. 13, 43, pl. 1, fig. 23.

Diagnosis: Test small, rapidly flaring, twisted biseral; rounded to somewhat truncated periphery; outline moderately lobate. Chambers slightly inflated, expanding rapidly in size; last pair of chambers could be broad, raised, and moderately inflated; chambers in latter part of test tend to slightly overlap previous ones, giving incised sutures. Sutures narrow and curved. Aperture a high asymmetrical arch with a prominent collar. Wall smooth and finely perforate.

Remarks: This species differs from *Streptochilus cubensis* in its slightly more compressed, twisted, rapidly flaring test, and asymmetrical aperture with a prominent collar.

Range: Middle Eocene–Middle Oligocene.

Streptochilus martini Pijpers, 1933
Plate 29, Figs. 4–6

Textularia martini Pijpers, 1933, Geology and Paleontology of Bonaire (Dutch West Indies). Geographische en Geologische Modedeelingen, Physiographische—Geologische Series, Utrecht, 8, 57, figs. 6–10.

Guembelina 5: SPDC Foraminiferal Type Collection.

Chiloguembelina martini Petters, 1983, Gulf of Guinea planktonic foraminiferal biochronology and geological history of the South Atlantic. J. Foramin. Res. 13, 45, pl. 1, fig. 25.

Chiloguembelina martini AGIP, 1988, Foraminifera of the Niger Delta, pl. 2, figs. 5a, b.

Diagnosis: Test broad, rapidly flaring, compressed fusiform and pointed at both ends, length almost double the width; ovate outline; narrow periphery, gently lobate; chambers rapidly expanding, narrow and strongly interlocking; reniform ultimate chamber, terminally raised; sutures wide, strongly depressed, strongly oblique; aperture high arch, rimmed; wall smooth, perforate.

Remarks: Like Beckmann's (1957) hypotypes in the Smithsonian Institution (USNM p. 5760 a–c, p. 5761 a–e, p. 5762 a–d), the Nigerian specimens of *Streptochilus martini* vary from gradually to rapidly flaring; gentle to moderately lobulate periphery. Both sets of specimens have strongly inclined sutures with interlocking chambers in the median part; smooth and perforate walls and reniform last pair of chambers. *Streptochilus martini* resembles *Chiloguembelina parallela*, but is less flared.

Range: Late Early–Late Eocene.

Streptochilus cf. *multicellaris* Hussey, 1949
Plate 29, Figs. 7 and 8

Guembelina multicellaris Hussey, 1949, Louisiana Cane River Eocene Foraminifera. J. Paleontol. 23, 130, pl. 27, fig. 10.

Chiloguembelina cf. *multicellaris* Beckmann, 1957, *Chloguembelina* Loeblich and Tappan and related foraminifera from the Lower Tertiary of Trinidad, B.W.I. US Natl. Museum Bull. 215, 91, pl. 21, fig. 17.

Diagnosis: Test small, slender, narrow, almost 3 times as long as wide, slightly compressed especially in the early portion of test; narrow periphery, even outline; chambers narrow, many, with very slow size increase, closely appressed, last pair of chambers globular; sutures faint depressions to indistinct at initial end, horizontal; aperture wide arch bordered by short thick rim; wall strongly perforate.

Range: Middle–Late Eocene.

Streptochilus victoriana Beckmann, 1957
Plate 29, Figs. 9 and 10

Chiloguembelina victoriana Beckmann, 1957, *Chiloguembelina* Loeblich and Tappan and related foraminifera from the Lower Tertiary of Trinidad, B.W.I. US Natl. Museum Bull. 215, 91, pl. 21. figs. 19, 20, text fig. 15 (46–48).

Chiloguembelina victoriana Petters, 1983, Gulf of Guinea planktonic foraminiferal biochronology and geological history of the South Atlantic. J. Foramin. Res. 13, 45, pl. 1, fig. 26.

Chiloguembelina cubensis AGIP, 1988, Foraminifera of the Niger Delta, pl. 2, figs. 1–3.

Diagnosis: Test small, elongate, slender, slightly compressed flaring; periphery rounded, gently lobulate; chambers broader than high, slightly inflated, increasing rapidly in size; sutures straight, wide, and depressed, oblique, initially but fairly horizontal in later part of test; aperture wide, circular, with short lip and oblique with one part of the lip more projecting; wall coarsely perforate, striate, and sometimes pustulose.

Remarks: Some of Beckmann's (1957) paratypes (USNM p. 5791 a–c) are less elongate and moderately flaring like the Niger Delta forms. The Nigerian specimens, like those figured by Poore (1979), show coarsely perforate and pustulose to striate ornamentation except in the last pair of chambers (Petters, 1983, pl. 1, fig. 26; AGIP, 1988, pl. 2, figs. 1–3). However, under the light microscope *S. victoriana* appears smooth, with morphologic variability of length and the moderately twisted basal portion of some tests. Beckmann (1957) observed that *Chiloguembelina victoriana* differs from *Chiloguembelina cubensis* (Palmer) in its higher and narrower aperture, smooth wall surface, and somewhat less inflated chambers.

Range: Early–Middle Miocene.

Genus: TURBOROTALIA Cushman and Bermudez, 1949
Turborotalia anfracta Parker, 1967
Plate 29, Fig. 11

Globorotalia anfracta Parker, 1967, Late Tertiary biostratigraphy (planktonic foraminifera) of tropical Indo-Pacific deep-sea cores. Bull. Am. Paleontol. 52, 175, pl. 28, figs. 3–8.

Globorotalia anfracta Petters, 1983, Gulf of Guinea planktonic foraminiferal biochronology and geological history of the South Atlantic. J. Foramin. Res. 13, 50, pl. 6, figs. 21, 24.

Diagnosis: Test small, circular; moderately high spiral side; slightly compressed; outline distinctly lobate; chambers four to five in final whorl, increased rapidly as added, slightly inflated; sutures incised, radial on both sides; umbilicus shallow, eccentric with aperture; aperture umbilical–extraumbilical, bordered by a small but distinct lip; wall glassy, thin, and sparsely perforate.

Remarks: Nigerian forms are mostly five-chambered and very smooth like the holotype in the Smithsonian Collection (USNM 643076). However, the holotype exhibits slightly greater extension of its ultimate chamber over the umbilicus. Cifelli and Scott (1986) observed that the smooth wall, spherical profile with broadly radial chambers found in *Turborotalia anfracta* may have been the ideal morphology and wall texture that was modified early in the evolution of Neogene turborotalids into the coarsely perforate thickened walls of the latter forms.

Range: Late Pliocene–Pleistocene.

Turborotalia cerroazulensis Cole, 1928
Plate 29, Figs. 12–14

Globigerina cerroazulensis Cole, 1928, A foraminiferal fauna from the Guayabal Formation in Mexico. Bull. Am. Paleontol. 14, 17, pl. 1, figs. 11–13.

Globorotalia cerroazulensis cerroazulensis Stainforth et al., 1975, Cenozoic planktonic foraminiferal zonation and characteristics of index forms. Paleontological Contribution, University of Kansas, Article 62, p. 256, figs. 7, 107.

Emended Diagnosis: Stainforth et al. (1975) stated the following which is valid for Niger Delta specimens of *Turborotalia cerroazulensis*: "Test large for the genus, a regular, trochospire approaching planoconvex with 4 to 5 chambers in last whorl. Spiral profile subcircular, flattened at sutures; side profile shows distinct but bluntly rounded shoulder between flattened spiral and inflated umbilical faces. Sutures distinct; on umbilical side radial, on spiral side recurved, defining lozenge to crescent-shaped chambers. Aperture a curved slit or low arch from umbilicus to periphery. Shell surface smooth, finely perforate, commonly glossy, but may be rugose in umbilical area. The side profile of this subspecies distinguishes it from the preceding *Turborotalia pomeroli* (evenly rounded, no shoulder) and the ensuing *T. cocoaensis* (angulate shoulder to acute margin). There is also an evolutionary tendency toward more precisely formed and sharply defined chambers, with *T. cerroazulensis cerroazulensis* in mid-position."

Range: Late Eocene.

Turborotalia continuosa Blow, 1959
Plate 29, Figs. 15 and 16

Globorotalia opima continuosa Blow, 1959, Age, correlation, and biostratigraphy of the upper Tocuyo (San Lorenzo) and Poron Formations, Eastern Falcon, Venezuela. Bull. Am. Paleontol. 39, 218–219, pl. 19, figs. 125 a–c.

Globorotalia opima continuosa Fayose, 1970, Stratigraphical paleontology of Afowo-1 well, Southwestern Nigeria. J. Min. Geol. 5 (1 and 2), 60, pl. 11, fig. 2.

Globorotalia continuosa Petters, 1983, Gulf of Guinea planktonic foraminiferal biochronology and geological history of the South Atlantic. J. Foramin. Res. 13, 50, pl. 6, figs. 12, 13.

Diagnosis: Test small, compact, subcircular to ovate, slightly inflated; outline lobate; equatorial periphery rounded; chambers subglobular; four chambers in final whorl; earlier chambers less distinct on flattened spiral side; sutures broad, depressed, radial on spiral and umbilical sides; umbilicus wide, shallow; aperture high, lipped, umbilical–extraumbilical arch; wall cancellate.

Remarks: There is close resemblance between *Turborotalia continuosa* and *T. nana* in test shape, size, number of chambers, and wall texture. However, there is a faster rate of chamber size increase in *T. continuosa.* Bolli and Saunders (1985) observed this similarity. They remarked: "This small, 4-chambered species was regarded by its author as closely related to the ancestral *G. opima nana* and to be itself the ancestor of *G. acostaensis*, which appears at the base of the Late Miocene." Bolli and Saunders also noted that *G. continuosa* appears to have developed gradually from the *G. opima nana* stock through the *Globigerina ciperoensis ciperoensis* Zone to the *Globigerinatella insueta* Zone, becoming more lobate and gradually acquiring a higher arched aperture. From *G. mayeri* the species is distinguished by its smaller size and by possessing only four instead of five to six chambers in the last whorl.

Range: Early–Middle Miocene.

Turborotalia danvillensis Howe and Wallace, 1932
Plate 29, Figs. 17 and 18

Globigerina danvillensis Howe and Wallace. In: Fayose, 1970, Stratigraphical paleontology of Afowo-1 well, Southwestern Nigeria. J. Min. Geol. 5 (1 and 2), pl. 11, fig. 3.

Description: Test small, low trochospiral, subovate; strongly lobate periphery; rounded axial periphery; chambers inflated, four in final whorl, increased rapidly in size; sutures depressed, radial; umbilicus narrow, shallow; aperture very low umbilical–extraumbilical arch; wall finely cancellate to sparingly pustulose on spiral side.

Range: Upper Eocene–Middle Miocene.

Turborotalia fohsi Cushman and Ellisor, 1939
Plate 29, Figs. 19 and 20; Plate 30, Figs. 1–3

Globorotalia fohsi fohsi Cushman and Ellisor, 1939, New species of Foraminifera from the Oligocene and Miocene. Contrib. Cushman Lab. Foramin. Res. 15, 12, pl. 2, fig. 6.

Globorotalia 52: SPDC Foraminiferal Type Collection.

Globorotalia fohsi fohsi Stainforth et al., 1975, Cenozoic planktonic foraminiferal zonation and characteristics of index forms. Paleontological Contribution, University of Kansas, Article 62, p. 274, fig. 117.

Diagnosis: Test discoid, lenticular trochospiral; more convex umbilically than on spiral side; more compressed in early part of test; somewhat inflated in later part; equatorial periphery smooth to gently lobate; axial periphery acute, angled; sutures slightly incised, radial on umbilical side, curved on spiral part; chambers five to seven in final whorl, compressed; increased rapidly in size; ultimate chamber much larger than penultimate; aperture low slit at base of ultimate chamber; wall smooth, coarsely perforate.

Remarks: This is an evolutionary early morphogroup in the *Turborotalia fohsi* series. It is distinguishable from the ancestral *T. peripheroronda* by its' sharper periphery.

Range: Middle Miocene.

Turborotalia increbescens Bandy, 1949
Plate 30, Figs. 4–6

Globigerina increbescens Bandy, 1949, Eocene and Oligocene forminifera from Little Stave Creek, Clarke County, Alabama. Bull. Am. Paleontol. 32, 120, pl. 23, fig. 3a–c.

Globorotalia increbescens Stainforth et al., 1975, Cenozoic planktonic foraminiferal zonation and characteristics of index forms. Paleontological Contribution, University of Kansas, Article 62, p. 283, fig. 123.

Emended Diagnosis: Stainforth et al. (1975) stated as follows: "Test a rather small, compact trochospire. Spiral side shows coil of subelliptical chambers increasing regularly in size, initially 5 but reducing to 4 per whorl. Umbilical side shows 4 inflated chambers whose sutures form a right-angled cross. Spiral outline squarish with rounded corners, indented at sutures; side profile bluntly ovate, thickness about two-thirds of maximum diameter. All sutures distinct, depressed. Aperture a conspicuous arch from axis almost to periphery, directed straightforward along curve of coiling. Surface distinctly and often coarsely cancellate on initial chambers, but becoming progressively less so to the extent that final chamber may be smooth and shiny."

Remarks: Niger Delta forms exhibit the previously mentioned characteristics, in addition to displaying the morphologic transition to *Globigerina ampliapertura*, the descendant of *T. increbescens*. This transition includes a more umbilically directed primary aperture in *G. ampliapertura*.

Range: Late Eocene–Early Oligocene.

Turborotalia inflata d'Orbigny, 1839; Jenkins, 1971
Plate 30, Figs. 7 and 8

Globigerina inflata d'Orbigny, 1839, Foraminiferes des Iles Canaries. In: Barker-Webb, P., Berthelot, S., Historie Naturelle des Canaries. Paris, vol. 2, Part 2, p. 134, pl. 2, figs. 7–9.

Globorotalia (Turborotalia) inflata Jenkins, 1971, New Zealand Cenozoic planktonic foraminifera. Paleontol. Bull. Wellington, 42, 116, pl. 11, figs. 282–287.

Globorotalia inflata Stainforth et al., 1975, Cenozoic planktonic foraminiferal zonation and characteristics of index forms. Paleontological Contribution, University of Kansas, Article 62, p. 360, figs. 171, 172.

Diagnosis: Test low trochospiral, tumid, inflated especially on umbilical side, almost biconvex; equatorial periphery moderately lobate; axial periphery broadly rounded; spiral side nearly flattened; chambers inflated, four to five in final whorl, increased rapidly in size; sutures depressed, curved on spiral side, almost radial umbilically; umbilicus narrow, deep; aperture broad, high arch, umbilical–extraumbilical, lacking a lip; wall perforate, granular in early part of test on spiral side, later smooth.

Remarks: *Turborotalia inflata* characterizes temperate to cooler oceanic water masses. Its presence in the Niger Delta suggests the influence of a cold water mass.

Range: Late Pliocene–Holocene.

Turborotalia kugleri Bolli, 1957
Plate 30, Figs. 9–11

Globorotalia kugleri Bolli, 1957a–c, Planktonic foraminifera from the Oligocene–Miocene Cipero and Lengua Foramation of Trinidad, etc. (mansion). In: Bolli, H.M., Saunders, J.B., Perch-Nielsen, K. (Eds.), 1985, Plankton Stratigraphy, vol. 1, Planktonic Foraminifera, Calcareous Nannofossils and Calpionellids, p. 118, pl. 28, figs. 5a,b.

Globorotalia 26: SPDC Foraminiferal Type Collection.

Globorotalia kugleri Stainforth et al., 1975, Cenozoic planktonic foraminiferal zonation and characteristics of index forms. Paleontological Contribution, University of Kansas, Article 62, p. 289, fig. 126.

Globorotalia kugleri Petters, 1983, Gulf of Guinea planktonic foraminiferal biochronology and geological history of the South Atlantic. J. Foramin. Res. 13, 52, pl. 6, figs. 16, 17.

Diagnosis: Test small, discoid, trochospire; flat to slightly raised spiral side; umbilical side broadly convex; compressed axial profile, smooth equatorial periphery; chambers six to eight in the last whorl, broadly angled; sutures incised, radial, straight on umbilical side; spiral sutures depressed, slanting backward, straight to recurved; aperture arched, umbilical–extra-umbilical; wall cancellate with dense, coarse perforations.

Remarks: *Turborotalia kugleri* resembles *T. peripheroronda* Blow and Banner, but the latter has a faster rate of chamber size increase, and has fewer (5) chambers in the final whorl.

Range: Late Oligocene–Earliest Miocene.

Turborotalia lobata Bermudez, 1949
Plate 30, Figs. 12 and 13

Globorotalia fohsi lobata Bermudez, 1949, Tertiary smaller Foraminifer of the Dominican Republic. Cushman Laboratory for Foraminiferal Research Special Publication No. 28, p. 286, pl. figs. 15–17.

Globorotalia lobata Postuma, 1971, Manual of Planktonic Foraminifera. Elsevier Publishing Co., Amsterdam, pp. 328–329.

Globorotalia fohsi lobata Stainforth et al., 1975, Cenozoic planktonic foraminiferal zonation and characteristics of index forms. Paleontological Contribution, University of Kansas, Article 62, p. 276, fig. 118.

Diagnosis: Test lenticular, low trochospiral, ovate, biconvex, compressed; equatorial periphery in latter part of test strongly lobate; axial periphery subacute to fairly rounded in initial part of test, latter acute and keeled; chambers compressed, six to eight in final whorl, increasing rapidly in size; sutures strongly inclined on spiral side, slightly curved to radial on umbilical side, broadly depressed, indistinct in early portion of test; aperture low interiomarginal arch, umbilical–extraumbilical bordered by lip where preservation is good; wall smooth, finely and densely perforate.

Remarks: This species is distinguishable from others in the *Turborotalia fohsi* series by its well-developed keel and strongly lobate outline and compressed test.

Range: Late–Middle Miocene.

Turborotalia mayeri Cushman and Ellisor, 1939
Plate 30, Fig. 14

Globorotalia mayeri Cushman and Ellisor, 1939, New species of Foraminifera from the Oligocene and Miocene. Contrib. Cushman Lab. Foramin. Res. 15, 11, pl. 2, figs. 4a–c.

Globigerina 56: SPDC Foraminiferal Type Collection.

Globorotalia mayeri Petters, 1983, Gulf of Guinea planktonic foraminiferal biochronology and geological history of the South Atlantic. J. Foramin. Res. 13, 52, pl. 6, figs. 6, 10, 11.

Diagnosis: Test small, circular, spiral-side flat, umbilical-side slightly convex; equatorial periphery gently lobate; axial periphery rounded; tightly coiled, compact test; chambers five to six in final whorl, increased gradually and regularly in size, wedge shaped especially on spiral side; sutures depressed, slightly curved on both sides; more strongly on spiral side; aperture very low slit bordered by pronounced lip; wall texture cancellate and coarsely perforate.

Remarks: This is a controversial species in spite of its stratigraphic importance and relevance to the radiation of Neogene globorotalids. The problem is to decide whether to separate forms referred to as *Globorotalia siakensis* LeRoy or consider this as a synonym of *Turborotalia mayeri*. Pertinent arguments summarized by Bolli and Saunders (1985) hinge on comparisons of shape of intercameral sutures and periphery. These authors concluded "that neither on morphologic nor on stratigraphic distribution criteria can a clear distinction between *T.* (*G*). *mayeri* and *G. siakensis* be maintained and that they fall within the variability of one species." Stainforth et al. (1975) also synonymized both taxa but preferred the name *G. siakensis*.

However, forms illustrated by Blow (1969) and Berggren and Amdrurer (1973) from the deep Atlantic Ocean as *Globorotalia siakensis* are sufficiently morphologically distinguishable to warrant the retention of *G. siakensis* as a separate species, an alternative taxonomic position summarized by Bolli and Saunders (1985) as follows: "contrary to findings in low latitudes, some workers in higher latitudes, such as Cita et al. (1978) and Borsetti et al. (1979) believe that they are in a position to distinguish clearly a *Globorotalia siakensis* type, i.e. forms with more radially arranged spiral sutures, more open umbilicus and more distinctly lobate equatorial outline, and a *T.* (*G.*) *mayeri* type which, compared with *G. siakensis*, possesses more oblique to slightly curved sutures on the spiral side, a very narrow umbilical area and a less lobate equatorial outline. " It isprecisely on the basis of the shape of the test outline, the umbilicus, and the sutures that *G. siakensis* is distinguished in the present study.

Range: Late Oligocene–Middle Miocene.

Turborotalia nana Bolli, 1957
Plate 30, Figs. 15–18

Globorotalia opima nana Bolli, 1957a–c, Planktonic foraminifera from the Oligocene–Miocene Cipero and Lengua Foramation of Trinidad, etc. (mansion). In: Bolli, H.M., Saunders, J.B., Perch-Nielsen, K. (Eds.), 1985, Plankton Stratigraphy, vol. 1. Planktonic Foraminifera, Calcareous Nannofossils and Calpionellids, p. 118, pl. 28, figs a–c.

Globorotalia 20: SPDC Foraminiferal Type Collection.

Globorotalia nana Petters, 1983, Gulf of Guinea planktonic foraminiferal biochronology and geological history of the South Atlantic. J. Foramin. Res. 13, 52, pl. 6, figs. 5, 9.

Diagnosis: Test small, tightly coiled, nearly a planispiral trochospire; rounded outline, faintly lobate; chambers rounded, thickness nearly equals chamber diameter, four chambers in final whorl; sutures slightly depressed, but distinct, radiate on umbilical side; aperture curved slit with rim extending toward periphery; wall cancellate.

Remarks: Stainforth et al. (1975) noted that "the smaller size of *Globorotalia opima nana* is supposedly its main distinction from *G. opima opima*, but in topotype material we could distinguish it also by its smoother test; initial chambers of *G. opima opima* are appreciably pustulose. The Miocene *Globorotalia continuosa* is grossly similar to *G. opima nana* but is more loosely coiled and has a larger, comma-shaped aperture."

Range: Late Eocene–Late Oligocene.

Turborotalia obesa Bolli, 1957
Plate 30 Figs. 19 and 20; Plate 31, Fig. 1

Globorotalia obesa Bolli, 1957a–c, Planktonic foraminifera from the Oligocene–Miocene Cipero and Lengua Foramation of Trinidad, etc. (mansion). In: Bolli, H.M., Saunders, J.B., Perch-Nielsen, K. (Eds.), 1985, Plankton Stratigraphy, vol. 1, Planktonic Foraminifera, Calcareous Nannofossils and Calpionellids, p. 119, pl. 29, figs. 2a–3.

Globorotalia obesa Petters, 1983, Gulf of Guinea planktonic foraminiferal biochronology and geological history of the South Atlantic. J. Foramin. Res. 13, 52, pl. 7, figs. 5, 6.

Diagnosis: Test small, low trochospire, quadrate; strongly lobate periphery; axial periphery rounded; chambers globular, four in the final whorl; sutures broad, deeply incised appearing star shaped, radiating evenly from the narrow deep, nearly closed umbilicus; straight on spiral side; broad sutures confer loose overall profile on test; aperture a low arch umbilical–extraumbilical, with distinct lip; wall rugose with minute pores.

Remarks: Four chambers loosely arranged; highlighting a quadrate profile; deep, small umbilicus; protruding ultimate chamber, and deep radiating sutures on the umbilical side are characteristics that distinguish *T. obesa* from similar cooccurring species of *Globigerina*, notably *G. praebulloides*, *G. occlusa*, and *G. foliata*.

Range: Early–Middle Miocene.

Turborotalia opima Bolli, 1957
Plate 31, Figs. 2–5

Globorotalia opima opima Bolli, 1957a–c, Planktonic foraminifera from the Oligocene–Miocene Cipero and Lengua Foramation of Trinidad, etc. In: Bolli, H.M., Saunders, J.B., Perch-Nielsen, K. (Eds.), 1985, Plankton Stratigraphy, vol. 1, Planktonic Foraminifera, Calcareous Nannofossils and Calpionellids. p. 117–118, pl. 28, figs. 1a-2.

Globorotalia 20: SPDC Foraminiferal Type Collection.

Globorotalia opima Stainforth et al., 1975, Cenozoic planktonic foraminiferal zonation and characteristics of index forms. Paleontological Contribution, University of Kansas, Article 62, p. 300, fig. 132.

Globorotalia opima Petters, 1985, J. Foramin. Res, 13, 52, pl. 6, figs. 4, 8.

Diagnosis: Test trochospiral, with depressed concave spiral side and bulbous chambers; initial part of test nearly planispiral; outline strongly lobate; nearly circular; chambers four to five in the last whorl, increasing gradually in size, sometimes rapidly; sutures initially obscure, later more distinct, broad and slightly incised; umbilicus wide and shallow; aperture a low arched slit umbilical extending to periphery, sometimes with lip; wall surface pustulose on umbilical side, but distinctly cancellate to rugose on spiral side.

Remarks: *Turborotalia opima* is grossly similar to *Turborotalia mayeri*, but differs in its more inflated bulbous, chunky, ovate test; and lesser number of chambers in the final whorl; whereas *T. mayeri* has five to six chambers in the last whorl.

Range: Oligocene.

Turborotalia peripheroacuta Blow and Banner, 1966
Plate 31, Figs. 6 and 7

Globorotalia (Turborotalia) peripheroacuta Blow and Banner, 1966, The morphology, taxonomy and biostratigraphy of *Globorotalia barisanensis* LeRoy, *Globorotalia fohsi* Cushman and Ellisor, and related taxa. Micropaleontology, 12, 294, pl. 1, fig. 2; pl. 2, figs. 4, 5, 13.

Globorotalia peripheroacuta Postuma, 1971, Manual of Planktonic Foraminifera. Elsevier Publishing Corp., Amsterdam, p. 346–347.

Globorotalia fohsi peripheroacuta Stainforth et al., 1975, Cenozoic planktonic foraminiferal zonation and characteristics of index forms. Paleontological Contribution, University of Kansas, Article 62, p. 276.

Description: Test low biconvex trochospire, compressed, equatorial periphery slightly lobate; axial periphery subacute to acute, more rounded in early part of test; chambers compressed six to seven in the last whorl, increasing gradually and regularly in size; sutures curved and, slightly depressed to flush on spiral side; depressed, curved, radial on umbilical side; aperture interiomarginal arch, low, umbilical–extraumbilical, with distinct lip; wall coarsely perforate to honeycomb.

Remarks: Berggren and Amdrurer (1971) illustrated specimens of *T. (G.) peripheroacuta* with two additional morphological features notably, fused early spiral sutures which form a solid calcite mass, and a distinctly cancellate wall type. Both characteristics are exhibited by Niger Delta forms.

Stainforth et al. (1975) suggested that "for those who choose to follow Blow and Banner (1966) in classifying the *Globorotalia fohsi* lineage on degree of development of a keel, this is the second stage of evolution, characterized by a more acute periphery than the primitive subspecies *T.* (*G.*) *peripheroronda* but lacking any sign of a keel. "

Range:Early–Middle Miocene.

Turborotalia peripheroronda Blow and Banner, 1966
Plate 31, Figs. 8–12

Globorotalia (Turborotalia) peripheroronda Blow and Banner, 1966, The morphology, taxonomy and biostratigraphy of *Globorotalia barisanensis* LeRoy, *Globorotalia fohsi* Cushman and Ellisor, and related taxa. Micropaleontology 12, 294, pl. 1, fig. 1; pl. 2, figs. 1–3.

Globorotalia 16: SPDC Foraminiferal Type Collection.

Globorotalia peripheroronda Petters, 1983, Gulf of Guinea planktonic foraminiferal biochronology and geological history of the South Atlantic. J. Foramin. Res. 13, 52, pl. 6, figs. 1–3.

Description: Test discoid trochospiral; umbilical side more convex than flat spiral side; ovate and tightly coiled; equatorial periphery moderately lobate; axial periphery broad and angled toward spiral side, umbilical side overhung by ultimate chamber; chambers five to six in final whorl; sutures distinct, lightly incised, radial to slightly slanting in latter part of test; on spiral surface sutures less incised; indistinct on initial part of test; latter slanting; aperture umbilical–extraumbilical, with rim.

Remarks: Bolli and Saunders (1985) stressed the fact that *T. peripheroronda* "is the earliest and most primitive subspecies of the (*Turoborotalia*) *fohsi* lineage. It possesses 5-6 chambers in the last whorl, is slightly elongate in equatorial view, is typical for all (*T.*) *fohsi* subspecies except for the end-form (*T.*) *robusta*, which tends to be more circular in outline. The characteristic for the subspecies is the bluntly rounded periphery."

Range: Early–Middle Miocene.

Turborotalia pomeroli Toumarkine and Bolli, 1970
Plate 31, Figs. 13 and 14

Globorotalia cerroazulensis pomeroli Toumarkine and Bolli, 1970, Evolution de *Globorotalia cerroazulensis* (Cole) dans l'Eocene moyen et superieur de Possagno (Italie). Rev. Micropaleontol. 13, 140, pl. 1, figs. 10–18; pl. 2, figs. 1, 2, 11–19.

Globorotalia cerroazulensis Petters, 1983, Gulf of Guinea planktonic foraminiferal biochronology and geological history of the South Atlantic. J. Foramin. Res. 13, 50, pl. 3, figs. 6, 7.

Turborotalia cerroazulensis pomeroli Tourmakine and Luterbacher, 1985, Paleocene and Eocene planktic foraminifera. In: Bolli, H.M., Saunders, J.B., Perch-Nielsen, K. (Eds.), 1985, Plankton Stratigraphy, Cambridge University Press, Cambridge, pp. 87–154. p1. 37, figs. 34.9, 35.4–9.

Description: Test circular, globose trochospiral, lenticular; almost plano-convex, but with slightly raised earlier whorl on the spiral side; umbilical side convex; lobate periphery; axial periphery somewhat compressed in early part of final whorl; rounded margin of ultimate chamber; chambers four to five in final whorl, slightly globose; sutures depressed slightly oblique; umbilicus shallow to closed; aperture low arch umbilical–extraumbilical; wall perforate, smooth.

Remarks: Slight increase in chamber size, test compression, and smooth, less perforate wall distinguish *T. pomeroli* from *T. cerroazulensis.*

Range: Middle–Late Eocene.

Turborotalia pseudopima Blow, 1969
Plate 31, Figs. 15–17

Globorotalia pseudopima Blow, 1969, Late Middle Eocene to Recent planktonic foraminiferal biostigraphy. In: Bronnimann, P., Renz, H.H. (Eds.), Proceedings of the First International Conference on Planktonic Microfossils, vol. 1. E.J. Brill, Geneva, p. 387, pl. 35, figs. 1–3.

Globorotalia pseudopima Bolli and Saunders, 1985, Oligocene to Holocene low latitutude plankton stratigraphy. In: Bolli, H.M., et al., (Eds.), 1985, Plankton Stratigraphy. Cambridge University Press, Cambridge, p. 211, figs. 27.7.

Diagnosis: Test low trochospire, umbilical side convex; ovate to slightly elongate profile; periphery moderately lobate; axial periphery rounded; chambers inflated, four to five in final whorl, increased moderately and regularly in size; sutures depressed, radial on both spiral and umbilical sides; aperture very low interiomarginal, umbilical–extraumbilical in narrow almost closed umbilicus; wall densely perforate.

Remarks: Bolli and Saunders (1985) remarked that compared with four-chambered *G. acostaensis,* the test of *T.* (*G.*) *pseudopima* is distinctly larger and the wall more robust and with coarser pores. The umbilicus is nearly closed in typical specimens. Forms transitional toward *Neoquadrina dutertrei,* display a more open umbilicus and a tendency for the apertures of the last chamber to become more umbilical. Except for their distinctly different stratigraphic ranges and different accompanying fauna, *T. pseudopima* and the Oligocene *T. opima opima* are difficult to distinguish from each other. Both possess usually four, more rarely four to five robust and inflated chambers in the last whorl, with about equally sized and spaced pores.

Range: Late Miocene–Recent.

Turborotalia siakensis LeRoy, 1939
Plate 31, Figs. 18 and 19

Globorotalia siakensis LeRoy, 1939, Some small foraminifera, ostracoda and otoliths from the Neogene ("Miocene") of the Rokan-Tapanoeli area, Central Sumatra. Natuurkwetenschappelijk Tijdschrift voor Nederlandsch Indie 99 (6), 39–40, pl. 3, figs. 30, 31.

Globorotalia siakensis Postuma, 1971, Manual of Planktonic Foraminifera. Elsevier Publishing Co., Amsterdam, p. 359.

Globorotalia siakensis Stainforth et al., 1975, Cenozoic planktonic foraminiferal zonation and characteristics of index forms. Paleontological Contribution, University of Kansas, Article 62, p. 317–320, fig. 143.

Diagnosis: Test small, low trochospire, flat spiral side; convex umbilical part; rounded to ovate outline; equatorial periphery strongly lobate; axial periphery rounded; chambers five to six in final whorl; sutures widely depressed and radiate on umbilical side, slightly curved spirally; aperture low arch, umbilical–extraumbilical, with lip; wall cancellate. See also discussion under *Turborotalia mayeri.*

Range: Late Oligocene–Middle Miocene.

Turborotalia subcretacea Lomnicki
Plate 31, Fig. 20

Globorotalia subcretacea (Lomnicki) Bolli, 1970, Initial Reports of the Deep Sea Drilling Project, p. 640, pl. 8, figs. 16–18.

Diagnosis: Test small, quadrate, low trochospire, flattened spiral side, slightly inflated; deeply lobate periphery; axial periphery rounded; chambers subglobular, four to five in the last whorl, increased rapidly in size, somewhat axially elongate; the test, therefore, appearing stellate; sutures wide, strongly depressed, radially aligned, umbilicus narrow, shallow, somewhat closed; aperture very low arch, umbilical–extraumbilical with lip; wall distinctly cancellate to honeycomb.

Remarks: *Turborotalia subcretacea* is quite distinctive because of its quadrate shape and inflated chambers; and especially the strongly cancellate wall texture, with pore pits well developed. Bolli (1970) figured similar materials from deep sea cores from Leg 4 of the Deep Sea Drilling Project along the northern margins of South America. Bolli's forms are also distinctly cancellate and quadrate, and were likened to *Globorotaloides hexagona*, which have paratypes in the Smithsonian Collection (USNM 5655 a–d) that are like Nigerian forms. It is likely that both the Niger Delta and Bolli's forms could be *G. hexagona,* but some of the paratypes are not typical *Globorotaloides* in that they have no bullae; but like Nigerian forms have thin lips, umbilical–extraumbilical primary apertures, low spiral sides, and are coarsely cancellate.

Range: Miocene–Pleistocene.

Genus: TENUITELLA Fleisher, 1974
Tenuitella munda Jenkins, 1966
Plate 32, Fig. 1

Globorotalia (Turborotalia) munda Jenkins, 1966, Planktonic foraminiferal zones and new taxa from the Danian to Lower Miocene of New Zealand. NZ J. Geol. Geophys. 8, 1121–1122, fig. 14, no. 126–133.

Globorotalia munda Petters, 1983, Gulf of Guinea planktonic foraminiferal biochronology and geological history of the South Atlantic. J. Foramin. Res. 13, 52, pl. 4, figs. 14, 26.

Globorotalia munda Carbonnel et al., 1996, Microfaunes et milieux de depots des niveaux oligocenes surmontant la 'discordance oligocene' sur la partite terrestre du bassin sedimentaire cotier du Benin. Actes Colloques Angers 1994, p. 240, pl. 1, fig. 4.

Diagnosis: Test small, low trochospire, quadrilobate equatorial periphery, rounded axial periphery; chambers subovate, slightly compressed, four to five in the final whorl; sutures curved on spiral side, depressed; radially depressed on umbilical side; umbilicus shallow; aperture low arch, interiomarginal, umbilical–extraumbilical, with thin lip; wall hispid to cancellate in final chamber.

Remarks: Jenkins (1985) stated that *Tenuitella munda* is characterized by four chambers in the final whorl and a low-lipped, slit aperture. However, Berggen (1969) showed five-chambered (final whorl) specimens that are similar to Nigerian forms. Jenkins (1985) further observed that there is some evidence that *T.* (*G.*) *munda* may have evolved from *Globigerina gemma*, and itself gave rise to the *Globigerina juvenilis* populations in the Late Oligocene *Globigerina euapertura* Zone, which eventually evolved into *G. glutinata* and possibly *G. bradyi.*

Range: Early–Middle Oligocene.

Genus: GLOBIGERINA d'Orbigny, 1826

Globigerina ampliapertura Bolli, 1959
Plate 32, Figs. 2 and 3

Globigeriana ampliapertura Bolli, 1957c, Planktonic foraminifera from the Eocene Navet and San Fernando formations of Trinidad, B.W.I., US Natl. Museum Bull. 215, 108, pl. 22, figs. 4a–7b.

Turborotalia ampliapertura AGIP, 1988, Foraminifera of the Niger Delta, pl. 3, figs. 8a–9b.

Globigerina ampliapertura Stainforth et al., 1975, Cenozoic planktonic foraminiferal zonation and characteristics of index forms. Paleontological Contribution, University of Kansas, Article 62, p. 248, fig. 102.

Emended Diagnosis: Stainforth et al. (1975) stated as follows about *G. ampliapertura*: "Test a regular trochospire of inflated chambers which maintain their shape while increasing rapidly in size. Consistently, 4 chambers in last whorl. Spiral outline subquadrate to subcircular; overall shape globose, spiral side flattened. Sutures distinct, depressed to incised; on umbilical side radial, on spiral side tangential to spire and giving it a rectangular aspect. Aperture a smooth-rimmed arch aligned obliquely across umbilical region, usually conspicuous though somewhat concealed by projecting anterior margin of last chamber, embracing 3 penultimate chambers."

Remarks: Niger Delta forms exhibit typically spirally flattened robust four-chambered test; with morphologic intergrades with *Turborotalia increbescens*.

Range: Late Eocene–Early Oligocene.

Globigerina angiporoides Hornibrook, 1961
Plate 32, Fig. 4

Globigerina angipora Stache Hornibrook, 1961, Tertiary foraminifera from the Oamuru District (N.Z.); Pt. 1-Systematics and distribution. NZ Geol. Surv. Paleontol. Bull. 34, 145, fig. 3.

Globigerina angiporoides Berggren and Amdrurer, 1971, Late Paleogene (Oligocene) and Neogene planktonic foraminiferal biostratigraphy of the Atlantic Ocean (Lat. 30°N to Lat. 30°S). Rev. Ital. Paleontol. 79, 374, pl. 25, figs. 1–4.

Globigerina angiporoides Stainforth et al., 1975, Cenozoic planktonic foraminiferal zonation and characteristics of index forms. Paleontological Contribution, University of Kansas, Article 62, p. 250, fig. 103.

Diagnosis: Test small, compact, rounded trochospire; periphery pronouncedly lobate; chambers often three in last whorls, hemispherical, increased in size rapidly and regularly, except the ultimate chamber; sutures distinct, depressed, radiate on umbilical side; umbilicus almost completely closed; aperture a very low interiomarginal arch or slit, rimmed by a thick lip, lying across the umbilicus; wall cancellate.

Remarks: This species is characterized by its coiled, circular outline; the alignment of the three chambers in the form of a cross; and the closed umbilicus.

Range: Late Eocene–Early Oligocene.

Globigerina angulio fficinalis Blow, 1969
Plate 32, Fig. 5

Globigerina angulio fficinalis Blow, 1969, Late Middle Eocene to Recent planktonic foraminiferal biostigraphy. In: Bronnimann, P., Renz, H.H. (Eds.), Proceedings of the First International Conference on Planktonic Microfossils, vol. 1. E.J. Brill, Geneva, p. 379, pl. 11, figs. 1–5.

Original Diagnosis: The small test consists of about 11 chambers arranged in a low trochospire with 4 chambers comprising the last whorl. The chambers are slightly to moderately inflated, closely appressed but separated by deeply incised ventral intercameral sutures. The intercameral sutures are marked by radial bands of less coarsely perforate wall and are distinctly U-shaped as in *G. angulisuturalis* Bolli. The distinct ventral intercameral sutures are formed by a sharp in-turning of the chamber walls and this in-turning gives a moderately lobate appearance to the equatorial profile of the test. The umbilicus is small, but deep and open and is surrounded by an imperforate area which grades into the imperforate lip of the final chamber. The aperture of the final chamber is a low arch, somewhat asymmetrical placed with respect to the umbilicus. On the dorsal side the intercameral sutures are less deeply incised less "U shaped" than on the ventral side. The spire is low, depressed and the early chambers indistinctly separated one from another. The wall is relatively densely perforate and the pores open into distinct pits between small rugosities.

Remarks: According to Blow (1969), *G. angulio fficinalis* differs from *G. angulisuturalis* in having a less highly arched aperture, a more closed umbilicus, less inflated chambers, and less deeply incised intercameral sutures especially over the equatorial–peripheral marginal area and over the dorsal–peripheral shoulders of the chambers. Frequently, four-chambered forms show the presence of abortive, small final chambers which are less coarsely perforate, a feature also evident in the Niger Delta materials; and suggestive of closeness to *G. angustiumbilicata*.

Range: Uppermost Eocene–Oligocene.

Globigerina angulisuturalis Bolli, 1957
Plate 32, Fig. 6

Globigerina angulisuturalis Bolli, 1957c, Planktonic foraminifera from the Eocene Navet and San Fernando formations of Trinidad, B.W.I., US Natl. Museum Bull. 215, 109, pl. 22, figs. 11a–c.

Globigerina 42: SPDC Foraminiferal Type Collection.

Globigerina angulisuturalis Postuma, 1971, Manual of Planktonic Foraminifera. Elsevier Publishing Co., Amsterdam, pp. 260–261.

Globigerina angulisuturalis Petters, 1983, Gulf of Guinea planktonic foraminiferal biochronology and geological history of the South Atlantic. J. Foramin. Res. 13, 54, pl. 6, figs. 15, 20.

Diagnosis: Test small, low trochospire; circular outline; strongly lobate periphery with deep angular, U-shaped sutures between chambers; rounded axial periphery; chambers consistently five in final whorl (except where broken as in the figured specimen in this study); chambers spherical, increased moderately in size; sutures broadly depressed radial on umbilical and spiral sides; deep, wide umbilicus; aperture interiomarginal umbilical; low arch bordered by lip; wall cancellate, with pustulose masses of calcite on umbilical shoulders of the V-shaped chambers on the umbilical side.

Range: Late Oligocene–Earliest Miocene.

Globigerina aff. *angulisuturalis* Berggren, 1969
Plate 32, Fig. 7–9

Globigerina aff. *angulisuturalis* Berggren, 1969, Paleogene biostratigraphy and planktonic Foraminifera of northern Europe. In: Bronnimann, P., Renz, H.H. (Eds.), Proceedings of the First International Conference on Planktonic Microfossils, vol. 1. E.J. Brill, Geneva, p. 159, pl. VII, figs. 4–8, 9–12, 13–16.

Diagnosis: Test small, low trochospiral, ovate, moderately inflated; strongly lobate equatorial periphery; axial periphery rounded; chambers subcircular, inflated, arranged in about two whorls; consistently five chambers in the last whorl; sutures strongly depressed, wide, radial on umbilical side; not so wide on spiral side, depressed, slightly curved; aperture very low, interiomarginal, umbilical to slightly extraumbilical; wall pustulose to cancellate.

Remarks: In his study of the Late Paleogene of Northern Europe, Berggren (1969) observed as follows: "Some samples examined from Astrup (Northern Germany) contained *Globorotalia opima* and some specimens which appear transitional from *G. opima nana* to *G. opima opima* as well as some specimens which are called here *Globigerina* sp. aff. *G. angulisuturalis* with its sharply incised and separated radial sutures. Similar forms were encountered in large numbers in the Niger Delta, in which the sutures are very close to those of *G. angulisuturalis*; with the overall test outline; number of chambers in the final whorl; rate of chamber size increase; the shape of the umbilicus; the position and shape of the aperture; and wall ornamentation also displaying closeness to *G. angulisuturalis*. Equally, the quadrate profile and other features ally these forms with *G. angustiumbilicata*." These similarities led Berggren (1969) to the conclusion that intermediate forms exist between *G. angustiumbilicata* and *G. angulisuturalis*, hence the former was the ancestor of the latter. The Niger Delta materials support this ancestry for *G. angulisuturalis*, and not *G. anguliofficinalis* which is very rare in the Niger Delta.

Range: Oligocene.

Globigerina angustiumbilicata Bolli, 1957
Plate 32, Figs. 10–12

Globigerina ciperoensis angustiumbilicata Bolli, 1957a–c, Planktonic foraminifera from the Eocene Navet and San Fernando formations of Trinidad, B.W.I., US Natl. Museum Bull. 215, 109, pl. 22, figs. 12 a–c.

Globigerina 41: SPDC Foraminiferal Type Collection.

Globigerina angustiumbilicata Petters, 1983, Gulf of Guinea planktonic foraminiferal biochronology and geological history of the South Atlantic. J. Foramin. Res. 13, 54, pl. 7, figs. 9, 10.

Diagnosis: Test small, low to moderate trochospire, quadrate; equatorial periphery strongly lobate; axial periphery rounded; chambers circular, inflated, increased slowly in size; usually five chambers in final whorl; sutures depressed; umbilicus very small or closed; aperture often hard to distinguish, but a very low slit, umbilical to slightly extraumbilical; wall pustulose; smooth ultimate chamber.

Remarks: This is a highly variable, generalized taxon with the basic morphology that could have been modified in the course of evolution into *G. ciperoensis*, *G. anguliofficinalis*, and *G. angulisuturalis*.

Range: Late Eocene–Early Miocene.

Globigerina aff. *angustiumbilicata*
Plate 32, Figs. 13 and 14

Diagnosis: Test small, ovate, low trochospiral; equatorial periphery strongly lobate; axial periphery rounded; chambers globular arranged in 2 to 2½ whorls; 4–5 chambers in the last whorl; chambers increasing gradually in size; sutures wide, depressed, radial on both umbilical and spiral sides; much wider, almost u-shaped umbilical sutures; umbilicus very shallow, merging imperceptibly with sutural depressions, somewhat quadrate; aperture very low slit, umbilical, and slightly extraumbilical; wall completely hispid.

Remarks: In gross morphology, this form is similar to *G. angustiumbilicata* but differs in the fact that its test is more compact, though the sutures are about the same in depth and width. *Globigerina* aff. *angustiumbilicata* also had a slower rate of chamber size increase in the final whorl. Besides, its test is uniformly and densely spinose and it is imperforate unlike *G. angustiumbilicata* s.s. The wall texture of this form shows a departure from characteristic Oligocene planktonic foraminiferal wall structures, and allies more with typical *Globigerina* wall structure of the Late Tertiary.

Range: Oligocene–Miocene.

Globigerina bulloides d'Orbigny, 1826
Plate 32, Figs. 15 and 16

Globigerina bulloides d'Orbigny, 1826, Tableau methodique de la classe des cephalopodes. Ann. Sci. Nat. Paris, ser. 1, 7, 277, no. 1.

Globigerina bulloides Bolli, 1957a–c, Planktonic foraminifera from the Eocene Navet and San Fernando formations of Trinidad, B.W.I., US Natl. Museum Bull. 215, 31, pl. 4, figs. 1a–c.

Globgerina bulloides bulloides Rögl and Bolli, 1973, Holocene to Pleistocene Planktonic Foraminifera of Leg 15, Site 147 (Cariaco Basin (Trench), Caribbean Sea) and Their Climatic Interpretation. Initial Reports of the Deep Sea Drilling Project, vol. XV, p. 562, pl. 1, figs. 1–9, 11; pl. 12, figs. 1, 8.

Diagnosis: Test small, low trochospire, ovate; equatorial periphery strongly lobate; axial periphery rounded; chambers inflated, ovate; four in the last whorl; sutures depressed, radial on both sides; umbilicus wide; aperture a low interiomarginal slit, umbilical; wall with tiny regularly distributed pores and slender spines.

Remarks: *Globgerina bulloides* is the type species of the genus *Globigerina.* There is much variation which led Rögl and Bolli (1973) to define a *Globigerina bulloides* group as follows: The test consists basically of a fairly low trochospire with 4 globular chambers in the last whorl.

The variability of this group is considerable and gave rise to the erection of a number of separate but obviously closely related species.

According to Rögl and Bolli (1973), temperature appears to determine the morphology of *G. bulloides*, the more compact *G. bulloides bulloides* with thick wall and small aperture being mostly restricted to colder waters, while morphotypes with delicate walls and more loosely arranged chambers and larger apertures are indicative of warm conditions. In the Niger Delta, compact *G. bulloides* is the dominant type, an indication of the influence of the cold Benguela Current.

Range: Miocene–Recent.

Globigerina ciperoensis Bolli, 1954
Plate 32, Figs. 17–20

Globigerina ciperoensis Bolli, 1957c, Planktonic foraminifera from the Eocene Navet and San Fernando formations of Trinidad, B.W.I., US Natl. Museum Bull. 215, 109, pl. 22, figs. 10a,b.

Globigerina ciperoensis AGIP, 1988, Foraminifera of the Niger Delta, pl. 3, figs. 6a,b.

Globigerina ciperoensis Petters, 1983, Gulf of Guinea planktonic foraminiferal biochronology and geological history of the South Atlantic. J. Foramin. Res. 13, 54, pl. 7, figs. 4, 15.

Diagnosis: Test small, low to medium trochospiral; equatorial periphery almost circular, lobate, almost pentagonal; rounded axial periphery; chambers spherical, consistently five in the last whorl, increased moderately in size, ultimate chamber sometimes smaller than penultimate; sutures depressed and radial on both spiral and umbilical sides; umbilicus conspicuously wide, open; aperture with a fairly high arch, interiomarginal, umbilical; bordered by a faint lip; wall finely perforate to cancellate, and may appear hispid in places.

Remarks: Stainforth et al. (1973) observed that "*Globigerina ciperoensis* is consistent in size and morphology. Somewhat similar species are *Globigerina angulisuturalis* which has conspicuous U-shaped channels between its chambers and *G. angustiumbilicata* which has a small or closed umbilicus."

Range: Oligocene–Earliest Miocene.

Globigerina eamsi Blow, 1959
Plate 33, Figs. 1–3

Globigerina eamsi Blow, 1959, Age, correlation, and biostratigraphy of the upper Tocuyo (San Lorenzo) and Poron Formations, Eastern Falcon, Venezuela. Bull. Am. Paleontol. 39, 176, pl. 9, figs. 39a–c.

Globigerina eamsi Blow, 1969, Late Middle Eocene to Recent planktonic foraminiferal biostigraphy, In: Bronnimann, P., Renz, H.H. (Eds.), Proceedings of the First International Conference on Planktonic Microfossils, vol. 1. E.J. Brill, Geneva, p. 318, pl. 15, figs. 1–4.

Globigerina eamsi Petters, 1983, Gulf of Guinea planktonic foraminiferal biochronology and geological history of the South Atlantic. J. Foramin. Res. 13, 54, pl. 7, fig. 16.

Diagnosis: Test moderate in size, high, spiroconvex trochospire; umbilical side inflated; circular outline; equatorial periphery strongly lobate, axially rounded; chambers 3½ to 4 in the final whorl surrounding a narrow, deep to closed umbilicus; chambers of the final whorl increased very slowly in size; globular chambers; sutures broad, depressed, radial in umbilical view; on spiral side less distinct, slightly curved; aperture low very small interiomarginal slit; wall smooth, coarsely pitted to cancellate.

Range: Late Miocene–Pliocene.

Globigerina euapertura Jenkins, 1960
Plate 33, Figs. 4 and 5

Globigerina euapertura Jenkins, 1960, Planktonic foraminifera from the Lakes Entrance oil shaft, Victoria, Australia. Micropaleontology 6, 351, pl. 1, figs. 8a–c.

Globigerina euapertura Bolli and Saunders, 1985, Oligocene to Holocene low latitutude plankton stratigraphy. In: Bolli, H.M., et al., (Eds.), 1985, Plankton Stratigraphy. Cambridge University Press, Cambridge, pp. 155–262. pl. 80, fig. 13.21.

Description: Test small, circular, inflated, compact, high trochospire; umbilical side less inflated; chambers inflated, circular, 3 to 3½ in final whorl, increased regularly and rapidly in size, arranged in 3 whorls which are clearly visible on the inflated spiral side; sutures broad, depressed, radial; aperture low arch, small, confined to an equally small umbilicus; wall texture distinctly cancellate.

Remarks: Blow (1969) characterized *G. euapertura* by the laterally restricted nature of the primary aperture which has a lip. Jenkins (1960) also noted the "coarsely pitted" wall, which is actually cancellate. Bolli and Saunders (1985) remarked that "*G. euapertura* is distinguished from *G. ampliapertura* by having the final chamber tending to embrace more of the earlier test resulting in the test becoming more globular. The aperture in *G. euapertura* is a lower arch compared with the almost circular opening of *G. ampliapertura*."

Range: Late Eocene–Early Miocene.

Globigerina falconensis Blow, 1959
Plate 33, Figs. 6–8

Globigerina falconensis Blow, 1959, Age, correlation, and biostratigraphy of the upper Tocuyo (San Lorenzo) and Poron Formations, Eastern Falcon, Venezuela. Bull. Am. Paleontol. 39, 177, pl. 9, figs. 40–41.

Globigerina falconensis Jenkins, 1971, New Zealand Cenozoic planktonic foraminifera. Paleontol. Bull. Wellington 42, 148, pl. 16, figs. 463–465.

Diagnosis: Test, small, compact, moderately high trochospire, globular and distinctly oblate–quadrate profile; umbilical side inflated; equatorial view strongly lobate; axial periphery rounded; chambers high inflated, globular; four in the final whorl, increasing slowly in size, the ultimate chamber slightly reduced or equal to penultimate in size; sutures widely depressed; straight, radial on umbilical side; short and slightly curved on spiral side; umbilicus narrow, quadrate outline, with sutures connecting the four umbilical corners; aperture very low arch bordered by a short lip; wall hispid.

Remarks: In overall test shape, forms referred here to *G. falconensis* are similar to *G. ouachitaensis* in the oblate–quadrate profile and in the shape of the umbilicus. The difference lies in the wider umbilicus in *G. ouachitaensis* and the cancellate wall texture of this species.

Globigerina falconensis in the Niger Delta also resembles *G. eamsi* in its umbilical view, the four-chambered final whorl in which the chambers increased slowly, and in the oblate test shape; but *G. eamsi* is more spiroconvex and cancellate than *G. falconensis*.

Range: Early Miocene–Recent.

Globigerina foliata Bolli, 1957
Plate 33, Figs. 9–13

Globigerina foliata Bolli, 1957c, Planktonic foraminifera from the Eocene Navet and San Fernando formations of Trinidad, B.W.I., US Natl. Museum Bull. 215, 111, pl. 24, figs. 1a–c.

Globigerina foliata Blow, 1969, Late Middle Eocene to Recent planktonic foraminiferal biostigraphy. In: Bronnimann, P., Renz, H.H. (Eds.), Proceedings of the First International Conference on Planktonic Microfossils, vol. 1. E.J. Brill, Geneva, p. 319, pl. 16, figs. 2, 3.

Globigerina foliata Fayose, 1970, Stratigraphical paleontology of Afowo-1 well, Southwestern Nigeria. J. Min. Geol. 5 (1 and 2), 43, pl. 12, fig. 4.

Diagnosis: Test low trochospiral, elongate, flat to concave spiral side; equatorial periphery strongly lobate; axially rounded; chambers spherical; 8–10 arranged in 2 whorls, 4 chambers in final whorl which increase very rapidly in size with the ultimate chamber pronouncedly elongate; sutures deeply incised, radial on both sides; small umbilicus; aperture low arch with a lip, interiomarginal, umbilical; wall hispid to finely cancellate.

Remarks: This species is distinguished by its elongate final chamber, otherwise it is quite similar to *Globigerina praebulloides* from which it probably evolved in the Early Miocene.

Range: Early–Middle Miocene.

Globigerina galvisi Bermudez, 1961
Plate 33, Figs. 14 and 15

Globigerina galvisi Bermudez, 1961, Contribucion al estudio de las Globigerinidea de la region Caribe-Antillana (Paleocene–Reciente). Memoria de III Congreso Geologies Venezolano, Caracas, vol. 3. Bolatin de Geologia, Publicacion especial No. 3, 1960, p. 1, 183, pl. 4, fig. 3.

Globigerina galvisi Blow, 1969, Late Middle Eocene to Recent planktonic foraminiferal biostigraphy. In: Bronnimann, P., Renz, H.H. (Eds.), Proceedings of the First International Conference on Planktonic Microfossils, vol. 1. E.J. Brill, Geneva, p. 319, pl. 5, figs. 1–3; pl.16, figs. 4, 5.

Globigerina galvisi Berggren and Amdrurer, 1973, Late Paleogene (Oligocene) and Neogene planktonic foraminiferal biostratigraphy of the Atlantic Ocean (Lat. 30°N to Lat. 30°S). Rev. Ital. Paleontol. 79, 376, pl. 26, figs. 1, 2.

Diagnosis: Test small, inflated, high trochospiral, loosely arranged chambers in final whorl, ovate; strongly lobate equatorial periphery; rounded axial periphery; chambers circular, inflated, four chambers in final whorl, increased moderately in size, until highly enlarged ultimate chamber which constitutes half of the test; sutures depressed, wide, distinct on umbilical side, slightly curved; so also on spiral side; umbilicus small, closed, occupied by primary aperture, a small low arch with a distinct lip; wall hispid, pustulose.

Remarks: Although Bolli and Saunders (1985) considered *Globigerina galvisi* to be the same species as *Globigerina yeguaensis*, Blow (1969) had earlier argued that both are morphologically distinct in spite of their almost identical stratigraphic distribution. From the refigured holotypes of *G. galvisi* and *G. yeguaensis* (Blow, 1969; pl. 5, figs. 1–3; pl. 3 figs. 12–14, respectively), the main difference lies in the relatively larger and higher ultimate chamber in *G. galvisi*, a distinction also evident in the forms illustrated by Berggren and Amdrurer (1973); and also applied here for identifying the Niger Delta *G. galvisi*. Another useful morphologic feature in the Niger Delta is the wall texture. Like the forms illustrated with the scanning electron microscope by Berggren and Amdrurer (1971), the Niger Delta specimens of *G. galvisi* are spinose, whereas *G. yeguaensis* has a distinctly cancellate wall structure.
Range: Late Eocene–Early Oligocene.

Globigerina gnaucki Blow and Banner, 1962
Plate 33, Figs. 16 and 17

Globigerina ouachitaensis gnaucki Blow and Banner, 1962, The Mid-Tertiary (Upper Eocene to Aquitanian) Globigerinacea; In: Eames, F.E., et al., Fundamentals of Mid-Tertiary Stratigraphical Correlation. Cambridge University Press, New York, NY, p. 91, pl. IX, figs. L–N.
Globigerina ouachitaensis gnaucki Bolli and Saunders, 1985, Oligocene to Holocene low latitude plankton stratigraphy. In: Bolli, H.M., et al., (Eds.), 1985, Plankton Stratigraphy, Cambridge University Press, Cambridge, p. 178, figs. 13, 16a–c.

Diagnosis: Test small, subquadrate trochospire, high spiral side, umbilical side concave, ovate shape; strongly lobate periphery; chambers inflated, four in the final whorl, expanding rapidly in size; sutures widely depressed accentuating lobate periphery and loose chamber arrangement; radial sutures umbilically and spirally; umbilicus open, deep; aperture low arched, umbilical; wall texture strongly cancellate.
Remarks: This species is distinguished from others in the *Globigerina ouachitaensis* group by its loose coil, rapid rate of chamber expansion, and broad sutural depressions which ally *G. gnaucki* with *G. angulisuturalis*. *Globigerina ouachitaensis* evolved into the short-ranging form *G. gnaucki* close to the Eocene–Oligocene boundary.
Range: Late Eocene–Early Oligocene.

Globigerina juvenelis Bolli, 1957
Plate 33, Figs. 18 and 19

Globigerina juvenelis Bolli, 1957c, Planktonic foraminifera from the Eocene Navet and San Fernando formations of Trinidad, B.W.I., US Natl. Museum Bull. 215, 110, pl. 24, figs. 5a–c.
Globigerina 34: SPDC Foraminiferal Type Collection.
Globigerina juvenelis Petters, 1983, Gulf of Guinea planktonic foraminiferal biochronology and geological history of the South Atlantic. J. Foramin. Res. 13, 54, pl. 7, figs. 22, 23.

Original Diagnosis: "Shape of test moderately to distinctly trochospiral; equatorial periphery distinctly lobate; wall calcareous, perforate, surface smooth to very finely pitted; chambers spherical to ovate; about 12, arranged in about 3 whorls; the 3–4 chambers of the last whorl increase rapidly in size; sutures on spiral side curved to radial in the early stage, radial in last whorl, depressed; on umbilical side radial, depressed; umbilicus small; aperture a low elongate slit, often with a thin lip, interiomarginal, umbilical; coiling random."

Remarks: The distinctly pustulose test and the greater number of chambers visible on the high spiral side, as well as the rapid rate of chamber expansion distinguish *G. juvenelis* from *G. bradyi*.

Range: Oligocene–Recent.

Globigerina leroyi Blow and Banner, 1962
Plate 34, Figs. 1 and 2

Globigerina praebulloides leroyi Blow and Banner, 1962, The Mid-Tertiary (Upper Eocene to Aquitanian) Globigerinacea. In: Eames, F.E., et al., Fundamentals of Mid-Tertiary Stratigraphical Correlation. Cambridge University Press, New York, NY, p. 93, pl. 9, figs. R–T.

Globigerina praebulloides leroyi Kiesel, Lotsch, and Trumper, 1969, Planktonic foraminifera from the Oligo/Miocene of the German Democratic Republic (GDR). In: Bronnimann, P., Renz, H.H. (Eds.), Proceedings of the First International Conference on Planktonic Microfossils, vol. II. E.J. Brill, Geneva, p. 342, pl. 1, figs 4a–c.

Diagnosis: Test small, low trochospiral, quadrate; strongly lobate, rounded axial periphery; chambers inflated, four in the final whorl, increased rapidly in size, ultimate chamber nearly half of test; sutures wide, depressed radial on umbilical side, slightly curved on spiral side; umbilicus small; aperture very low, interiomarginal, umbilical, with thin lip; wall cancellate.

Remarks: The authors cited previously consider *G. leroyi* as a subspecies of *G. praebulloides*, a common Oligocene–Miocene taxon. Bolli and Saunders (1985) compared the three subspecies in the *G. praebulloides* group, namely *Globigerina praebulloides praebulloides*, *G. praebulloides occlusa*, and *G. praebulloides leroyi* and noted that the criteria for separating them are indeed subtle. Blow and Banner (1962) who erected the three subspecies distinguished them as follows: "*G. praebulloides leroyi* is distinguished from *G. praebulloides praebulloides* principally by its smaller umbilicus, lower and more symmetrical aperture, slower rate of chamber enlargement and greater degree of embrace between its chambers. *G. praebulloides occlusa* differs from *G. praebulloides* in possessing a smaller, shallower umbilicus, a smaller lower aperture which lacks a lip or rim and possessing a slightly thicker, rougher and more coarsely perforate wall. *G. praebulloides occlusa* is distinguished from *G. praebulloides leroyi* in possessing a shallower umbilicus and an asymmetrical aperture which lacks a lip. The chambers of *G. praebulloides occlusa* are more embracing than those of *G. praebulloides praebulloides*, but less tightly embracing than those of *G. praebulloides leroyi*. "

Range: Late Eocene–Middle Miocene.

Globigerina occlusa Blow and Banner, 1962
Plate 34, Figs. 3–5

Globigerina praebulloides occlusa Blow and Banner, 1962, The Mid-Tertiary (Upper Eocene to Aquitanian) Globigerinacea. In: Eames, F.E., et al., Fundamentals of Mid-Tertiary Stratigraphical Correlation. Cambridge University Press, New York, NY, p. 95–96, pl. 9 u–w, figs. 14 i,ii.

Globigerina occlusa Petters, 1983. Gulf of Guinea planktonic foraminiferal biochronology and geological history of the South Atlantic. J. Foramin. Res. 13, 54, pl. 7, Figs. 7, 8.

Diagnosis: Test low to moderate trochospire, ovate to quadrate, inflated, loosely coiled; strongly lobate equatorial periphery; axial periphery rounded; chambers, inflated, circular, arranged in 2½ to 3 whorls; 4 chambers in the final whorl, which increased very rapidly in size; sutures widely depressed, radial on umbilical side, slightly bent on spiral side, depressed; aperture low interiomarginal, umbilical, without a lip; wall coarsely cancellate.

Remarks: *Globigerina occlusa* exhibits the coarsely cancellate wall texture of Late Eocene–Oligocene Globigerininids hence, as suggested by Bolli and Saunders (1985), it is probably the ancestor to the *G. praebulloides* stock. Its umbilicus is narrower than in *G. praebulloides*.

Range: Middle Eocene–Middle Miocene.

Globigerina officinalis Subbotina, 1953
Plate 34, Figs. 6 and 7

Globigerina officinalis Subbotina, 1953, Iskopaemye Foraminifery SSR. Globigerinidy, Khantkeninidy i Globorotaliidy (Fossil Foraminifera of the USSR. Globigerinidae, Hantkeninidae and Globorotaliidae). Trudy Vsesoyuznogo Neftyanogo Nauchnoissledovatel skogo Geologorazvedochnogo Instituta (VNIGRI), Ser. 76, p. 78, pl. 11, figs. 1–7.

Globigerina officinalis Petters, 1983, Gulf of Guinea planktonic foraminiferal biochronology and geological history of the South Atlantic. J. Foramin. Res. 13, 56, pl. 4, figs. 8, 17.

Diagnosis: Test small, low trochospire, tightly coiled, ovate shape, strongly lobate; rounded axial periphery chambers inflated, circular, last whorl with 3½ to 4 globular chambers which increase gradually in size; sutures depressed, wide on both sides accentuating the strongly indented equatorial periphery; umbilicus very small, shallow; aperture a low umbilical arch bordered with a distinct lip; wall smooth, cancellate.

Remarks: *Globigerina officinalis* differs from *G. ouachitaensis* in its rapidly increasing chambers of its final whorl; sizeable ultimate chamber comprising one-third of the entire test; the tighter coil; and in the reduction in the size of the umbilicus.

Range: Late Middle Eocene–Oligocene.

Globigerina ouachitaensis Howe and Wallace, 1932
Plate 34, Figs. 8–10

Globigerina ouachitaensis Howe and Wallace, 1932. A Monograph of the foraminiferal family Nonionidae, No. 191, Louisiana. Louisiana Department of Conservation, Geological Bulletin 2, p. 74, pl. 10, figs. 7a,b.

Globigerina 66: SPDC Foraminiferal Type Collection.

Globigerina ouachitaensis Postuma, 1981, Manual of Planktonic Foraminifera. Elsevier Publishing Co., Amsterdam, pp. 152–153.

Globigerina ouachitaensis Petters, 1983, Gulf of Guinea planktonic foraminiferal biochronology and geological history of the South Atlantic. J. Foramin. Res. 13, 56, pl. 7, figs. 1, 2, 11.

Diagnosis: Test small, medium to high trochospiral, spiroconvex; equatorial periphery distinctly lobate, axial periphery rounded; chambers subspherical; 2½ to 3 whorls; 4 chambers in the last whorl juxtaposed in two pairs of opposite chambers; regular but moderate rate of chamber size increase in final whorl; sutures depressed, radial on umbilical side, on spiral side curved to radial; umbilicus moderate; aperture distinct low interiomarginal arch, bordered by lip; wall distinctly cancellate.

Remarks: The moderate centralized umbilicus; quadrate outline; four inflated chambers in the final whorl; and cancellate wall characterize *G. ouachitaensis*. *Globigerina gnaucki* differs in the rapid rate of chamber size increase and the more elongate test profile; while *G. ciperoensis* can be distinguished by its five chambers in the last whorl and hispid umbilical ornamentation. Bolli and Saunders (1985) observed that "problems may arise in distinguishing *G. ouachitaensis* s.l. from *G. praebulloides* s.l., especially when intermediate forms are present. The *G. praebulloides* subspecies differs from those of *G. ouachitaensis* in the more rapidly increasing size in particular of the ultimate chamber; resulting in a distinctly elongated equatorial outline instead of a more quadrate one in *G. ouachitaensis*. Furthermore, the umbilicus in *G. praebulloides* does not possess the quadrate outline that is typical for *G. ouachitaensis*."

Range: Middle Eocene–Middle Oligocene.

Globigerina cf. *ouachitaensis*
Plate 34, Figs. 11–13

Diagnosis: Test small, moderately high trochospire; circular outline, inflated; equatorial periphery strongly lobate, axial periphery rounded; chambers circular, globular; four chambers in the last whorl; increasing gradually in size, with slight reduction in size of ultimate chamber; three whorls in the test; sutures wide, depressed on both umbilical and spiral sides, radial on both sides; umbilicus wide, deep; quadrate umbilical outline; aperture very low basal slit, umbilical to slightly extraumbilical; wall hispid, especially on the umbilical side, moderately cancellate, but more so on spiral side; last chamber smooth, perforate to lightly hispid.

Remarks: *Globigerina* cf. *ouachitaensis* exhibits the gross test morphology of the *G. ouachitaensis* group, differing mainly in its more loosely arranged chambers; reduction in the size of the final chamber; lower trochospire; but most especially in wall surface ornamentation. Whereas *G. ouachitaensis* is coarsely cancellate with pores surrounded with cylindrical ridges, the wall of *G.* cf. *ouachitaensis* shows obliteration, and crowding of pore pits and ridges by spines. This species represents subsequent evolutionary adaptation of *G. ouachitaensis* at the higher stratigraphic levels.

Range: Oligocene–Miocene.

Globigerina praebulloides Blow, 1959
Plate 34, Figs. 14–16

Globigerina praebulloides Blow, 1959, Age, correlation, and biostratigraphy of the upper Tocuyo (San Lorenzo) and Poron Formations, Eastern Falcon, Venezuela. Bull. Am. Paleontol. 39, 180, pl. 8, figs. 47 a–c; pl. 9, fig. 48.

Globigerina praebulloides Postuma, 1971, Manual of Planktonic Foraminifera. Elsevier Publishing Co., Amsterdam, pp. 268–269.

Globigerina praebulloides Petters, 1983, Gulf of Guinea planktonic foraminiferal biochronology and geological history of the South Atlantic. J. Foramin. Res. 13, 56, pl. 7, fig. 3.
Globigerina praebulloides AGIP, 1988, Foraminifera of the Niger Delta, pl. 2, fig. 12.

Diagnosis: Test low trochospire, equatorial periphery distinctly lobate, axial periphery rounded; chambers inflated, arranged in about 2½ whorls; 4 chambers in final whorl with moderate size increase; sutures on spiral side depressed, radial to slightly slanting; radial on umbilical side, depressed; umbilicus, wide and quadrate in shape; aperture moderately high interiomarginal arch; wall cancellate to weakly pustulose.
Range: Late Eocene–Middle Miocene.

Globigerina prasaepis Blow, 1979
Plate 34, Figs. 17–19

Globigerina prasaepis Blow, 1979, Danian to Oligocene planktonic foraminiferal biostratigraphy. The Cainozoic Globigerinida. A Study of the Morphology, Taxonomy, Evolutionary Relationships and the Stratigraphical Distribution of Some Globigerinida (Mainly Globigerinacea). E.J. Brill, Leiden, p. 382, pl. 10, fig. 13, pl. 18, figs. 3–7.

Diagnosis: Test globose, low trochospire; equatorial periphery circular; axial periphery rounded; chambers moderately inflated, subspherical, closely appressed, particularly the ultimate chamber; four chambers in the final whorl; sutures incised, radial on umbilical side; umbilicus narrow, deep; aperture low interiomarginal arch, extending across umbilicus; wall distinctly cancellate.
Range: Uppermost Eocene–Late Oligocene.

Globigerina quinqueloba Natland, 1938
Plate 35, Figs. 1–3

Globigerina quinqueloba Natland, 1938. New Species of Foraminifera From Off the West Coast of North America and From the Later Tertiary of the Los Angeles Basin. California, Technical Series, vol. 4, p. 149, figs. 6, 7.
Globigerina quinqueloba Rögl and Bolli, 1973, Initial Reports of the Deep Sea Drilling Project, p. 564, pl. 4, figs. 10, 12.

Emended Diagnosis: Rögl and Bolli (1973) emended the diagnosis of *G. quinqueloba* as follows: "Shape of test very low trochospiral; 5 chambers in the last whorl. Chambers subglobular, last chamber radially elongate and extending over the umbilicus. The umbilicus and the umbilical–extraumbilical aperture also may be partially covered by an apertural flap. Blow, (1969, p. 373, no. 203) places the species in *Turborotalia,* which appears justified considering the shape of test and last chamber. However, living *Globigerina quinqueloba* possesses very thin, long and flexible spines (Hemleben, 1969, pl. 17, fig. 4) which are not preserved in specimens from the sediment. With the possession of spines, *Globigerina quinqueloba* belongs to the family Globigerinidae and not Globorotaliidae and *Turborotalia.* The position of *Globigerina* looks questionable based on shape of test. The wall is thin, vitreous transparent, almost smooth, with few but large pores and only few and small bases of spines. The apertural flap is missing in globular final chambers."

Remarks: Niger Delta forms of *G. quinqueloba* are similar to those figured in Rögl and Bolli (1973) except that the ultimate chambers do not lie completely across the umbilicus, hence their apertures are more visible and globigerine in shape and position.
Range: Middle Miocene–Recent.

Globigerina yeguaensis Weinzien and Applin, 1929
Plate 35, Figs. 4 and 5

Globigerina yeguaensis Weinzien and Applin, 1929. The Clairborne formation on the coastal domes. J. Paleontol. 3, p. 408, pl. 43, figs. 1a,b.
Globigerina yeguaensis Postuma, 1971, Manual of Planktonic Foraminifera. Elsevier Publishing Co., Amsterdam, pp. 162–163.

Diagnosis: Test small, low, rounded trochospiral; slightly convex spiral side; convex umbilically; strongly lobate periphery; broadly rounded axial periphery; chambers inflated, subspherical, arranged in three whorls; three to four chambers in the final whorl, increased fairly rapidly in size; sutures radial on umbilical side, depressed, slightly curved on spiral side, depressed; small umbilicus because of highly compact test; aperture very low, umbilical arch, interiomarginal with thin lip; wall cancellate.
Remarks: There is close similarity between *G. yeguanensis* and *G. galvisi*, but in the latter the chambers increased more rapidly and the wall texture is hispid.
Range: Late Eocene–Early Oligocene.

Globigerina utilisindex Jenkins and Orr, 1973
Plate 35, Fig. 6

Globigerina utilisindex Jenkins and Orr, 1973, J. Foramin. Res. 3, 133–135, pl. 1, figs. 1–6; pl. 2, figs. 1–9; pl. 3, figs. 1–3.

Diagnosis: Test medium in size, inflated rectangular shape, very low trochospire, the last whorl markedly enlarged on the umbilical side; equatorial periphery smooth to only slightly indented, rounded axial periphery, chambers inflated, squarish shape; three enlarged chambers in the last whorl, the ultimate chamber having a flattish termination; the last chamber constitutes over half of the test; sutures wide, slightly depressed; umbilicus closed; aperture very low slit of the base of ultimate chamber, indistinct where poorly preserved; wall pustulose and coarsely perforate.
Remarks: This species is distinctive on account of its rectangular inflated shape and flat-topped ultimate chamber.
Range: Upper Eocene–Oligocene.

Globigerina nepenthes Todd, 1957
Plate 35, Figs. 19 and 20

Globigerina nepenthes Todd, 1957, Geology of Saipan, Mariana Island, Part 3, Paleontology. Chapter H, Smaller Foraminifera, US Geol. Surv. Prof. Paper, 280H, p. 301, pl. 78, fig. 7.

Globigerina (Zeoglobigerina) nepenthes Todd, Kennett, and Srinivasan, 1983, Taxonomic notes 1983 (www-odp.tamu.edu/publications/165_SR/chap_02/c2_9.htm), p. 49, pl. 9, figs. 1–3.

Globigerina nepenthes Todd, 1957, Bolli and Saunders, 1985, Oligocene to Holocene low latitutude plankton stratigraphy. In: Bolli, H.M., et al., (Eds), 1985, Plankton Stratigraphy, Cambridge University Press, Cambridge, p. 201, fig. 25; 1–4.

Diagnosis: Text compactly coiled except for the last-formed cap-shaped chamber that has a large aperture usually bordered by a thickened rim, four chambers in the final whorl; sutures obscured on earlier chambers due to test thickening, radial to slightly curved on later chambers, slightly depressed; surface cancellate with distinct pore pits and interpore ridges.

Remarks: This species can show considerable morphological variation in closeness of coiling and thickness of test wall, which has led to a number of subspecies being proposed; for example, *G. nepenthes delicatula* Bronnimann and Resig (1971).

Range: Miocene–Pliocene.

Genus: GLOBOROTALOIDES Bolli, 1957
Globorotaloides suteri Bolli, 1957
Plate 35, Figs. 7–9

Globorotaloides suteri Bolli, 1957a–c, Planktonic foraminifera from the Eocene Navet and San Fernando formations of Trinidad, B.W.I., US Natl. Museum Bull. 215, 117, pl. 27, figs. 9a–13b.

Globorotaloides suteri Stainforth et al., 1975., Cenozoic planktonic foraminiferal zonation and characteristics of index forms. Paleontological Contribution, University of Kansas, Article 62, p. 322–323, fig. 146.

Globorotaloides suteri Petters, 1983, Gulf of Guinea planktonic foraminiferal biochronology and geological history of the South Atlantic. J. Foramin. Res. 13, 57, pl. 6, figs. 25, 26.

Diagnosis: Test small subcircular, slightly compressed low trochospiral; lobate periphery. Chambers five to six in the final whorl, increased rapidly in size, last chamber inflated; sutures depressed, radial umbilically and spirally; aperture extraumbilical slit bordered by distinct lip; umbilicus partially closed, may possess bulla; wall distinctly cancellate.

Remarks: This tiny species often shows no bulla in Niger Delta forms. Stainforth et al. (1975) observed that individual specimens vary greatly, and small ones may have bullas whereas large ones may retain slit apertures. Nonbullate specimens with only slit apertures are common in the Niger Delta.

Range: Middle Eocene–Middle Miocene.

Genus: GLOBOQUADRINA Finlay, 1947
Globoquadrina altispira Cushman and Jarvis, 1936
Plate 35, Figs. 10 and 11

Globigerina altispira Cushman and Jarvis, 1936, Three new Foraminifera from the Miocene Bowden Marl of Jamaica. Contrib. Cushman Lab. Foramin. Res. 12, 3–5, pl. 1.

Globoquadrina altispira altispira Stainforth et al., 1975, Cenozoic planktonic foraminiferal zonation and characteristics of index forms. Paleontological Contribution, University of Kansas, Article 62, p. 245, fig. 100.

Diagnosis: Test large, high conical trochospire, nearly same test height as diameter; spiral outline subcircular, slight indentation at sutures; ovate to subtriangular side profile; chambers closely packed; initial chambers globular, and rapidly enlarging, later in much closer contact; about four to six chambers in final whorl; sutures depressed and distinct; aperture umbilical; umbilicus, wide and deep, showing apertural teeth of earlier chambers; wall finely cancellate.

Range: Early Miocene–End Middle Pliocene.

Globoquadrina dehiscens Chapman, Parr, and Collins, 1934
Plate 35, Figs. 12 and 13

Globorotalia dehiscens Chapman, Parr, and Collins, 1934, Tertiary Foraminifera of Victoria, Australia. The Balcombian deposits of Port Phillip, Part, 3. Zool. J. Linnean Soc. 38, 569, pl. 11, figs 36a–c.

Globoquadrina dehiscens Bolli, 1957a–c, US Natl. Museum Bull. 215, 31, pl. 5, figs. 4a–6.

Diagnosis: Test large, low trochospiral, flat spiral side; umbilical side convex; equatorial periphery slightly lobate; outline almost quadrate; axial periphery rounded with chamber shoulders near umbilicus; chambers spherical in initial part of test; later compressed; four chambers in the last whorl, increased rapidly in size; sutures depressed; radial on spiral side, slightly curved; radial on umbilical side; umbilicus wide and deep; aperture intraumbilical with a flap; wall cancellate.

Remarks: This species can be differentiated from *Dentoglobigerina baroemeonsis* by its narrower umbilicus and enlarged ultimate chamber; apart from its cancellate wall.

Range: Lower–Middle Late Miocene.

Globoquadrina globosa Bolli, 1957
Plate 35, Figs. 14 and 15

Globoquadrina altispira globosa Bolli, 1957a–c, Planktonic foraminifera from the Oligocene–Miocene Cipero and Lengua Foramation of Trinidad, etc. (mansion). In: Bolli, H.M., Saunders, J.B., Perch-Nielsen, K. (Eds.), 1985, Plankton Stratigraphy, vol. 1, Planktonic Foraminifera, Calcareous Nannofossils and Calpionellids, p. 111, pl. 24, figs 9a–10c.

Globoquadrina globosa Petters, 1983, Gulf of Guinea planktonic foraminiferal biochronology and geological history of the South Atlantic. J. Foramin. Res. 13, 56, pl. 8, figs. 22, 27.

Diagnosis: Test large, inflated, moderately high to low trochospire, convex umbilical side; moderately lobate equatorial periphery; axial periphery rounded; chambers globular, circular, four to five in the final whorl; sutures deep and radial on umbilical side; less depressed and narrower on spiral side, straight; umbilicus large, deep; aperture intraumbilical, covered by shamber flaps; wall finely cancellate to pustulose, especially on umbilical view.

Range: Early Miocene–Pliocene.

Globoquadrina praedehiscens Blow and Banner, 1962
Plate 35, Fig. 16

Globoquadrina dehiscens praedehiscens Blow and Banner, 1962, The Mid-Tertiary (Upper Eocene to Aquitanian) Globigerinacea. In: Eames, F.E., et al., Fundamentals of Mid-Tertiary Stratigraphical Correlation. Cambridge University Press, New York, NY, p. 185, pl. 11, figs. 57a–c.

Diagnosis: Test large, low, compact trochospire, convex umbilical side, highly inflated; subtriangular outline; slightly lobate periphery; chambers inflated, laterally compressed; four in final whorl, increased rapidly in size; ultimate chamber comprising one-third of the test; sutures broadly depressed, radial on umbilical side; umbilicus narrow; aperture very low interiomarginal slit; wall coarsely perforate.

Remarks: This species differs from *G. dehiscens* in its highly compact test, narrow umbilicus, and enlarged ultimate chamber. Blow (1996) suggested the evolution of *G. praedehiscens* from *Globigerina tripartita*, a three-chambered, inflated Oligocene species that has a tighter coil than *G. praedehiscens.*

Range: Late Oligocene–Early Miocene.

Globoquadrina venezuelana Hedberg, 1937
Plate 35, Figs. 17–20

Globigerina venezuelana Hedberg, 1937, Foraminifera of the middle Tertiary Carapita Formation of northeastern Venezuela. J. Paleontol. 11 (8), 681, pl. 92, figs. 7a,b.

Globoquadrina venezuelana Petters, 1983, Gulf of Guinea planktonic foraminiferal biochronology and geological history of the South Atlantic. J. Foramin. Res. 13, 56, pl. 8, figs. 12–13.

Diagnosis: Test large, inflated, quadrate, medium to low trochospire; strongly inflated umbilical side; moderately lobate equatorial periphery; chambers spherical, those of last whorl four, laterally compressed, increased rapidly in size; but ultimate chamber could be reduced in size; sutures, depressed, radial distinct; deep umbilicus of variable size depending on degree of chamber axial prolongation; aperture intraumbilical, very low, may have a small blunt triangular lip; wall cancellate.

Range: Late Oligocene–Middle Pliocene.

Genus: DENTOGLOBIGERINA Blow, 1979
Dentoglobigerina baroemoensis LeRoy, 1939
Plate 36, Figs. 1–3

Globoquadrina baroemoensis LeRoy, 1939, Some small foraminifera, ostracoda and otoliths from the Neogene ("Miocene") of the Rokan-Tapanoeli area, Central Sumatra. Natuurkwetenschappelijk Tijdschrift voor Nederlandsch Indie, 99 (6), 263, pl. 6, figs. 1–2; pl. 27, figs. 4, 8.

Dentoglobigerina cf. *baroemoensis* Carbonnel et al., 1996, Microfaunes et milieux de depots des niveaux oligocenes surmontant la 'discordance oligocene' sur la partite terrestre du bassin sedimentaire cotier du Benin. Actes Colloques Angers 1994, 240, pl. 1, figs. 5.

Diagnosis: Test medium, quadrate, low trochospiral, flat spiral side; strongly umbilico-convex; moderately lobate equatorial periphery; axial periphery rounded; chamber subcircular laterally compressed and angulate, 4 to 4½ in the final whorl; expanding rapidly in size; sutures depressed, deep and wide on umbilical side, slanting on spiral side producing chamber overlap; umbilicus deep and very wide and excavated; aperture interiomarginal, umbilical with triangular projecting tooth; wall coarsely perforate.

Remarks: This species is distinguished from cooccurring *Globoquadrina* species by its wide umbilicus, otherwise it can be confused with *G. dehiscens.*

Range: Middle Oligocene–Late Miocene.

Genus: NEOGLOBOQUADRINA Brady, Frerichs, and Vincent, 1967
Neogloboquadrina acostaensis Blow, 1959
Plate 36, Figs. 4–7

Globorotalia acostaensis Blow, 1959, Age, correlation, and biostratigraphy of the upper Tocuyo (San Lorenzo) and Poron Formations, Eastern Falcon, Venezuela. Bull. Am. Paleontol. 39, 208–210, pl. 17, figs. 106a–c, 107.

Neogloboquadrina acostaensis Petters, 1983, Gulf of Guinea planktonic foraminiferal biochronology and geological history of the South Atlantic. J. Foramin. Res. 13, 56, pl. 6, figs. 18–19.

Diagnosis: Test small, thick, low biconvex, trochospire, circular outline; strongly lobate equatorial periphery; axial periphery slightly compressed; chambers, ovate to spherical, four to six in the final whorl, increased gradually and regularly in size; sutures straight, radial, more prominent on umbilical side; umbilicus largely covered by apertural lip; aperture a low arch, interiomarginal.

Remarks: About this distinctive marker species Stainforth et al. (1975) stated as follows: "*Globorotalia acostaensis* is regarded as intermediate between *Globorotalia continuosa* Blow (1969) (more compact quadrate form) and *Globorotalia humerosa* Takayanagi and Saito (1962) (larger, looser coil, more distinct umbilicus). It is grossly similar to *Globorotalia siakensis* LeRoy (1939) but has more inflated chambers, a more concealed aperture, and a more rapidly opening spire. Within the lower part of the *Globorotalia acostaensis* Zone, this species coiled almost exclusively to the left, whereas in the upper part of the zone dominantly right and random coiling predominates. This horizon of coiling change has been observed in the South China Sea, in the Caribbean and Gulf of Mexico regions, and in Italy within the lower part of the Messinian stratotype of Selli (1971) . . . the datum, therefore, approximates closely the base of the Messinian stage. "

Range: Late Miocene–Pliocene.

Neogloboquadrina dutertrei d'Orbigny, 1939
Plate 36, Figs. 8–10

Globigerina dutertrei d'Orbigny, 1839, Foraminiferes des Iles Canaries. In: Barker-Webb, P., Berthelot, S., Historie Naturelle des Canaries. Paris, vol. 2, Part 2, p. 84, pl. 4, figs. 19–21.

Neogloboquadrina dutertrei Bolli and Saunders, 1985, Oligocene to Holocene low latitutude plankton stratigraphy. In: Bolli, H.M., et al., (Eds.), 1985 Plankton Stratigraphy, Cambridge University Press, Cambridge, p. 208, figs. 27.3–4.

Diagnosis: Test thick, circular, inflated, high trochospire; lobate periphery; axial periphery rounded; chambers five to six in the final whorl, globular, expanded gradually and regularly; inflated ovate chambers, elongated parallel to spiral axis; sutures depressed, radial, umbilicus, quadrate, deep; aperture umbilical, low arch; wall coarsely perforate, pustulose around umbilicus.

Remarks: Bolli and Saunders (1985) figured forms with mainly four chambers and a small fifth chamber in the final whorl, like the Niger Delta specimens illustrated here. The main differences between *N. dutertrei* and *N. humerosa* are the higher spiral side and the tighter coiling and hence a more circular outline in *N. dutertrei*. As for the generic status of *Neogloboquadrina* Bolli and Saunders (1985) stated that "The type species of *Neogloboquadrina* is *Globigerina dutertrei* d'Orbigny. As regards the arrangement of chambers, position of apertures and the tooth-like umbilical flaps, *Neogloboquadrina* combines features of the three genera *Globigerina*, *Globorotalia* and *Globoquadrina*." *Neogloboquadrina dutertrei* is regarded as the end form of the bioseries *N. acostaensis–N. humerosa–N. dutertrei*.

Range: Pliocene–Holocene.

Neogloboquadrina humerosa Tagayanagi and Saito, 1962
Plate 36, Figs. 11 and 12

Globorotalia humerosa Takayanagi and Saito, 1962, Planktonic Foraminifera from the Mobori Formation, Shikoku, Japan. Tohoku Univ. Sci. Rep. Ser.2 (Geol.), 5, 78, pl. 26, figs 1–2.

Neogloboquadrina humerosa Petters, 1983, Gulf of Guinea planktonic foraminiferal biochronology and geological history of the South Atlantic. J. Foramin. Res. 13, 56, pl. 6, figs 7, 14.

Diagnosis: Test low trochospire, circular, inflated, spiral side almost flat; equatorial periphery slightly lobate; axial periphery rounded but somewhat compressed; chambers moderately inflated, five to seven in final whorl; slow rate of chamber size increase in final whorl; sutures depressed, radial on both sides; umbilicus wide, deep; primary aperture low umbilical–extraumbilical arch, sometimes bordered by lip; wall coarsely perforate, granular umbilical region.

Range: Latest Miocene–Early Pleistocene.

Genus: GLOBIGERINOIDES Cushman, 1927
Globigerinoides bisphericus Todd, 1954
Plate 36, Fig. 13

Globigerinoides bispherica Todd, 1954, Appendix. In: Todd, R., Cloud Jr., P.E., Low, D., Schmidt, R.G. (Eds.), Probable Occurrence of Oligocene on Saipan. Am. J. Sci. 252, 681, pl. 1, figs. 1a–c.

Globigerinoides bisphericus Bolli and Saunders, 1985, Oligocene to Holocene low latitutude plankton stratigraphy. In: Bolli, H.M., et al., (Eds.), 1985, Plankton Stratigraphy, Cambridge University Press, Cambridge, p. 199, figs. 24.8.

Diagnosis: Test moderate, globular, very low trochospiral, equatorial periphery lobate; axial periphery rounded; chambers in early whorl concealed; last whorl shows two chambers, the ultimate chamber occupying two-thirds of the test; sutures only distinct in later part of the test, wide and depressed; aperture and umbilicus covered by the enlarged ultimate chamber; wall cancellate.

Remarks: Bolli and Saunders (1985) described this as "a short-lived species differing from its probable ancestor, *G. trilobus* (Reuss), in its more enveloping last chamber which almost completely conceals the umbilicus. Only 2 apertures are present along the suture between the last and earliest chambers, though more apertures may remain open on the spire."

Range: Early Miocene.

Globigerinoides bulloideus Crescenti, 1966
Plate 36, Figs. 14–16

Globigerinoides bulloideus Crescenti, 1966, Sulla biostratigrafia del Miocene affiorante al confine marchigiano 'abruzzese. Geol. Romana, 5, 43, pl. 9, figs. 9a–c.

Diagnosis: Test small, very high trochospire, triangular outline; equatorial periphery strongly lobate; axial periphery rounded; chambers circular, inflated; four in the final whorl; increased rapidly in size even in earlier whorl on the spiral side; ultimate chamber diminutive, elongate in axis of coiling; sutures wide, depressed, radial on umbilicus shallow, narrow; primary aperture moderately high arch, with short lip; single supplementary aperture on spiral side; wall cancellate.

Remarks: Bolli and Saunders (1985) remarked that: "This species is characterised by a wide primary aperture and only one small supplementary aperture. In its shape, arrangement of chambers and in its size, it is very close to *Globigerina bulloides.* It differs from *Globigerinoides bolli* in possessing a wider primary aperture, a more lobate equatorial profile and less embracing chambers. *G. altiaperturus* differs from *G. bulloideus* in the higher and distinctly larger supplementary aperture. Compared with *G. primordius, G. bulloideus* possesses a larger primary aperture and a less lobate outline."

Range: Middle–Late Miocene.

Globigerinoides extremus Bolli and Bermudez, 1965
Plate 36, Figs. 17 and 18

Globigerinoides obliquus (Bolli) var. or subsp. *extremus* Bolli and Bermudez, 1965, Zonation based on Planktonic Foraminifera of Middle Miocene to Pliocene warm-water sediments. Assoc. Venezolana Geol. Min. Petr. Bol. Informativo, 8 (5), 139, pl. 1, figs. 10–12.

Diagnosis: Test medium to high trochospire, all chambers of the final whorl distinctly compressed, the four chambers of the final whorl increasing regularly in size as added, last chamber somewhat reduced in size, and distinctly flattened; sutures on both sides radial to slightly curved and depressed; surface distinctly pitted; primary aperture interiomarginal, umbilical, a distinct arch of medium height; chambers of the final whorl with one supplementary aperture opposite the primary one.

Range: Late Miocene–Late Pliocene.

Globigerinoides immaturus LeRoy, 1939
Plate 37, Figs. 1–3

Globigerinoides sacculiferus (Brady) var. *immatura* LeRoy, 1939, Some small foraminifera, ostracoda and otoliths from the Neogene ("Miocene") of the Rokan-Tapanoeli area, Central Sumatra. Natuurkwetenschappelijk Tijdschrift voor Nederlandsch Indie, 99 (6), 263, pl. 3, figs. 19–21.

Globigerinoides immatura Petters, 1983, Gulf of Guinea planktonic foraminiferal biochronology and geological history of the South Atlantic. J. Foramin. Res. 13, 56, pl. 8, figs. 8, 21.

Diagnosis: Test trochospiral, ovate, elongate, inflated equally on both sides; equatorial periphery lobate; axial periphery broadly rounded; chambers spherical, arranged in about three whorls; three chambers in the last whorl which increased very rapidly in size, the ultimate chamber constituting one-half of the entire test; sutures broadly depressed to incised, radial on umbilical side, slightly curved on spiral side; umbilicus very narrow to nearly closed; aperture interiomarginal, umbilical, low to medium with a lip; on spiral side there is one secondly aperture almost opposite the primary aperture; wall finely cancellate.

Remarks: This species is similar to *G. inusitatus*, but the latter has a wider umbilicus and higher primary aperture.

Range: Lower Miocene–Recent.

Globigerinoides inusitatus Jenkins, 1966
Plate 37, Figs. 4 and 5

Globigerinoides inusitatus Jenkins, 1966, Planktonic foraminiferal zones and new taxa from the Danian to Lower Miocene of New Zealand. NZ J. Geol. Geophys. 8, 1108, fig. 9, no. 72–80.

Original Diagnosis: Jenkins (1966) described the holotype of this species as follows: "Test free, dextrally coiled, low trochospiral, equatorial periphery lobulate, axial periphery rounded. Wall calcareous, perforate, surface finely perforate. Chambers spherical in outline, 13 forming the test, arranged in about three whorls, with five in the first whorl, four in the second and final whorls, the final chamber slightly depressed; proloculus diameter 0.02 mm. Sutures on the spiral side radial to slightly curved, depressed; on the umbilical side radial, depressed. Umbilicus, small open. Primary aperture a very low arch; interiomarginal, umbilical; supplementary aperture at the base of the final chamber on the spiral side, a high arch with a thick rim."

Remarks: Typical forms are present in the Niger Delta which show a considerably fast rate of increase in the size of the chambers in the final whorl right from the first chamber of the last whorl, as is the case with forms figured in Jenkins (1971, pl. 19, figs. 562–570). As a result the final chamber in *G. inusitatus* does not appear as enlarged as in *G. trilobus*.

Range: Oligocene–Early Miocene.

Globigerinoides obliquus Bolli, 1957
Plate 37, Figs. 6–9

Globigerinoides obliqua Bolli, 1957b, Planktonic foraminifera from the Oligocene–Miocene Cipero and Lengua Foramation of Trinidad, etc. (mansion). Plankton Stratigraphy: Volume 1, Planktonic Foraminifera, calcareous nannofossils and calpionellids. Edited by Hans M. Bolli, John B. Saunders, Katharina Perch-Nielsen p. 113, pl. 25, figs. 9a–c.

Globigerinoides obliquus Postuma, 1971, Manual of Planktonic Foraminifera. Elsevier Publishing Co., Amsterdam, pp. 296–297.

Diagnosis: Test inflated, trochospire, biconvex, more convex umbilically; equatorial periphery strongly lobate, axial periphery rounded; chambers inflated, spherical; three to four in the final whorl, ultimate chamber compressed and laterally oblique; sutures depressed, curved on umbilical and spiral sides; umbilicus very narrow almost entirely occupied by primary aperture; primary aperture umbilical, interiomarginal, high arch bordered by a rim; small secondary apertural suture on spiral side; wall coarsely cancellate.

Remarks: This species differs from *G. extremus* in the less oblique ultimate chamber, hence the primary aperture is not as asymmetrical as in *G. extremus*.

Range: Early Miocene–Early Pliocene.

Globigerinoides primordius Blow and Banner, 1962
Plate 37, Figs. 10–12

Globigerinoides quadrilobatus primordius Blow and Banner, 1962, The Mid-Tertiary (Upper Eocene to Aquitanian) Globigerinacea. In: Eames, F.E., et al., Fundamentals of Mid-Tertiary stratigraphical correlation. Cambridge University Press, New York, NY, p. 115, pl. 9, figs. Dd–Ff.

Globigerinoides primordius Blow and Banner, 1971, Manual of Planktonic Foraminifera, Elsevier, vol. 1, pp. 298–299.

Diagnosis: Test low trochospire, unequally biconvex; equatorial periphery strongly lobate; axial periphery rounded; chambers globular, inflated; four chambers in final whorl, increasing rapidly in size, ultimate chamber constituting up to half the size of the test; sutures depressed, shallow, radial on both umbilical and spiral sides; umbilicus narrow; primary aperture low, interiomarginal; one secondary aperture on spiral side; wall coarsely perforate.

Remarks: Bolli and Saunders (1985) offered very illuminating comments on the evolution of *G. primordius*: "This species is the immediate ancestor of *Globigerinoides trilobus* stock, with overall morphological features already very close, in particular to *G. trilobus immaturus*. Typical *G. primordius* specimens differ in having more delicately built chamber walls, in the primary aperture being higher arched, and in possessing only a single secondary opening, situated on the spiral side of the last chamber. It is generally agreed that the immediate ancestor of this first *Globigerinoides* species is a *Globigerina*. According to Blow and Banner (1962) it is very close to *G. praebulloides occlusa* from which it differs primarily only in the presence of a spiral secondary aperture."

Range: Late Oligocene–Earliest Miocene.

Globigerinoides quadrilobatus d'Orbigny, 1846
Plate 37, Figs. 13 and 14

Globigerinoides quadrilobatus d'Orbigny, 1846, Foraminiferes fossiles du Bassin Tertiare de Vienne (Autriche). Gide et Comp. Paris, p. 303, pl. 21.

Globigerinoides quadrilobatus Bolli and Saunders, 1985, Oligocene to Holocene low latitude plankton stratigraphy. In: Bolli, H.M., et al., (Eds.), 1985, Plankton Stratigraphy. Cambridge University Press, Cambridge, p. 192, fig. 20.18.

Diagnosis: Test moderate in size, ovate, low trochospire, inflated especially on umbilical side, trilobate; equatorial periphery strongly lobate; axial periphery rounded; chambers spherical, inflated; three in final whorl, increase rapidly in size; ultimate chamber occupies about half of the test; sutures incised, radial on umbilical side; slightly curved on spiral side;

umbilicus small; primary aperture low interiomarginal arch at base of ultimate chamber, occupies entire umbilicus sitting across penultimate and antepenultimate chambers; secondary sutural aperture on spiral side single, opposite main aperture; wall distinctly cancellate.

Remarks: The Niger Delta *G. quadrilobatus* was identified using Bolli and Saunders' (1985) illustration of the lectotype; the only difference being the fact that the lectotype has 3½ chambers in the final whorl. On the taxonomic status of *G. quadrilobatus*, Blow (1969) and Stainforth et al. (1975) considered this as the group that is generally more primitive, hence subspecies status was accorded to forms similar to this root stock.

Range: Miocene–Recent.

Globigerinoides ruber d'Orbigny, 1839
Plate 37, Figs. 15 and 16

Globigerinoides ruber d'Orbigny, 1839, Foraminiferes des Iles Canaries. In: Barker-Webb, P. and S. Berthelot, Historie Naturelle des Canaries. Paris, vol. 2, Part 2, pp. 82–83, pl. 4, figs. 12–14.

Globigerinoides ruber Postuma, 1971, Manual of Planktonic Foraminifera. Elsevier Publishing Co., Amsterdam, pp. 300–301.

Diagnosis: Test inflated, subovate, high trochospire, convex umbilical side; equatorial periphery distinctly lobate; axial periphery rounded; chambers highly inflated, spherical; initially closely touching, later loosely arranged; 3½ chambers in final whorl; ultimate chamber constitutes ½ to 1/3rd of the test; sutures strongly depressed, incised; radial on umbilical side; primary aperture large, high umbilical arch, semicircular sitting atop penultimate and antepenultimate chambers; two supplementary apertures on spiral side; wall coarsely perforated and spinose.

Remarks: So varied is this group that it forms the cores of a morphologic plexus, differentiated on the basis of the relative elevation of the trochospire. Forms in this group include *Globigerinoides seiglei* (high-spired) *G. pyramidalis* (lofty-spired) and *G. mitra* (very large, aberrant with a colossal spire like the bishop's mitre). *Globigerinoides ruber* evolved from *G. quadrilobatus altiapertura*, though the latter is smaller, and less elevated.

Range: Early Miocene–Recent.

Globigerinoides sacculifer Brady, 1877
Plate 37, Figs. 17–20

Globigerina sacculifer Brady, 1877, Supplementary note on the Foraminifera of the Chalk (?) of the New Britain group. Geological Magazine, 4, 535.

Globigerinoides sacculifera Postuma, 1971, Manual of Planktonic Foraminifera. Elsevier Publishing Co., Amsterdam, pp. 302–303.

Globigerinoides sacculifer Petters, 1983, Gulf of Guinea planktonic foraminiferal biochronology and geological history of the South Atlantic. J. Foramin. Res. 13, 56, pl. 8, figs. 6, 16.

Diagnosis: Test modestly high to low trochospire, biconvex; equatorial periphery lobate; axial periphery rounded to slightly flattened in the ultimate chamber; chambers spherical, three to four in the last whorl which increase moderately in size; the ultimate chamber is usually elongate sac-like, and may be reduced in size; sutures on spiral side slightly curved, depressed; radial, depressed on umbilical side; umbilicus fairly narrow; primary aperture low arch, interiomarginal, umbilical, bordered by a distinct lip; only one sutural supplementary aperture on spiral side opposite primary aperture.

Range: Early Miocene–Recent.

Globigerinoides seiglei Bermudez and Bolli, 1969
Plate 38, Figs. 1 and 2

Globigerinoides rubra (d'Orbigny) subspecies *seiglei* Bermudez and Bolli, 1969, Consideraciones sobre los sedimentos del Mioceno Medio al Reciente de las costas central oriental de Venezuela. Boletin de Geologia, Venezuela, v. 10, p. 137–223, p. 164, pl. 8, figs 10–12.

Globigerinoides seigliei Bolli and Saunders, 1985, Oligocene to Holocene low latitutude plankton stratigraphy. In: Bolli, H.M., et al., (Eds.), 1985, Plankton Stratigraphy. Cambridge University Press, Cambridge, p. 196, fig. 20.5.

Diagnosis: Test large, very high conical, compact trochospire, inflated, convex umbilical side; triangular umbilical profile; strongly lobate equatorial periphery; axial periphery rounded; chambers spherical, inflated; three in the final whorl, increased slowly and regularly in size; ultimate chamber reduced in size; sutures depressed, deep and radial on umbilical side, curved on spiral side; primary aperture high arch, umbilical, astride penultimate and antepenultimate chambers; secondary spiral sutural apertures two; a larger one opposite the primary aperture; the smaller between the last three chambers before the ultimate chamber; wall perforate, finely spinose.

Range: Late Miocene–Middle Pliocene.

Globigerinoides sicanus de Stefani, 1950
Plate 38, Fig. 3

Globigerinoides sicanus de Stefani, 1950, Sualcune manifestazioni di idrocarburi in provincia de Palermo e descrizione di formaminiferi nuovi. Plinia (Palermo, Italy), 3, nota 4, 9, fig. 6.

Globigerinoides sicanus Postuma, 1971, Manual of Planktonic Foraminifera. Elsevier Publishing Co., Amsterdam, pp. 304–305.

Diagnosis: Test very low trochospire, ovoid, spherical; divided into two by the ultimate chamber which constitutes over half of the test, and is hemispherical in shape; the other part is the initial coil where chambers increased normally and gradually in size; supplementary apertures are slits or triangular openings on spiral side; surface cancellate.

Range: Early–Middle Miocene transition.

Globigerinoides subquadratus Bronniman, 1954
Plate 38, Figs. 4 and 5

Globigerinoides subquadratus Bronnimann, 1954, Appendix: descriptions of new species. In: Todd, R., Cloud, P.E., Low, D., Schmidt, R.G. (Eds.), Probable Occurrence of Oligocene in Japan, Am. J. Sci., 252, 680, pl. 1, figs. 8a–c.

Globigerinoides subquadrata Postuma, 1971, Manual of Planktonic Foraminifera. Elsevier Publishing Co., Amsterdam, pp. 306–307.

Globigerinoides subquadratus Bolli and Saunders, 1985, Oligocene to Holocene low latitude plankton stratigraphy. In: Bolli, H.M., et al., (Eds.), 1985, Plankton Stratigraphy. Cambridge University Press, Cambridge, p. 192, fig. 20.6.

Description: Test low trochospiral, inflated, subquadrate; equatorial periphery lobate; axial periphery broadly rounded; chambers spherical; 3 to 3½ in the last whorl; increasing rapidly in size; sutures depressed, radial on both sides; umbilicus narrow; primary aperture interiomarginal, umbilical, medium arch; spiral side displays two distinct sutural apertures; wall cancellate.

Remarks: This species differs from *G. primordius*, *G. immaturus*, and *G. inusitatus* in the more quadrate profile and the presence of a pair of sutural supplementary apertures on the spiral side. The shape distinguishes it from *G. ruber*.

Range: Early–Middle Miocene.

Globigerinoides triloba Reuss, 1850
Plate 38, Figs. 6 and 7

Globigerinoides triloba Reuss, 1850, Neues Foraminiferen aus den Schichten des Osterreichischen Tertiarbeckens. Denkschriften der Kaiserlichen Akademie der Wissenschaften, Mathematisch Naturwissenschaftliche Classe 1, p. 347, pl. 47, fig. 11.

Diagnosis: Test low trochospire, inflated and convex umbilical side; subtriangular profile; equatorial periphery moderately lobate; axial periphery rounded; chambers spherical, inflated; three chambers in final whorl, expanded markedly in size; last chamber constitutes, two-thirds of the test; sutures curved on spiral side, depressed; on the umbilical side also depressed, radial; umbilicus, indistinct, concealed by ultimate chamber; primary aperture umbilical, interiomarginal, low arch bordered by rim; one secondary aperture on spiral suture opposite primary aperture; wall cancellate.

Remarks: Postuma (1971) drew the following comparisons between *G. triloba* and other *Globigerinoides*. "This species is closely related to several others. *Globigerinoides trilobus* (Reuss) differs from *G. immaturus* LeRoy in having a final chamber that is larger than all earlier chambers combined. *G. altiaperturus* Bolli is distinguished from *G. triloba* in having a high arched primary aperture."

Range: Early Miocene–Recent.

Genus: PRAEORBULINA Olsson, 1964
Praeorbulina circularis Blow, 1956
Plate 38, Fig. 8

Globigerinoides glomerosa circularis Blow, 1956, Origin and evolution of the foraminiferal genus *Orbulina* d'Orbigny. Micropaleontology, 2, 65, text figs. 2–4.

Diagnosis: Test spherical, single-chambered, with ultimate chamber enveloping entire test; aperture consist of round pores along base of ultimate chamber; outline of the globigerine initial part of test preserved as relict in the initial part of the test; wall finely perforate.

Remarks: According to Bolli and Saunders (1985) "*P. glomerosa circularis* differs from *P. glomerosa glomerosa* in having many circular, pore-like apertures between the last chamber and earlier ones rather than slit-like apertures of the latter subspecies. It is almost completely spherical with sutures hardly incised. It differs from *Orbulina suturalis* in having all the apertures confined to the sutures."

Range: Early Miocene–Middle Miocene boundary.

Praeorbulina curva Blow, 1956
Plate 38, Figs. 9–11

Globigerinoides glomerosa curva Blow, 1956, Origin and evolution of the foraminiferal genus *Orbulina* d'Orbigny. Micropaleontology, 2, 65, text-figs. 1, 10, 11.

Praeorbulina glomerosa curva Bolli and Saunders, 1985, Oligocene to Holocene low latitude plankton stratigraphy. In: Bolli, H.M., et al., (Eds.), 1985, Plankton Stratigraphy. Cambridge University Press, Cambridge, p. 199, figs. 23.5; 24.6.

Diagnosis: Test spherical, globular; early trochospiral part a short projection like the pilot's cage at the base of a dirigible balloon; in which case the ultimate chamber comprises nearly 90% of the test; wall cancellate; apertures are openings around the base of the ultimate chamber.

Remarks: According to Bolli and Saunders (1985), "*P. glomerosa curva* differs from *P. sicana* in having 7 instead of 4 apertures around the base of the last chamber in the holotype. It also shows greater envelopment of the test by the last chamber."

Range: Late Early Miocene.

Praeorbulina glomerosa Blow, 1956
Plate 38, Figs. 12–14

Globigerinoides glomerosa Blow, 1956, Origin and evolution of the foraminiferal genus *Orbulina* d'Orbigny. Micropaleontology, 2, 64–65, figs. 1, 2.

Praeorbulina glomerosa glomerosa Stainforth et al., 1975, Cenozoic planktonic foraminiferal zonation and characteristics of index forms. Paleontological Contribution, University of Kansas, Article 62, p. 281, fig. 121.

Praeorbulina glomerosa Petters, 1983, Gulf of Guinea planktonic foraminiferal biochronology and geological history of the South Atlantic. J. Foramin. Res. 13, 57, pl. 8, figs. 23, 24.

Diagnosis: Test globular, spherical, early part trochospiral; periphery nearly circular; adult test comprises mostly one single chamber which envelopes one-half to three-fourths of the initial globigerine part; aperture evenly spaced slits restricted to the suture of the final chamber, varying from short slits to large circular pores; surface finely to coarsely cancellate.

Remarks: Praeorbulina glomerosa differs from *P. curva* in having more supplementary apertures around the base of the ultimate chamber and in having a test which is nearly spherical in outline as a result of the large extent of enveloping by the ultimate chamber.

Range: Late Early Miocene.

Genus: ORBULINA d'Orbigny, 1839
Orbulina bilobata d'Orbigny, 1846
Plate 38, Fig. 15

Globigerina bilobata d'Orbigny, 1846, Foraminiferes fossiles du Bassin Tertiare de Vienne (Autriche). Gide et Comp. Paris, p. 164, p. 9, figs. 11–14.

Biorbulina bilobata Blow, 1969, Late Middle Eocene to Recent planktonic foraminiferal biostratigraphy. In: Bronnimann, P., Renz, H.H. (Eds.), Proceedings of the First International Conference on Planktonic Microfossils, vol. 1. E.J. Brill, Geneva, p. 334, pl. 23, figs. 5, 6.

Biorbulina bilobata Petters, 1983, Gulf of Guinea planktonic foraminiferal biochronology and geological history of the South Atlantic. J. Foramin. Res. 13, 57, pl. 8, fig. 7.

Diagnosis: Test bilobate, inflated; early trochospiral part hardly noticeable on most adults; two spherical chambers in the last whorl, the ultimate above and the penultimate below; apertures are scattered all over the surface of the test; wall finely perforate.

Range: Middle Miocene–Pliocene.

Orbulina suturalis Bronnimann, 1951
Plate 38, Fig. 16

Orbulina suturalis Bronnimann, 1951, *Globigerinita naparimaensis* n. gen, n. sp. from the Miocene of Trinidad, B.W.I. Contrib. Cushman Found. Foramin. Res. 2, 135, text-fig. IV; figs. 15, 16, 20.

Diagnosis: Test spherical; early trochospiral part barely visible. Final chamber has enveloped and almost completely resorbed the trochospiral stage. Aperture at adult stage consists of areal small openings which are scattered all over the ultimate chamber; with small sutural supplementary openings along the sutures separating the last chamber from the penultimate and earlier chambers.

Remarks: Bolli and Saunders (1985) noted that: "Areal as well as sutural apertures distinguish this species from *Praeobulina glomerosa circularis*. It differs from the closely related *O. universa* in having the earlier chambers of the test breaking the outline of the sphere."

Range: Early Middle Miocene–Recent.

Orbulina universa d'Orbigny, 1839
Plate 38, Fig. 19

Orbulina universa d'Orbigny, 1839, Foraminiferes des Iles Canaries. In: Barker-Webb, P., Berthelot, S., Historie Naturelle des Canaries. Paris, vol. 2, Part 2, p. 119–146. pl. 1, fig. 1.

Orbulina universa Bolli, Saunders, and Perch-Nelsen, 1985, Oligocene to Holocene low latitude plankton stratigraphy. In: Bolli, H.M., et al. (Eds.), 1985, Plankton Stratigraphy, Cambridge University Press, Cambridge, p. 200, figs. 23.2, 24.3.

Diagnosis: Test spherical, shows the early chambers internally when broken, apertural pores widely scattered across the surface of the last chamber.

Range: Miocene–Recent.

Genus: GLOBOROTALIA Cushman, 1927
Globorotalia archeomenardii Bolli, 1957
Plate 38, Figs. 17 and 18

Globorotalia archeomenardii Bolli, 1957a–c, US Natl. Museum Bull. No. 215, p. 119, pl. 28, figs. 11a–c.

Globorotalia 27: SPDC Foraminiferal Type Collection.

Globorotalia archeomenardii Postuma, 1971, Manual of Planktonic Foraminifera. Elsevier Publishing Co., Amsterdam, pp. 316–317.

Original Diagnosis: "Shape of test low trochospiral, compressed; equatorial periphery slightly lobate; axial periphery acute with a thin but distinct keel; wall calcareous, very finely perforate, surface smooth; chambers angular rhomboid, strongly compressed; 12–15, arranged in about 3 whorls, the 4–5 chambers of the last whorl increase fairly rapidly in size; sutures on spiral side strongly curved; on umbilical side radial to slightly curved, depressed; umbilicus small, fairly shallow; aperture a low slit, interiomarginal, umbilical–extraumbilical". Coiling random.

Range: Late Early Miocene–Early Middle Miocene.

Globorotalia crassaformis Galloway and Wissler, 1927
Plate 39, Figs. 1–7

Globigerina crassaformis Galloway and Wissler, 1927, Pleistocene Foraminifera from the Lomita Quarry, Palos Verdes Hills, California. J. Paleontol. 1, 41, pl. 7, fig. 2.

Globorotalia 50: SPDC, Foraminiferal Type Collection.

Globorotalia crassaformis Postuma, 1971, Manual of Planktonic Foraminifera. Elsevier Publishing Co., Amsterdam, pp. 318–319.

Globorotalia crassaformis Petters, 1983, Gulf of Guinea planktonic foraminiferal biochronology and geological history of the South Atlantic. J. Foramin. Res. 13, 50, pl. 5, figs. 21, 26.

Diagnosis: Test medium, low trochospiral, flat spiral side with raised inner whorl; moderately convex umbilical side; equatorial periphery lobate to quadrate; axial section planoconvex, subacute to surrounded; chambers compressed; 4 to 4½ in the final whorl, increased rapidly in size, ultimate chamber constituting ½ to 1/3rd of the test; sutures on spiral side distinctly curved, depressed; depressed, radial to slightly sigmoidal on umbilical side; umbilicus narrow to closed; aperture a long slit umbilical–extraumbilical, bordered by lip; wall finely perforate, sparsely rugose on umbilical side.

Remarks: Niger Delta forms scarcely develop keels unlike in other regions of the world (Caribbean-Gulf of Mexico, Italy, New Zealand) where this cool- to warm-water species thrives alike.

Range: Middle Pliocene–Holocene.

Globorotalia cultrata d'Orbigny, 1839
Plate 39, Fig. 8–13

Rotalina cultrata d'Orbigny, 1839, Foraminiferes des Iles Canaries. In: Barker-Webb, P., Berthelot, S., Historie Naturelle des Canaries. Paris, vol. 2, Part 2, p. 76, pl. 5, figs. 7–9.

Globorotalia cultrata Banner and Blow, 1960, The taxonomy, morphology and affinities of the genera included in the subfamily Hastigerininae. Micropaleontology, 6, 24, pl. 6, figs. 1a–c.

Diagnosis: Test low trochospiral, lenticular, ovate outline; biconvex, slightly concave umbilical side; smoothly convex spirally; equatorial periphery lobulate; axial periphery narrow, keeled; chambers strongly compressed, wedge-shaped on umbilical side, rhomb-shaped on spiral side; five to six chambers in final whorl, increase in size rapid, flaring test in which the final chamber is considerably enlarged, doubling the size of the penultimate; sutures depressed and radially arranged on umbilical side, flush, curved on spiral side with considerable thickening near the junction with peripheral keel; umbilicus very narrow; aperture a low slit umbilical–extraumbilical rimmed with short lip; wall densely and finely perforate.

Remarks: This form can be easily distinguished from *G. menardii* by its more open coil, flaring test, and a more rapid increase in chamber size. Bolli and Saunders (1985) in their comparison of both forms concluded: "Our investigations, mainly in the tropical Caribbean/Atlantic province, have led to the distinction on the Quaternary–Holocene of two *G. menardii* types, a robust, thick-walled form, fairly circular in equatorial outline and with a thick peripheral keel, and a more delicate thinner-walled form, with a less marked peripheral keel and a more elongate outline in equatorial view. When comparing these two types with Banner and Blow (1960a), the robust form fits well with their lectotype of *G. menardii menardii* and the delicate, more elongate form with their neotype of *G. menardii cultrata.*"

Range: Pliocene–Recent.

Globorotalia flexuosa Koch, 1923
Plate 39, Fig. 14

Pulvinulina tumida flexuosa Koch, 1923, Die jung-tertiare Foraminifere fauna von Kabu (Res. Surabaja, Java). Eclogae Geol. Helvet. 18, 357, figs. 9, 10.

Globorotalia flexuosa Stainforth et al., 1975, Cenozoic planktonic foraminiferal zonation and characteristics of index forms. Paleontological Contribution, University of Kansas, Article 62, p. 355, fig. 168.

Description: Stainforth et al. (1975) diagnosis and discussion of this rare species is instructive: "Test a large, heavily keeled, biconvex trochospire. Identical in all respects with *Globorotalia tumida* (Brady, 1877) except that anterior rim of last chamber is bent sharply toward umbilicus, in extreme cases perpendicular to plane of coiling."

Remarks: Probably this form is only a morphologic variant of *Globorotalia tumida*—possibly it is a teratoid form resulting from paleotemperature changes; for instance the distribution chart of Bolli (1966) clearly shows coincidence of its appearance with abrupt reversals of coiling direction, and Lamb and Beard (1972) noted coincidence of its appearance with interglacial intervals. Regardless of such subjective considerations, *Globorotalia flexuosa* is a conspicuous form, well suited for biostratigraphic use.

Range: Pliocene–Pleistocene.

Globorotalia margaritae Bolli and Bermudez, 1965
Plate 39, Figs. 15 and 16

Globorotalia margaritae Bolli and Bermudez, 1965, Zonation based on Planktonic Foraminifera of Middle Miocene to Pliocene warm-water sediments. Assoc. Venezolana Geol. Min. Petr. Bol. Informativo, 8 (5), 139, figs. 16–18.

Globorotalia margaritae Stainforth et al., 1975, Cenozoic planktonic foraminiferal zonation and characteristics of index forms. Paleontological Contribution, University of Kansas, Article 62, p. 363, figs. 175–176.

Diagnosis: Test lenticular, low trochospire, weak keel; equatorial periphery ovate, moderately lobate; axial periphery slightly convex spirally, also inflated toward umbilicus; chambers compressed toward periphery, four to five in last whorl, increased slowly in size, regular in shape, but more loosely attached later; crescentic shape on spiral side; umbilicus almost closed; aperture a low arched slit, extending from umbilicus to periphery, bordered by thin rim; wall smooth, perforate, granular on early part of test.

Remarks: Coiling in this species is mostly sinistral.

Range: Early–Middle Pliocene.

Globorotalia menardii d'Orbigny, 1826
Plate 39, Figs. 17 and 18

Rotalia menardii d'Orbigny, 1826, Tableau methodique de la classe des cephalopodes. Ann. Sci. Nat. Paris, ser. 1, 7, 273.

Globorotalia menardii menardii Bolli and Saunders, 1985, Oligocene to Holocene low latitude plankton stratigraphy. In: Bolli, H.M., et al., (Eds.), 1985, Plankton Stratigraphy, Cambridge University Press, Cambridge, p. 226, figs. 32.4, 34.5–7.

Diagnosis: Test very low trochospiral, biconvex, compressed; lobate equatorial periphery; axial periphery narrow with prominent keel; chambers strongly compressed, five to seven in final whorl increased moderately and regularly in size; sutures on spiral side strongly limbate, curved; on umbilical side depressed, radial; umbilicus moderately wide, shallow; aperture interiomarginal, umbilical–extraumbilical, low slit, bordered by a distinct lip; wall finely perforate, slightly rugose in early part, later wall smooth.

Remarks: Bolli and Saunder (1985) distinguished *G. menardii menardii* from *G. cultrata* offering the following remarks: "Our investigations, mainly in the tropical Caribbean/Atlantic province, have led to the distinction on the Quaternary–Holocene of two *G. menardii* types, a robust, thick-walled form, fairly circular in equatorial outline and with a thick peripheral keel, and a more delicate, thinner-walled form, with a less marked peripheral keel and a more elongate outline in equatorial view. The robust form fits well with the lectotype of *G. menardii menardii* and the delicate, more elongate form with the neotype of *G. menardii cultrata*." Both forms are readily distinguishable in the Niger Delta.

Range: Late Miocene–Recent.

Globorotalia merotumida Blow and Banner, 1965
Plate 39, Figs. 19 and 20

Globorotalia (Globorotalia) merotumida Blow and Banner, 1965, In: Banner and Blow, 1965, Nature, 207, 1352, figs. 1a–c.

Globorotalia merotumida Bolli and Saunders, 1985, Oligocene to Holocene low latitutude plankton stratigraphy. In: Bolli, H.M., et al., (Eds.), 1985, Plankton Stratigraphy, Cambridge University Press, Cambridge, p. 227, fig. 33.7.

Diagnosis: Test very low trochospiral; flat spiral side; umbilical side moderately convex; equatorial periphery subcircular to rounded, faintly lobate; axial periphery acute with a thick keel; chambers compressed, angular rhomboid; five to six in the final whorl, increasing rapidly in size, the ultimate chamber considerably enlarged; sutures incised and curved on umbilical side; limbate, curved on spiral side, much thickened peripherally at junction with the keel; umbilicus mostly closed; aperture low slit at base of ultimate chamber, umbilical–extra-umbilical, bordered by a lip; wall densely and finely perforate.

Remarks: The main, practically discernible difference between *G. merotumida*, the ancestor and its successor, *G. plesiotumida* is the more circular outline of *G. merotumida*, with the ultimate chamber in *G. plesiotumida* somewhat erect or raised. The evolutionary succession in this bioseries actually started with the *G. menardii* ancestor to *G. merotumida*, through *G. plesistumida* to *G. tumida*.

Range: Late Miocene–Early Pliocene.

Globorotalia plesiotumida Blow and Banner, 1965
Plate 40, Fig. 1

Globorotalia (*G*) *tumida* Brady *plesiotumida* Blow and Banner, 1965, Blow and Banner, 1965, In: Banner and Blow, 1965, Nature, 207, 1353, figs. 2a–c.

Globorotalia plesiotumida Stainforth et al., 1975, Cenozoic planktonic foraminiferal zonation and characteristics of index forms. Paleontological Contribution, University of Kansas, Article 62, p. 396, fig. 194.

Original Diagnosis: Stainforth et al. (1975) presented the most appropriate diagnosis of this species: "Test a lenticular trochospire rimmed by a blunt keel. Generally 6 chambers in last whorl, increasing somewhat irregularly in size as added, especially in radial length. Equatorial profile ear-shaped; slightly lobate, last chamber projecting noticeably; axial profile biconvex, slightly more inflated on umbilical than on spiral side. Sutures on spiral side limbate, initially flush, later depressed to sunken, oblique to recurved, coalescing into keel; on umbilical side incised, radial, almost straight. Umbilicus almost closed, umbilical shoulders developed only on last 3 chambers. Aperture a low arch from umbilicus almost to keel, carrying narrow lip along whole length. Wall smooth, densely and uniformly perforate except for granular area adjacent to the umbilicus."

Remarks: The narrow, elongate, raised ultimate chamber in published illustrations of *G. plesiotumida* (Blow, 1969, pl. 8, figs. 7–9; Bolli and Saunders, 1985, figs. 5 a–c; Stainforth et al., 1975, figs. 194, 1–9) is perhaps, the most practical means of differentiating *G. plesiotumida* from *G. merotumida* and *G. tumida*. This stable biocharacter was applied to the Niger Delta, although only the spiral view of *G. plesiotumida* is illustrated here.

Range: Latest Miocene–Early Pliocene.

Globorotalia scitula Brady, 1882
Plate 40, Figs. 2 and 3

Pulvinulina scitula Brady, 1882, Report on the Foraminifera. Proc. R. Soc. Edinb., 11, 716.
Globorotalia 21: SPDC Foraminiferal Type Collection.

Globorotalia scitula Petters, 1983, Gulf of Guinea planktonic foraminiferal biochronology and geological history of the South Atlantic. J. Foramin. Res. 13, 313, fig. 140.

Diagnosis: The following diagnosis was given for this species: "Test a discoidal trochospire of chambers which increase regularly in size and maintain constant shape. Final whorl of 4 to 5 chambers. Profile in vertical aspect subcircular to pentagonal, may be slightly lobate; in side profile spiral and umbilical faces form an acute angle but periphery is bluntly rounded. Sutures on spiral side very strongly recurved, resulting in elongate crescentic chambers; sutures on umbilical side almost straight, radial. Umbilicus closed or narrow and deep. Surface smooth, finely perforate, but may be pustulose near aperture."

Range: Middle Miocene–Recent.

Globorotalia tumida Brady, 1877
Plate 40, Figs. 4–9

Pulvinulina menardii tumida Brady, 1877, Supplementary note on the Foraminifera of the Chalk of the New Britain group. Geol. Mag. 4, 535.

Globorotalia 63: SPDC Foraminiferal Type Collection.

Globorotalia tumida Petters, 1983, Gulf of Guinea planktonic foraminiferal biochronology and geological history of the South Atlantic. J. Foramin. Res. 13, 52, pl. 5, figs. 5, 25.

Diagnosis: Stainforth et al. (1975) gave the following concise characterization of *Globorotalia tumida*: "Test a large, biconvex, axially tumid trochospire rimmed by heavy, blunt keel. Chambers 5 to 6 in last whorl increase irregularly in size as added, especially in radial length. Equatorial profile ear-shaped, slightly lobate; axial profile rhomboid to drop-shaped. Sutures on spiral side limbate, curved to hooked obliquely backwards, coalescing into thick keel; on umbilical side sinuously radial. Umbilicus narrow, deep. Aperture a slit or low arch between umbilicus and keel generally bordered by rim or flap. Wall finely perforate, smooth except for pustulose area centered on umbilicus."

Remarks: Stainforth et al. (1975) stated that "the type species of the genus *Globorotalia* has the same gross morphology as many other Late Neogene species (i.e., the *menardi* form group). Distinguishing features are its large size, exceptionally heavy keel, biconvex tumidity, and rapidly opening spire. *Globorotalia tumida* is a warm-water species. Abrupt and abundant appearance of large specimens coincided with the end of the latest glacial interval so this species is a natural zonal index for sediments of Holocene age."

Range: Pliocene–Recent.

Genus: SPHAEROIDINELLOPSIS Banner and Blow, 1959
Sphaeroidinellopsis disjuncta Finlay, 1940
Plate 40, Figs. 10–12

Sphaeroidinella disjuncta Finlay, 1940, New Zealand Foraminifera: key species in stratigraphy—No. 4. Trans. R. Soc. NZ 69, 469, pl. 67, figs. 224–228.

Sphaeroidinella disjuncta Jenkins, 1971, New Zealand Cenozoic planktonic foraminifera. Paleontol. Bull. Wellington 42, 171, pl. 17, figs. 536–538.

Sphaeroidinella disjuncta Bolli and Saunders, 1985, Oligocene to Holocene low latitutude plankton stratigraphy. In: Bolli, H.M., et al., (Eds.), 1985, Plankton Stratigraphy, Cambridge University Press, Cambridge, p. 241, figs. 38.18–38.22.

Diagnosis: Jenkins (1971) redescribed the holotype of *Sphaeroidinella disjuncta* as follows: "Test free, trochospiral; equatorial periphery quadrilobate; axial periphery rounded. Wall calcareous, thick, finely perforate, interpore areas thickened, rounded, with minute pustules. Chamber outlines spherical, early chambers obscured by test thickening, 4 in the final whorl, the first 3 of which show a slow increase in size but the final chamber smaller than the penultimate and positioned more towards the umbilicus. Sutures on the umbilical side deeply cut, radial; on the spiral side, radial to slightly curved, depressed. Umbilicus small, deep but unfilled. Aperture interiomarginal, umbilical, bordered by a fairly thick fimbrate margin."

Remarks: The morphology of *S. disjuncta* can be easily confused with those of *S. multiloba* and *S. seminulina*, hence Bolli and Saunders (1985) clarified it as follows: "*S. disjuncta* is the stratigraphically earliest species of the lineage (*Sphaeroidinellopsis disjuncta–S. multiloba–S. seminulina–S. sphaeroides–Sphaeroidinella*). In the early part of its range specimens are smaller compared with later ones, mostly with 3 chambers in the last whorl, occasionally 4. The average test size increases gradually during the *Globorotalia fohsi* Zone; 4-chambered specimens become more frequent and occasionally even 5-chambered forms appear. The shape of its chambers is subglobular, the last chambers have a tendency to become radially somewhat elongated. Sutural incisions in 3-chambered forms are slight; they are deeper in 4-chambered specimens. The aperture of the last chamber leading into the umbilicus is often protected by a distinct rounded lip, particularly in the larger specimens. Higher in the lineage, beginning with *S. multiloba,* the rounded lips gradually take on the flange-like shape typical for the forms in the later part of the lineage (*S. seminulina, S. sphaeroides).* The walls are massive and distinctly pitted. Increased cortex formation that eventually leads to the smooth glassy surface characteriszing *S. multiloba* gradually develops through transitional forms from *S. disjuncta.*" From the foregoing, it is evident that important morphological stages in the *Sphaeroidinellopsis* bioseries are represented in the Niger Delta.

Range: Middle–Late Miocene.

Sphaeroidinellopsis multiloba LeRoy, 1944
Plate 40, Figs. 13–15

Sphaeroidinella multiloba LeRoy, 1944. Miocene foraminifera from Sumatra and Java, Netherlands East Indies, Part. 1. Miocene foraminifera of Central Sumatra, Netherlands East Indies. Colorado Sch. Mines Q. 39, 91, pl. 4, figs. 7–9.

Sphaeroidinellopsis multiloba Bolli and Saunders, 1985, Oligocene to Holocene low latitutude plankton stratigraphy. In: Bolli, H.M., et al., (Eds.), 1985, Plankton Stratigraphy, Cambridge University Press, Cambridge, p. 241, figs. 38.14–38.17.

Diagnosis: Test small, low trochospire, elongate outline; inflated trilobate test; equatorial periphery strongly lobate; axial periphery rounded; chambers surrounded, greatly inflated; three to four in final whorl, nearly of uniform size, the ultimate chamber slightly reduced in size; chambers axially elongate; sutures wide, depressed, radial; umbilicus narrow, shallow or closed beneath the cortex; aperture a low arch, bordered by thick rim; wall smooth, glassy, with heavy cortex, through which are minute widely spaced pores; cortex may form a crenulated edge around umbilicus and also smoothen the sutures.

Remarks: Distinctive characteristics which allow differentiation of *S. multiloba* from *S. disjuncta* and *S. seminulina* are its trilobate, highly inflated test, rounded chambers, and loosely arranged chambers. Bolli and Saunders (1985) also noted that "as in *S. disjuncta*, forms with a smaller, radially extended, occasionally sacculiferid end-chamber are often present."

Range: Middle–Early Late Miocene.

Sphaeroidinellopsis sphaeroides Lamb, 1969
Plate 40, Figs. 16 and 17

Sphaeroidinellopsis sphaeroides Lamb, 1969, Planktonic foraminiferal datums and Late Neogene epoch boundaries in the Mediterranean Caribbean, and Gulf of Mexico. Trans. Gulf Coast Assoc. Geol. Soc. 19, 571, 578, pl. 1, fig. 1.

Sphaeroidinellopsis sphaeroides Bolli and Saunders, 1985, Oligocene to Holocene low latitude plankton stratigraphy. In: Bolli, H.M., et al., (Eds.), 1985, Plankton Stratigraphy, Cambridge University Press, Cambridge, p. 244, figs. 38.4–38.5.

Diagnosis: Test inflated almost spherical; equatorial profile rounded, only slightly indented but not lobate; axial periphery lobate; chambers somewhat globular, but quite indistinct, with only a vague outline of what was a three-chambered test in the adult form; sutures indistinct, masked by cortex, represented by faint wide depressions between chambers; aperture faint, as relicts of slits faintly traceable across the middle of the test; wall heavily covered by the cortex which has obliterated most of the surface features and resulted in a roughly spherical test.

Remarks: In shape, relict apertural slits and heavy secondary calcification (cortex) *S. sphaeroides* encountered in the Niger Delta is quite typical; since forms illustrated by Bolli and Saunders (1985) share the previously mentioned characteristics but are not perfect spheres after all. Bolli and Saunders concluded that "with a test shape very close or identical to *Sphaeroidinella dehiscens* and with the apertural features of *Sphaeroidinellopsis seminulina* (one aperture only), *S. sphaeroides* is regarded as a transitional form between these two taxa. Its stratigraphic range is not significantly different from that of *S. seminulina*, except for a trend to become more typically developed and more frequent in the upper part of the Early Miocene, that is, immediately before the first appearance of *Sphaeroidinella*."

Range: Latest Miocene–End of Middle Pliocene.

Sphaeroidinellopsis subdehiscens Blow, 1959
Plate 40, Figs. 18 and 19

Sphaeroidinella dehiscens subdehiscens Blow, 1959, Age, correlation, and biostratigraphy of the upper Tocuyo (San Lorenzo) and Poron Formations, Eastern Falcon, Venezuela. Bull. Am. Paleontol. 39, 67–251, pl. 12, fig. 71.

Sphaeroidinella dehiscens subdehiscens Postuma, 1971, Manual of Planktonic Foraminifera. Elsevier Publishing Co., Amsterdam, pp. 388–389.

Sphaeroidinellopsis subdehiscens Petters, 1983, Gulf of Guinea planktonic foraminiferal biochronology and geological history of the South Atlantic. J. Foramin. Res. 13, 57, pl. 8, figs. 1–2.

Diagnosis: Test low trochospire, compact to ovoid shape; equatorial periphery ovate, lobate; axial periphery rounded; chambers subglobular to somewhat radially elongate; three chambers in the final whorl, increased slowly in size, may be reduced especially in ultimate chamber; sutures straight, flush on spiral side, depressed on umbilical side; umbilicus irregular in shape, reduced or closed being covered by cortex; aperture an elongate slit, umbilical, rimmed by thickened and irregular margins of the chambers of the final whorl; may appear as a low arch because of shell thickening at the base of the ultimate chamber; wall consists of an initial coarsely perforate globigerine test covered by cortex of secondary calcite layers deposited on the adult test, thereby causing a restriction of pores, and produces a smooth glassy test.

Remarks: Typical forms are found in the Niger Delta. According to Stainforth et al. (1975): "this species differs from *Sphaeroidinellopsis seminulina* (Schwager, 1866) primarily in having 3 (rather than 4 or more) chambers in the last whorl and in the heavier cortex. It differs from *Sphaeroidinellopsis paenedehiscens* Blow (1969) mainly in lacking a completely crenulate umbilical aperture and in being less smoothly ovate."

Range: Middle–Late Miocene.

Genus: PULLENIATINA Cushman, 1927.

Pulleniatina obliquiloculata Parker and Jones, 1865

Plate 40, Fig. 20

Pullenia sphaeroides (d'Orbigny) var. *obliquiloculata* Parker and Jones, 1865, On the foraminifera from North Atlantic and Artic Oceans, including Davis Straits and Baffin's Bay. Royal Society of London, Philosophical Transactions, 155, In: Carpenter, 183 (*nomen nudum*) Parker and Jones, 1865, p. 365–368, pl. 19, figs. 4a,b.

Pulleniatina obliquiloculata (Parker and Jones, 1865), Bolli and Saunders, 1985, Oligocene to Holocene low latitutude plankton stratigraphy. In: Bolli, H.M., et al., (Eds.), 1985, Plankton Stratigraphy, Cambridge University Press, Cambridge, pp. 247–248, Figs. 40.4, 41.9–12, 7, 10, 11.

Diagnosis: Test globose, initially trochospiral, later streptospiral almost involute, axial periphery broadly rounded, four to five broadly rounded chambers in the final whorl; sutures almost flush; surface smooth, granular in apertural areas; umbilicus covered; aperture a low arch, extending from umbilical area to periphery and onto spiral side.

Range: Early Pliocene–Recent.

Pulleniatina primalis Banner and Blow, 1967

Plate 41, Fig. 1

Pulleniatina primalis Banner and Blow, 1967, The origin, evolution and taxonomy of the foraminiferal genus *Pulleniatina* Cushman, 1927. Micropaleontology, 13 (2), 142, pl. 1, figs. 3–8.

Diagnosis: Test tumid, but with a low, partially resorbed trochospire; equatorial periphery subrounded to subpentagonal; ovate axial periphery, flattened spiral side; chambers slightly inflated, broadly rounded, four to five in final whorl, increased slowly in size; sutures distinct, depressed, radial, slightly curved; umbilicus shallow, wide; aperture a low arch extending from umbilicus toward periphery; wall smooth, pustulose in umbilical region.

Remarks: Stainforth et al. (1975), observed that *P. primalis* evolved from *Turborotalia acostaensis* in the Late Miocene and gave rise to *P. obliquiloculata* in the Middle Pliocene.

Range: Late Miocene to end of Middle Pliocene.

Genus: CATAPSYDRAX Bolli, Loeblich and Tappan, 1957
Catapsydrax martini Blow and Banner, 1962
Plate 41, Figs. 2 and 3

Globigerinita martini martini Blow and Banner, 1962, The Mid-Tertiary (Upper Eocene to Aquitanian) Globigerinacea. In: Eames, F.E., et al., (Eds.), Fundamentals of Mid-Tertiary Stratigraphical Correlation. Cambridge University Press, New York, NY, pp. 110–111, pl. XIV, fig. 10.

Diagnosis: Test small, biconvex subquadrate trochospire; spiral side moderately high, umbilical side dominated by bulla; equatorial periphery lobate; axial periphery rounded; chambers four in final whorl, increased gradually in size, globular, circular in shape; sutures wide, depressed, radial; umbilicus large, entirely occupied by bulla; aperture totally covered by bulla; wall cancellate on chambers and bulla; bulla large and bulging.

Remarks: A combination of the quadrate, four-chambered compact test, cancellate wall and fairly large, bulging bulla which occupies the whole umbilical area sets this species apart from other forms of *Catapsydrax*.

Range: Late Eocene–Middle Oligocene.

Catapsydrax cf. *howei*
Plate 41, Fig. 4

Catapsydrax unicavus Petters, 1983, Gulf of Guinea planktonic foraminiferal biochronology and geological history of the South Atlantic. J. Foramin. Res. 13, 57, pl. 8, figs. 10, 15.

Diagnosis: Test small, low trochospire, subtriangular, circular, inflated; slightly lobate periphery; chambers globular, 3½ to 4 in the last whorl; increased rapidly in size; sutures depressed, wide; umbilical area covered by a very large and highly inflated bulla, covering entire umbilical area and sometimes more than half of the test; aperture not visible because of bulla; wall cancellate; bulla is coarsely perforate.

Remarks: Petters (1983) presented this form under *C. unicavus*, and considered it to be a teratoid form of this species, which normally has a small umbilical bulla. But the common occurrence of the form reported here in the Niger Delta, and its extremely pronouned bulla justifies separating it from *C. unicavus sensu stricto*. Furthermore, the wall texture of *C. unicavus* shows a striking superposition of spines upon the basically cancellate wall. Nonetheless, the cancellate wall which seems to be fundamental to the genus *Catapsydrax*, is retained in C. cf. *howei* without modification, except in the bulla which is coarsely perforate and not cancellate. It is possible that C. cf. *howei* is a stable morphologic offshoot from *C. unicavus*, rather than an odd morphological development. The spiral views of *C. unicavus* and C. cf. *howei* are, however, basically similar, being a low, flattened trochospire. The partially stratigraphically overlapping species, *Catapsydrax dissimilis*, which also occurs in the Niger Delta (Petters, 1983), differs in that it has a high spiral view.

Range: Miocene.

Catapsydrax stainforthi Bolli, Loeblich and Tappan, 1957
Plate 41, Fig. 5

Catypsydrax stainforthi Bolli, Loeblich and Tappan, 1957, Planktonic foraminifera of Paleocene and Early Eocene age from the Gulf and Atlantic Coastal Plains. US Natl. Museum Bull. 215, 37, pl. 7, Figs 11a–c.

Diagnosis: The diagnosis of Stainforth et al. (1975) best suited the Niger Delta materials: "Test small, quadrate-discoidal. On spiral side a flat embryonic coil, almost flush with surface, is followed by a very low spire of ellipsoidal chambers which increase slightly in size as added. Usually 4 chambers in final whorl (5 in holotype). Spiral profile mildly lobate, side profile bluntly ovate. Sutures initially obscure, later distinct and lightly incised; spiral suture lobate, intercameral sutures radial. On umbilical side final chambers surround an umbilicus completely covered by a bulla. Primary aperture concealed by bulla. Infralaminal apertures in the form of small arched openings on bulla above each intercameral sutures, may be ahead of tubules aligned with suture, may have delicate flaring lip. Wall distinctly cancellate."

Range: Early Miocene.

Catapsydrax unicavus Bolli, Loeblich and Tappan, 1957
Plate 41, Figs. 6 and 7

Catapsydrax unicavus Bolli, Loeblich and Tappan, 1957, Planktonic foraminifera of Paleocene and Early Eocene age from the Gulf and Atlantic Coastal Plains. US Natl. Museum Bull. 215, 37, pl. 7, fig. 9.

Globigerinita unicava unicava Blow, 1969, Late Middle Eocene to Recent planktonic foraminiferal biostratigraphy. In: Bronnimann, P., Renz, H.H. (Eds.), Proceedings of the First International Conference on Planktonic Microfossils, vol. 1. E.J. Brill, Geneva, p. 330, pl. 24, figs. 8, 9.

Diagnosis: Test small, quadrate, inflated, low trochospire; equatorial periphery slightly lobate; axial periphery squared; chambers inflated, circular, commonly four in final whorl, increased rapidly in size; spiral sutures indistinct; radial, wide depressions around bulla; umbilicus covered by large, circular, protruding smooth bulla; aperture covered by bulla, sometimes may appear as a crescentic arch above bulla with faint lip, wall cancellate, coarsely hispid; bulla smooth, different in wall texture from the chambers.

Range: Late Middle Eocene–Early Miocene.

Genus: GLOBIGERINITA Bronnimann, 1951
Globigerinita incrusta Akers, 1955
Plate 41, Fig. 8

Globigerinita incrusta Akers, 1955, Some planktonic Foraminifera of the American Gulf Coast and suggested correlations with the Caribbean Tertiary. J. Paleontol., 29, 655, pl. 65, fig. 2 (part).

Globigerinita incrusta Stainforth et al., 1975, Cenozoic planktonic foraminiferal zonation and characteristics of index forms. Paleontological Contribution, University of Kansas, Article 62, p. 286, fig. 124.

Diagnosis: Test small, high trochospire, inflated, quadrate to trilobate strongly lobate periphery; chambers subglobular; three to four in the final whorl; sutures depressed, distinct; umbilicus covered by bulla, often slightly inflated; similar in wall texture to the rest of the test; aperture infralaminal, four little semicircular openings could be preserved at the ends of bulla extensions along umbilical sutures; wall texture pustulose.

Remarks: Stainforth et al. (1975) remarked as follows: "*Globigerinita incrusta* and *Catapsydrax stainforthi* have bullas of similar shape but are distinct in other features, especially their shell texture. *Globigerinita incrusta* appears to be a primitive form from which, by developing variability of the bulla, arose the group of forms placed under *Globigerinita naparimaensis*."

Range: Miocene–Recent.

Globigerinita naparimaensis Bronnimann, 1951.
Plate 41, Figs. 9 and 10

Globigerinita naparimaensis Bronnimann, 1951, *Globigerinita naparimaensis* n. gen, n. sp. from the Miocene of Trinidad, B.W.I. Contrib. Cushman Found. Foramin. Res. 2, 16–18, figs. 1–14.

Globigerinita naparimaensis Postuma, 1971, Manual of Planktonic Foraminifera. Elsevier Publishing Co., Amsterdam, pp. 282–283.

Diagnosis: Test small, ovate, trochospiral middle of spiral side elevated; umbilical side inflated; equatorial periphery strongly lobate; axial periphery rounded; chambers spherical to subglobular; three to four in the final whorl, increased rapidly in size; sutures broadly depressed indistinct; umbilicus closed by bulla, which is slightly inflated with tubulose openings along the sutures; aperture covered by bulla; wall pustulose.

Range: Middle Miocene–Recent.

Genus: HASTIGERINA Thomson, 1876
Hastigerina aequilateralis Brady, 1879
Plate 41, Fig. 11

Globigerina aequilateralis Brady, 1879, Notes on some of the reticularian Rhizopoda of the'Challenger' Expedition: Part I, on new or little-known arenaceous types. Q. J. Microsc. Sci. 19, 9, pl. 80, figs. 18–21.

Hastigerina aequilateralis Postuma, 1971, Manual of Planktonic Foraminifera. Elsevier Publishing Co., Amsterdam, pp. 366–367.

Hastigerina aequilateralis Petters, 1983, Gulf of Guinea planktonic foraminiferal biochronology and geological history of the South Atlantic. J. Foramin. Res. 13, 48, pl. 8, fig. 3.

Diagnosis: Test medium, planispiral evolute; equatorial periphery lobate; axial periphery rounded; chambers globular to subglobular; five to six chambers in the last whorl, chambers increased rapidly in size; sutures distinct, incised, radial; aperture equatorial, interiomarginal, broad low arch; wall hispid, sparingly spinose.

Range: Late Middle Miocene–Recent.

Hastigerina siphonifera d'Orbigny, 1839
Plate 41, Fig. 12

Globigerina siphonifera d'Orbigny, 1839, Foraminiferes des Iles Canaries. In: Barker-Webb, P., Berthelot, S., Historie Naturelle des Canaries. Paris, vol. 2, Part 2, p. 83, pl. 4, figs. 15–18.

Hastigerina siphonifera Banner and Blow, 1960, The taxonomy, morphology and affinities of the genera included in the subfamily Hastigerininae. Micropaleontology, 6, 22, figs. 2a–c.

Diagnosis: Test medium, planispiral involute, circular outline, biumbonate; equatorial periphery strongly lobate; axial periphery broad, rounded; chambers circular to semicircular, inflated; four to five chambers in the final whorl which increased rapidly in size, the ultimate chambers constituting half of the entire test and overhanging both sides of the test; sutures depressed, incised; radial on both sides; aperture a low gaping arch; wall hispid, finely perforate.

Remarks: Niger Delta specimens are very similar to those of the tropical Atlantic Caribbean region figured by Rögl and Bolli (1973, pl. 3, figs. 15–18). It differs from *H. aequilateralis* in the enlargement of the enveloping ultimate chamber.

Range: Middle Miocene–Recent.

Plates 1–41

1
2
3
4
5
6
8
9
10
11
12
13
14
15
16
17
18
19
20

EXPLANATION OF PLATE 1

FIGURE 1	*Saccammina complanata*. Utapate-1 well, 8200 ft., ×125.
FIGURE 2	*Saccammina complanata*. Utapate-1 well, 8240–8260 ft., ×200.
FIGURE 3	*Saccammina sphaerica*. Ameena-1 well, 1950 m, ×330.
FIGURE 4	*Saccammina atlanticus*. Ameena-1 well, 1910 m, ×45.
FIGURE 5	*Ammodiscus glabratus*. Utapate-1 well, 8240–8280 ft., ×200.
FIGURE 6	*Glomospira charoides*. Ameena-1 well, 2093 m, ×200.
FIGURE 7	*Glomospira gordialis*. Ameena-1 well, 2093 m, ×250.
FIGURE 8	*Reophax* sp. 1. Utapate-1 well, 8120–8180 ft., ×150.
FIGURE 9	*Reophax morrisoni*. Utapate-1 well, 8120–8180 ft., ×150.
FIGURE 10	*Arenoturrispirillina* sp. Ekanga-1 well, 1530 m, umbilical view, ×275.
FIGURE 11	*Arenoturrispirillina* sp. Ekanga-1 well, 1530 m, spiral view, ×300.
FIGURE 12	*Ammobaculites* sp. 1. Ibotio-1 well, 7360 ft., ×125.
FIGURE 13	*Ammobaculites strathearnensis*. Abo North-1 well, 2460 m, ×100.
FIGURE 14	*Ammobaculites strathearnensis*. Bosi-6 well, 2530 m, ×300.
FIGURE 15	*Ammobaculites strathearnensis*. Banga-SW2 well, 6700 ft., ×150.
FIGURE 16	*Ammobaculites strathearnensis*. Ibotio1 well, 8490–8550 ft., ×125.
FIGURE 17	*Ammobaculites agglutinans*. Utapate-1 well, 7800 ft., ×100.
FIGURE 18	*Ammobaculites agglutinans*. Utonana-1 well, 8080–8110 ft., ×100.
FIGURE 19	*Ammobaculites coprolithiformis*. Ihuo-1 well, 9950 ft., ×100.
FIGURE 20	*Ammoscalaria pseudospiralis*. Obeaja-1 well, 7360 ft., ×380.

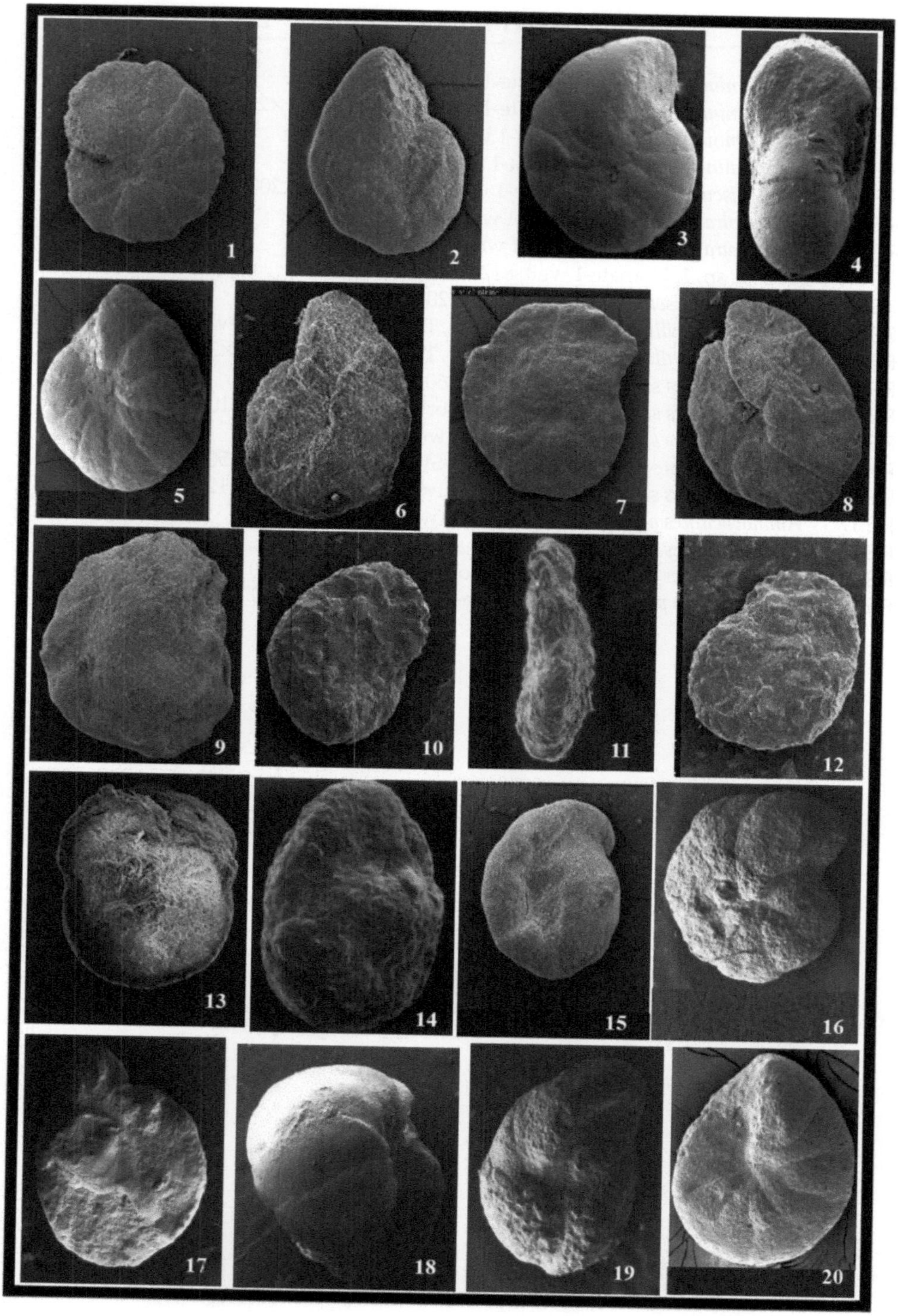
1
2
3
4
5
6
7
8
9
10
11
12
13
14
15
16
17
18
19
20

EXPLANATION OF PLATE 2

FIGURE 1	*Cyclammina cancellata*. HB-1 well, 9740–9800 ft., side view, ×50.
FIGURE 2	*Cyclammina cancellata*. HB-1 well, 9740-9800 ft., apertural view, ×125.
FIGURE 3	*Cyclammina cancellata*. Ameena-1 well, 1950 m, side view, ×45.
FIGURE 4	*Cyclammina cancellata*. Ameena-1 well, 2295 m, apertural view, ×100.
FIGURE 5	*Cyclammina cancellata*. Ameena-1 well, 2295 m, ×65.
FIGURE 6	*Cyclammina minima*. Andoni-1 well, 8400–8440 ft., ×100.
FIGURE 7	*Cyclammina minima*. Utapete-1 well, 8360–8420 ft., ×150.
FIGURE 8	*Haplophragmoides* sp. 1. Akamba-1 well, 1350–1360 ft., ×100.
FIGURE 9	*Haplophragmoides compressa*. Utapate-1 well, 8240–8280 ft., ×100.
FIGURE 10	*Haplophragmoides excavate*. Gilli Gilli-1 well, 8430 ft., umbilical view, ×300.
FIGURE 11	*Haplophragmoides excavate*. Gilli Gilli-1 well, 8430 ft., apertural view, ×300.
FIGURE 12	*Haplophragmoides excavate*. Gilli Gilli-1 well, 8430 ft., dorsal view, ×300.
FIGURE 13	*Haplophragmoides narivaensis*. Yomene-1 well, 4880–492 ft., ×225.
FIGURE 14	*Haplophragmoides narivaensis*. Gilli Gilli-1 well, 10,790 ft., ×460.
FIGURE 15	*Haplophragmoides obliquicameratus*. Utapate-1 well, 8120–8180 ft., ×75.
FIGURE 16	*Haplophragmoides obliquicameratus*. Ameena-1 well, 2135 m, ×60.
FIGURE 17	*Haplophragmoides obliquicameratus*. Ameena-1 well, 2355 m, ×60.
FIGURE 18	*Haplophragmoides bradyi*. Aparo-1 well, 8760 ft., edge view, ×100.
FIGURE 19	*Haplophragmoides bradyi*. Aparo-1 well, 8880 ft., umblical view, ×100.
FIGURE 20	*Haplophragmoides canariensis*. Ameena-1 well, 1950 m, ×230.

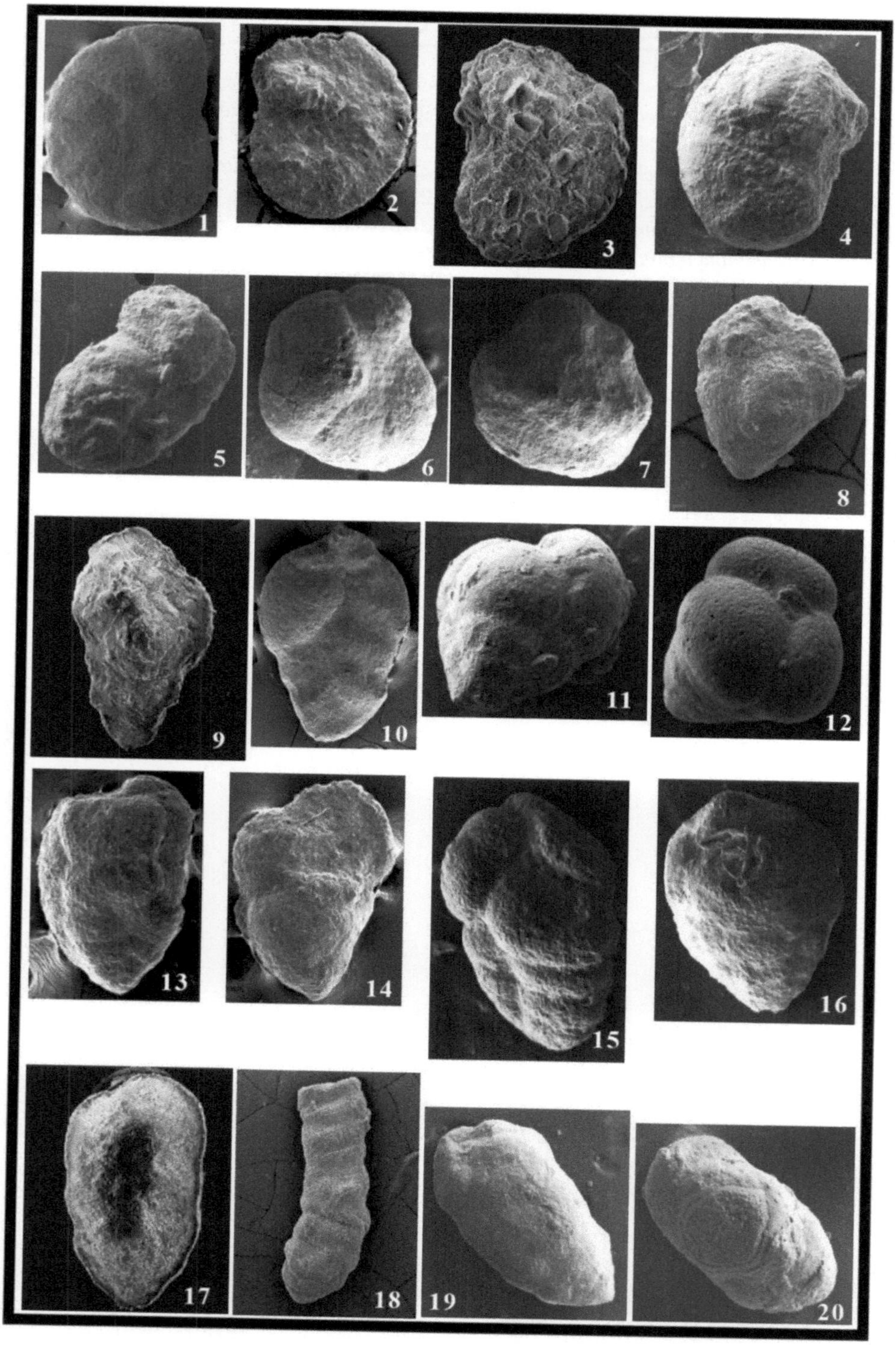
1
2
3
4
5
6
7
8
9
10
11
12
13
14
15
16
17
18
19
20

EXPLANATION OF PLATE 3

FIGURE 1 *Haplophragmoides canariensis*. Ameena-1 well, 1950 m, side view, ×80.
FIGURE 2 *Alveolophragmium crassum*. Ameena-1 well, 2295 m, ×330.
FIGURE 3 *Alveolophragmium crassum*. Adoni-1 well, 8600–8640 ft., ×150.
FIGURE 4 *Alveolophragmium scitulum*. Ameena-1 well, 2395 m, side view, ×200.
FIGURE 5 *Alveolophragmium scitulum*. Ameena-1 well, 2395 m, side view, ×170.
FIGURE 6 *Alveolophragmium subglobosum*. Ameena-1 well, 2395 m, side view, ×60.
FIGURE 7 *Alveolophragmium subglobosum*. Ameena-1 well, 2395 m, apertural view, ×100.
FIGURE 8 *Eggerella* sp. 1. Yomene-1 well, 4880–4920 ft., ×225.
FIGURE 9 *Eggerella scabra*. Adoni-1 well, 4880–4920 ft., side view, ×275.
FIGURE 10 *Eggerella bradyi*. Ameena-1 well, 2295 m, side view, ×110.
FIGURE 11 *Eggerella bradyi*. Aparo-1 well, 7680 ft., side view, ×100.
FIGURE 12 *Eggerella bradyi*. Aparo-1 well, 7680 ft., apertural view, ×200.
FIGURE 13 *Eggerella forestensis*. Ameena-1 well, 2156 m, side view, ×190.
FIGURE 14 *Eggerella forestensis*. Ameena-1 well, 2435 m, side view, ×220.
FIGURE 15 *Eggerella propinqua*. Aparo-1 well, 7860 ft., side view, ×170.
FIGURE 16 *Gravellina narivaensis*. Ameena-1 well, 1970 m, side view, ×170.
FIGURE 17 *Karreriella siphonella*. Ibotio-1 well, 4690–4750 ft., side view, ×165.
FIGURE 18 *Karreriella subcylindrica*. Utapate-1 well, 8360–8420 ft., side view, ×75.
FIGURE 19 *Karreriella bradyi*. Ameena-1 well, 2114 m, side view, ×100.
FIGURE 20 *Karreriella gaudryinoides*. Ameena-1 well, 2415 m, apertural view, ×170.

1
2
3
4
5
6
7
8
9
10
11
12
13
14
15
16
17
18
19
20

EXPLANATION OF PLATE 4

FIGURE 1	*Karreriella gaudryinoides*. Ameena-1 well, 2415 m, edge view, ×200.
FIGURE 2	*Karreriella gaudryinoides*. Ameena-1 well, 2415 m, side view, ×200.
FIGURE 3	*Valvulina flexilis*. Utapete-1well, 8360–8420 ft., side view, ×75.
FIGURE 4	*Valvulina flexilis*. Ameena-1 well, 2295 m, side view, ×70.
FIGURE 5	*Valvulina flexilis*. Ameena-1 well, 2295 m, side view, ×65.
FIGURE 6	*Valvulina pennatula*. Aparo-4 well, 12,120 ft., side view, ×60.
FIGURE 7	*Martinottiela communis*. Ameena-1 well, 1950 m, 2415 m, side view, ×200.
FIGURE 8	*Poritextularia aff. p*. Obeaja-1 well, 6160–6190 ft., side view, ×100.
FIGURE 9	*Poritextularia aff. p*. Akamba-1 well, 8340–8386 ft., side view, ×50.
FIGURE 10	*Spiroplectammina wrightii*. Utonana-1 well, 8320–8350 ft., 8320–8350 ft., side view, ×225.
FIGURE 11	*Spiroplectammina wrightii*. Mosogar-1 well, 5200 ft., side view, ×250.
FIGURE 12	*Textularia elegans*. Utapete-1 well, 8360–8420 ft., side view, ×250.
FIGURE 13	*Textularia soldanii*. Anyala-8 well, 3220–3310 ft., side view, ×100.
FIGURE 14	*Textularia parvula*. Ameena-1 well, 1770 m, side view, ×250.
FIGURE 15	*Textularia gramen*. Ameena-1 well, 1655 m, side view, ×200.
FIGURE 16	*Textularia sagittula*. Abo North-2 well, 1460 m, side view, ×200.
FIGURE 17	*Textularia sagittula*. Abo North-2 well, 1460 m, side view, ×300.
FIGURE 18	*Textularia sagittula*. Abo North-2 well, 1480 m, side view, ×300.
FIGURE 19	*Textulariella barretti*. Abo North-2 well, side view, ×150.
FIGURE 20	*Textulariella barretti*. Abo North-2 well, umbilical view, ×150.

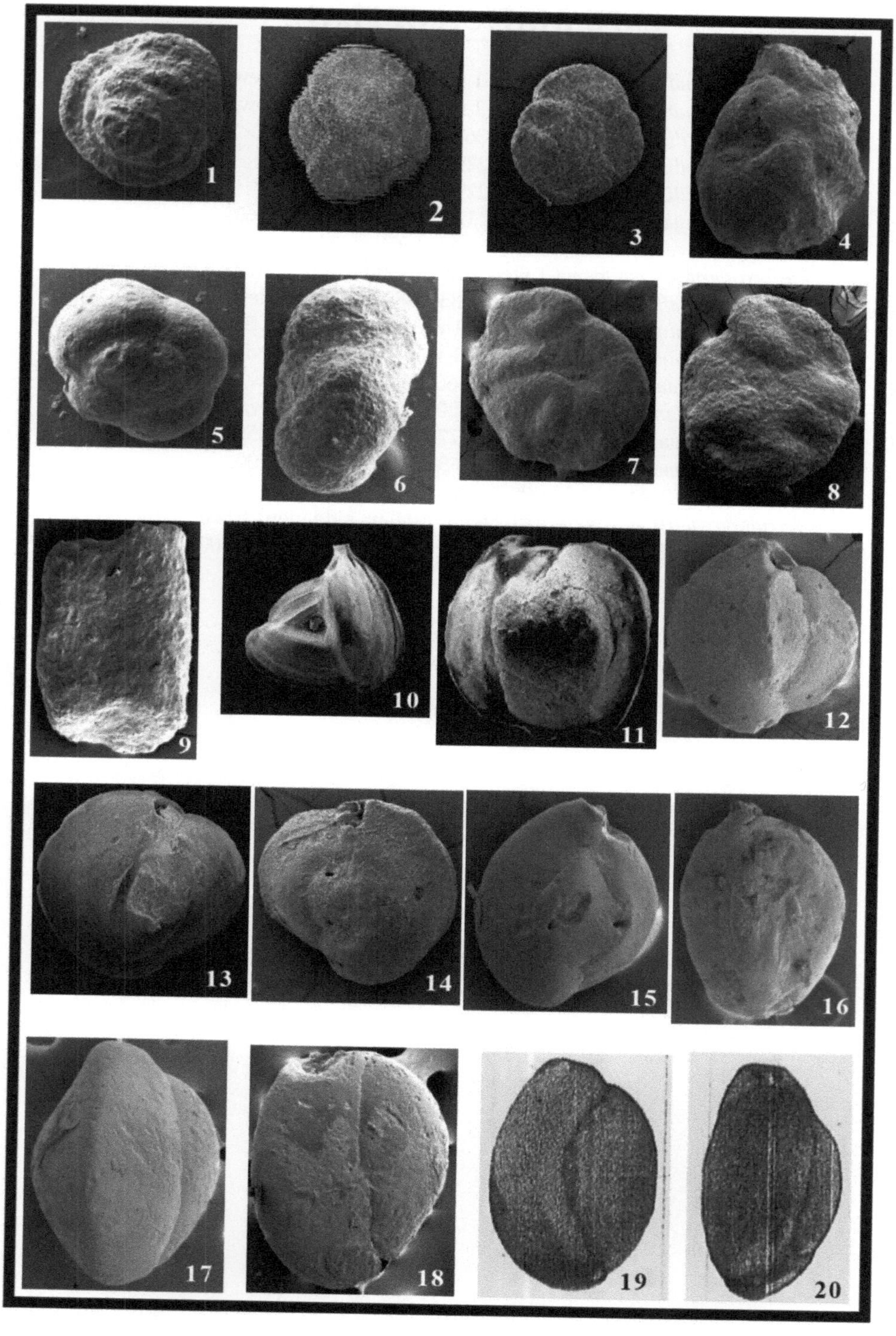
1
2
3
4
5
6
7
8
9
10
11
12
13
14
15
16
17
18
19
20

EXPLANATION OF PLATE 5

FIGURE 1 *Textulariella barretti*. Abo North-2 well, 2040 m, spiral view, ×150.
FIGURE 2 *Trochammina* sp. 1. Utapete-1 well, 8280–8340 ft., umbilical view, ×125.
FIGURE 3 *Trochammina* sp. 1. Utapete-1 well, 8280–8340 ft., side view, ×175.
FIGURE 4 *Trochammina proteus*. Ameena-1 well, 1495 m, side view, ×190.
FIGURE 5 *Trochammina globigeriniformis*. Ameena-1 well, 2355 m, spiral view, ×150.
FIGURE 6 *Trochammina globigeriniformis*. Ameena-1 well, 2355 m, umbilical view, ×150.
FIGURE 7 *Recurvoides deformis*. Ameena-1 well, 1950 m, spiral view, ×60.
FIGURE 8 *Recurvoides deformis*. Ibotio-1 well, 4720 ft., spiral view, ×80.
FIGURE 9 *Bathysiphon* sp. Ameena-1 well, 2435 m, side view, ×130.
FIGURE 10 *Ptychomiliola polygonia*. Akamba-1 well, 150–170 m, side view, ×75.
FIGURE 11 *Quinqueloculina lamarckiana*. Andoni-1 well, 6520–6580 ft., side view, ×200.
FIGURE 12 *Quinqueloculina lamarckiana*. Anyala-8 well, 1600 ft., side view, 150.
FIGURE 13 *Quinqueloculina microcostata*. Anyala-4 well, 1392–1416 ft., side view, ×100.
FIGURE 14 *Quinqueloculina microcostata*. Anyala-8 well, 1570–1660 ft., side view, ×150.
FIGURE 15 *Quinqueloculina microcostata*. Obeaja-1, well, 6286 ft., side view, ×160.
FIGURE 16 *Quinqueloculina seminulum*. Agberi North-1 well, 3232 ft., side view, ×140.
FIGURE 17 *Quinqueloculina seminulum*. Anyala-8 well, 1720 ft., edge view, ×180.
FIGURE 18 *Quinqueloculina seminulum*. Agberi North-1 well, 3200 ft., side view, ×270.
FIGURE 19 *Quinqueloculina seminulum*. Anyala-8 well, 3040–3130 ft., side view, ×250.
FIGURE 20 *Quinqueloculina seminulum*. Anyala-8 well, 3040–3130 ft., edge view, ×250.

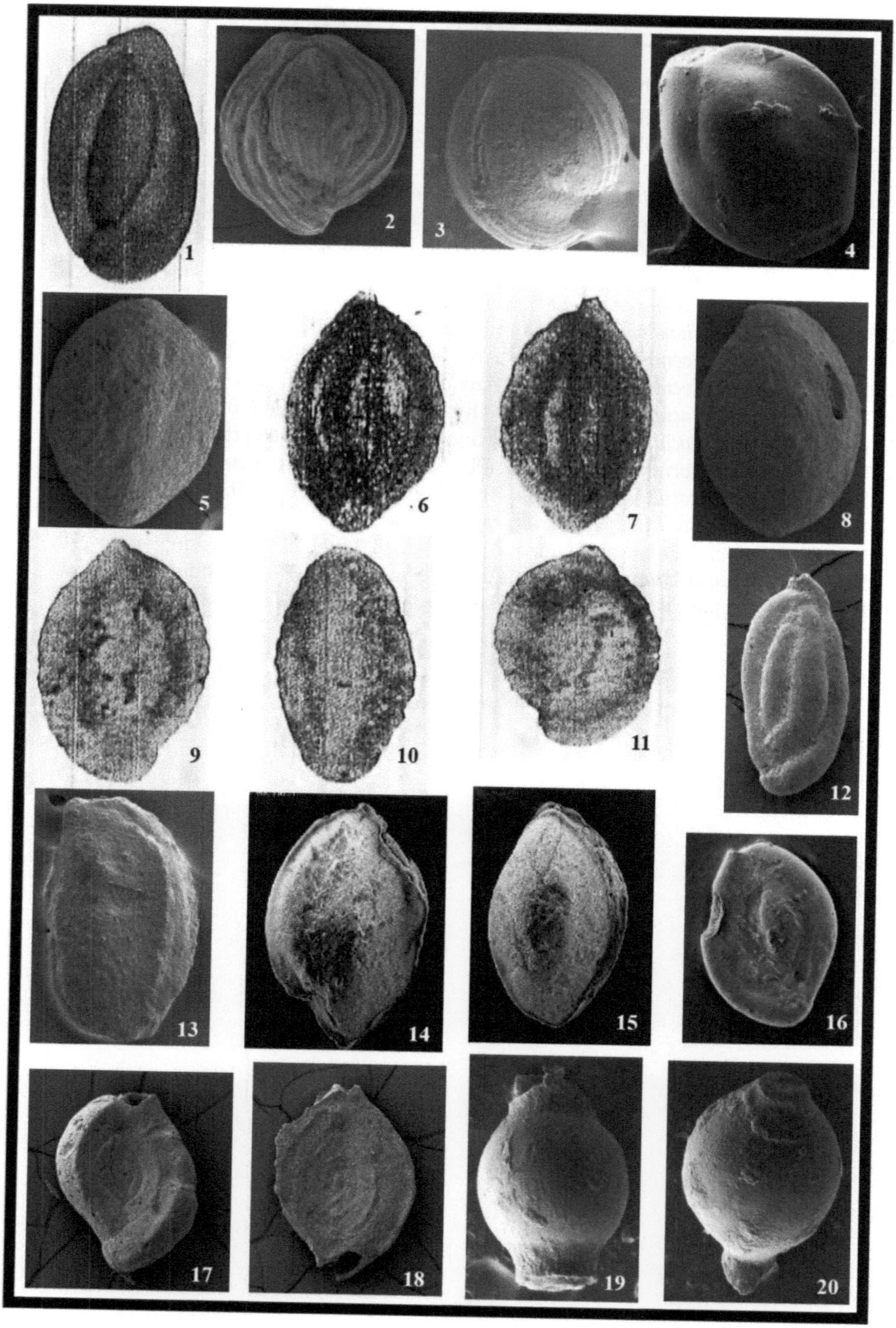
1
2
3
4
5
6
7
8
9
10
11
12
13
14
15
16
17
18
19
20

EXPLANATION OF PLATE 6

FIGURE 1 *Quinqueloculina seminulum*. Anyala-8 well, 3040–3130 ft., side view, ×250.
FIGURE 2 *Quinqueloculina rhodiensis*. Agberi North-1 well, 3248 ft., side view, ×140.
FIGURE 3 *Quinqueloculina bicornis*. Aparo-4 well, 7260 ft., side view, ×100.
FIGURE 4 *Quinqueloculina vulgaris*. Aparo-1 well, 7260 ft., side view, ×250.
FIGURE 5 *Sigmoilopsis schlumbergeri*. Ameena-1 well, 2253 m, side view, ×170.
FIGURE 6 *Sigmoilopsis schlumbergeri*. Kolo Creek-1 well, 12,640–12,680 ft., side view, ×275.
FIGURE 7 *Sigmoilopsis schlumbergeri*. Kolo Creek-1 well, 12,640–12,680 ft., side view, ×275.
FIGURE 8 *Sigmoilopsis celata*. Ameena-1 well, 2030 m, side view, ×200.
FIGURE 9 *Sigmoilopsis celata*. Anyala-1 well, 3040–3130 ft., side view, ×250.
FIGURE 10 *Sigmoilopsis celata*. Anyala-1 well, 3040–3130 ft., edge view, ×250.
FIGURE 11 *Sigmoilopsis celata*. Anyala-1 well, 3040–3130 ft., side view, ×250.
FIGURE 12 *Spirosigmoilopsis tenuis*. Anyala-8-1 well, 3040–3130 ft., side view, ×275.
FIGURE 13 *Spirosigmoilopsis tenuis*. Obeaja-1 well, 6799 ft., side view, ×150.
FIGURE 14 *Spirosigmoilopsis oligocaenica*. Kolo Creek-1 well, 12,640–12,680 ft., side view, ×325.
FIGURE 15 *Spirosigmoilopsis oligocaenica*. Kolo Creek-1 well, 12,640–12,680 ft., side view, ×275.
FIGURE 16 *Spirosigmoilopsis oligocaenica*. Fragbene-1 well, 8340 ft., side view, ×450.
FIGURE 17 *Spiroloculina depressa*. Anyala-1 well, 7780–7870 ft. side view, ×100.
FIGURE 18 *Spiroloculina depressa*. Anyala-1 well, 7780–7870 ft. side view, ×100.
FIGURE 19 *Nodosaria radicula*. Abo North-2 well, 1660 m, side view, ×200.
FIGURE 20 *Nodosaria radicula*. Abo North-2 well, 1660 m, apertural view, ×200.

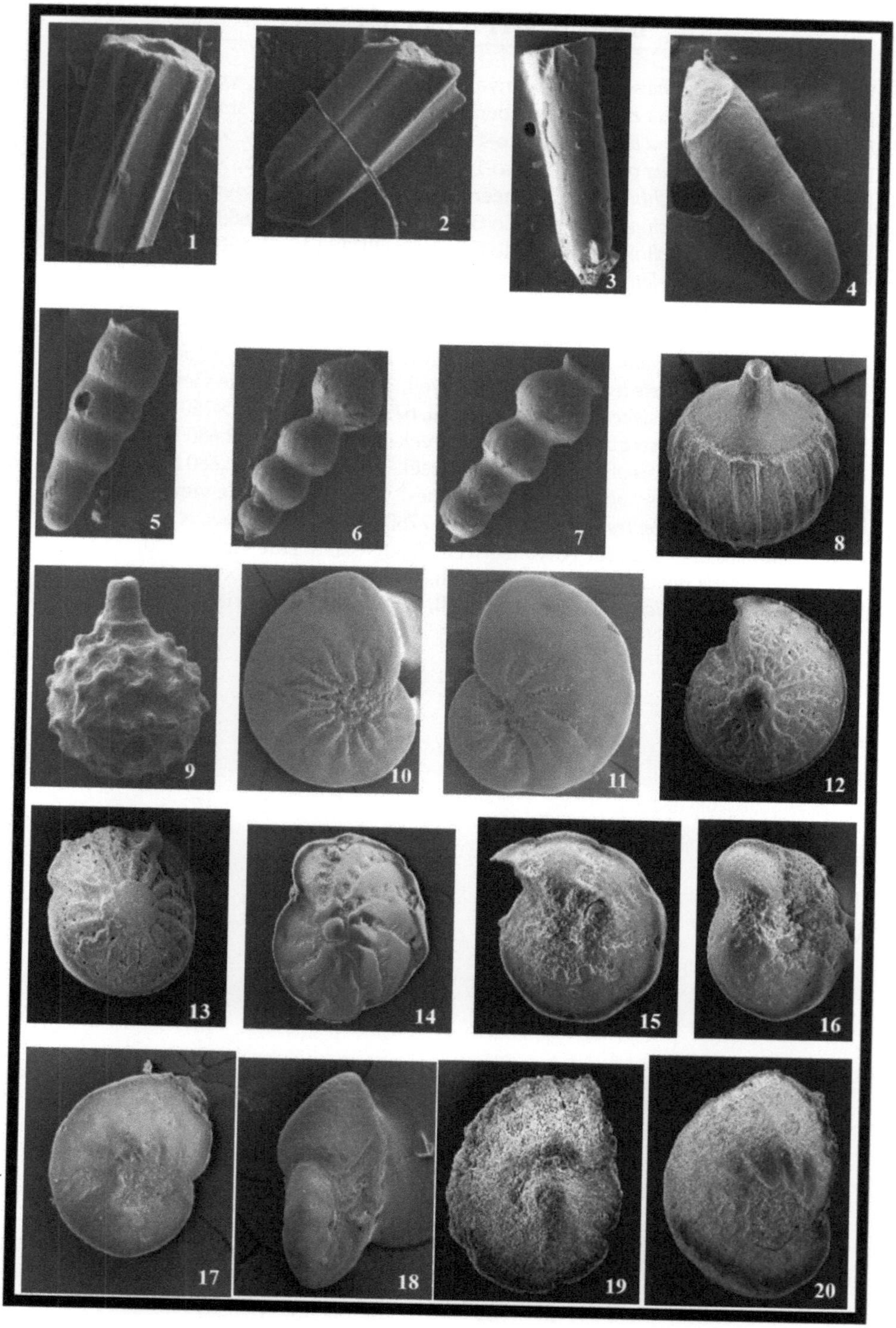
1
2
3
4
5
6
7
8
9
10
11
12
13
14
15
16
17
18
19
20

EXPLANATION OF PLATE 7

FIGURE 1 *Nodosaria raphanistrum*. Aparo-4 well, 7500 ft., side view ×150.
FIGURE 2 *Nodosaria raphanistrum*. Aparo-4 well, 7500 ft., side view ×150.
FIGURE 3 *Dentalina leguminiformis*. Abo North-2 well, 1910 m, side view, ×150.
FIGURE 4 *Dentalina leguminiformis*. Abo North-2 well, 1910 m, side view, ×150.
FIGURE 5 *Stilostomella advena*. Ameena-1 well, 1495 m, side view, ×250.
FIGURE 6 *Stilostomella monilis*. Ameena-1 well, 1435 m, side view, ×250.
FIGURE 7 *Stilostomella monilis*. Ameena-1 well, 1435 m, side view, ×200.
FIGURE 8 *Lagena striata*. Anyala-4 well, 5250–5310 ft., side view, ×400.
FIGURE 9 *Oolina faveolata*. Ameena-1 well, 530 m, side view, ×300.
FIGURE 10 *Elphidium incertum* Anyala-8 well, 1540 ft., side view, ×160.
FIGURE 11 *Elphidium incertum* Anyala-8 well, 1540 ft., side view, ×190.
FIGURE 12 *Elphidium flexuosum*. Akamba-1 well, 940 m, side view, ×100.
FIGURE 13 *Elphidium flexuosum*. Akamba-1 well, 940 m, side/apertural view, ×100.
FIGURE 14 *Elphidium advenum*. Ekanga-2 well, 650 m, side view, ×250.
FIGURE 15 *Cribroelphidium decipiens*. Akamba-1 well, 1140 m, side view, ×300.
FIGURE 16 *Cribroelphidium decipiens*. Akamba-1 well, 1140 m, side/apertural view, ×300.
FIGURE 17 *Florilus atlanticus*. HB-1 well. 10,040–10,100 ft., side view, ×175.
FIGURE 18 *Florilus atlanticus*. HB-1 well. 10,040–10,100 ft., apertural view, ×250.
FIGURE 19 *Florilus costiferum* (*Nonion*-4). Andoni-1 well, 6520–6580 ft., side view, ×300.
FIGURE 20 *Florilus costiferum* (*Nonion*-6). Utapete-1 well, 9840–9900 ft., side view, ×200.

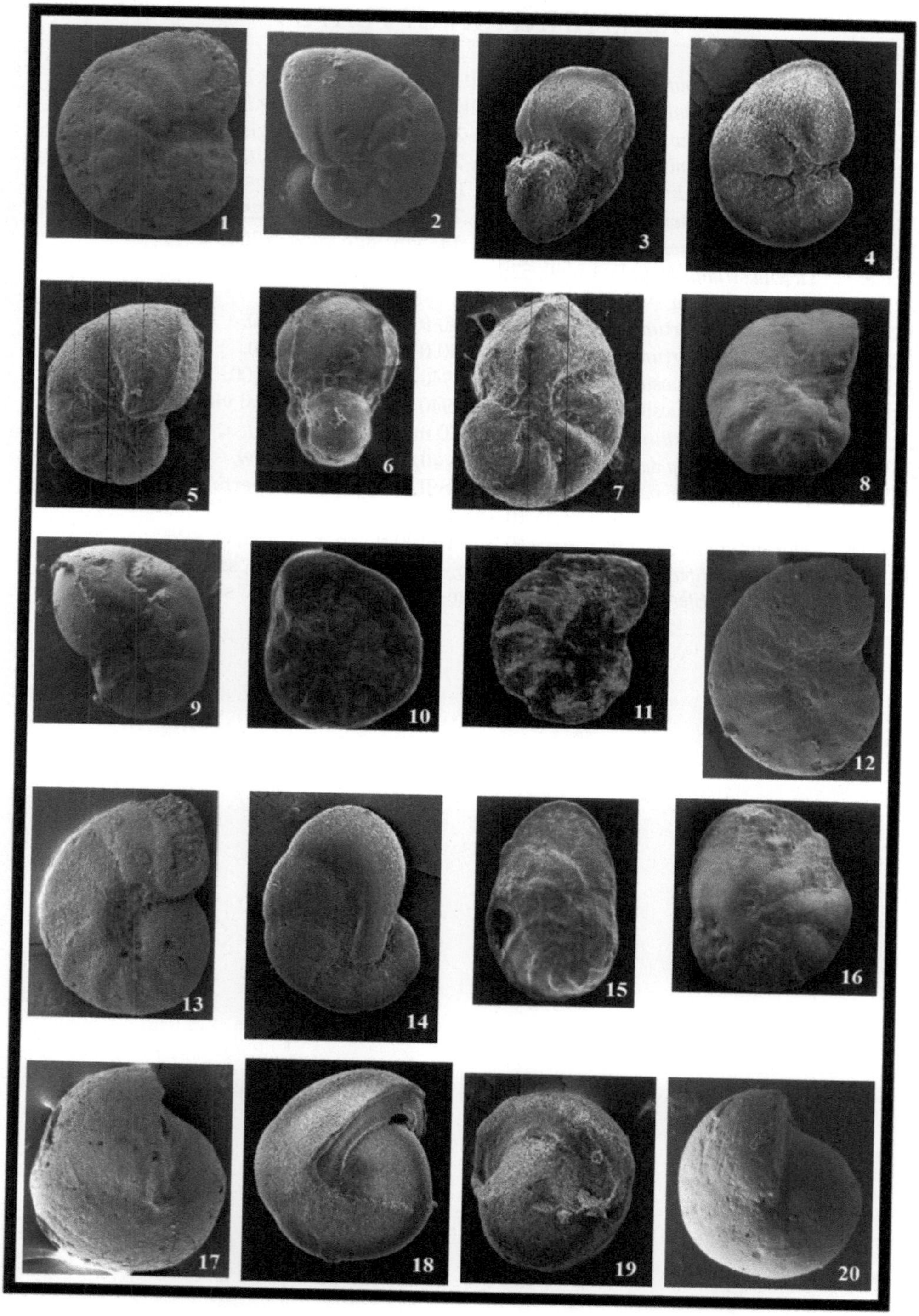
1
2
3
4
5
6
7
8
9
10
11
12
13
14
15
16
17
18
19
20

EXPLANATION OF PLATE 8

FIGURE 1 *Florilus boueanum*. Anyala-8 well, 1660 ft., side view, ×270.
FIGURE 2 *Florilus insectus*. Abo North-2 well, 1560 m, side view, ×200.
FIGURE 3 *Nonion centrosulcatum*. Benin River-3 well, 6550–6640 ft., apertural view, ×325.
FIGURE 4 *Nonion centrosulcatum*. Benin River-3 well, 6550–6640 ft., side view, ×275.
FIGURE 5 *Nonion centrosulcatum*. Fragbene-1 well, 8010 ft., side view, ×600.
FIGURE 6 *Nonion centrosulcatum*. Fragbene-1 well, 8010 ft., apertural view, ×600.
FIGURE 7 *Nonion centrosulcatum*. Fragbene-1 well, 8010 ft., side view, ×600.
FIGURE 8 *Nonion oyae*. Gilli Gilli-1 well, 9730 ft., side view, ×820.
FIGURE 9 *Nonion ecuadoranum*. Aparo-4 well, 6435 ft., side view, ×200.
FIGURE 10 *Cribrononion texana*. Gilli Gilli well, 11,230 ft., side view, ×800.
FIGURE 11 *Nonionella* sp. 1. Jesse-1 well, 10,995 ft., side view, ×800.
FIGURE 12 *Nonionella auris*. Mosogar-1 well, 4835 ft., spiral view, ×300.
FIGURE 13 *Nonionella auris*. Obeaja-1 well, 6672 ft., spiral view, ×270.
FIGURE 14 *Nonionella danvillensis*. Amakubu-1 well, 4880–4940 ft., umbilical view, ×300.
FIGURE 15 *Nonionella danvillensis*. Gilli Gilli-1 well, 10,520 ft., spiral view, ×420.
FIGURE 16 *Nonionella danvillensis*. Gilli Gilli-1 well, 10,520 ft., Umbilical view, ×440.
FIGURE 17 *Pullenia bulloides*. Anyala-8 well. 6610 ft., side view, ×400.
FIGURE 18 *Pullenia bulloides*. Anyala-4 well, 5250–5310 ft., side view, ×300.
FIGURE 19 *Pullenia bulloides*. Anyala-4 well, 5250–5310 ft., apertural view, ×325.
FIGURE 20 *Pullenia quinqueloba*. Aparo-4 well, 7260 ft., side view, ×250.

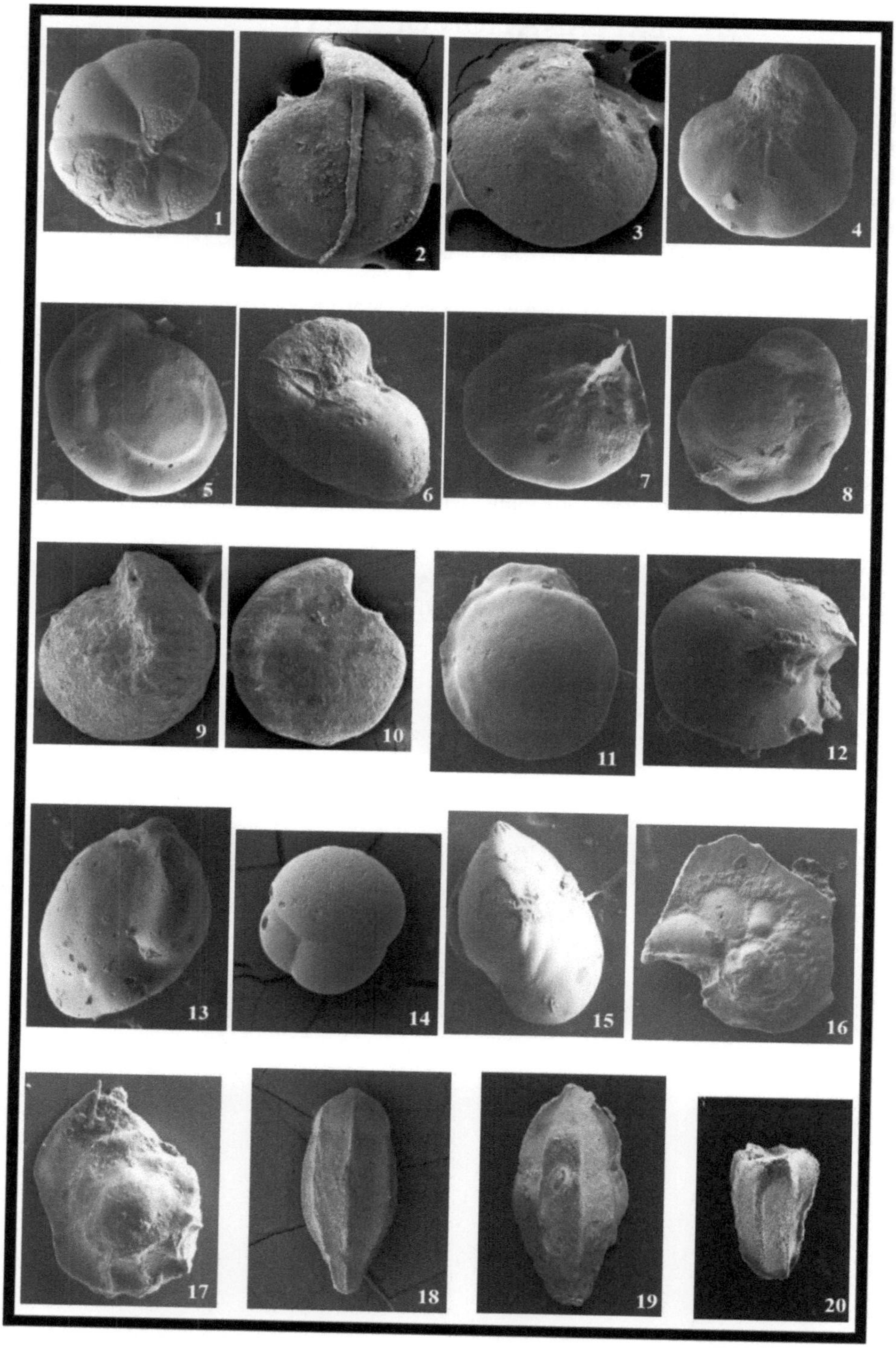
1
2
3
4
5
6
7
8
9
10
11
12
13
14
15
16
17
18
19
20

EXPLANATION OF PLATE 9

FIGURE 1	*Pullenia salisburyi*. Aparo-4 well, 13,680 ft., side view, ×200.
FIGURE 2	*Gyroidina soldanii*. Ameena-1 well, 1495 m.
FIGURE 3	*Gyroidina soldanii*. Ameena-1 well, 1495 m, spiral view, ×200.
FIGURE 4	*Gyroidina soldanii*. Ameena-1 well, 1495 m, umbilical view, ×270.
FIGURE 5	*Gyroidina orbicularis*. Aparo-4 well, 7560 ft., spiral view, ×150.
FIGURE 6	*Gyroidina orbicularis*. Aparo-4 well, 7560 ft., apertural view, ×150.
FIGURE 7	*Gyroidina orbicularis*. Aparo-4 well, 7560 ft., umbilical view, ×150.
FIGURE 8	*Gyroidina orbicularis*. Aparo-4 well, 7560 ft., spiral view, ×150.
FIGURE 9	*Gyroidinoides neosoldanii*. Ameena-1 well, 1930 m, umbilical view, ×200.
FIGURE 10	*Gyroidinoides neosoldanii*. Ameena-1 well, 1930 m, 2375 m, spiral view, ×200.
FIGURE 11	*Oridorsalis umbonatus*. Opuekeba-3 well, 8320 ft., dorsal view, ×100.
FIGURE 12	*Oridorsalis umbonatus*. Opuekeba-3 well, 8320 ft., ventral view, ×150.
FIGURE 13	*Oridorsalis umbonatus*. Opuekeba-3 well, 8320 ft., edge view, ×150.
FIGURE 14	*Sphaeroidina bulloides*. Anyala-8 well. 1570–1660 ft., side view, ×200.
FIGURE 15	*Saracenaria nuttalli*. Abo North-2 well, 1540 m, side view, ×100.
FIGURE 16	*Laticarinina pauperata*. Aparo-4 well, 7260 ft., ventral view, ×150.
FIGURE 17	*Laticarinina pauperata*. Aparo-4 well, 7260 ft., dorsal view, ×150.
FIGURE 18	*Rectoglandulina comatula*. HB-1 well, 9620–9680 ft., side view, ×125.
FIGURE 19	*Rectoglandulina comatula*. Anyala-8 well, 1570–1660 ft., side view, ×125.
FIGURE 20	*Marginulina costata*. Akamba-1 well, 1380–1400 ft., side view, ×50.

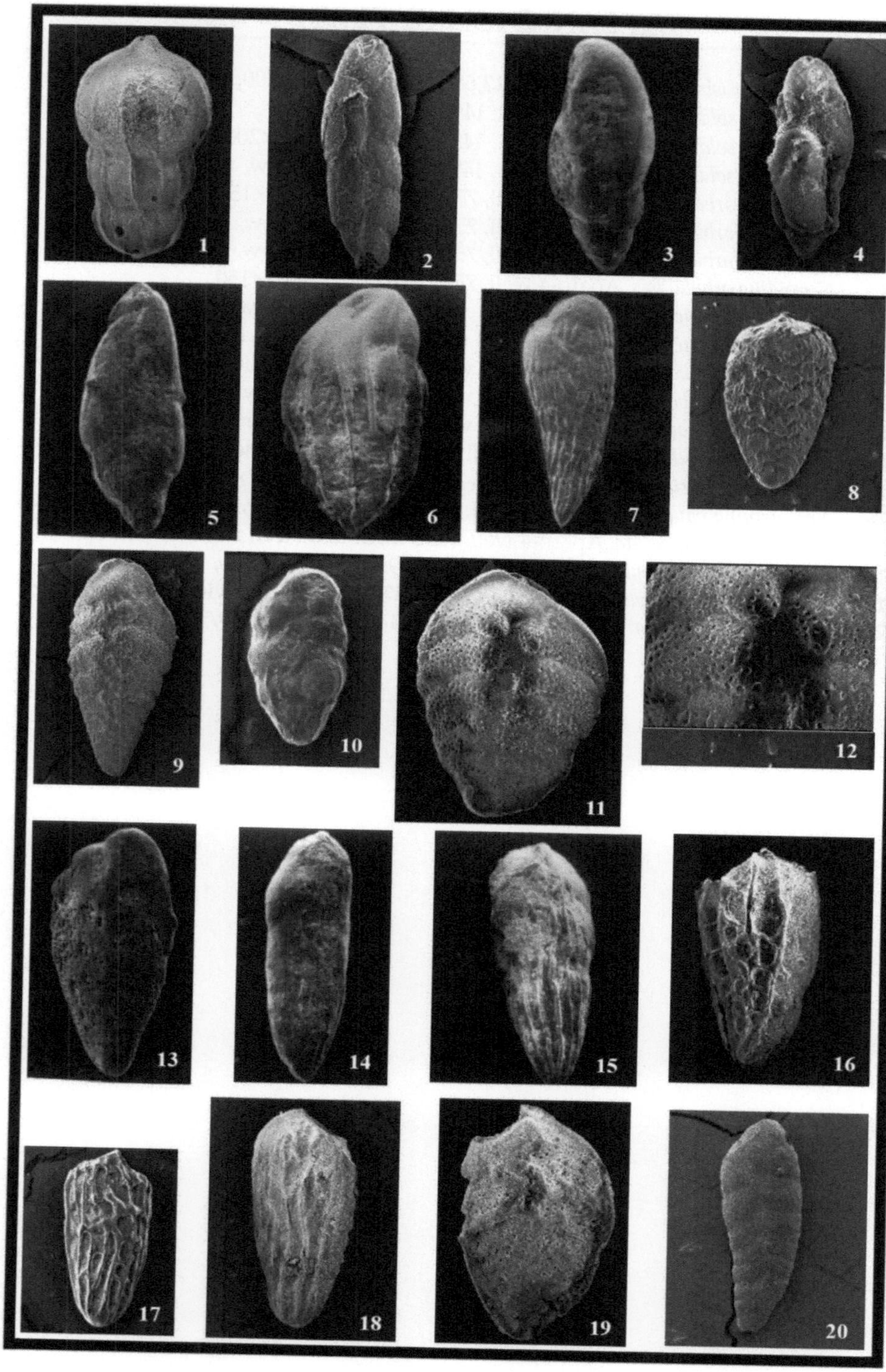
1
2
3
4
5
6
7
8
9
10
11
12
13
14
15
16
17
18
19
20

EXPLANATION OF PLATE 10

FIGURE 1 *Amphicoryna scalaris caudate*. Beniboye North-1 well, 1308–1332 m, side view, ×225.
FIGURE 2 *Altistoma tenuis*. Amakubu-1 4790–4850 well, side view, ×325.
FIGURE 3 *Altistoma tenuis*. Gilli Gilli-1 well, 10,280 ft., side view, ×500.
FIGURE 4 *Altistoma scalaris*. Amakubu-1 well, 4790–4850 ft., side view, ×200.
FIGURE 5 *Virgulina* sp. Gilli Gilli-1 well, 10,280 ft., side view, ×460.
FIGURE 6 *Bolivina* sp. Gilli Gilli-1 well, 8430 ft., side view, ×500.
FIGURE 7 *Bolivina afra*. Ihuo-1 well, 10,860 ft., side view, ×300.
FIGURE 8 *Bolivina africana*. Gilli Gilli-1 well, 11,390 ft., side view, ×600.
FIGURE 9 *Bolivina beyrichi*. Fragbene-1 well, 8320–8350 ft., side view, ×420.
FIGURE 10 *Bolivina denticulocameratus*. Gilli Gilli-1 well, 9850 ft., side view, ×500.
FIGURE 11 *Bolivina dertonensis*. Benin River-3 well, 6490–6550 ft., side view, ×325.
FIGURE 12 *Bolivina dertonensis*. Benin River-3 well, 6490–6550 ft., side view (enlarged), ×1000.
FIGURE 13 *Bolivina dertonensis*. Fragbene-1 well, 9110 ft., side view, ×420.
FIGURE 14 *Bolivina ihuoensis*. Pologbene-1 well, 9205 ft., ×500.
FIGURE 15 *Bolivina ihuoensis*. Jesse-1 well, 11,390 ft., side view, ×200.
FIGURE 16 *Bolivina interjuncta*. Kolo Creek-1 well, 10,320–10,400 ft., side view, ×80.
FIGURE 17 *Bolivina mandoroveensis*. Utonana-1 well, 8320–8340 ft., side view, ×75.
FIGURE 18 *Bolivina mandoroveensis*. Amakubu-1 well, 14,670 ft., side view, ×300.
FIGURE 19 *Bolivina miocenica*. Utapete-1 well, 9760–9820 ft., side view, ×250.
FIGURE 20 *Bolivina punctata*. Anyala-8 well, 2620–2680 ft., side view, ×225.

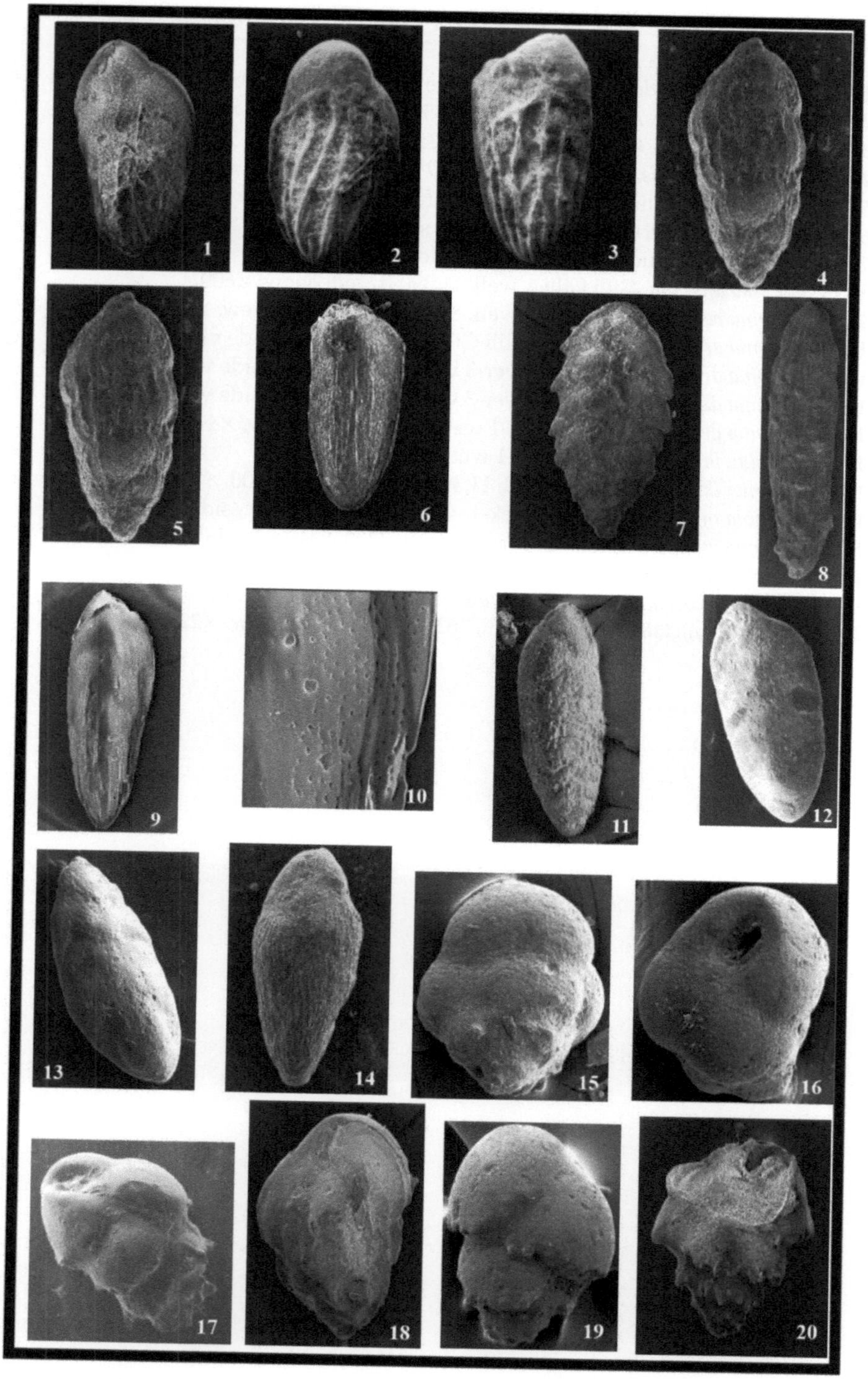
1
2
3
4
5
6
7
8
9
10
11
12
13
14
15
16
17
18
19
20

EXPLANATION OF PLATE 11

FIGURE 1	*Bolivina scalprata*. Benin River-3 well, 7240–7330 ft., side view, ×125.
FIGURE 2	*Bolivina scalprata*. Amakubu-1 well, 14,670 ft., side view, ×250.
FIGURE 3	*Bolivina simplex*. Ihuo-1 well, 16,850 ft., side view, ×200.
FIGURE 4	*Bolivina tenuicostata*. Gilli Gilli-1 well, 11,290 ft., side view, ×800.
FIGURE 5	*Bolivina tenuicostata*. Gilli Gill-1 well, 19,470 ft., side view, ×500.
FIGURE 6	*Brizalina imperatrix*. Benin River-3 well, 7090–7150 ft., side view, ×225.
FIGURE 7	*Brizalina spinata*. Fragbene-1 well, 7440–7470 ft., side view, ×200.
FIGURE 8	*Brizalina striatula spinata*. Adagbassa Creek-1 well, 4960 ft., side view, ×85.
FIGURE 9	*Brizalina aenariensis*. Ekanga-1 well, 8430 ft., side view, ×200.
FIGURE 10	*Brizalina aenariensis*. Ekanga-1 well, 8430 ft., side view (enlarged), ×1000.
FIGURE 11	*Brizalina isidroensis*. HB-1 well, 4760 ft., side view, ×250.
FIGURE 12	*Brizalina dilatata*. Gbokoda-4 well, 9040 ft., side view, ×300.
FIGURE 13	*Brizalina dilatata*. Gbokoda-4 well, 9040 ft., side view, ×300.
FIGURE 14	*Bulimina* sp. 1. Gilli Gilli-1 well, 19,470 ft., side view, ×500.
FIGURE 15	*Bulimina aculeata*. Ameena-1 well, 1455 m, 1950 m, side view, ×330.
FIGURE 16	*Bulimina aculeata*. Ameena-1 well, 1455 m, 1950 m, apertural view, ×370.
FIGURE 17	*Bulimina aculeata*. Abo North-2 well, 1249 m, side/apertural view, ×150.
FIGURE 18	*Bulimina costata*. HB-1 well, 19,470 ft., side view, ×500.
FIGURE 19	*Bulimina marginata*. Anyala-8 well, 6670 ft., side view, ×270.
FIGURE 20	*Bulimina marginata*. Anyala-4 well, 5250–5310 ft., apertural view, ×250.

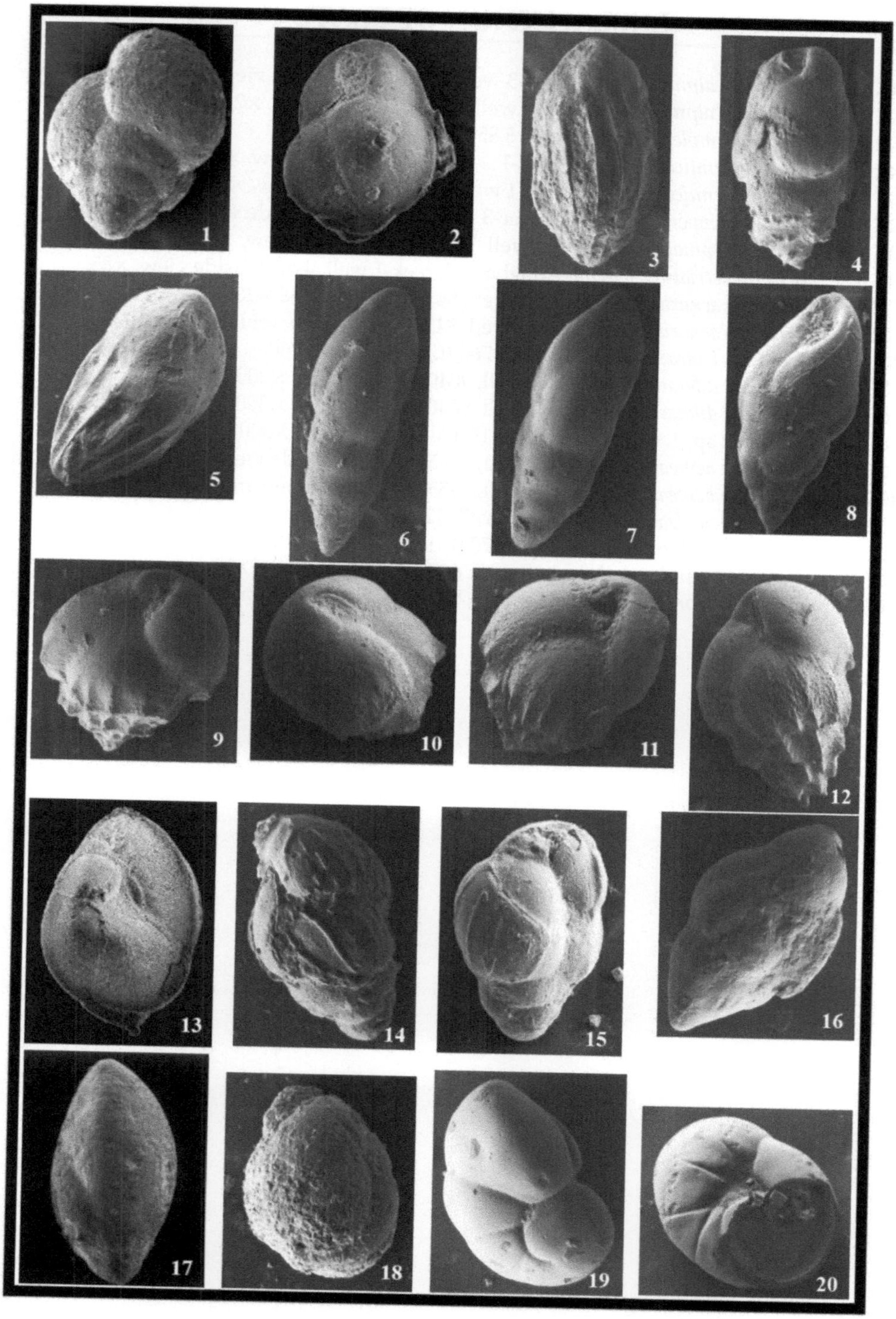
1
2
3
4
5
6
7
8
9
10
11
12
13
14
15
16
17
18
19
20

EXPLANATION OF PLATE 12

FIGURE 1	*Bulimina minima*. Ameena-1 well, 1495 m, side view, ×330.
FIGURE 2	*Bulimina minima*. Inanga-1 well, 13,340–13,400 ft., side view, ×350.
FIGURE 3	*Bulimina alazanensis*. Abo North-2 well, 1249 m, side view, ×250.
FIGURE 4	*Bulimina basisipinosa*. Abo North-2 well, 1480 m, side view, ×200.
FIGURE 5	*Bulimina buchiana*. Aparo-4 well, 7560 ft., side view, ×200.
FIGURE 6	*Bulimina exilis*. Aparo-4 well, 7380 ft., side view, ×200.
FIGURE 7	*Bulimina exilis*. Aparo-4 well, 7380 ft., edge view, ×200.
FIGURE 8	*Bulimina exilis*. Aparo-4 well, 7380 ft., apertural view, ×200.
FIGURE 9	*Bulimina inflate*. Aparo-4 well, 7260 ft., side view/apertural view, ×170.
FIGURE 10	*Bulimina inflate*. Aparo-4 well, 7260 ft., apertural view, ×170.
FIGURE 11	*Bulimina spicata*. Abo North-2, 1620 m, apertural view, ×150.
FIGURE 12	*Bulimina spicata*. Abo North-2, 1620 m, side view, ×170.
FIGURE 13	*Praeglobobulimina ovate*. Utapete-1 well, 9760–9820 ft., side view, ×200.
FIGURE 14	*Praeglobobulimina pupoides*. Meji-30 well, 10,090 ft., side view, ×250.
FIGURE 15	*Praeglobobulimina pupoides*. Abiteye-25 well, 10,460 ft., apertural view, ×250.
FIGURE 16	*Praeglobobulimina pupoides*. Abiteye-25 well, 10,760, side view, ×250.
FIGURE 17	*Globobulimina* sp. 1. Obeaja-1 well, 7540 ft., side view, ×50.
FIGURE 18	*Globobulimina ovula*. Ameena-1 well, 1595 m, side view, ×200.
FIGURE 19	*Ceratobulimina pacifica*. Aparo-4 well, 7440 ft., dorsal, side view, ×200.
FIGURE 20	*Ceratobulimina pacifica*. Aparo-4 well, 7440 ft., ventral view, ×100.

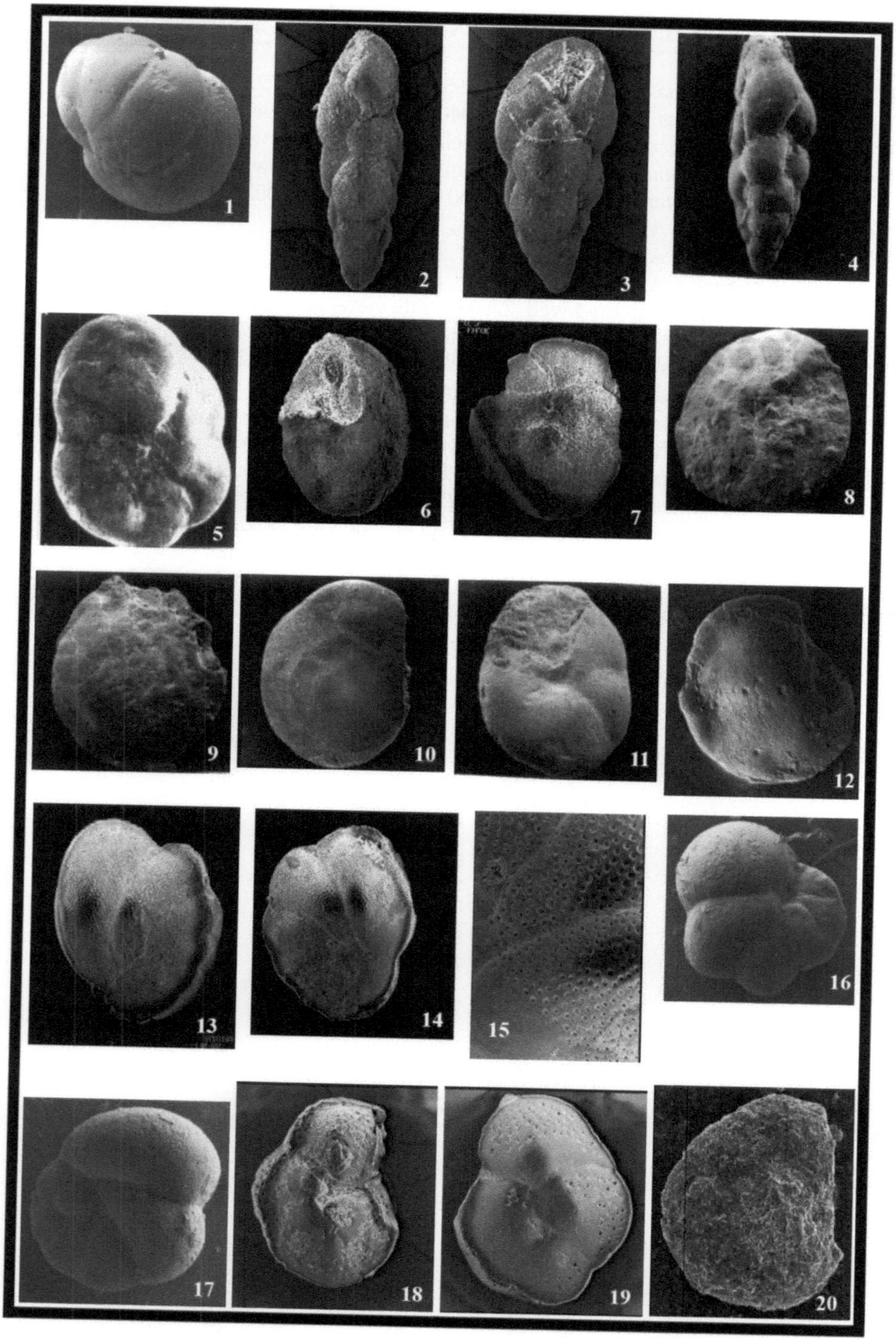
1
2
3
4
5
6
7
8
9
10
11
12
13
14
15
16
17
18
19
20

EXPLANATION OF PLATE 13

FIGURE 1 *Ceratobulimina pacifica*. Aparo-4, well, 7440 ft., edge, ×200.
FIGURE 2 *Buliminella subfusiformis*. Obeaja-1 well, 7300–7330 ft., apertural view, ×250.
FIGURE 3 *Buliminella subfusiformis*. Obeaja-1 well, 7300–7330 ft., apertural view, ×350.
FIGURE 4 *Buliminella subfusiformis*. Gilli Gilli-1 well, 7800 ft., side view, ×450.
FIGURE 5 *Discorbis* sp. 1. Gilli Gilli-1 well, 11,520 ft., umbilical view, ×300.
FIGURE 6 *Epistominella pontoni*. Obeaja-1 well, 7480 ft., umbilical view, ×300.
FIGURE 7 *Epistominella pontoni*. Adagbassa Creek-1 well, 4890 ft., spiral view, ×250.
FIGURE 8 *Epistominella pontoni*. Obeaja-1 well, 7360 ft., umbilical view, ×800.
FIGURE 9 *Epistominella pontoni*. Obeaja-1 well, 7360 ft., spiral view, ×650.
FIGURE 10 *Epistominella vitrea*. Obeaja-1 well, 7360 ft., spiral view, ×500.
FIGURE 11 *Epistominella vitrea*. Obeaja-1 well, 7360 ft., spiral view, ×560.
FIGURE 12 *Epistominella pacifica*. Gbokoda-4 well, 9040 ft., umbilical view, ×200.
FIGURE 13 *Cancris auriculus*. Akamba-1 well, 170–190 m, spiral view, ×150.
FIGURE 14 *Cancris auriculus*. Akamba-1 well, 170–190 m, umbilical view, ×575.
FIGURE 15 *Cancris auriculus*. Akamba-1 well, 170–190 m, umbilical view (enlarged) view, ×1500.
FIGURE 16 *Cancris turgidus*. Aparo-4 well, 13,680 ft., umbilical view, ×200.
FIGURE 17 *Cancris turgidus*. Aparo-4 well, 13,680 ft., spiral view, ×200.
FIGURE 18 *Rosalina floridana*. Anyala-4 well, 5250–5331 ft., umbilical view, ×400.
FIGURE 19 *Rosalina floridana*. Anyala-4 well, 5250–5331 ft., spiral view, ×410.
FIGURE 20 *Valvulineria* sp. 1. Gilli Gilli-1 well, 10,910 ft., spiral view, ×480.

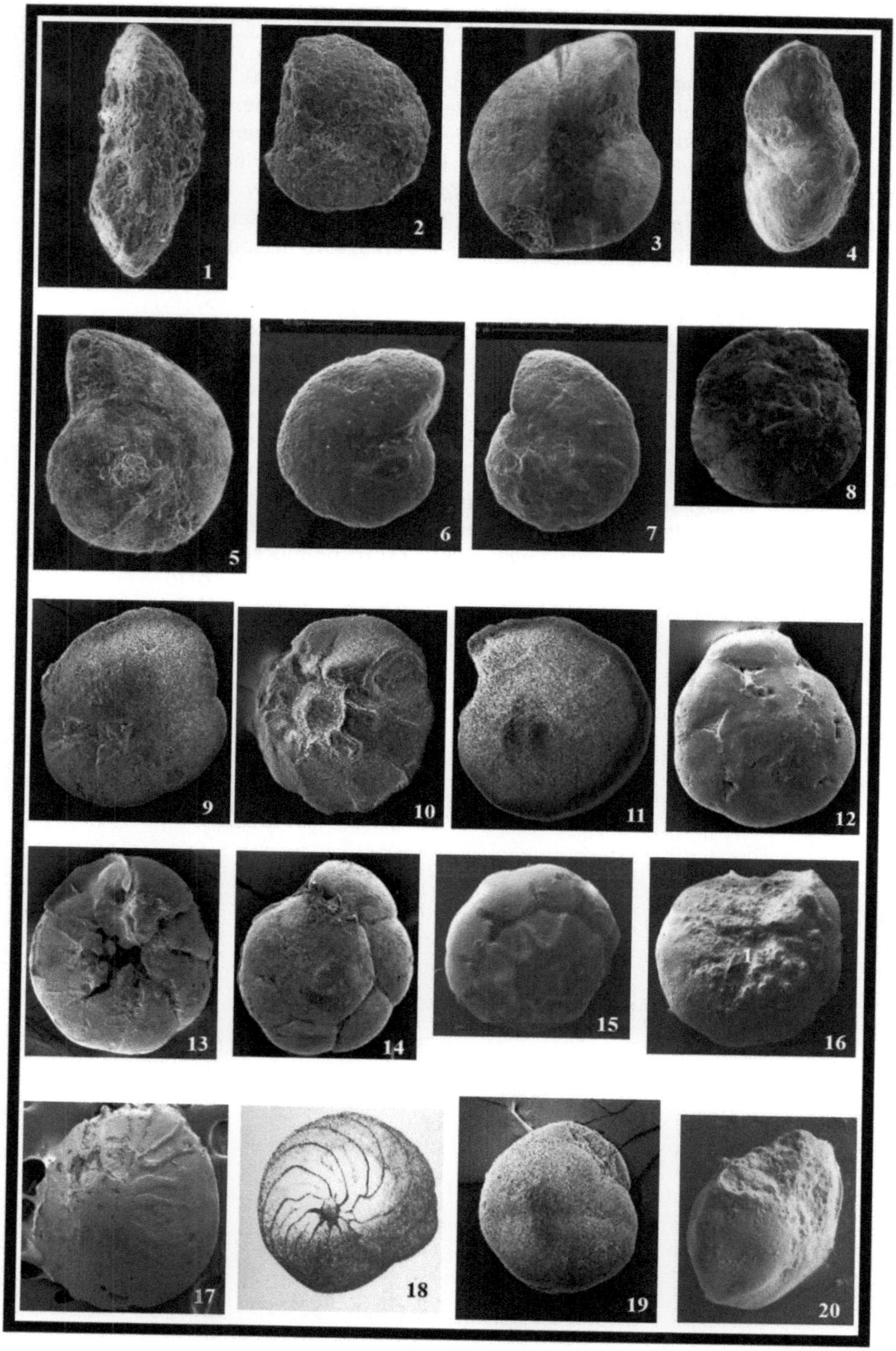
1
2
3
4
5
6
7
8
9
10
11
12
13
14
15
16
17
18
19
20

EXPLANATION OF PLATE 14

FIGURE 1	*Valvulineria* sp. 1. Gilli Gilli-1 well, 10,910 ft., apertural view, ×480.
FIGURE 2	*Valvulineria* sp. 1. Gilli Gilli-1 well, 10,910 ft., umbilical view, ×480.
FIGURE 3	*Valvulineria jacksonensis*. Gilli Gilli-1 10,910 ft., umbilical view, ×480.
FIGURE 4	*Valvulineria jacksonensis*. Gilli Gilli-1 10,910 ft., apertural view, ×480.
FIGURE 5	*Valvulineria jacksonensis*. Gilli Gilli-1 10,910 ft., spiral view, ×480.
FIGURE 6	*Valvulineria martinezensis*. Ihuo-1 well, 8800 ft., umbilical view, ×400.
FIGURE 7	*Valvulineria martinezensis*. Ihuo-1 well, 8800 ft., spiral view, ×275.
FIGURE 8	*Valvulineria trochoidea*. Gilli Gilli-1 well, 9850 ft., spiral view, ×250.
FIGURE 9	*Valvulineria wilcoxensis*. Obeaja-1 well, 7300–7330 ft., umbilical view, ×225.
FIGURE 10	*Ammonia beccarii*. Inanga-1 well, 3095 ft., umbilical view, ×275.
FIGURE 11	*Ammonia beccarii*. Inanga-1 well, 3095 ft., spiral view, ×275.
FIGURE 12	*Ammonia beccarii*. Ameena-1 well, 650 m, spiral view, ×300.
FIGURE 13	*Ammonia beccarii*. Ameena-1 well, 650 m, umbilical view, ×300.
FIGURE 14	*Ammonia beccarii*. Ameena-1 well, 650 m, spiral view, ×330.
FIGURE 15	*Ammonia tepida*. Abo North-2 well, 1520 m, spiral view, ×200.
FIGURE 16	*Ammonia tepida*. Abo North-2 well, 1520 m, umbilical view, ×200.
FIGURE 17	*Amphistegina lessonii*. Anyala-8 well, 1600 ft., side view, ×55.
FIGURE 18	*Amphistegina lessonii*. HB-1 well, 4760 ft., side view, ×50.
FIGURE 19	*Cassidulina crassa*. HB-1 well, 3320–3380 ft., side view, ×425.
FIGURE 20	*Cassidulina laevigata*. Gbokoda-4 well, 4090 ft., edge view, ×400.

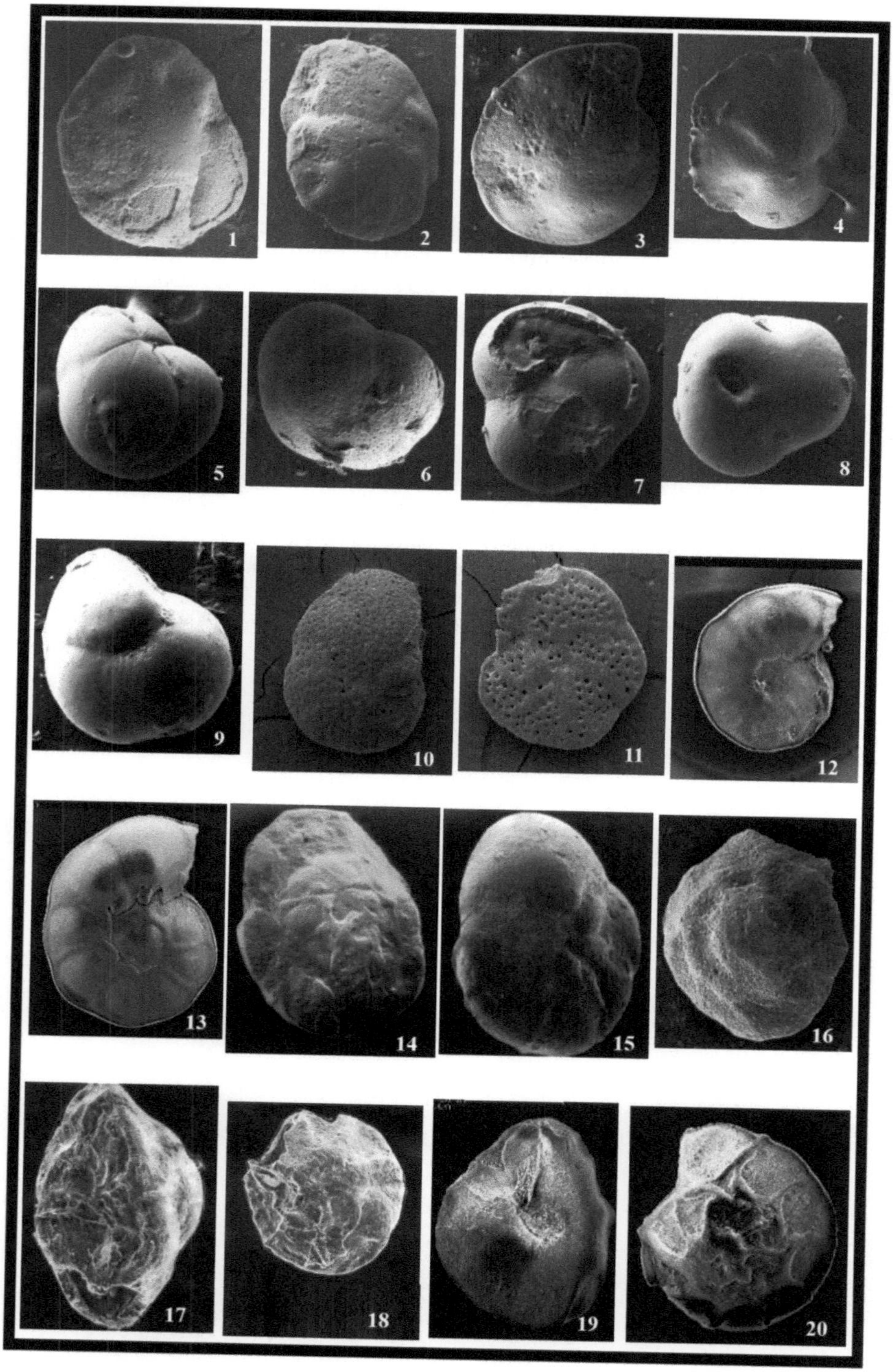
1
2
3
4
5
6
7
8
9
10
11
12
13
14
15
16
17
18
19
20

EXPLANATION OF PLATE 15

FIGURE 1 *Cassidulina laevigata*. Gbokoda-4 well, 4150 ft., side view, ×400.
FIGURE 2 *Cassidulina laevigata*. Gbokoda-4 well, 4150 ft., apertural view, ×200.
FIGURE 3 *Cassidulina neocarinata*. Opuekeba-3 well, 7840 m, side view, ×300.
FIGURE 4 *Cassidulina neocarinata*. Bonga SW-2 well, 6700 ft., side view, ×150.
FIGURE 5 *Globocassidulina subglobosa*. Aparo-4 well, 7260 ft., umbilical view, ×150.
FIGURE 6 *Globocassidulina subglobosa*. Aparo-4 well, 7260 ft., spiral view, ×200.
FIGURE 7 *Globocassidulina subglobosa*. Aparo-4 well, 7260 ft., apertural view, ×150.
FIGURE 8 *Globocassidulina oblonga*. Bonga SW-2 well, 7060 ft., apertural view, ×250.
FIGURE 9 *Globocassidulina oblonga*. Bonga SW-2 well, 7060 ft., side view, ×250.
FIGURE 10 *Cibicides lobatulus*. Anyala-8 well, 2680–2740 ft., umbilical view, ×150.
FIGURE 11 *Cibicides lobatulus*. Anyala-8 well, 2680–2740 ft., spiral view, ×225.
FIGURE 12 *Hyalinea balthica*. Ekanga-2 well, 650 m, umbilical view, ×250.
FIGURE 13 *Hyalinea balthica*. Anyala-4 well, 870–930 ft., spiral view, ×200.
FIGURE 14 *Eponides africana*. Gilli Gilli-1 well, 9350 ft., spiral view, ×250.
FIGURE 15 *Eponides africana*. Gilli Gilli-1 well, 9350 ft., umbilical view, ×300.
FIGURE 16 *Eponides berthelotianus*. Amakubu-1 well, 9380 ft., spiral view, ×700.
FIGURE 17 *Eponides berthelotianus*. Amakubu-1 well, 9380 ft., apertural view, ×700.
FIGURE 18 *Eponides berthelotianus*. Amakubu-1 well, 9380 ft., umbilical view, ×700.
FIGURE 19 *Eponides eshira*. Kolo Creek-1 well, 12,640–126,80 ft., umbilical view, ×175.
FIGURE 20 *Eponides eshira*. Kolo Creek-1 well, 12,640–126,80 ft., spiral view, ×175.

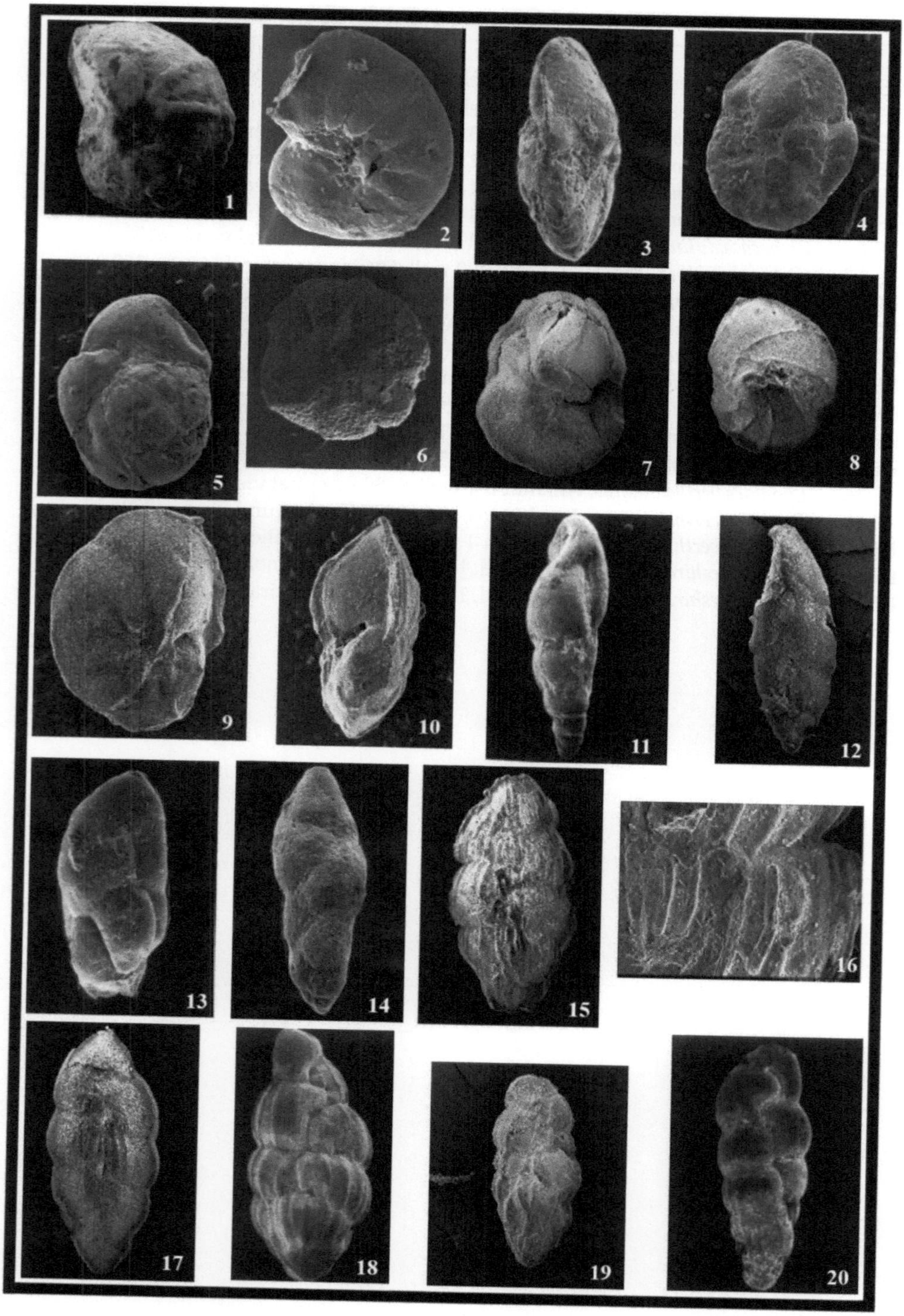
1
2
3
4
5
6
7
8
9
10
11
12
13
14
15
16
17
18
19
20

EXPLANATION OF PLATE 16

FIGURE 1 *Eponides eshira*. Gilli Gill-1 well, 18,430 ft., umbilical view, ×175.
FIGURE 2 *Eponides iojimaensis*. Mosogar-2 well, 8305 ft., umbilical view, ×300.
FIGURE 3 *Eponides pseudoelevatus*. Ihuo-1, 7780 ft., apertural view, ×240.
FIGURE 4 *Eponides pseudoelevatus*. Ihuo-1, 7780 ft., umbilical view, ×240.
FIGURE 5 *Eponides pseudoelevatus*. Ihuo-1, 7780 ft., spiral view, ×240.
FIGURE 6 *Eponides ornatus*. Gbokoda-4 well, 7330 ft., umbilical view, ×200.
FIGURE 7 *Cibicorbis inflata*. Benin River-3 well, 8110–8170 ft., umbilical view, ×75.
FIGURE 8 *Cibicorbis inflata*. Benin River-3 well, 8110–8170 ft., spiral view, ×75.
FIGURE 9 *Cibicorbis inflata*. Gilli Gilli-1 well, 9380 ft., umbilical view, ×320.
FIGURE 10 *Cibicorbis inflata*. Gilli Gilli-1 well, 9380 ft., apertural view, ×320.
FIGURE 11 *Fursenkoina howei*. Amakubu-1 well, 9590 ft., side view, ×320.
FIGURE 12 *Fursenkoina punctata*. Amakubu-1 well, 4790–4850 ft., side view, ×325.
FIGURE 13 *Fursenkoina schreibersiana*. Gilli Gilli-1 well, 8430 ft., side view, ×650.
FIGURE 14 *Fursenkoina wilcoxensis*. Gilli Gilli-1 well, 8430 ft., side view, ×650.
FIGURE 15 *Hopkinsina bononiensis*. Fragbene-1 well, 6150–6210 ft., side view, ×175.
FIGURE 16 *Hopkinsina bononiensis*. Fragbene-1 well, 6150–6210 ft., side view (enlarged), ×1000.
FIGURE 17 *Hopkinsina bononiensis*. Fragbene-1 well, 6150–6210 ft., side view, ×200.
FIGURE 18 *Hopkinsina semiornata*. Mosogar-1 well, 18,270 ft., side view, ×210.
FIGURE 19 *Hopkinsina semiornata*. Tuomo West-1 well, 4132–4140 m, side view, ×100.
FIGURE 20 *Hopkinsina danvillensis*. Gilli Gilli-1 well, 19,730 ft., side view, ×260.

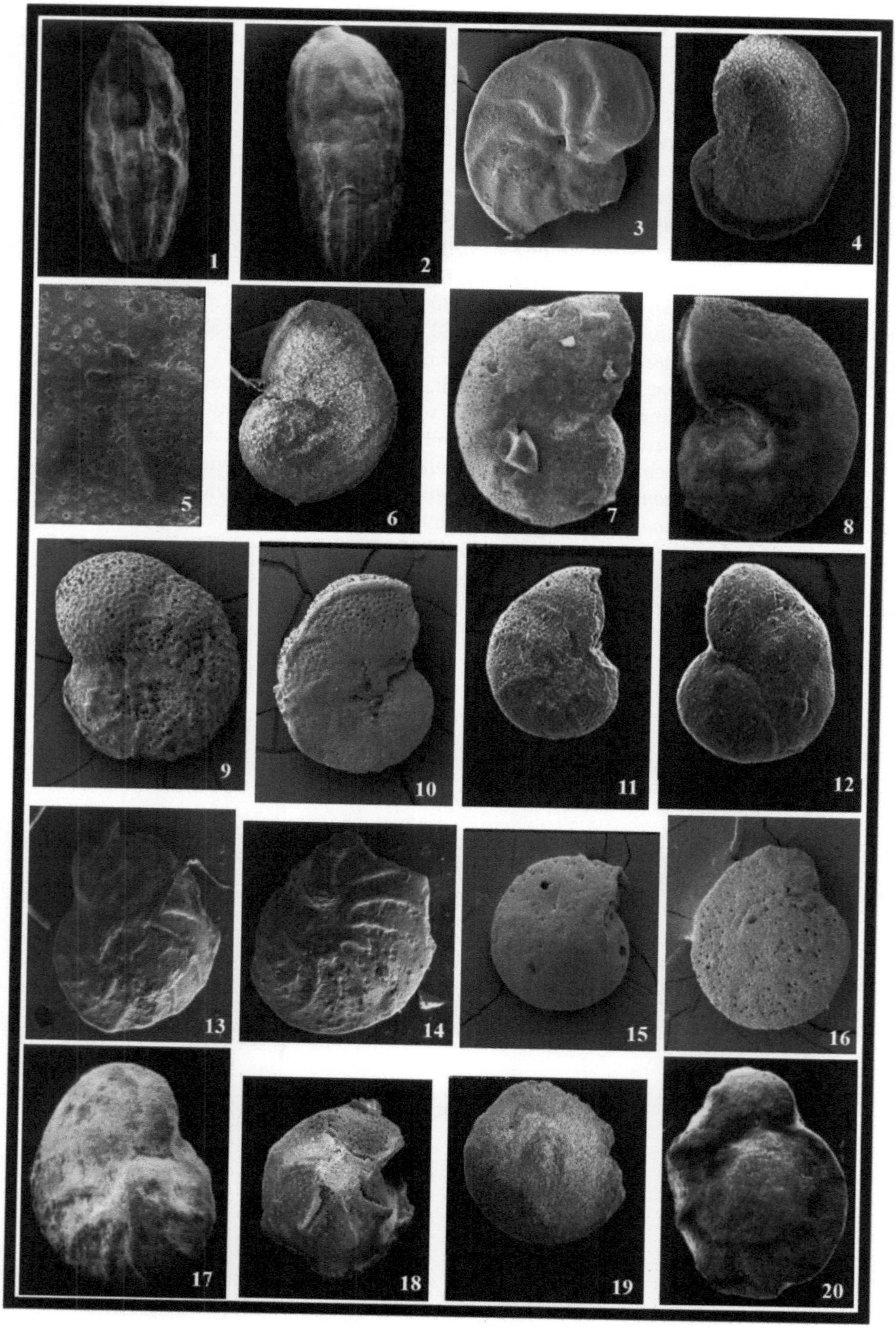
1
2
3
4
5
6
7
8
9
10
11
12
13
14
15
16
17
18
19
20

EXPLANATION OF PLATE 17

FIGURE 1	*Hopkinsina hourqi*. Amakubu-1 well, 1991 ft., side view, ×200.
FIGURE 2	*Hopkinsina hourqi*. Amakubu-1 well, 1991 ft., side view, ×80.
FIGURE 3	*Hanzawaia mantaensis*. Ameena-1 well, 1435 m, ventral view, ×230.
FIGURE 4	*Hanzawaia concentrica*. Adagbassa Creek-1 well, 4980–5000 ft., dorsal view, ×175.
FIGURE 5	*Hanzawaia concentrica*. Adagbassa Creek-1 well, 4980–5000 ft., dorsal view (enlarged), ×775.
FIGURE 6	*Hanzawaia concentrica*. Adagbassa Creek-1 well, 4980–5000 ft., ventral view, ×225.
FIGURE 7	*Hanzawaia concentrica*. Obeaja-1 well, 7360 ft., dorsal view, ×500.
FIGURE 8	*Hanzawaia concentrica*. Obeaja-1 well, 7360 ft., ventral view, ×500.
FIGURE 9	*Hanzawaia strattoni*. Anyala-8 well, 3430–3520 ft., dorsal view, ×155.
FIGURE 10	*Hanzawaia strattoni*. Anyala-8 well, 3430–3520 ft., ventral view, ×155.
FIGURE 11	*Hanzawaia strattoni*. Obeaja-1 well, 17,300 ft., dorsal view, ×500.
FIGURE 12	*Hanzawaia strattoni*. Obeaja-1 well, 17,300 ft., ventral view, ×500.
FIGURE 13	*Hanzawaia isidroensis*. Aparo-4 well, 13,980 ft., dorsal view, ×150.
FIGURE 14	*Hanzawaia isidroensis*. Aparo-4 well, 13,980 ft., ventral view, ×150.
FIGURE 15	*Heterolepa floridana*. Anyala-4 well, 1416–1422 ft., umbilical view, ×150.
FIGURE 16	*Heterolepa floridana*. Anyala-4 well, 1416–1422 ft., spiral view, ×175.
FIGURE 17	*Heterolepa floridana*. Amakubu-1 well, 4550 ft., umbilical view, ×500.
FIGURE 18	*Heterolepa pseudoungeriana*. Anyala-4 well, 1416–1422 ft., umbilical view, ×150.
FIGURE 19	*Heterolepa pseudoungeriana*. Anyala-4 well, 1416–1422 ft., spiral view, ×125.
FIGURE 20	*Heterolepa pseudoungeriana*. Amakubu-1 well, 4550 ft., spiral view, ×420.

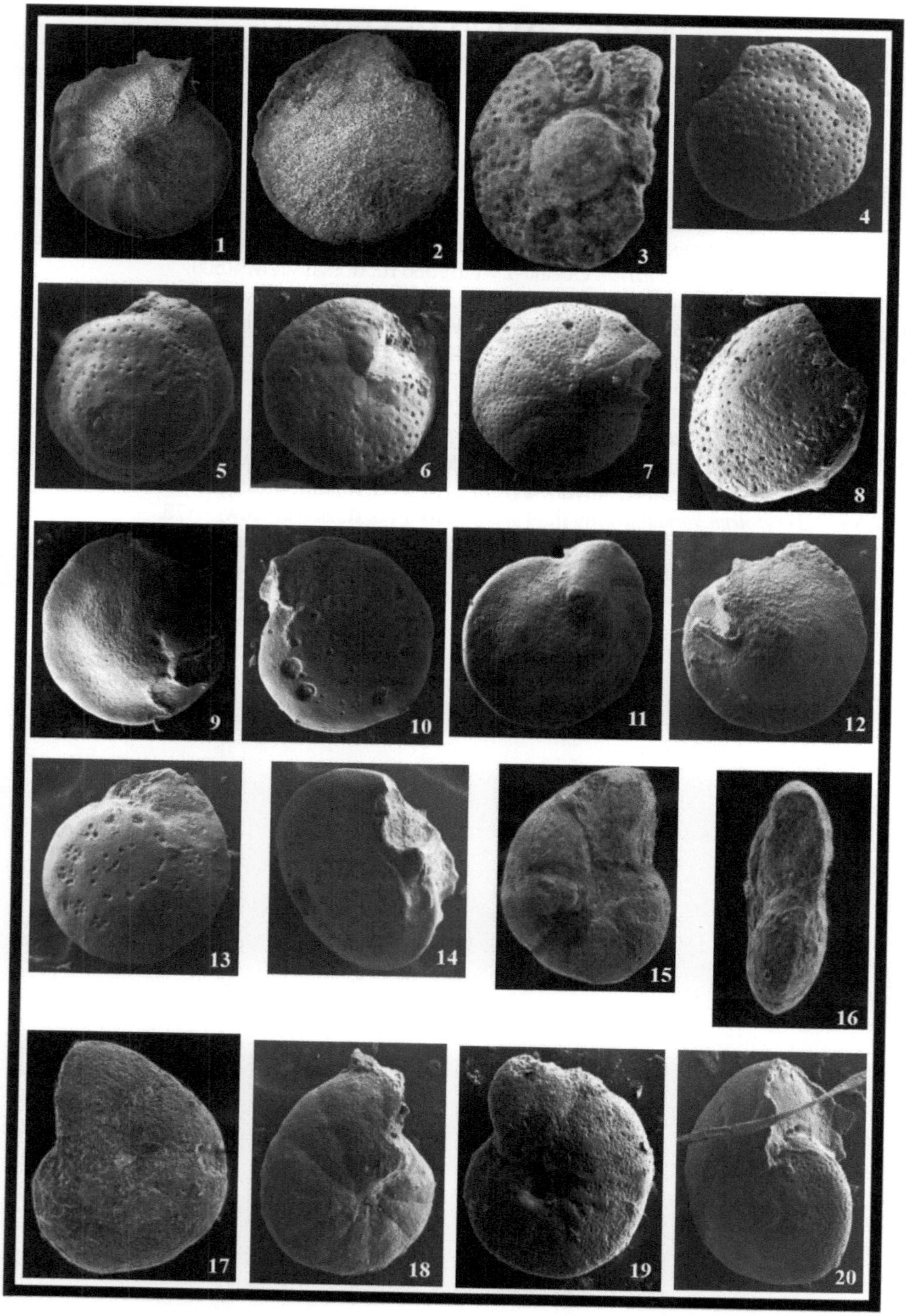
1
2
3
4
5
6
7
8
9
10
11
12
13
14
15
16
17
18
19
20

EXPLANATION OF PLATE 18

FIGURE 1	*Heterolepa mckannai*. Fragbene-1 well, 5610–5670 ft., umbilical view, ×175.
FIGURE 2	*Heterolepa mckannai*. Fragbene-1 well, 5610–5670 ft., spiral view, ×175.
FIGURE 3	*Heterolepa mckannai*. Amakubu-1 well, 4550 ft., spiral view, ×700.
FIGURE 4	*Heterolepa bellincioni*. Aparo-4 well, 7140 ft., spiral view, ×150.
FIGURE 5	*Heterolepa bellincioni*. Aparo-4 well, 7140 ft., spiral view, ×200.
FIGURE 6	*Heterolepa bellincioni*. Aparo-4 well, 7140 ft., umbilical view, ×200.
FIGURE 7	*Cibicidoides pseudoungerianus*. Ameena-1 well, 650 m, ventral view, ×150.
FIGURE 8	*Cibicidoides pseudoungerianus*. Ameena-1 well, 650 m, dorsal view, ×200.
FIGURE 9	*Cibicidoides crebbsi*. Aparo-4 well, 7320 ft., ventral view, ×150.
FIGURE 10	*Cibicidoides crebbsi*. Aparo-4 well, 7320 ft., dorsal view, ×150.
FIGURE 11	*Cibicidoides crebbsi*. Aparo-4 well, 7320 ft., apertural view, ×150.
FIGURE 12	*Cibicidoides pachyderma*. Aparo-4 well, 7560 ft., ventral view, ×100.
FIGURE 13	*Cibicidoides pachyderma*. Aparo-4 well, 7560 ft., dorsal view, ×100.
FIGURE 14	*Cibicidoides pachyderma*. Aparo-4 well, 7560 ft., apertural view, ×150.
FIGURE 15	*Anomalinoides madrugaensis*. Ihuo-1 well, 10870 ft., umbilical view, ×150.
FIGURE 16	*Anomalinoides madrugaensis*. Ihuo-1 well, 10870 ft., apertural view, ×50.
FIGURE 17	*Anomalinoides madrugaensis*. Ihuo-1 well, 10870 ft., spiral view, ×50.
FIGURE 18	*Melonis padanum*. Abo North-2 well, 2400 m, side view, ×150.
FIGURE 19	*Melonis affinis*. Abo North-2 well, 1560 m, side view, ×150.
FIGURE 20	*Melonis affinis*. Abo North-2 well, 1560 m, apertural view, ×300.

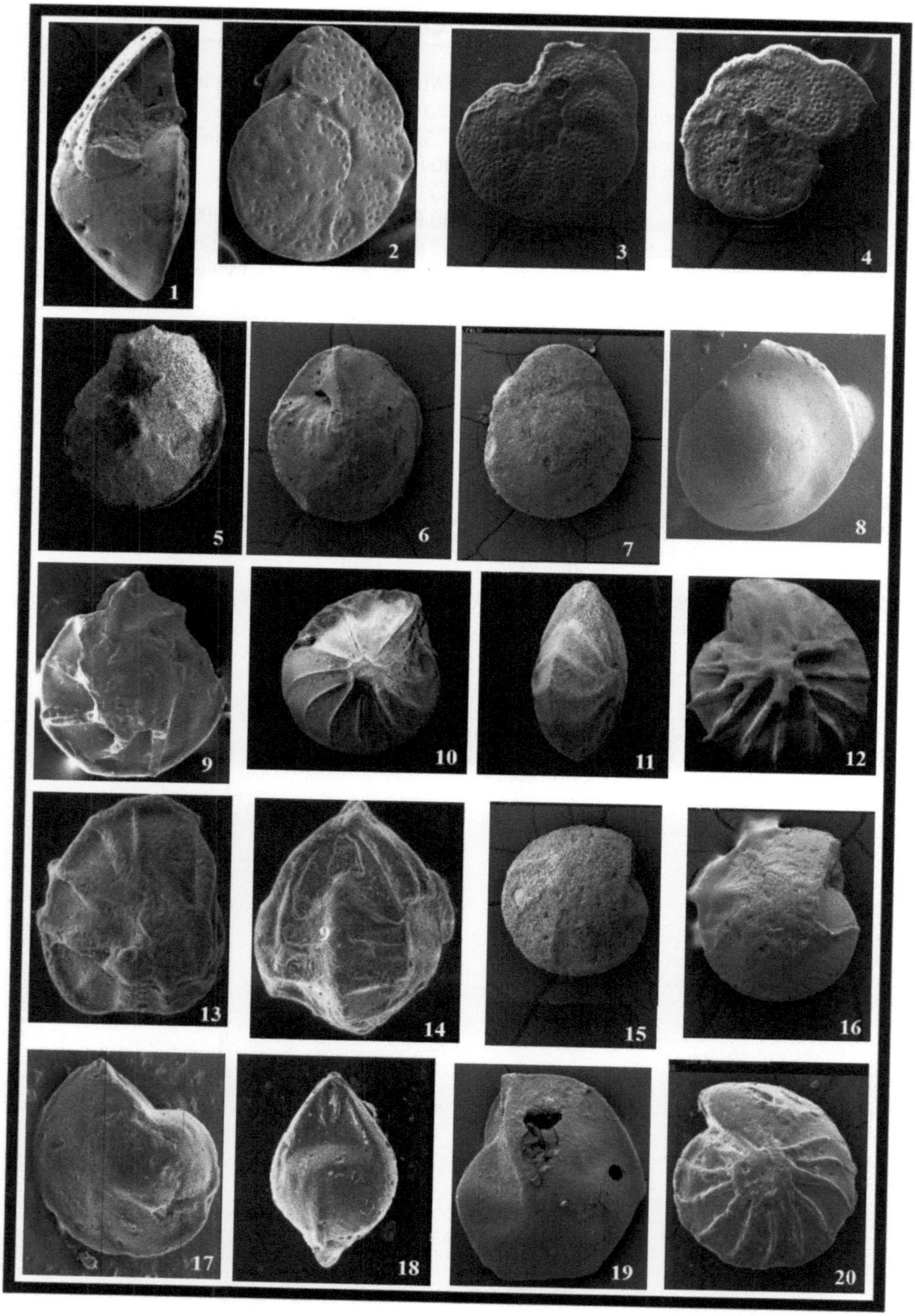
1
2
3
4
5
6
7
8
9
10
11
12
13
14
15
16
17
18
19
20

EXPLANATION OF PLATE 19

FIGURE 1 *Planulina wuellerstorfi*. Ameena-1 well, 1475 m, apertural view, ×370.
FIGURE 2 *Planulina wuellerstorfi*. Ameena-1 well, 1475 m, ventral view, ×200.
FIGURE 3 *Planulina ariminensis*. Abo North-2 well, 1249 m, dorsal view, ×100.
FIGURE 4 *Planulina ariminensis*. Abo North-2 well, 1249 m, ventral view, ×100.
FIGURE 5 *Heterostegina* sp. 1. Utonana-1 well, 4260–4290 ft., side view, ×100.
FIGURE 6 *Hoeglundina elegans*. Anyala-8 well, 3220–3310 ft., apertural view, ×75.
FIGURE 7 *Hoeglundina elegans*. Anyala-8 well, 3220–3310 ft., spiral view, ×75.
FIGURE 8 *Hoeglundina elegans*. Aparo-4 well, 7140 ft., spiral view, ×150.
FIGURE 9 *Lenticulina calcar*. Ameena-1 well, 1455 m, side view, ×120.
FIGURE 10 *Lenticulina grandis*. Benin River-3 well, 7240–7330 ft., side view, ×75.
FIGURE 11 *Lenticulina grandis*. Benin River-3 well, 7240–7330 ft., apertural view, ×100.
FIGURE 12 *Lenticulina grandis*. Benin River-3 well, 7240–7330 ft., side view, ×100.
FIGURE 13 *Lenticulina grandis*. Obeaja-1 well, 7510 ft., side view, ×500.
FIGURE 14 *Lenticulina grandis*. Obeaja-1 well, 7510 ft., apertural view, ×500.
FIGURE 15 *Lenticulina inornata*. Utapete-1 well, 9760–9820 ft., side view, ×75.
FIGURE 16 *Lenticulina inornata*. Utapete-1 well, 9760–9820 ft., apertural view, ×100.
FIGURE 17 *Lenticulina inornata*. Utapete-1 well, 9760–9820 ft., side view, ×400.
FIGURE 18 *Lenticulina inornata*. Utapete-1 well, 9760–9820 ft., apertural view, ×400.
FIGURE 19 *Lenticulina rotulata*. Ameena-1 well, 1950 m, side view, ×170.
FIGURE 20 *Lenticulina pseudomammiligerus*. Gilli Gilli-1 well, 10,910 ft., side view, ×450.

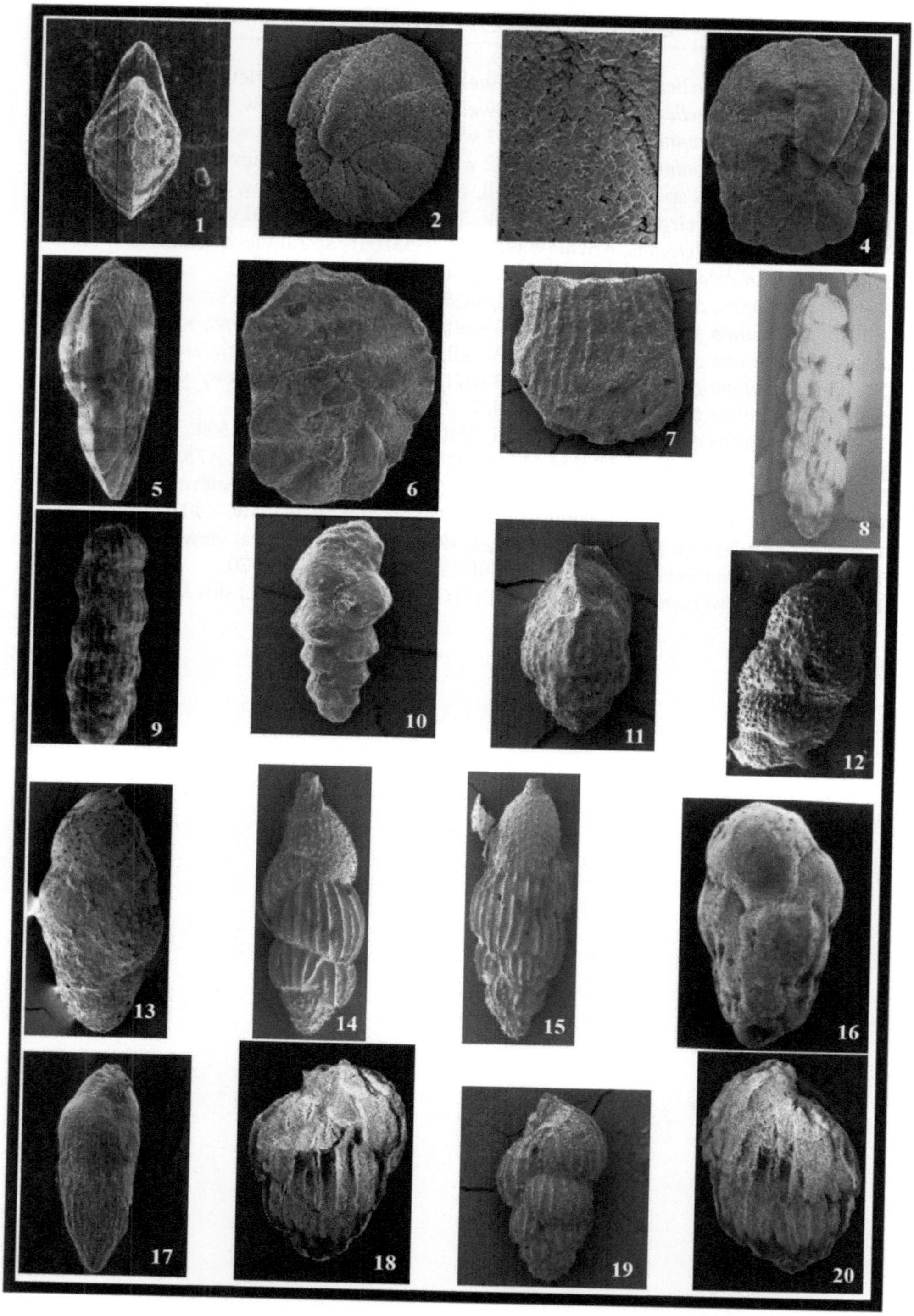
1
2
4
5
6
7
8
9
10
11
12
13
14
15
16
17
18
19
20

EXPLANATION OF PLATE 20

FIGURE 1	*Lenticulina pseudomammiligerus*. Gilli Gilli-1 well, 10,910 ft., apertural view, ×300.
FIGURE 2	*Megastomella africana*. Fragbene-1 well, 6690–6720 ft., umbilical view, ×175.
FIGURE 3	*Megastomella africana*. Fragbene-1 well, 6690–6720 ft., umbilical view (enlarged), ×775.
FIGURE 4	*Megastomella africana*. Fragbene-1 well, 8010 ft., umbilical view, ×400.
FIGURE 5	*Megastomella africana*. Fragbene-1 well, 8010 ft., apertural view, ×400.
FIGURE 6	*Megastomella africana*. Fragbene-1 well, 8010 ft., spiral view, ×400.
FIGURE 7	*Plectofrondicularia* sp. 1. Anyala-8 well, 4960–5020 ft., side view, ×175.
FIGURE 8	*Rectuvigerina multicostata*. HB-1 well, 3320–3380 ft., side view, ×225.
FIGURE 9	*Rectuvigerina tenuicostata*. Gilli Gilli-1 well, 9730 ft., side view, ×400.
FIGURE 10	*Sagrina circumspinosa*. Ihuo-1 well, 6955 ft., side view, ×500.
FIGURE 11	*Trifarina reussi*. Anyala-8 well, 2680–2770 ft., side view, ×300.
FIGURE 12	*Siphouvigerina auberiana*. Abo North-2 well, 1249 m, side view, ×200.
FIGURE 13	*Uvigerina auberiana*. Ameena-1 well, 1950 m, side view, ×100.
FIGURE 14	*Uvigerina pygmaea*. Anyala-8 well, 1660 ft., side view, ×160.
FIGURE 15	*Uvigerina pygmaea*. Anyala-8 well, 1660 ft., side view, ×190.
FIGURE 16	*Uvigerina gallowayi*. Gilli Gilli-1 well, 1042 ft., side view, ×380.
FIGURE 17	*Uvigerina havanensis*. Gilli Gilli-1 well, 9670 ft., side view, ×500.
FIGURE 18	*Uvigerina peregrina*. Akamba-1 well, 620–670 m, side view, ×150.
FIGURE 19	*Uvigerina peregrina*. Anyala-8, well, 2680–2770 ft., side view, ×200.
FIGURE 20	*Uvigerina subperegrina*. Ibotio-1 well, 4680–4780 ft., side view, ×175.

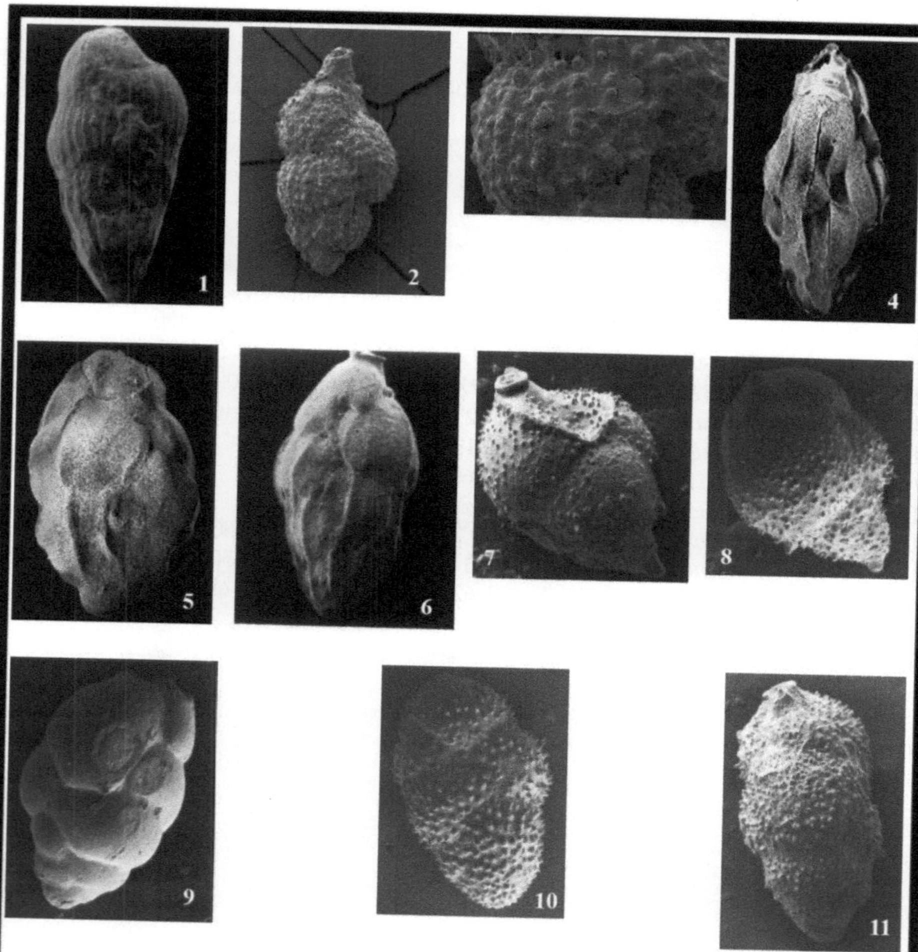
1
2
4
5
6
7
8
9
10
11

EXPLANATION OF PLATE 21

FIGURE 1 *Uvigerina subperegrina*. Amakubu-1 well, 5030 ft., side view, ×500.
FIGURE 2 *Uvigerina proboscidea*. Anyala-8 well, 2680–2770 ft., side view, ×225.
FIGURE 3 *Uvigerina proboscidea*. Anyala-8 well, 2680–2770 ft., side view (enlarged), ×775.
FIGURE 4 *Uvigerina sparsicostata*. Utonana-1 well, 4540–4570 ft., side view, ×150.
FIGURE 5 *Uvigerina sparsicostata*. Utonana-1 well, 4540–4570 ft., side view, ×150.
FIGURE 6 *Uvigerina sparsicostata*. Obeaja-1 well, 7510 ft., side view, ×260.
FIGURE 7 *Uvigerina asperula*. Abo North-2 well, 1400 m, side view, ×150.
FIGURE 8 *Uvigerina asperula*. Abo North-2 well, 1400 m, side view, ×150.
FIGURE 9 *Uvigerina mantaensis*. Utonana-9 ST well, 8140 ft., side view, ×150.
FIGURE 10 *Uvigerina hispida*. Abo North-2 well, 1400 m, side view, ×100.
FIGURE 11 *Uvigerina hispida*. Abo North-2 well, 1520 m, apertural view, ×100.

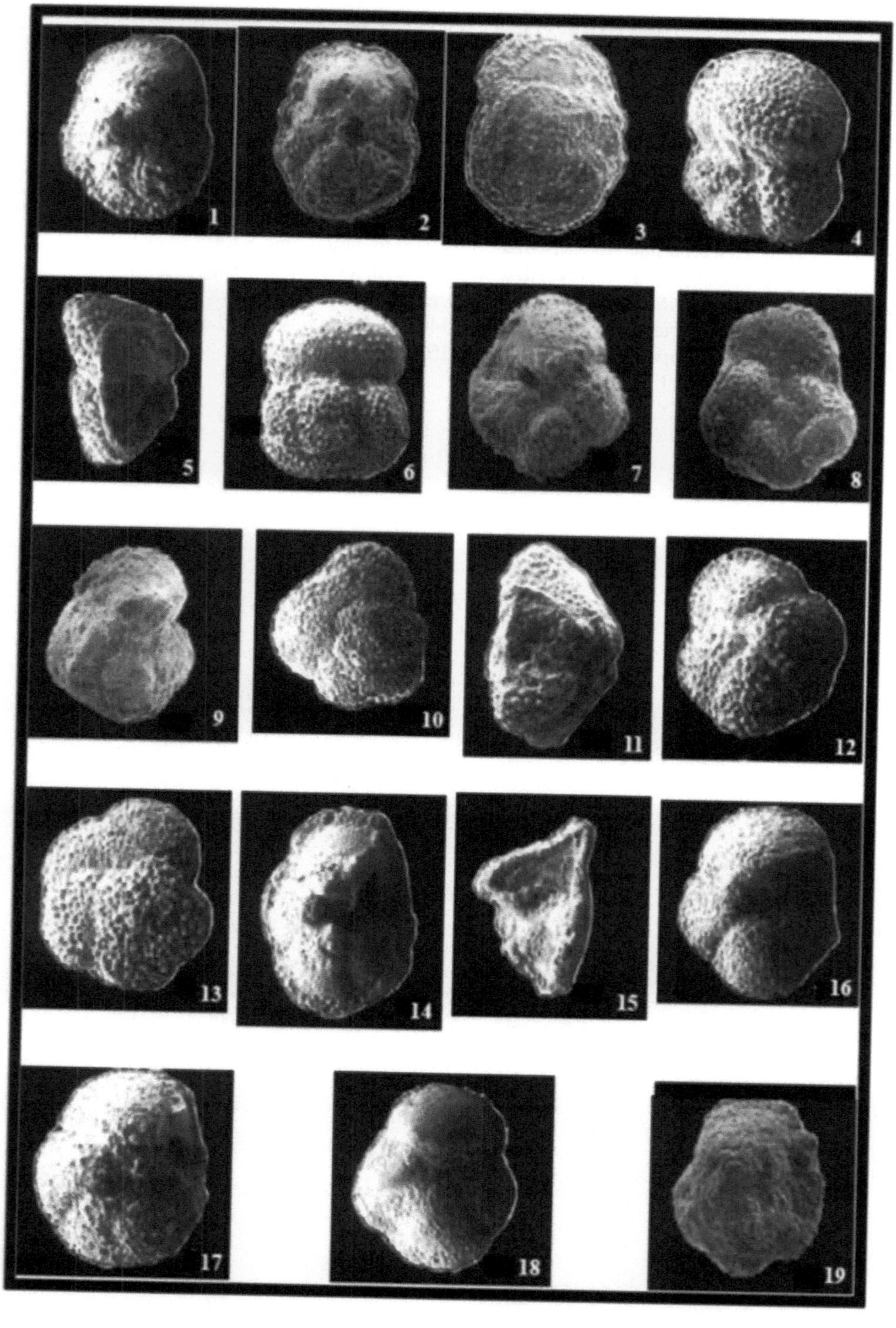
1
2
3
4
5
6
7
8
9
10
11
12
13
14
15
16
17
18
19

EXPLANATION OF PLATE 22

FIGURE 1 *Morozovella acuta* (Toulmin). Gbekebo-1 well, 2420 ft., umbilical view ×175.

FIGURE 2 *Morozovella acutispira* (Bolli and Cita). East Obe-1 well, 7776–7821 ft., umbilical view, ×152.

FIGURE 2 *Morozovella acutispira* (Bolli and Cita). East Obe-1 well, 7776–7821 ft., spiral view, ×152.

FIGURE 4 *Morozovella aequa* (Cushman and Renz). Gbekebo-1 well, 2370 ft., umbilical view, ×234.

FIGURE 5 *Morozovella aequa* (Cushman and Renz). Gbekebo-1 well, 2370 ft., edge view, ×195.

FIGURE 6 *Morozovella aequa* (Cushman and Renz). Gbekebo-1well, 2370 ft., spiral view, ×195.

FIGURE 7 *Morozovella angulata* (White). Baba-1 well, 8070–8130 ft. umbilical view, ×167.

FIGURE 8 *Morozovella angulata* (White). Baba-1 well, 8070–8130 ft., spiral view, ×123.

FIGURE 9 *Morozovella angulata* (White). Baba-1 well, 8070–8130 ft., umbilical view, ×126.

FIGURE 10 *Morozovella conicotruncata* (Subbotina). Araromi-1 well, 985 ft., spiral view, ×162.

FIGURE 11 *Morozovella conicotruncata* (Subbotina). Araromi-1 well, 985 ft., edge view, ×253.

FIGURE 12 *Morozovella conicotruncata* (Subbotina). Araromi-1 well, 985 ft., umbilical view, ×162.

FIGURE 13 *Morozovella edgari* (Premoli Silva and Bolli). Gbekebo-1 well, 2420 ft., spiral view, ×260.

FIGURE 14 *Morozovella gracilis* (Cushman). Araromi-1, 1090 ft., umbilical view, ×146.

FIGURE 15 *Morozovella gracilis* (Cushman). Araromi-1, 1090 ft., edge view, ×162.

FIGURE 16 *Morozovella lensiformis* (Subbotina). Gbekebo-1 well, 2420 ft., umbilical view, ×162.

FIGURE 17 *Morozovella lensiformis* (Subbotina). Gbekebo-1 well, 2420 ft., umbilical view, ×195.

FIGURE 18 *Morozovella lensiformis* (Subbotina). Araromi-1 well, 1350 ft., umbilical view, ×162.

FIGURE 19 *Morozovella marginodentata* (Subbotina). Baba-1 well, 8070–8130 ft., umbilical view, ×116.

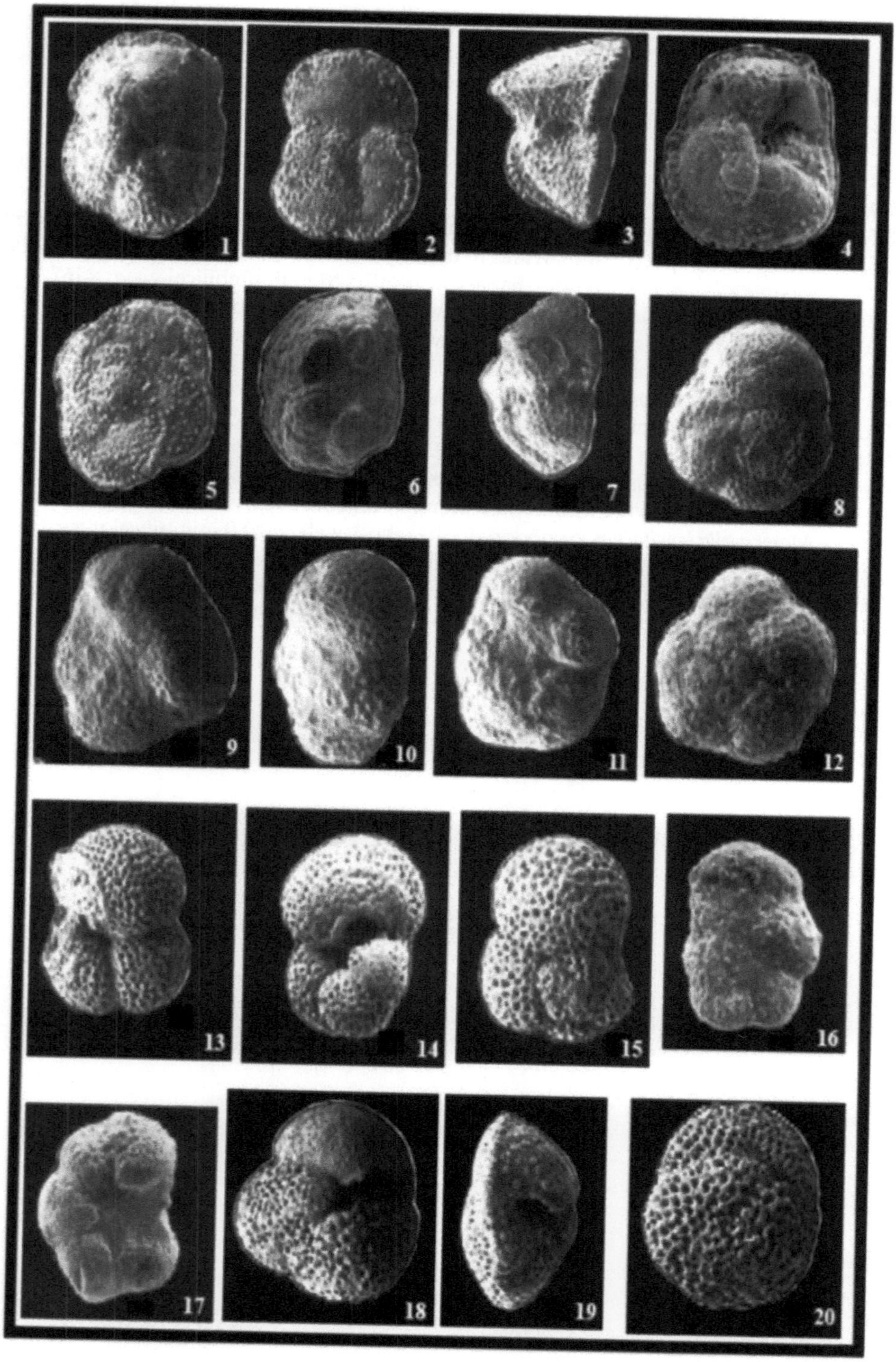
1
2
3
4
5
6
7
8
9
10
11
12
13
14
15
16
17
18
19
20

EXPLANATION OF PLATE 23

FIGURE 1 *Morozovella quetra* (Bolli). Araromi-1 well, 1350 ft., umbilical view, ×162.
FIGURE 2 *Morozovella quetra* (Bolli). Araromi-1 well, 1350 ft., spiral view, ×162.
FIGURE 3 *Morozovella quetra* (Bolli). Araromi-1 well, 1350 ft., edge view, ×162.
FIGURE 4 *Morozovella subbotinae* (Morozova). Araromi-1 well, 1264 ft., umbilical view, ×142.
FIGURE 5 *Morozovella velascoensis* (Cushman). Stuk-2 well, 3710 ft., spiral view, ×208.
FIGURE 6 *Morozovella velascoensis* (Cushman). Baba-1 well, 8070–8130 ft., umbilical view, ×153.
FIGURE 7 *Morozovella velascoensis* (Cushman). Ituk-2 well, 3710 ft., edge view; ×162.
FIGURE 8 *Subbotina corpulenta* (Subbotina). Gilli Gilli-1 well, 8295 ft., spiral view, ×280.
FIGURE 9 *Subbotina corpulenta* (Subbotina). Gilli Gilli-1 well, 8295 ft., umbilical view, ×188.
FIGURE 10 *Subbotina corpulenta* (Subbotina). Gilli Gilli-1 well, 8295 ft., edge view, ×215.
FIGURE 11 *Subbotina inconstans* (Subbotina). Gilli Gilli-1 well, 9020 ft., umbilical view, ×182.
FIGURE 12 *Subbotina inconstans* (Subbotina). Awaizombe-1 well, 9580–9600 ft., spiral view, ×182.
FIGURE 13 *Subbotina triangularis* (White). Araromi-1 well, 1045 ft., umbilical view, ×227.
FIGURE 14 *Subbotina triangularis* (White). Araromi-1 well, 1045 ft., edge view, ×260.
FIGURE 15 *Subbotina triangularis* (White). Araromi-1 well, 1045 ft., spiral view, ×260.
FIGURE 16 *Praemurica uncinata* (Bolli). Baba-1 well, 8070–8130 ft., spiral view, ×215.
FIGURE 17 *Praemurica uncinata* (Bolli). Baba-1 well, 8070–8130 ft., umbilical view, ×173.
FIGURE 18 *Igorina albeari* (Cushman and Bermudez). Gbekebo-1 well, 2360 ft., umbilical view, ×227.
FIGURE 19 *Igorina albeari* (Cushman and Bermudez). Gbekebo-1 well, 2360 ft., edge view, ×234.
FIGURE 20 *Igorina albeari* (Cushman and Bermudez). Gbekebo-1 well, 2360 ft., spiral view, ×260.

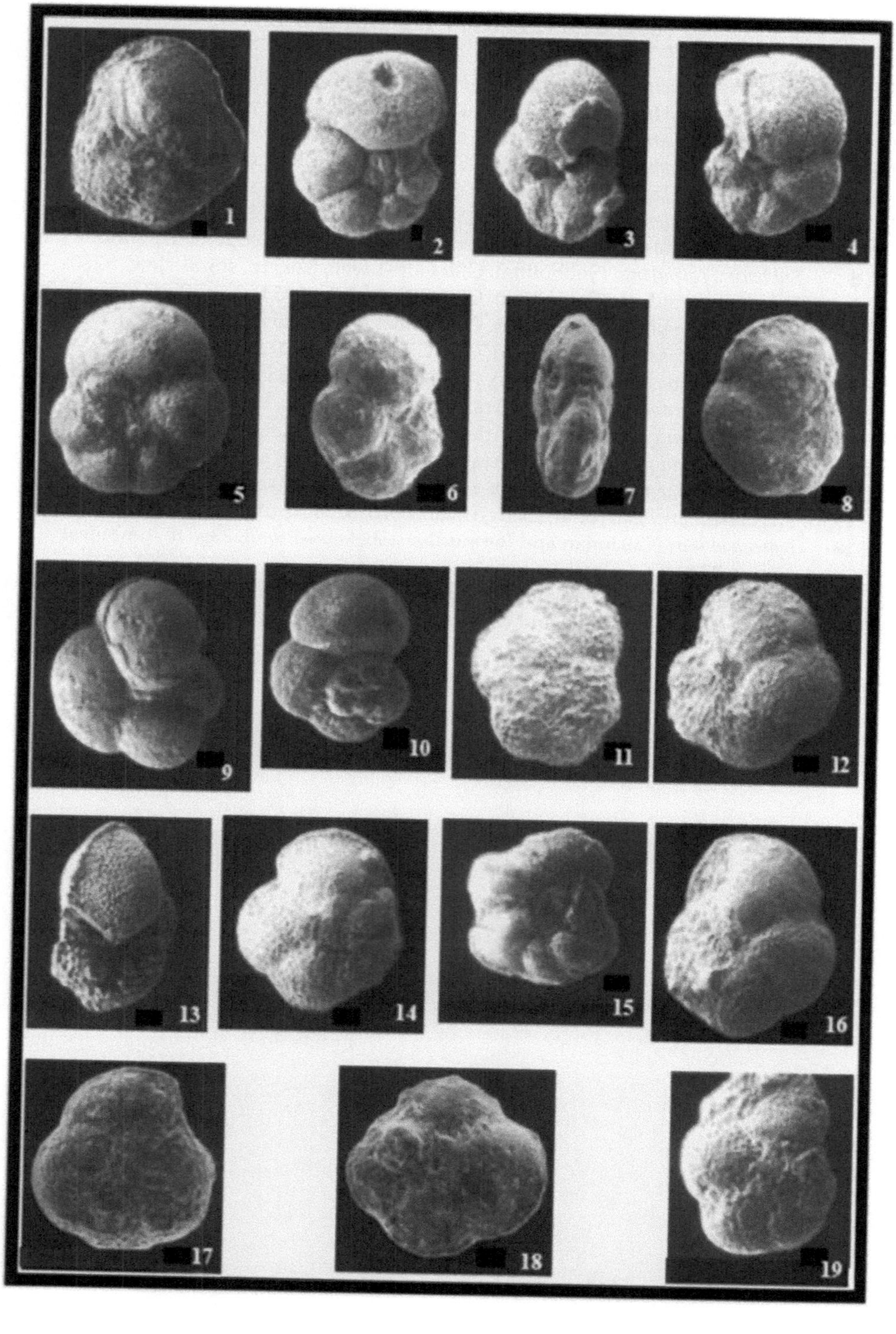
1
2
3
4
5
6
7
8
9
10
11
12
13
14
15
16
17
18
19

EXPLANATION OF PLATE 24

FIGURE 1	*Igorina pusilla* (Bolli). Ituk-2 well, 3710 ft., umbilical view, ×292.
FIGURE 2	*Globanomalina chapmani* (Parr). Araromi-1 well, 985 ft., spiral view, ×226.
FIGURE 3	*Globanomalina chapmani* (Parr). Araromi-1 well, 985 ft., edge view, ×208.
FIGURE 4	*Globanomalina chapmani* (Parr). Araromi-1 well, 985 ft., umbilical view, ×227.
FIGURE 5	*Globanomalina compressa* (Bolli). Gilli Gilli-1 well, 6939 ft., spiral view, ×240.
FIGURE 6	*Globanomalina compressa* (Bolli). Gilli Gilli-1 well, 6939 ft., umbilical view, ×200.
FIGURE 7	*Globanomalina compressa* (Bolli). Gilli Gilli-1 well, 6939 ft., edge view, ×195.
FIGURE 8	*Globanomalina compressa* (Bolli). Gilli Gilli-1 well, 6939 ft., umbilical view, ×197.
FIGURE 9	*Globanomalina imitata* (Subbotina). Gbekebo-1 well, 2330 ft., umbilical view, ×195.
FIGURE 10	*Globanomalina imitata* (Subbotina). Gbekebo-1 well, 2330 ft., spiral view, ×358.
FIGURE 11	*Globanomalina planoconica* (Subbotina). Gbekebo-1 well, 2370 ft., spiral view, ×325.
FIGURE 12	*Globanomalina planoconica* (Subbotina). Gbekebo-1 well, 2370 ft., umbilical view, ×260.
FIGURE 13	*Globanomalina pseudomenardii* (Bolli). Ikono-1 well, 2860 ft., edge view, ×260.
FIGURE 14	*Globanomalina pseudomenardii* (Bolli). Ikono-1 well, 2860 ft., umbilical view, ×260.
FIGURE 15	*Globanomalina pseudomenardii* (Bolli). Ikono-1 well, 2860 ft., spiral view, ×195.
FIGURE 16	*Globanomalina renzi* (Bolli). Gbekebo-1 well, 2370 ft., umbilical view, ×325.
FIGURE 17	*Globanomalina renzi* (Bolli). Gbekebo-1 well, 2370 ft., spiral view, ×296.
FIGURE 18	*Globanomalina renzi* (Bolli). Tonjor-1 well, 7500 ft., umbilical view, ×276.
FIGURE 19	*Globanomalina renzi* (Bolli). Tonjor-1 well, 7500 ft., spiral view, ×358.

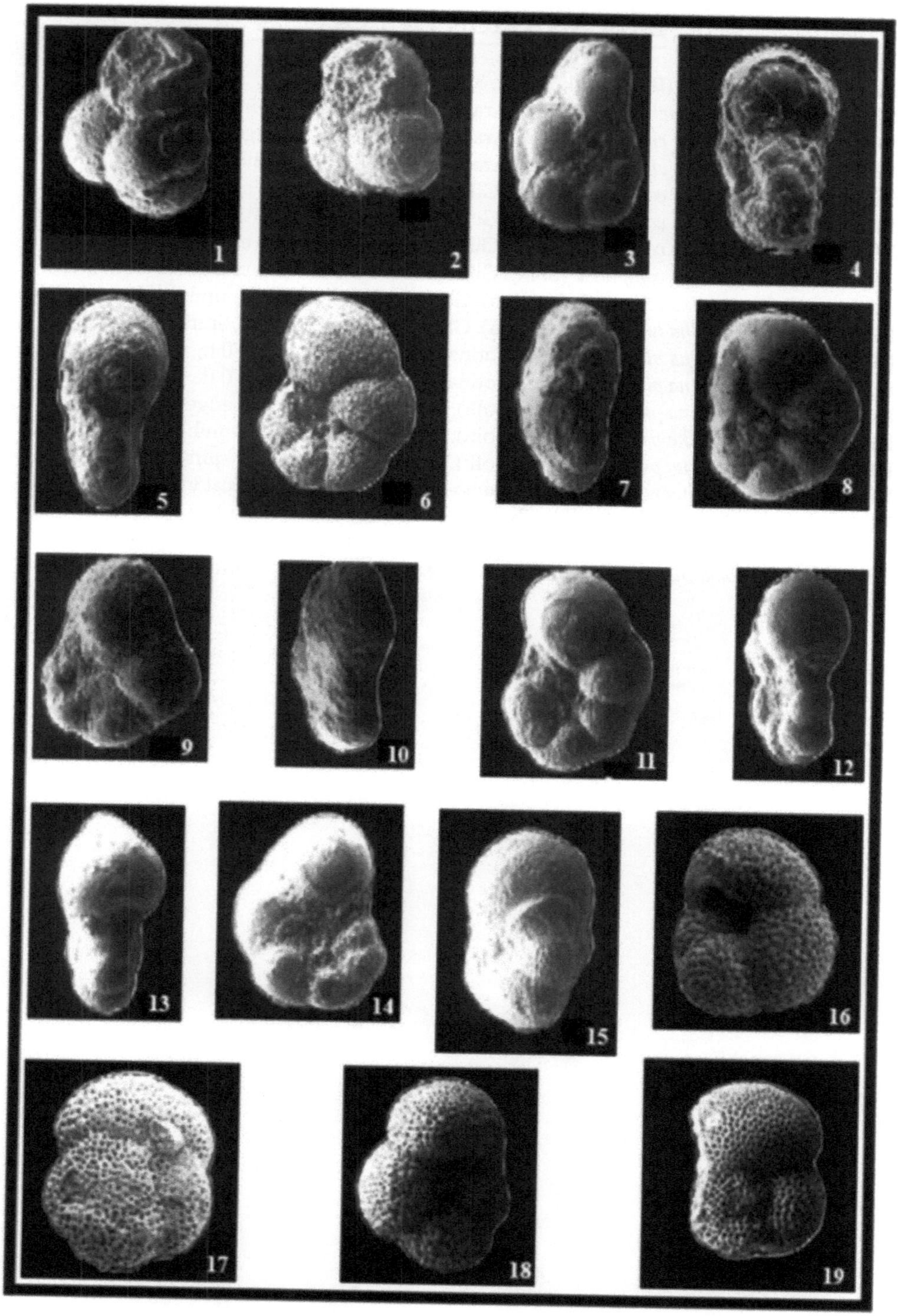
1
2
3
4
5
6
7
8
9
10
11
12
13
14
15
16
17
18
19

EXPLANATION OF PLATE 25

FIGURE 1	*Globoconusa daubjergensis* (Bronnimann). Araromi-1 well, 780 ft., spiral view, ×391.
FIGURE 2	*Globoconusa daubjergensis* (Bronnimann). Araromi-1 well, 780 ft., umbilical view, ×391.
FIGURE 3	*Pseudohastigerina barbadoensis* (Blow). Tonjor-1 well, 8110 ft., side view, ×227.
FIGURE 4	*Pseudohastigerina barbadoensis* (Blow). Gilli Gilli-1 well, 6939 ft., edge view, ×227.
FIGURE 5	*Pseudohastigerina danvillensis* (Howe and Wallace). Ihuo-1 well, 6017 ft., edge view, ×299.
FIGURE 6	*Pseudohastigerina danvillensis* (Howe and Wallace). Ihuo-1 well, 6017 ft., side view, ×273.
FIGURE 7	*Pseudohastigerina micra* (Cole). Gilli Gilli-1 well, 8820 ft., edge view, ×325.
FIGURE 8	*Pseudohastigerina micra* (Cole). Gilli Gilli-1 well, 8820 ft., side view, ×762.
FIGURE 9	*Pseudohastigerina* cf. *micra*. Gilli Gilli-1 well, 8820 ft., side view, ×195.
FIGURE 10	*Pseudohastigerina* cf. *micra*. Gilli Gilli-1 well, 8820 ft., edge view, ×279.
FIGURE 11	*Pseudohastigerina wilcoxensis* (Cushman and Ponton), Tonjor-1 well, 8110 ft., side view, ×292.
FIGURE 12	*Pseudohastigerina wilcoxensis* (Cushman and Ponton), Tonjor-1 well, 8110 ft., edge view, ×312.
FIGURE 13	*Pseudohastigerina* cf. *wilcoxensis*. Tonjor-1 well, 8110 ft., edge view, ×292.
FIGURE 14	*Pseudohastigerina* cf. *wilcoxensis*. Tonjor-1 well, 8110 ft., side view, ×325.
FIGURE 15	*"Hastigerina" bolivariana* (V. Petters). Awaizombe-1 well, 9780–9790 ft., edge view, ×169.
FIGURE 16	*Acarinina bullbrooki* (Bolli). Araromi-1 well, 852 ft., umbilical view, ×174.
FIGURE 17	*Acarinina bullbrooki* (Bolli). Araromi-1 well, 852 ft., spiral view, ×170.
FIGURE 18	*Acarinina esnaensis* (LeRoy). Benin West-1 well, 1079 ft., umbilical view, ×162.
FIGURE 19	*Acarinina esnaensis* (LeRoy). Benin West-1 well, 1079 ft., spiral view, ×162.

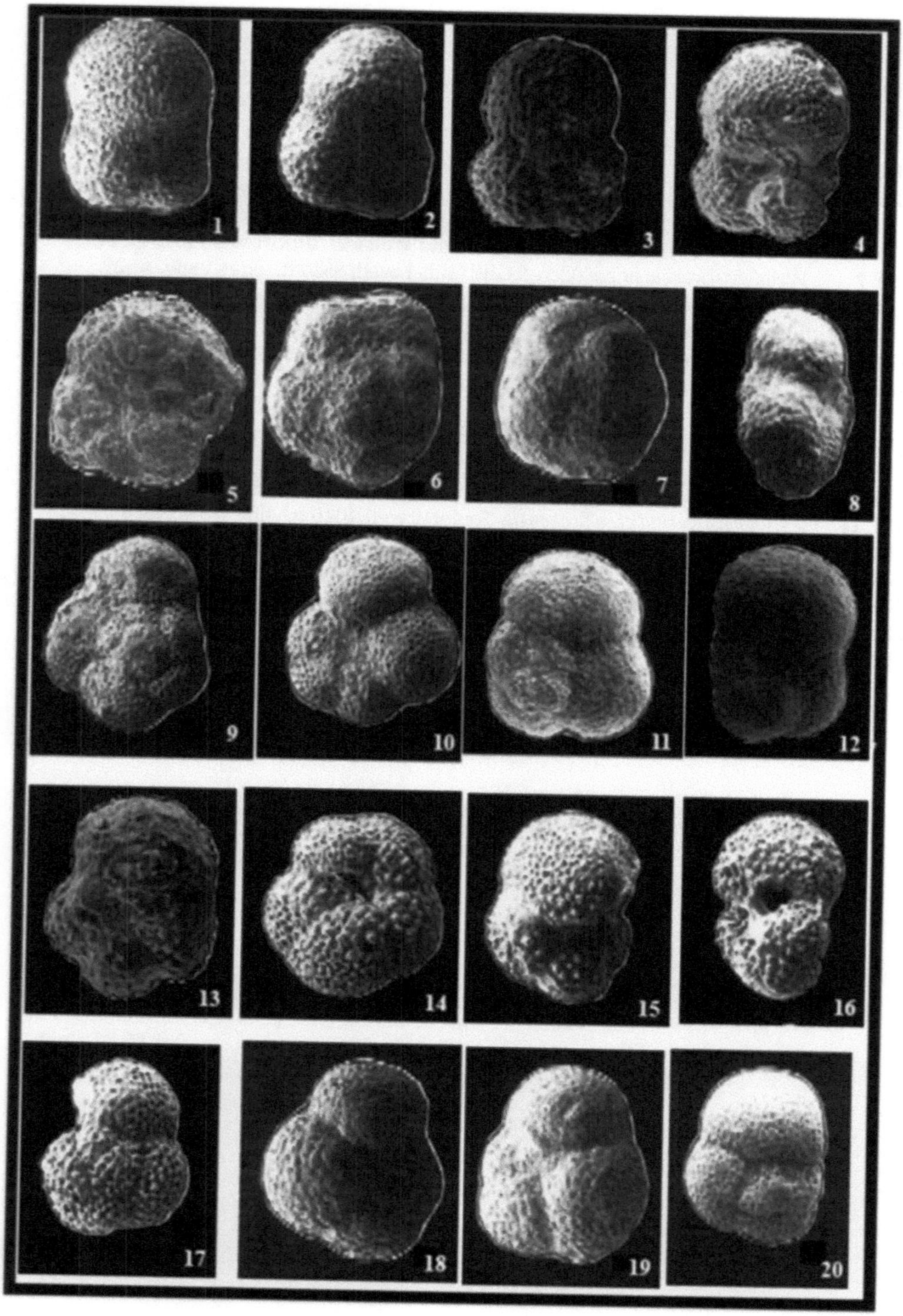
1
2
3
4
5
6
7
8
9
10
11
12
13
14
15
16
17
18
19
20

EXPLANATION OF PLATE 26

FIGURE 1	*Acarinina frontosa* (Subbotina). Araromi-1 well, 695 ft., umbilical view, ×195.
FIGURE 2	*Acarinina frontosa* (Subbotina). Araromi-1 well, 695 ft., spiral view, ×182.
FIGURE 3	*Acarinina inaequispira* (Subbotina). Ore-1 well, 372 ft., spiral view, ×190.
FIGURE 4	*Acarinina inaequispira* (Subbotina). Ore-1 well, 372 ft., umbilical view, ×169.
FIGURE 5	*Acarinina linaperta* (Subbotina). Ore-1 well, 372 ft., umbilical view, ×160.
FIGURE 6	*Acarinina mckannai* (White). Gbekebo-1 well, 2370 ft., spiral view, ×195.
FIGURE 7	*Acarinina mckannai* (White). Gbekebo-1 well, 2370 ft., umbilical view, ×162.
FIGURE 8	*Acarinina pentacamerata* (Subbotina). Ikono-1 well, 3060 ft., edge view, ×247.
FIGURE 9	*Acarinina pentacamerata* (Subbotina). Ikono-1 well, 3060 ft., spiral view, ×240.
FIGURE 10	*Acarinina pentacamerata* (Subbotina). Ikono-1 well, 3060 ft., umbilical view, ×240.
FIGURE 11	*Acarinina primitiva* (Finlay). Araromi-1 well, 924 ft., spiral view, ×239.
FIGURE 12	*Acarinina primitiva* (Finlay). Araromi-1 well, 924 ft., oblique umbilical view, ×187.
FIGURE 13	*Acarinina pseudotopilensis* (Subbotina). Ore-1 well, 372 ft., umbilical view, ×171.
FIGURE 14	*Acarinina strabocela* (Loeblich and Tappan). Gbekebo-1 well, 2370 ft., umbilical view, ×188.
FIGURE 15	*Acarinina tribulosa* (Loeblich and Tappan). Iju-1 well, 435 ft., umbilical view, ×195.
FIGURE 16	*Acarinina tribulosa* (Loeblich and Tappan). Iju-1 well, 435 ft., edge view, ×195.
FIGURE 17	*Acarinina tribulosa* (Loeblich and Tappan). Iju-1 well, 435 ft., spiral view, ×195.
FIGURE 18	*Acarinina wilcoxensis* (Cushman and Ponton). Ikang-1 well, 2980 ft., umbilical view, ×260.
FIGURE 19	*Acarinina wilcoxensis* (Cushman and Ponton). Ikang-1 well, 2980 ft., umbilical view, ×279.
FIGURE 20	*Acarinina wilcoxensis* (Cushman and Ponton). Ikang-1 well, 2980 ft., spiral view, ×260.

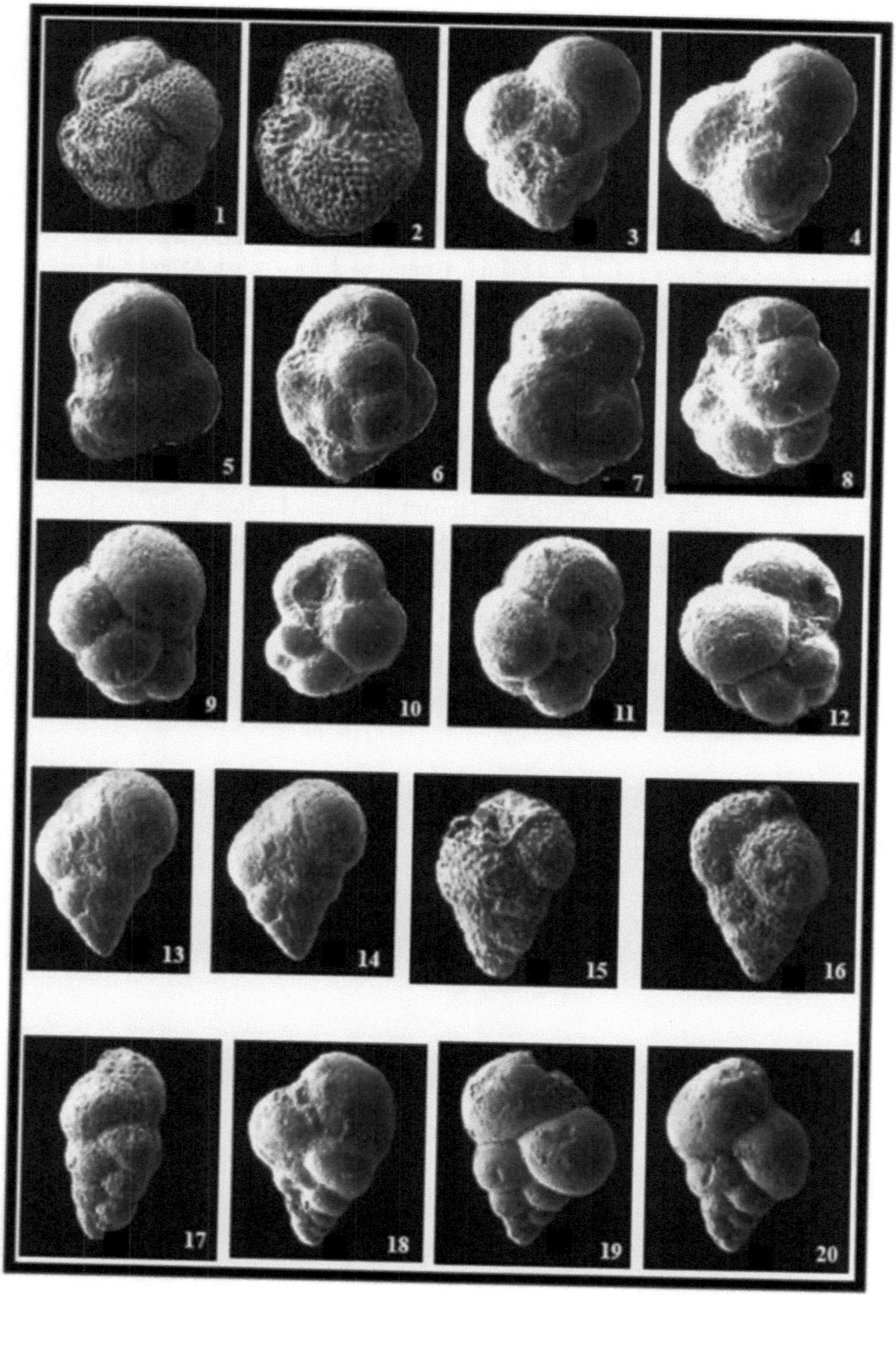
1
2
3
4
5
6
7
8
9
10
11
12
13
14
15
16
17
18
19
20

EXPLANATION OF PLATE 27

FIGURE 1 *Truncorotaloides rori* (Bronnimann and Bermudez). Gbekebo-1 well, 2370 ft., spiral view, ×162.

FIGURE 2 *Truncorotaloides rori* (Bronnimann and Bermudez). Gbekebo-1 well, 2370 ft., edge view, ×195.

FIGURE 3 *Guembelitria columbiana* (Howe). Tonjor-1 well, 8110 ft., umbilical view, ×486.

FIGURE 4 *Guembelitria columbiana* (Howe). Tonjor-1 well, 8110 ft., umbilical view, ×486.

FIGURE 5 *Guembelitria columbiana* (Howe). Tonjor-1 well, 8110 ft., spiral view, ×486.

FIGURE 6 *Cassigerinelloita amekiensis* (Stolk). Tonjor-1 well, 8110 ft., spiral view, ×389.

FIGURE 7 *Cassigerinelloita amekiensis* (Stolk). Tonjor-1 well, 8110 ft., umbilical view, ×455.

FIGURE 8 *Cassigerinella boudecensis* (Pokorny). Akata-1 well, 9703–9768 ft., side view, ×325.

FIGURE 9 *Cassigerinella boudecensis* (Pokorny). Akata-1 well, 9703–9768 ft., umbilical view, ×325.

FIGURE 10 *Cassigerinella chipolensis* (Cushman and Ponton). Obeaja-1 well, 5574 ft., umbilical view, ×422.

FIGURE 11 *Cassigerinella chipolensis* (Cushman and Ponton). Obeaja-1 well, 5574 ft., umbilical view, ×455.

FIGURE 12 *Cassigerinella chipolensis* (Cushman and Ponton). Obeaja-1 well, 5574 ft., oblique umbilical view, ×422.

FIGURE 13 *Chiloguembelina crinita* (Glaessner). Tonjor-1 well, 8020 ft., side view, ×240.

FIGURE 14 *Chiloguembelina crinita* (Glaessner). Tonjor-1 well, 8020 ft., side view, ×234.

FIGURE 15 *Chiloguembelina midwayensis* (Cushman). Tonjor-1 well, 8020 ft., side view, ×299.

FIGURE 16 *Chiloguembelina midwayensis* (Cushman). Tonjor-1 well, 8020 ft., side view, ×260.

FIGURE 17 *Chiloguembelina midwayensis* (Cushman). Tonjor-1 well, 8020 ft., side view, ×364.

FIGURE 18 *Chiloguembelina parallela* (Beckmann). Tonjor-1 well, 8110 ft., edge view, ×318.

FIGURE 19 *Chiloguembelina parallela* (Beckmann). Tonjor-1 well, 8110 ft., side view, ×305.

FIGURE 20 *Chiloguembelina parallela* (Beckmann). Tonjor-1 well, 8110 ft., side view, ×358.

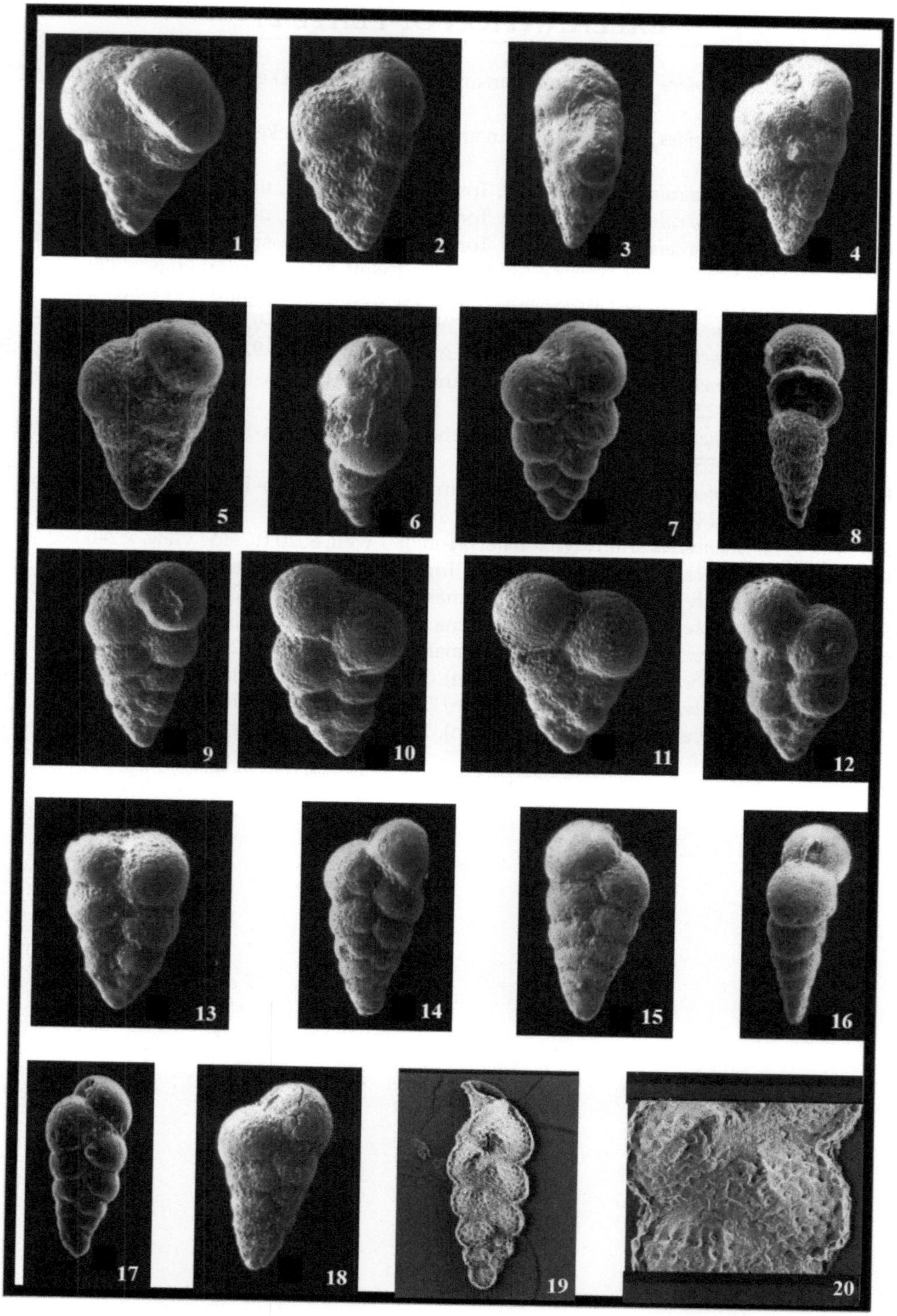
1
2
3
4
5
6
7
8
9
10
11
12
13
14
15
16
17
18
19
20

EXPLANATION OF PLATE 28

FIGURE 1 *Chiloguembelina strombiformis* (Beckmann). Tonjor-1 well, 8020 ft., side view, ×227.

FIGURE 2 *Chiloguembelina subtriangularis* (Beckmann). Baba-1 well, 8970–9030 ft., side view, ×325.

FIGURE 3 *Chiloguembelina subtriangularis* (Beckmann). Baba-1 well, 8970–9030 ft., edge view, ×325.

FIGURE 4 *Chiloguembelina subtriangularis* (Beckmann). Baba-1 well, 8970–9030 ft., side view, ×286.

FIGURE 5 *Chiloguembelina wilcoxensis* (Cushman and Ponton). Tonjor-1 well, 8110 ft., side view, ×455.

FIGURE 6 *Chiloguembelina wilcoxensis* (Cushman and Ponton). Tonjor-1 well, 8110 ft., edge view, ×358.

FIGURE 7 *Chiloguembelina wilcoxensis* (Cushman and Ponton). Tonjor-1 well, 8110 ft., side view, ×393.

FIGURE 8 *Chiloguembelina wilcoxensis* (Cushman and Ponton). Tonjor-1 well, 8110 ft., side view, ×390.

FIGURE 9 *Chiloguembelina wilcoxensis* (Cushman and Ponton). Tonjor-1 well, 8110 ft., side view, ×422.

FIGURE 10 *Streptochilus cubensis* (Palmer). Pologbene-1 well, 6090–6540 ft., side view, ×422.

FIGURE 11 *Streptochilus cubensis* (Palmer). Pologbene-1 well, 6090–6540 ft., side view, ×292.

FIGURE 12 *Streptochilus cubensis* (Palmer). Pologbene-1 well, 6090–6540 ft., side view, ×364.

FIGURE 13 *Streptochilus cubensis* (Palmer). Pologbene-1 well, 6090–6540 ft., side view, ×325.

FIGURE 14 *Streptochilus dubeyi* (Petters). Batan-1 well, 5100 ft., side view, ×279.

FIGURE 15 *Streptochilus dubeyi* (Petters). Batan-1 well, 5100 ft., side view, ×292.

FIGURE 16 *Streptochilus dubeyi* (Petters). Batan-1 well, 5100 ft., edge view, ×292.

FIGURE 17 *Streptochilus dubeyi* (Petters). Etelebou-1 well, 2080–2097 m, side view, ×260. Depths are in meters.

FIGURE 18 *Streptochilus dubeyi* (Petters). Batan-1 well, 5100 ft., side view, ×325.

FIGURE 19 (?)*Streptochilus* sp. 1. Ekanga-2 well, 670 m, side view, ×250.

FIGURE 20 (?)*Streptochilus* sp. 1. Ekanga-2 well, 670 m, same specimen, side view (enlarged), ×800.

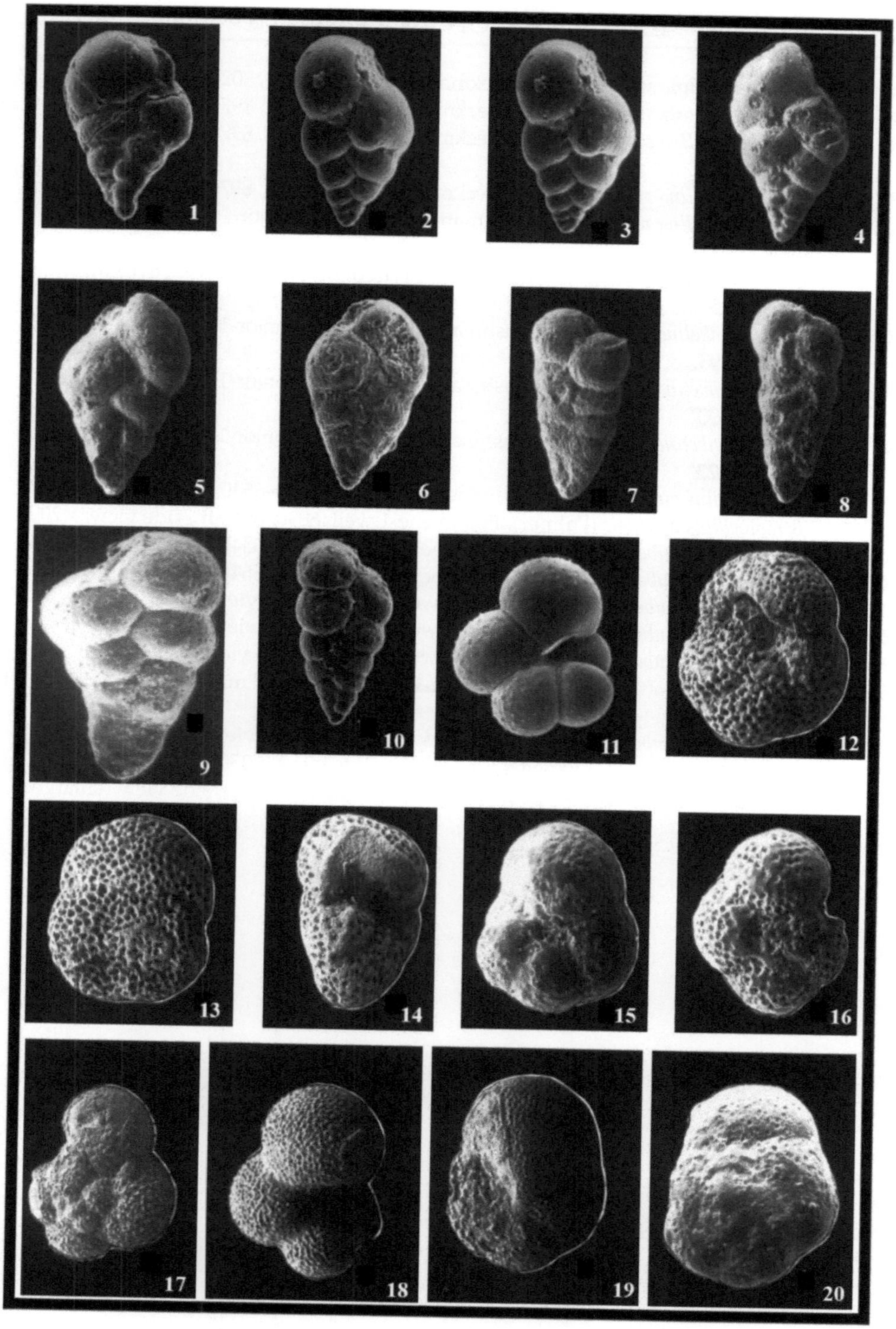
1
2
3
4
5
6
7
8
9
10
11
12
13
14
15
16
17
18
19
20

EXPLANATION OF PLATE 29

FIGURE 1 *Streptochilus heterostoma* (Bermudez). Tonjor-1 well, 8110 ft., side view, ×227.
FIGURE 2 *Streptochilus heterostoma* (Bermudez). Tonjor-1 well, 8110 ft., side view, ×209.
FIGURE 3 *Streptochilus heterostoma* (Bermudez). Tonjor-1 well, 8110 ft., side view, ×209.
FIGURE 4 *Streptochilus martini* (Pijpers). Tonjor-1 well, 8110 ft., side view, ×279.
FIGURE 5 *Streptochilus martini* (Pijpers). Tonjor-1 well, 8110 ft., side view, ×292.
FIGURE 6 *Streptochilus martini* (Pijpers). Tonjor-1 well, 8110 ft., side view, ×292.
FIGURE 7 *Streptochilus* cf. *multicellaris* (Hussey). Tonjor-1 well, 8020 ft., side view, ×286.
FIGURE 8 *Streptochilus* cf. *multicellaris* (Hussey). Tonjor-1 well, 8020 ft., side view, ×325.
FIGURE 9 *Streptochilus victoriana* (Beckman). Tuomo-2 well, 8045 ft., side view, ×358.
FIGURE 10 *Streptochilus victoriana* (Beckman). Pologbene-1 well, 6090–6540 ft., side view, ×325.
FIGURE 11 *Turborotalia anfracta* (Parker). Meren-1 well, 333 ft., umbilical view, ×326.
FIGURE 12 *Turborotalia cerroazulensis* (Cole). Tonjor-1 well, 8450 ft., umbilical view, ×227.
FIGURE 13 *Turborotalia cerroazulensis* (Cole). Tonjor-1 well, 8450 ft., spiral view, ×240.
FIGURE 14 *Turborotalia cerroazulensis* (Cole). Tonjor-1 well, 8450 ft., edge view, ×227.
FIGURE 15 *Turborotalia continuosa* (Blow). Baba-1 well, 2025–2079 ft., umbilical view, ×195.
FIGURE 16 *Turborotalia continuosa* (Blow). Baba-1 well, 2025–2079 ft., spiral view, ×208.
FIGURE 17 *Turborotalia danvillensis* (Howe and Wallace). Ewan-1 well, 6900–6960 ft., spiral view, ×208.
FIGURE 18 *Turborotalia danvillensis* (Howe and Wallace). Ewan-1 well, 6900–6960 ft., umbilical view, ×169.
FIGURE 19 *Turborotalia fohsi* (Cushman and Ellisor). Ewan-1 well, 6120-6180 ft., umbilical view, ×182.
FIGURE 20 *Turborotalia fohsi* (Cushman and Ellisor). Etelebou-1 well, 6930 ft., spiral view, ×195.

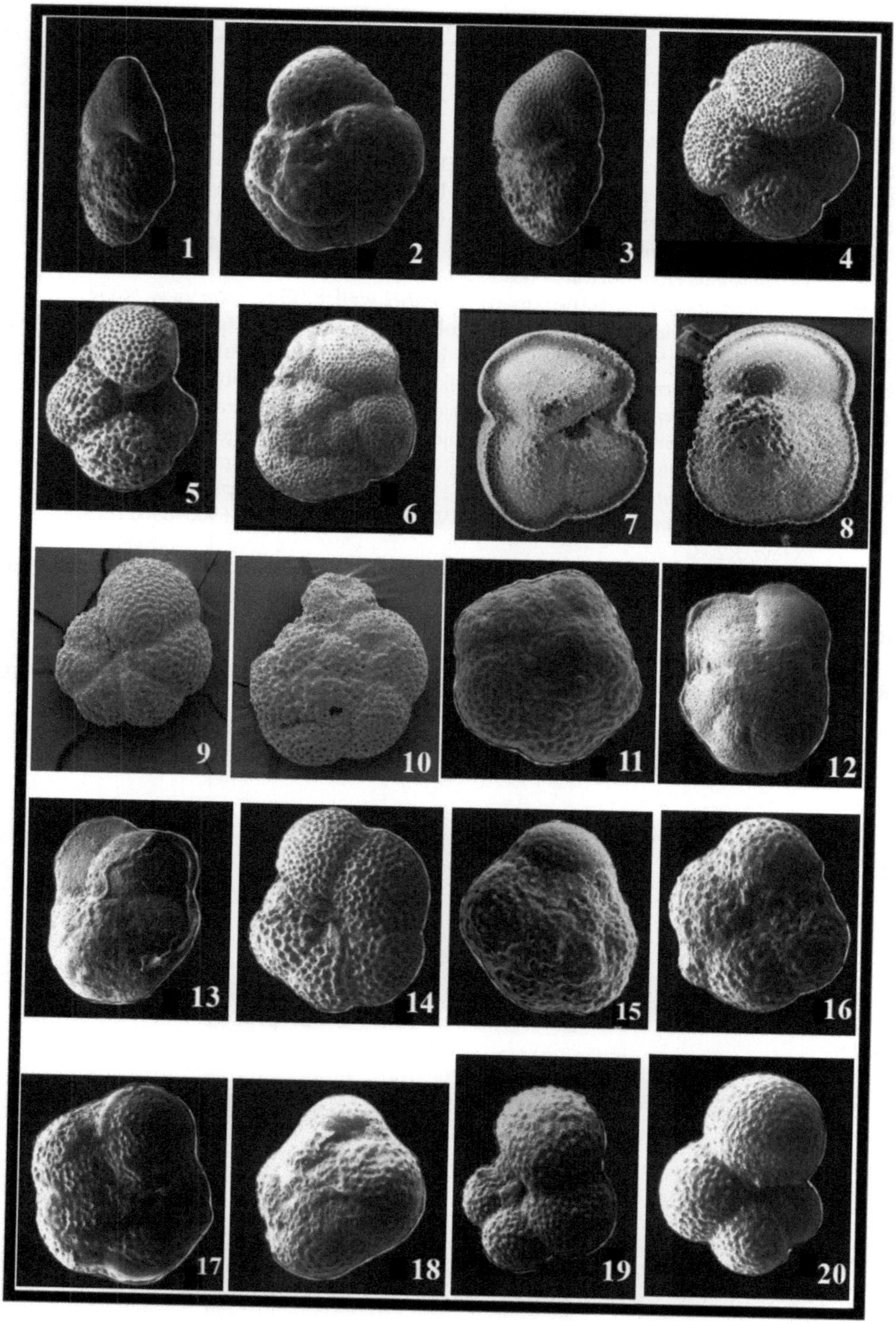
1
2
3
4
5
6
7
8
9
10
11
12
13
14
15
16
17
18
19
20

EXPLANATION OF PLATE 30

FIGURE 1 *Turborotalia fohsi* (Cushman and Ellisor). Ewan-1 well, 6120 ft., edge view, ×208.
FIGURE 2 *Turborotalia fohsi* (Cushman and Ellisor). Manuso-1 well, 6920 ft., spiral view, ×364.
FIGURE 3 *Turborotalia fohsi* (Cushman and Ellisor). Ewan-1 well, 6120–6180 ft., edge view, ×182.
FIGURE 4 *Turborotalia increbescens* (Bandy). Gbekebo-1 well, 1750 ft., umbilical view, ×160.
FIGURE 5 *Turborotalia increbescens* (Bandy). Gbekebo-1 well, 1750 ft., umbilical view, ×160.
FIGURE 6 *Turborotalia increbescens* (Bandy). Gbekebo-1 well, 1750 ft., spiral view, ×160.
FIGURE 7 *Turborotalia inflata* (d'Orbigny). Ekanga-2 well, 660 m, umbilical view, ×250.
FIGURE 8 *Turborotalia inflata* (d'Orbigny). Anyala-4 well, 870–930 ft., spiral view, ×300.
FIGURE 9 *Turborotalia kugleri* (Bolli). Benin River-3 well, 6550–6640, ft., umbilical view, ×300
FIGURE 10 *Turborotalia kugleri* (Bolli). Amakubu-1 well, 4970 ft., spiral view, ×350.
FIGURE 11 *Turborotalia kugleri* (Bolli). Gilli Gilli-1 well, 8820 ft., spiral view, ×303.
FIGURE 12 *Turborotalia lobata* (Bermudez). Bita-9 well, 9170 ft., umbilical view, ×190.
FIGURE 13 *Turborotalia lobata* (Bermudez). Bita-9 well, 9170 ft., spiral view, ×180.
FIGURE 14 *Turborotalia mayeri* (Cushman and Ellisor). Pologbene-1 well, 6090–6540 ft., umbilical view, ×260.
FIGURE 15 *Turborotalia nana* (Bolli). Gilli Gilli-1 well, 8820 ft., spiral view, ×312.
FIGURE 16 *Turborotalia nana* (Bolli). Meta-1 well, 9440–9500 ft., umbilical view, ×325.
FIGURE 17 *Turborotalia nana* (Bolli). Meta-1 well, 10,700–10,760 ft., umbilical view, ×208.
FIGURE 18 *Turborotalia nana* (Bolli). Gilli Gilli-1 well, 8820 ft., umbilical view, ×260.
FIGURE 19 *Turborotalia obesa* (Bolli). Pologbene-1 well, 6090–6540 ft., umbilical view, ×195.
FIGURE 20 *Turborotalia obesa* (Bolli). Pologbene-1 well, 6090–6540 ft., umbilical view, ×325.

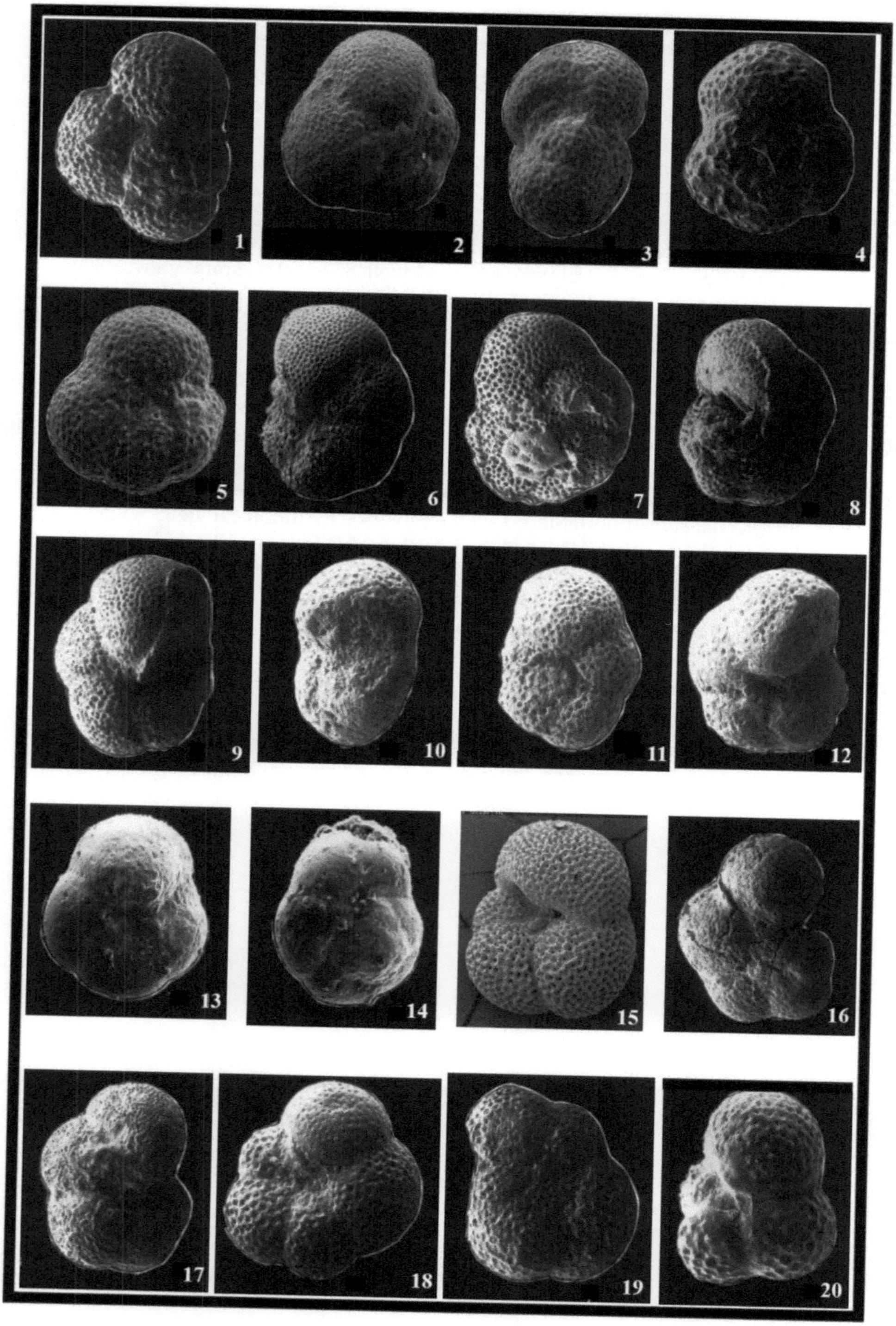
1
2
3
4
5
6
7
8
9
10
11
12
13
14
15
16
17
18
19
20

EXPLANATION OF PLATE 31

FIGURE 1	*Turborotalia obesa* (Bolli). Pologbene-1 well, 6090–6450 ft., spiral view, ×325.
FIGURE 2	*Turborotalia opima* (Bolli). Meta-1 well, 9440–9500 ft., umbilical view, ×162.
FIGURE 3	*Turborotalia opima* (Bolli). Gilli Gilli-1 well, 8775 ft., edge view, ×130.
FIGURE 4	*Turborotalia opima* (Bolli). Meta-1 well, 9440–9500 ft., spiral view, ×352.
FIGURE 5	*Turborotalia opima* (Bolli). Meta-1 well, 9440–9500 ft., umbilical view, ×260.
FIGURE 6	*Turborotalia peripheroacuta* (Blow and Banner). Ewan-1 well, 6120–6180 ft., umbilical view, ×162.
FIGURE 7	*Turborotalia peripheroacuta* (Blow and Banner). Meta-1 well, 10,220–10,280 ft., spiral view, ×324.
FIGURE 8	*Turborotalia peripheroronda* (Blow and Banner). Ewan-1 well, 6540–6600 ft., oblique umbilical view, ×195.
FIGURE 9	*Turborotalia peripheroronda* (Blow and Banner). Etelebou-1 well, 6930 ft., umbilical view, ×195.
FIGURE 10	*Turborotalia peripheroronda* (Blow and Banner). Utapate-1 well, 10,100 ft., edge view, ×260.
FIGURE 11	*Turborotalia peripheroronda* (Blow and Banner). Utapate-1 well, 10,100 ft., spiral view, ×195.
FIGURE 12	*Turborotalia peripheroronda* (Blow and Banner). Utapate-1 well, 10,100 ft., umbilical view, ×273.
FIGURE 13	*Turborotalia pomeroli* (Tourmarkine and Bolli). Baba-1 well, 8070–8130 ft., umbilical view, ×207.
FIGURE 14	*Turborotalia pomeroli* (Tourmarkine and Bolli). Baba-1 well, 8070–8130 ft., spiral view, ×207.
FIGURE 15	*Turborotalia pseudopima* (Blow). Akamba-1 well, 150–170 m, umbilical view, ×175.
FIGURE 16	*Turborotalia pseudopima* (Blow). Pologbene-1 well, 6090–6540 ft., spiral view, ×236.
FIGURE 17	*Turborotalia pseudopima* (Blow). Pologbene-1 well, 6090–6540 ft., umbilical view, ×236.
FIGURE 18	*Turborotalia siakensis* (LeRoy). Meta-1 well, 9440–9500 ft., umbilical view, ×162.
FIGURE 19	*Turborotalia siakensis* (LeRoy). Meta-1 well, 9440–9500 ft., spiral view, ×160.
FIGURE 20	*Turborotalia subcretacea* (Lomnicki). Opuama-1 well, 4224 ft., umbilical view, ×292.

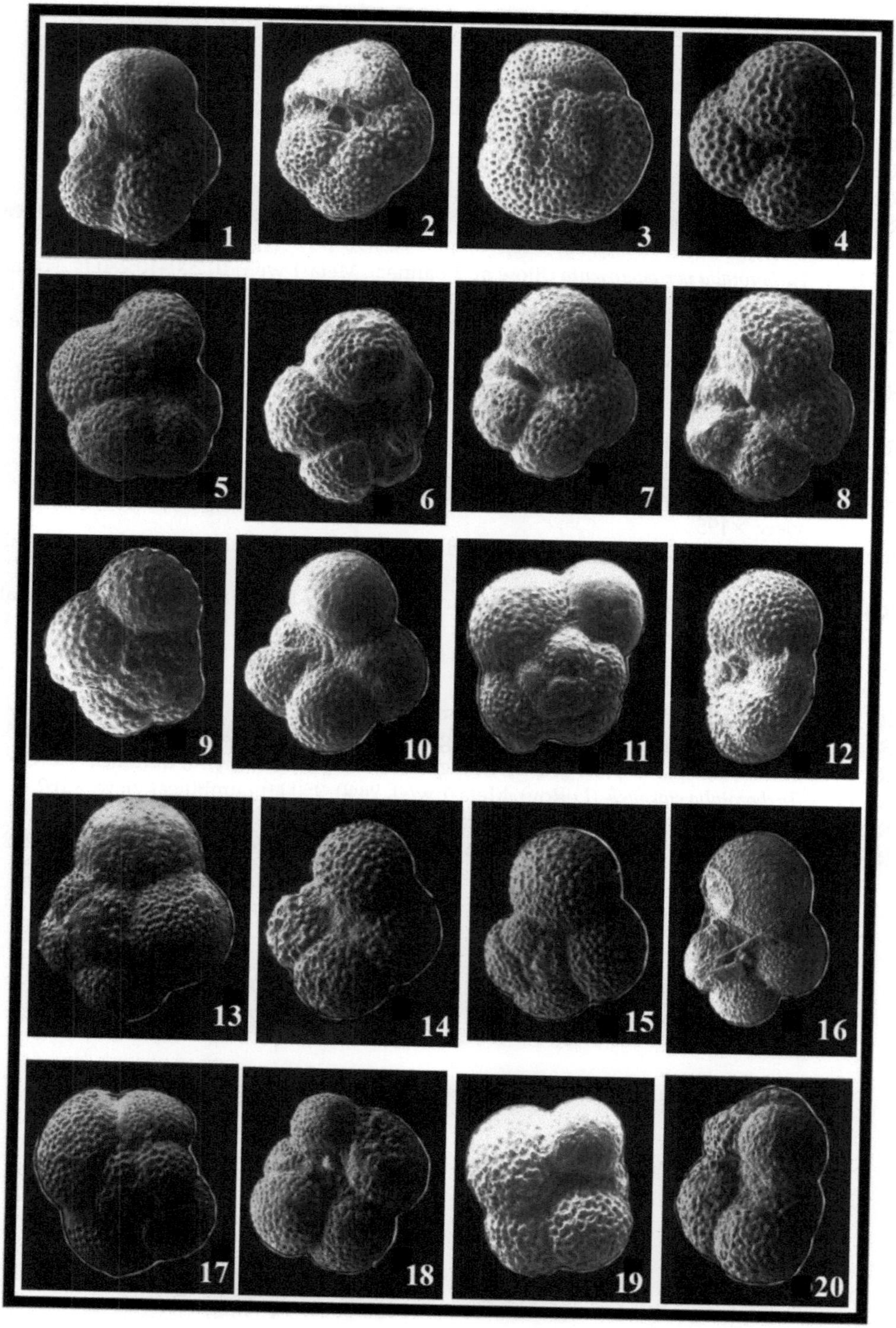
1
2
3
4
5
6
7
8
9
10
11
12
13
14
15
16
17
18
19
20

EXPLANATION OF PLATE 32

FIGURE 1 *Tenuitella munda* (Jenkins). Pologbene-1 well, 6090–6540 ft., umbilical view, ×201.
FIGURE 2 *Globigerina ampliapertura* (Bolli). Gbekebo-1 well, 2330 ft., umbilical view, ×162.
FIGURE 3 *Globigerina ampliapertura* (Bolli). Gbekebo-1 well, 2330 ft., spiral view, ×195.
FIGURE 4 *Globigerina angiporoides* (Hornibrook). Ewan-1, 6540–660 ft., umbilical view, ×227.
FIGURE 5 *Globigerina anguliofficinalis* (Blow). Ewan-1, 6900–6960 ft., umbilical view, ×201.
FIGURE 6 *Globigerina angulisuturalis* (Bolli). Ewan-1, 6900–6960 ft., umbilical view, ×201.
FIGURE 7 *Globigerina* aff. *angulisuturalis*. Obeaja-1 well, 7055 ft., umbilical view, ×358.
FIGURE 8 *Globigerina* aff. *angulisuturalis*. Obeaja-1 well, 7055 ft., umbilical view, ×358.
FIGURE 9 *Globigerina* aff. *angulisuturalis*. Obeaja-1 well, 7055 ft., spiral view, ×358.
FIGURE 10 *Globigerina angustumbilicata* (Bolli). Batan-1 well, 10,900 ft., umbilical view, ×273.
FIGURE 11 *Globigerina angustumbilicata* (Bolli). Batan-1 well, 10,900 ft., spiral view, ×260.
FIGURE 12 *Globigerina angustumbilicata* (Bolli). Batan-1 well, 10,900 ft., edge view of same sample, ×260.
FIGURE 13 *Globigerina* aff. *angustumbilicata*. Ewan-1 well, 7020–7080 ft., spiral view, ×279.
FIGURE 14 *Globigerina* aff. *angustumbilicata*. Ewan-1 well, 7020–7080 ft., umbilical view, ×279.
FIGURE 15 *Globigerina bulloides* (d'Orbigny). Akata-1 well, 9703–9768 ft., spiral view, ×162.
FIGURE 16 *Globigerina bulloides* (d'Orbigny). Akata-1 well, 9703–9768 ft., umbilical view, ×149.
FIGURE 17 *Globigerina ciperoensis* (Bolli). Opuekeba-1 well, 7640–7740 ft., spiral view, ×227.
FIGURE 18 *Globigerina ciperoensis* (Bolli). Pologbene-1 well, 6090–6540 ft., umbilical view, ×312.
FIGURE 19 *Globigerina ciperoensis* (Bolli). Ewan-1 well, 6090–6960 ft., umbilical view, ×325.
FIGURE 20 *Globigerina ciperoensis* (Bolli). Opuekeba-1 well, 7640–7700 ft., edge view, ×227.

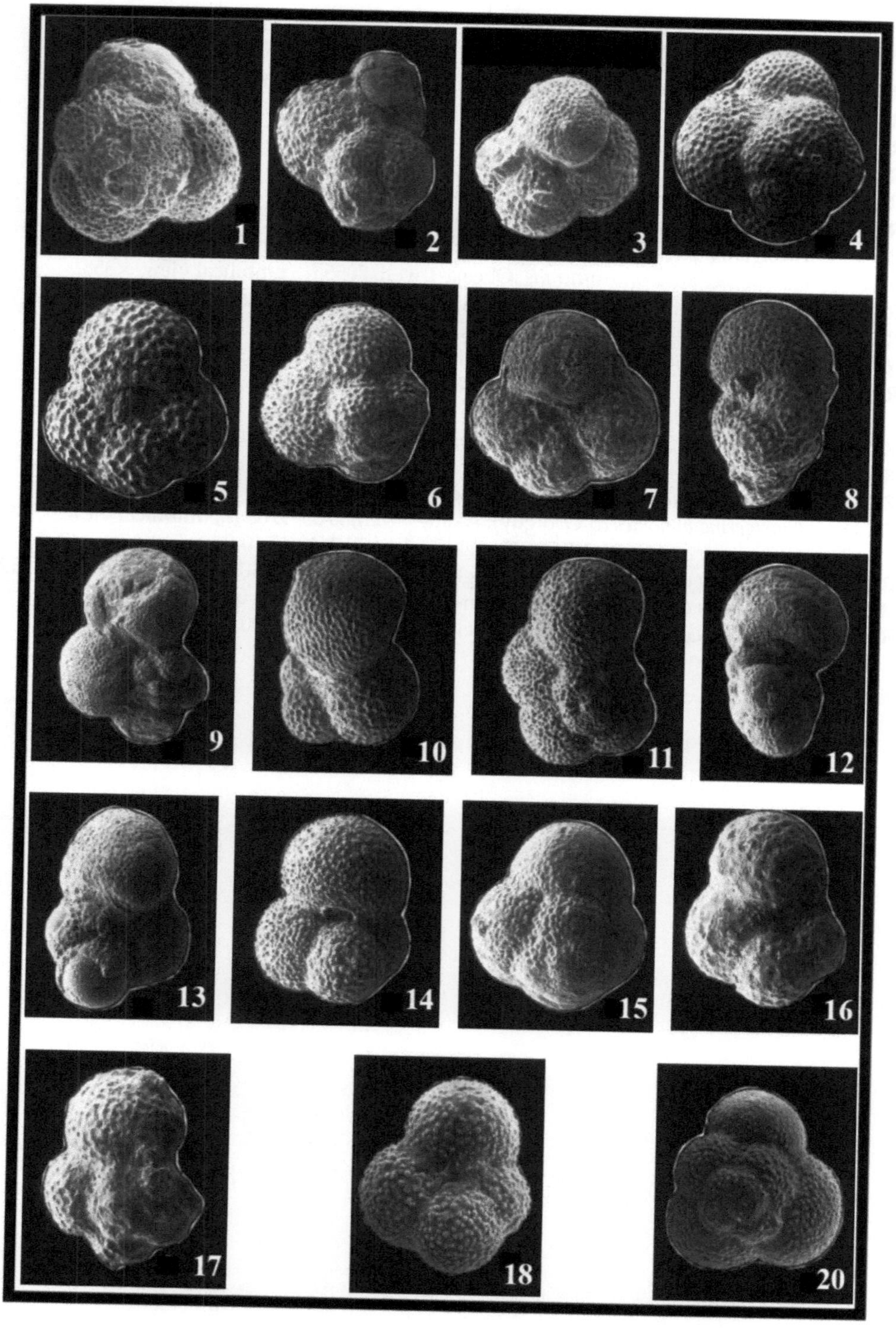
1
2
3
4
5
6
7
8
9
10
11
12
13
14
15
16
17
18
20

EXPLANATION OF PLATE 33

FIGURE 1	*Globigerina eamsi* (Blow). Meta-1 well, 5373–5409 ft., spiral view, ×157.
FIGURE 2	*Globigerina eamsi* (Blow). Akata-1 well, 9703–9768 ft., spiral view, ×149.
FIGURE 3	*Globigerina eamsi* (Blow). Akata-1 well, 9703–9768 ft., umbilical view, ×162.
FIGURE 4	*Globigerina euapertura* (Jenkins). Ewan-1 well, 6540–6600 ft., spiral view, ×182.
FIGURE 5	*Globigerina euapertura* (Jenkins). Ewan-1 well, 6540–6600 ft., umbilical view, ×305.
FIGURE 6	*Globigerina falconensis* (Blow). Akata-1 well, 9703–9768 ft., spiral view, ×195.
FIGURE 7	*Globigerina falconensis* (Blow). Akata-1 well, 9703–9768 ft., umbilical view, ×292.
FIGURE 8	*Globigerina falconensis* (Blow). Akata-1 well, 9703–9768 ft., edge view, ×279.
FIGURE 9	*Globigerina foliata* (Bolli). Opuekeba-1 well, 9870–9930 ft., spiral view, ×149.
FIGURE 10	*Globigerina foliata* (Bolli). Opuekeba-1 well, 9870–9930 ft., oblique umbilical view, ×162.
FIGURE 11	*Globigerina foliata* (Bolli). Opuama-1 well, 4234 ft., spiral view, ×188.
FIGURE 12	*Globigerina foliata* (Bolli). Opuekeba-1 well, 9870–9930 ft., edge view, ×182.
FIGURE 13	*Globigerina foliata* (Bolli). Opuama-1 well, 4224 ft., umbilical view, ×175.
FIGURE 14	*Globigerina galvis* (Bermudez). Akata-1 well, 9703–9768 ft., umbilical view, ×389.
FIGURE 15	*Globigerina galvis* (Bermudez). Akata-1 well, 9703–9768 ft., spiral view, ×488.
FIGURE 16	*Globigerina gnaucki* (Blow and Banner). Idu-2 well, 9450 ft., umbilical view, ×292.
FIGURE 17	*Globigerina gnaucki* (Blow and Banner). Idu-2 well, 9450 ft., spiral view, ×292.
FIGURE 18	*Globigerina juvenelis* (Bolli). Malu-1 well, 6912–6948 ft., umbilical view, ×337.
FIGURE 19	*Globigerina juvenelis* (Bolli). Malu-1 well, 6912–6948 ft., spiral view, ×337.

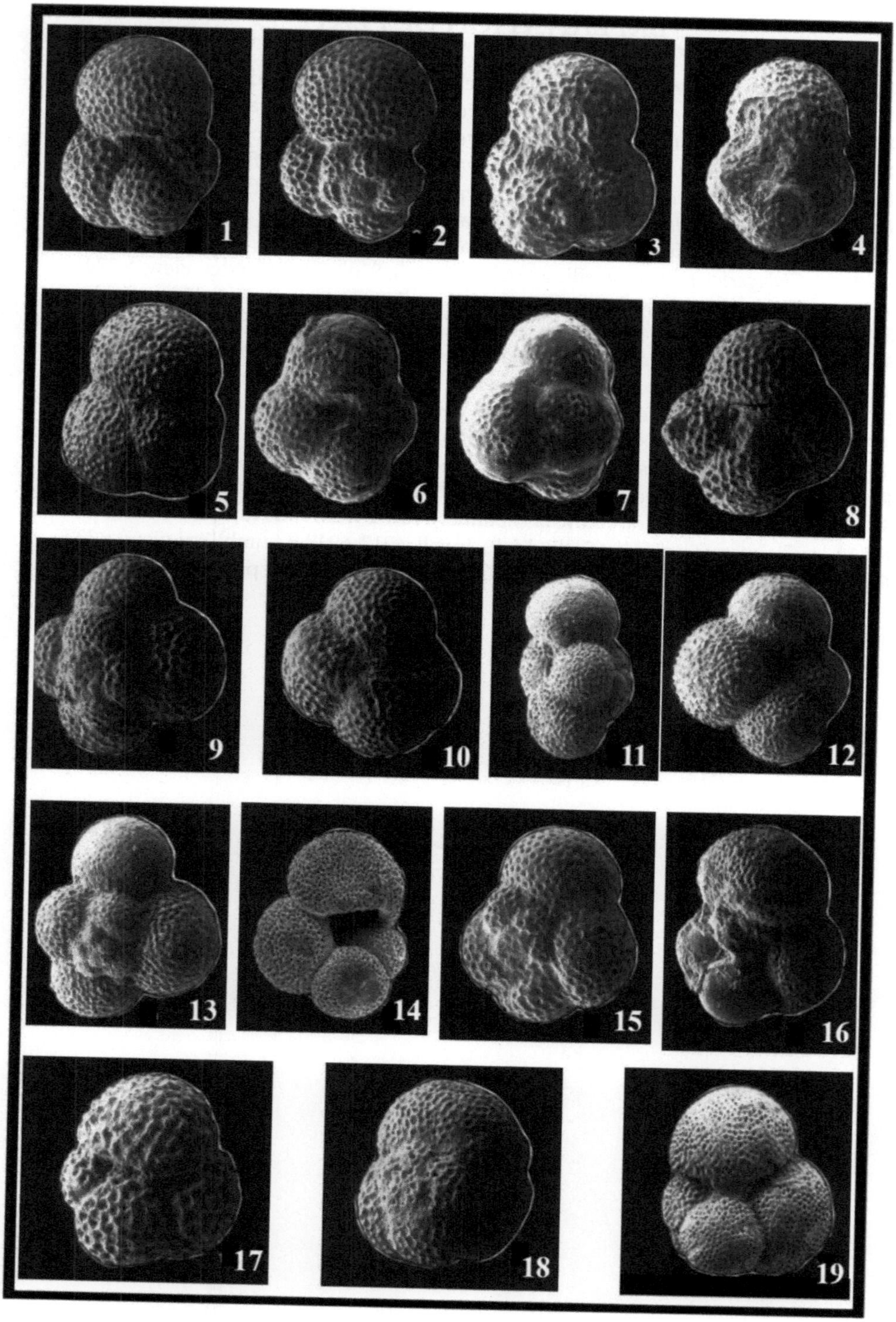
1
2
3
4
5
6
7
8
9
10
11
12
13
14
15
16
17
18
19

EXPLANATION OF PLATE 34

FIGURE 1	*Globigerina leroyi* (Blow and Banner). Batan-1 well, 10,900 ft., umbilical view, ×169.
FIGURE 2	*Globigerina leroyi* (Blow and Banner). Batan-1 well, 10,900 ft., spiral view, ×181.
FIGURE 3	*Globigerina occlusa* (Blow and Banner). Pologbene-1 well, 6090–6540 ft., umbilical view, ×325.
FIGURE 4	*Globigerina occlusa* (Blow and Banner). Obiafu-1 well, 8370 ft., umbilical view, ×344.
FIGURE 5	*Globigerina occlusa* (Blow and Banner). Ewan-1 well, 6900–6960 ft., spiral view, ×182.
FIGURE 6	*Globigerina officinalis* (Subbotina). Parabe-1 well, 7340 ft., umbilical view, ×234.
FIGURE 7	*Globigerina officinalis* (Subbotina). Parabe-1 well, 7340 ft., spiral view, ×234.
FIGURE 8	*Globigerina ouachitaensis* (Howe and Wallace). Ewan-1 well, 6900–6960 ft., umbilical view, ×208.
FIGURE 9	*Globigerina ouachitaensis* (Howe and Wallace). Ewan-1 well, 6120–6180 ft., spiral view, ×247.
FIGURE 10	*Globigerina ouachitaensis* (Howe and Wallace). Ewan-1 well, 6120–6180 ft., umbilical view, ×188.
FIGURE 11	*Globigerina* cf. *ouachitaensis*. Obiafu-13 well, 8370 ft., edge view, ×214.
FIGURE 12	*Globigerina* cf. *ouachitaensis*. Obiafu-13 well, 8370 ft., umbilical view, ×195.
FIGURE 13	*Globigerina* cf. *ouachitaensis*. Obiafu-13 well, 8370 ft., spiral view, ×260.
FIGURE 14	*Globigerina praebulloides* (Blow). Sonam-9 well, 4194–4257 ft., umbilical view, ×143.
FIGURE 15	*Globigerina praebulloides* (Blow). Pologbene-1 well, 6090–6540 ft., spiral view, ×292.
FIGURE 16	*Globigerina praebulloides* (Blow). Opuama-1 well, 4224 ft., umbilical view, ×195.
FIGURE 17	*Globigerina prasaepis* (Blow). Ewan-1 well, 6900–6950 ft., umbilical view, ×292.
FIGURE 18	*Globigerina prasaepis* (Blow). Ewan-1 well, 6900–6950 ft., spiral view, ×156.
FIGURE 19	*Globigerina prasaepis* (Blow). Tuomo-1 well, 7290–7320 ft., umbilical view, ×180.

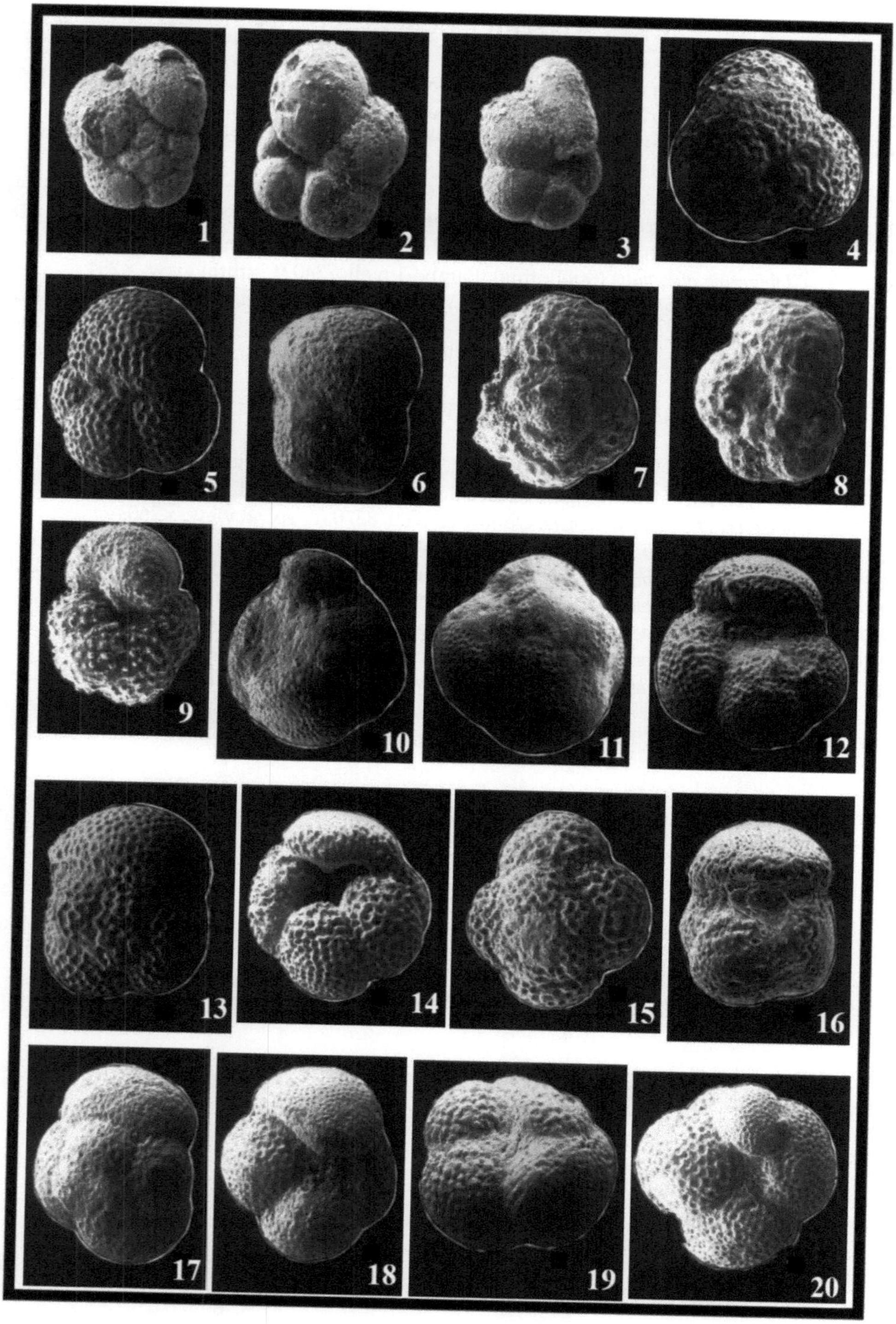
1
2
3
4
5
6
7
8
9
10
11
12
13
14
15
16
17
18
19
20

EXPLANATION OF PLATE 35

FIGURE 1	*Globigerina quinqueloba* (Natland). Ubulu-1 well, 7520 ft., spiral view, ×325.
FIGURE 2	*Globigerina quinqueloba* (Natland). Ubulu-1 well, 7520 ft., umbilical view, ×389.
FIGURE 3	*Globigerina quinqueloba* (Natland). Ubulu-1 well, 7520 ft., umbilical view, ×332.
FIGURE 4	*Globigerina yeguaensis* (Weinzie and Applin). Ewan-1 well, 6540–6600 ft., spiral view, ×221.
FIGURE 5	*Globigerina yeguaensis* (Weinzie and Applin). Ewan-1 well, 6540–6600 ft., umbilical view, ×201.
FIGURE 6	*Globigerina utilisindex* (Jenkins and Orr). Baba-1 well, 8610–8670 ft., umbilical view, ×116
FIGURE 7	*Globorotaloides suteri* (Bolli). Obeaja-1 well, 7055 ft., spiral view; ×488.
FIGURE 8	*Globorotaloides suteri* (Bolli). Obeaja-1 well, 7055 ft., spiral view; ×455.
FIGURE 9	*Globorotaloides suteri* (Bolli). Obeaja-1, well 7055 ft., umbilical view; ×455.
FIGURE 10	*Globoquadrina altispira* (Bermudez). Malu-2 well, 8650–8680 ft., umbilical view, ×130.
FIGURE 11	*Globoquadrina altispira* (Bermudez). Malu-2 well, 8650–8680 ft., spiral view, ×130.
FIGURE 12	*Globoquadrina dehiscens* (Chapman). Ewan-1 well, 6120–6180 ft., umbilical view, ×136.
FIGURE 13	*Globoquadrina dehiscens* (Chapman). Ewan-1 well, 6120–6180 ft., spiral view, ×201.
FIGURE 14	*Globoquadrina globosa* (Bolli). Meta-1 well, 9440–9500 ft., umbilical view, ×175.
FIGURE 15	*Globoquadrina globosa* (Bolli). Ewan-1 well, 6540–6600 ft., spiral view, ×336.
FIGURE 16	*Globoquadrina praedehiscens* (Blow and Banner). Ewan-1 well, 6120–6180 ft., umbilical view, ×110.
FIGURE 17	*Globoquadrina venezuelana* (Hedberg). Opuekeba-1 well, 7160–7220 ft., spiral view, ×208.
FIGURE 18	*Globoquadrina venezuelana* (Hedberg). Opuekeba-1 well, 7160–7220 ft., umbilical view, ×175.
FIGURE 19	*Globoquadrina venezuelana* (Hedberg). Baba-1 well, 2670–2850 ft., umbilical view, ×195.
FIGURE 20	*Globoquadrina venezuelana* (Hedberg). Baba-1 well, 2670–2850 ft., umbilical view, ×162.

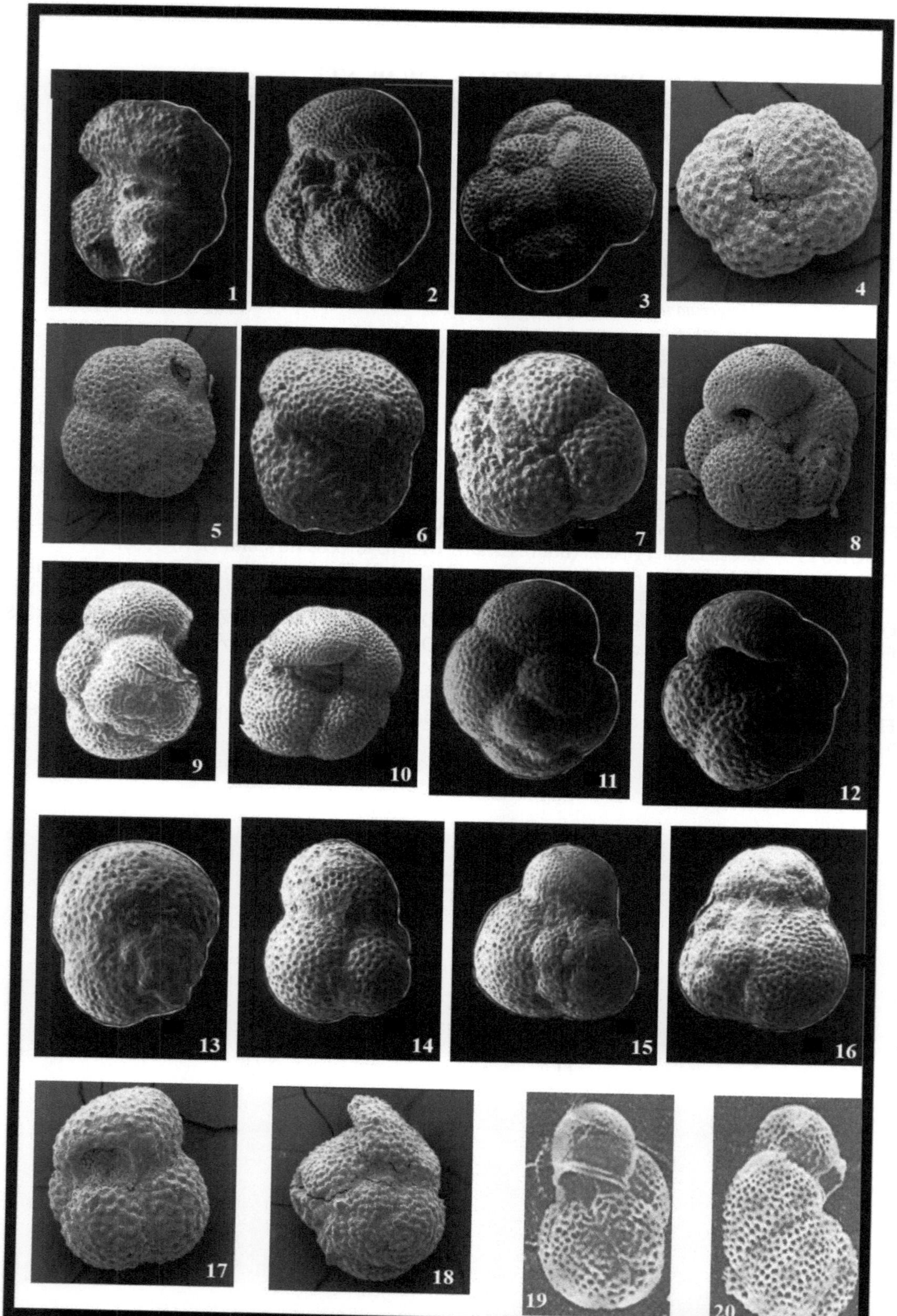
1
2
3
4
5
6
7
8
9
10
11
12
13
14
15
16
17
18
19
20

EXPLANATION OF PLATE 36

FIGURE 1	*Dentoglobigerina baroemoensis* (LeRoy). Etelebou-1 well, 9050 ft., umbilical view, ×130.
FIGURE 2	*Dentoglobigerina baroemoensis* (LeRoy). West Isan-1 well, 4360 ft., umbilical view, ×130.
FIGURE 3	*Dentoglobigerina baroemoensis* (LeRoy). West Isan-1 well, 4360 ft., spiral view, ×110.
FIGURE 4	*Neogloboquadrina acostaensis* (Blow, 1959). HB-1 well, 3320–3380 ft., umbilical view, ×300.
FIGURE 5	*Neogloboquadrina acostaensis* (Blow, 1959). HB-1 well, 3320–3380 ft., spiral view, ×225.
FIGURE 6	*Neogloboquadrina acostaensis* (Blow, 1959). Bita-1 well, 3710 ft., umbilical view, ×227.
FIGURE 7	*Neogloboquadrina acostaensis* (Blow, 1959). Bita-1 well, 3710 ft., umbilical view, ×227.
FIGURE 8	*Neogloboquadrina dutertrei* (d'Orbigny). Akamba-1 well, 150–170 m, umbilical view, ×150.
FIGURE 9	*Neogloboquadrina dutertrei* (d'Orbigny). Akamba-1 well, 150–170 m, another specimen, spiral view, ×110.
FIGURE 10	*Neogloboquadrina dutertrei* (d'Orbigny). Akamba-1 well, 150–170 m, same specimen, umbilical view, ×97.
FIGURE 11	*Neogloboquadrina humerosa* (Takayanagi and Saito). Baba-1 well, 1010–1025 ft., spiral view, ×149.
FIGURE 12	*Neogloboquadrina humerosa* (Takayanagi and Saito). Baba-1 well, 1010–1025 ft., umbilical view, ×149.
FIGURE 13	*Globigerinoides bisphericus* (Todd). Tuomo-1 well, 3432–3440 ft., spiral view, ×162.
FIGURE 14	*Globigerinoides bulloideus* (Crescent). Uquo-1 well, 6400 ft., umbilical view, ×130.
FIGURE 15	*Globigerinoides bulloideus* (Crescent). Uquo-1 well, 6400 ft., spiral view, ×130.
FIGURE 16	*Globigerinoides bulloideus* (Crescent). Uquo-1 well, 6400 ft., spiral view, ×130.
FIGURE 17	*Globigerinoides extremus* (Bolli and Bermudez, 1965). HB-1 well, 10,040–10,100 ft., umbilical view, ×300.
FIGURE 18	*Globigerinoides extremus* (Bolli and Bermudez, 1965). HB-1 well, 10,040–10,100 ft., spiral view, ×200.
FIGURE 19	*Globigerina nepenthes* (Todd, 1957). Bonga SW-2 well, 6820–7420 ft., umbilical view, ×100.
FIGURE 20	*Globigerina nepenthes* (Todd, 1957). Bonga SW-2 well, 6820–7420 ft., side view, ×100.

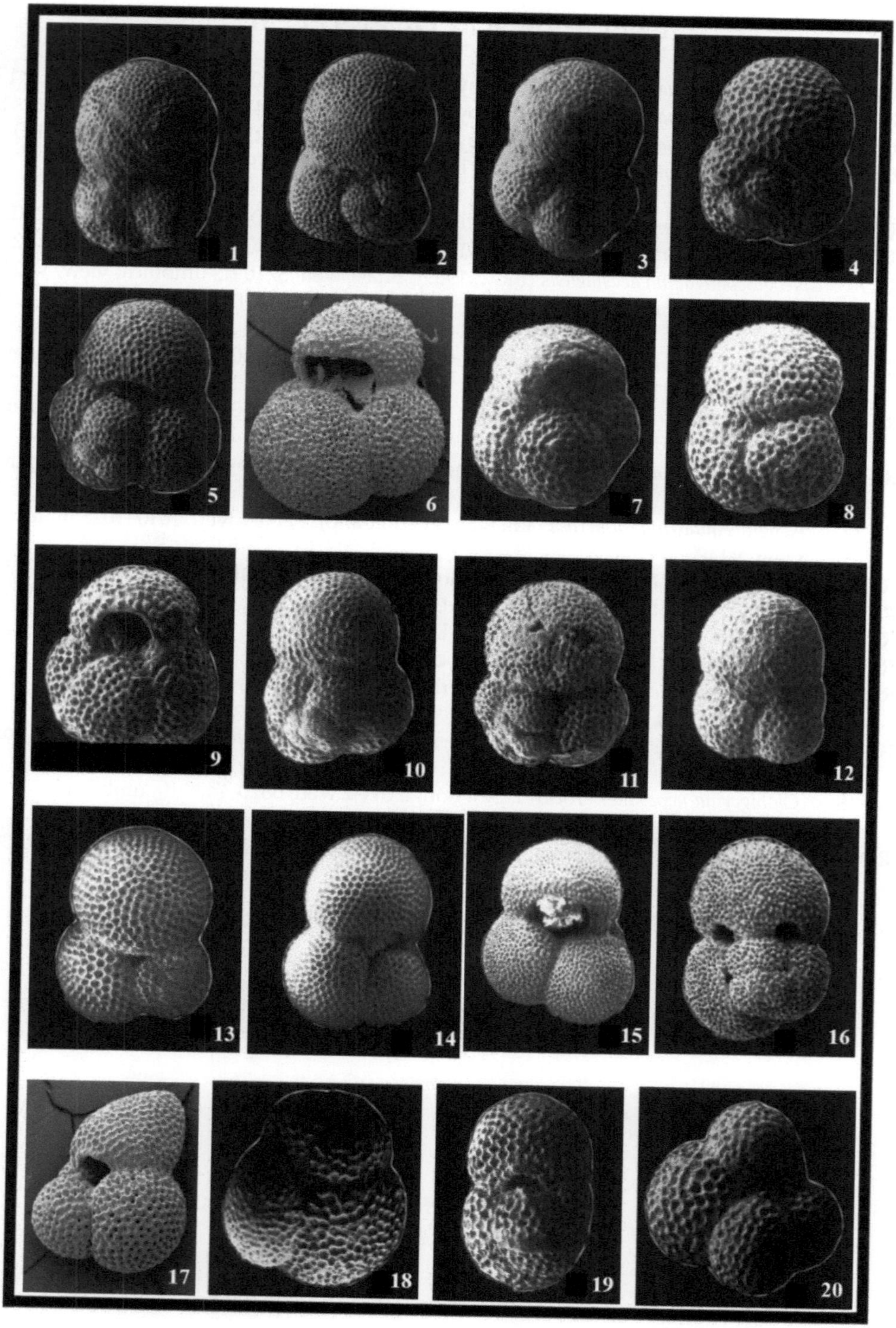
1
2
3
4
5
6
7
8
9
10
11
12
13
14
15
16
17
18
19
20

EXPLANATION OF PLATE 37

FIGURE 1	*Globigerinoides immaturus* (LeRoy). Ewan-1 well, 6120–6180 ft., umbilical view, ×162.
FIGURE 2	*Globigerinoides immaturus* (LeRoy). Ewan-1 well, 6120–6180 ft., spiral view, ×175.
FIGURE 3	*Globigerinoides immaturus* (LeRoy). Ewan-1 well, 6900–6960 ft., umbilical view, ×136.
FIGURE 4	*Globigerinoides inusitatus* (Jenkins). West Isan-1 well, 4360–4420 ft., spiral view, ×130.
FIGURE 5	*Globigerinoides inusitatus* (Jenkins). Ewan-1 well, 6900–6960 ft., spiral view, ×156.
FIGURE 6	*Globigerinoides obliquus* (Bolli). Akamba-1 well, 150–170 m, umbilical view, ×200.
FIGURE 7	*Globigerinoides obliquus* (Bolli). Meren-1 well, 3300 ft., umbilical view, ×195.
FIGURE 8	*Globigerinoides obliquus* (Bolli). Benin West-1 well, 1079 ft., spiral view, ×175.
FIGURE 9	*Globigerinoides obliquus* (Bolli). Benin West-1 well, 1079 ft., umbilical view, ×182.
FIGURE 10	*Globigerinoides primordius* (Blow and Banner). Okan-4 well, 9510–9570 ft., spiral view, ×156.
FIGURE 11	*Globigerinoides primordius* (Blow and Banner). Tuomo-1 well, 2430–2440 ft., spiral view, ×149.
FIGURE 12	*Globigerinoides primordius* (Blow and Banner). Utapate-1 well, 10,100 ft., umbilical view, ×143.
FIGURE 13	*Globigerinoides quadrilobatus* (d'Orbigny). Akamba-1 well, 510–570 ft., spiral view, ×130.
FIGURE 14	*Globigerinoides quadrilobatus* (d'Orbigny). Akamba-1 well, 510–570 ft., umbilical view, ×91.
FIGURE 15	*Globigerinoides ruber* (d'Orbigny). Akamba-1 well, 510–570 ft., umbilical view, ×97.
FIGURE 16	*Globigerinoides ruber* (d'Orbigny). East Obe-1 well, 1836–1890 ft., spiral view, ×97.
FIGURE 17	*Globigerinoides sacculifer* (Brady). Akamba-1 well, 150–170 m, umbilical view, ×170.
FIGURE 18	*Globigerinoides sacculifer* (Brady). Ewan-1 well, 6540–6600 ft., spiral view, ×195.
FIGURE 19	*Globigerinoides sacculifer* (Brady). Okan-4, 3880–3940 ft., edge view, ×150.
FIGURE 20	*Globigerinoides sacculifer* (Brady). Ewan-1, 6540–6600 ft., umbilical view, ×208.

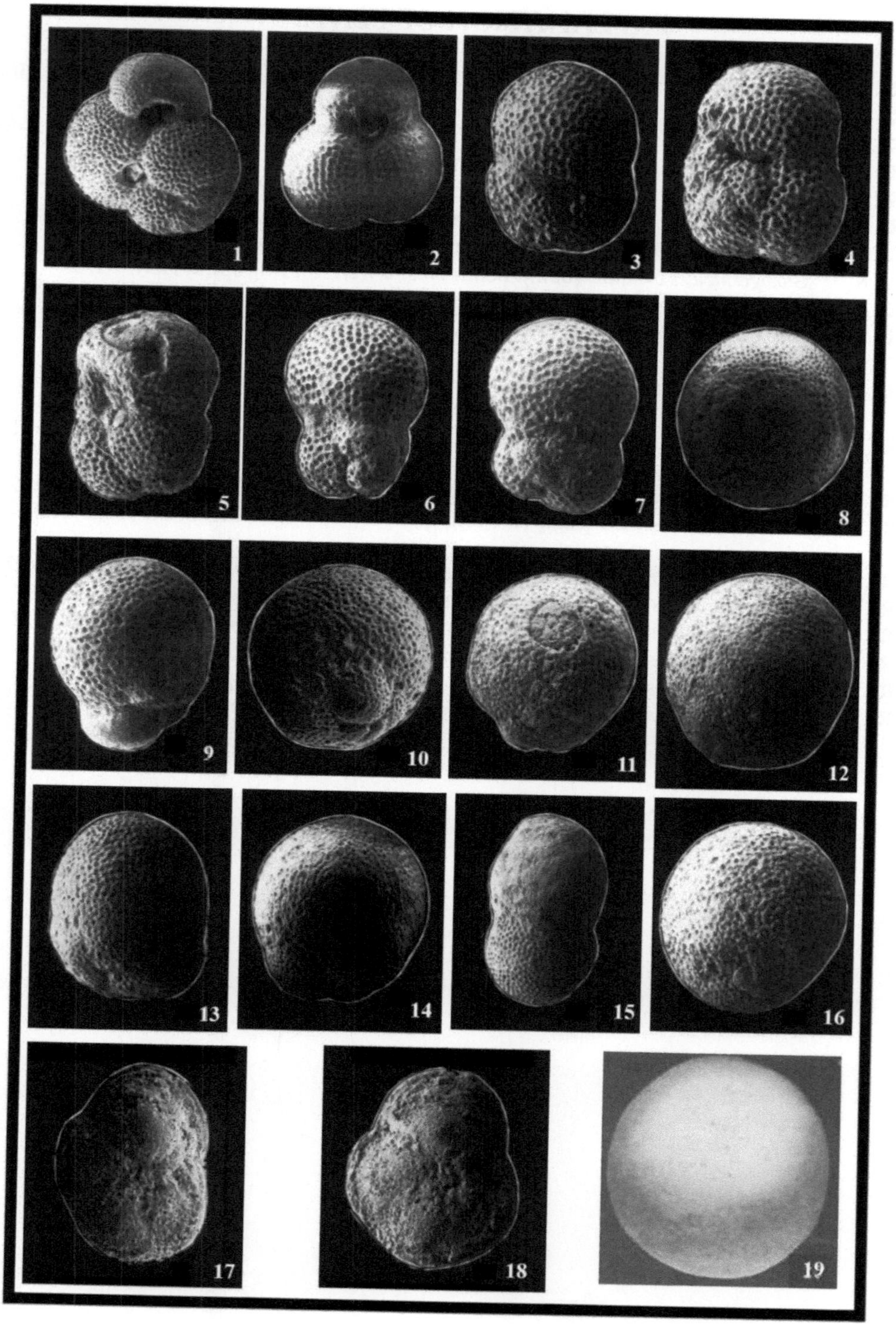
1
2
3
4
5
6
7
8
9
10
11
12
13
14
15
16
17
18
19

EXPLANATION OF PLATE 38

FIGURE 1 *Globigerinoides seiglei* (Bermudez and Bolli). Akamba-1 well, 510–570 ft., spiral view, ×130.

FIGURE 2 *Globigerinoides seiglei* (Bermudez and Bolli). Akamba-1 well, 510–570 ft., umbilical view, ×130.

FIGURE 3 *Globigerinoides sicanus* (de Stefani). Ireti-9 well, 7320 ft., spiral view, ×156.

FIGURE 4 *Globigerinoides subquadratus* (Bronnimann). Utapate-1 well, 10,100 ft., spiral view, ×143.

FIGURE 5 *Globigerinoides subquadratus* (Bronnimann). Utapate-1 well, 10,100 ft., spiral view, ×143.

FIGURE 6 *Globigerinoides triloba* (Reuss). Okan-1 well, 9510–9570 ft., spiral view, ×169.

FIGURE 7 *Globigerinoides triloba* (Reuss). Okan-1 well, 9510–9570 ft., spiral view, ×169.

FIGURE 8 *Praeorbulina circularis* (Blow). Manuso-1 well, 677 ft., side view, ×130.

FIGURE 9 *Praeorbulina curva* (Blow). Utapate-1 well, 10,300 ft., side view, ×162.

FIGURE 10 *Praeorbulina curva* (Blow). Opuama-1 well, 3801 ft., side view, ×162.

FIGURE 11 *Praeorbulina curva* (Blow). Utapate-1 well, 10,300 ft., side view, ×162.

FIGURE 12 *Praeorbulina glomerosa* (Blow). Utapate-1 well, 9800 ft., side view, ×130.

FIGURE 13 *Praeorbulina glomerosa* (Blow). Delta South-1 well, 9870–9930 ft., side view, ×169.

FIGURE 14 *Praeorbulina glomerosa* (Blow). Delta South-1 well, 9870–9930 ft., side view, ×162.

FIGURE 15 *Orbulina bilobata* (d'Orbigny). Delta South-1 well, 9870–9930 ft., side view, ×150.

FIGURE 16 *Orbulina suturalis* (Bronimann). Utapate-1 well, 9800 ft., side view, ×162.

FIGURE 17 *Globorotalia archemenardi* (Bolli). Ruta-9 well, 4918–5094 ft., umbilical view, ×162.

FIGURE 18 *Globorotalia archemenardi* (Bolli). Ruta-9 well, 4918–5094 ft., umbilical view, ×162.

FIGURE 19 *Orbulina universa* (d'Orbigny). Utapate-1 well, 9800 ft., side view, ×150.

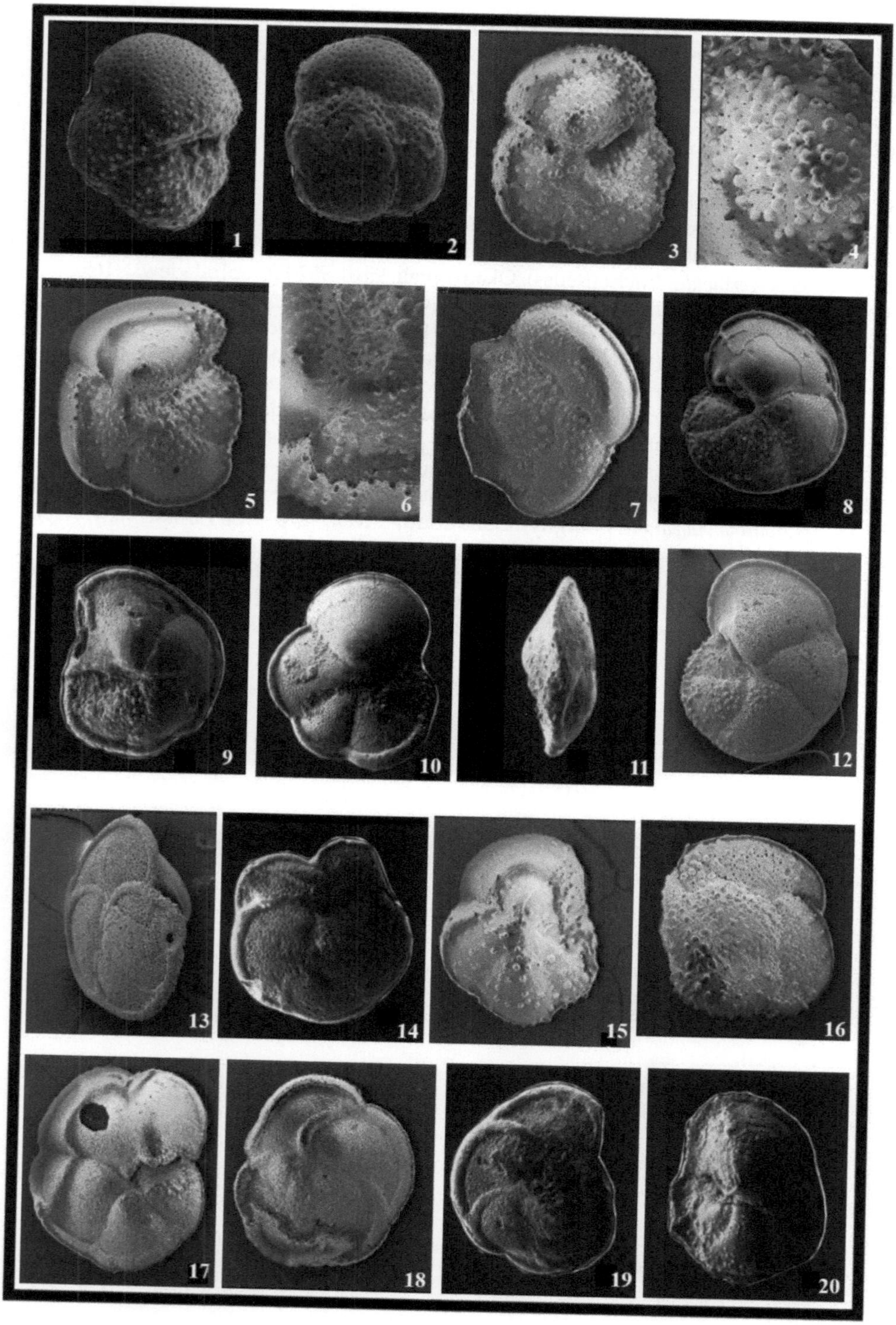
1
2
3
4
5
6
7
8
9
10
11
12
13
14
15
16
17
18
19
20

EXPLANATION OF PLATE 39

FIGURE 1 *Globorotalia crassaformis* (Galloway and Wissler). Baba-1 well, 855–909 ft., umbilical view, ×236.

FIGURE 2 *Globorotalia crassaformis* (Galloway and Wissler). Baba-1 well, 855–909 ft., spiral view, ×230.

FIGURE 3 *Globorotalia crassaformis* (Galloway and Wissler). Ekanga-2 well, 660 m, umbilical view, ×250.

FIGURE 4 *Globorotalia crassaformis* (Galloway and Wissler). Ekanga-2 well, 660 m, umbilical view of same specimen (Fig. 3) enlarged, ×1000.

FIGURE 5 *Globorotalia crassaformis* (Galloway and Wissler). Anyala-4 well, 810–870 ft., umbilical view, ×200.

FIGURE 6 *Globorotalia crassaformis* (Galloway and Wissler). Anyala-4 well, 810–870 ft., umbilical view of same sample (Fig. 5) enlarged, ×800.

FIGURE 7 *Globorotalia crassaformis* (Galloway and Wissler). Anyala-4 well, 810–870 ft., spiral view, ×200.

FIGURE 8 *Globorotalia cultrata* (d'Orbigny, 1826). Ruta-9 well, 4914–5094 ft., umbilical view, ×145.

FIGURE 9 *Globorotalia cultrata* (d'Orbigny, 1826). Ruta-9 well, 4914–5094 ft., umbilical view, ×143.

FIGURE 10 *Globorotalia cultrata* (d'Orbigny, 1826). Ruta-9 well, 4918–5094 ft., umbilical view, ×130.

FIGURE 11 *Globorotalia cultrata* (d'Orbigny, 1826). Ruta-9 well, 4918–5094 ft., edge view, ×117.

FIGURE 12 *Globorotalia cultrata* (d'Orbigny, 1826). Akamba-1 well, 600–620 m, umbilical view, ×200.

FIGURE 13 *Globorotalia cultrata* (d'Orbigny, 1826). Akamba-1 well, 600–620 m, spiral/side view, ×200.

FIGURE 14 *Globorotalia flexuosa* (Koch). Ruta-9 well, 4918–5094 ft., spiral view, ×182.

FIGURE 15 *Globorotalia margaritae* (Bolli and Bermudez, 1965). Ekanga-2 well, 650 m, umbilical view, ×300

FIGURE 16 *Globorotalia margaritae* (Bolli and Bermudez, 1965). Ekanga-2 well, 650 m, spiral view, ×250.

FIGURE 17 *Globorotalia menardii* (d'Orbigny, 1826). Anyala-4 well, 810–870 ft., umbilical view, ×90.

FIGURE 18 *Globorotalia menardii* (d'Orbigny, 1826). Anyala-4 well, 810–870 ft., spiral view, ×125.

FIGURE 19 *Globorotalia merotumida* (Blow and Banner). Ruta-9 well, 4914–5094 ft., spiral view, ×143.

FIGURE 20 *Globorotalia merotumida* (Blow and Banner). Ruta-9 well, 4914–5094 ft., umbilical view, ×162.

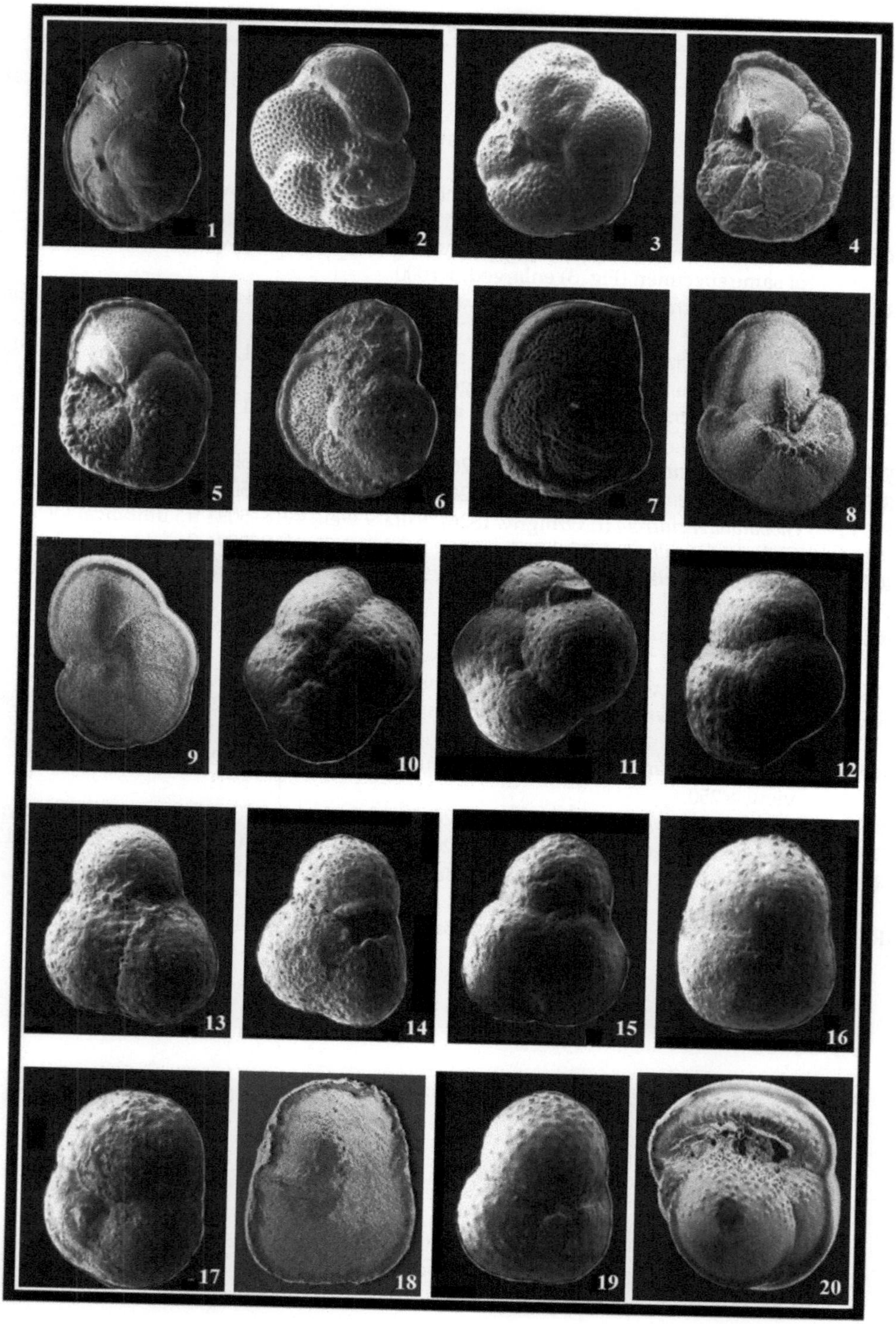
1
2
3
4
5
6
7
8
9
10
11
12
13
14
15
16
17
18
19
20

EXPLANATION OF PLATE 40

FIGURE 1	*Globorotalia plesiotumida* (Blow). Ruta-9 well, 4918–5094 ft., spiral view, ×117.
FIGURE 2	*Globorotalia scitula* (Banner and Blow). Opuama-1 well, 4320 ft., spiral view, ×201.
FIGURE 3	*Globorotalia scitula* (Banner and Blow). Opuama-1 well, 4320 ft., umbilical view, ×253.
FIGURE 4	*Globorotalia tumida* (Brady). Ruta-1 well, 3030–3060 ft., umbilical view, ×143.
FIGURE 5	*Globorotalia tumida* (Brady). Sonam-1 well, 4660–4670 ft., umbilical view, ×195.
FIGURE 6	*Globorotalia tumida* (Brady). Sonam-1 well, 4660–4670 ft., spiral view, ×195.
FIGURE 7	*Globorotalia tumida* (Brady). Ruta-9 well, 4914–5094 ft., spiral view, ×162.
FIGURE 8	*Globorotalia tumida* (Brady). Anyala-4 well, 930–990 ft., umbilical view, ×80.
FIGURE 9	*Globorotalia tumida* (Brady). Anyala-4 well, 930–990 ft., spiral view, ×85.
FIGURE 10	*Sphaeroidinellopsis disjuncta* (Finlay). Meta-1 well, 9440–9500 ft., umbilical view, ×130.
FIGURE 11	*Sphaeroidinellopsis disjuncta* (Finlay). Meta-1 well, 9440–9500 ft., umbilical view, ×130.
FIGURE 12	*Sphaeroidinellopsis disjuncta* (Finlay). Malu-2 well, 8650–8980 ft., spiral view, ×208.
FIGURE 13	*Sphaeroidinellopsis multiloba* (LeRoy). Malu-2 well, 8650–8980 ft., umbilical view, ×136.
FIGURE 14	*Sphaeroidinellopsis multiloba* (LeRoy). Malu-2 well, 8650–8980 ft., umbilical view, ×130.
FIGURE 15	*Sphaeroidinellopsis multiloba* (LeRoy). Malu-2 well, 8650–8980 ft., umbilical view, ×154.
FIGURE 16	*Sphaeroidinellopsis sphaeroides* (Lamb). Ruta-9 well, 4918–5094 ft., side view, ×221.
FIGURE 17	*Sphaeroidinellopsis sphaeroides* (Lamb). Ruta-9 well, 4918–5094 ft., umbilical view, ×185.
FIGURE 18	*Sphaeroidinellopsis subdehiscens* (Blow). Ekanga-2 well, 910 m, ×300.
FIGURE 19	*Sphaeroidinellopsis subdehiscens* (Blow). Ruta-9 well, 4918–5094 ft., umbilical view, ×169.
FIGURE 20	*Pulleniatina obliquiloculata*. Anyala-4 well, 870–980 ft., ×200.

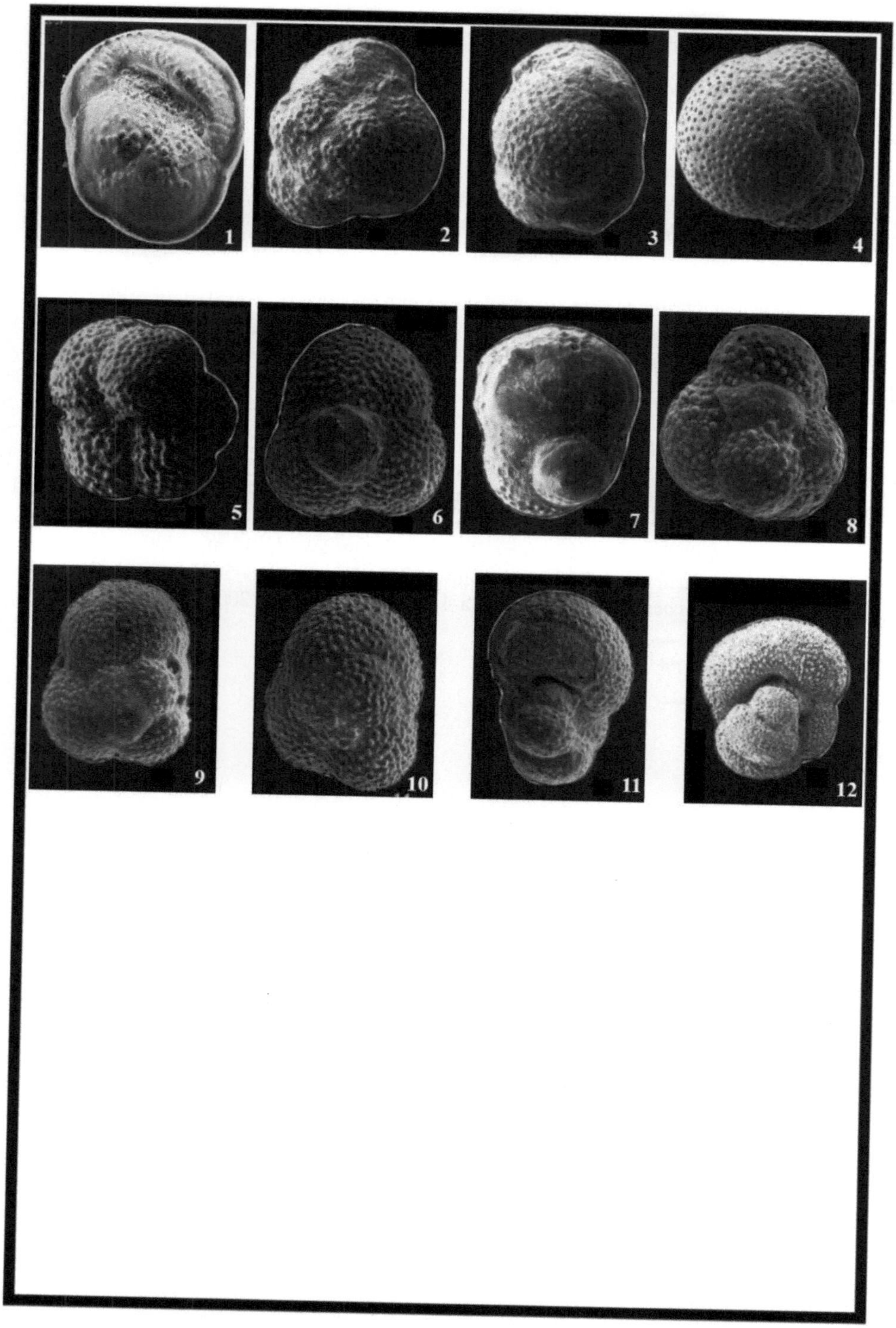
1
2
3
4
5
6
7
8
9
10
11
12

EXPLANATION OF PLATE 41

FIGURE 1 *Pulleniatina primalis* (Banner and Blow, 1967). Anyala-4 well, 870-930 ft., ×200.

FIGURE 2 *Catapsydrax martini* (Blow and Banner, 1967). Parabe–1 well, 9500 ft., spiral view, ×208.

FIGURE 3 *Catapsydrax martini* (Blow and Banner, 1967). Parabe-1 well, 9500 ft., umbilical view, ×214.

FIGURE 4 *Catapsydrax* cf. *howei*. Opuama-1 well, 4224 ft., umbilical view, ×286.

FIGURE 5 *Catapsydrax stainforthi* (Bolli, Loeblich, and Tappan). Ewan-1 well, 7390–7720 ft., umbilical view, ×214.

FIGURE 6 *Catapsydrax unicavus* (Bolli, Loeblich, and Tappan). Okan–1 well, 3880–3940 ft., umbilical view, ×201.

FIGURE 7 *Catapsydrax unicavus* (Bolli, Loeblich, and Tappan). Okan-1 well, 3880–3940 ft., umbilical view, ×162.

FIGURE 8 *Globigerinita incrusta* (Akers). Meta-1 well, 5373–5490 ft., umbilical view, ×347.

FIGURE 9 *Globigerinita naparimaensis* (Bronnimann). Meta-1 well, 5373–5490 ft., edge view, ×325.

FIGURE 10 *Globigerinita naparimaensis* (Bronnimann). Meta-1 well, 5373–5490 ft., spiral view, ×237.

FIGURE 11 *Hastigerina aequilateralis* (Brady). Baba-1 well, 855–909 ft., edge view, ×171.

FIGURE 12 *Hastigerina siphonifera* (d'Orbigny). Akamba-1 well, 510–570 ft., edge view, ×97.

PART II

CENOZOIC CALCAREOUS NANNOFOSSIL BIOSTRATIGRAPHY OF THE NIGER DELTA

A.S. Oyebamiji, O.S. Adegoke, J.J. Edet, P.L. Osterloff & O.K. Ulu

CHAPTER

1

Introduction

OUTLINE

1.1 ORIGIN AND CONCEPT OF THE STRATIGRAPHIC COMMITTEE OF THE NIGER DELTA (STRATCOM)

The need to set up the Stratigraphic Committee (formerly Commission) of the Niger Delta was mooted by Shell Petroleum Development Company of Nigeria (SPDC) through the auspices of Dr. T.J.A. Reijers (Chairman), with Mr. A.A. Adesida as Secretary in 1995. The proposal was warmly accepted and supported by other operating Joint Venture companies. The idea was then presented to the Department of Petroleum Resources (DPR) on September 5, 1995, the exploration OPTS (October 4, 1995), Nigerian Association of Petroleum Explorationists (NAPE) (November 17, 1995), management of OPTS (March 20, 1996), and National Petroleum Investment Services (NAPIMS) (March 22, 1996). All these bodies endorsed and welcomed the concept.

The Committee was formally inaugurated on October 20, 1995. Membership was made up of senior representatives of the DPR, NAPIMS, SPDC, Chevron Nigeria Limited (CNL), Nigerian Petroleum Development Company (NPDC), Elf Petroleum Nigeria Limited (EPNL) (now Total EandP Nig. Ltd.), Nigerian Agip Oil Company Ltd. (NAOC), Mobil Producing Nigeria (MPN) Unlimited (now ExxonMobil), Texaco Overseas (Nig.) Petroleum

Cenozoic Foraminifera and Calcareous Nannofossil Biostratigraphy of the Niger Delta. http://dx.doi.org/10.1016/B978-0-12-812161-0.00007-7

Company Unlimited (TOPCON) (now defunct after merger with CNL), and Ashland (now Addax) distributed in four technical subcommittees. The DPR is Chairman of StratCom with NAPIMS and the Geological Survey of Nigeria taking Vice Chairmanship positions. The structure includes an Editorial Board that is to take care of intermediate and final publications. The Committee is to be aided by the Advisory Board in which representatives of four Nigerian Universities and at least six major service companies are represented. In addition, NAPE and Nigerian Mining and Geosciences Society (NMGS) are represented in the Committee.

The concept at inception was to pull together and make full use of the vast but diverse, array of exploration and production data on the Niger Delta which, until then, was exclusively in the archives of the various international operating companies and the DPR. In doing this, the major operating companies hoped to standardize the data and thereby develop a unified geological, structural, and chronostratigraphic scheme that can be applied throughout the Niger Delta.

The objectives of the proposed StratCom project include:

1. Formalization of a hierarchically organized computer-compatible lithostratigraphy.
2. Establishment of chronostratigraphic and sequence stratigraphic standards for the entire Niger Delta.
3. Identification, naming, and harmonization of the nomenclature of stratigraphically relevant main structural elements of the Niger Delta.
4. Linking genetic sequences through seismostratigraphy.

The scope of work as originally designed was to:

1. Carry out detailed lithostratigraphic, biostratigraphic, and sequence stratigraphic analysis and reanalysis of about 150 selected wells.
2. Carry out detailed structural and seismosequence stratigraphic interpretation of at least nine dip lines and four strike lines released for this purpose by the operators.

Though this work scope was modified over time, the requirements initially defined still form the lodestone for the effort of StratCom. To achieve the laudable objectives set by StratCom, the then six major operators of Joint Ventures with the Nigerian National Petroleum Corporation (NNPC) entered into a working agreement on the March 22, 1996. The overall objective was to complete all defined studies and publish the results before the beginning of the new millennium (2000).

1.2 THE ORGANIZATION THAT OVERSEES STRATCOM ACTIVITIES

As set up, the activities of StratCom are to be overseen jointly by DPR and NAPIMS, with DPR as Chairman. In its place as Chairman of the Committee, the DPR steers and coordinates the activities of StratCom. NAPIMS acts as a supervisor of the Joint Venture partners composed of most of the participating companies, and in its capacity, the first Vice Chairman assures that the yearly StratCom budgetary allocations contained in the studies budget of each of the participating companies are justified and accounted for at the end of each year.

1.3 THE COMMITTEES OF STRATCOM

The various Technical Committees set up by StratCom for the implementation of its assignments include an Advisory Board and four multidisciplinary committees as listed:

1. The Advisory Board
2. Lithostratigraphic Committee
3. Biostratigraphic Committee
4. Seismosequence Stratigraphic Committee
5. Structural Committee

1.4 THE CALCAREOUS NANNOFOSSIL CONSORTIUM

The Agreement for the Calcareous Nannofossil Taxonomy Project of the Niger Delta, (Contract No. W03265B dated July, 1999), was made between StratCom comprising the following companies:

1. SPDC
2. CNL
3. EPNL
4. MPN
5. TOPCON
6. NAOC

with Global Energy Company Limited (representing a Consortium comprising of Mosunmolu Limited and Global Energy Company Limited) as Contractor. With the departure of the nannofossil expert, Mr. O. Ogunjobi from Global Energy Company Limited soon after the commencement of the project, Mosunmolu Limited, with Mr. Ajibola Oyebamiji as General Manager and Team Lead assumed full responsibility for project implementation.

The project objectives as set out in the Agreement include the following:

- Harmonization of the taxonomic nomenclature and zonal schemes of all the member companies of the Stratigraphic Committee of the Niger Delta.
- Recognition of new Maximum Flooding Surfaces (MFS).
- Characterization of the MFS using chronostratigraphically significant Calcareous Nannofossil species.

The project deliverables are:

- A Final Report containing the Niger Delta Calcareous Nannofossil zonal scheme and the detailed taxonomic description of the chronostratigraphically and paleoenvironmentally significant species.
- A Photocatalogue of key Niger Delta Calcareous Nannofossil marker species.

The project, as initially defined, involved the study of 700 samples from selected wells along the StratCom transect lines (Fig. 1.1). During subsequent revisions and updates of the original report, in response to the need for improved understanding of the nannofossil

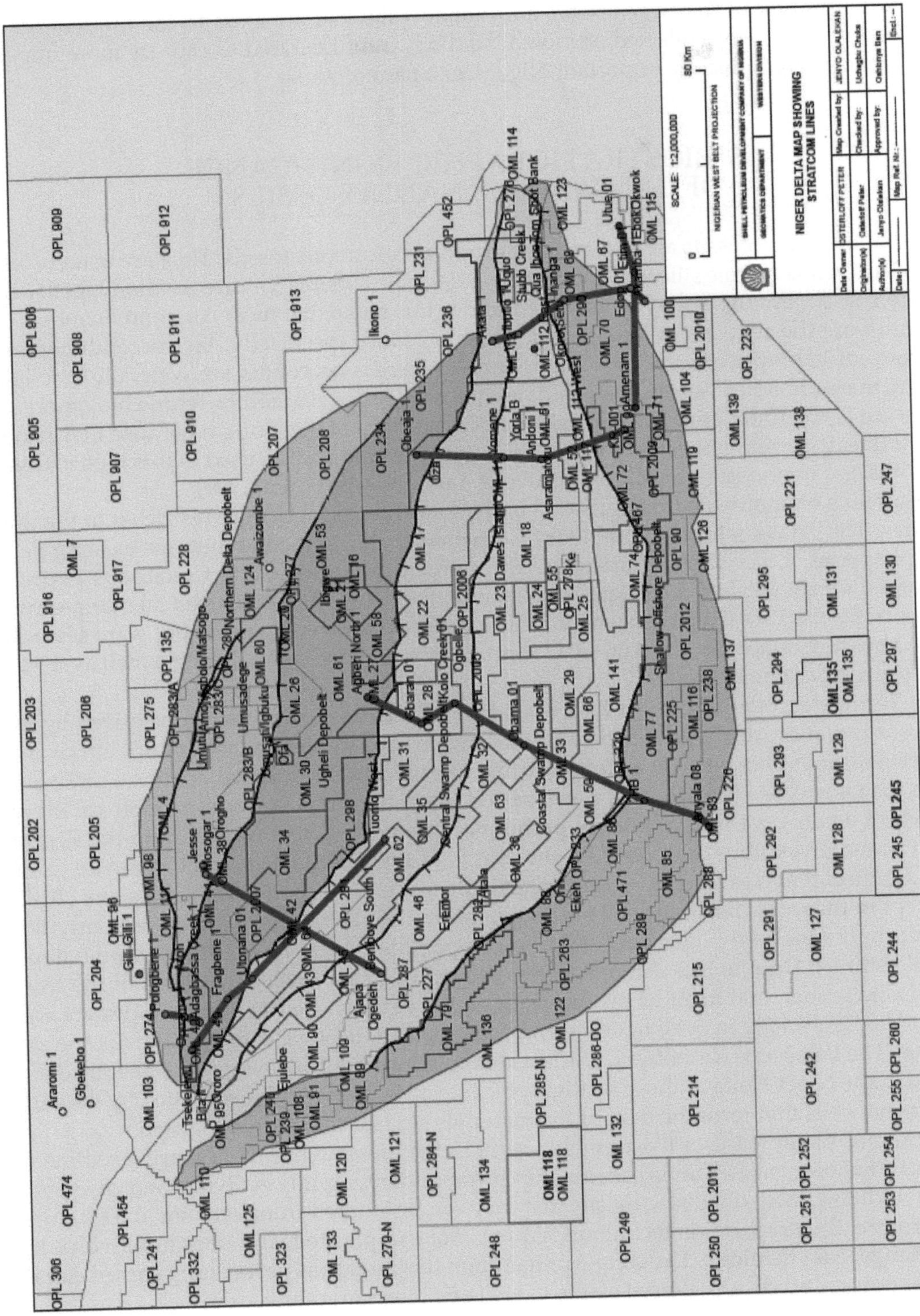

FIGURE 1.1 **Map of Niger Delta showing the StratCom Lines in the Project Area.**

chronostratigraphy of the Niger Delta, additional wells that are close to the original Strat-Com transect lines were selected, sampled, studied, and illustrated to capture more important chronostratigraphically important Niger Delta nannofossils.

1.5 BIOSTRATIGRAPHIC SIGNIFICANCE OF CALCAREOUS NANNOFOSSILS

Calcareous nannofossils are the smallest group of calcareous fossils. They are usually less than 30 μm in size. While siliceous ascidian spicules, calcispheres (calcareous dinoflagellates), and juvenile foraminifera may be included under this name, the most common members of the group are the remains of haptophytic algae or, more specifically, the coccolith-bearing subgroup of haptophytes, the coccolithophores. Whereas coccoliths are generally disc-like in form, there are other variously shaped nannofossils called nannoliths whose biological affinities are less certain. They are generally presumed to be of algal origin because they show comparable distribution with coccoliths. Thus, by definition and, as used in this report, fossil coccoliths and nannoliths constitute calcareous nannofossils (Bown, 1999).

Calcareous nannofossils first appeared in the Late Triassic and were restricted to the low latitudes. The Triassic/Jurassic boundary extinction event claimed all but one existing nannofossil species, *Crucirhabdulus primulus*. Throughout the Jurassic and Cretaceous periods, nannofossils reestablished their dominance within the oceans and colonized all marine environments. By the Late Cretaceous, nannofossils attained a diversity peak, but were severely affected by the Cretaceous/Tertiary (K/T) boundary extinction event during which over 90% of the species became extinct.

The Tertiary recovery from the K/T boundary extinction event was characterized by the Paleocene radiation, during which time different coccolith families and new nannolith groups were established. This influx was followed by a significant decline that continued into and within the Oligocene. The Miocene was a period of significant increase in nannofossil diversity, which again gave way to a more pronounced increase that continued into the Recent. The Miocene diversity was doused by the effects of the Middle Miocene Carbonate Crash (MMCC) effect that was recently documented in the Niger Delta by Fadiya and Salami (2012).

Except in the most distal, prodelta settings, calcareous nannofossils are more poorly represented in the modern Niger Delta than in comparable typical low-latitude marine depositional settings. Despite this difference, the pattern of nannofossil occurrence in the fossil Niger Delta is such that most of the biostratigraphic datums used in low latitudes are well documented especially in the Oligocene and Miocene and, to some extent, in the Eocene and Pliocene. The Paleocene nannofossil worldwide radiations are, however, yet to be fully documented in the Niger Delta as these stratigraphic intervals were rarely penetrated by the early exploration wells that form the bulk of the materials studied for this monograph.

Calcareous nannofossils are extremely useful in the biostratigraphy of marine sediments of Jurassic to Recent ages. This is because they are abundant, rapidly evolving, and cosmopolitan. In addition, their small size means that they can be studied from very small rock chips. Furthermore, the preparation technique is quick, not complicated, and does not involve the use of dangerous chemicals. The simple preparation technique has particularly earned calcareous nannofossil study a preferred status in well-site biostratigraphy.

1.6 PREVIOUS CALCAREOUS NANNOFOSSIL BIOSTRATIGRAPHIC WORK IN THE NIGER DELTA

Routine biostratigraphic studies of ditch-cutting samples from drilled wells have been an important correlation tool in the Niger Delta over the last three decades. Such studies were initially carried out on foraminifera (benthonics and planktonics) and palynomorphs (pollen, spores, and dinoflagellate cysts). Biostratigraphers based their benthonic zonation and paleoenvironmental interpretation on in-house schemes that were previously established by SPDC, MPN, EPNL, and other companies, as well as published works by Adegoke et al. (1976), Petters (1982), among others. The planktonic foraminiferal zonations used in the Niger Delta have initially been based on correlation with the worldwide zonation scheme of Blow (1969a,b, 1979), and more recently those of Berggren et al. (1998a,b), Hardenbol et al. (1998), and Gradstein et al. (2004a,b, 2010, 2014).

The use of nannofossils in routine biostratigraphy in the Niger Delta had a humble beginning in the early 1990s, coinciding with the period when sequence stratigraphy was initially applied to hydrocarbon exploration. The need for more comprehensive and better refined zones at scales of less than 1 Ma necessitated the use of nannofossils as a complement to foraminifera and palynomorphs. At present, nannofossils have become one of the more important tools used in biostratigraphy especially in the deep offshore of the Gulf of Guinea where deltaic influence is less pronounced. Recent intense exploration activities in the Niger Delta deepwater have revealed the presence of appreciable quantities of nannofossils that are capable of generating huge in-house databases, especially within the Neogene and Paleogene, for the use of the oil industry.

At the commencement of this project in 1995, there were virtually no published works on the nannofossil biostratigraphy of the Niger Delta. The pioneering Nigerian nannofossil studies were the unpublished Master of Science degree dissertations of Ogunjobi (1994) and Oyebamiji (1997) from University College London. Both authors were sponsored by the indigenous oil and gas service companies, Global Energy Ltd. and Mosunmolu Limited, their employers. The successful completion of these Master of Science degree theses marked the beginning of nannofossil manpower development and use in Nigeria. It was not until 2007 and later, that the nannofossil published works on Niger Delta samples appeared in the literature, initially as forms from unnamed Niger Delta deepwater wells (Boboye and Fowora, 2007; Ehinola et al., 2009; Boboye and Adeleye, 2009; Fadiya, 2012, 2014; Fadiya and Salami, 2012; Osterloff et al., 2013). As with other microfossil groups, many unpublished, proprietary reports on calcareous nannofossils from wells in the Niger Delta exist in the archives of the oil companies operating in Nigeria. The contents of some of these unpublished reports were consulted for this project.

The chronostratigraphic interpretation of the nannofossil data carried out in the Niger Delta is based on the worldwide zonation schemes of Martini (1971) and Okada and Bukry (1980). The present study has also benefited from these and several additional references especially Gallagher (1990), Rio et al. (1990), Young (1990), Young et al. (1994), Raffi et al. (1995), Varol (1998), and the globally calibrated works of Gradstein et al. (2004a,b, 2012).

1.7 STUDY MATERIALS

This StratCom calcareous nannofossil study of the Niger Delta was initially based on the analysis of over 700 ditch-cutting samples from 40 wells that were selected along the StratCom Transect Lines (Fig. 1.1). As the study of nannofossils became more attractive in the

Niger Delta, and to enhance the quality of the report, more than 800 additional samples were selected from new wells that are close to the StratCom transect lines. The samples studied were, in most cases, selected from shales sections that are rich in planktonic and/or benthonic foraminifera, most of which are known from existing biostratigraphic reports to be marine-condensed sections. The studied wells are listed in Table 1.1.

TABLE 1.1 Stratcom Wells Studied For Calcareous Nannofossils

Well	Line	Operator	OML
Jesse-1	H	SPDC	38
Ovhor-10	—	SPDC	38
Opuama-1	L	SPDC	40
Amakubu-1	I	SPDC	40
Umutu-1	J	SPDC	38
Zarama-7	—	SPDC	28
Igodo-1	—	SPDC	43
Pologbene-1	I	SPDC	1
Gilli Gilli-1	J	SPDC	96
Mosogar-2	—	SPDC	38
Akata-1	B1	SPDC	13
Ibotio-1	B1	SPDC	13
Obeaja-1	C	SPDC	11
Yomene-1	C	SPDC	11
Adagbassa Creek-1	I	SPDC	40
Utapate-1	M	SPDC	13
Andoni-1	C	SPDC	11
HB-1	F	SPDC	77
Batan-1	H	SPDC	42
Obama North-1	F	SPDC	29
KR-1	C	SPDC	72
Koronama-3	—	SPDC	72
Amakubu-1	I	SPDC	40
Pologbene-1	I	SPDC	40
Afam-1	C	SPDC	11
Tonjor-1	I	SPDC	1
Bomu-1	C	SPDC	11
Gbaran-2	F	SPDC	28
Kolo Creek-39	—	SPDC	28
Anyala-4 and Anyala-8	F	CNL	83

TABLE 1.1 Stratcom Wells Studied For Calcareous Nannofossils (*cont.*)

Well	Line	Operator	OML
Benin River-3	—	CNL	49
Utonana-1	L	CNL	42
Okubie-8	—	CNL	86
Fragbene-1	L	CNL	49
Umuoru-1	F	NAOC	61
Agberi North-1	F	NAOC	61
Nemomai-1	—	NAOC	OPL 275
Beniboye North-1	H	NAOC	62
Tuomo West-1	L	NAOC	62
Inyi Creek-1	—	NAOC	OPL 135
Adofi-1	—	NAOC	OPL 275
Ukpokiti-1	H	Elf	57
Akamba-1	B1	Elf	97
Edop-33a	B1	MPN	67
Etim-1	O	MPN	67
Utue-17a	O	MPN	67
Inanga-1	B1	MPN	69
Otobo North-1	—	MPN	OPL 221

CNL, Chevron Nigeria Limited; MPN, Mobil Producing Nigeria; NAOC, Nigerian Agip Oil Company Ltd.; OML, Oil Mining License; OPL, Oil Prospecting License; SPDC, Shell Petroleum Development Company.

Data from the Early–Middle Paleocene sequences are not included in this report largely because sedimentary sections of this age have not been penetrated by oil wells and have thus not been studied in great detail.

It is evident that some of the oil well samples used in this study may have been affected by poor storage conditions over the years. This has resulted in the recovery of etched and sometimes dissolved nannofossil tests at some horizons. This was particularly noticed in samples from the Early Eocene and Paleocene in which poor preservation and low abundances were the rule rather than the exception.

1.8 SAMPLE PREPARATION TECHNIQUES

For all the samples used in this project, slides were made from selected portions of the ditch cuttings and were studied under Olympus BH2 and Zeiss Polarizing microscopes at magnifications between ×1000 and ×1250. The preparation of cuttings for analysis followed standard procedures. Typically, several small chips of cuttings were disaggregated in distilled water in a beaker to make a suspension. A small part of this suspension was decanted into a vial and diluted in distilled water. About 1 mL of the diluted solution was pipetted onto a cover slip (22 × 40 mm) and dried on a hot plate. The cover slip was then affixed face down, on a glass slide with Canada Balsam or Norland Optical Adhesive (RI = 1.56) as the mounting medium. Other methods of preparation used occasionally included standard simple smear, gravity settling, and short-centrifuging techniques.

1.9 DATA COLLECTION AND PRESENTATION

A semiquantitative method of counting nannofossils was adopted in this project. All the nannofossils encountered in not less than two sweeps of the longer side of the 22 × 40 mm slide were counted. This is the counting method used in routine nannofossil biostratigraphic studies by many authors, as it allows for faster quantification of individual species.

CHAPTER

2

Calcareous Nannofossil Biozonation

OUTLINE

2.1 INTRODUCTION

Seventeen (17) Paleogene and twenty one (21) Neogene Calcareous Nannofossil Zones were recognized and described in this study based on the calcareous nannofossil distribution in the selected Niger Delta transect wells (Enclosure 2.1). Each proposed Zone is described stating its boundary criteria, as well as its characteristic assemblage composition, using features that can be easily recognized under a Transmitted Light Microscope. The following criteria were employed for the recognition of the proposed Niger Delta Zones:

- Taxon Acme,
- First and Last Occurrences (FO and LO),
- First and Last Common Occurrences (FCO and LCO),
- Frequent occurrences, and
- The ratio of key marker taxa. This refers to the ratio of the quantitative abundance of any two marker species, for example, the ratio of *Discoaster berggrenii:Discoaster quinqueramus* is 2:1 within the proposed NND11 Zone and 1:2 in the NND10 Zone (Enclosure 2.2).

The recognized zones were correlated with the low latitude calcareous nannofossil schemes that have been established by several authors, particularly the worldwide zonation schemes of Martini (1971) and Okada and Bukry (1980), among others.

The more recent detailed study of the NN5–NN4 zone of the Niger Delta by Osterloff et al. (2013) and the Niger Delta Neogene zonation sheme of Fadiya and Salami (2012), show that

Cenozoic Foraminifera and Calcareous Nannofossil Biostratigraphy of the Niger Delta. http://dx.doi.org/10.1016/B978-0-12-812161-0.00008-9

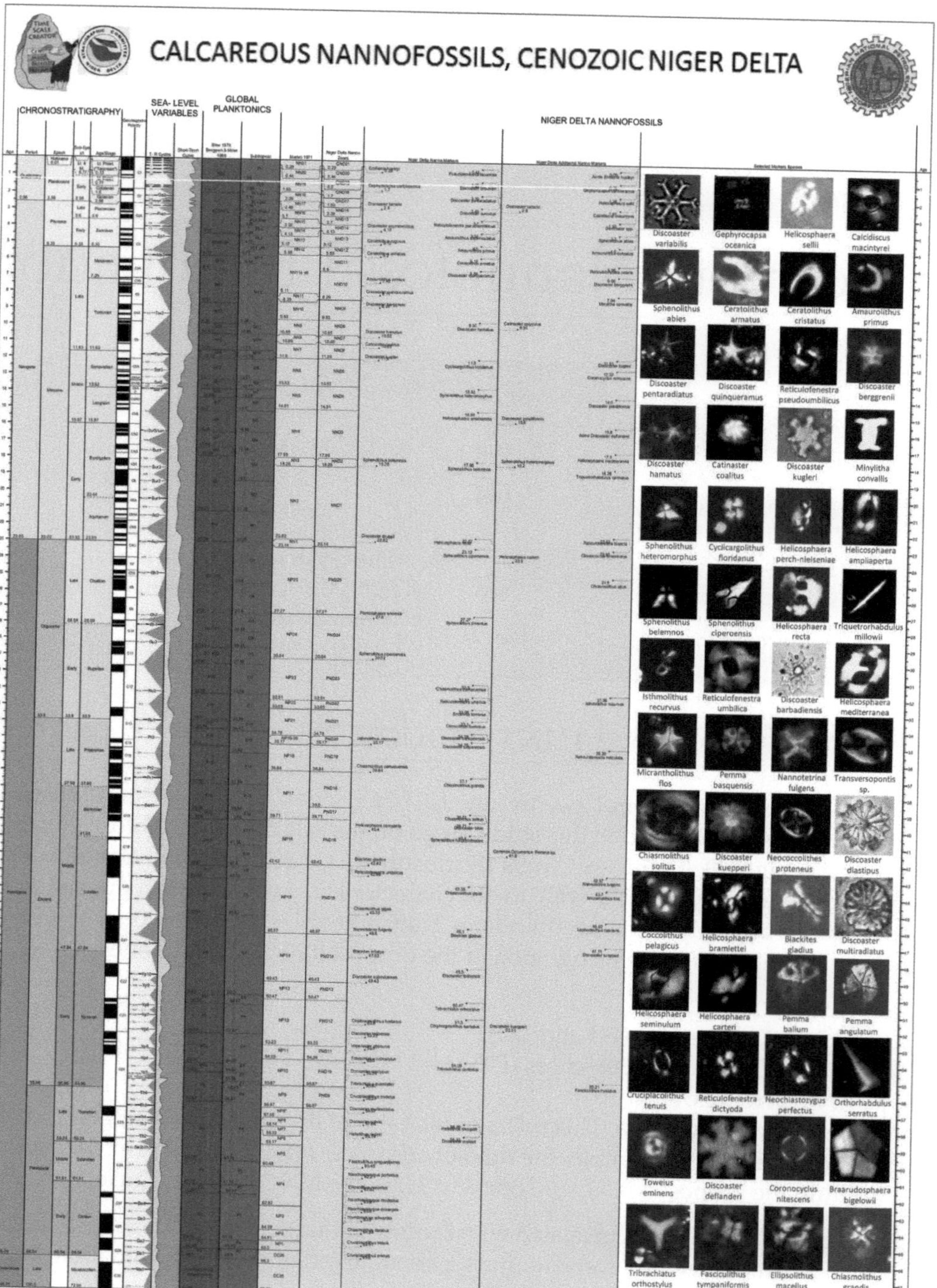

ENCLOSURE 2.1 **Calcareous nannofossil zonation of the Cenozoic Niger Delta.**

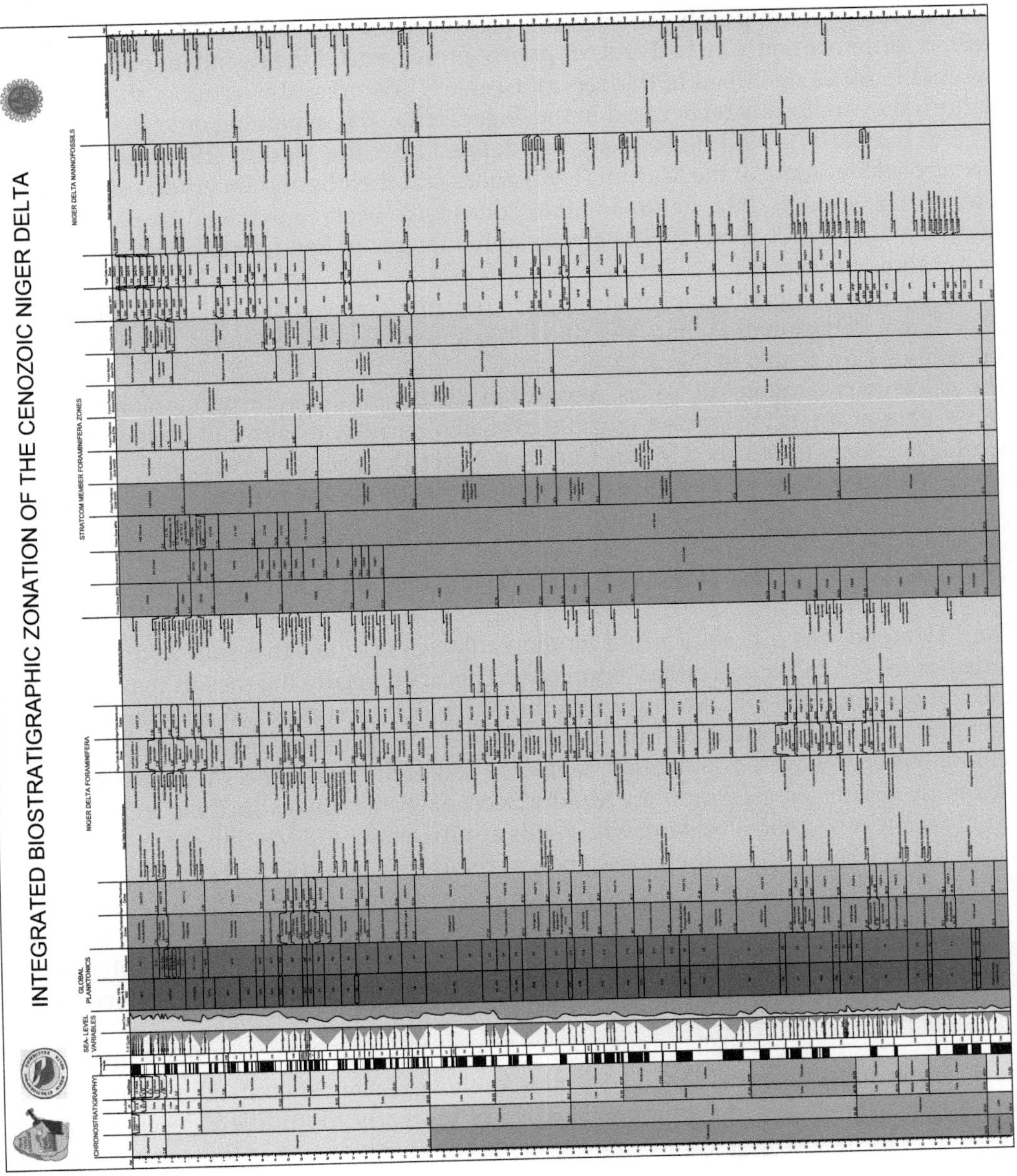

ENCLOSURE 2.2 **Integrated biostratigraphic zonation of the Cenozoic Niger Delta.**

other criteria can be used to attain an equally effective refined zonation of the Niger Delta sequence.

The abundance of calcareous nannofossils in the Paleogene–Neogene strata of the Niger Delta varies markedly. Nannofossils are common at some horizons, few and very rare at others. In some old and poorly stored wells, nannofossil data are generally scanty, and the recovered forms are badly etched and/or poorly preserved. Of the twenty one (21) Neogene Nannofossils zones described in this report, twelve (12) represent new zones that are based on datums that are locally recognized in the Niger Delta. The remaining nine zones, particularly from the Early to Middle Miocene, are defined after the Martini (1971) zonal scheme. Where applicable, some of the Martini (1971) zones and their boundary criteria are retained. Where there is enough data, datums within zones have been constrained by dated marine flooding events (see Chapter 4). The stratigraphic ranges of key nannofossils in the Niger Delta are shown in Fig. 2.1.

The stratigraphic distribution of the species recorded, as well as the significant datums, zonal and age determination, and identified Maximum Flooding Surfaces (MFS) of five selected wells are presented in Fig. 2.2A–E.

The calcareous nannofossil zones recognized in this study were harmonized with the in-house data of the listed operating companies and correlated with the foraminiferal and palynological zones established for the Cenozoic Niger Delta (Enclosure 2.2 and Part 1 of this volume), to demonstrate an integrated chronostratigraphy for the region.

2.2 PALEOGENE BIOZONATION

The Paleocene was a time of rapid nannoplankton evolution that followed the decline recorded at the Cretaceous/Tertiary boundary. This is especially the case in the subtropical latitudinal province (10–35°S) within which the Niger Delta is situated. Within the time interval of approximately 12 Ma constituting the Paleocene epoch, over 25 new genera evolved. Several of these, however, disappeared before the end of the Paleocene. By the Early Eocene, the diversity again increased until the Middle Eocene when many extinctions were recorded with only a few new genera existing within this approximately 18 Ma time span. As a result, many of the markers used in the biozonation of the Eocene are species belonging to genera that that had their origin in the Paleocene, ranging up through and into the Eocene.

Calcareous nannofossil diversity is poor in the earliest Oligocene times but they gradually blossomed with the appearance of new species especially members of the family Sphenolithaceae toward the latter part of the Early Oligocene. Another floral decline was observed in the earliest part of the Late Oligocene.

Of the over 40 StratCom wells studied, only a few penetrated Oligocene to Middle Eocene sections, while only one, the Gilli Gilli-1 well is known to have penetrated Late Paleocene to Early Eocene sediments. The nannofossil record of the Early to Middle Paleocene was not included in this study largely because no nannofossil-bearing section of this age has yet been studied in detail. This is probably because true delta development probably began only during the Late Paleocene/Early Eocene when sediments began to build out beyond the horsts and grabens, which existed between the basement rocks that formed the northern flanks of the emerging present-day Niger Delta (Adegoke et al., 1987; Osterloff et al., 2015).

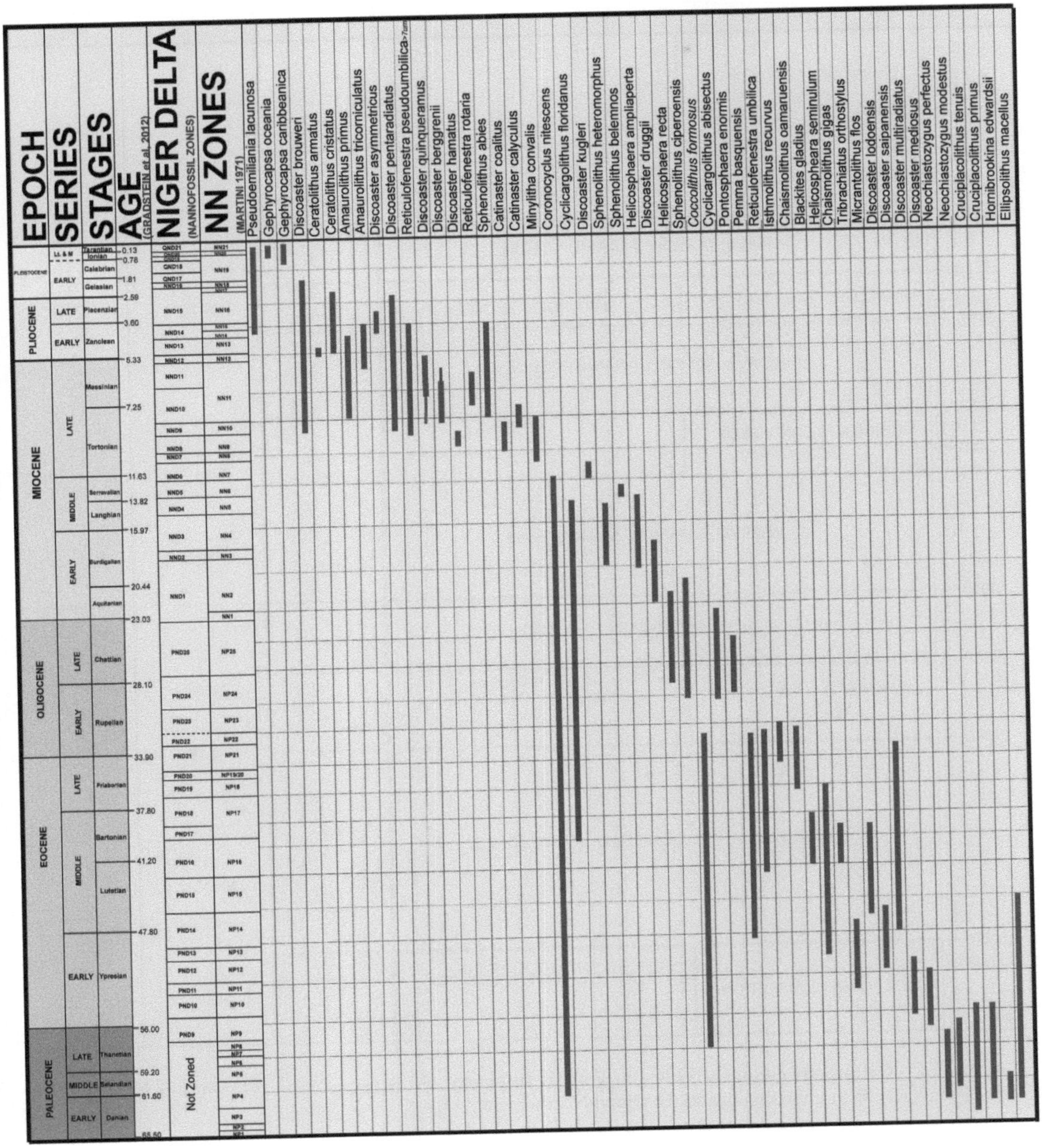

FIGURE 2.1 **Stratigraphic distribution of key calcareous nannofossils in the Niger Delta.**

(A)

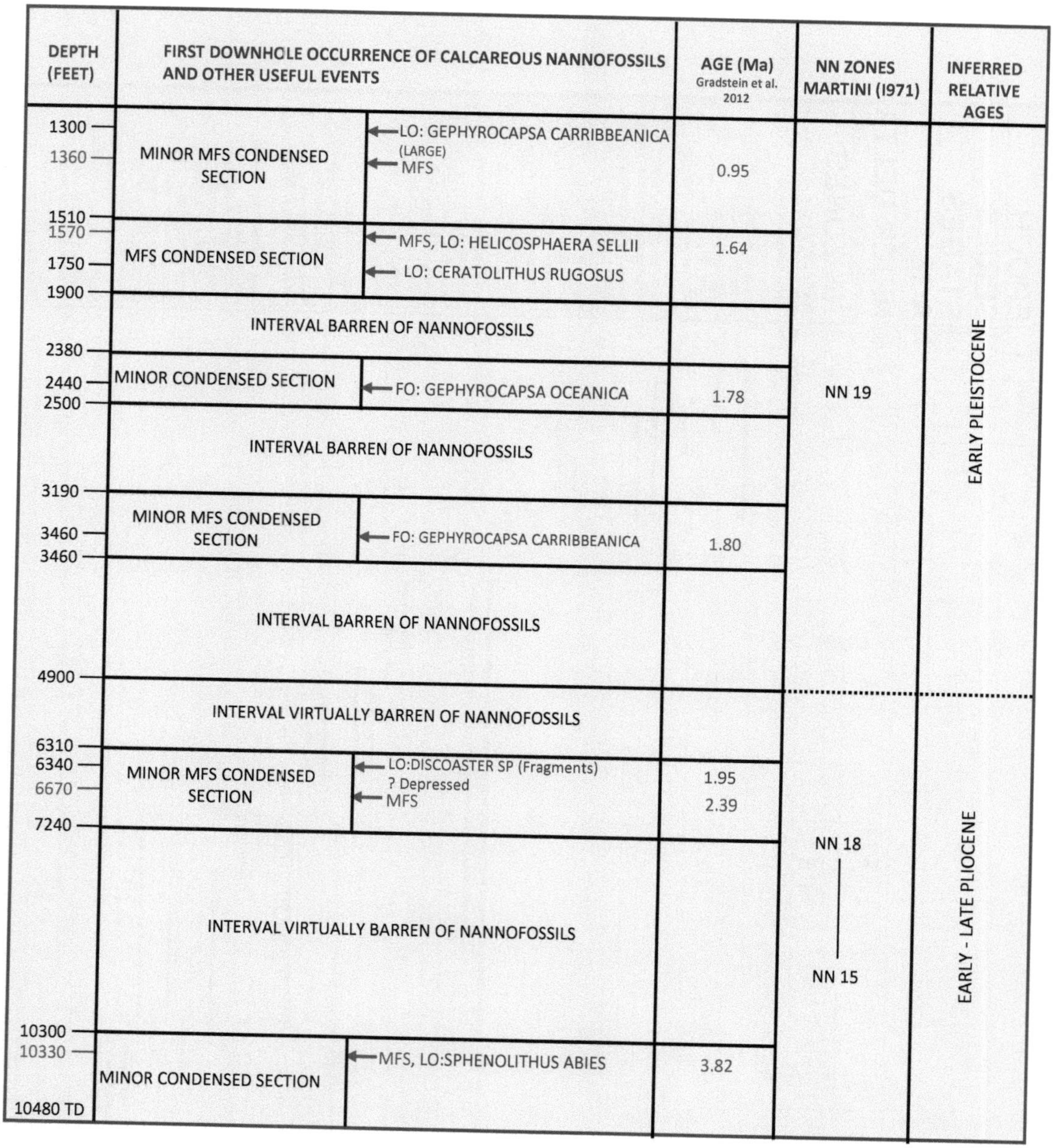

FIGURE 2.2 **Calcareous nannofosil biostratigraphic summary.** (A) Anyala-8 well, (B) Gbaran-2 well, (C) Kolo Creek-39 well, (D) Nemomai-1 well, and (E) Otobo North-1 well.

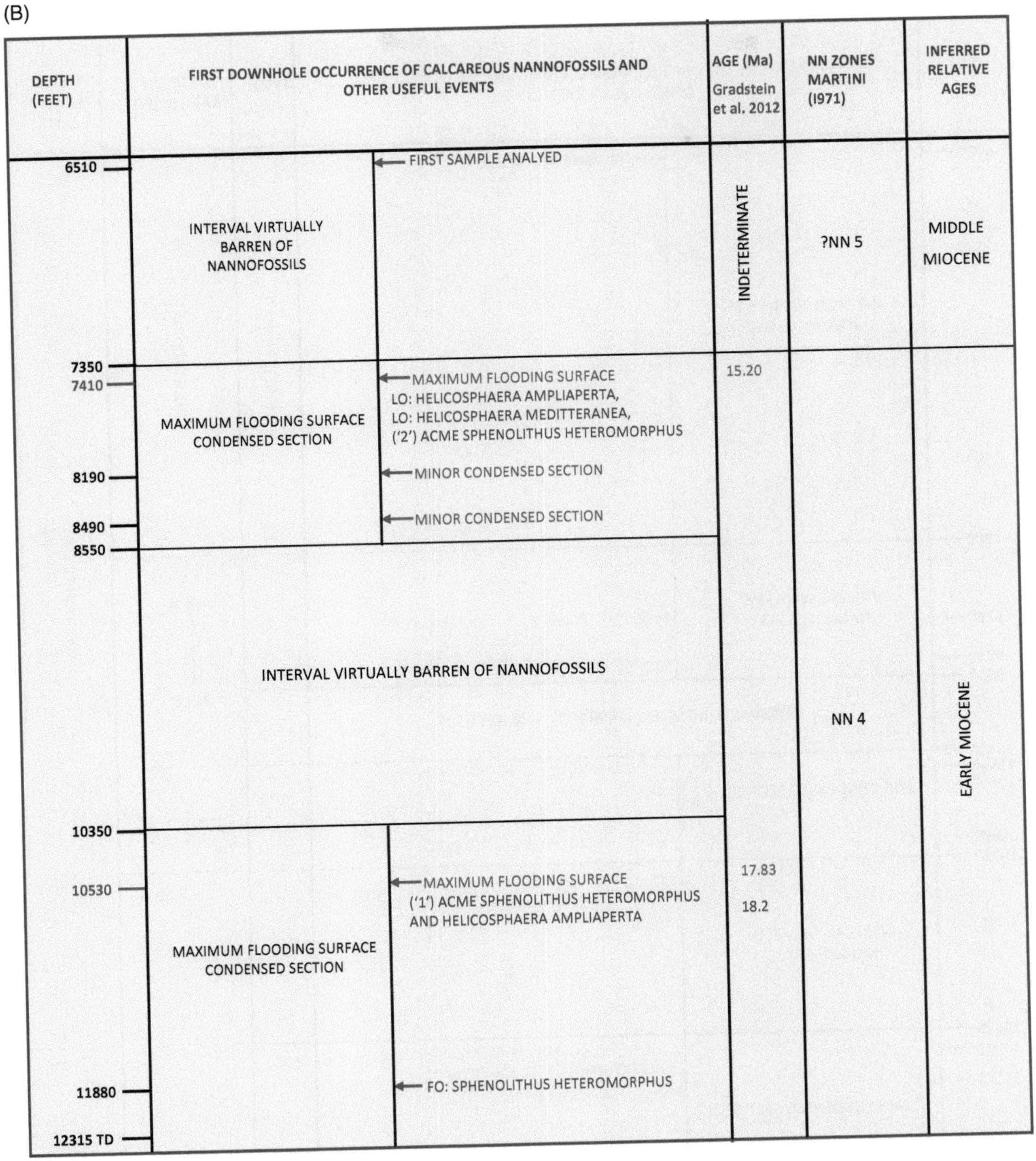

FIGURE 2.2 **(Cont.)**

(C)

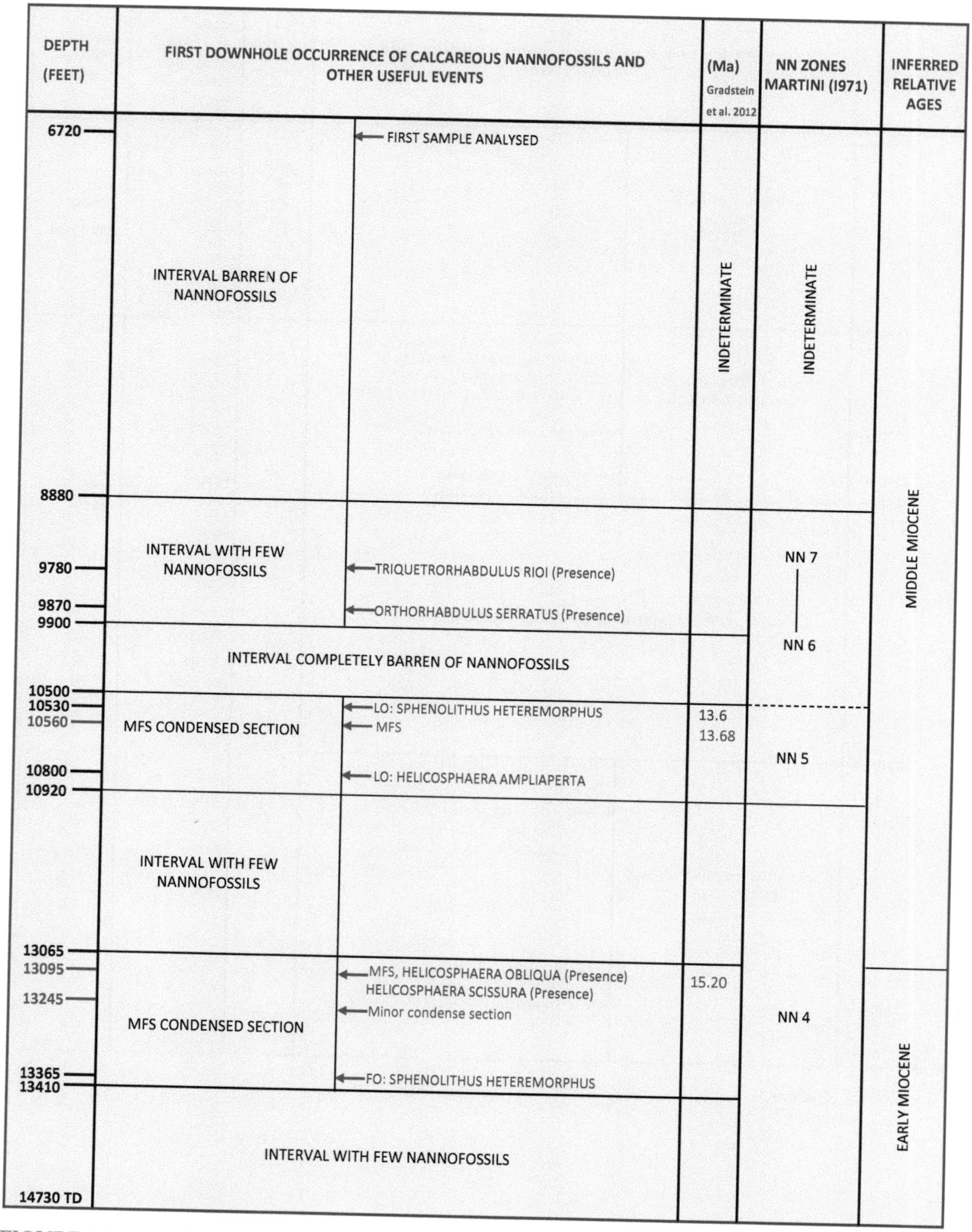

FIGURE 2.2 **(Cont.)**

(D)

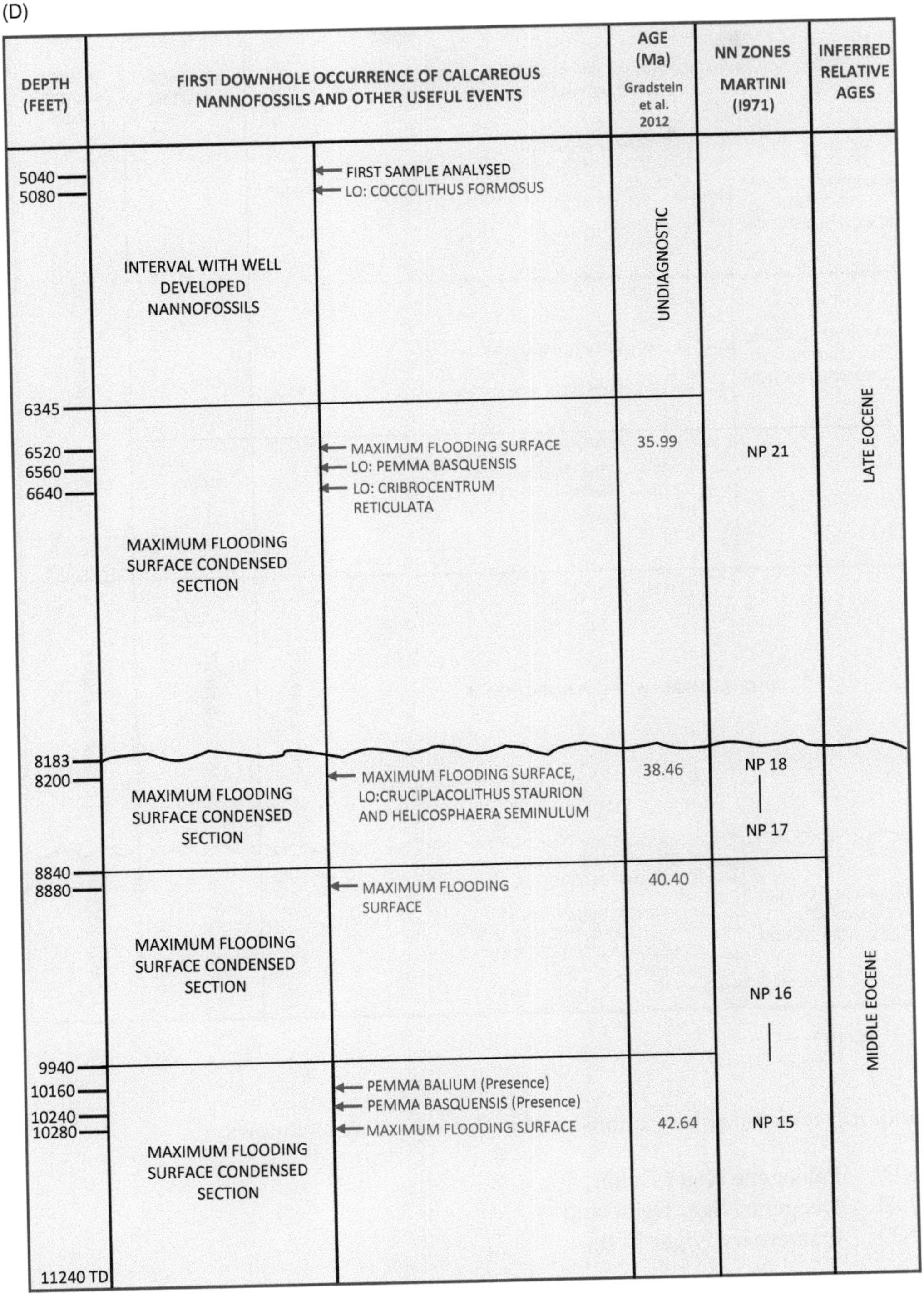

FIGURE 2.2 ***(Cont.)***

(E)

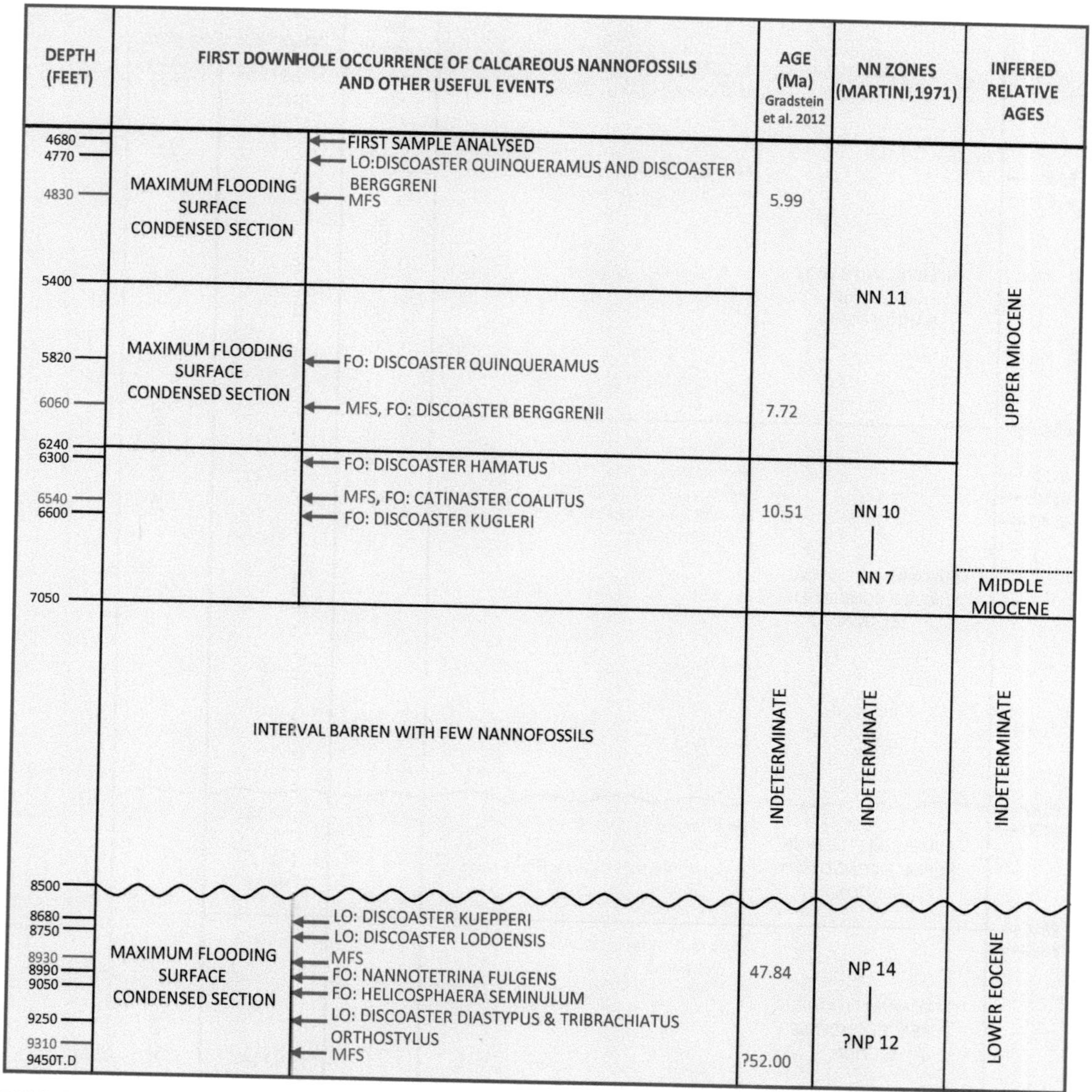

FIGURE 2.2 (*Cont.*)

The standardized zonal abbreviations used in this report are as follows:

1. PND Paleogene Niger Delta,
2. NND Neogene Niger Delta, and
3. QND Quaternary Niger Delta.

The zones established are described in detail below from the oldest age to youngest.

2.2.1 The Paleogene Niger Delta (PND9–25) Calcareous Nannofossil Zones

The Paleogene sections in the Niger Delta are yet to be studied in as much detail as their Neogene and Pleistocene counterparts. As a result, the zones described in this section are basically founded on the zonation scheme of Martini (1971). They have been supplemented, where appropriate, by events and datums recognized within and outside the StratCom scope of work in the Niger Delta. Most of these supplementary events have been recognized in zonation schemes proposed for other provinces in the world. The few zones that are new have been constrained in age by foraminifera and palynomorph data that are available for several of the intervals where they were observed. The Paleogene Niger Delta Calcareous Nannofossil Zones are coded PND Zones 9–25 as detailed further.

PND9 ZONE
This Zone is equivalent to the *Discoaster multiradiatus* Zone of Bramlette and Sullivan (1961) and NP9 Zone of Martini (1971).
Definition: FO (Base) of *D. multiradiatus* to FO (Base) *Tribrachiatus bramlettei* or any of the youngest species of *Fasculithus.*
Age: Late Paleocene/Early Eocene.
Occurrence: This zone was recorded only in the Gilli Gilli-1 and Ihuo-1 wells.
Remarks: This zone has also been defined by earlier authors to range from the FO (Base) of *D. multiradiatus* to the FO (Base) of *T. bramlettei*. Within this Zone, there are several species of *Rhomboaster* and the FO (Base) of several *Discoaster* spp. Where *Tribrachiatus* is absent or very rare, *Fasciculithus* can be used to approximately define the top of NP9 and, by extension, the Paleocene/Eocene boundary.

PND10 ZONE
This Zone is equivalent to the *Tribrachiatus contortus* Zone of Hay (1964) and Bukry (1973a), the NP10 Zone of Martini (1971), and the CP9a Zone of Okada and Bukry (1980).
Definition: Interval from the FO (Base) of *T. bramlettei* (and/or *Discoaster diastypus)* to the LO (Top) of *T. contortus.*
Age: Early Eocene.
Occurrence: Gilli Gilli-1 and Deep Offshore of Nigeria's Otobo North-1 wells.
Remarks: When *Tribrachiatus* spp. is present, it is very easy to recognize the top and base of this Zone in the Niger Delta. In sections where *T. contortus* is very rare and thus it's LO (Top) is difficult to find, the FO (Base) of *Tribrachiatus orthostylus*, which usually occurs shortly before the LO (Top) of *T. contortus* can be used as an approximation of the NP10/NP11 boundary. Other important events within the Zone are the LO of *Discoaster multiradiatus* and *Ellipsolithus macellus.*

PND11 ZONE
This Zone is equivalent to the *Discoaster binodosus* Zone of Mohler and Hay, In: Hay et al. (1967), the NP11 Zone of Martini (1971), and the CP9b Zone of Okada and Bukry (1980).
Definition: Interval from the LO (Top) of *T. contortus* to the FO (Base) of *Discoaster lodoensis.*
Age: Early Eocene.
Occurrence: Gilli Gilli-1 and Deep Offshore Nigeria, Otobo North-1 wells.

Remarks: *D. binodosus* may or may not be present in the Zone, which is defined by the absence of the marker species for the zones that are above and below. *Chiasmolithus grandis*, *Imperiaster obscurus*, *Discoaster distinctus*, *Sphenolithus editus*, and *Sphenolithus radians* are important nannofossils constituting the assemblage within this Zone.

PND12 ZONE

This Zone is equivalent to the *T. orthostylus* Zone of Bronnimann and Stradner (1960) and Bukry (1973b), NP12 Zone of Martini (1971), and CP10 of Okada and Bukry (1980).
Definition: Interval from the FO (Base) of *D. lodoensis* to the LO (Top) of *T. orthostylus*.
Age: Early Eocene.
Occurrence: Gilli Gilli-1 and Deep Offshore Nigeria's Otobo North-1 wells.
Remarks: In assemblages where *D. lodoensis* and *T. orthostylus* are poorly represented, the presence of *Discoaster kuepperi* or *Rhabdosphaera truncata* (the FO of which approximates the FO of *D. lodoensis*) can still be used to define the zone. Species of *Tribrachiatus* and *I. obscurus* dominate this Zone.

PND13 ZONE

This Zone is equivalent to the *D. lodoensis* Zone of Bronnimann and Stradner (1960) and Bukry (1973a), the NP13 Zone of Martini (1971), and CP11 of Okada and Bukry (1980).
Definition: Interval from the LO (Top) of *T. orthostylus* to the FO (Base) of *Discoaster sublodoensis*.
Age: Early Eocene.
Occurrence: Deep Offshore Nigeria Otobo North-1 and Gilli Gilli-1 wells.
Remarks: The base of the Zone is easily determined in the Niger Delta by the LO (Top) of *T. orthostylus*. *D. diastypus* has it's LO (Top) in NP13, while *Sphenolithus conspicuus* and *S. editus* disappear near the base of the PND13 zone.

PND14 ZONE

This Zone is equivalent to the *D. sublodoensis* zone of Hay (1964) and Bukry (1973a), NP14 Zone of Martini (1971), and the CP12 Zone of Okada and Bukry (1980).
Definition: Interval from the FO (Base) of *D. sublodoensis* to the FO (Base) of *Nannotetrina* spp.
Age: Early/Middle Eocene.
Occurrence: Deep Offshore Nigeria (Otobo North-1) and Gilli Gilli-1 wells.
Remarks: *D. lodoensis* overlaps with *D. sublodoensis* in the lower part of the Zone. The top of the Zone has sometimes been approximated by using the FO of any of the species of *Nannotetrina* instead of *Nannotetrina fulgens*, which can be very rare in some sections where other species of *Nannotetrina* have been found to be consistently present.

PND15 ZONE

This Zone is equivalent to part of the *N. fulgens* Zone of Hay, In: Hay et al. (1967), emended by Martini (1970), Bukry (1973a), and the NP15 Zone of Martini (1971).
Definition: Interval from the FO (Base) of *N. fulgens* to either the LO (Top) of *N. fulgens* or the LO of *Blackites gladius*.
Age: Middle Eocene
Occurrence: Nemomai-1, Deep Offshore Nigeria Otobo North-1, and Aroh-1 wells.
Remarks: The Zone is recognized as the total range of *Nannotetrina* spp. It is more convenient to use the presence of this genus rather than the different species in the

Niger Delta. *Chiasmolithus gigas*, although rare, has been found to cooccur with the zonal marker. The FO of *Reticulofenestra reticulata* (*Cribrocentrum reticulatum*) also occurs within the zone, while the FO (Base) of *Reticulofenestra umbilicus* occurs near the top of the zone.

PND16 ZONE

This Zone is equivalent to part of the *Discoaster tanii nodifer* Zone (Hay et al., 1967), emend. Martini (1971), and the NP16 Zone of Martini (1971).
Definition: Interval from LO (Top) of *N. fulgens* to the LO (Top) of *Chiasmolithus solitus*.
Age: Middle Eocene.
Occurrence: Deep Offshore Nigeria Otobo North-1 and Aroh-1 wells.
Remarks: The FO (Base) of *Reticulofenestra umbilicus*, is slightly older but approximates this event. It has been used in place of the rare *B. gladius*. The LO (Top) of *Nannotetrina* spp. also approximates the PND16/PND15 zonal boundary. The FO (Base) of *Helicosphaera compacta* and the LO (Top) of *Sphenolithus furcatolithoides* is found within this Zone. The LCO of *Pemma* spp. or the influx of *Pemma papillatum* is a major characteristic of this Zone.

PND17 ZONE

The Zone is equivalent to the lower part of the *Discoaster saipanensis* Zone of Martini (1971) and the lower part of the NP17 Zone of Martini (1971).
Definition: Interval from the LO (Top) of *C. solitus* to the LO (Top) of *Transversopontis* spp.
Age: Middle Eocene.
Occurrence: Deep Offshore Nigeria (Otobo North-1 well), Nemomai-1, Adofi-1, and Gilli Gilli-1 wells.
Remarks: This is a relatively short duration zone that is recognized only through close sampling.

PND18 ZONE

This Zone is equivalent to the "upper" part of the NP17 Zone of Martini (1971).
Definition: Interval from the LOs of either of *Transversopontis* spp. or *Discoaster bifax* to the FO (Base) of *C. oamaruensis*.
Age: Middle/Late Eocene.
Occurrence: Gilli Gilli-1 and Nemomai-1 wells.
Remarks: In the Niger Delta, this Zone has been recognized in several deep offshore sections.

PND19 ZONE

The PND19 Zone is equivalent to the *Chiasmolithus oamaruensis* Zone of Martini (1970) and the NP18 Zone of Martini (1971).
Definition: Interval from the FO (Base) of *C. oamaruensis* to the FO (Base) of *Ismolithus recurvus*.
Age: Late Eocene.
Occurrence: Mosogar-2, Nemomai-1, and Jesse-1 wells.
Remarks: This Zone can be difficult to recognize in low latitudes. *R. reticulata* is a reliable species, which occurs throughout the PND19 Zone. *Helicosphaera bramlettei* and *Sphenolithus predistentus*, as well as the LO (Top) of *Cruciplacolithus* spp., *Helicosphaera seminulum*, and *Neococcolith* spp. are found frequently occurring within this Zone.

PND20 ZONE

This is a composite Zone that is equivalent to the NP19 and NP20 zones of Martini (1971).
Definition: Interval between the FO of *Ismolithus recurvus* to the LO of disc-shaped *Discoaster* species.
Age: Late Eocene.
Occurrence: Mosogar-2 and Jesse-1 wells. *Remarks*: If *S. radians* or the bigger *Sphenolithus pseudoradians* are found, then this Zone can be divided into the NP20 and NP19 zones (Martini, 1971). *R. reticulata* (*C. reticulatum*) and *I. recurvus* are common within the upper part of this composite zone.

PND21 ZONE

This Zone is equivalent to the *Ericsonia subdistichus* Zone of Roth and Hay In: Hay et al. (1967), emended by Martini and Worsley (1970), and the NP21 Zone of Martini (1971).
Definition: Interval between the LO (Top) of disc-shaped discoasters to the LO (Top) of *Coccolithus formosus.*
Age: Late Eocene/Early Oligocene.
Occurrence: Amakubu-1, Iheoma-1, Nemomai-1, and Pologbene-1 wells.
Remarks: Commonly found ranging into this Zone are *Discoaster barbadiensis, D. saipanensis*, and the LO (Top) of *R. reticulata* (*C. reticulatum*). The acme of *Clausioccocus* spp. and the LO of *Clausiococcus subdistichus* marks the Eocene/Oligocene boundary (33.90 Ma). In the absence of *Clausioccocus* spp., the presence of disc-shaped *Discoaster* species is taken to approximate this boundary.

PND22 ZONE

This Zone is equivalent to the *R. reticulata* Zone of Bramlette and Wilcoxon (1967a), emended by Martini and Worsley (1970), and the NP22 Zone of Martini (1971).
Definition: Interval between the LO (Top) of *C. formosus* to the LO (Top) of *R. umbilica.*
Age: Early Oligocene.
Occurrence: Amakubu-1, Opuama-1, Benin River-3, and Iheoma-1 wells.
Remarks: The LO (Top) of *R. umbilicus* and *Reticulofenestra hillae* is an approximate alternative to *C. formosus. Lanternithus minutus*, which is commonly recognized in other provinces and which is recorded in many Early Oligocene zones, has not been found in the Niger Delta. This is mainly due to the poorly fossiliferous nature of the well sections so far analyzed in the Niger Delta.

PND23 ZONE

This Zone is equivalent to the *S. predistentus* Zone of Bramlette and Wilcoxon (1967a), emended by Martini (1970), and the NP23 Zone of Martini (1971).
Definition: Interval between the LO (Top) of *R. umbilica* and the FO of *Sphenolithus ciperoensis* [and/or FO (Base) of *Cyclicargolithus abisectus*].
Age: Early Oligocene.
Occurrence: Amakubu-1, Opuama-1, Benin River-3, and Iheoma-1 wells.
Remarks: *Helicosphaera recta* FO (Base) is an alternative marker for the top of this Zone. *R. umbilicus* is distinguishable from other reticulofenestrid nannofossils by its larger size and corresponding larger central area (see Appendix 1). *I. recurvus* and *C. oamaruensis* commonly have their LO's (Top) within this Zone. Also if the samples are rich, it is often easy to find the FO (Base) of *Sphenolithus distentus* at the lower part of the Zone.

The PND23 Zone can sometimes be difficult to recognize especially where specimens of sphenoliths are very small and rare, and when *S. ciperoensis* is difficult to identify. On the other hand, the base is very easy to recognize because of the presence of the ubiquitous *R. umbilicus*.

PND24 ZONE
This Zone is equivalent to the *S. distentus* Zone of Bramlette and Wilcoxon (1967a) and the NP24 Zone of Martini (1971).
Definition: Interval between the FO (Base) of *S. ciperoensis* and the LO (Top) of *S. distentus*.
Age: Early/Late Oligocene.
Occurrence: Pologbene-1 and Fragbene-1 wells.
Remarks: In the Niger Delta, the FO's (Base) of *H. recta* and *C. abisectus* commonly occur with *S. ciperoensis*. The FO (Base) of *Pontosphaera enormis* has also shown to be a useful event for the subdivision of the Upper Oligocene and is an approximation of the NP24/NP25 boundary (Martini, 1981a). Other important nannofossils within the zone include *H. bramlettei* and *Helicosphaera euphratis*.

PND25 ZONE
This Zone is equivalent to the *S. ciperoensis* Zone of Bramlette and Wilcoxon (1967a) as emended by Martini (1976) and the NP25 Zone Martini (1971).
Definition: Interval between the LO (Top) of *S. distentus* to LO (Top) *S. ciperoensis* and/or *Reticulofenestra bisectus* and *H. recta*.
Age: Late Oligocene.
Occurrence: Nkali-1 well.
Remarks: *Triquetrorhabdulus milowii*, *Triquetrorhabdulus carinatus*, *C. abisectus*, *Cyclicargolithus floridanus*, *Coccolithus pelagicus*, *Coccolithus miopelagicus*, *Chiasmolithus altus*, and *Helicosphaera truempyi* are important taxa observed within the PND25 Zone. *Discoaster deflandrei* with seven rays has been recorded in this Zone and continues to exist in older strata.

2.3 NEOGENE BIOZONATION

The Miocene, Pliocene, and Pleistocene Epochs represent much of the time interval during which the River Niger deposited enormous volumes of sediments in the Niger Delta. Large volumes of nannofossil-rich shales and silts were deposited distally. These horizons have been drilled in both the onshore and offshore portions of the Niger Delta. The operating companies have thus carried out several high-resolution biostratigraphic analytical projects many of which included nannofossil studies. The validity of the results have been confirmed by the analysis of over 700 samples selected across major Condensed Sections from 40 of the Strat-Com wells provided for study. For ease of comparison, the equivalent NN Zones of Martini (1971) and CN Zones of Okada and Bukry (1980) are indicated further.

The Late Pleistocene sequences of the Niger Delta are yet to be studied in detail. Samples available for study are recovered from a few wells, while much of the least known sections are within the non-nannofossil bearing fluvial/alluvial, very proximal marine sediments of the Benin Formation.

2.3.1 The Neogene Niger Delta (NND1–16) Calcareous Nannofossil Zones

Sixteen Neogene Calcareous Nannofossil Zones recognized in the Niger Delta are described in detail.

NND1 ZONE

The NND1 Zone is equivalent to the *T. carinatus* and *Discoaster druggii* zones of Bramlette and Wilcoxon (1967a) and the NN1 and NN2 zones of Martini and Worsley (1970).
Definition: Interval between the LO (Top) of *H. recta* (and/or FO (Base) of *S. ciperoensis*) to the LO (Top) of *Sphenolithus belemnos* and/or LO (Top) *T. carinatus*.
Age: Late Oligocene/Early Miocene
Occurrence: Obeaja-1, Opuama-1, Amara-1, Fragbene-1, Benin River-3, Amakubu-1 Adagbassa Creek-1, Zarama-7, and Agberi North-1 wells.
Remarks: This zone is relatively thick and is the oldest zone represented in the Miocene. *C. floridanus* and *T. carinatus* are usually present throughout the Zone, the former being especially common. In poorly preserved material and in samples with common detritus, it can be difficult to distinguish poorly preserved *T. carinatus* from other elongated calcite particles. Also *D. druggii* can be difficult to distinguish from other discoasters when in overgrown specimens.

The base of this Zone is taken to approximate the Oligocene/Miocene boundary, as it is easy to define based on the disappearance of typical Oligocene species, such as *S. ciperoensis*, *Reticulofenestra bisecta*, and *Reticulofenestra scrippsae*. Other significant events recorded within the Zone include the FO of *D. druggii* (very rare) and *Helicosphaera carteri* (acme).

The Oligocene–Miocene turnover in nannofossils is characterized by the LO (Top) of *H. recta*, LCO of *R. bisecta*, LO (Top) of *Zygrhablithus bijugatus*, and LO (Top) of *S. ciperoensis*. The first indication of the Neogene is often easily diagnosed by the continuous occurrence of *H. carteri*. The FO (Base) of *H. carteri* has been found to be gradational but useful in the Niger Delta.

NND2 ZONE

The Zone is equivalent to the *S. belemnos* Zone of Bramlette and Wilcoxon (1967a), the NN3 Zone of Martini (1971), and CN2 Zone of Okada and Bukry (1980).
Definition: Interval between the LO (Top) of *T. carinatus* and/or the FO (Base) of *S. belemnos* to the LO (Top) of *S. belemnos*.
Age: Early Miocene.
Occurrence: Nkali-1 and Zarama-7 wells.
Remarks: The nannofossil assemblage of this Zone is often dominated by *D. deflandrei* and *C. floridanus*. Besides, *Sphenolithus moriformis*, *Sphenolithus dissimilis*, *S. belemnos*, *C. miopelagicus*, and *T. milowii* are also common. In poorly preserved material, it can be difficult to distinguish *Sphenolithus heteromorphus* from *Sphenolithus conicus*, especially if one compares the forms to the holotypes of the two species. The FO (Base) of *S. heteromorphus* can also be used as the base of the zone in sections with poor records of *S. belemnos*.

NND3 ZONE

The Zone is equivalent to *Helicosphaera ampliaperta* Zone of Bramlette and Wilcoxon (1967a) and NN4 Zone of Martini (1971).

Definition: Interval between the LO (Top) of *S. belemnos* or LO (Base) of *S. heteromorphus* and the LO (Top) of *H. ampliaperta*.
Age: Early/Middle Miocene.
Occurrence: Nkali-1, Opuama-1, Zarama-7, Gbaran-2, and Kolo Creek-39 wells.
Remarks: As in the NND4 Zone, this Zone is easy to identify because the nominate taxon, *H. ampliaperta* is abundant in the Niger Delta. Both the LCO and the acme of *D. deflandrei* are found within the Zone. The FO (Base) of *Discoaster petaliformis* is commonly found in association with abundant *D. deflandrei*. The cooccurrence of *S. heteromorphus* and *H. ampliaperta* is seen throughout the zone. The Early/Middle Miocene boundary is placed above the acme of *D. deflandrei*.

NND4 ZONE
This Zone is equivalent to the *S. heteromorphus* Zone of Bramlette and Wilcoxon (1967a), the NN5 Zone of Martini (1971), and the CN4 zone of Okada and Bukry (1980).
Definition: Interval between the LO (Top) of *H. ampliaperta* and LO (Top) of *S. heteromorphus*.
Age: Middle Miocene.
Occurrence: Beniboye North-1, Tuomo West-1, Gbaran-2, and Nkali-1 wells.
Remarks: *S. heteromorphus*, *Discoaster sanmiguelensis*, and *D. deflandrei* are important markers within this zone. The zone is thus differentiated from the preceding NND3 Zone by the absence of *H. ampliaperta*. The Zone is thus easy to identify. The LO of *Helicosphaera meditteranea* is close to the top of the Zone.

NND5 ZONE
This Zone is equivalent to the *Discoaster exilis* Zone of Hay (1970), the NN6 Zone of Martini (1971), and the CN5a Subzone of Okada and Bukry (1980).
Definition: Interval between the LO (Top) of *S. heteromorphus* to the FO (Base) of *Discoaster kugleri* and/or LO (Top) of *C. floridanus*.
Age: Middle Miocene.
Occurrence: Kolo Creek-39, Beniboye North-1, and Utapate-1 wells.
Remarks: This zone and the succeeding NND6 and NND7 as seen in the Niger Delta are characterized by paucity of calcareous nannofossils. This observation was earlier made by Fadiya and Salami (2012) and linked to the Middle Miocene carbonate crash observed in the eastern equatorial Pacific by King et al. The event is now regarded as worldwide, as it has been documented from many oceans around the world where it spanned the Middle to Late Miocene and are characterized by lower calcium carbonate and poor preservation of calcium carbonate microfossils. This zone, as in the succeeding zones, are also characterized by the presence of reticulofenestrids. The LO (Top) of *Orthorhabdulus serratus* is rare near the boundary of NND4/NND5.

NND6 ZONE
This Zone is equivalent to the *D. kugleri* Zone of Bramlette and Wilcoxon (1967a); the NN7 Zone of Martini (1971), and the CN5a Zone of Okada and Bukry (1980).
Definition: Interval between the FO (Base) of *D. kugleri* or LO (Top) *C. floridanus*, and the FO (Base) of *Catinaster coalitus*.
Age: Middle/Late Miocene.
Occurrence: Beniboye North-1 and Kolo Creek-39 wells.

Remarks: Where *Discoaster* and *C. coalitus* are rare or missing, it will be difficult to characterize the NND6 Zone because of the poor distribution of nannofossils in this interval. Fornaciari et al. (1996) suggest that the LCO of *Calcidiscus premacintyrei*, which occurs just before the FO (Base) of *D. kugleri* is the best alternative.

NND7 ZONE

The NND7 Zone is equivalent to the *C. coalitus* Zone of Bramlette and Wilcoxon (1967a), the NN8 of Martini (1971), and the *C. coalitus* Zone (CN6) of Okada and Bukry (1980).
Definition: Interval between the FO (Base) of *C. coalitus* to the FO (Base) of *D. hamatus*.
Age: Late Miocene.
Occurrence: Beniboye North-1 well.
Remarks: The Zone is characterized by the cooccurrence of *Discoaster bollii* and *C. coalitus*. The influx of five-rayed *Discoaster* and the FCO of *Sphenolithus abies* normally marks the top of the Zone and the presence of NND8 Zone above. This interval is also characterized by paucity of nannofossils. *D. bollii*, *C. coalitus*, *H. carteri*, *S. moriformis*, and *Reticulofenestra* spp. are common nannofossils within the zone.

NND8 ZONE

This Zone is equivalent to the *D. hamatus* Zone of Bramlette and Wilcoxon (1967a), the NN9 Zone of Martini (1971), and the *D. hamatus* Zone (CN7) of Okada and Bukry (1980).
Definition: Interval between the FO (Base) and LO (Top) of *D. hamatus*.
Age: Late Miocene.
Occurrence: Beniboye North-1 well.
Remarks: Five-rayed *Discoaster* species notably *D. hamatus* and *Discoaster prepentaradiatus* have their FOs within the Zone.

The FCO of *S. abies* has been consistently observed within this Zone in studies of the NND8 Zone in the Niger Delta. This high-abundance occurrence is such that *S. moriformis* replaces *S. abies* as the dominant species of *Sphenolithus*.

NND9 ZONE

This Zone is equivalent to *Discoaster calcaris* NN10 Zone of Martini (1969) and the *Discoaster neohamatus* Zone (CN8) of Okada and Bukry (1980).
Definition: Interval between the LO of *D. hamatus* (9.53 Ma) to the FO (Base) of *D. quinqueramus* (8.11 Ma) or *D. berggrenii* (8.29 Ma).
Age: Late Miocene.
Occurrence: Utapete-1, Beniboye North-1, Ibotio-1, Andoni-1, and Gbadudu-2 (7500–9816 ft.) wells.
Remarks: The important bioevents observed within this Zone include the LOs (Tops) of *D. bollii*, *D. prepentaradiatus*, *C. coalitus*, and *Helicosphaera walbersdorfensis*, as well as the abundant occurrence of small reticulofenestrids.

NND10 ZONE

The Zone is equivalent to the lower part of the *D. quinqueramus* Zone of Gartner (1969) and lower part of the NN11a Zone of Martini (1971), and the *D. quinqueramus* Subzone (CN9a) of Okada and Bukry (1980).
Definition: Interval from the FO (Base) of *D. berggrenii* or *D. quinqueramus* to the FO of any species of *Ceratolithus*, *Amaurolithus*, or any intermediate form.

Age: Late Miocene.
Occurrence: The zone has been recorded in the HB-1, Igodo-1, Koronama-3, Yomene-1, Etim-1, Utue-17a, Inanga-1 (9,930–11,000 ft.), Andoni-1, and Gbadudu-2 wells (6,611–7,500 ft.).
Remarks: The abundance ratio of species of *Discoaster* in this Zone is such that the *D. quinqueramus/D. berggrenii* ratio is approximately 1:2. The LO (Top) of *Minylitha convallis* and *Catinaster mexicanus* is an important datum within this Zone. The recorded FO (Base) of *Amaurolithus primus*, the first horseshoe-shaped calcareous nannofossils in the Neogene is also within this Zone

NND11 ZONE
This Zone is equivalent to the upper *D. quinqueramus* Zone of Gartner (1969), the upper part of the NN11b Zone of Martini (1971), and the *D. quinqueramus* Zone (CN9b) of Okada and Bukry (1980).
Definition: Interval from the FO of *Amaurolithus* spp. to the LCO of *D. quinqueramus* (5.59 Ma).
Age: Late Miocene.
Occurrence: Okubie-8, Igodo-1, HB-1, Yomene-1, Etim-1, Utue-17A, Inanga-1 (8010–9930 ft.), Andoni-1, and Gbadudu-2 wells (3361–6611 ft.).
Remarks: The abundance ratio of *D. quinqueramus/D. berggrenii* within this Zone is such that it is approximately 2:1. An acme of *S. abies* is also often found within this Zone. It is practical to recognize the base of this Zone at the FO of any species of *Amaurolithus* or any intermediate variants (see Appendix 1). This event occurs halfway into the NN11 Zone of Martini (1971).

NND12 ZONE
This Zone is equivalent to the *Amaurolithus tricorniculatus* Zone of Gartner (1969), the NN12 Zone of Martini (1971), and the lower and middle subzones of the *A. tricorniculatus* Zone (CN10) of Okada and Bukry (1980).
Definition: Interval from the LCO of *D. quinqueramus* (5.59 Ma) to the FO (Base) of *Ceratolithus rugosus* and/or LO (Top) *Ceratolithus armatus* at (5.12 Ma).
Age: Late Miocene/Early Pliocene.
Occurrence: Inanga-1 (8010 ft.), Akamba-1, Okubie-8, HB-1, and Gbadudu-2 (3361–1640 ft.) wells.
Remarks: The Miocene/Pliocene boundary is placed approximately at the FO (Base) of *Ceratolithus acutus* (often based on the presence of the species). *C. acutus* is very rare but forms a very useful datum when found. *C. armatus* also helps to mark the Miocene/Pliocene boundary, placing the boundary below the FO (Base) of the *C. armatus*. The FO (Base) of *C. acutus* and *C. armatus*, as well as the LO (Top) of *Triquetrorhabdulus rugosus* are associated with the boundary. Fadiya and Salami (2012) also used the FO of either *C. armatus* or *C. acutus* to mark the Miocene/Pliocene boundary.

NND13 ZONE
The NND13 Zone is equivalent to the *C. rugosus* Zone of Gartner (1969), the NN13 Zone of Martini (1971), and the lower part of the *C. rugosus* Subzone (CN10c) of Okada and Bukry (1980).

Definition: Interval from the FO (Base) of *C. rugosus* to the FCO of *Gephyrocapsa* spp. (2–4 μm), the FO (Base) of *C. rugosus* to FO (Base) of *Discoaster asymmetricus*, and the FO (Base) of *C. rugosus* and/or LO (Top) of *C. acutus* to FO (Base) of *D. asymmetricus*.
Age: Early Pliocene.
Occurrence: Inanga-1 (7000–8010 ft.), Akamba-1, HB-1, Okubie-8, Anyala-4 (7980–8340 ft.), and Gbadudu-2 (1640 ft.) wells.
Remarks: The coccolith assemblage within this zone is composed of *Gephyrocapsa* spp., *Discoaster pentaradiatus*, *Reticulofenestra pseudoumbilicus*, *Reticulofenestra* spp., *P. lacunosa*, *S. abies*, and *Helicosphaera sellii*. It constitutes roughly 50% of the nannofossil population. *C. rugosus* may be rare in the Zone particularly in onshore sections. The base may thus be difficult to determine in such sections. The equivalent base or FCO of *Ceratolithus cristatus* can also be used where present.

NND14 ZONE

This Zone is equivalent to the *D. asymmetricus* and the *R. pseudoumbilica* zones of Gartner (1969), the NN14 and NN15 zones of Martini (1971), and the CN10c–d and CN11 zones of Okada and Bukry (1980).
Definition: Interval from FCO of small *Gephyrocapsa* spp. (2–4 μm) and/or FO (Base) of *D. asymmetricus* to the LO (Top) of *R. pseudoumbilicus* (>7 μm) or *Sphenolithus* spp (3.70 Ma).
Age: Early Pliocene.
Occurrence: Inanga-1 (5490–7000 ft.), Akamba-1, HB-1, Okubie-8, Anyala-4 (1500–7980 ft.), and Gbadudu-2 (870–1640 ft.) wells.
Remarks: There is a sharp decrease in the nonbirefringent ceratoliths, the LO (Top) of *A. tricorniculatus* and/or the LO (Top) of *Amaurolithus delicatus* within this zone. In addition, medium to small *Gephyrocapsa* spp. and *P. lacunosa*, ceratoliths, and Pliocene-age helicosphaerids are common within the Zone.

NND15 ZONE

This Zone is equivalent to the *D. pentaradiatus* Zone of Martini and Worsley (1970) and the *Discoaster surculus* Zone of Hay and Schmidt, In: Hay et al. (1967), emended by Gartner (1969), the NN17 and NN16 zones of Martini (1971), and CN12a–c Zone of Okada and Bukry (1980).
Definition: Interval from the LO (Top) of large forms of *R. pseudoumbilica* (>7 μm) or *Sphenolithus* spp. to the LO (Top) of *D. pentaradiatus* (2.39 Ma).
Age: Early Pliocene/Late Pliocene/Early Pleistocene.
Occurrence: Anyala-4 (1,230–1,500 ft.), Anyala-8 (10,450 ft.), Okubie-8, and Etim-1 wells.
Remarks: The Pliocene/Pleistocene boundary is at the upper part of this zone. In good marine sections, an uphole increase of *D. pentaradiatus* is found in association with this zone. The LO (Top) of *D. surculus* (2.49 Ma), marks the boundary between the NN17 and NN16 zones of Martini (1971). *P. lacunosa* becomes well established in the lower part of this zone while *Discoaster challengeri* and *Discoaster variabilis* disappear.

NND16 ZONE

This Zone is equivalent to the *Discoaster brouweri* Zone of Boudreaux and Hay (1969), the NN18 Zone of Martini (1971), and the CN12d Zone of Okada and Bukry (1980).

Definition: Interval from the LO of *D. pentaradiatus* (2.39 Ma) to the LO of *D. brouweri* (1.93 Ma).
Age: Early Pleistocene.
Occurrence: Anyala-4 (1230 ft.).
Remarks: The LO of all species of *Discoaster* marks the top of this zone. Within the Zone and in poorly fossiliferous samples, the presence of broken rays of *Discoaster* often forms a reliable basis for the establishment of this zonal interval. The size of specimens of *Gephyrocapsa* increases toward the top of this zone.

2.4 THE QUATERNARY NIGER DELTA (QND17–21) CALCAREOUS NANNOFOSSIL ZONES

Only the QND17–21 Calcareous Nannofossil Zones are recognized in the Niger Delta as characterized further.

QND17 ZONE
This Zone is equivalent to the lower part of the NN19 *P. lacunosa* Zone of Martini (1971), the *Calcidiscus macintyrei* Zone of Gartner (1977), and the Lower part of the CN13 *Crenalithus doronicoides* Zone of Okada and Bukry (1980).
Definition: Interval from the LO (Top) of *D. brouweri* (1.93 Ma) to the FO of *Gephyrocapsa caribbeanica* (1.70 Ma) or/and the LO (Top) of *C. macintyrei.*
Age: Early Pleistocene.
Occurrence: Anyala-4 well (870–1140 ft.).
Remarks: This Zone is marked by the extinction of all *Discoaster* species, as well as, by the replacement of *Pseudoemiliania* by *Gephyrocapsa* as the dominant reticulofenestrid types.

QND18 ZONE
The QND18 Zone is equivalent to the small *Gephyrocapsa* Zone and the *H. sellii* Zone of Gartner (1977), the middle part of the NN19 *P. lacunosa* Zone of Martini (1971), and the upper part of the CN13 *C. doronicoides* Zone of Okada and Bukry (1980).
Definition: Interval from the LO (Top) *C. macintyrei* to the LO (Top) of dominantly small *Gephyrocapsa.*
Age: Early Pleistocene.
Occurrence: Anyala-4 well (780–870 ft.) and Anyala-8 well (8020 ft.).
Remarks: The zonal interval is rich in coccoliths particularly *G. caribbeanica* and *P. lacunosa.* The FO of *G. caribbeanica* is a very consistent datum. It is particularly easy to recognize because the species is abundant from the beginning of its FO.

QND19 ZONE
This Zone is equivalent to the upper part of the NN19 *P. lacunosa* Zone of Martini (1971), the *P. lacunosa* Zone of Gartner (1977), and the Lower part of the CN14 *Gephyrocapsa oceanica* Zone.
Definition: Interval from the LO (Top) of dominantly small *Gephyrocapsa* to the LO (Top) of *P. lacunosa* (0.44 Ma).
Age: Middle Pleistocene/Early Pleistocene.

Occurrence: Anyala-4 well (340–780 ft.) and Anyala-8 well (1600–6310 ft.).
Remarks: Small *Gephyrocapsa* spp. dominates this zonal assemblage. At least two nannofossil population peaks are observed within the Zone. The lowermost is characterized by the acme of small *Gephyrocapsa* spp., and has been dated 0.92 Ma. The LO (Top) of *P. lacunosa* occurs at 0.44 Ma (Hine, 1990; Gradstein et al., 2012). According to Takayama and Sato (1987) the FO (Base) of *Helicosphaera inversa* occurs at 0.48 Ma.

QND20 ZONE

This Zone is equivalent to the *G. oceanica* Zone of Boudreaux and Hay, In: Hay et al. (1967), the *Gephyrocapsa* Zone of Gartner (1969), and the NN20 Zone of Martini (1971), the upper part of the *G. oceanica* Zone (CN14), and the *C. cristatus* Subzone (CN14b) of Okada and Bukry (1980).
Definition: Interval from the LO of *P. lacunosa* (0.44 Ma) to the FO of *Emiliania huxleyi* (0.29 Ma).
Age: Middle Pleistocene.
Occurrence: This zone is not encountered in any of the studied StratCom wells.
Remarks: The LO (top) of *P. lacunosa* was considered a synchronous event by Berggren et al. (1980). Also the FO (Base) of *Gephyrocapsa ericsonii* and *Gephyrocapsa muellerae* approximates the NN20/NN21 boundary.

QND21 ZONE

This Zone is equivalent to the *E. huxleyi* Zone of Boudreaux and Hay, In: Hay et al. (1967), the NN21 Zone *E. huxleyi* Zone Martini (1971), *E. huxleyi* Zone of Gartner (1977), and CN15 *E. huxleyi* Zone of Okada and Bukry (1980).
Definition: Interval above the FO (Base) of *E. huxleyi* to Recent.
Age: Late Pleistocene.
Occurrence: This zone was not encountered in any of the wells.
Remarks: The acme of *E. huxleyi* (Gartner, 1977) is a very important bioevent observed within this Zone in the Niger Delta. The FO (Base) of *E. huxleyi* is estimated at 0.29 Ma (Thierstein et al., 1977; Gradstein et al., 2012). During the Late Pleistocene, the genera *Emiliania*, *Gephyrocapsa*, *Pseudoemiliania*, and *Reticulofenestra* of the Family Noelaerhabdaceae dominated the nannofossil assemblages. High rates of evolutionary appearances, extinctions and, most importantly, "acmes" of nannofossils of this age provided essential tools for the stratigraphic subdivision into smaller time units that are useful for high-resolution biostratigraphic studies.

CHAPTER

3

Calcareous Nannofossil Assemblages of Major Condensed Sections of the Niger Delta

OUTLINE

Cenozoic Foraminifera and Calcareous Nannofossil Biostratigraphy of the Niger Delta. http://dx.doi.org/10.1016/B978-0-12-812161-0.00009-0

3.1 INTRODUCTION

The identification and age dating of condensed sections using the contained calcareous nannofossils is an important component of the high-resolution calcareous nannofossil biostratigraphy and zonation of the Niger Delta. This is because the ability to fingerprint the pelagic shales using the nannofossil assemblages enhances local and fieldwide correlations with foraminiferal and palynological data and further enhances the application of the principles of sequence stratigraphy.

The calcareous nannofossils of the Niger Delta basin, as in other basins of the world, have been shown to have assemblages that characterize the major condensed sections. In this regard, some local events and the distribution patterns of some species have been found useful in characterizing the major condensed sections of the Tertiary Niger Delta sequences. Such characteristics include the biodatum First and Last Occurrences, Top and Base occurrence, size and shape variations of species within and across zones, and taxon quantitative distribution within the stratigraphic column. Other important criteria in the identification of the condensed sections include the ratio of taxa occurrences along the stratigraphic column and acme events.

It has also been observed that in the Niger Delta Tertiary sequences, especially in the Neogene (Early Miocene–Early Pliocene), many of the condensed sections were associated with the acme events of certain taxa especially some discoasters (see also Osterloff et al., 2015). Species of the same genus have also been found to show different abundances in different condensed sections. Some of these criteria have been used earlier for zonation in the deep offshore Niger Delta (Fadiya and Salami, 2012) and in this project. Data from other Niger Delta wells were integrated with those studied in this project for the condensed section descriptions. The condensed sections described here are the major ones with basinwide occurrence in most of the wells that penetrated the age in the Niger Delta. The condensed sections followed the Gradstein et al. (2012) Time Scale Chart. The condensed sections were correlated with equivalent ones in the Haq et al. (1987), Haq et al. (1988), and Hardenbol et al. (1998) sequence cycle charts. The applied condensed sections age were calibrated using dated bioevents from and Gradstein et al. (2012).

3.1.1 55.82 Ma Condensed Section

Age: Late Paleocene/Early Eocene (Thanetian/Ypresian).

Equivalent Condensed Sections: 54.80 Ma of Haq et al. (1987), 55.90 Ma of Haq et al. (1988), and the 54.84 Ma condensed section of Hardenbol et al. (1998).

Zone of Occurrence: PND9 Niger Delta Nannofossil Zone (NP9 Zone of Martini, 1971).

Characteristic Assemblage: *Rhomboaster bitrifida, Rhomboaster cuspis, Rhomboaster clacitrapa, Ellipsolithus macellus, Ellipsolithus distichus, Toweius eminens, Toweius* spp., *Tribrachiatus bramlettei, Chiasmoilithus solitus, Cruciplacolithus frequens, Neococcolithes dubius, Coccolithus formosus, Coccolithus pelagicus, Neochiostozygus* spp., *Discoaster multiradiatus, Discoaster barbadiensis, Cruciplacolithus primus*, and *Sphenolithus moriformis* (Fig. 3.1).

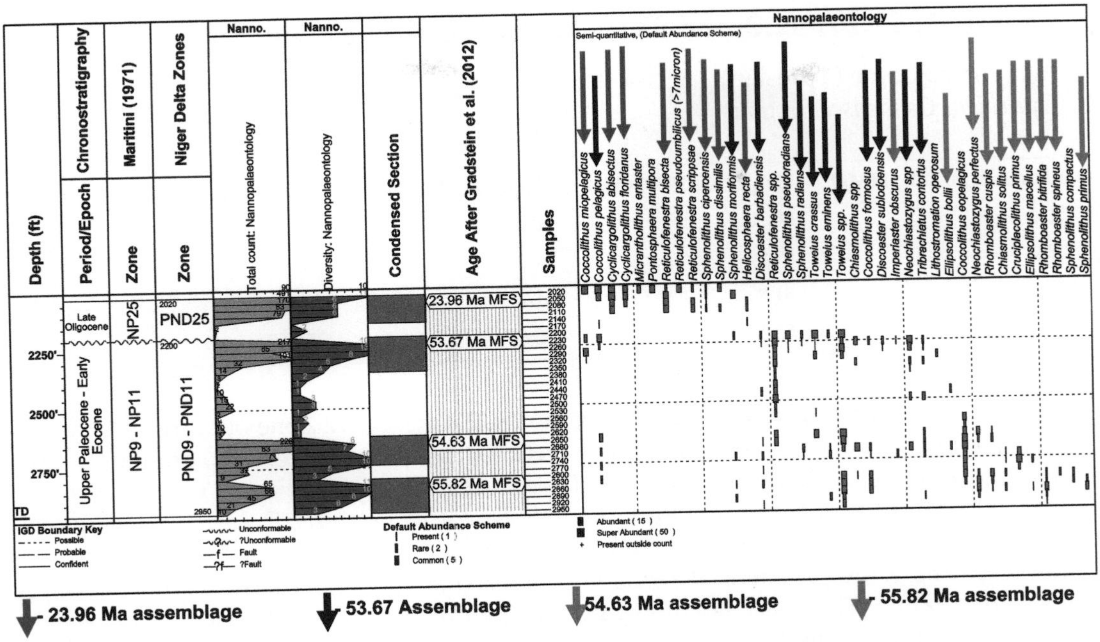

FIGURE 3.1 **Calcareous nannofossil assemblages of the 23.96, 53.67, 54.63, and 55.82 Ma condensed sections.**

3.1.2 54.63 Ma Condensed Section

Age: Early Eocene (Ypresian).

Equivalent Condensed Sections: 54.30 Ma of Haq et al. (1987), 54.6 Ma of Haq et al. (1988), and the 54.09 Ma condensed section of Hardenbol et al. (1998).

Zone of Occurrence: Niger Delta PND10 nannofossil zone (NP10 Zone of Martini, 1971).

Characteristic Assemblage: *Tribrachiatus contortus, Discoaster barbadiensis, Toweius eminens, Toweius* spp., *Cruciplacolithus* spp., *Rhomboaster cuspis, Neococcolithes dubius, Coccolithus formosus, Coccolithus pelagicus, Sphenolithus radians*, and *Sphenolithus primus*.

3.1.3 53.67 Ma Condensed Section

Age: Early Eocene (Ypresian).

Equivalent Condensed Sections: 53.5 Ma of Haq et al. (1987), 53.6 Ma of Haq et al. (1988), and the 53.24 Ma condensed section of Hardenbol et al. (1998).

Zone of Occurrence: Niger Delta PND11 nannofossil zone (NP11 Zone of Martini, 1971).

Characteristic Assemblage: *Imperiaster obscurus, Tribrachiatus orthostylus, T. contortus, T. eminens, Toweius* spp., *Chiasmolithus grandis, R. cuspis, C. formosus, C. pelagicus, S. radians, D. barbadiensis*, and *S. moriformis*.

3.1.4 50.51, 51.27, 52.0, and 53.01 Ma Condensed Sections

Age: Early Eocene (Ypresian).
Equivalent Condensed Sections:

Gradstein et al. (2012) (Ma)	Haq et al. (1987) (Ma)	Haq et al. (1988) (Ma)	Hardenbol et al. (1998) (Ma)
50.51	50.30	50.0	50.65
51.27	51.0	51.0	51.27
52.0	51.8	51.7	51.84
53.01	52.5	52.6	52.65

Zone of Occurrence: Niger Delta PND12 nannofossil zone (NP12 Zone of Martini, 1971).
Characteristic Assemblage: The condensed sections are characterized by abundant occurrence of *T. orthostylus*, *Discoaster kuepperi*, *Toweius callosus*, *Toweius* spp., *Toweius pertusus*, *C. grandis*, *C. solitus*, and *N. dubius*. Others include *Discoaster lodoensis*, *D. kuepperi*, *C. formosus*, *C. pelagicus*, *S. radians*, *D. barbadiensis*, and *S. moriformis*.
Remarks: These four condensed sections succeed one another rapidly within the Niger Delta PND12 Zone. The fossil assemblage is basically the same and the condensed sections are therefore, differentiated mainly by the sediment stacking patterns and the corresponding nannofossil abundance and diversity peaks (Figs. 3.1 and 3.2).

3.1.5 49.81 Ma Condensed Section

Age: Early/Middle Eocene (Ypresian/Lutetian).
Equivalent Condensed Sections: 49.80 Ma of Haq et al. (1987) and the 50.51 Ma condensed section of Hardenbol et al. (1998).

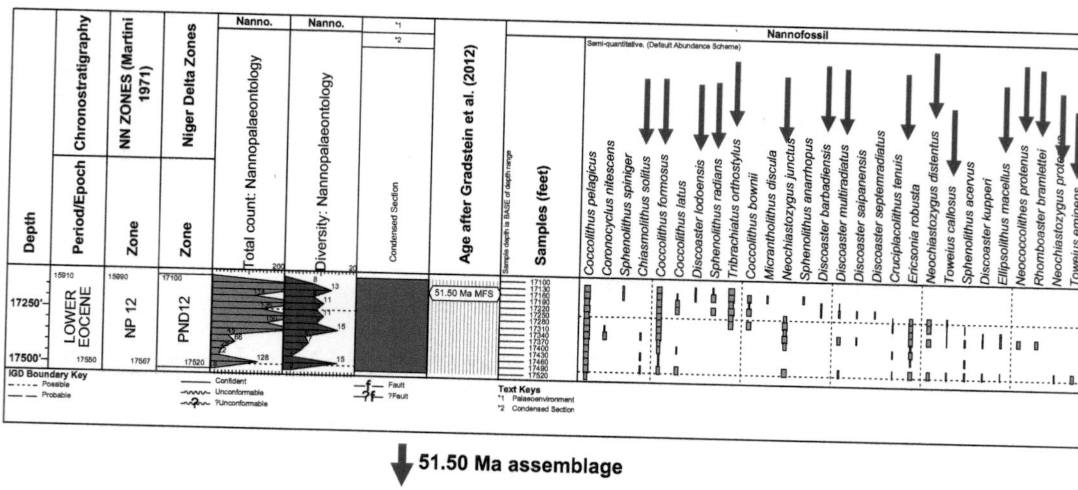

FIGURE 3.2 **Calcareous nannofossil assemblages of the 51.50 Ma condensed section.**

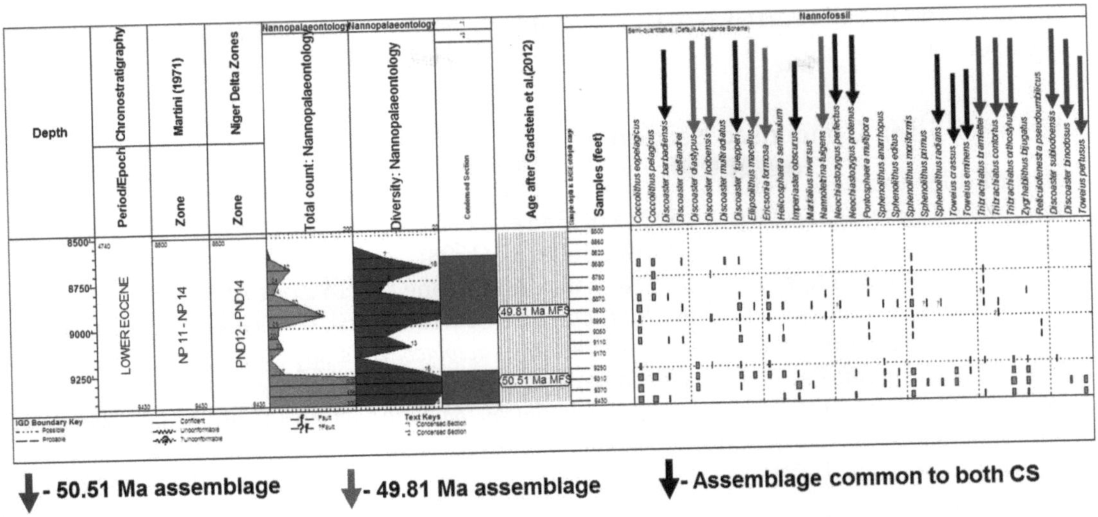

FIGURE 3.3 **Calcareous nannofossil assemblages of the 50.51 and 49.81 Ma condensed sections.**

Zone of Occurrence: Niger Delta PND13 nannofossil zone (NP13 Zone of Martini, 1971).

Characteristic Assemblage: *D. lodoensis*, *C. solitus*, *C. grandis*, *Pemma papiplatum*, *Micrantholithus bulbosus*, *Braarudosphaera bigelowii*, *Lophodolithus mochloporus*, *C. formosus*, *Helicosphaera lophota*, *Reticulofenestra* spp., *Neococcolithes* spp., *T. callosus*, and *Toweius* spp. (Fig. 3.3)

3.1.6 47.84 Ma Condensed Section

Age: Middle Eocene (Lutetian).

Equivalent Condensed Sections: 49.0 Ma of Haq et al. (1987), 48.9 Ma of Haq et al. (1988), and the 49.08 Ma condensed section of Hardenbol et al. (1998).

Zone of Occurrence: Niger Delta PND14 nannofossil zone (NP14 Zone of Martini, 1971).

Characteristic Assemblage: *D. lodoensis*, *Discoaster sublodoensis*, *C. solitus*, *C. grandis*, *Pemma balium*, *P. papillatum*, *M. bulbosus*, *B. bigelowii*, *L. mochloporus*, *C. formosus*, *Helicosphaera seminulum*, *H. lophota*, and some species of *Reticulofenestra*.

Remarks: This is the first condensed section within the PND14 Zone and the First Occurrence (Base) of *D. lodoensis* is commonly found within this condensed section.

3.1.7 46.82 Ma Condensed Section

Age: Middle Eocene (Lutetian).

Equivalent Condensed Sections: 48.0 Ma of Haq et al. (1987) and the 47.70 Ma condensed section of Hardenbol et al. (1998).

Zone of Occurrence: Niger Delta PND14 nannofossil zone (NP14 Zone of Martini, 1971).

Characteristic Assemblage: *Nannotetrina fulgens*, *Nannotetrina cristata*, *C. solitus*, *C. grandis*, *Pemma basquensis*, *P. papillatum*, *M. bulbosus*, *Cruciplacolithus* spp., *Chiasmolithus* spp., *Blackites* spp., *B. bigelowii*, *L. mochloporus*, *C. formosus*, *H. seminulum*, *H. lophota*, and some species of *Reticulofenestra*.

Remarks: The First Occurrence (Base) of *N. fulgens* is commonly associated with this condensed section.

3.1.8 44.75 Ma Condensed Section

Age: Middle Eocene (Lutetian).

Equivalent Condensed Sections: 45.50 Ma of Haq et al. (1987), 46.1 Ma of Haq et al. (1988), and the 45.15 Ma condensed section of Hardenbol et al. (1998).

Zone of Occurrence: Niger Delta PND14 nannofossil zone (NP14 Zone of Martini, 1971).

Characteristic Assemblage: *H. seminulum, Discoaster barbardiensis, Chiasmolithus gigas, N. fulgens, N. cristata, Chiasmolithus* spp., *C. pelagicus, Cruciplacolithus* spp., *C. formosus, Neococcolithes* spp., *Lophodolithus* spp., *L. mochloporus*, and *Zygodiscus* spp.

Remarks: The pentaliths (*P. basquensis, P. papillatum, M. bulbosus*, and *B. bigelowii*) show fewer numbers than in the succeeding 42.64 Ma condensed section.

3.1.9 42.64 Ma Condensed Section

Age: Middle Eocene (Lutetian)

Equivalent Condensed Sections: 43.0 Ma of Haq et al. (1987), 43.2 Ma of Haq et al. (1988), and the 42.78 Ma condensed section of Hardenbol et al. (1998).

Zone of Occurrence: Niger Delta PND15 nannofossil zone (NP15 Zone of Martini, 1971).

Characteristic Assemblage: The condensed section is characterized by the common occurrences of the pentaliths (*P. basquensis, P. papillatum, Pemma angulatum, M. bulbosus*, and *B. bigelowii*) and the species of *Nannotetrina* (*N. fulgens, Nannotetrina pappii*, and *N. cristata*). *D. barbadiensis, Discoaster saipanensis, Chiasmolithus* spp., *Cruciplacolithus* spp., *C. formosus, H. seminulum, Neococcolithus* spp., *Lophodolithus* spp., *L. mochloporus*, and *Zygodiscus* spp. formed part of the assemblage within the condensed section.

Remarks: The First Occurrences of *Reticulofenestra umbilica* and *Reticulofenestra reticulata* are found within the condensed section.

3.1.10 40.4 Ma Condensed Section

Age: Late Eocene (Bartonian).

Equivalent Condensed Sections: 40.0 Ma of Haq et al. (1987), 41.0 Ma of Haq et al. (1988), and the 40.38 Ma condensed section of Hardenbol et al. (1998).

Zone of Occurrence: Niger Delta PND16 nannofossil zone (NP16 Zone of Martini, 1971).

Characteristic Assemblage: The condensed section is characterized by the peak abundance of the pentaliths (*P. basquensis, P. papillatum, P. angulatum, M. bulbosus*, and *B. bigelowii*). The genus *Nannotetrina* typified by *N. fulgens, Nannotetrina quadrata*, and *N. cristata* are an important group within the condensed section. Other important nannofossils characterizing the condensed section include *C. solitus, Chiasmolithus* spp., *Helicosphaera wilcoxonii, Helicosphaera compacta, H. lophota, D. saipanensis, D. barbadiensis, R. reticulata, R. umbilica*, and *Reticulofenestra* spp.

Remarks: This is by far the most conspicuous condensed section within the Middle Eocene Bartonian age.

3.1.11 37.36 Ma Condensed Section

Age: Late Eocene (Bartonian/Priabonian).

Equivalent Condensed Sections: 38.8 Ma of Haq et al. (1987), 39.4 Ma of Haq et al. (1988), and the 38.1 Ma condensed section of Hardenbol et al. (1998).

Zone of Occurrence: Niger Delta PND18 nannofossil zone (NP17 Zone of Martini, 1971).

Characteristic Assemblage: Common occurrences of the pentaliths (*P. basquensis, M. bulbosus,* and *B. bigelowii*) are found within this condensed section. Other nannofossils here include *Helicosphaera bramlettei, H. wilcoxonii, R. reticulata, Reticulofenestra bisecta, Reticulofenestra scrippsae, Chiasmolithus oamaruensis, H. lophota, Helicosphaera euphratis, H. compacta, D. barbadiensis,* and *D. saipanensis.*

Remarks: The First Occurrence of *R. bisecta* is found within this condensed section, while the pentaliths shows a gradual reduction in quantity from the preceding 40.4 Ma condensed section.

3.1.12 35.99 Ma Condensed Section

Age: Late Eocene (Priabonian).

Equivalent Condensed Sections: 36.5 Ma of Haq et al. (1987), 35.9 Ma of Haq et al. (1988), and the 35.31 Ma condensed section of Hardenbol et al. (1998).

Zone of Occurrence: Niger Delta PND19 nannofossil zone (NP18 Zone of Martini, 1971).

Characteristic Assemblage: *R. reticulata, R. bisecta, R. scrippsae, R. umbilicus, C. oamaruensis, Ismolithus recurvus, Cruciplacolithus* spp., *H. bramlettei, H. wilcoxonii, Helicosphaera reticulata, H. lophota, H. euphratis, H. compacta, D. saipanensis,* and *D. barbadiensis* (Fig. 3.4)

Remarks: *Cyclicargolithus floridanus* is reduced drastically in number as it is close to its First Occurrence. The pentaliths (*M. bulbosus, B. bigelowii,* and *P. basquensis*) also occur in low numbers within the condensed section.

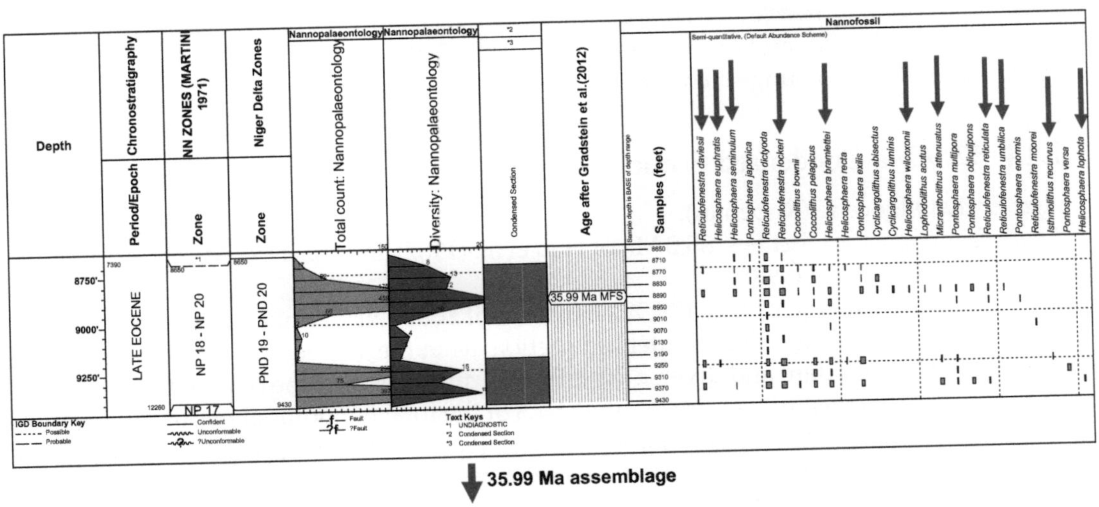

FIGURE 3.4 **Calcareous nannofossil assemblage of the 35.59 Ma condensed section.**

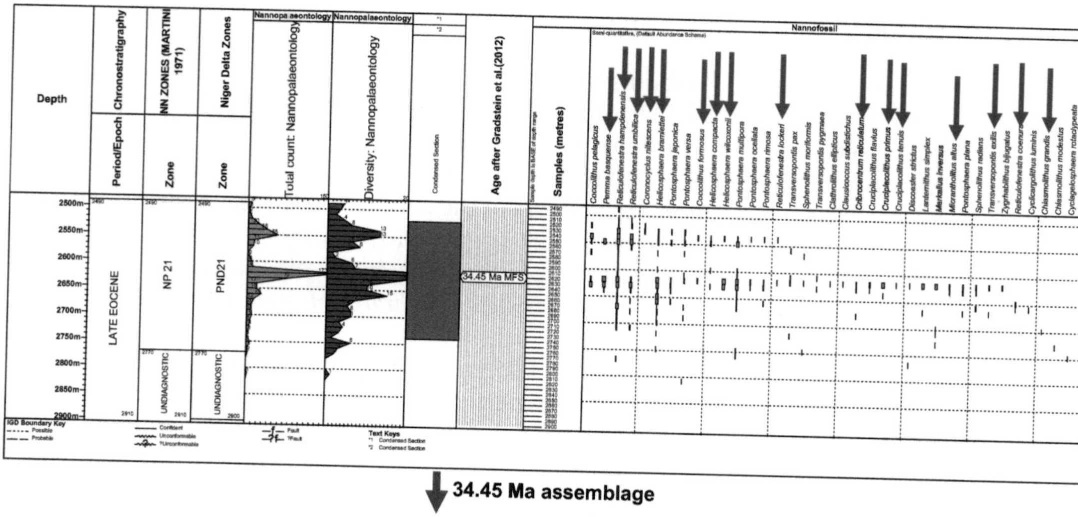

FIGURE 3.5 **Calcareous nannofossil assemblage of the 34.45 Ma condensed section.**

3.1.13 34.45 Ma Condensed Section

Age: Late Eocene (Priabonian)

Equivalent Condensed Sections: 35.0 Ma of Haq et al. (1987), 34.0 Ma of Haq et al. (1988), and the 33.70 Ma condensed section of Hardenbol et al. (1998).

Zone of Occurrence: Niger Delta PND21 nannofossil zone (NP21 Zone of Martini, 1971).

Characteristic Assemblage: This condensed section is characterized by the relatively high abundance of reticulofenestrids and *Sphenolithus*, which reached their peak in the 32.9 Ma condensed section. Of importance is the occurrence of *R. reticulata*, *D. barbardiensis*, and the acme of *Clausiococcus* spp. Other important nannofossils constituting the assemblage here include *R. umbilica*, *R. bisecta*, *R. scrippsae*, *H. compacta*, *H. bramlettei*, *C. floridanus*, *Coccolithus miopelagicus*, *C. pelagicus*, and *Discoaster deflandrei* (Fig. 3.5).

Remarks: This is the youngest condensed section in the Upper Eocene.

3.1.14 32.9 Ma Condensed Section

Age: Early Oligocene (Rupelian).

Equivalent Condensed Sections: 35.0 Ma of Haq et al. (1987) and the 32.47 Ma condensed section of Hardenbol et al. (1998).

Zone of Occurrence: Niger Delta PND22 nannofossil zone (NP22 Zone of Martini, 1971).

Characteristic Assemblage: This condensed section, as in the succeeding one, is also characterized by highly abundant and diverse assemblages dominated by the reticulofenestrids and *Sphenolithus*, which constitute approximately 80% of the total nannofossil population. Important nannofossils here include *H. compacta*, *R. scrippsae*, *R. bisecta*, *Sphenolithus predistentus*, *S. moriformis*, *H. bramlettei*, *C. floridanus*, *C. miopelagicus*, *C. pelagicus*, and *D. deflandrei*.

Remarks: The downhole increase of the reticulofenestrids and *Sphenolithus* reached their peak within this condensed section. The large reticulofenestrids finally attain their biggest size and transform to the *R. umbilica/hillae* complex, which distinguishes the condensed section from the younger 30.72 Ma Condensed Section earlier. Many of the medium-sized reticulofenestrids also approache *R. reticulata* of Upper Eocene age.

3.1.15 30.54 Ma Condensed Section

Age: Early Oligocene (Rupelian).

Equivalent Condensed Sections: 33.5 Ma of Haq et al. (1988) and the 30.58 Ma Condensed Section of Hardenbol et al. (1998).

Zone of Occurrence: Niger Delta PND21 nannofossil zone (NP21 Zone of Martini, 1971).

Characteristic Assemblage: *H. compacta*, *R. scrippsae*, *R. bisecta*, *S. predistentus*, *S. moriformis*, *H. bramlettei*, *C. floridanus*, *Reticulofenestra*, *Cyclicargolithus abisectus*, *Helicosphaera recta*, *C. miopelagicus*, *C. pelagicus*, and *D. deflandrei*.

Remarks: This condensed section is one of the most obvious within the Early Oligocene Rupelian age. The condensed section is characterized by the downhole increase in the reticulofenestrids. Large forms of *Reticulofenestra* spp. and abies-looking *Sphenolithus* spp. reached their peak within this condensed section.

3.1.16 28.64 Ma Condensed Section

Age: Early Oligocene (Rupelian).

Equivalent Condensed Sections: 32.0 Ma of Haq et al. (1987), 31.30 Ma of Haq et al. (1988), and the 29.7 Ma Condensed Section of Hardenbol et al. (1998).

Zone of Occurrence: Niger Delta PND24 nannofossil zone (NP24 Zone of Martini, 1971).

Characteristic Assemblage: *R. scrippsae*, *R. bisecta*, *Sphenolithus distentus*, *H. bramlettei*, *C. floridanus*, *C. abisectus*, *H. recta*, *C. miopelagicus*, *C. pelagicus*, and *D. deflandrei* constitute the nannofossil assemblage.

Remarks: As noted in the succeeding condensed section, large forms of *Reticulofenestra* spp. and *Sphenolithus* including the abies-looking *Sphenolithus* continue to form a significant population of the assemblage.

3.1.17 27.79 Ma Condensed Section

Age: Late Oligocene (Chattian).

Equivalent Condensed Sections: 29.0 Ma of Haq et al. (1987), 28.10 Ma of Haq et al. (1988), and the 28.0 Ma Condensed Section of Hardenbol et al. (1998).

Zone of Occurrence: Niger Delta PND24 nannofossil zone (NP24 Zone of Martini, 1971).

Characteristic Assemblage: *H. bramlettei*, *S. distentus*, *R. scrippsae*, *R. bisecta*, *C. floridanus*, *C. abisectus*, *H. recta*, *C. miopelagicus*, *C. pelagicus*, and *D. deflandrei*.

Remarks: An influx of typically large forms of *Reticulofenestra* and *Sphenolithus* is common within this condensed section and increases in number as one goes into older condensed sections. While many of the reticulofenestrids appear to be intermediate between *Reticulofenestra pseudoumbilicus* and *R. umbilicus/hillae*, the *Sphenolithus* are large with many resembling *Sphenolithus abies*.

3.1.18 26.2 Ma Condensed Section

Age: Late Oligocene (Chattian).

Equivalent Condensed Sections: 27.50 Ma of Haq et al. (1987), 26.20 Ma of Haq et al. (1988), and the 26.40 Ma Condensed Section of Hardenbol et al. (1998).

Zone of Occurrence: Niger Delta PND25 nannofossil zone (NP25 Zone of Martini, 1971).

Characteristic Assemblage: *R. scrippsae, R. bisecta, C. floridanus, C. abisectus, Sphenolithus ciperoensis, Helicosphaera truempyi, H. recta, C. miopelagicus, C. pelagicus, D. deflandrei, Transversopontis* spp., *H. euphratis, Micrantholithus attenuatus*, and some species of *Pontosphaera* and *Reticulofenestra.*

3.1.19 23.96 Ma Condensed Section

Age: Late Oligocene (Chattian).

Equivalent Condensed Sections: 26.0 Ma of Haq et al. (1987), 24.30 Ma of Haq et al. (1988), and the 24.57 Ma Condensed Section of Hardenbol et al. (1998).

Zone of Occurrence: Niger Delta PND25 nannofossil zone (NP25 Zone of Martini, 1971).

Characteristic Assemblage: *S. ciperoensis, R. scrippsae, R. bisecta, C. miopelagicus, C. pelagicus, C. floridanus, C. abisectus, H. truempyi, H. recta, Triquetrorhabdulus carinatus, Zygrhablithus bijugatus, Ilselithina fusa, D. deflandrei, Discoaster formosus, Transversopontis* spp., *H. euphratis* and *Pontosphaera enormis*. Other important nannofossils within the section include *D. deflandrei, S. moriformis*, and *Reticulofenestra* spp. (Fig. 3.6).

Remarks: This condensed section is a conspicuous condensed section and the youngest within the Late Oligocene. The condensed section is typified by the high abundance of

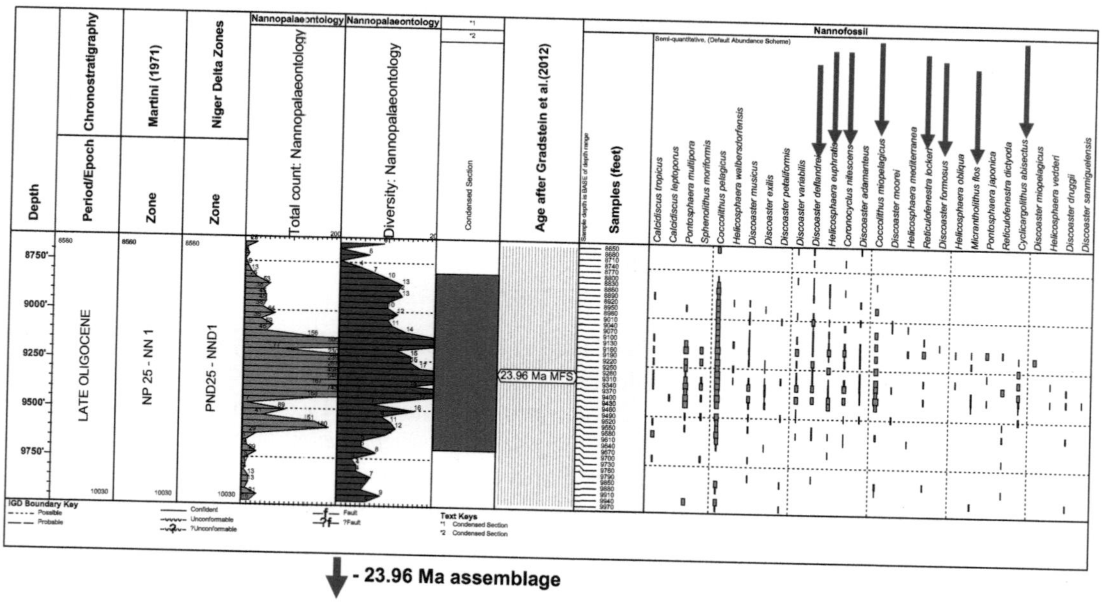

FIGURE 3.6 **Calcareous nannofossil assemblage of the 23.96 Ma condensed section.**

C. floridanus, *C. miopelagicus*, and *C. pelagicus*, which persisted into the Early Miocene. The reticulofenestrids are an important group within the condensed section.

3.1.20 22.24 Ma Condensed Section

Age: Early Miocene (Aquitanian).

Equivalent Condensed Sections: 24.80 Ma of Haq et al. (1987), 23.20 Ma of Haq et al. (1988), and the 22.90 Ma Condensed Section of Hardenbol et al. (1998).

Zone of Occurrence: Niger Delta NND1 nannofossil zone (NN1 Zone of Martini, 1971).

Characteristic Assemblage: *H. truempyi*, *H. recta*, *C. pelagicus*, *C. miopelagicus*, *C. floridanus*, *C. abisectus*, *D. deflandrei*, *Transversopontis* spp., *H. euphratis*, and *Micrantholithus* spp.

Remarks: This is the oldest condensed section within the NND1 Niger Delta nannofossil zone. The condensed section is typified by high abundance of *C. miopelagicus*, *C. pelagicus*, and *C. floridanus*. The reticulofenestrids are an important group within the condensed section.

3.1.21 20.94 Ma Condensed Section

Age: Early Miocene (Aquitanian).

Equivalent Condensed Sections: 21.50 Ma of Haq et al. (1987), 22.0 Ma of Haq et al. (1988), and the 21.30 Ma condensed section of Hardenbol et al. (1998).

Zone of Occurrence: Niger Delta NND1 nannofossil zone (NN1 and NN2 Zones of Martini, 1971).

Characteristic Assemblage: The condensed section is typified by the downhole increase in *C. miopelagicus* and *C. pelagicus*. Other important nannofossils within the section include *C. floridanus*, *D. deflandrei*, *S. moriformis*, and *Reticulofenestra* spp. Typically, large forms of *C. floridanus* (?*C. abisectus*) are found within this condensed section. The reticulofenestrids are an important group with a relatively high abundance within the condensed section (Fig. 3.7).

3.1.22 19.8 Ma Condensed Section

Age: Early Miocene (Burdigalian).

Equivalent Condensed Sections: 18.50 Ma of Haq et al. (1987), 20.50 Ma of Haq et al. (1988), and the 19.95 Ma Condensed Section of Hardenbol et al. (1998).

Zone of Occurrence: Niger Delta NND1 nannofossil zone (NN2 Zone of Martini, 1971).

Characteristic Assemblage: *T. carinatus*, *Discoaster druggii*, *D. deflandrei*, *C. miopelagicus*, *Helicosphaera ampliaperta*, *C. pelagicus*, *C. floridanus*, *Sphenolithus dissimilis*, *Sphenolithus conicus*, *S. moriformis*, and *Reticulofenestra* spp. (Fig. 3.8).

3.1.23 18.64 Ma Condensed Section

Age: Early Miocene (Burdigalian).

Equivalent Condensed Sections: 19.40 Ma of Haq et al. (1988) and the 19.10 Ma Condensed Section of Hardenbol et al. (1998).

Zone of Occurrence: Niger Delta NND2 nannofossil zone (NN3 Zone of Martini, 1971).

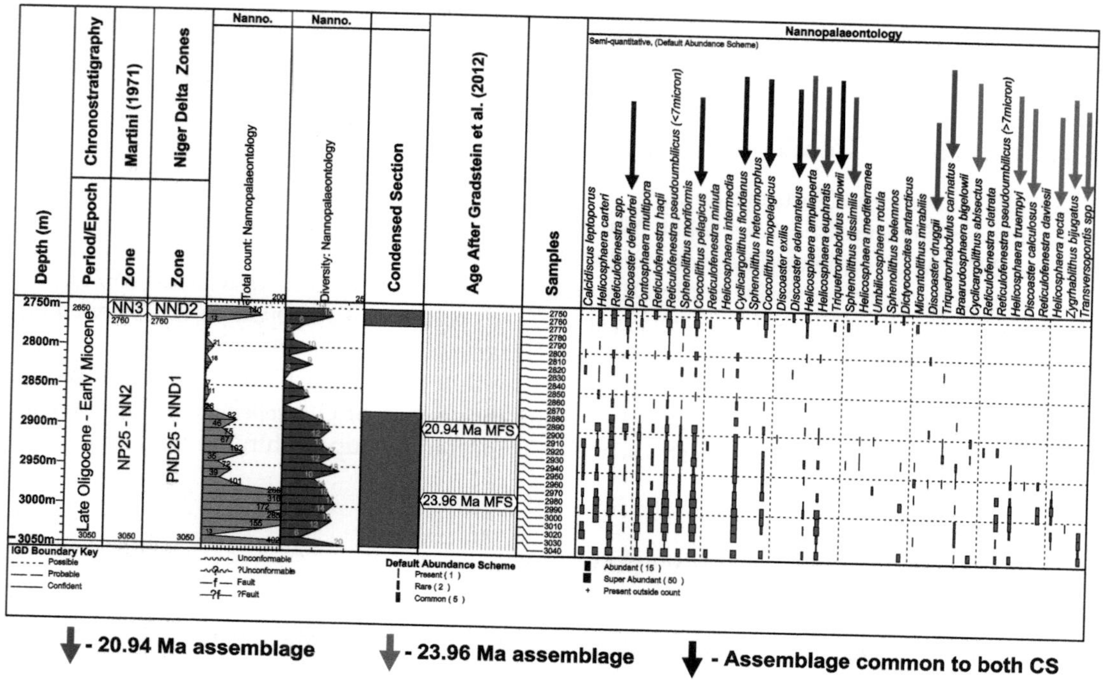

FIGURE 3.7 **Calcareous nannofossil assemblages of the 20.94 and 23.96 Ma condensed sections.**

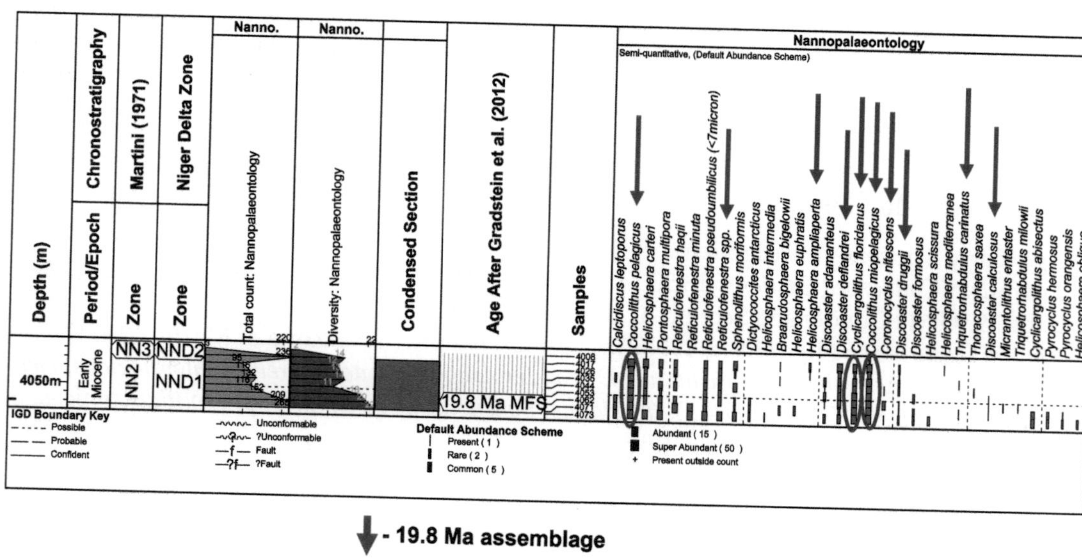

FIGURE 3.8 **Calcareous nannofossil assemblages of the 19.8 Ma condensed section.**

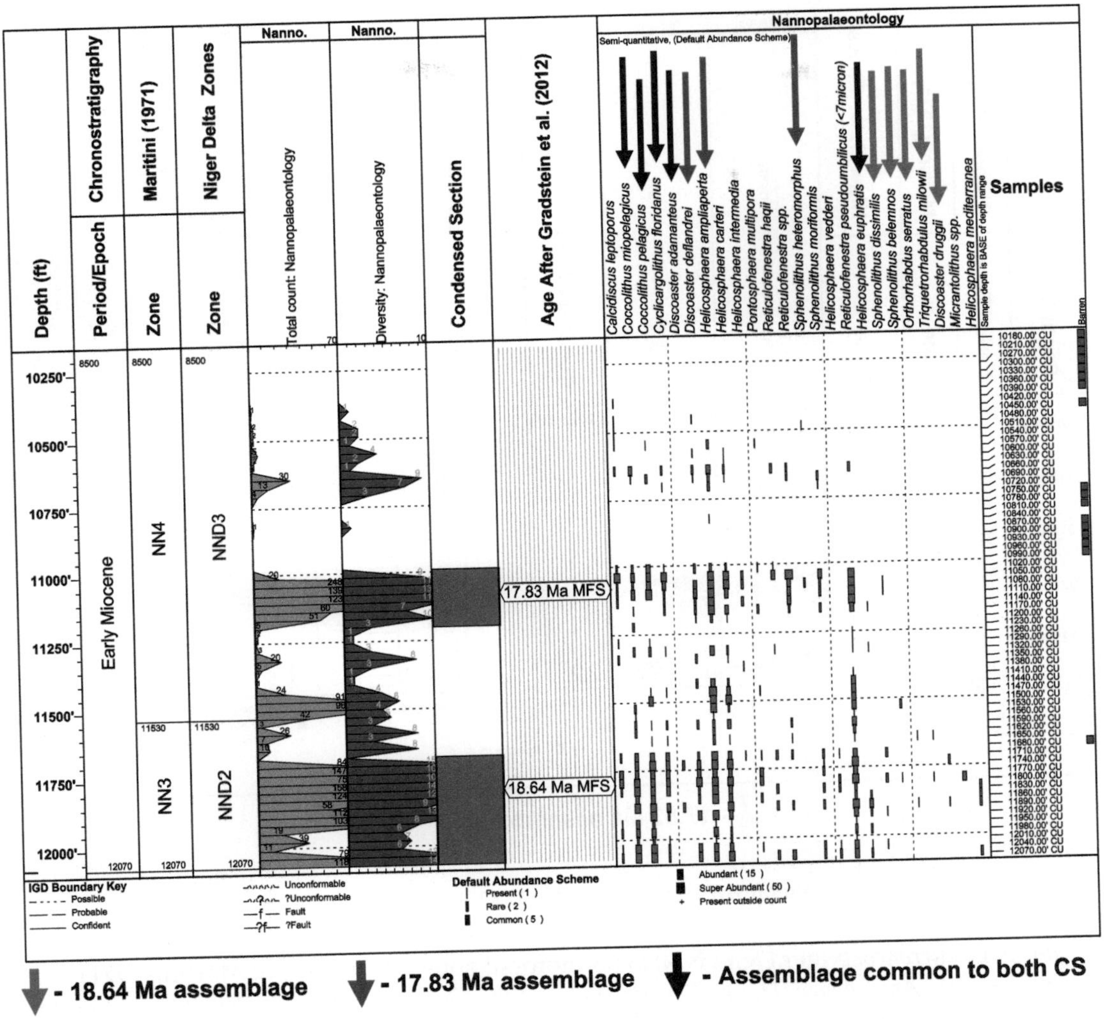

FIGURE 3.9 **Calcareous nannofossil assemblages of the 18.64 and 17.83 Ma condensed sections.**

Characteristic Assemblage: *H. ampliaperta*, *C. floridanus*, *Sphenolithus belemnos* (rare), *D. deflandrei*, *Discoaster calculosus*, *S. dissimilis*, *S. conicus*, *S. moriformis*, *Triquetrorhabdulus milowii*, *C. pelagicus*, *C. miopelagicus*, and *Reticulofenestra* spp.

Remarks: This Condensed Section is easily distinguished from those above it by the absence of *Sphenolithus heteromorphus* (Fig. 3.9).

3.1.24 17.83 Ma Condensed Section

Age: Early Miocene (Burdigalian).

Equivalent Condensed Sections: 17.40 Ma of Haq et al. (1988) and the 18.0 Ma condensed section of Hardenbol et al. (1998).

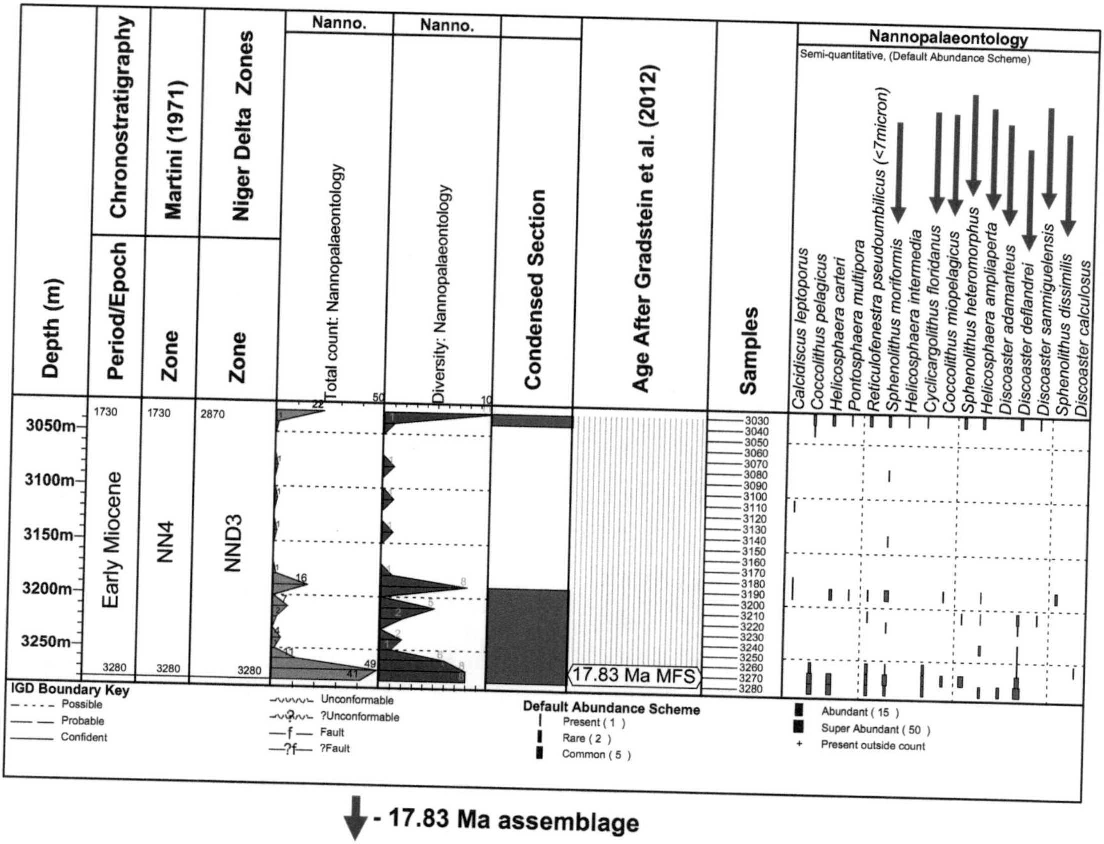

FIGURE 3.10 **Calcareous nannofossil assemblage of the 17.83 Ma condensed section.**

Zone of Occurrence: Niger Delta NND3 nannofossil zone (NN4 Zone of Martini, 1971).

Characteristic Assemblage: *H. ampliaperta*, *Sphenolithus heteromorphus*, *D. deflandrei*, *C. floridanus*, *Helicosphaera carteri*, *Pontosphaera multipora*, *S. moriformis*, *H. euphratis*, *C. pelagicus*, *C. miopelagicus*, *R. pseudoumbilicus* (<7 μm), *R. pseudoumbilicus* (>7 μm), *Reticulofenestra* spp., and *Reticulofenestra haqii*.

Remarks: The specimens of *H. ampliaperta* here consist of robust and oval-shaped forms (Figs. 3.9 and 3.10).

3.1.25 16.96 Ma Condensed Section

Age: Early Miocene (Burdigalian).

Equivalent Condensed Sections: 17.50 Ma of Haq et al. (1987) and the 16.86 Ma Condensed Section of Hardenbol et al. (1998).

Zone of Occurrence: Niger Delta NND3 nannofossil zone (NN4 Zone of Martini, 1971).

Characteristic Assemblage: *H. ampliaperta*, *S. heteromorphus*, *C. floridanus*, *D. deflandrei*, *Discoaster sanmiguelensis*, *Discoaster adamanteus*, *H. carteri*, *P. multipora*, *S. moriformis*, *C. miopelagicus*,

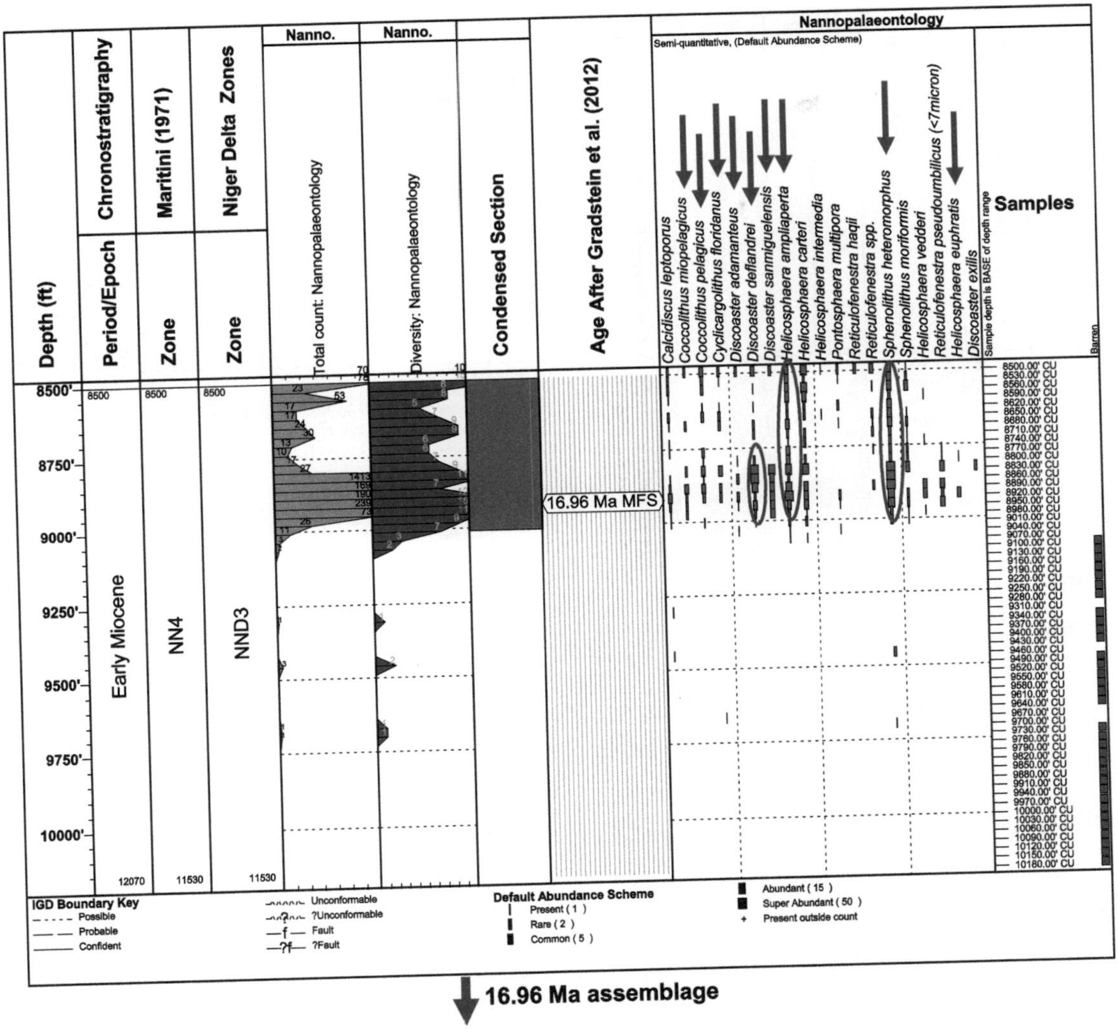

FIGURE 3.11 **Calcareous nannofossil assemblages of the 16.96 Ma condensed section.**

R. pseudoumbilicus (<7 μm), *R. pseudoumbilicus* (>7 μm), *Reticulofenestra* spp., and *R. haqii* (Fig. 3.11).

Remarks: This condensed section is characterized by the acme event of *D. deflandrei* and the downhole increase in *H. ampliaperta*.

3.1.26 15.2 Ma Condensed Section

Age: Middle Miocene (Langhian).

Equivalent Condensed Sections: 16.0 Ma of Haq et al. (1987), 16.0 Ma of Haq et al. (1988), and the 15.6 Ma Condensed Section of Hardenbol et al. (1998).

Zone of Occurrence: Niger Delta NND3 nannofossil zone (NN4 Zone of Martini, 1971).

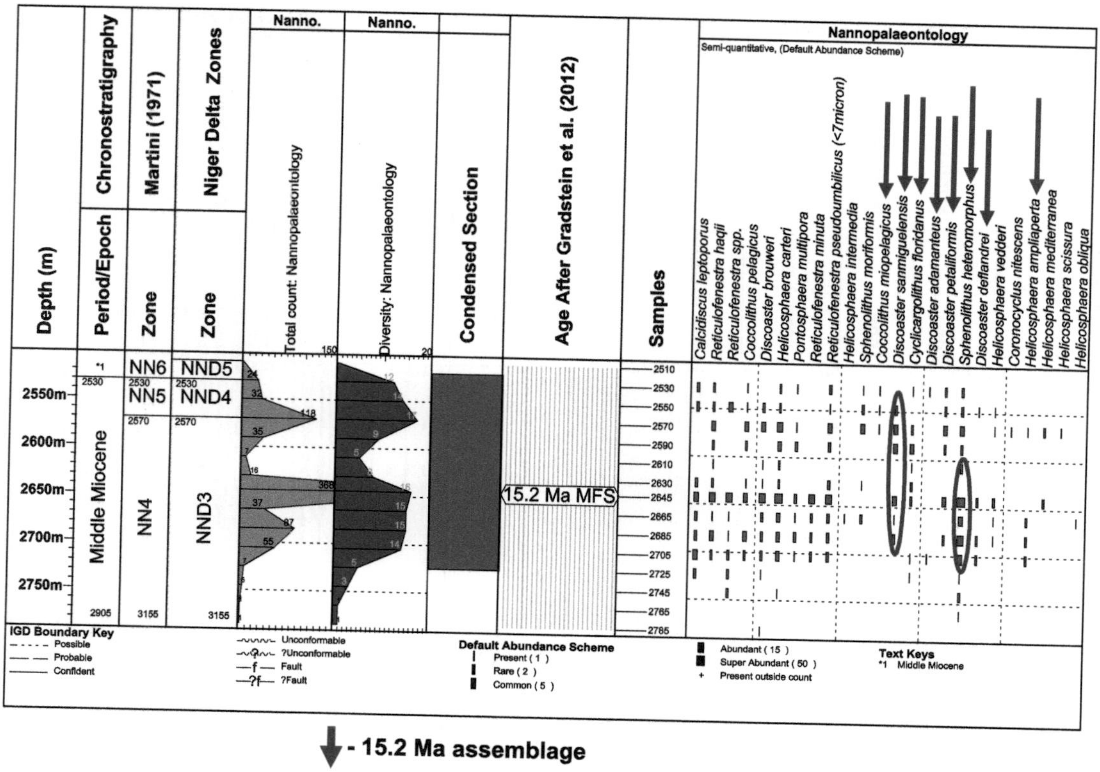

FIGURE 3.12 **Calcareous nannofossil assemblages of the 15.2 Ma condensed section.**

Characteristic Assemblage: *S. heteromorphus, H. carteri, C. floridanus, D. sanmiguelensis, D. adamanteus, C. miopelagicus*, reticulofenestrids [*R. pseudoumbilicus* (<7 μm), *R. pseudoumbilicus* (>7 μm), *R. haqii, R. minutula*, and *R. minuta*], and *P. multipora*.

Remarks: This condensed section is characterized by the acme of *S. heteromorphus* and *D. sanmiguelensis* (Fig. 3.12).

3.1.27 13.65 Ma Condensed Section

Age: Middle Miocene (Serravalian).

Equivalent Condensed Sections: 15.0 Ma of Haq et al. (1987), 15.0 Ma of Haq et al. (1988), and the 14.20 Ma Condensed Section of Hardenbol et al. (1998).

Zone of Occurrence: Niger Delta NND4 nannofossil zone (NN5 Zone of Martini, 1971).

Characteristic Assemblage: *S. heteromorphus, C. floridanus*, reticulofenestrids [*R. pseudoumbilicus* (<7 μm), *R. pseudoumbilicus* (>7 μm), *R. haqii, R. minutula*, and *R. minuta*], and *P. multipora*. Also noticeable within the interval is the abundant occurrence of *H. carteri* and *Calcidiscus leptoporous*.

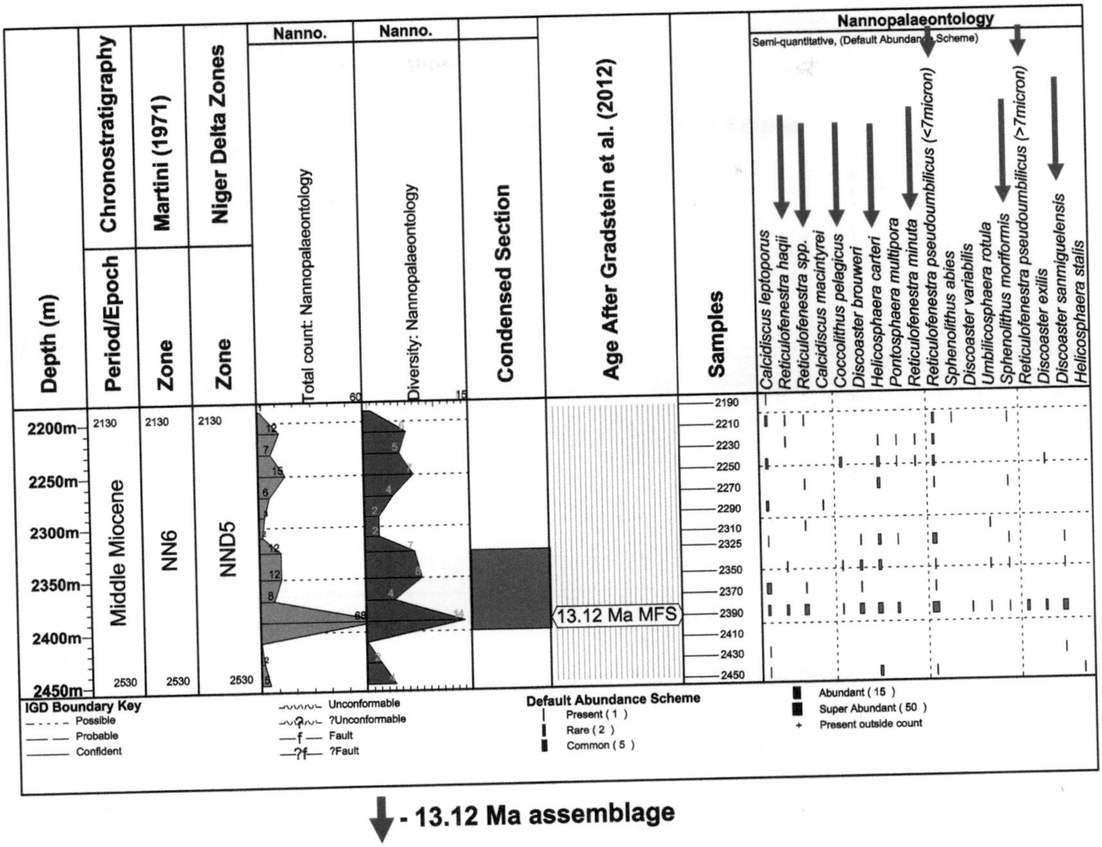

FIGURE 3.13 **Calcareous nannofossil assemblages of the 13.12 Ma condensed section.**

3.1.28 13.12 Ma Condensed Section

Age: Middle Miocene (Serravalian).

Equivalent Condensed Sections: 13.4 Ma of Haq et al. (1987), 13.4 Ma of Haq et al. (1988), and the 13.15 Ma condensed section of Hardenbol et al. (1998).

Zone of Occurrence: Niger Delta NND5 nannofossil zone (NN6 Zone of Martini, 1971).

Characteristic Assemblage: Reticulofenestrids [*R. pseudoumbilicus* (<7 μm), *R. pseudoumbilicus* (>7 μm), *R. haqii*, *R. minutula*, and *R. minuta*], and *C. floridanus*. Also noticeable within the interval is the abundant occurrence of *H. carteri* and *C. leptoporous* (Fig. 3.13).

Remarks: This condensed section is more conspicuous than the succeeding 12.26 Ma condensed section even though found within the "carbonate crash" sequence of the Middle Miocene Serravallian age.

Remarks: This condensed section is easily identifiable from the preceding 13.68 Ma condensed section by the absence of *S. heteromorphus*.

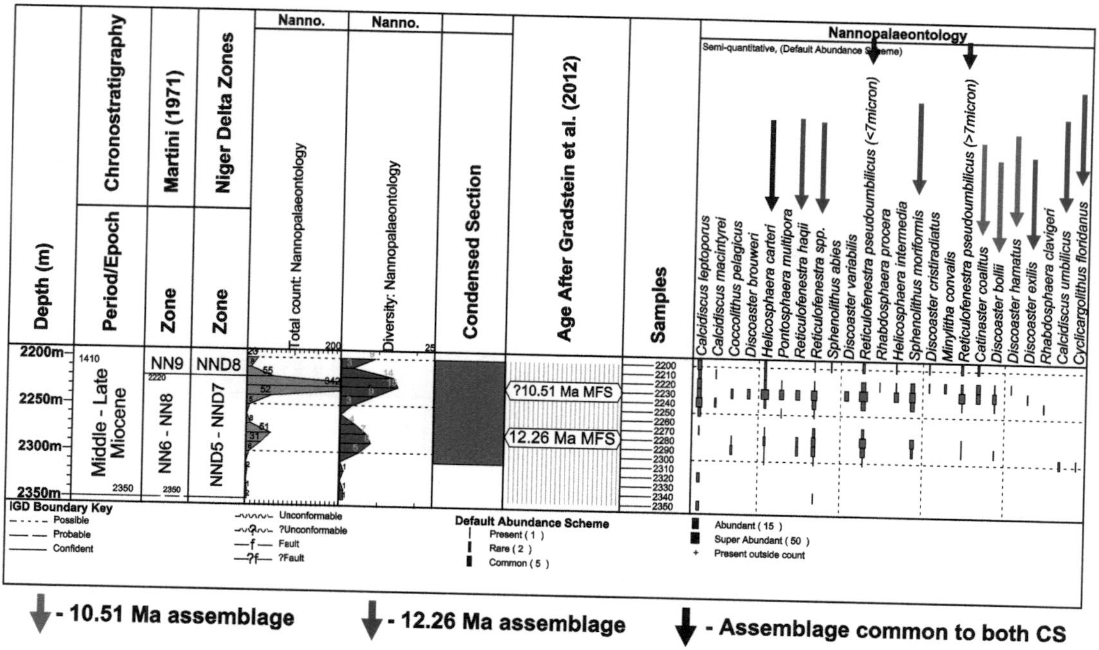

FIGURE 3.14 **Calcareous nannofossil assemblages of the 10.51 and 12.26 Ma condensed sections.**

3.1.29 12.26 Ma Condensed Section

Age: Middle Miocene (Serravalian)

Equivalent Condensed Sections: 11.6 Ma of Haq et al. (1987), 11.6 Ma of Haq et al. (1988), and the 12.18 Ma Condensed Section of Hardenbol et al. (1998).

Zone of Occurrence: Niger Delta NND5 nannofossil zone (NN6 Zone of Martini, 1971).

Characteristic Assemblage: The slight floral distribution is dominated by the reticulofenestrids [*R. pseudoumbilicus* (>7 μm), *R. haqii*, *R. minutula*, and *Reticulofenestra* spp.]. Others include *H. carteri*, *S. moriformis*, and *P. multipora* (Fig. 3.14).

Remarks: This condensed section is rarely observed in most Niger Delta sequences and at best, is represented by a slight floral distribution commonly believed as relics of the condensed section. This is due to paucity of nannofossils within the Middle Miocene Serravallian age. This low distribution of calcareous nannofossils, as well as other calcareous microfossils has been linked to the global "Middle Miocene Carbonate Crash" (Fadiya and Salami, 2012). This event that affected the better part of the Middle Miocene Serravallian age and the upper part of the Langhian is believed to have spanned approximately 4.36 Ma from the upper NN5 (NND4) to lower NN8 (NND7) nannofossil zones. Discoasters are very rare within the condensed section with spot occurrences of *D. sanmiguelensis*.

3.1.30 10.51 Ma Condensed Section

Age: Late Miocene (Tortonian).

Equivalent Condensed Sections: 9.2 Ma of Haq et al. (1987), 9.5 Ma of Haq et al. (1988), and the 9.5 Ma Condensed Section of Hardenbol et al. (1998).

Zone of Occurrence: Niger Delta NND8 nannofossil zone (NN9 Zone of Martini, 1971).

Characteristic Assemblage: *Discoaster bollii*, *Discoaster hamatus*, *Catinaster coalitus*, *Minylitha convalis*, *R. pseudoumbilicus* (>7 μm), *R. pseudoumbilicus* (<7 μm), *Discoaster neohamatus*, *S. moriformis*, *S. abies*, and *Discoaster bellus* (Fig. 3.15A–C).

Remarks: This is a major condensed section characterized by the acme event of both *D. bollii* and *D. hamatus*. In addition, *S. moriformis* dominates *S. abies* within this condensed section in contrast to the observation made in the 7.72 Ma condensed section.

3.1.31 8.8 Ma Condensed Section of Hardenbol et al. (1998)

Age: Late Miocene (Tortonian).

Equivalent Condensed Sections: Documented only by Hardenbol et al. (1998).

Zone of Occurrence: Niger Delta NND9 nannofossil zone (NN10 Zone of Martini, 1971).

Characteristic Assemblage: *D. neohamatus*, *Discoaster prepentaradiatus*, *R. pseudoumbilicus* (>7 μm), *Catinaster mexicanus*, *S. moriformis*, *S. abies*, *Discoaster neoerectus*, *Discoaster variabilis*, and *M. convalis* (Fig. 3.15A–C).

Remarks: The First Occurrences of *Discoaster pentaradiatus*, *Discoaster berggrenii*/*quinqueramus*, as well as the Last Occurrences of *D. prepentaradiatus*, *D. bollii*, and *R.*

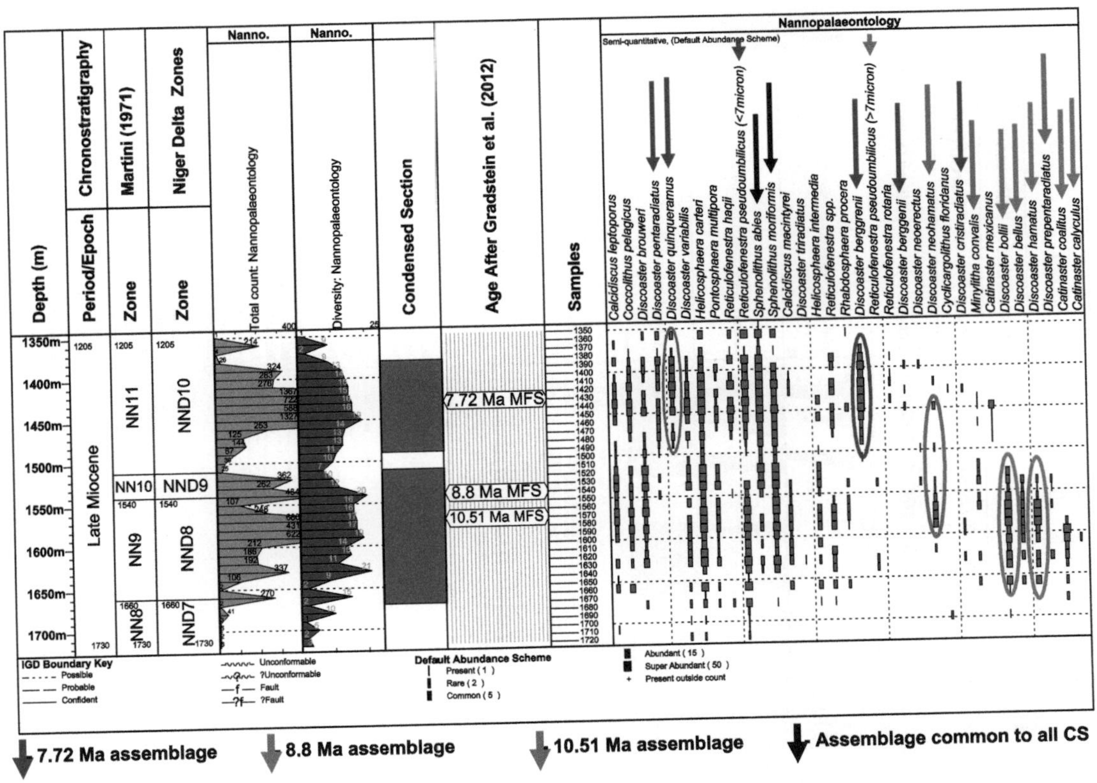

FIGURE 3.15 **Calcareous nannofossil assemblages of the 7.72 Ma condensed sections.**

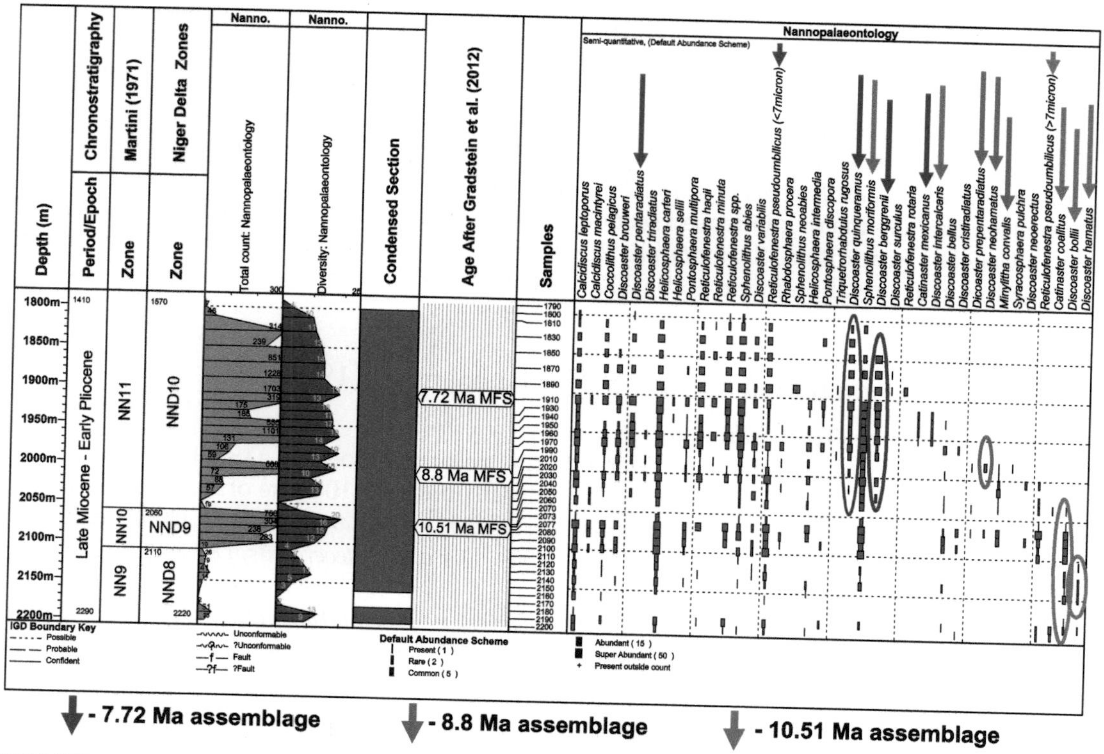

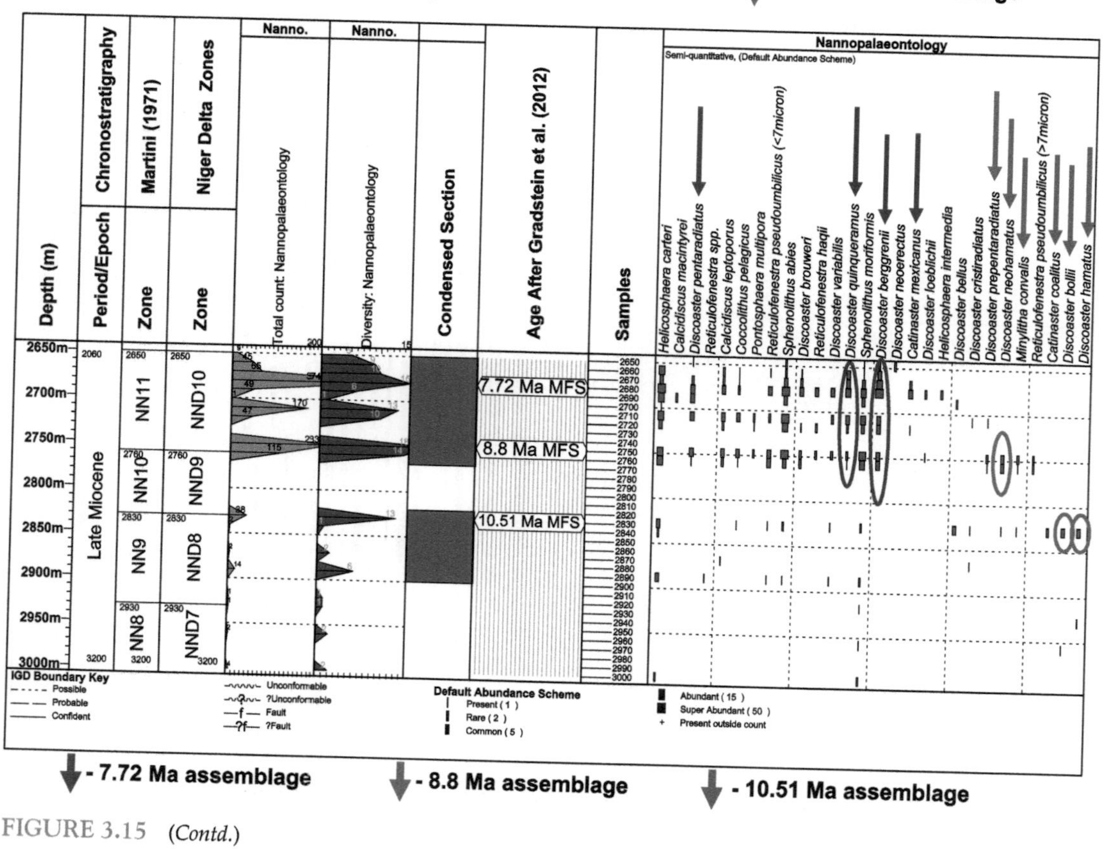

FIGURE 3.15 (*Contd.*)

pseudoumbilicus (>7 μm) were found to occur within this condensed section. The acme event of *D. neohamatus* also characterizes the condensed section. The condensed section, which is seen in many wells in the Niger Delta has only been documented by Hardenbol et al. (1998) and was missing in the Haq et al. (1987, 1988), and Gradstein et al. (2012) sequence cycle charts.

3.1.32 7.72 Ma Condensed Section

Age: Late Miocene (Tortonian).

Equivalent Condensed Sections: 7.30 Ma of Haq et al. (1987), 7.40 Ma of Haq et al. (1988), and the 7.30 Ma condensed section of Hardenbol et al. (1998).

Zone of Occurrence: Niger Delta NND10 nannofossil zone (NN11 Zone of Martini, 1971).

Characteristic Assemblage: *D. berggrenii, D. quinqueramus, S. abies, D. pentaradiatus, Discoaster brouweri, D. variabilis, S. moriformis*, reticulofenestrids [*R. pseudoumbilicus* (<7 μm), *R. pseudoumbilicus* (>7 μm) *Reticulofenestra* spp., *R. haqii*, and *R. minutula*]. The lower part of the condensed section is also characterized by the occurrence of *Discoaster berggenii* (a third morphotype of the group), *M. convalis, C. mexicanus*, and *Discoaster loeblichii* (Fig. 3.15A–C).

Remarks: The 7.72 Ma condensed section is characterized by the acme event of *D. berggrenii* where it also dominates over *D. quinqueramus* in abundance. As noticed *S. abies* also dominates over *S. moriformis* in abundance. It is important to note the dominance of *D. berggrenii over D. quinqueramus*, as well as the dominance of *S. abies* over *S. moriformis* even in sections with poor nannofossil distribution (Fig. 3.16A–B). Establishing the abundance proportion coupled with the presence of either of *C. mexicanus* or *M. convalis* is enough evidence for the 7.7 Ma condensed section.

3.1.33 5.99 Ma Condensed Section

Age: Late Miocene (Messinian).

Equivalent Condensed Sections: 5.8 Ma of Haq et al. (1987), 6.0 Ma of Haq et al. (1988), and the 6.0 Ma condensed section of Hardenbol et al. (1998).

Zone of Occurrence: Niger Delta NND11 nannofossil zone (NN11 Zone of Martini, 1971).

Characteristic Assemblage: *Discoaster quinqueramus, D. berggrenii, D. pentaradiatus, D. brouweri, D. variabilis, S. abies, S. moriformis*, reticulofenestrids [*R. pseudoumbilicus* (<7 μm), *R. pseudoumbilicus* (>7 μm) *Reticulofenestra* spp., *R. haqii*, and *R. minutula*], and *Amaurolithus delicatus*.

Remarks: The condensed section is characterized by the acme of *D. quinqueramus* and the dominance of *D. quinqueramus* over *D. berggrenii*, both of which are morphotypes of the same group. *D. quinqueramus* found within this condensed section are typically long-rayed in contrast to those found in the preceding 7.72 Ma condensed section. In addition, *S. abies* is conspicuously dominant over *S. moriformis* within the condensed section (Fig. 3.17). The species dominance of *D. quinqueramus* over *D. berggrenii* and *S. abies* over *S. moriformis* within the 5.99 Ma condensed section is also obvious in sections with poor nannofossil distribution (Fig. 3.18).

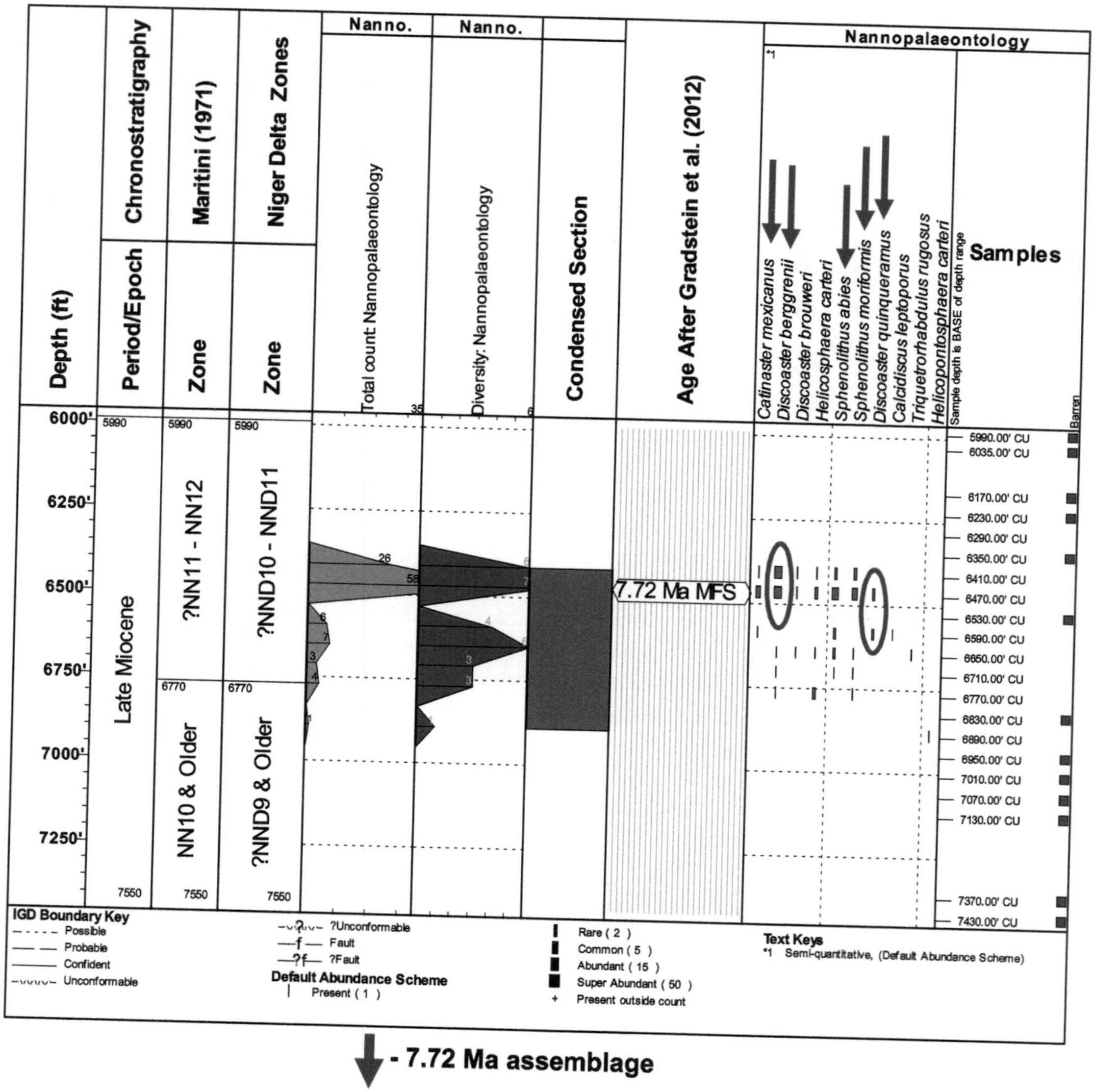

FIGURE 3.16 **Calcareous nannofossil assemblages of the 7.72 Ma condensed section (nannofossil-poor sections).**

3.1.34 5.47 Ma Condensed Section

Age: Late Miocene (Messinian).

Equivalent Condensed Sections: 5.0 Ma of Haq et al. (1987), 5.0 Ma of Haq et al. (1988), and the 5.05 Ma condensed section of Hardenbol et al. (1998).

Zone of Occurrence: Niger Delta NND12 nannofossil zone (NN12 Zone of Martini, 1971).

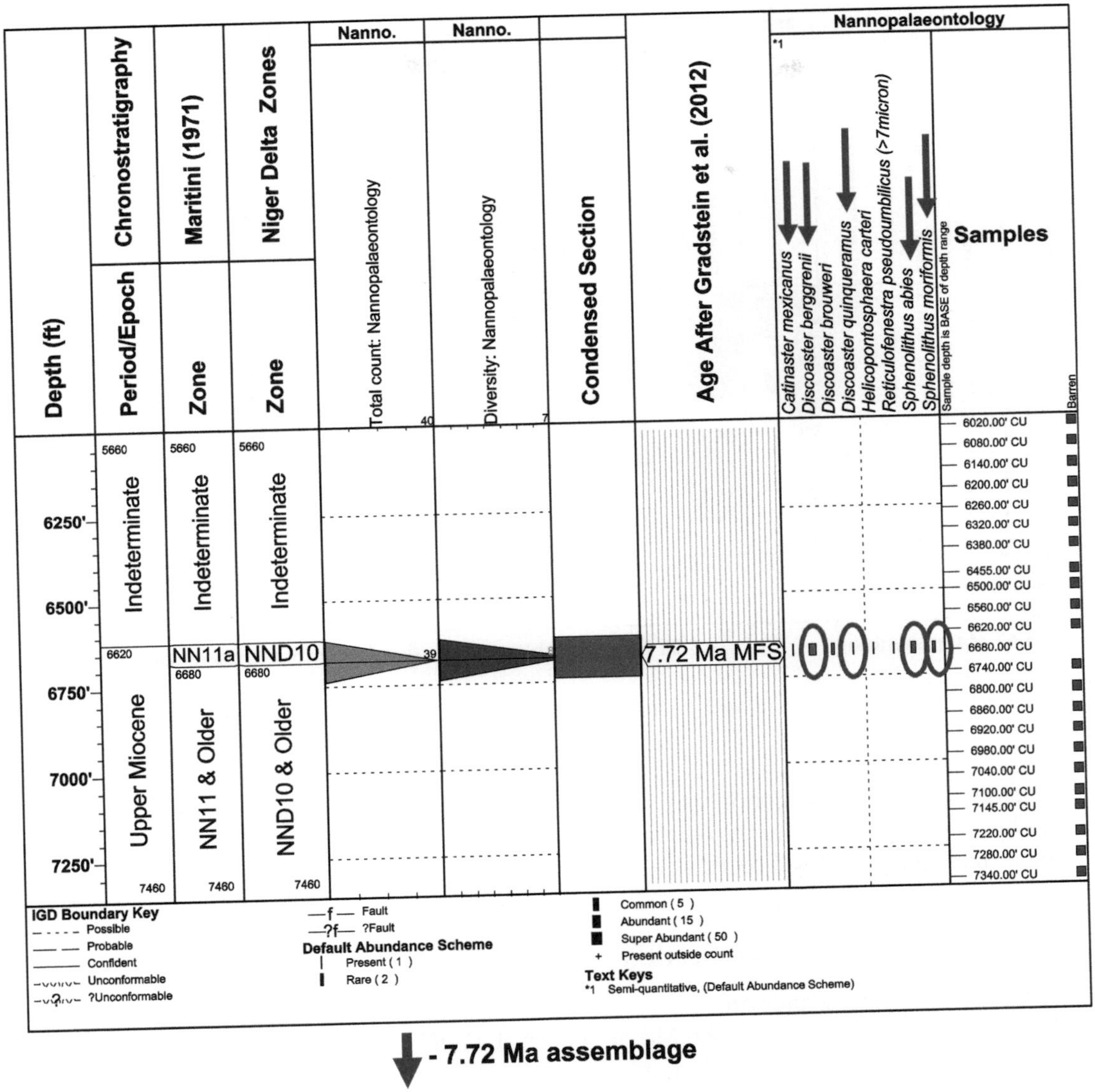

FIGURE 3.16 (*Contd.*)

Characteristic Assemblage: The condensed section is characterized by abundant and diverse nannofossil assemblages including the reticulofenestrids [*Reticulofenestra pseudoumbilicus* (<7 μm), *R. pseudoumbilicus* (>7 μm), *Reticulofenestra* spp., *R. haqii*, and *R. minutula*], *D. pentaradiatus*, *S. abies*, various species of *Scyphosphaera*, *Amaurolithus* (*A. delicatus* and *A. tricorniculatus*), ceratoliths, *Helicosphaera sellii*, *D. brouweri*, and *D. variabilis* (Fig. 3.19).

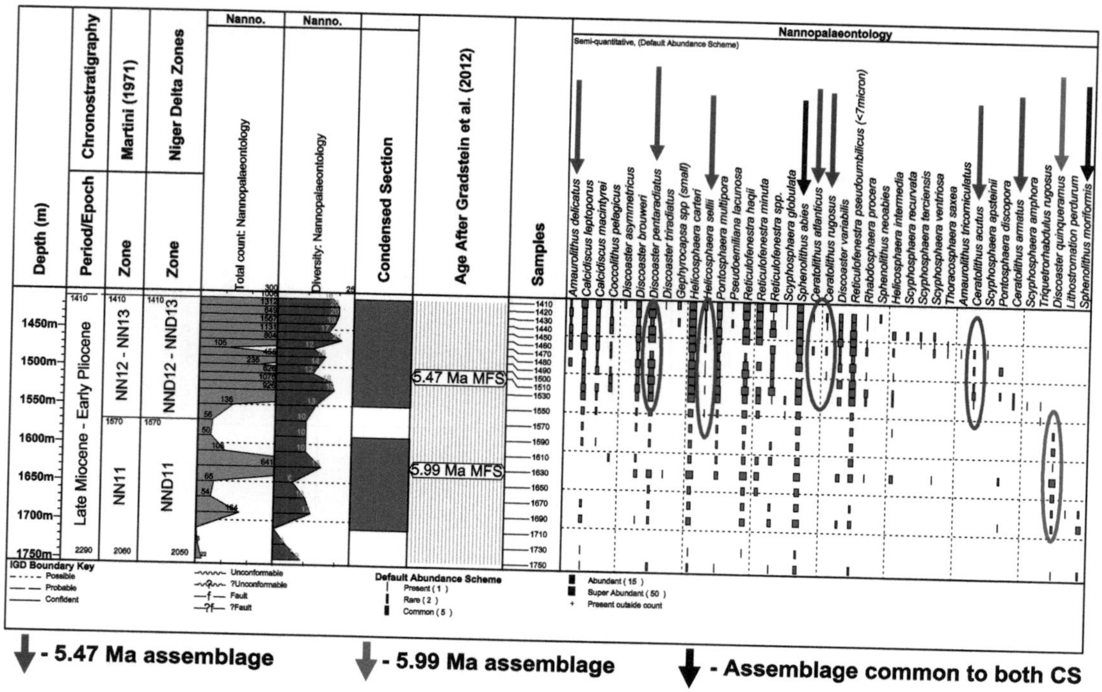

FIGURE 3.17 **Calcareous nannofossil assemblages of the 5.99 and 5.47 Ma condensed sections.**

Remarks: This is a major condensed section characterized by the acme occurrence of *D. pentaradiatus*. The appearance of the ceratoliths (*Ceratolithus armatus* and *Ceratolithus acutus*) also characterizes the condensed section.

3.1.35 4.16 Ma Condensed Section

Age: Early Pliocene (Zanclean).

Equivalent Condensed Sections: 4.0 Ma of Haq et al. (1987), 3.9 Ma of Haq et al. (1988), and the 4.0 Ma condensed section of Hardenbol et al. (1998).

Zone of Occurrence: Niger Delta NND13 nannofossil zone (NN13 Zone of Martini, 1971).

Characteristic Assemblage: This is a minor condensed section within the Early Pliocene and is characterized by the reticulofenestrids (*R. pseudoumbilicus*, *Reticulofenestra* spp., *R. haqii*, and *R. minutula*), *S. abies*, *Pseudoemiliania lacunosa*, *Gephyrocapsa* spp. (2–4 μm), *Ceratolithus cristatus*, *H. carteri*, *H. sellii*, and *D. pentaradiatus* (Fig. 3.19).

3.1.36 3.82 Ma Condensed Section

Age: Early Pliocene (Zanclean).

Equivalent Condensed Sections: 3.4 Ma of Haq et al. (1987), 3.4 Ma of Haq et al. (1988), and the 3.58 Ma Condensed Section of Hardenbol et al. (1998).

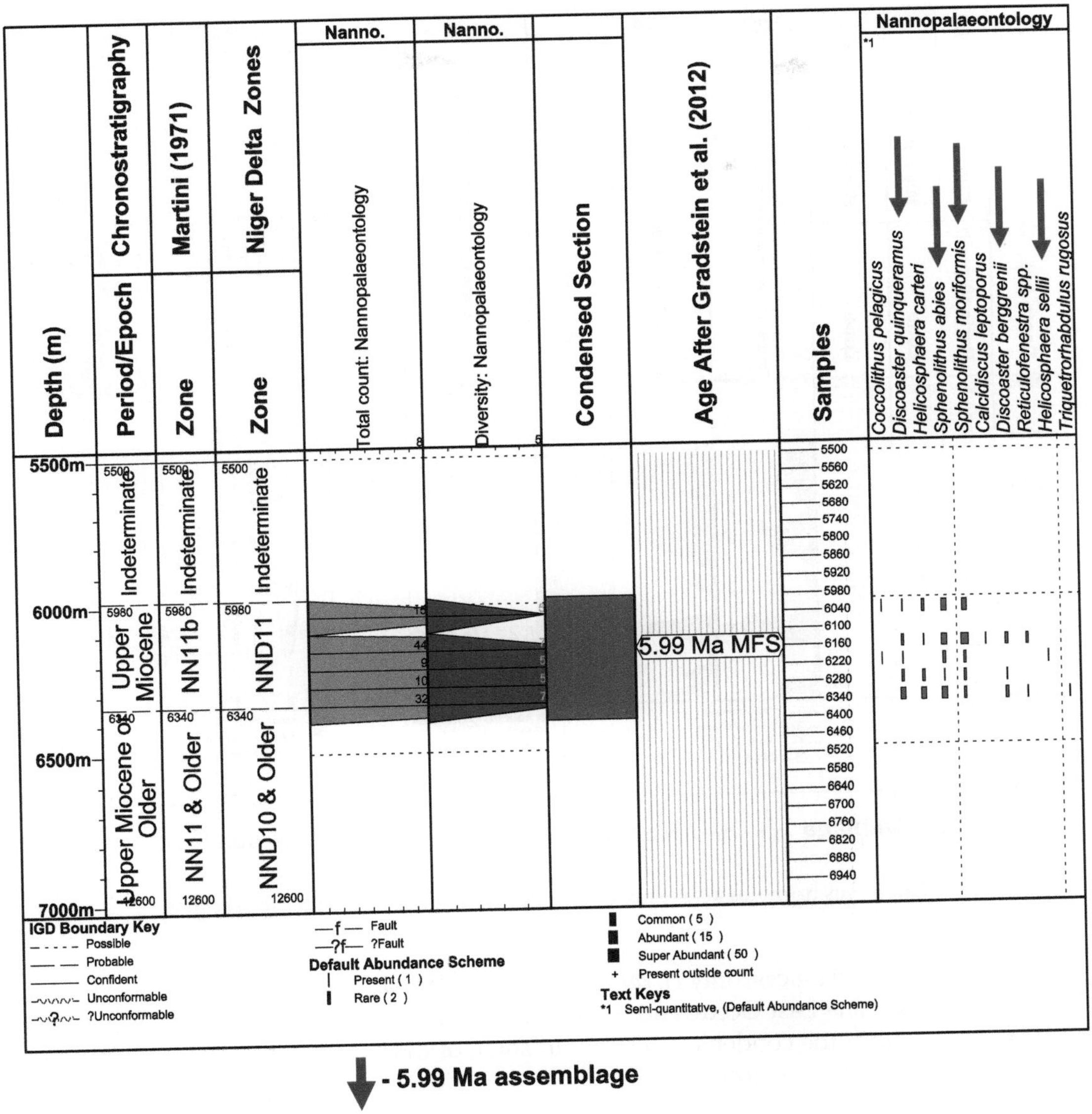

FIGURE 3.18 **Calcareous nannofossil assemblage of the 5.99 Ma condensed section (nannofossil-poor sections).**

Zone of Occurrence: Niger Delta NND14 nannofossil zone (NN15 Zone of Martini, 1971).

Characteristic Assemblage: *S. abies*, *Sphenolithus* spp., *R. pseudoumbilicus*, *C. pelagicus*, *R. minutula*, *R. haqii*, *P. lacunosa*, *D. pentaradiatus*, *Discoaster surculus*, *Gephyrocapsa* spp. (2–4 μm), and *P. multipora*. This condensed section within the Middle Pliocene is characterized by *C. pelagicus*, *R. minutula*, *R. haqii*, *P. lacunosa*, *D. pentaradiatus*, *D. surculus*, *Gephyrocapsa* spp. (2–4 μm), and *P. multipora* constitute the assemblage together with abundant records of *H. carteri* and *H. sellii* (Fig. 3.20).

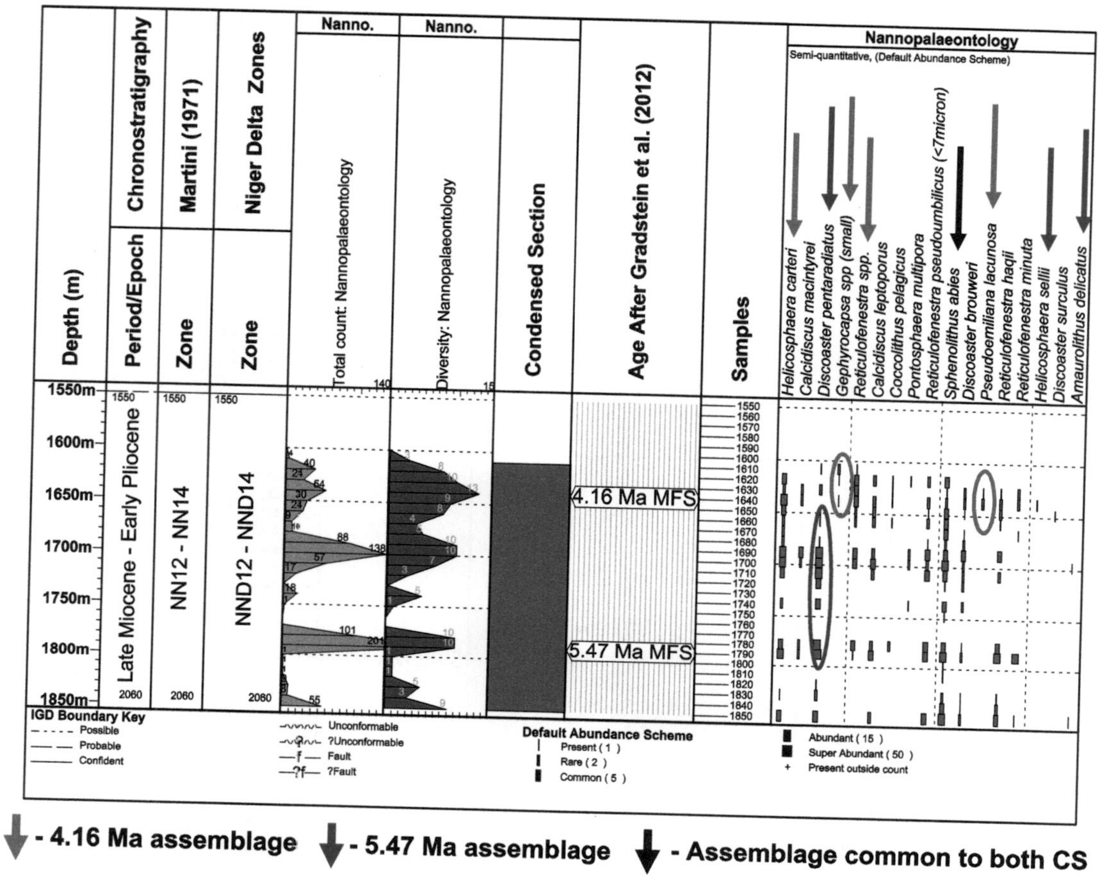

FIGURE 3.19 **Calcareous nannofossil assemblages of the 5.47 and 4.16 Ma condensed sections.**

Remarks: In many instances, only *H. carteri*, *H. sellii*, *P. lacunosa*, and *R. haqii* are found in the condensed section. The Last Occurrence of *S. abies*, *Sphenolithus* spp., and *R. pseudoumbilicus* are also found within the condensed section. In some of the condensed sections of this age, the discoasters are absent entirely.

3.1.37 2.92 Ma Condensed Section

Age: Late Pliocene (Piancenzian).

Equivalent Condensed Sections: 3.4 Ma of Haq et al. (1987), 3.4 Ma of Haq et al. (1988), and the 3.21 Ma condensed section of Hardenbol et al. (1998).

Zone of Occurrence: Niger Delta NND15 nannofossil zone (NN16 Zone of Martini, 1971).

Characteristic Assemblage: *D. pentaradiatus*, *D. brouweri*, *Reticulofenestra* spp., *Gephyrocapsa* spp. (2–4 μm), *P. lacunosa*, and *Calcidiscus macintyrei*.

Remarks: This represents the last condensed section with discoasters in the Upper Pliocene of the Niger Delta. In many instances, the discoasters are absent.

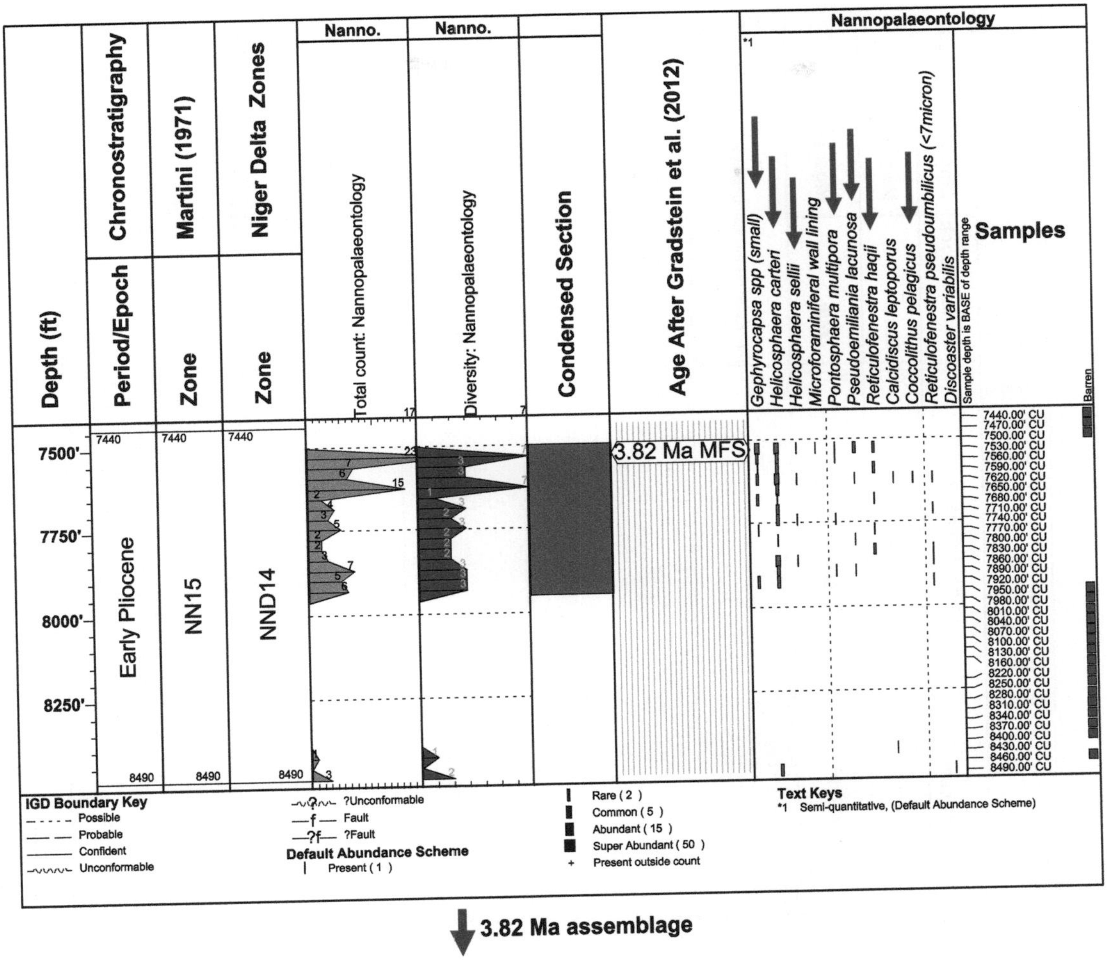

FIGURE 3.20 **Calcareous nannofossil assemblage of the 3.82 Ma condensed section.**

3.1.38 1.76 Ma Condensed Section

Age: Early Pleistocene (Calabrian).

Equivalent Condensed Sections: 2.0 Ma of Haq et al. (1987), 2.0 Ma of Haq et al. (1988), and the 1.59 Ma condensed section of Hardenbol et al. (1998).

Zone of Occurrence: Niger Delta QND17 nannofossil zone (NN19 Zone of Martini, 1971).

Characteristic Assemblage: The assemblage includes abundant reticulofenestrids dominated by *Gephyrocapsa caribbeanica*, *Gephyrocapsa oceanica*, *Gephyrocapsa parallela*, *Gephyrocapsa* spp. (2–4 μm), and *P. lacunosa*. Other important nannofossils within the condensed section include *C. macintyrei*, *H. sellii*, *Ceratolithus telesmus*, and *Helicosphaera inversa* (Fig. 3.21).

Remarks: This is the most obvious and clearly defined condensed section in the Early Pleistocene. The most conspicuous characteristic of this condensed section is the abundant and

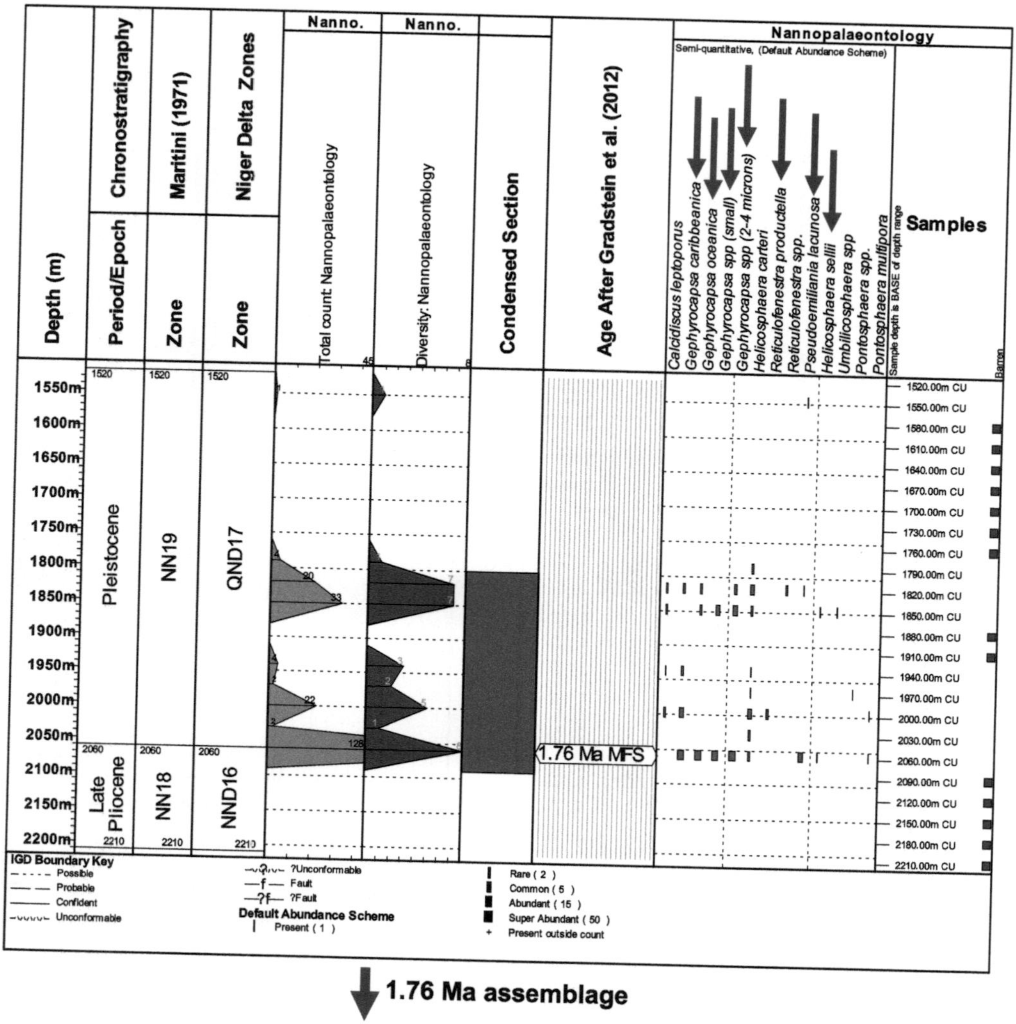

FIGURE 3.21 **Nannofossil assemblage of the 1.76 Ma condensed section.**

dominant occurrence of the genus *Gephyrocapsa*. Discoasters are completely absent within the condensed section.

This data show conclusively that different calcareous nannofossil assemblages can be used to characterize the different condensed sections in the Niger Delta and the data can be applied in the recognition and correlation of cap rocks and sealing shales.

CHAPTER

4

Systematic Paleontology: Niger Delta Calcareous Nannofossils

The task of unifying and harmonizing the fossil taxonomic concepts used by the different operating companies in the Niger Delta is one of the key objectives of the Calcareous Nannofossil Taxonomy project. However, unlike palynomorphs and foraminifera, the nannofossil taxonomic concepts employed by the various companies operating in the Niger Delta were based on those published in standard texts, such as Perch-Nielsen (1985a,b), Martini (1971), Okada and Bukry (1980), and others. The systematic taxonomy follows the recommendations of the International Code of Botanical Nomenclature. This chapter, therefore, concentrates on the description of the morphologic features of the marker species commonly encountered during routine analysis as observed under the light transmitting microscope (LTM). These features are described with emphasis on their stratigraphic importance. No new species are described.

The summary of the stratigraphic distribution of the species recorded, as well as the significant datums, zonal, and age determinations and the identified Maximum Flooding Surfaces (MFS) in selected study wells are presented in Fig. 2.2A–E. The calcareous nannofossil assemblages that characterize the major condensed sections that have so far been identified in the Niger Delta are discussed in details in Chapter 3 and summarized graphically in Figs. 3.1–3.21.

Family: DISCOASTERACEAE Tan, 1927
Genus: DISCOASTER Tan, 1927b

Discoaster sp.
Plate 1, Fig. 12

Note: This is an unidentified species of *Discoaster* found in the studied Niger Delta samples.

Discoaster adamanteus Bramlette and Wilcoxon, 1967a,b
Plate 5, Fig. 16

Note: Six-rayed *Discoaster* with no bifurcation. The rays are joined almost at their total length. Species are usually small with conspicuous ridges at the central area (CA).

Cenozoic Foraminifera and Calcareous Nannofossil Biostratigraphy of the Niger Delta. http://dx.doi.org/10.1016/B978-0-12-812161-0.00010-7

Stratigraphic Range: Late Oligocene PND23–Early Miocene NND2 Zones.

Discoaster asymmetricus Gartner, 1969
Plate 3, Fig. 17; Plate 5, Figs. 3 and 4

Note: This species is similar to *D. pentaradiatus* but with five asymmetric rays and no bifurcation. *D. asymmetricus* is rare in the Niger Delta.
Stratigraphic Range: NND14–NND15 Zones.

Discoaster berggrenii Bukry, 1971
Plate 1, Figs. 9 and 20; Plate 2, Fig. 1; Plate 3, Figs. 7–12, 18–20

Note: This species is similar to *D. quinqueramus* but with shorter arms and relatively more imposing proximal knob. The population of *D. berggrenii* decreases at the expense of *D. quinqueramus* in the upper part of their cooccuring ranges. Variants with far shorter arms, for example, *D. berggrenii* have also been reported within their range. The variants are here included under *D. berggrenii.*
Stratigraphic Range: Restricted to the NND10–NND11 Zones.

Discoaster bollii Martini and Bramlette, 1963
Plate 2, Fig. 2

Note: CA is generally small but with characteristic large distal boss, which is higher than the proximal one. This feature is particularly observed when the specimen lies on its side. Bifurcation is present on the six rays. Species is usually $< 10\ \mu m$.
Stratigraphic Range: Late Miocene NND7–NND9 Zones.

Discoaster brouweri Tan, 1927
Plate 1, Figs. 8, 10, 13, 14; Plate 4, Figs. 1–6

Note: *Discoaster* species with six nonbifurcating rays, symmetrical, and often with proximal extensions. Its last occurrence (LO) was in the Late Pliocene NND16 Zone at/or close to the Pliocene–Pleistocene boundary. *D. brouweri* is the most common six-rayed Miocene/Pliocene *Discoaster* species without bifurcation found in the Niger Delta.
Stratigraphic Range: NND6–NND16 Zones.

Discoaster deflandrei Bramlette and Riedel, 1954
Plate 2, Fig. 11

Note: *Discoaster deflandrei* has six rays with a wide featureless CA and wide bifurcations at the tip of the short rays. Average size is 7–8 μm. *D. deflandrei* is very common in the Early Miocene NND4 Zone. Slight morphological modifications have been observed within the species, for example, rare seven-rayed variants have been observed occurring from the Oligocene to Early Miocene.
Stratigraphic Range: Paleogene–Middle Miocene NND4 Zone.

Discoaster druggii Bramlette and Wilcoxon, 1967a,b
Plate 5, Fig. 20

Note: Size about 15 μm. Similar but larger than *Discoaster deflandrei* with which it cooccurs in the Early Miocene NND1 Zone. This form is very rare in the studied well sections.

Stratigraphic Range: Early Miocene NND1.

Discoaster exilis Martini and Bramlette, 1963
Plate 2, Figs. 17–20

Note: Six-rayed *Discoaster* species with tapering arms, which bear well-developed bifurcations at the tip and a stellate knob in the CA. Rays are characteristically more slender than those of other *Discoaster* species of the same age.

Stratigraphic Range: NND3–NND8 Zones.

Discoaster hamatus Martini and Bramlette, 1963
Plate 5, Fig. 15

Note: *Discoaster hamatus* is distinguished from all other five-rayed *Discoaster* species by the clockwise pointing of its arm in proximal view. The CA is small and the arms are long and slender and bent proximally. The six-rayed variant of the species, *D. neohamatus*, has an extended range and cooccurs with *D. quinqueramus* in the ND10 Zone.

Stratigraphic Range: Restricted to the NND8 Zone.

Discoaster kugleri Martini and Bramlette, 1963
Plate 1, Figs. 1, 5, 16

Note: Six-rayed *Discoaster* species with large and entirely featureless distal CA. The inter-ray area is broad, rays are short and free with notched ends, and mostly thickest at their tips. This sometimes appears like a small bifurcation. The proximal CA bears a low-stellate central knob.

Stratigraphic Range: Restricted to the NND6 Zone.

Discoaster kuepperi Stradner, 1959
Plate 5, Fig. 17

Note: Rays are of equal length, usually 8–12 in number, and CA is filled with a distinct boss. *Discoaster kuepperi* is common in the Early Eocene of the Niger Delta.

Stratigraphic Range: Early Eocene PND12–PND14 Zones.

Discoaster multiradiatus Bramlette and Riedel, 1954
Plate 2, Fig. 16

Note: *Discoaster multiradiatus* is rosette shaped with rays numbering from 25 to 29 that are joined along most of their length. It has a central knob that appears raised. It shows a slight birefringence in cross-polarized light. The features are more discernible in phase contrast.

Stratigraphic Range: Late Paleocene PND9–Early Eocene PND10 Zones.

Discoaster pentaradiatus Tan, 1927 emend. Bramlette and Riedel, 1954
Plate 1, Figs. 4 and 5; Plate 2, Figs. 3 and 4; Plate 3, Fig. 6

Note: *Discoaster pentaradiatus* has five thin, tapering symmetrical arms, with terminal thin delicate bifurcations, which are rarely preserved in most samples. The form shows slightly high birefringence when viewed between crossed nicols, as well as characteristic sutural extinction lines in the CA that make *D. pentaradiatus* identifiable even when the arms are not preserved. With these features and a lack of a boss in the CA, *D. pentaradiatus* is separable from *D. berggrenii* even when badly preserved.

Stratigraphic Range: NND9–Upper NND15 Zones.

Discoaster quinqueramus Gartner, 1969
Plate 1, Figs. 1–3, 11, 19; Plate 2, Figs. 7–10; Plate 3, Fig. 5; Plate 5, Figs. 5–12

Note: *Discoaster quinqueramus* has five slender arms with nonbifurcating tips. It exhibits a pronounced proximal knob and distal sutural ridges that make it distinguishable from other discoasters. The proportion of the CA to the arms varies, so does the overall size. Specimens with three to four arms have not been observed in the Niger Delta, although are reported from other basins worldwide.

Stratigraphic Range: Restricted to the NND10–NND11 Zones.

Discoaster surculus Martini and Bramlette, 1963
Plate 4, Figs. 15–20

Note: Six symmetrical-rayed *Discoaster* species, with ray tips that trifurcate into tongue-like protrusions. This feature makes *D. surculus* easily discernible.

Stratigraphic Range: NND9–NND15 Zones.

Discoaster variabilis Martini and Bramlette, 1963
Plate 1, Figs. 6, 7, 17, 18; Plate 2, Fig. 5; Plate 3, Figs. 1–4; Plate 4, Figs. 7–12

Note: This species has well-developed symmetric bifurcation on the six rays. The CA has ridges on either side and sometimes with very weak bosses.

Stratigraphic Range: Late Miocene NND8–Pliocene NND15 Zones.

Genus: CATINASTER Martini and Bramlette, 1963

Diagnosis: These are medium-sized, basket-shaped nannoliths that are formed as a result of fusion of the bifurcations of small discoasters (Young, 1998). They appear under the LTM like birefringent *Discoaster* species, but with the arms connected together at their tips.

Catinaster calyculus Martini and Bramlette, 1963
Plate 12, Fig. 11

Note: This species possesses six curved rays, which extend beyond the confines of the basket structure.

Stratigraphic Range: Late Miocene NND8–NND9 Zones.

Catinaster coalitus Martini and Bramlette, 1963
Plate 12, Figs. 12–14, 16

Note: This species possesses six rays that are confined to the basket. It is easy to identify as the forms display high birefringence in cross-polarized light.

Stratigraphic Range: Late Miocene NND7–NND9 Zones.

Family: FASCICULITHACEAE Hay and Mohler, 1967
Genus: FASCICULITHUS Bramlette and Sullivan, 1961

Diagnosis: These are cylindrical-shaped nannoliths consisting of a proximal column, disk (lateral elements), and a distal cone. They are very useful in Paleocene biostratigraphy.

Fasciculithus tympaniformis Hay and Mohler, 1967
Plate 16, Fig. 10

Note: Highly birefringent (usually yellow), with a simple proximal column which tapers distally.

Stratigraphic Range: Paleocene NP5–NP9 Zones.

Family: SPHENOLITHACEAE Deflandre, 1952
Genus: SPHENOLITHUS Deflandre In: Grassé, 1952

Diagnosis: Sphenoliths are nannoliths with a proximal column, one or more tiers of lateral elements, and an elongated apical structure, which sometimes bifurcates in some species. Many of the species have short stratigraphic ranges and are widely used as stratigraphic markers, especially within Oligocene to Early Miocene sequences.

Sphenolithus abies Deflandre, 1954
Plate 8, Figs. 3, 4, 15, 19, 20

Note: *S. abies* is the most common sphenolith in the Niger Delta. It occurs with *S. moriformis* in the "early" Late Miocene but it possesses a more cuspate outline, more elevated with apical spine than *S. moriformis*. Intermediate morphologic forms exist between the two forms in the "early" Late Miocene. *Sphenolithus neoabies* (Bukry and Bramlette, 1969) are considered smaller variants of *S. abies*.

Stratigraphic Range: NND8–NND14 Zones.

Sphenolithus belemnos Bramlette and Wilcoxon, 1967a,b
Plate 8, Fig. 9

Note: Dart-shaped sphenolith, with a well-developed apical spine, with sturdy proximal column that is relatively higher than the lateral element. The proximal column is higher in *S. belemnos* than in *S. heteromorphus* with which it can be confused. This species is rare in the Niger Delta.

Stratigraphic Range: Restricted to the Early Miocene NND2 Zone.

Sphenolithus ciperoensis Bramlette and Wilcoxon, 1967a,b
Plate 8, Fig. 10

Note: This form is rather small in size, but can be identified by the X-shaped extinction lines formed between the basal column and the apical spine when the stage is rotated to 45° under crossed nicols. *S. ciperoensis* possesses a well-developed apical spine which terminates in an X-shaped bifurcation.
Stratigraphic Range: Late Oligocene PND24–PND25 Zones.

Sphenolithus heteromorphus Deflandre, 1953
Plate 8, Figs. 1, 2, 5, 6, 8

Note: *S. heteromorphus* is very common in the Early Miocene of the Niger Delta. It is easily identifiable with its well-developed apical spine, which can be exceptionally high in some forms. The proximal column is of equal height with the lateral elements. This feature forms the main difference between *S. heteromorphus* and other sphenoliths that possess high apical spines, for example, *S. belemnos.*
Stratigraphic Range: Early–Middle Miocene NND2–NND4 Zones.

Sphenolithus moriformis Bronnimann and Stradner, 1960
Plate 8, Figs. 7, 11, 16–18

Note: Form with hemispherical outline, similar to *S. abies*, but with no visible apical spine. The entire structure often shows higher birefringence than *S. abies*. *Sphenolithus compactus* Martini, 1965 is considered a synonym of *S. moriformis.*
Stratigraphic Range: Early Eocene–Late Miocene NND9 Zone.

Sphenolithus primus Perch-Nielsen, 1971
Plate 8, Figs. 13 and 14

Note: Dome-shaped, oldest form of *Sphenolithus*, commonly found in the Early Paleocene. It consists, like other sphenoliths, of a proximal column and lateral elements.
Stratigraphic Range: Early Paleocene NP4–Early Eocene PND11 Zones.

Family: ZYGODISCACEAE Hay and Mohler, 1967
Genus: ISMOLITHUS Deflandre, 1954

Ismolithus recurvus Deflandre, 1954
Plate 20, Fig. 2

Note: Long elliptical to rhomboid outline with two bridges parallel to the axis each made up of two parts joining along the major axis which thickens slightly. *Ismolithus recurvus* is stratigraphically important in the Late Eocene–Early Oligocene interval. It is rare in the Niger Delta.
Stratigraphic Range: PND20–PND22 Zones.

Genus: NEOCHIASTOZYGUS Perch-Nielsen, 1971

Diagnosis: This genus is defined as elliptical coccoliths with two walls of inclined and/or vertical elements with a central cross (Perch-Nielsen, 1985a,b).

Neochiastozygus sp.

Note: Same description as *N. perfectus* but with a more pronounced central cross.

Neochiastozygus perfectus Perch-Nielsen, 1971
Plate 19, Figs. 3 and 14

Note: Species of this genus are relatively small, characterized by a rim consisting of two well-developed thin walls. The central cross is narrow and symmetrical with a higher inner wall. Nearly all species of *Neochiastozygus* are useful in the biozonation of the Paleocene.

Stratigraphic Range: Early Paleocene NP4–NP6 Zones.

Genus: ZYGODISCUS Bramlette and Sullivan, 1961

Diagnosis: The genus *Zygodiscus* consists of coccoliths with an I-shaped bridge, elliptical in outline, and two thin walls.

Zygodiscus bramlettei Perch-Nielsen, 1981
Plate 19, Fig. 15

Note: *Zygodiscus bramlettei* is elliptical and has a very thin outer wall and thick inner walls with openings surrounded by a ring of elements. This is an important species in the biozonation of the Paleocene.

Stratigraphic Range: Early Paleocene NP5–Late Paleocene NP9 Zones.

Family: HELICOSPHAERACEAE Black, 1971
Genus: HELICOSPHAERA Kamptner, 1954

Diagnosis: The genus *Helicosphaera* is structurally made up of three closely appressed units: the blankets, the flange, and the proximal plate. Between crossed nicols, all *Helicosphaera* species are birefringent. The optical feature of the blanket and that of the bridge or bar are used for the classification of the genus.

Helicosphaera ampliaperta Bramlette and Wilcoxon, 1967a,b
Plate 7, Figs. 1, 2–4

Note: *Helicosphaera* species with wide-opened central opening and a weak wing. The general outline of the species varies from elliptical to roughly triangular. The outline of the central opening corresponds to that of the periphery of the flange. *Helicosphaera scissura* Miller, 1981 is similar to *H. ampliaperta* with the exception of the narrow outline of the central opening. In the Niger Delta, both *H. ampliaperta* and *H. scissura* have been observed in the Early Miocene having approximately the same vertical distribution.

Stratigraphic Range: Early Miocene NND1–NND3 Zones.

Helicosphaera carteri Wallich, 1877; Kamptner, 1954
Plate 6, Figs. 1, 2, 4–8, 13–15; Plate 7, Figs. 5–8

Note: This name is used for any *Helicosphaera* that is medium to large in size and which possesses two uninclined small pores in the CA. *H. carteri* is the most common species of *Helicosphaera* in the Niger Delta.
Stratigraphic Range: NND1–NND21 Zones.

Helicosphaera carteri var. *wallichi* Lohmann, 1902; Theodoridis, 1984
Plate 6, Figs. 17 and 18

Notes: This species is similar to *H. carteri* but with closed or sometimes inclined central openings. The species was recently referred to *H. carteri carteri* [International Nannoplankton Association (INA) Discussion Forum].
Stratigraphic Range: Early Miocene NND1–Recent.

Helicosphaera euphratis Haq, 1966
Plate 18, Fig. 18

Note: Large and symmetrically elliptical helicolith with a bridge, which seals its central opening completely. *H. euphratis* resembles *H. intermedia* but differs in having an inclined bridge that covers the entire central opening of the proximal plate. It is common in the western Niger Delta.
Stratigraphic Range: Paleogene–Early Miocene NND3 Zone.

Helicosphaera recta Haq, 1966; Jafar and Martini, 1975
Plate 18, Fig. 17

Note: This form is relatively large. The CA has two large pores that are separated by a narrow conjunct bar. The distinguishing feature is the wing that terminates in a spur.
Stratigraphic Range: PND24–NND1 Zones.

Helicosphaera sellii Bukry and Bramlette, 1969; Jafar and Martini, 1975
Plate 6, Figs. 9 and 10; Plate 7, Figs. 19 and 20

Note: This species is similar to *H. carteri* but has large, slightly inclined pores in the CA.
Stratigraphic Range: NND11–NND18 Zones.

Helicosphaera seminulum Bramlette and Sullivan, 1961
Plate 7, Fig. 11; Plate 17, Figs. 10–12, 20

Note: This species is similar in outline and size to *H. truempyi* but has an increased flange with an acute notch.
Stratigraphic Range: PND12–PND16 Zones.

Helicosphaera stalis Muller, 1974
Plate 6, Fig. 16

Note: *Helicosphaera stalis* is a very small species of *Helicosphaera* with a lenticular or rhomboid outline and a broad, straight bar, which is inclined normally. *H. stalis* is common in the western Niger Delta.

Stratigraphic Range: NND5–NND11 Zones.

Helicosphaera truempyi Biolzi and Perch-Nielsen, 1982
Plate 18, Fig. 20

Note: Large species with open CA spanned by horizontal, disjunct bar. Sometimes the bar is missing in badly preserved forms in which case it may resemble *H. ampliaperta*. The difference is often recognized by the flange present in the *H. truempyi* that is missing in *H. ampliaperta*.

Stratigraphic Range: PND25–NND1 Zones.

Helicosphaera walbersdorfensis Muller, 1974
Plate 6, Fig. 20

Note: Small species, triangular to symmetrically elliptical helicoliths with a distinct, narrow central opening. This species occurs up to the Late Miocene NND9 Zone in the Niger Delta.

Stratigraphic Range: NND4–NND9 Zones.

Family: PONTOSPHAERACEAE Lemmermann, 1908
Genus: PONTOSPHAERA Lohmann, 1902

Diagnosis: The genus *Pontosphaera* contains murolith coccoliths having variable numbers of perforations in the CA (Young, 1998).

Pontosphaera discopora Schiller, 1925
Plate 13, Figs. 1 and 4

Note: *Pontosphaera discopora* is similar to *P. multipora* but differs with an elevated rim and lesser pores in the CA.

Stratigraphic Range: Paleogene–QND21 Zone.

Pontosphaera multipora Kamptner, 1948
Plate 13, Figs. 2, 3, 6–8

Note: *Pontosphaera multipora* has three or more cycles of large pores in the CA with the CA protruding above the rim.

Stratigraphic Range: Paleogene–QND21 Zone.

Genus: TRANSVERSOPONTIS Hay, Mohler, and Wade 1966

Diagnosis: The genus *Transversopontis* is here described under the Family *Pontosphaeraceae* (Perch-Nielsen, 1985a,b). They are usually characterized by two large openings in the CA.

Transversopontis pulcheroides Sullivan, 1964a,b
Plate 19, Fig. 6

Note: *Transversopontis pulcheroides* is elliptical in shape, with two large openings in the CA spanned by a broader bridge. *Transversopontis* is common to abundant in Middle Eocene NP15/NP16 Zones in the Niger Delta.

Stratigraphic Range: Early Eocene PND13–Late Oligocene PND24 Zones.

Genus: SYRACOSPHAERA Lohmann, 1902

The genus *Syracosphaera* is usually dithecate with exothecal coccoliths of various shapes. Endothecal coccoliths are muroliths with one, two, or three flanges often differentiated into body and circum-flagellar coccolits and sometimes antapical coccoliths. Exothecal coccoliths are planolithic muroliths or dome shaped.

Syracosphaera sp.
Plate 20, Fig. 14

Note: Similar to *Syracosphaera pulchra* but without radial laths in the CA. This species was found in the Anyala field.

Syracosphaera pulchra Lohmann, 1902
Plate 20, Fig. 6

Note: *Syracosphaera pulchra* is elliptical in outline, with a birefringent rim and central radial laths.

Stratigraphic Range: Paleogene–QND21 Zone.

Genus SCYPHOSPHAERA Lohman 1902 Diagnosis: The genus *Scyphosphaera* is similar to *Pontosphaera* but with rims elevated into vase-like structures (Young, 1998)

Scyphosphaera pulcherrima Deflandre, 1942
Plate 20, Fig. 7

Note: *Scyphosphaera pulcherrima* has a wide opening and a well-developed collar. The species is common in the Pliocene–Pleistocene sequences of the Niger Delta.

Stratigraphic Range: NND6–QND19 Zones.

Family: COCCOLITHACEAE Poche, 1913
Genus: COCCOLITHUS Schwarz, 1894

Coccolithus formosus Kamptner, 1963
Plate 16, Figs. 11 and 12

Note: Large elliptical coccolith, with open CA and highly birefringent (yellow). They are useful within the Early Oligocene–Early Eocene.

Stratigraphic Range: Early Eocene PND9–Early Oligocene PND22 Zones.

Coccolithus pelagicus Wallich, 1877; Schiller, 1930
Plate 14, Figs. 9–11, 13–16, 19, 20

Note: This species is common from the Eocene to Pleistocene. It is round to slightly elliptical and consists of two visible cycles. The inner cycle shows higher birefringence than the outer cycle. A small opening in the center is characteristic. Sizes ranging from 5 to 16 μm are common. The large size ranges have been the basis of definition of different species: *Coccolithus eopelagicus* (>16 μm) (Bramlette and Riedel, 1954) Bramlette and Sullivan, 1961, *Coccolithus miopelagicus* (14–16 μm) Bukry, 1971, and *Coccolithus pliopelagicus* Wise, 1973 (small CA, with no bar). None of these species are used as stratigraphic marker in the Niger Delta.

Stratigraphic Range: Eocene–QND 21.

Coccolithus subpertusus Kamptner, 1963
Plate 14, Fig. 17

Note: *Coccolithus subpertusus* differs from *Coccolithus formosus* in size. *Coccolithus formosus* is usually >10 μm, while *Coccolithus subpertusus* is between 7 and 10 μm.

Stratigraphic Range: Restricted to the entire Paleocene interval.

Genus: SOLIDOPONS Theodoridis, 1984
Solidopons petrae Theodoridis, 1984
Plate 19, Fig. 13

Note: *Solidopons petrae* is elliptical in outline, with a narrow rim and a wide CA usually with a disjunct, transverse bridge (Young, 1998). *Solidopons petrae* is rare in the shallow offshore areas but common in the deep water of the Niger Delta.

Stratigraphic Range: Early Miocene NND2–Late Miocene NND8 Zones.

Genus: UMBILICOSPHAERA Lohmann, 1902
Umbilicosphaera rotula Kamptner, 1956; Varol, 1982
Plate 20, Fig. 10

Note: Circular to subcircular form with wide CA and narrow rim. Medium-size range, often less than 5 μm.

Stratigraphic Range: *U. rotula* occurs from Early Miocene NND1–Late Pliocene NND15 Zones in the Niger Delta.

Genus: CRUCIPLACOLITHUS Hay and Mohler In: Hay et al., 1967
Cruciplacolithus primus Perch-Nielsen, 1977
Plate 19, Figs. 8 and 17

Note: Small coccolith, size range between 3 and 5 μm, with thin characteristic central cross. The first occurrence (FO) is useful in defining the Cretaceous/Tertiary boundary.

Stratigraphic Range: Early Paleocene NP1–NP2 Zones.

Cruciplacolithus tenuis Stradner, 1961
Plate 19, Fig. 10

Note: Similar to *Cruciplacolithus primus* but differs in the structure of the CA cross. The CA cross in *Cruciplacolithus tenuis* has small feet, in *Cruciplacolithus primus*, the feet are absent.

Stratigraphic Range: Paleocene NP3–NP9 Zones.

Family: CALCIDISCACEAE Young and Bown, 1997
Genus: CALCIDISCUS Kamptner, 1950

Calcidiscus tropicus
Plate 10, Figs. 11, 12, 15–17

Note: *Calcidiscus tropicus* is similar to *Calcidiscus macintyrei*, but differs in size. *Calcidiscus tropicus* is usually 7–10 μm in size, while *Calcidiscus macintyrei* is often >10 μm.
Stratigraphic Range: Early Miocene NND3–Late Miocene NND9 Zones.

Calcidiscus leptoporus Murray and Blackman, 1898
Plate 10, Figs. 1–6

Note: This species is usually <10 μm and broadly elliptical to circular with a closed CA. The proximal shield is often more birefringent than the distal. The number of elements in the shields is less than in *Calcidiscus macintyrei.*
Stratigraphic Range: Neogene–Recent.

Calcidiscus macintyrei Bukry and Bramlette, 1969
Plate 10, Figs. 7–9

Note: *C. macintyrei* differs from *C. leptoporus* in size. *C. macintyrei* is often less than 10 μm, with higher number of elements in the CA. The CA in *C. leptoporus* is closed, while the CA is open in *C. macintyrei.*
Stratigraphic Range: NND6–QND19 Zones.

Genus: CHIASMOLITHUS Hay, Mohler, and Wade, 1966
Chiasmolithus altus Bukry and Percival, 1971
Plate 19, Fig. 16

Note: Forms with small CA, central cross meets at right angles and of equal length.
Stratigraphic Range: Early Oligocene PND22–PND25 Zones.

Chiasmolithus bidens Bramlette and Sullivan, 1961; Hay and Mohler, 1967
Plate 21, Figs. 7 and 8

Note: This species is an elliptical coccolith with two cycles, inner and outer, and a central structure representing a cross that is not aligned with the axes of the coccolith and which occupies a considerable portion of the CA. The space left within the CA is small compared to the portion occupied by the structural features. The inner cycle is observed to be more bi-refringent than the outer cycle. The outer cycle consists of prominent elements, which builds the crest. The average sizes of the forms seen are 7–9 μm.
Stratigraphic Range: PND10–PND11 Zones.

Chiasmolithus solitus Bramlette and Sullivan, 1961, Locker, 1968
Plate 17, Figs. 13, 16–19

Note: Coccoliths with two cycles of elements in the proximal shield; CA with a cross and narrow distal shield. *Chiasmolithus solitus* differs from *Chiasmolithus bidens* in the absence of "teeth" (Perch-Nielsen, 1985a,b).

Stratigraphic Range: Eocene PND10–PND17 Zones.

Family: CALSIOSOLENIACEAE Kamptner, 1937
Genus: CALSIOSOLENIA Gran, 1912

Diagnosis: Rhomboid shaped muroliths, usually small and similar to *Syracosphaera* (Bown, 1999).

Calsiosolenia murrayi Gran, 1912
Plate 20, Figs. 12 and 15

Note: Very small scapholiths, usually less than 3 μm. Identification is sometimes difficult.

Stratigraphic Range: Paleogene–Recent.

Family: NOELAERHABDACEAE Jerkovic, 1970 emend. Young and Bown, 1997
Genus: CYCLICARGOLITHUS Bukry, 1971

Diagnosis: This genus is defined as a circular to subcircular coccolith with a small CA (Young, 1998).

Cyclicargolithus abisectus Muller, 1970
Plate 15, Figs. 13 and 16

Note: This form is similar to but larger than *C. floridanus*. The form is often larger than 11 μm.

Stratigraphic Range: Oligocene PND24–PND25 Zones.

Cyclicargolithus floridanus Roth and Hay, In: Hay et al., 1967; Bukry, 1971
Plate 15, Figs. 5–7

Note: Subcircular form with a birefringent distal shield and a relatively small central opening. The extinction lines between the wall and the shield are curved and continuous. The species is similar to *Cyclicargolithus abisectus* but with a size of <10 μm.

Stratigraphic Range: Paleogene Middle Miocene NND6 Zone.

Cyclicargolithus lumina Sullivan, 1965; Bukry, 1971
Plate 17, Fig. 1

Note: Medium-sized, circular discoidal placolith, consisting of two very closely appressed plates. CA slightly depressed, less than half the total diameter.

Genus: PSEUDOEMILIANIA Gartner, 1969

Definition: Circular or round, rarely elliptical coccoliths, with a variable number of slits in the distal shield.

Pseudoemiliania lacunosa Kamptner, 1963; Gartner, 1969
Plate 12, Figs. 17–19

Remarks: *P. lacunosa* is circular to subcircular, with more than 12 slits in the distal shield. The size is medium, often less than 4 μm. Although, identification of the slits under the LTM is possible mainly in phase contrast, the form is easily identified by the circular outline of the coccolith. It is similar to *Reticulofenestra rotaria*, but the latter usually occurs in NND11 Zone in association with *D. quinqueramus*.

Stratigraphic Range: With other species of *Pseudoemiliania*, *P. lacunosa* occurs from Early Pliocene NND14–Early Pleistocene QND19 Zones in the Niger Delta.

Genus: GEPHYROCAPSA Kamptner, 1943

Diagnosis: These are forms similar to *Reticulofenestra* but with a bridge in the CA. They have been observed in the Niger Delta from the Early Pliocene. Two species of *Gephyrocapsa* are important markers in the biostratigraphy of the Niger Delta, namely, small *Gephyrocapsa* spp. in the Pliocene and *G. caribbeanica* in the Pleistocene.

Small *Gephyrocapsa* spp.

Note: These are smaller *Gephyrocapsa* species not larger than 3 μm. They are common to abundant from the Early Pliocene NND14 Zone. Sporadic or rare occurrences of them are recorded around the Miocene/Pliocene boundary. The bridge spanning the CA has variable angles. The roundness and pore width are also variable.

Stratigraphic Range: QND14 Zone.

Gephyrocapsa caribbeanica Boudreaux and Hay, 1967
Plate 21, Figs. 2 and 3

Note: This is largest species of the genus *Gephyrocapsa* observed in the Niger Delta. The bridge spanning the CA has variable angles. The roundness and pore width are also variable. *G. caribbeanica* is referred to as "Large" *Gephyrocapsa* spp. in many nannofossil reports made available for this project. The lower size limit is 4–5 μm.

Stratigraphic Range: Restricted to QND18 Zone.

Gephyrocapsa oceanica Kamptner, 1943
Plate 20, Fig. 5

Note: Elliptical placolith with a central bridge usually at an angle less than 45°. *Geophyrocapsa oceanica* is common in the Pleistocene sediments of the Niger Delta. The size varies between 3 and 4 μm.

Stratigraphic Range: Early Pleistocene QND19–Recent.

Genus: HORNIBROOKINA Edwards, 1973

Hornibrookina teuriensis Edwards, 1973
Plate 19, Fig. 5

Note: *Hornibrookina* as a genus is phylogenetically linked to the Cretaceous genus *Biscutum*. It seems to have survived the Cretacous–Tertiary mass extinction. *Hornibrookina teuriensis* has a wide CA covered by more or less radially oriented elements. Nearly all the species of *Hornibrookina* are restricted to the Early Paleocene.

Stratigraphic Range: Paleocene NP3–NP5 Zones.

Genus: TOWEIUS Hay and Mohler, 1967

Toweius crassus Bramlette and Sullivan, 1961
Plate 19, Fig. 4

Note: Medium-size *Toweius* species with a small CA opening. The species is highly birefringent. The shields are bright with no clear perforations in the CA.

Stratigraphic Range: *Toweius crassus* is restricted to the Early Eocene PND13 Zone.

Toweius eminens Bramlette and Sullivan, 1961
Plate 19, Fig. 9

Note: The genus *Toweius* refers to elliptical coccoliths with a distal shield, double-proximal shield, and two rows of elements in the CA covered by a relatively fine-meshed grid. *Toweius eminens* has a high (yellow) birefringence, four large perforations in the CA, and the perforations are situated on the distal side of the coccolith. It differs from other species by the number and size of perforations in the CA.

Stratigraphic Range: *Toweius eminens* is rare in the Paleocene NP7 Zone but common between Early NP9 and Paleocene NP10 Zones.

Toweius ovae Perch-Nielsen, 1977
Plate 19, Fig. 11

Note: *Toweius ovae* is a very large form of *Toweius* with six large perforations in the CA. It is highly birefringent when viewed between crossed nicols with the six perforations appearing as dark holes (Perch-Nielsen, 1985a,b).

Stratigraphic Range: *Toweius ovae* is restricted to the Early Eocene PND9 Zone.

Toweius selandianus Perch-Nielsen, 1977
Plate 19, Fig. 1

Note: Very small form, usually less than 4 μm in size, with relatively large perforations in the CA surrounded by two walls. The species has a very weak birefringence.

Stratigraphic Range: *Toweius selandianus* is restricted to the Middle Paleocene NP5 Zone in the Niger Delta.

Genus: RETICULOFENESTRA Hay, Mohler, and Wade 1966 (?1967)

Description: The genus *Reticulofenestra* includes elliptical, circular to subcircular coccoliths with an open or closed CA. Species are grouped based on the CA opening, type of structure in the CA, and size variation.

Reticulofenestra haqii Backman, 1978
Plate 11, Figs. 10–12, 19

Note: Form with size range 3–5 μm, elliptical, and birefringent.
Stratigraphic Range: (?)Early Miocene–(?)Late Pliocene NND1–NND14 Zones.

Reticulofenestra minuta Roth, 1970
Plate 11, Fig. 14

Note: Form with size less than 3 μm, elliptical, and weakly birefringent. Usage includes small species of *Dictyococites*, with CA closed by plates formed from inner tube cycle (Black, 1967) found in many nannofossil reports from operators in the Niger Delta.
Stratigraphic Range: Paleogene–QND21.

Reticulofenestra minutula Gartner, 1967; Haq and Berggren, 1978
Plate 20, Fig. 18

Note: Form similar to *Pseudoemiliania lacunosa* but usually 3–5 μ in size. The species is common in the Late Pliocene.
Stratigraphic Range: Early Pliocene NND14–Late Pliocene NND16 Zones.

Reticulofenestra pseudoumbilicus Gartner, 1967
Plate 11, Figs. 1–9, 18

Note: Similar to *R. haqii* but with size >5 μm. *R. pseudoumbilicus* of size >7 μm have been observed in the NND5 Zone of the Niger Delta.
Stratigraphic Range: Early Miocene–Late Pliocene NND1–NND15 Zones.

Reticulofenestra rotaria Theodoridis, 1984
Plate 21, Figs. 14–16

Note: Very distinctive species with characteristic circular outline with wide CA. In cross-polarized light, this form is similar in appearance to *Pseudoemiliania lacunosa* (Kamptner, 1963; Gartner, 1969). *R. rotaria* occurs in the Late Miocene, while *P. lacunosa* is an Early Pliocene/Pleistocene form.

Stratigraphic Range: Restricted to NND11 Zone. Noted as a key marker with the LO of its range near top *Discoaster quinqueramus*/*Discoaster berggrenii*.

Reticulofenestra umbilica Bramlette and Wilcoxon, 1966b
Plate 17, Figs. 8 and 9

Note: This species has a large elliptical form with a high wall bordering a medium to large CA. The CA is often open without the fine net that is typical of very well-preserved forms. Size range between 15 μ and 16 μm is very common. Both the FO and the LO of *R. umbilica* are useful in the Eocene.

Stratigraphic Range: PND 15–PND 22 Zones.

Family: RHABDOSPHAERACEAE Haeckel, 1894a,b.
Genus: RHABDOSPHAERA Haeckel, 1894a,b

Rhabdosphaera claviger Murray and Blackman, 1898
Plate 20, Fig. 13 (SEM)

Note: *Rhabdosphaera claviger* has straight narrow spire and a symmetrical base. It is the most common rhabdolith in the Neogene of the Niger Delta.

Stratigraphic Range: Paleogene–QND21 Zone.

Genus: DICTYOCOCCITES Black, 1967
Reticulofenestra scrippsae Bukry and Percival, 1971
Plate 11, Fig. 20

Note: Elliptical, small-sized coccolith with bent extinction line (Perch-Nielsen, 1985a,b). LO useful in defining the top of the Late Oligocene PND25 Zone.

Stratigraphic Range: Middle Eocene PND16–Late Oligocene PND25 Zones.

Genus: EMILIANIA Hay and Mohler, In: Hay et al., 1966
Emiliania huxleyi Lohmann, 1902

Note: This species occurs as a small coccolith often about 2–3 μm in LTM. They dominate the assemblage in the Late Pleistocene.

Stratigraphic Range: Late Pleistocene NN21 Zone.

Family: CERATOLITHACEAE Norris, 1965

Remarks: The family embraces the Neogene horseshoe-shaped nannoliths. Two genera are identified, strongly birefringent *Ceratolithus* and the weakly to nonbirefringent *Amaurolithus*. Both are stratigraphically useful in the Upper Miocene to Pliocene of the Niger Delta.

Genus: AMAUROLITHUS Gartner and Bukry, 1975a,b

Note: These consist of the horseshoe-shaped nannoliths, which are weakly birefringent or nonbirefringent between crossed nicols. These appear this way because the *c*-axis of the microscope lies perpendicular to the plane of the horseshoe.

Amaurolithus delicatus Gartner and Bukry, 1975a,b
Plate 9, Figs. 1, 2, 10

Note: Simple and delicate horseshoe-shaped species with a rounded apex and no apical spine or extension. Intermediate forms occurring within the range of *A. delicatus* are common. These exhibit features of *A. delicatus* but are strongly birefringent like *Ceratolithus*. *A. delicatus* is the most commonly observed ceratolithaceae in the Niger Delta.

Stratigraphic Range: Late Miocene NND11–Early Pliocene NND14 Zones.

Genus: CERATOLITHUS Kamptner, 1950

Diagnosis: *Ceratolithus* contains species of horseshoe-shaped nannoliths, which show strong birefringence in cross-polarized light.

Ceratolithus acutus Gartner and Bukry, 1974
Plate 21, Figs 1 and 2

Note: When found during analysis, *C. acutus* is the nannofossil that best approximates the Miocene/Pliocene boundary. It often occurs associated with the 5.0 Ma MFS condensed section of Haq et al. (1988). The species possesses a characteristic apical structure that terminates in an acute angle to a sharp point. The two arms are unequal.

Stratigraphic Range: NND12–(?)NN13 Zones.

Ceratolithus cristatus Kamptner, 1950
Plate 9, Figs. 4 and 5

Note: Horseshoe-shape, birefringent with variable size, and some degree of ornamentation. The two arms are equal with no apical structure.

Stratigraphic Range: Pliocene NND13–Pleistocene NND21 Zones.

Ceratolithus rugosus Bukry and Bramlette, 1968
Plate 20, Fig. 17

Note: This species does not possess apical structures, but has characteristic rows of nodes on the arm. They have been observed sometimes heavily calcified.

Stratigraphic Range: Early Pliocene NND13 Zone.

Family: TRIQUETRORHABDULACEAE Lipps, 1969
Genus: TRIQUETRORHABDULUS Martini, 1965

Diagnosis: These are nannoliths formed by a single calcite crystal with a variable *c*-axis orientation. They are elongated in shape and formed of three blades. They show bright birefringence in crossed nicols.

Triquetrorhabdulus carinatus Martini, 1965
Plate 9, Figs. 11 and 12

Note: These are elongated forms, with *c*-axis parallel to length of the nannolith and all the three blades are identical.
Stratigraphic Range: Paleocene–Early Miocene NND1 Zone.

Triquetrorhabdulus milowii Bukry, 1971
Plate 21, Fig. 13

Note: This species is relatively short and broad, blades are curved but similar. It shows a very bright birefringence between crossed nicols.
Stratigraphic Range: NND1–NND5 Zones.

Triquetrorhabdulus rugosus Bramlette and Wilcoxon, 1967a,b
Plate 21, Figs. 7 and 8

Note: This species is characterized by a lateral blade that is straight or slightly curved and often wider than other species of *Triquetrorhabdulus*.
Stratigraphic Range: NND5–NND12 Zones.

Family: BRAARUDOSPHAERACEAE Deflandre, 1947

Remarks: This family includes calcareous pentaliths whose genera are distinguished by the form of individual pentagonal nannoliths, the presence of a hole in each segment, as well as ornaments. The genera within the family include: *Braarudosphaera*, *Pemma*, *Pentaster*, and *Micrantholithus*.

Genus: BRAARUDOSPHAERA Deflandre, 1947
Braarudosphaera bigelowii Gran and Braarud, 1935; Deflandre, 1947
Plate 9, Fig. 20

Note: Formed from 12 pentaliths, each with 5 crystal units showing tangential *c*-axis orientation. *B. bigelowii* is a pentagonal nannolith with a tiny hole at the center. It exhibits high birefringence under crossed polars with one of the pentaliths going into extinction, while the other remains bright when the stage is rotated. The species is commonly observed within condensed intervals.
Stratigraphic Range: Cretaceous–Recent.

Genus: BIANTHOLITHUS Bramlette and Martini, 1964
Biantholithus hughesii Bramlette and Martini, 1964
Plate 20, Fig. 4

Note: This species of *Biantholithus* is similar to *Braarudosphaera* but without pentaliths. *B. hughesii* is radial in shape with a wavy outline. The FO of the type species, *Biantholithus sparsus* is useful in the delineation of the Cretaceous/Paleocene (Tertiary) boundary.
Stratigraphic Range: Paleogene NP3–?NP8 Zones.

Genus: MICRANTHOLITUS Deflandre, 1954

Diagnosis: Species of the genus *Micrantholithus* are differentiated by having an individual pentagonal nannolith and by the presence or absence of a hole in each segment. The genus *Micrantholithus* is distinguished from other genera by the triangular segments composing the pentaliths. Species of *Micrantholithus* are usually common in the Middle Eocene of the Niger Delta.

Micrantholithus entaster Bramlette and Sullivan, 1961
Plate 16, Fig. 19

Note: *Micrantholithus entaster* include pentagonal nannoliths with triangular segments composing the pentalith. In the Niger Delta, *Micrantholithus* species, though common in the Middle Eocene, are not generally used for biozonation in the Cenozoic.

Stratigraphic Range: Middle Eocene PND16–Late Miocene NND8 Zones.

Genus: PEMMA Martini, 1959

Pemma angulatum Martini, 1959
Plate 13, Fig. 18

Note: The genus *Pemma* is similar to *Braarudosphaera* but with perforations in each segment. This form (*Pemma angulatum*) is common in the western Niger Delta especially in the Middle Eocene PND15/PND16 Zones.

Stratigraphic Range: Eocene–(?)Early Oligocene PND21 Zone.

Genus MINYLITHA Bukry, 1973b

Diagnosis: This monospecific genus is described as thick, plate-like individual calcite element with polygonal outline, depressed CA, and raised rim.

Minylitha convalis Bukry, 1973b
Plate 20, Fig. 11

Note: *Minylitha convalis*, the only species of the genus *Minylitha* is a rhomboid-shaped nannolith with elevated rim. The species is often found particularly in NND10–NND11 Zones within condensed intervals.

Stratigraphic Range: Late Miocene NND9–NND11 Zones.

Genus: ELLIPSOLITHUS Sullivan, 1964a,b

Diagnosis: This genus is characterized by the bright distal shield when viewed between crossed nicols, elliptical to long-elliptical shape with an elongated CA. Species within the genus are differentiated based on the structure/organization of elements in the CA.

Ellipsolithus distichus Bramlette and Sullivan, 1961; Sullivan, 1964a,b
Plate 19, Fig. 7

Note: Elliptical, with wider CA than other species, relatively large, elongated with radially oriented openings or perforations in the CA.

Stratigraphic Range: Early Paleocene NP4–Late Paleocene NP9 Zones.

Ellipsolithus macellus Bramlette and Sullivan, 1961; Sullivan, 1964a,b
Plate 19, Fig. 2

Note: CA fully covered. Species is elliptical in outline.

Stratigraphic Range: Early Paleocene NP4–NP12 Zones.

Family: CALYPTROPHAERACEAE Boudreaux and Hay, 1969
Genus: ZYGRHABLITHUS Deflandre, 1959

Zygrhablithus bijugatus Deflandre, 1959
Plate 9, Figs. 18 and 19

Note: *Zygrhablithus bijugatus* are rod-shaped holococcoliths with a small-basal disc (Young, 1998). They form an important component of the calcareous nannofossils of the Early Miocene Niger Delta.

Stratigraphic Range: Paleocene–Early Miocene NND1 Zones.

COCCOLITH *Incertae sedis*
Genus: CORONOCYCLUS Hay et al., 1966

Coronocyclus nitescens Kamptner, 1963
Plate 14, Figs. 1–3

Note: *Coronocyclus nitescens* is ring-shaped, with open CA, and spines on the distal side. It is highly birefringent. The species is abundant in the Eocene of the Niger Delta.

Stratigraphic Range: Paleogene–Middle Miocene NN5 Zone.

Genus: NANNOTETRINA Achuthan and Stradner, 1969

Diagnosis: Genus *Nannotetrina* is discussed here under "*Incertae sedis*," a category that includes families, which are not related to the other well-established families in the Tertiary. The genus *Nannotetrina* consists of all calcareous nannofossils with a central cross either formed by four straight or with bent arms.

Nannotetrina cristata Martini, 1958
Plate 19, Fig. 19

Note: *Nannotetrina cristata* is a relatively small species of *Nannotetrina* with a massive central cross and a basket that is almost as wide as the span of the arms of the cross.

Stratigraphic Range: PND14–PND15 Zones.

Nannotetrina fulgens Stradner, 1960
Plate 19, Fig. 18

Note: *Nannotetrina fulgens* has a small basket in the CA with long and tapering arms.
Stratigraphic Range: Restricted to the Middle Eocene PND15 Zone.

Genus: TRIBRACHIATUS Heckel, 1968

Diagnosis: Species of this genus are useful in the Eocene of the Niger Delta and elsewhere. Three arms form the basic unit of the nannofossils. Two sets of these three arms with about equal angles between them form the basic unit of some species. Specimens of *Tribrachiatus* are very easy to identify under the LTM by their peculiar morphology.

Tribrachiatus bramlettei Heckel, 1968
Plate 19, Fig. 20

Note: Species is characterized by two sets of three arms with about equal angles between them. No bifurcation is apparent at the tips of the arms. Species show weak to bright birefringence in crossed nicols.
Stratigraphic Range: Restricted to Early Eocene PND10 Zone.

Tribrachiatus contortus Bronnimann and Stradner, 1960
Plate 20, Fig. 1

Note: Very distinctive Eocene age nannofossils, weakly birefringent in between crossed nicols, three armed, at different angles to each other, and with the larger angle between two arms of different layers equal to 90°. The genus *Tribrachiatus* forms an important constituent of Early Eocene calcareous nannofossils assemblages of the Niger Delta.
Stratigraphic Range: Restricted to the Early Eocene PND10 Zone.

Tribrachiatus orthostylus Romein, 1979
Plate 16, Figs. 15 and 16

Note: This species has two sets of the basic three arms with the larger angle between two arms of different layers being 90°. Slight bifurcation is observed on the tips of the arms. *T. orthostylus* shows weak birefringence between crossed nicols.
Stratigraphic Range: Early Eocene PND10–PND11 Zones.

Genus: TETRALITHOIDES Theodoridis, 1984, emend. Jordan et al., 1993
Tetralithoides symeonidesii Theodoridis, 1984
Plate 20, Figs. 3 and 9

Note: Forms with elliptical outline, narrow rim with four plates filling the CA opening. Rare occurrences were recorded in the shallow offshore but common in deepwater in the Niger Delta.
Stratigraphic Range: NND1–Early Pliocene NND12 Zones.

Plates 1–21

Cenozoic Foraminifera and Calcareous Nannofossil Biostratigraphy of the Niger Delta. http://dx.doi.org/10.1016/B978-0-12-812161-0.00013-2

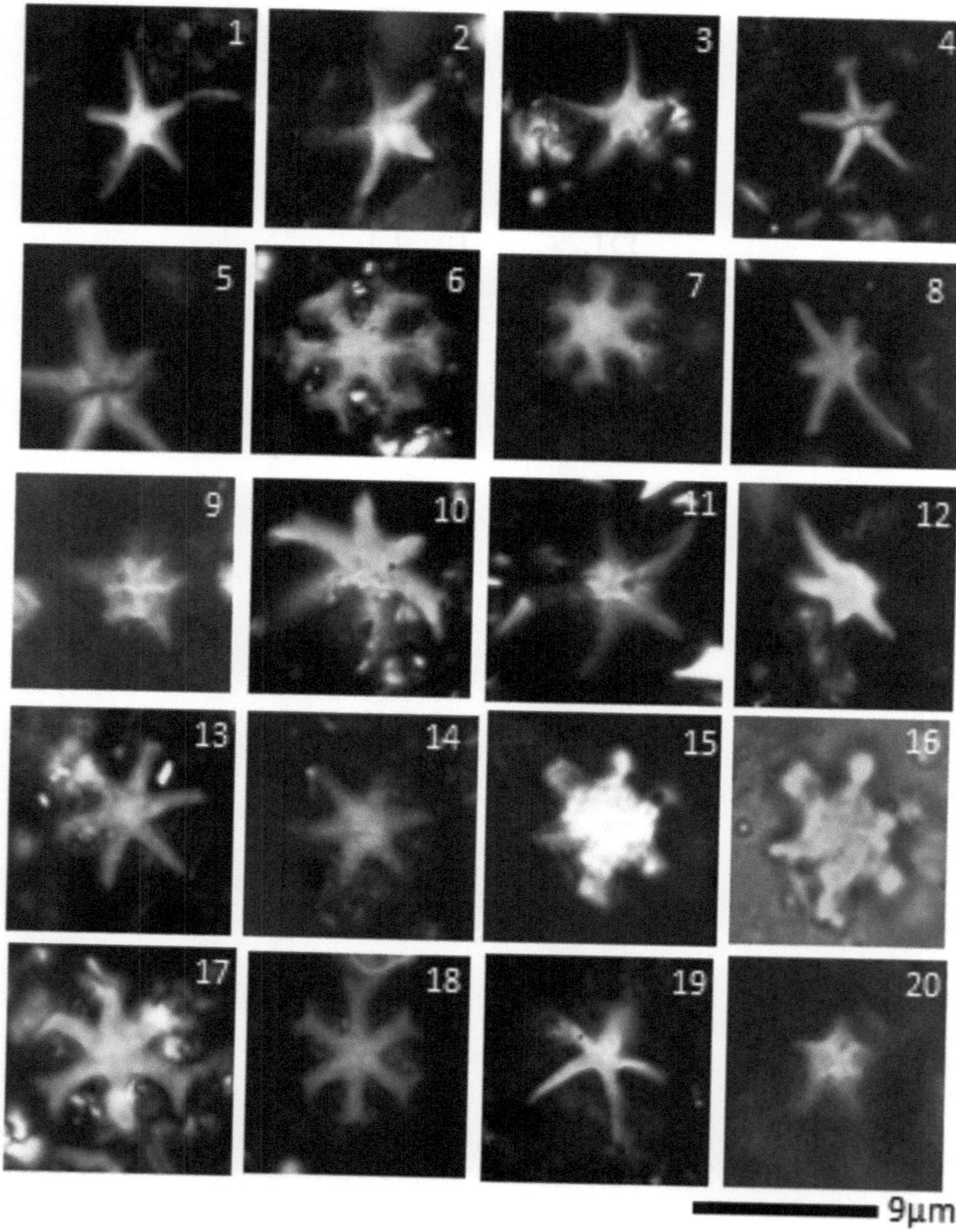
1
2
3
4
5
6
7
8
9
10
11
12
13
14
15
16
17
18
19
20
9μm

EXPLANATION OF PLATE 1

FIGURE 1	*Discoaster quinqueramus*. Koronama-3 well, 9260 ft.
FIGURE 2	*Discoaster quinqueramus*. Koronama-3 well, 9260 ft.
FIGURE 3	*Discoaster quinqueramus*. Igodo-1 well, 4400 ft.
FIGURE 4	*Discoaster pentaradiatus*. Etim-1 well, 1000–1050 ft.
FIGURE 5	*Discoaster pentaradiatus*. Etim-1 well, 3140–3180 ft.
FIGURE 6	*Discoaster variabilis*. Utapate-1 well, 8050 ft.
FIGURE 7	*Discoaster variabilis*. Utapate-1 well, 8050 ft.
FIGURE 8	*Discoaster brouweri*. Anyala-8 well, 8470 ft.
FIGURE 9	*Discoaster berggrenii*. Inanga-1 well, 9530 ft.
FIGURE 10	*Discoaster brouweri*. Anyala-8 well, 8470 ft.
FIGURE 11	*Discoaster quinqueramus* (broken). Inanga-1 well, 9530 ft.
FIGURE 12	*Discoaster* sp. Gilli Gilli-1 well, 11,030 ft.
FIGURE 13	*Discoaster brouweri*. Anyala-8 well, 8470 ft.
FIGURE 14	*Discoaster brouweri*. Anyala-8 well, 8470 ft.
FIGURE 15	*Discoaster kugleri*. Beniboye North-1 well, 1910 m.
FIGURE 16	*Discoaster kugleri*. Beniboye North-1 well, 1910 m.
FIGURE 17	*Discoaster variabilis*. Anyala-8 well, 8410 ft.
FIGURE 18	*Discoaster variabilis*. Anyala-8 well, 8410 ft.
FIGURE 19	*Discoaster quinqueramus*. Inanga-1 well, 9530 ft.
FIGURE 20	*Discoaster berggrenii*. Inanga-1 well, 9530 ft.

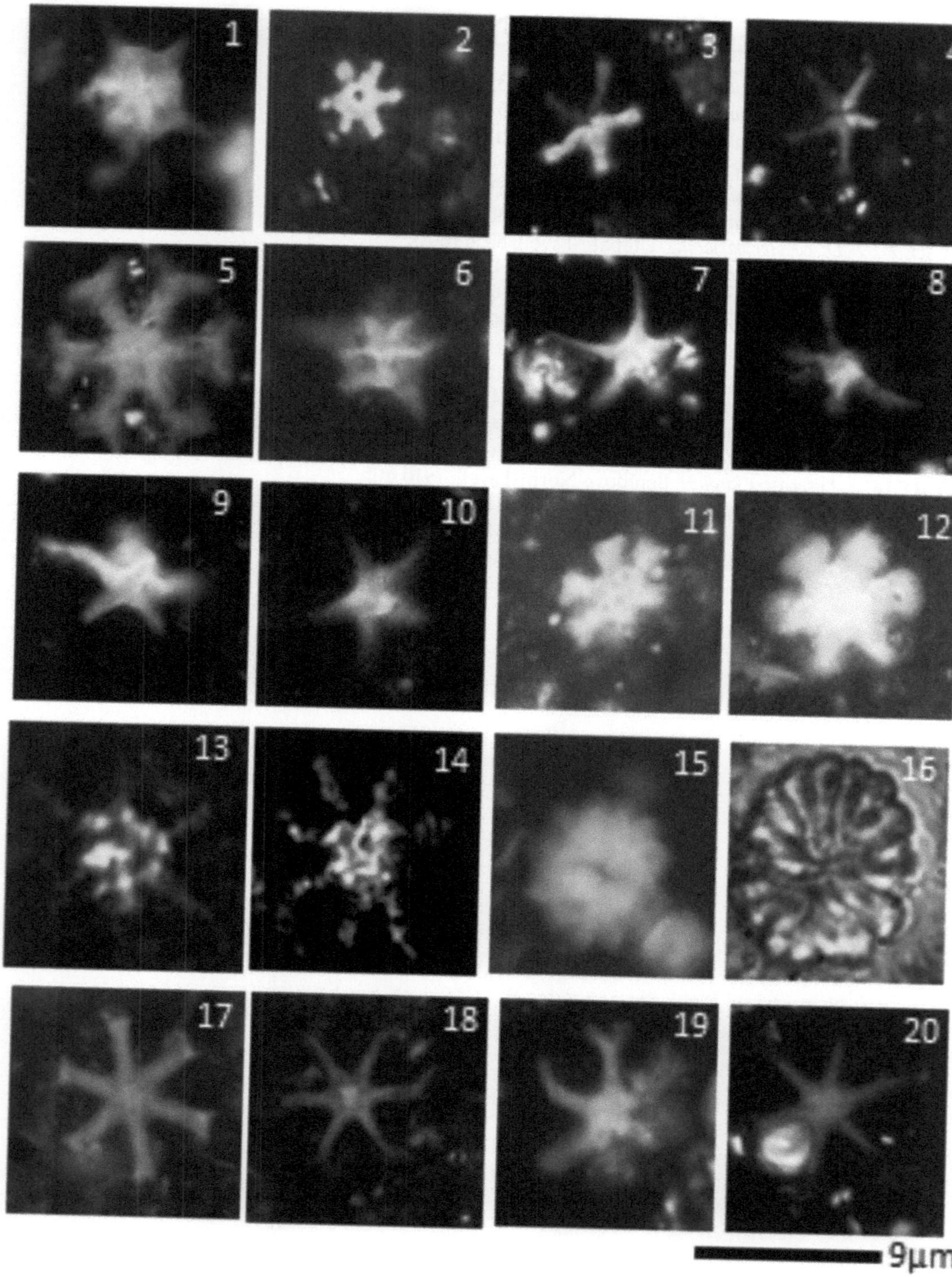
1
2
3
4
5
6
7
8
9
10
11
12
13
14
15
16
17
18
19
20
9μm

EXPLANATION OF PLATE 2

FIGURE 1 *Discoaster berggrenii*. Inanga-1 well, 9530 ft.
FIGURE 2 *Discoaster bollii*. Utapate-1 well, 8050 ft.
FIGURE 3 *Discoaster pentaradiatus*. Etim-1 well, 3140–3180 ft.
FIGURE 4 *Discoaster pentaradiatus*. Etim-1 well, 3140–3180 ft.
FIGURE 5 *Discoaster variabilis*. Anyala-8 well, 8410 ft.
FIGURE 6 *Discoaster berggrenii*. Inanga-1 well, 9530 ft.
FIGURE 7 *Discoaster quinqueramus*. Inanga-1 well, 8570 ft.
FIGURE 8 *Discoaster quinqueramus*. Inanga-1 well, 8570 ft.
FIGURE 9 *Discoaster quinqueramus*. Inanga-1 well, 8570 ft.
FIGURE 10 *Discoaster quinqueramus*. Inanga-1 well, 8570 ft.
FIGURE 11 *Discoaster deflandrei*. Akata-1 well, 7830 ft.
FIGURE 12 *Discoaster deflandrei*. Akata-1 well, 7830 ft.
FIGURE 13 *Discoaster mahmoudi*. Nemomai-1 well, 5200 ft.
FIGURE 14 *Discoaster mediosus*. Nemomai-1 well, 5720 ft.
FIGURE 15 *Discoaster multiradiatus*. Nemomai-1 well, 6680 ft.
FIGURE 16 *Discoaster multiradiatus*. Nemomai-1 well, 6680 ft.
FIGURE 17 *Discoaster deflandrei*. Akata-1 well, 7830 ft.
FIGURE 18 *Discoaster exilis*. Gbaran-2 well, 7410 ft.
FIGURE 19 *Discoaster exilis*. Gbaran-2 well, 7410 ft.
FIGURE 20 *Discoaster brouweri*. Gbaran-2 well, 8490 ft.

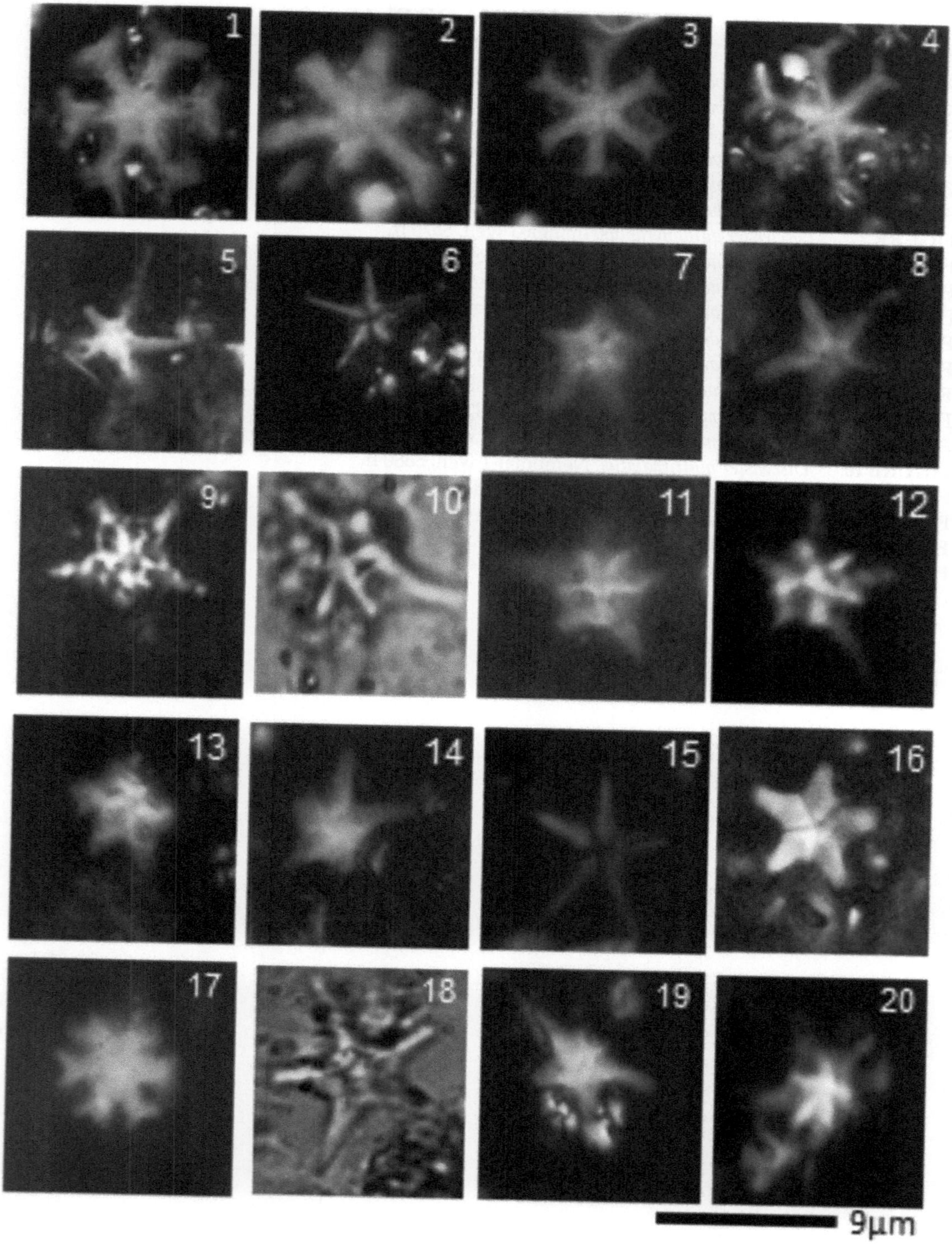
1
2
3
4
5
6
7
8
9
10
11
12
13
14
15
16
17
18
19
20
9μm

EXPLANATION OF PLATE 3

FIGURE 1	*Discoaster variabilis*. Utapate-1 well, 8050 ft.
FIGURE 2	*Discoaster variabilis*. Utapate-1 well, 8050 ft.
FIGURE 3	*Discoaster variabilis*. Utapate-1 well, 8050 ft.
FIGURE 4	*Discoaster variabilis*. Utapate-1 well, 8050 ft.
FIGURE 5	*Discoaster quinqueramus*. Inanga-1 well, 8570 ft.
FIGURE 6	*Discoaster pentaradiatus*. Akamba-1 well, 1160 m.
FIGURE 7	*Discoaster berggrenii*. Koronama-3 well, 8900 ft.
FIGURE 8	*Discoaster berggrenii*. Koronama-3 well, 8900 ft.
FIGURE 9	*Discoaster berggrenii* (note large central knob). Koronama-3 well, 8960 ft.
FIGURE 10	*Discoaster berggrenii* (note large central knob). Koronama-3 well, 8960 ft.
FIGURE 11	*Discoaster berggrenii* (note large central knob). Koronama-3 well, 8960 ft.
FIGURE 12	*Discoaster berggrenii* (note large central knob). Koronama-3 well, 8960 ft.
FIGURE 13	*Discoaster berggrenii*. Inanga-1 well, 9530 ft.
FIGURE 14	*Discoaster berggrenii*. Inanga-1 well, 9530 ft.
FIGURE 15	*Discoaster pentaradiatus*. Etim-1 well, 3140–3180 ft.
FIGURE 16	*Discoaster pentaradiatus*. Etim-1 well, 3140–3180 ft.
FIGURE 17	*Discoaster deflandrei*. Anyala-8 well, 8470 ft.
FIGURE 18	*Discoaster berggrenii*. Inanga-1 well, 9530 ft.
FIGURE 19	*Discoaster berggrenii*. Inanga-1 well, 9530 ft.
FIGURE 20	*Discoaster berggrenii*. Inanga-1 well, 9530 ft.

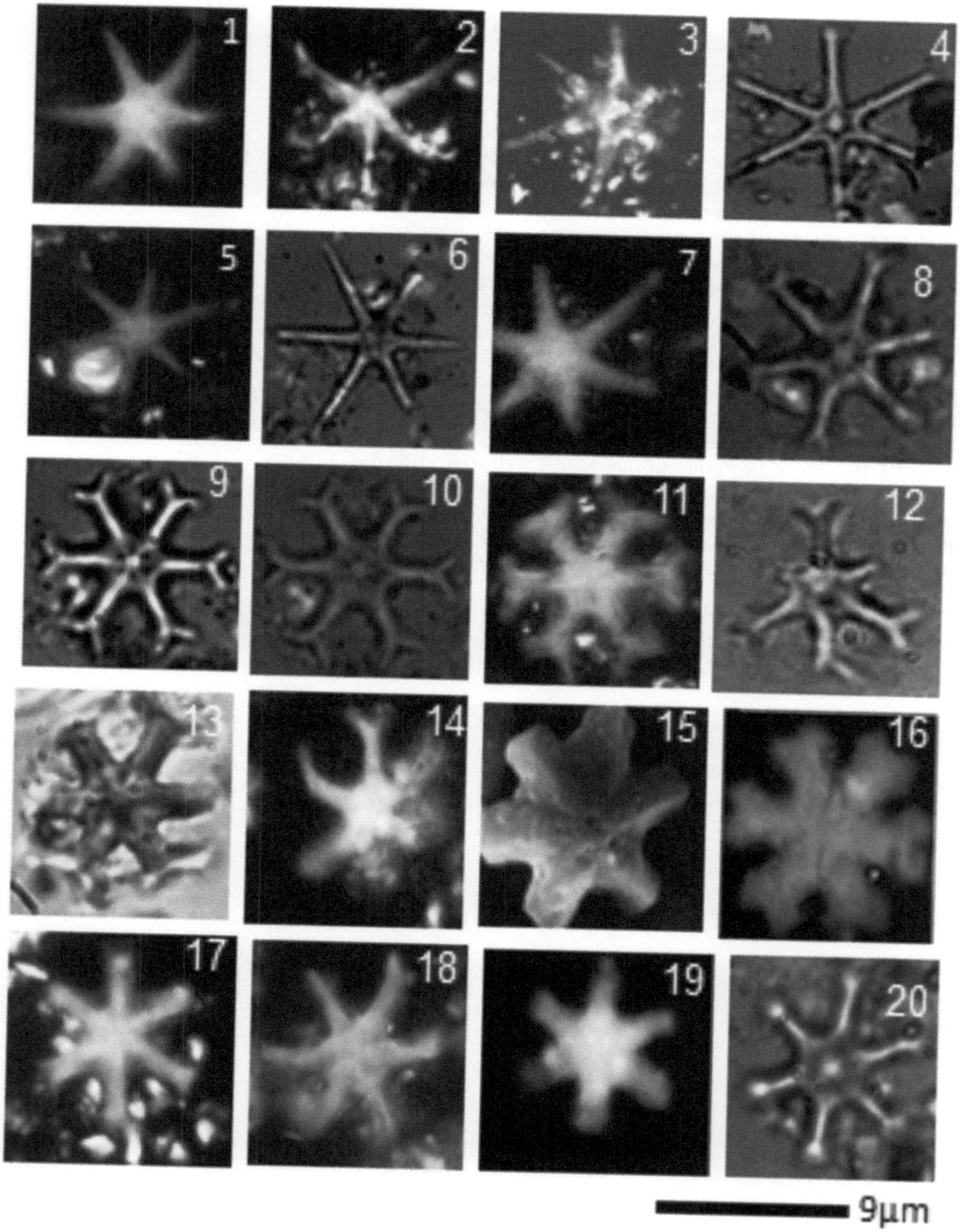
1
2
3
4
5
6
7
8
9
10
11
12
13
14
15
16
17
18
19
20
9µm

EXPLANATION OF PLATE 4

FIGURE 1 *Discoaster brouweri*. Anyala-8 well, 8470 ft.
FIGURE 2 *Discoaster brouweri*. Anyala-8 well, 8470 ft.
FIGURE 3 *Discoaster brouweri*. Anyala-8 well, 8470 ft.
FIGURE 4 *Discoaster brouweri*. Anyala-8 well, 8470 ft.
FIGURE 5 *Discoaster brouweri*. Anyala-8 well, 8470 ft.
FIGURE 6 *Discoaster brouweri*. Anyala-8 well, 8470 ft.
FIGURE 7 *Discoaster brouweri*. Anyala-8 well, 8470 ft.
FIGURE 8 *Discoaster exilis*. Anyala-8 well, 8410 ft.
FIGURE 9 *Discoaster variabilis*. Anyala-8 well, 8410 ft.
FIGURE 10 *Discoaster variabilis*. Anyala-8 well, 8410 ft.
FIGURE 11 *Discoaster variabilis*. Anyala-8 well, 8410 ft.
FIGURE 12 *Discoaster variabilis*. Anyala-8 well, 8410 ft.
FIGURE 13 *Discoaster challengeri*. Anyala-8 well, 8410 ft.
FIGURE 14 *Discoaster* sp. Anyala-8 well, 8410 ft.
FIGURE 15 *Discoaster* sp. Anyala-8 well, 8410 ft.
FIGURE 16 *Discoaster deflandrei*. Anyala-8 well, 8410 ft.
FIGURE 17 *Discoaster* sp. (six rays). Anyala-8 well, 8410 ft.
FIGURE 18 *Discoaster* sp. (six rays). Anyala-8 well, 8410 ft.
FIGURE 19 *Discoaster* sp. (six rays). Anyala-8 well, 8410 ft.
FIGURE 20 *Discoaster surculus*. Anyala-8 well, 8470 ft.

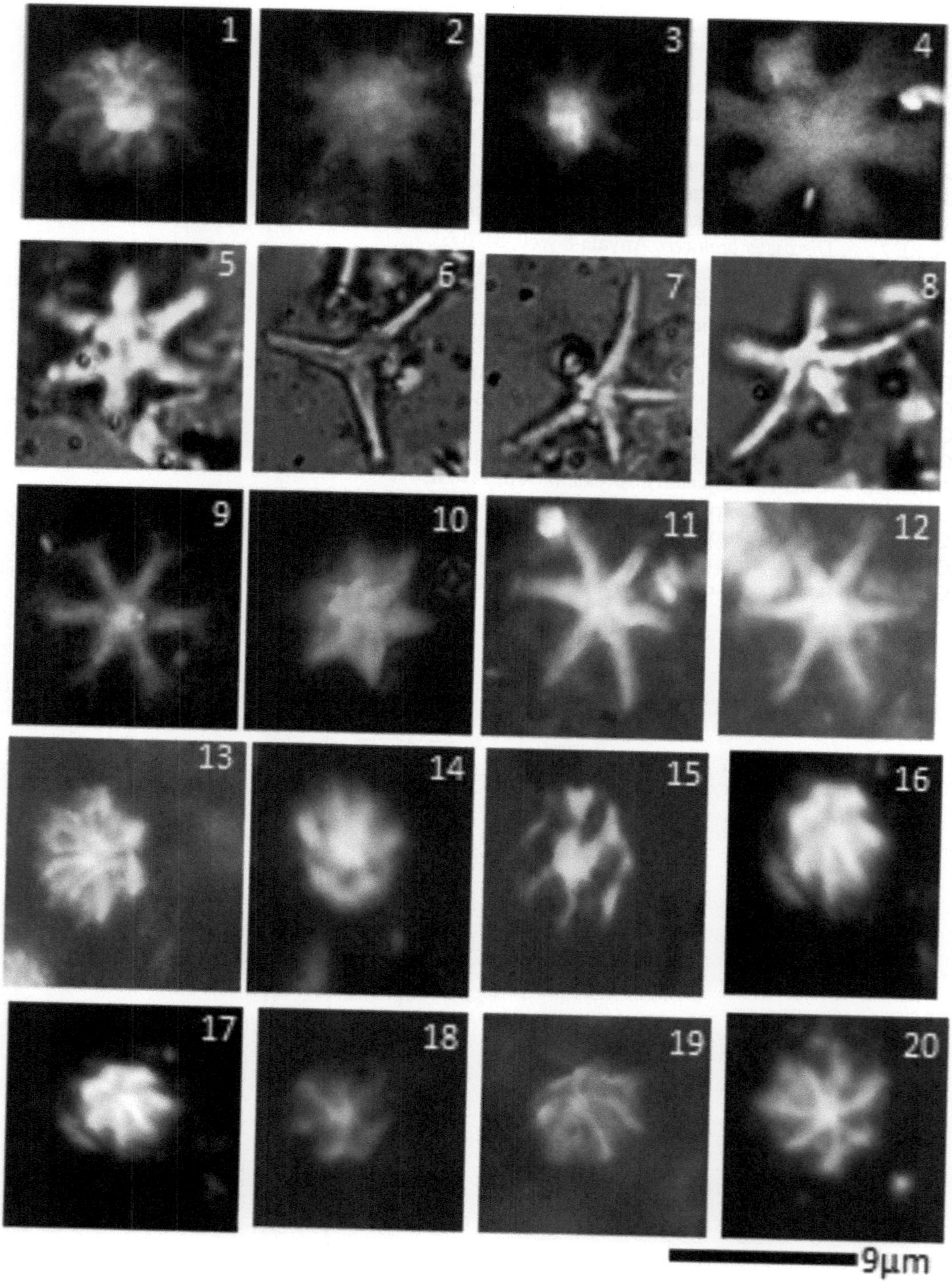
1
2
3
4
5
6
7
8
9
10
11
12
13
14
15
16
17
18
19
20
9μm

EXPLANATION OF PLATE 5

FIGURE 1 *Discoaster kuepperi*. Otobo North-1 well, 9020 ft.
FIGURE 2 *Discoaster mediosus*. Otobo North-1 well, 8840 ft.
FIGURE 3 *Discoaster binodosus*. Otobo North-1 well, 8840 ft.
FIGURE 4 *Discoaster druggii*. Benin River-3 well, 7990 ft.
FIGURE 5 *Discoaster sanmiguelensis*. Utonana-1 well, 5180–5200 ft.
FIGURE 6 *Discoaster triradiatus*. Anyala-8 well, 8470 ft.
FIGURE 7 *Discoaster quinqueramus*. Inanga-1 well, 9530 ft.
FIGURE 8 *Discoaster quinqueramus*. Inanga-1 well, 9530 ft.
FIGURE 9 *Discoaster variabilis*. Otobo North-1 well, 8840 ft.
FIGURE 10 *Discoaster* sp. Otobo North-1 well, 6600 ft.
FIGURE 11 *Discoaster brouweri*. Otobo North-1 well, 8840 ft.
FIGURE 12 *Discoaster brouweri*. Otobo North-1 well, 8840 ft.
FIGURE 13 *Catinaster mexicanus*. Beniboye North-1 well, 1910 ft.
FIGURE 14 *Catinaster mexicanus*. Beniboye North-1 well, 1910 ft.
FIGURE 15 *Catinaster coalitus*. Beniboye North-1 well, 2030 ft.
FIGURE 16 *Catinaster coalitus*. Beniboye North-1 well, 2030 ft.
FIGURE 17 *Catinaster coalitus*. Beniboye North-1 well, 2030 ft.
FIGURE 18 *Catinaster mexicanus*. Otobo North-1 well, 5820 ft.
FIGURE 19 *Catinaster mexicanus*. Otobo North-1 well, 5820 ft.
FIGURE 20 *Catinaster mexicanus*. Otobo North-1 well, 5820 ft.

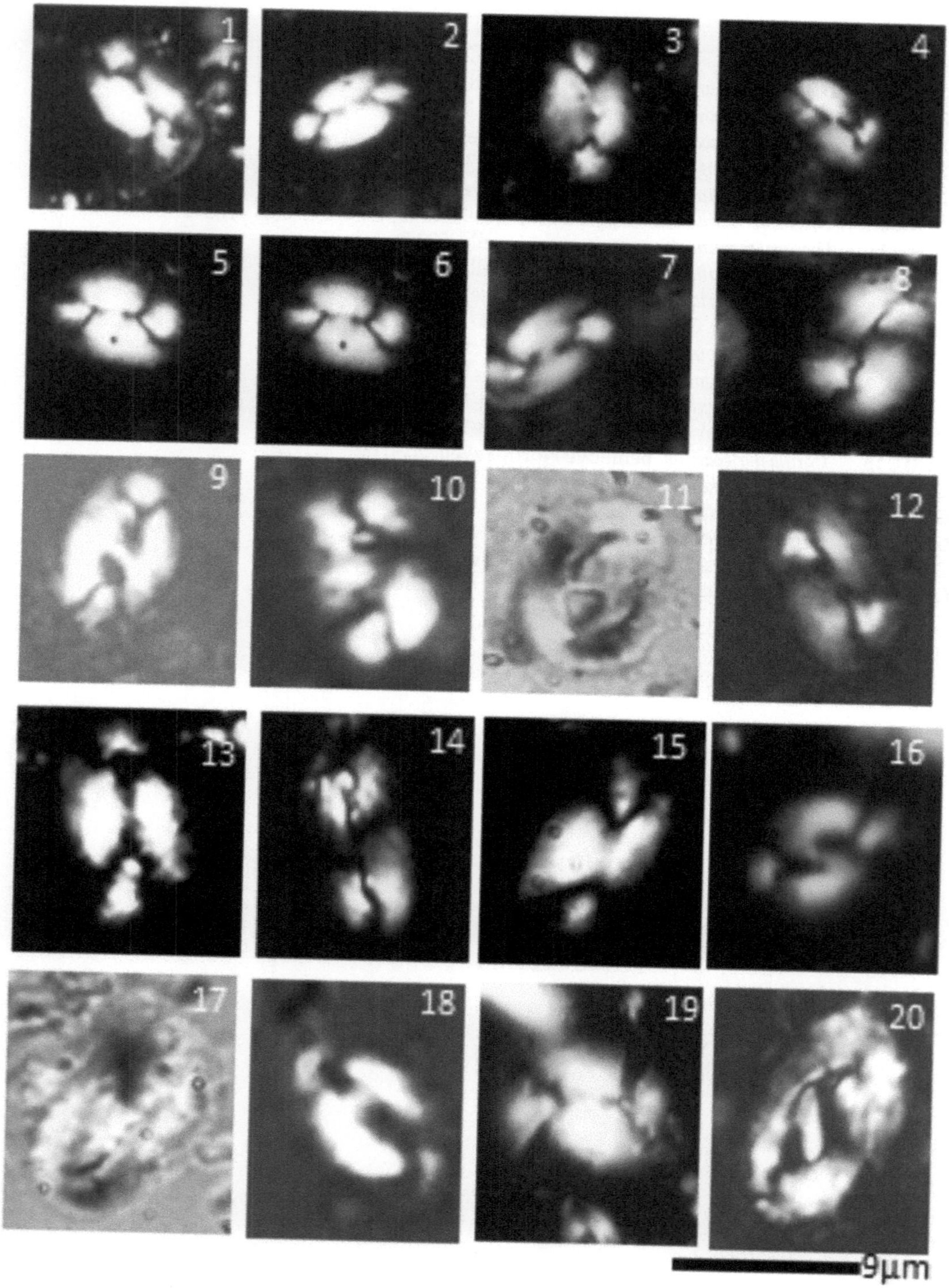
1
2
3
4
5
6
7
8
9
10
11
12
13
14
15
16
17
18
19
20
9µm

EXPLANATION OF PLATE 6

FIGURE 1	*Helicosphaera carteri*. Gbaran-2 well, 8190 ft.
FIGURE 2	*Helicosphaera carteri*. Gbaran-2 well, 8190 ft.
FIGURE 3	*Helicosphaera carteri*. Gbaran-2 well, 8190 ft.
FIGURE 4	*Helicosphaera carteri*. Gbaran-2 well, 8190 ft.
FIGURE 5	*Helicosphaera carteri*. Gbaran-2 well, 8190 ft.
FIGURE 6	*Helicosphaera carteri*. Gbaran-2 well, 8190 ft.
FIGURE 7	*Helicosphaera carteri*. Gbaran-2 well, 8190 ft.
FIGURE 8	*Helicosphaera carteri*. Gbaran-2 well, 8190 ft.
FIGURE 9	*Helicosphaera sellii*. Kolo Creek-39 well, 13,170 ft.
FIGURE 10	*Helicosphaera* sp. Kolo Creek-39 well, 7110 ft.
FIGURE 11	*Helicosphaera inversa*. Inanga-1 well, 8670 ft.
FIGURE 12	*Helicosphaera carteri*. Ovhor-10 well, 8070 ft.
FIGURE 13	*Helicosphaera carteri*. Ovhor-10 well, 8070 ft.
FIGURE 14	*Helicosphaera carteri*. Ovhor-10 well, 8070 ft.
FIGURE 15	*Helicosphaera carteri*. Ovhor-10 well, 8070 ft.
FIGURE 16	*Helicosphaera stalis*. Inanga-1 well, 8670 ft.
FIGURE 17	*Helicosphaera carteri*. Kolo Creek-39 well, 13,245 ft.
FIGURE 18	*Helicosphaera waltrans*. Tuomo West-1 well, 3844–3852 ft.
FIGURE 19	*Helicosphaera carteri*. Gbaran-2 well, 7410 ft.
FIGURE 20	*Helicosphaera intermedia*. Beniboye North-1 well, 1300 m.

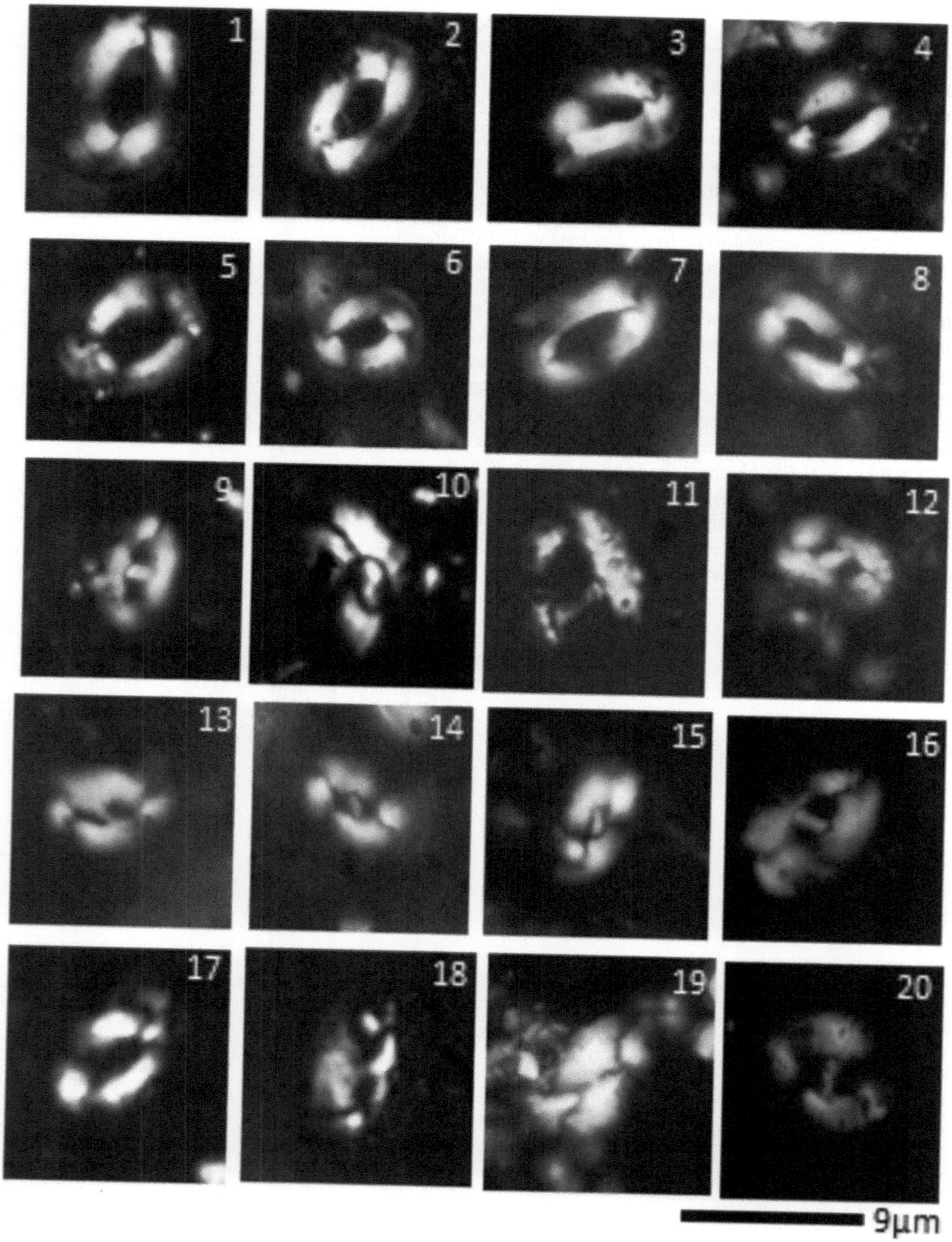
1
2
3
4
5
6
7
8
9
10
11
12
13
14
15
16
17
18
19
20
9µm

EXPLANATION OF PLATE 7

FIGURE 1	*Helicosphaera ampliaperta*. Tuomo West-1 well, 2906–2914 ft.
FIGURE 2	*Helicosphaera ampliaperta*. Tuomo West-1 well, 2906–2914 ft.
FIGURE 3	*Helicosphaera ampliaperta*. Tuomo West-1 well, 2906–2914 ft.
FIGURE 4	*Helicosphaera ampliaperta*. Tuomo West-1 well, 2906–2914 ft.
FIGURE 5	*Helicosphaera ampliaperta*. Tuomo West-1 well, 2906–2914 ft.
FIGURE 6	*Helicosphaera ampliaperta*. Tuomo West-1 well, 2906–2914 ft.
FIGURE 7	*Helicosphaera ampliaperta*. Tuomo West-1 well, 2906–2914 ft.
FIGURE 8	*Helicosphaera ampliaperta*. Tuomo West-1 well, 2906–2914 ft.
FIGURE 9	*Helicosphaera seminulum*. Nemomai-1 well, 8510 ft.
FIGURE 10	*Helicosphaera bramlettei*. Nemomai-1 well, 8600 ft.
FIGURE 11	*Helicosphaera seminulum*. Nemomai-1 well, 8600 ft.
FIGURE 12	*Helicosphaera seminulum*. Nemomai-1 well, 8600 ft.
FIGURE 13	*Helicosphaera* sp. Utonana-1 well, 5180–5200 ft.
FIGURE 14	*Helicosphaera intermedia*. Fragbene-1 well, 9090 ft.
FIGURE 15	*Helicosphaera intermedia*. Fragbene-1 well, 9090 ft.
FIGURE 16	*Helicosphaera seminulum*. Fragbene-1 well, 7050 ft.
FIGURE 17	*Helicosphaera ampliaperta*. Opuama-1 well, 4200 ft.
FIGURE 18	*Helicosphaera perch-nielseniae*. Fragbene-1 well, 9480 ft.
FIGURE 19	*Helicosphaera carteri*. Kolo Creek-39 well, 13,245 ft.
FIGURE 20	*Helicosphaera seminulum*. Fragbene-1 well, 9090 ft.

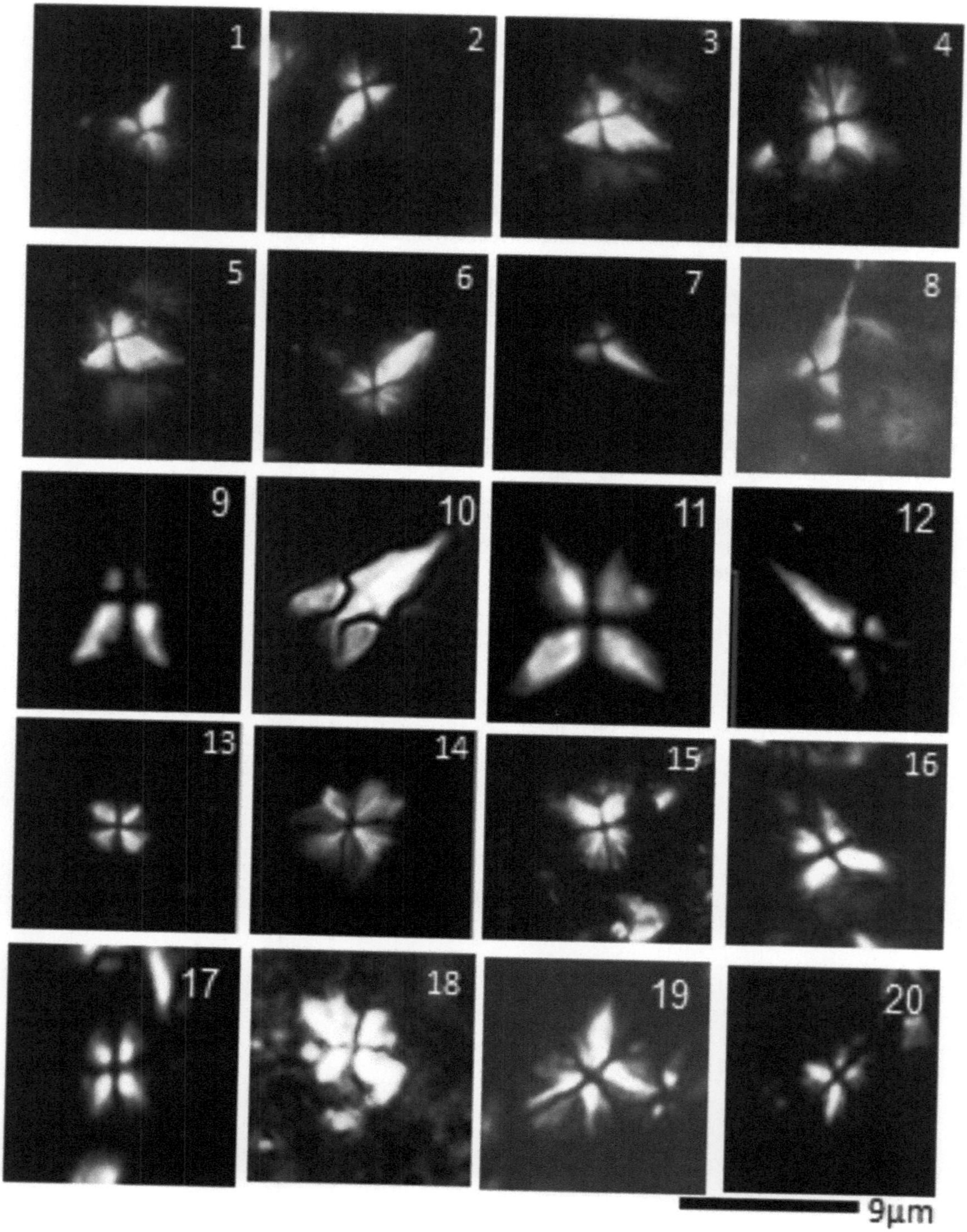

EXPLANATION OF PLATE 8

FIGURE 1 *Sphenolithus heteromorphus*. Gbaran-2 well, 7410 ft.
FIGURE 2 *Sphenolithus heteromorphus*. Gbaran-2 well, 7410 ft.
FIGURE 3 *Sphenolithus heteromorphus*. Gbaran-2 well, 7410 ft.
FIGURE 4 *Sphenolithus abies*. Okubie-8 well, 3660 ft.
FIGURE 5 *Sphenolithus heteromorphus*. Gbaran-2 well, 8190 ft.
FIGURE 6 *Sphenolithus heteromorphus*. Gbaran-2 well, 8190 ft.
FIGURE 7 *Sphenolithus heteromorphus*. Gbaran-2 well, 8190 ft.
FIGURE 8 *Sphenolithus heteromorphus*. Gbaran-2 well, 8190 ft.
FIGURE 9 *Sphenolithus belemnos*. Tuomo West-1 well, 2906–2914 ft.
FIGURE 10 *Sphenolithus ciperoensis*. Benin River-3 well, 9300 ft.
FIGURE 11 *Sphenolithus moriformis*. Tuomo West-1 well, 2906–2914 ft.
FIGURE 12 *Sphenolithus delphix*. Benin River-3 well, 9300 ft.
FIGURE 13 *Sphenolithus primus*. Gilli Gilli-1 well, 9980 ft.
FIGURE 14 *Sphenolithus primus*. Gilli Gilli-1 well, 9980 ft.
FIGURE 15 *Sphenolithus abies*. Igodo-1 well, 4460 ft.
FIGURE 16 *Sphenolithus moriformis*. Kolo Creek-39 well, 13,110 ft.
FIGURE 17 *Sphenolithus moriformis*. Kolo Creek-39 well, 13,110 ft.
FIGURE 18 *Sphenolithus moriformis*. Kolo Creek-39 well, 13,110 ft.
FIGURE 19 *Sphenolithus abies*. Koronama-3 well, 8540 ft.
FIGURE 20 *Sphenolithus abies*. Koronama-3 well, 8540 ft.

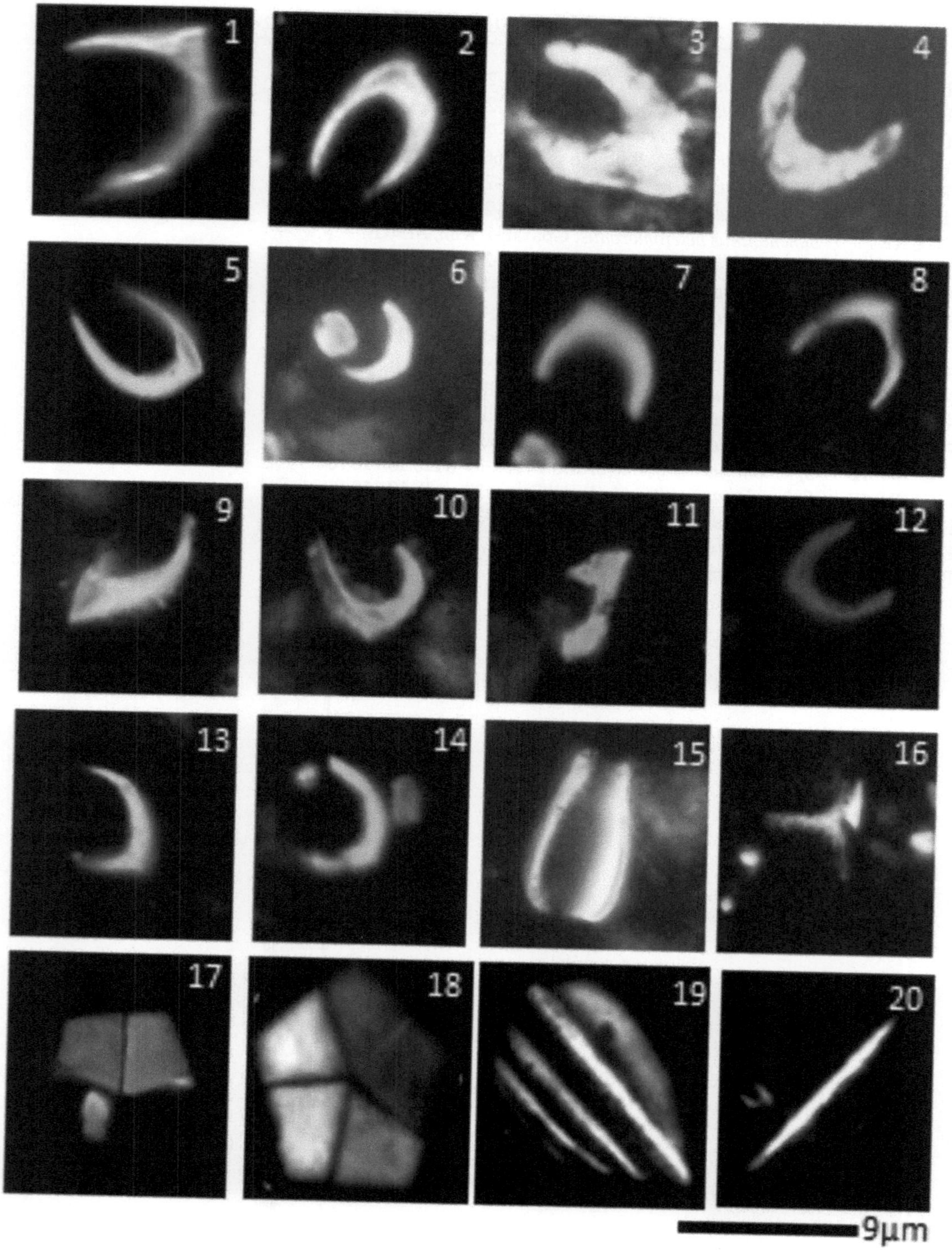
1
2
3
4
5
6
7
8
9
10
11
12
13
14
15
16
17
18
19
20
9µm

EXPLANATION OF PLATE 9

FIGURE 1 *Amaurolithus acutus*. Inanga-1 well, 9840 ft.
FIGURE 2 *Ceratolithus cristatus*. Anyala-8 well, 3250 ft.
FIGURE 3 *Ceratolithus armatus*. Etim-1 well, 2450–2500 ft.
FIGURE 4 *Amaurolithus delicatus*. Inanga-1 well, 9840 ft.
FIGURE 5 *Ceratolithus cristatus*. Anyala-8 well, 3250 ft.
FIGURE 6 *Ceratolithus cristatus*. Anyala-8 well, 3250 ft.
FIGURE 7 *Ceratolithus cristatus*. Anyala-8 well, 3250 ft.
FIGURE 8 *Ceratolithus cristatus*. Anyala-8 well, 3250 ft.
FIGURE 9 *Ceratolithus armatus*. Anyala-8 well, 3250 ft.
FIGURE 10 *Ceratolithus cristatus*. Inanga-1 well, 9840 ft.
FIGURE 11 *Ceratolithus armatus*. Etim-1 well, 2500–2550 ft.
FIGURE 12 *Ceratolithus cristatus*. Inanga-1 well, 9870 ft.
FIGURE 13 *Ceratolithus armatus*. Etim-1 well, 2500–2550 ft.
FIGURE 14 *Amaurolithus primus*. Inanga-1 well, 9840 ft.
FIGURE 15 *Scyphosphaera apsteinii*. Otobo North-1 well, 5820 ft.
FIGURE 16 *Rhabdosphaera clavigera*. Kolo Creek-39 well, 13,245 ft.
FIGURE 17 *Braarudosphaera bigelowii*. Anyala-8 well, 1660 ft.
FIGURE 18 *Braarudosphaera bigelowii*. Anyala-8 well, 1660 ft.
FIGURE 19 *Triquetrorhabdulus challengeri*. Fragbene-1 well, 5170 ft.
FIGURE 20 *Triquetrorhabdulus carinatus*. Fragbene-1 well, 5170 ft.

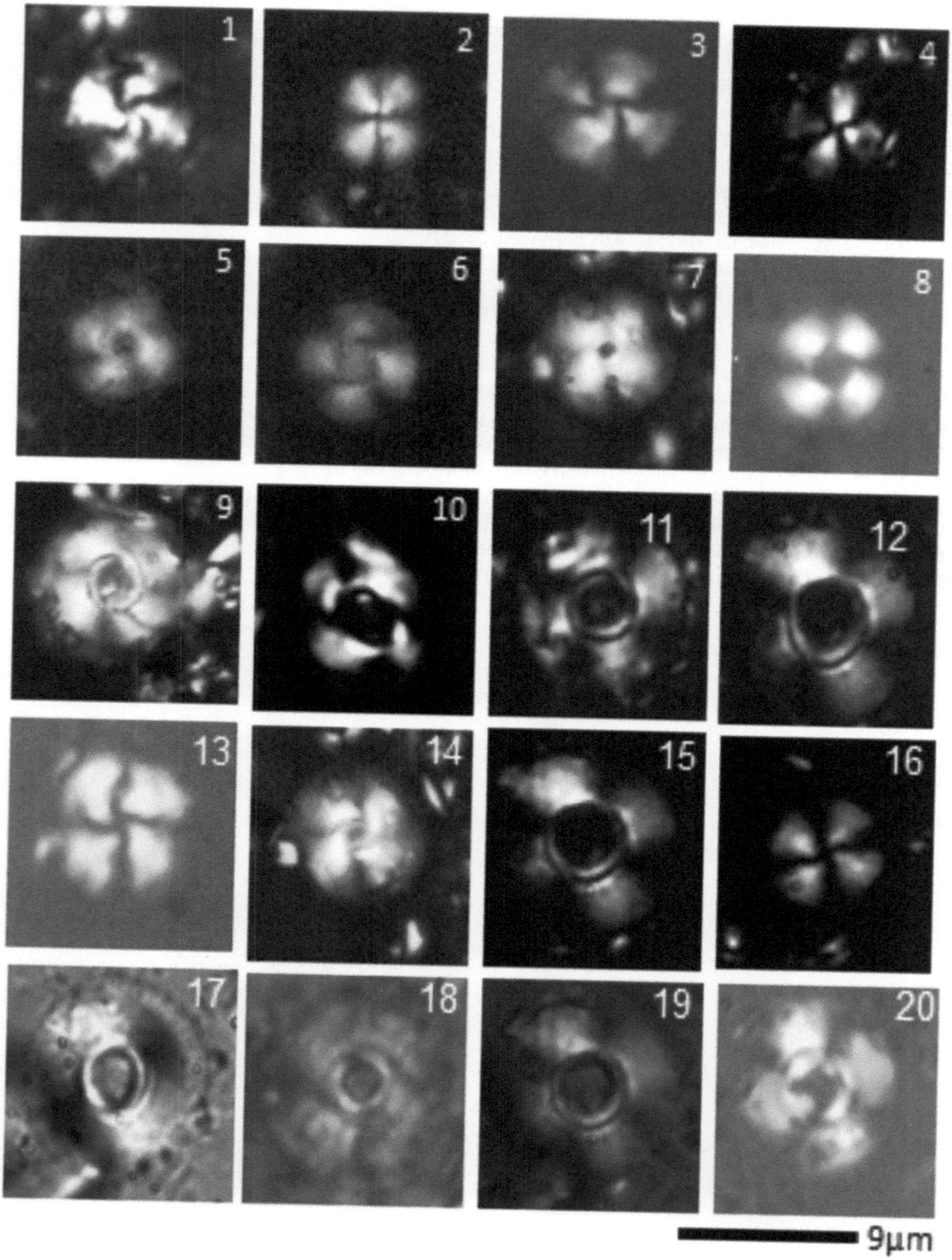
1
2
3
4
5
6
7
8
9
10
11
12
13
14
15
16
17
18
19
20
9μm

EXPLANATION OF PLATE 10

FIGURE 1 *Calcidiscus leptoporus*. Zarama-7 well, 5630 ft.
FIGURE 2 *Calcidiscus leptoporus*. Zarama-7 well, 5630 ft.
FIGURE 3 *Calcidiscus leptoporus*. Zarama-7 well, 5630 ft.
FIGURE 4 *Calcidiscus leptoporus*. Zarama-7 well, 5630 ft.
FIGURE 5 *Calcidiscus leptoporus*. Zarama-7 well, 5630 ft.
FIGURE 6 *Calcidiscus leptoporus*. Zarama-7 well, 5630 ft.
FIGURE 7 *Calcidiscus leptoporus*. Zarama-7 well, 5630 ft.
FIGURE 8 *Calcidiscus leptoporus*. Zarama-7 well, 5630 ft.
FIGURE 9 *Calcidiscus leptoporus*. Zarama-7 well, 5630 ft.
FIGURE 10 *Calcidiscus premacintyrei*. Bosi-6 well, 4170 m.
FIGURE 11 *Calcidiscus tropicus*. Anyala-4 well, 430 ft.
FIGURE 12 *Calcidiscus tropicus*. Anyala-4 well, 430 ft.
FIGURE 13 *Calcidiscus leptoporus*. Kolo Creek-39 well, 13,245 ft.
FIGURE 14 *Calcidiscus macintyrei*. Koronama-3 well, 9380 ft.
FIGURE 15 *Calcidiscus tropicus*. Anyala-4 well, 340 ft.
FIGURE 16 *Calcidiscus leptoporus*. Anyala-4 well, 340 ft.
FIGURE 17 *Calcidiscus tropicus*. Anyala-4 well, 430 ft.
FIGURE 18 *Calcidiscus tropicus*. Anyala-4 well, 430 ft.
FIGURE 19 *Calcidiscus tropicus*. Anyala-4 well, 430 ft.
FIGURE 20 *Calcidiscus premacintyrei*. Bosi-6 well, 4170 m.

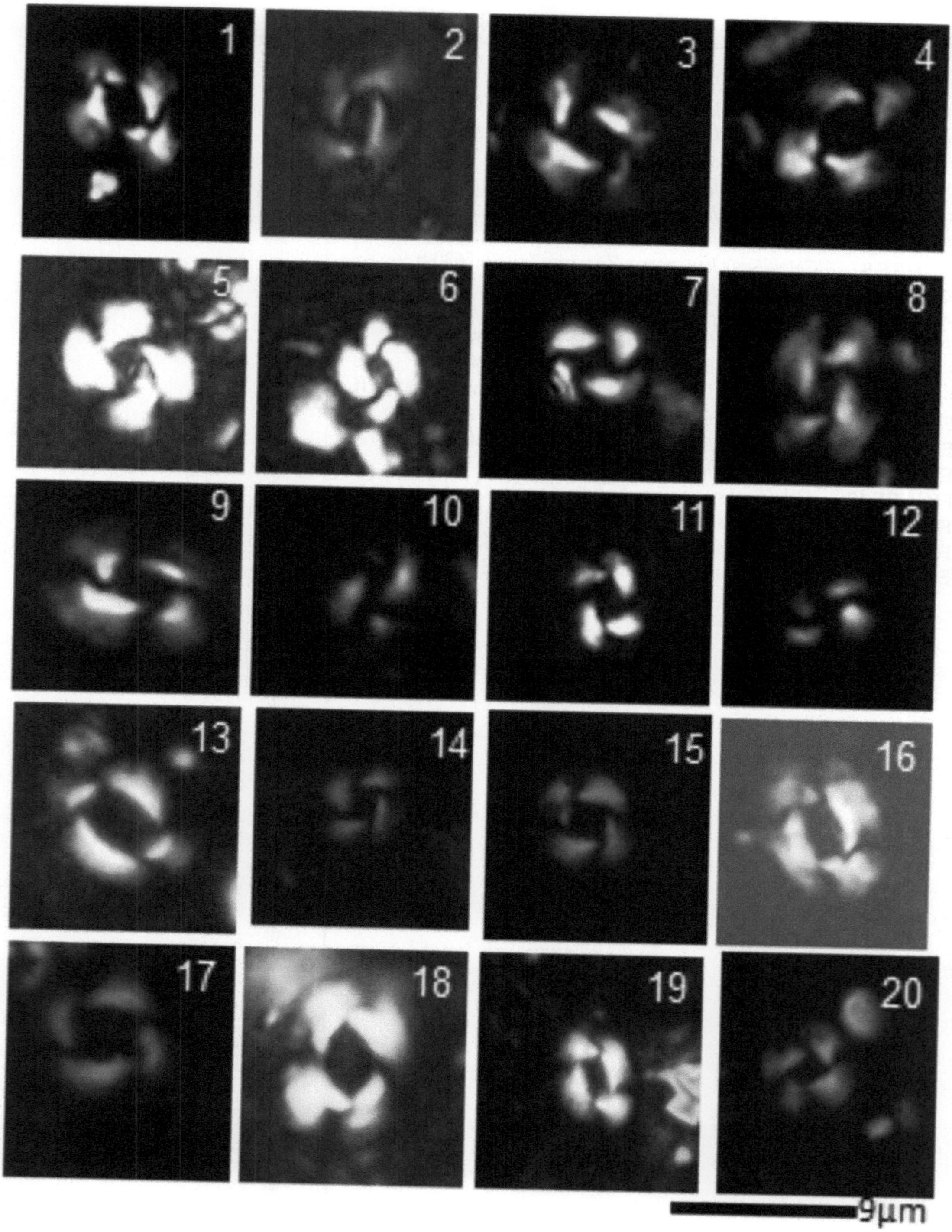
1
2
3
4
5
6
7
8
9
10
11
12
13
14
15
16
17
18
19
20
9µm

EXPLANATION OF PLATE 11

FIGURE 1 *Reticulofenestra pseudoumbilicus* (5–7 μm). Gbaran-2 well, 7410 ft.
FIGURE 2 *Reticulofenestra pseudoumbilicus* (5–7 μm). Gbaran-2 well, 7410 ft.
FIGURE 3 *Reticulofenestra pseudoumbilicus* (5–7 μm). Gbaran-2 well, 7410 ft.
FIGURE 4 *Reticulofenestra pseudoumbilicus* (>7 μm). Zarama-7 well, 5690 ft.
FIGURE 5 *Reticulofenestra pseudoumbilicus* (5–7 μm). Ovhor-10 well, 8070 ft.
FIGURE 6 *Reticulofenestra pseudoumbilicus* (5–7 μm). Ovhor-10 well, 8070 ft.
FIGURE 7 *Reticulofenestra pseudoumbilicus* (5–7 μm). Koronama-3 well, 9320 ft.
FIGURE 8 *Reticulofenestra pseudoumbilicus* (5–7 μm). Koronama-3 well, 9320 ft.
FIGURE 9 *Reticulofenestra pseudoumbilicus* (5–7 μm). Koronama-3 well, 9320 ft.
FIGURE 10 *Reticulofenestra haqii* (3–5 μm). Gbaran-2 well, 7410 ft.
FIGURE 11 *Reticulofenestra haqii* (3–5 μm). Gbaran-2 well, 7410 ft.
FIGURE 12 *Reticulofenestra haqii* (3–5 μm). Gbaran-2 well, 7410 ft.
FIGURE 13 *Reticulofenestra pseudoumbilicus* (5–7 μm). Zarama-7 well, 6290 ft.
FIGURE 14 *Reticulofenestra minuta*. Inanga-1 well, 5490 ft.
FIGURE 15 *Reticulofenestra haqii* (3–5 μm). Fragbene-1 well, 7380 ft.
FIGURE 16 *Reticulofenestra pseudoumbilicus* (5–7 μm). Koronama-3 well, 9320 ft.
FIGURE 17 *Reticulofenestra pseudoumbilicus* (5–7 μm). Koronama-3 well, 9320 ft.
FIGURE 18 *Reticulofenestra pseudoumbilicus* (5–7 μm). Ovhor-10 well, 8070 ft.
FIGURE 19 *Reticulofenestra haqii* (3–5 μm). Zarama-7 well, 6290 ft.
FIGURE 20 *Reticulofenestra scrippsae*. Benin River-3 well, 8170 ft.

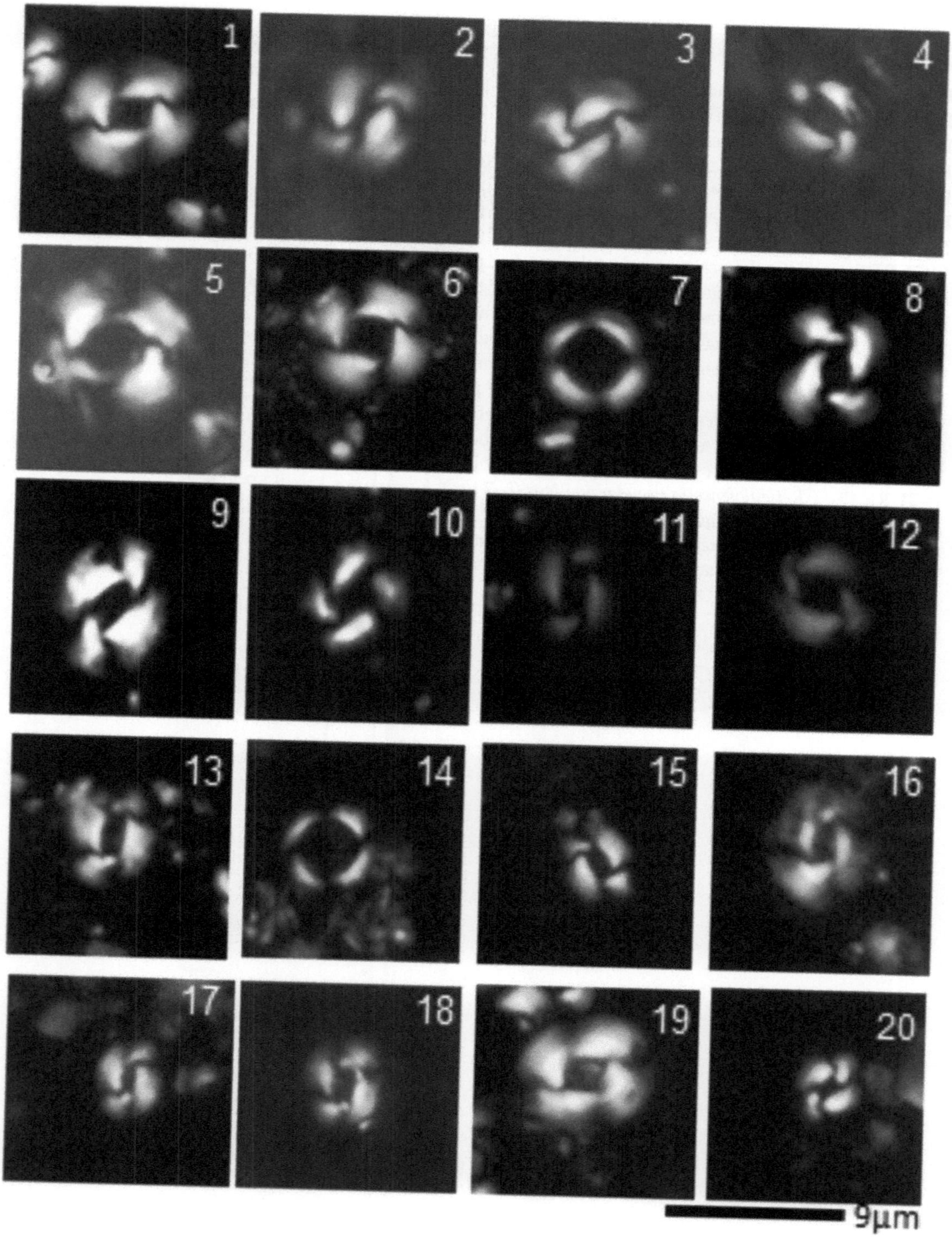
1
2
3
4
5
6
7
8
9
10
11
12
13
14
15
16
17
18
19
20
9µm

EXPLANATION OF PLATE 12

FIGURE 1	*Reticulofenestra pseudoumbilicus* (5–7 μm). Zarama-7 well, 6290 ft.
FIGURE 2	*Reticulofenestra haqii* (3–5 μm). Ovhor-10 well, 8010 ft.
FIGURE 3	*Reticulofenestra haqii* (3–5 μm). Ovhor-10 well, 8010 ft.
FIGURE 4	*Reticulofenestra haqii* (3–5 μm). Ovhor-10 well, 8010 ft.
FIGURE 5	*Reticulofenestra pseudoumbilicus* (>7 μm). Inanga-1 well, 5490 ft.
FIGURE 6	*Reticulofenestra haqii* (5–7 μm). Ovhor-10 well, 8010 ft.
FIGURE 7	*Reticulofenestra rotaria*. Ovhor-10 well, 8010 ft.
FIGURE 8	*Reticulofenestra pseudoumbilicus* (5–7 μm). Tuomo West-1 well, 3644–3652 ft.
FIGURE 9	*Reticulofenestra pseudoumbilicus* (5–7 μm). Tuomo West-1 well, 3644–3652 ft.
FIGURE 10	*Reticulofenestra pseudoumbilicus* (5–7 μm). Fragbene-1 well, 7050 ft.
FIGURE 11	*Reticulofenestra pseudoumbilicus* (5–7 μm). Fragbene-1 well, 7050 ft.
FIGURE 12	*Reticulofenestra pseudoumbilicus* (5–7 μm). Fragbene-1 well, 7050 ft.
FIGURE 13	*Reticulofenestra pseudoumbilicus* (5–7 μm). Fragbene-1 well, 7050 ft.
FIGURE 14	*Reticulofenestra rotaria*. Otobo North-1 well, 5400 ft.
FIGURE 15	*Reticulofenestra haqii* (3–5 μm). Otobo North-1 well, 5400 ft.
FIGURE 16	*Reticulofenestra haqii* (5–7 μm). Otobo North-1 well, 5400 ft.
FIGURE 17	*Reticulofenestra haqii* (3–5 μm). Otobo North-1 well, 6440 ft.
FIGURE 18	*Reticulofenestra haqii* (3–5 μm). Otobo North-1 well, 6440 ft.
FIGURE 19	*Reticulofenestra haqii* (>7 μm). Otobo North-1 well, 5400 ft.
FIGURE 20	*Reticulofenestra minuta*. Ovhor-10 well, 8010 ft.

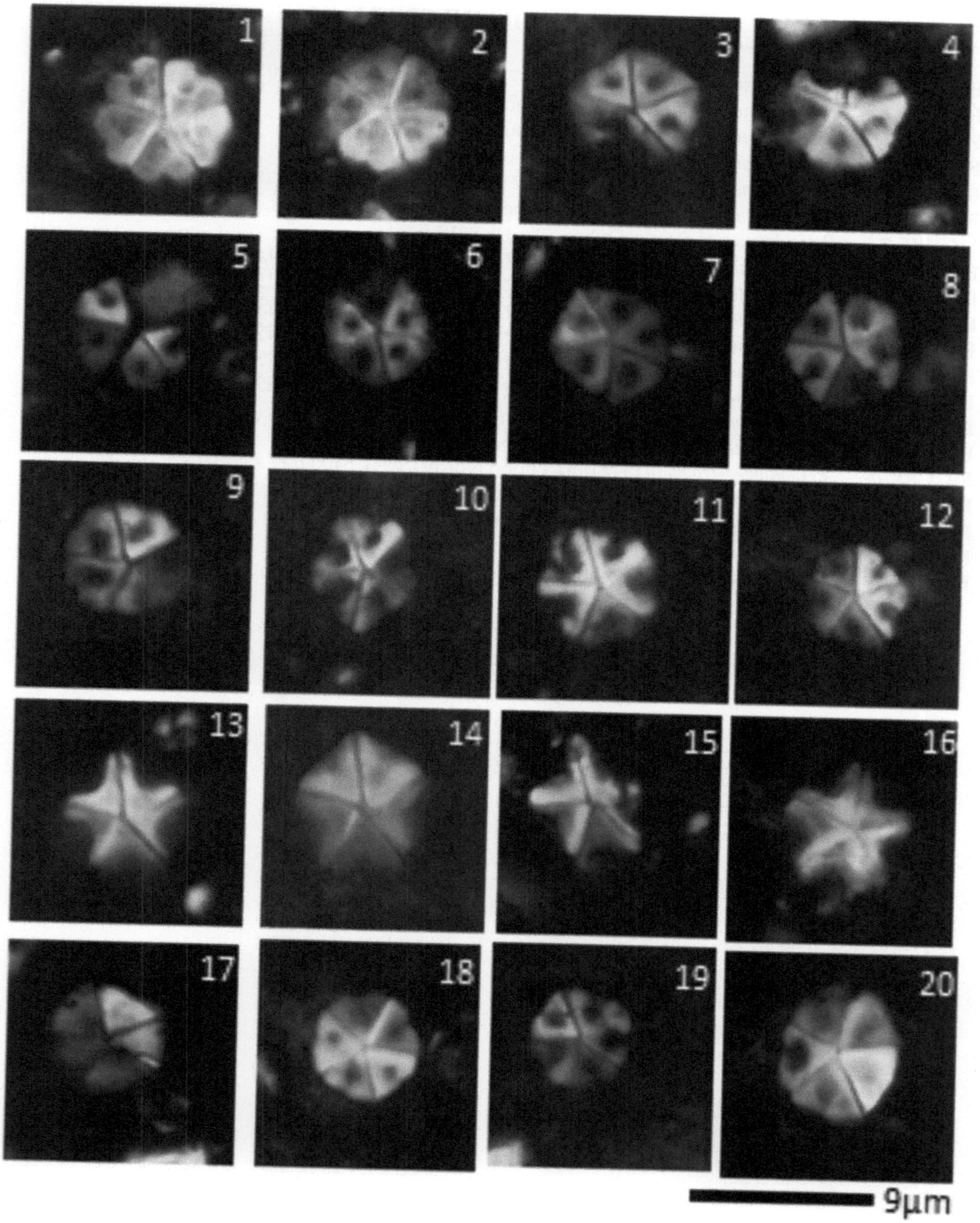
1
2
3
4
5
6
7
8
9
10
11
12
13
14
15
16
17
18
19
20
9μm

EXPLANATION OF PLATE 13

FIGURE 1 *Pemma basquensis*. Inyi Creek-1 well, 9,980–10,010 ft.
FIGURE 2 *Pemma basquensis*. Inyi Creek-1 well, 9,980–10,010 ft.
FIGURE 3 *Pemma basquensis*. Inyi Creek-1 well, 9,980–10,010 ft.
FIGURE 4 *Pemma basquensis*. Inyi Creek-1 well, 9,980–10,010 ft.
FIGURE 5 *Pemma basquensis*. Nemomai-1 well, 10,160 ft.
FIGURE 6 *Pemma basquensis*. Nemomai-1 well, 10,160 ft.
FIGURE 7 *Pemma basquensis*. Nemomai-1 well, 10,160 ft.
FIGURE 8 *Pemma basquensis*. Nemomai-1 well, 10,160 ft.
FIGURE 9 *Pemma basquensis*. Nemomai-1 well, 10,160 ft.
FIGURE 10 *Micrantholithus bulbosus*. Inyi Creek-1 well, 10,250–10,280 ft.
FIGURE 11 *Micrantholithus bulbosus*. Inyi Creek-1 well, 10,250–10,280 ft.
FIGURE 12 *Micrantholithus bulbosus*. Inyi Creek-1 well, 10,250–10,280 ft.
FIGURE 13 *Micrantholithus astrum*. Aroh-1 well, 12,400 ft.
FIGURE 14 *Micrantholithus breviradiatus*. Aroh-1 well, 12,400 ft.
FIGURE 15 *Micrantholithus astrum*. Creek-1 well, 10,250–10,280 ft.
FIGURE 16 *Micrantholithus astrum*. Creek-1 well, 10,250–10,280 ft.
FIGURE 17 *Pemma basquensis*. Inyi Creek-1 well, 10,250–10,280 ft.
FIGURE 18 *Pemma basquensis*. Inyi Creek-1 well, 10,250–10,280 ft.
FIGURE 19 *Micrantholithus bulbosus*. Inyi Creek-1 well, 11,210–11,230 ft.
FIGURE 20 *Pemma basquensis*. Inyi Creek-1 well, 11,210–11,230 ft.

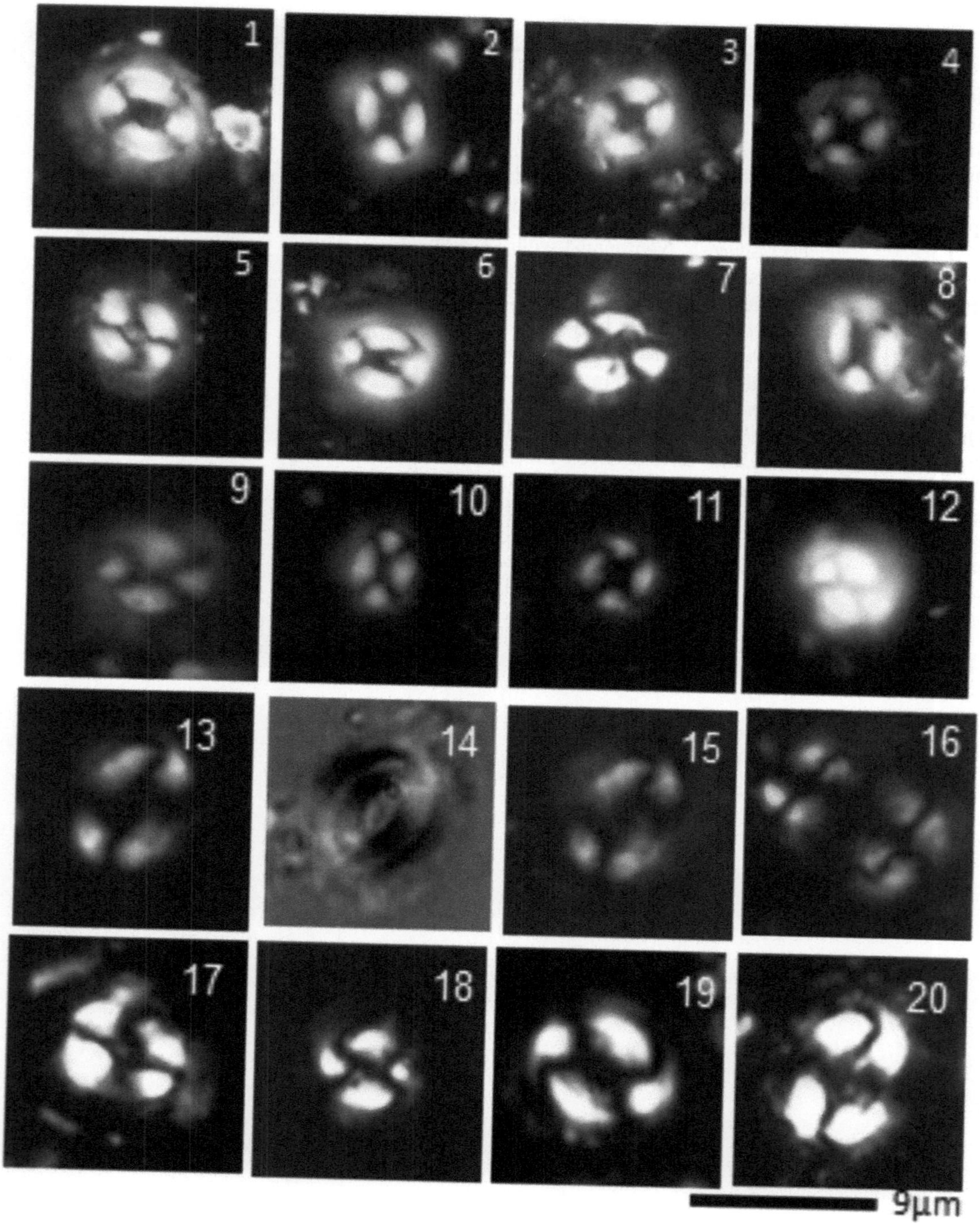
1
2
3
4
5
6
7
8
9
10
11
12
13
14
15
16
17
18
19
20
9µm

EXPLANATION OF PLATE 14

FIGURE 1 *Coccolithus miopelagicus*. Adofi-1 well, 9065 ft.
FIGURE 2 *Coccolithus pelagicus*. Adofi-1 well, 9095 ft.
FIGURE 3 *Coccolithus pelagicus*. Adofi-1 well, 9095 ft.
FIGURE 4 *Coccolithus pelagicus*. Fragbene-1 well, 7860 ft.
FIGURE 5 *Coccolithus pelagicus*. Fragbene-1 well, 7860 ft.
FIGURE 6 *Coccolithus pelagicus*. Fragbene-1 well, 7050 ft.
FIGURE 7 *Coccolithus pelagicus*. Fragbene-1 well, 7050 ft.
FIGURE 8 *Coccolithus pelagicus*. Fragbene-1 well, 9480 ft.
FIGURE 9 *Coccolithus pelagicus*. Ovhor-10 well, 8070 ft.
FIGURE 10 *Coccolithus pelagicus*. Nemomai-1 well, 5120 ft.
FIGURE 11 *Coccolithus pelagicus*. Nemomai-1 well, 5120 ft.
FIGURE 12 *Coccolithus pelagicus*. Tuomo West-1 well, 2914–2922 ft.
FIGURE 13 *Coccolithus pelagicus*. Zarama-7 well, 10,730 ft.
FIGURE 14 *Coccolithus pelagicus*. Zarama-7 well, 10,730 ft.
FIGURE 15 *Coccolithus pelagicus*. Zarama-7 well, 10,730 ft.
FIGURE 16 *Coccolithus pelagicus*. Zarama-7 well, 10,970 ft.
FIGURE 17 *Coccolithus pelagicus*. Zarama-7 well, 11,270 ft.
FIGURE 18 *Coccolithus pelagicus*. Zarama-7 well, 11,270 ft.
FIGURE 19 *Coccolithus pelagicus*. Zarama-7 well, 11,270 ft.
FIGURE 20 *Coccolithus pelagicus*. Zarama-7 well, 11,270 ft.

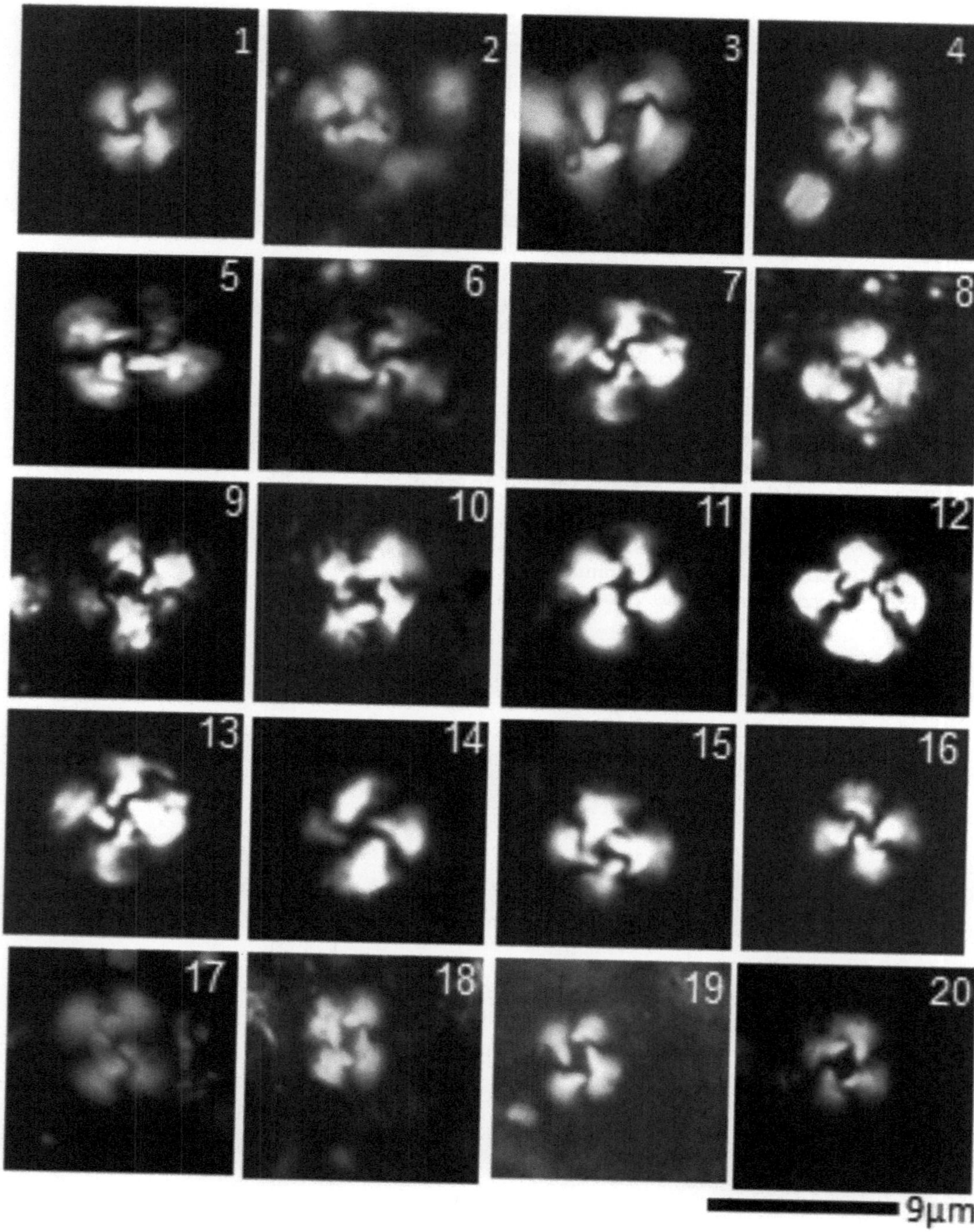
1
2
3
4
5
6
7
8
9
10
11
12
13
14
15
16
17
18
19
20
9μm

EXPLANATION OF PLATE 15

FIGURE 1 *Cyclicargolithus floridanus*. Zarama-7 well, 10,940 ft.
FIGURE 2 *Cyclicargolithus floridanus*. Zarama-7 well, 10,940 ft.
FIGURE 3 *Cyclicargolithus floridanus*. Zarama-7 well, 10,940 ft.
FIGURE 4 *Cyclicargolithus floridanus*. Zarama-7 well, 10,940 ft.
FIGURE 5 *Cyclicargolithus floridanus*. Zarama-7 well, 10,910 ft.
FIGURE 6 *Cyclicargolithus floridanus*. Zarama-7 well, 10,910 ft.
FIGURE 7 *Cyclicargolithus floridanus*. Zarama-7 well, 10,910 ft.
FIGURE 8 *Cyclicargolithus floridanus*. Fragbene-1 well, 9090 ft.
FIGURE 9 *Cyclicargolithus floridanus*. Fragbene-1 well, 9090 ft.
FIGURE 10 *Cyclicargolithus floridanus*. Fragbene-1 well, 9400 ft.
FIGURE 11 *Cyclicargolithus floridanus*. Fragbene-1 well, 9400 ft.
FIGURE 12 *Cyclicargolithus floridanus*. Fragbene-1 well, 7050 ft.
FIGURE 13 *Cyclicargolithus floridanus*. Fragbene-1 well, 7050 ft.
FIGURE 14 *Cyclicargolithus floridanus*. Fragbene-1 well, 7860 ft.
FIGURE 15 *Cyclicargolithus floridanus*. Fragbene-1 well, 7860 ft.
FIGURE 16 *Cyclicargolithus floridanus*. Fragbene-1 well, 7860 ft.
FIGURE 17 *Cyclicargolithus abisectus*. Nemomai-1 well, 8200 ft.
FIGURE 18 *Cyclicargolithus floridanus*. Otobo North-1 well, 8650 ft.
FIGURE 19 *Cyclicargolithus floridanus*. Otobo North-1 well, 8650 ft.
FIGURE 20 *Cyclicargolithus floridanus*. Adofi-1 well, 9065 ft.

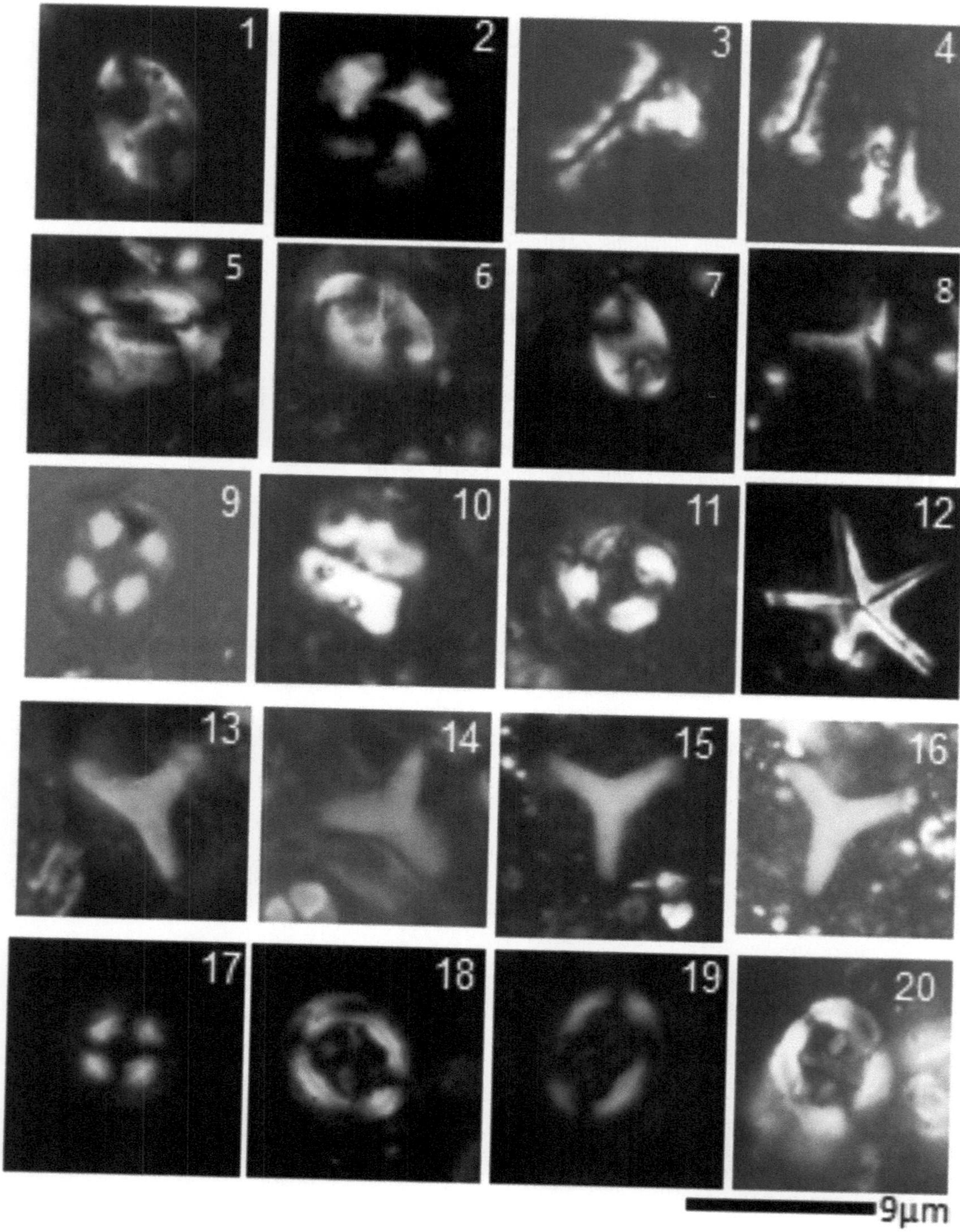
1
2
3
4
5
6
7
8
9
10
11
12
13
14
15
16
17
18
19
20
9μm

EXPLANATION OF PLATE 16

FIGURE 1	*Pontosphaera exilis*. Nemomai-1 well, 6640 ft.
FIGURE 2	*Reticulofenestra dictyoda*. Nemomai-1 well, 8200 ft.
FIGURE 3	*Zygrhablithus bijugatus*. Inyi Creek-1 well, 9,980–10,010 ft.
FIGURE 4	*Zygrhablithus bijugatus*. Inyi Creek-1 well, 9,980–10,010 ft.
FIGURE 5	*Ellipsolithus macellus*. Inyi Creek-1 well, 10,250–10,280 ft.
FIGURE 6	*Neochiastozygus* sp. Inyi Creek-1 well, 11,180–11,210 ft.
FIGURE 7	*Pontosphaera exilis*. Inyi Creek-1 well, 11,180–11,210 ft.
FIGURE 8	*Blackite* sp. Inyi Creek-1 well, 11,210–11,230 ft.
FIGURE 9	*Coccolithus subpertusus*. Nemomai-1 well, 6160 ft.
FIGURE 10	*Fasciculithus tympaniformis*. Gilli Gilli-1 well, 1290 ft.
FIGURE 11	*Coccolithus formosus*. Nemomai-1 well, 5350 ft.
FIGURE 12	*Micrantholithus attenuatus*. Inyi Creek-1 well, 11,180–11,210 ft.
FIGURE 13	*Tribrachiatus orthostylus*. Otobo North-1 well, 9340 ft.
FIGURE 14	*Tribrachiatus orthostylus*. Otobo North-1 well, 9340 ft.
FIGURE 15	*Tribrachiatus orthostylus*. Otobo North-1 well, 9340 ft.
FIGURE 16	*Tribrachiatus orthostylus*. Otobo North-1 well, 9340 ft.
FIGURE 17	*Coccolithus formosus*. Nemomai-1 well, 5320 ft.
FIGURE 18	*Cruciplacolithus tenuis*. Nemomai-1 well, 5080 ft.
FIGURE 19	*Cruciplacolithus* sp. Okagba-1 well, 5050 ft.
FIGURE 20	*Cruciplacolithus tenuis*. Nemomai-1 well, 5120 ft.

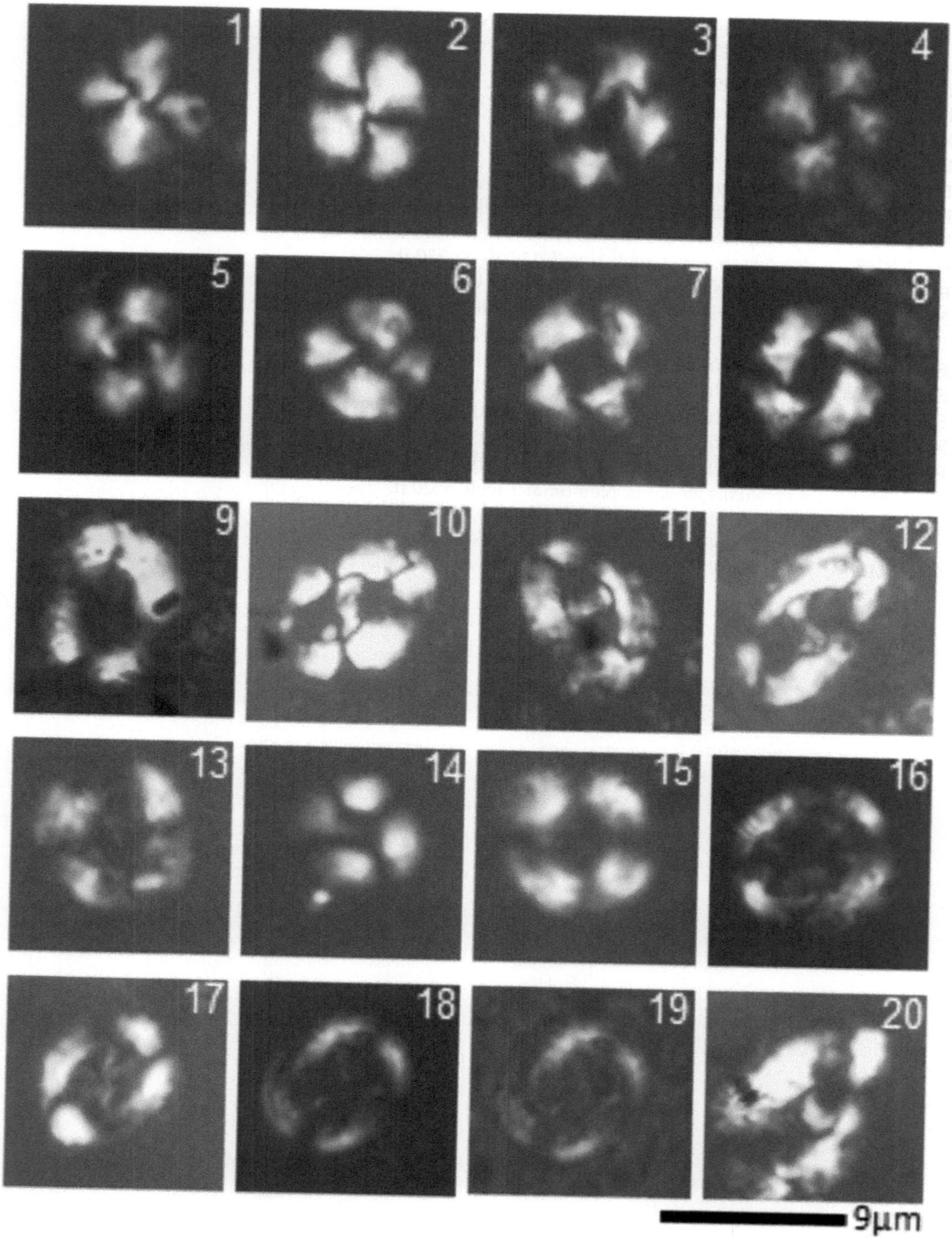
1
2
3
4
5
6
7
8
9
10
11
12
13
14
15
16
17
18
19
20
9µm

EXPLANATION OF PLATE 17

FIGURE 1 *Cyclicargolithus luminis*. Inyi Creek-1 well, 9,980–10,010 ft.
FIGURE 2 *Cyclagelosphaera margerelii*. Inyi Creek-1 well, 9,980–10,010 ft.
FIGURE 3 *Reticulofenestra dictyoda*. Inyi Creek-1 well, 9,980–10,010 ft.
FIGURE 4 *Reticulofenestra dictyoda*. Inyi Creek-1 well, 9,980–10,010 ft.
FIGURE 5 *Reticulofenestra dictyoda*. Inyi Creek-1 well, 9,980–10,010 ft.
FIGURE 6 *Reticulofenestra reticulata*. Inyi Creek-1 well, 9,980–10,010 ft.
FIGURE 7 *Reticulofenestra dictyoda*. Inyi Creek-1 well, 10,250–10,280 ft.
FIGURE 8 *Reticulofenestra umbilica*. Inyi Creek-1 well, 10,250–10,280 ft.
FIGURE 9 *Reticulofenestra umbilica*. Inyi Creek-1 well, 10,950–10,980 ft.
FIGURE 10 *Helicosphaera seminulum*. Inyi Creek-1 well, 11,210–11,230 ft.
FIGURE 11 *Helicosphaera seminulum*. Inyi Creek-1 well, 11,210–11,230 ft.
FIGURE 12 *Helicosphaera seminulum*. Inyi Creek-1 well, 11,210–11,230 ft.
FIGURE 13 *Chiasmolithus solitus*. Inyi Creek-1 well, 9,980–10,010 ft.
FIGURE 14 *Coccolithus formosus*. Inyi Creek-1 well, 9,980–10,010 ft.
FIGURE 15 *Coccolithus formosus*. Inyi Creek-1 well, 9,980–10,010 ft.
FIGURE 16 *Chiasmolithus solitus*. Inyi Creek-1 well, 10,070–10,100 ft.
FIGURE 17 *Chiasmolithus solitus*. Inyi Creek-1 well, 10,250–10,280 ft.
FIGURE 18 *Chiasmolithus solitus*. Inyi Creek-1 well, 11,210–11,230 ft.
FIGURE 19 *Chiasmolithus solitus*. Inyi Creek-1 well, 11,210–11,230 ft.
FIGURE 20 *Helicosphaera seminulum*. Inyi Creek-1 well, 9,980–10,010 ft.

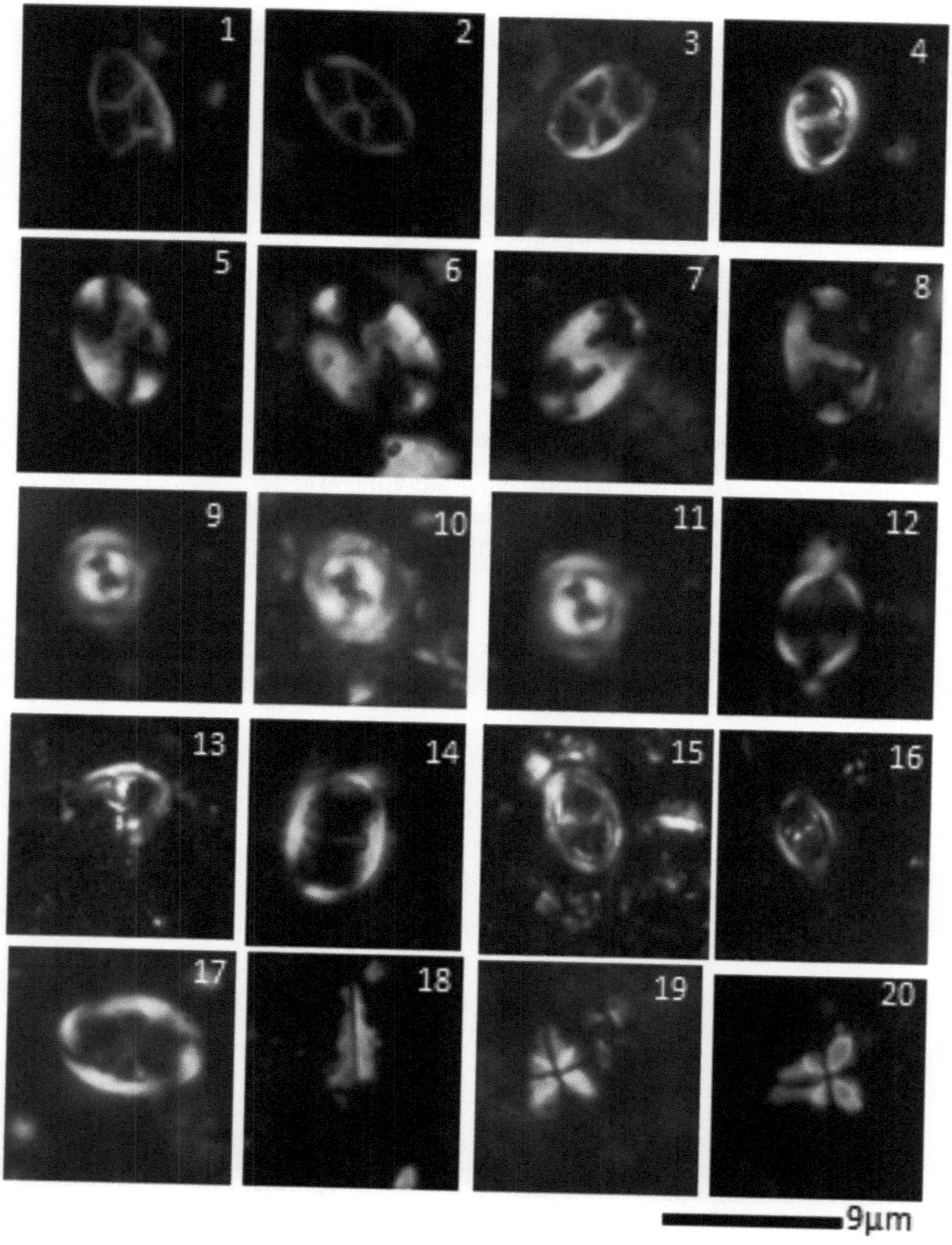
1
2
3
4
5
6
7
8
9
10
11
12
13
14
15
16
17
18
19
20
9μm

EXPLANATION OF PLATE 18

FIGURE 1 *Neococcolithes dubius*. Inyi Creek-1 well, 10,070–10,100 ft.
FIGURE 2 *Neococcolithes protenus*. Inyi Creek-1 well, 11,180–11,210 ft.
FIGURE 3 *Neococcolithes protenus*. Inyi Creek-1 well, 11,180–11,210 ft.
FIGURE 4 *Zeugrhabdotus sigmoides*. Inyi Creek-1 well, 9,980–10,010 ft.
FIGURE 5 *Transversopontis pulcher*. Inyi Creek-1 well, 9,980–10,010 ft.
FIGURE 6 *Transversopontis* pulcheroides. Inyi Creek-1 well, 10,070–10,100 ft.
FIGURE 7 *Transversopontis* pulcheroides. Inyi Creek-1 well, 10,070–10,100 ft.
FIGURE 8 *Pontosphaera exilis*. Inyi Creek-1 well, 10,920–10,950 ft.
FIGURE 9 *Toweius eminens*. Inyi Creek-1 well, 10,920–10,950 ft.
FIGURE 10 *Toweius occultatus*. Inyi Creek-1 well, 11,180–11,210 ft.
FIGURE 11 *Toweius occultatus*. Inyi Creek-1 well, 11,180–11,210 ft.
FIGURE 12 *Pontosphaera versa*. Inyi Creek-1 well, 10,210–10,230 ft.
FIGURE 13 *Zeugrhabdotus sigmoides* (broken). Inyi Creek-1 well, 10,350–10,380 ft.
FIGURE 14 *Lophodolithus nascens*. Inyi Creek-1 well, 11,180–11,210 ft.
FIGURE 15 *Zeugrhabdotus sigmoides*. Adofi-1 well, 9065 ft.
FIGURE 16 *Zeugrhabdotus sigmoides*. Adofi-1 well, 9065 ft.
FIGURE 17 *Neochiastozygus junctus*. Benin River-3 well, 7990 ft.
FIGURE 18 *Blackite* sp. Benin River-3 well, 7990 ft.
FIGURE 19 *Sphenolithus spiniger*. Benin River-3 well, 9930 ft.
FIGURE 20 *Sphenolithus spiniger*. Benin River-3 well, 9930 ft.

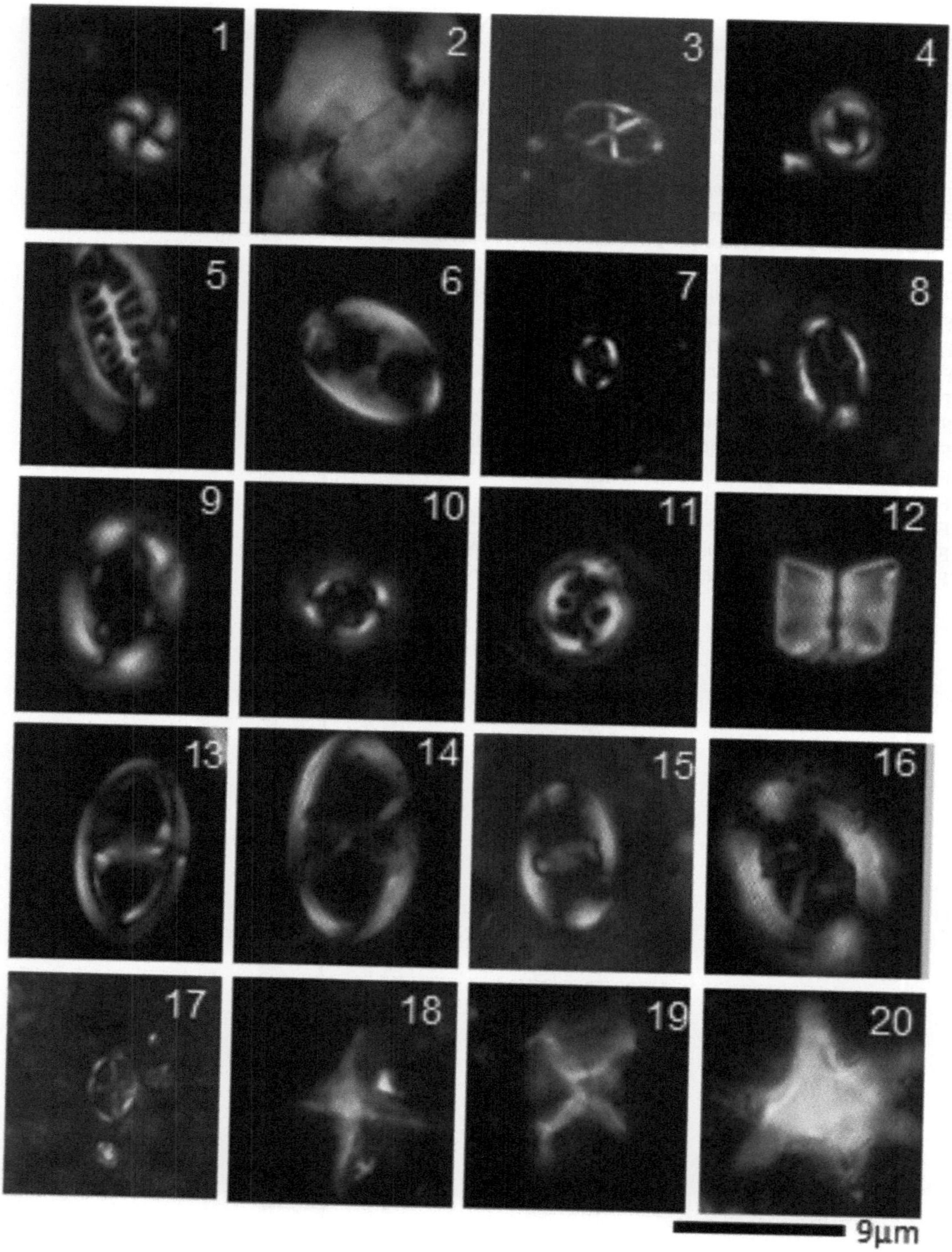
1
2
3
4
5
6
7
8
9
10
11
12
13
14
15
16
17
18
19
20
9μm

EXPLANATION OF PLATE 19

FIGURE 1	*Toweius rotundus*. Gilli Gilli-1 well, 11,450 ft.
FIGURE 2	*Ellipsolithus macellus*. Gilli Gilli-1 well, 9670 ft.
FIGURE 3	*Neococcolithes protenus*. Gilli Gilli-1 well, 10,280 ft.
FIGURE 4	*Toweius crassus*. Otobo North-1 well, 9330 ft.
FIGURE 5	*Ellipsolithus distichus*. Gilli Gilli-1 well, 11,450 ft.
FIGURE 6	*Transversopontis pulcheroides*. Gilli Gilli-1 well, 10,520 ft.
FIGURE 7	*Reticulofenestra minutula*. Utip-1 well, 2500 ft.
FIGURE 8	*Cruciplacolithus primus*. Gilli Gilli-1 well, 11,450 ft.
FIGURE 9	*Cruciplacolithus* sp. Gilli Gilli-1 well, 10,520 ft.
FIGURE 10	*Cruciplacolithus tenuis*. Gilli Gilli-1 well, 11,030 ft.
FIGURE 11	*Toweius tovae*. Gilli Gilli-1 well, 10,520 ft.
FIGURE 12	*Fasciculithus tympaniformis*. Gilli Gilli-1 well, 11,290 ft.
FIGURE 13	*Zeugrhabdotus sigmoides*. Fragbene-1 well, 4750 ft.
FIGURE 14	*Neochiastozygus* cf. *perfectus*. Otobo North-1 well, 9400 ft.
FIGURE 15	*Zygodiscus plectopons*. Otobo North-1 well, 9400 ft.
FIGURE 16	*Chaismolithus altus*. Otobo North-1, 8650 ft.
FIGURE 17	*Cruciplacolithus primus*. Gilli Gilli-1 well, 11,450 ft.
FIGURE 18	*Nannotetrina fulgens*. Otobo North-1 well, 8720 ft.
FIGURE 19	*Nannotetrina cristata*. Otobo North-1 well, 8720 ft.
FIGURE 20	*Rhomboaster bramlettei*. Otobo North-1 well, 9400 ft.

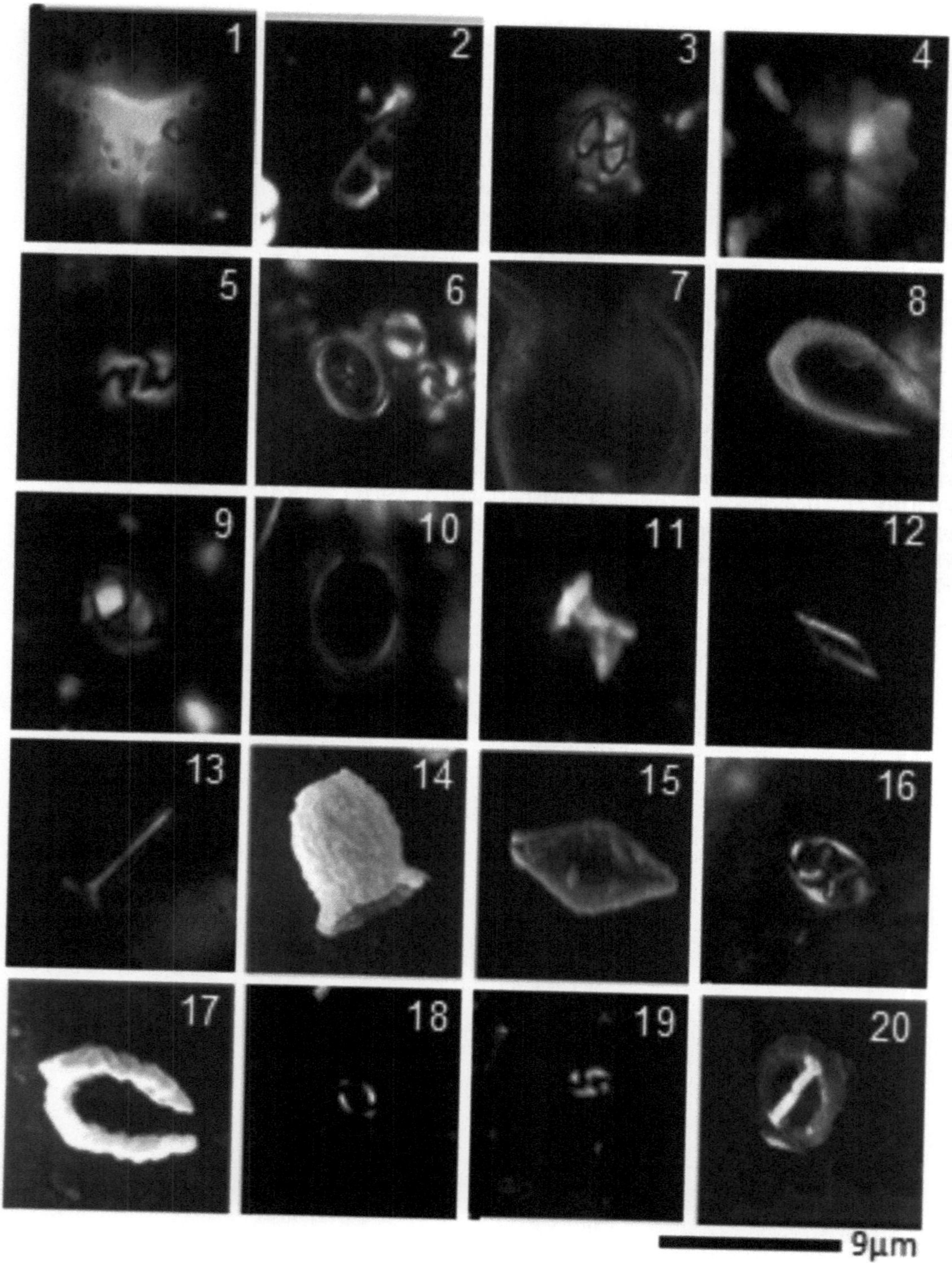
1
2
3
4
5
6
7
8
9
10
11
12
13
14
15
16
17
18
19
20
9μm

EXPLANATION OF PLATE 20

FIGURE 1	*Rhomboaster* sp. Otobo North-1 well, 9400 ft.
FIGURE 2	*Ismolithus recurvus*. Amakubu-1 well, 9120 ft.
FIGURE 3	*Tetralithoides symeonidesii*. Opuama-1 well, 4237 ft.
FIGURE 4	*Biantholithus hughesii*. Gilli Gilli-1 well, 1450 ft.
FIGURE 5	*Gephyrocapsa oceanica*. Anyala-8 well, 1720 ft.
FIGURE 6	*Syracosphaera pulchra*. Anyala-4 well, 3250 ft.
FIGURE 7	*Syracosphaera pulcherrima*. Adaka-1 well, 5950 ft.
FIGURE 8	*Ceratolithus telesmus*. Anyala-8 well, 3250 ft.
FIGURE 9	*Tetralithoides symeonidesii*. Benin River-3 well, 8760 ft.
FIGURE 10	*Umbilicosphaera rotula*. Anyala-8 well, 6420 ft.
FIGURE 11	*Minylitha convallis*. Inanga-1 well, 10,730 ft.
FIGURE 12	*Calsiosolenia murrayi*. Anyala-8 well, 1600–1720 ft.
FIGURE 13	*Rhabdosphaera clavigeri*. Anyala-4 well, 330 ft.
FIGURE 14	*Scyphosphaera* sp. Anyala-8 well, 8410 ft.
FIGURE 15	*Calsiosolenia murrayi*. Anyala-8 well, 8410 ft.
FIGURE 16	*Neoccolithus* sp. Otobo North-1 well, 9400 ft.
FIGURE 17	*Ceratolithus rugosus*. Anyala-8 well, 8410 ft.
FIGURE 18	*Reticulofenestra minutula*. Utip-1 well, 2100 ft.
FIGURE 19	*Gephyrocapsa* sp. (small). Anyala-8 well, 4930 ft.
FIGURE 20	*Gephyrocapsa* sp. Anyala-8 well, 8410 ft.

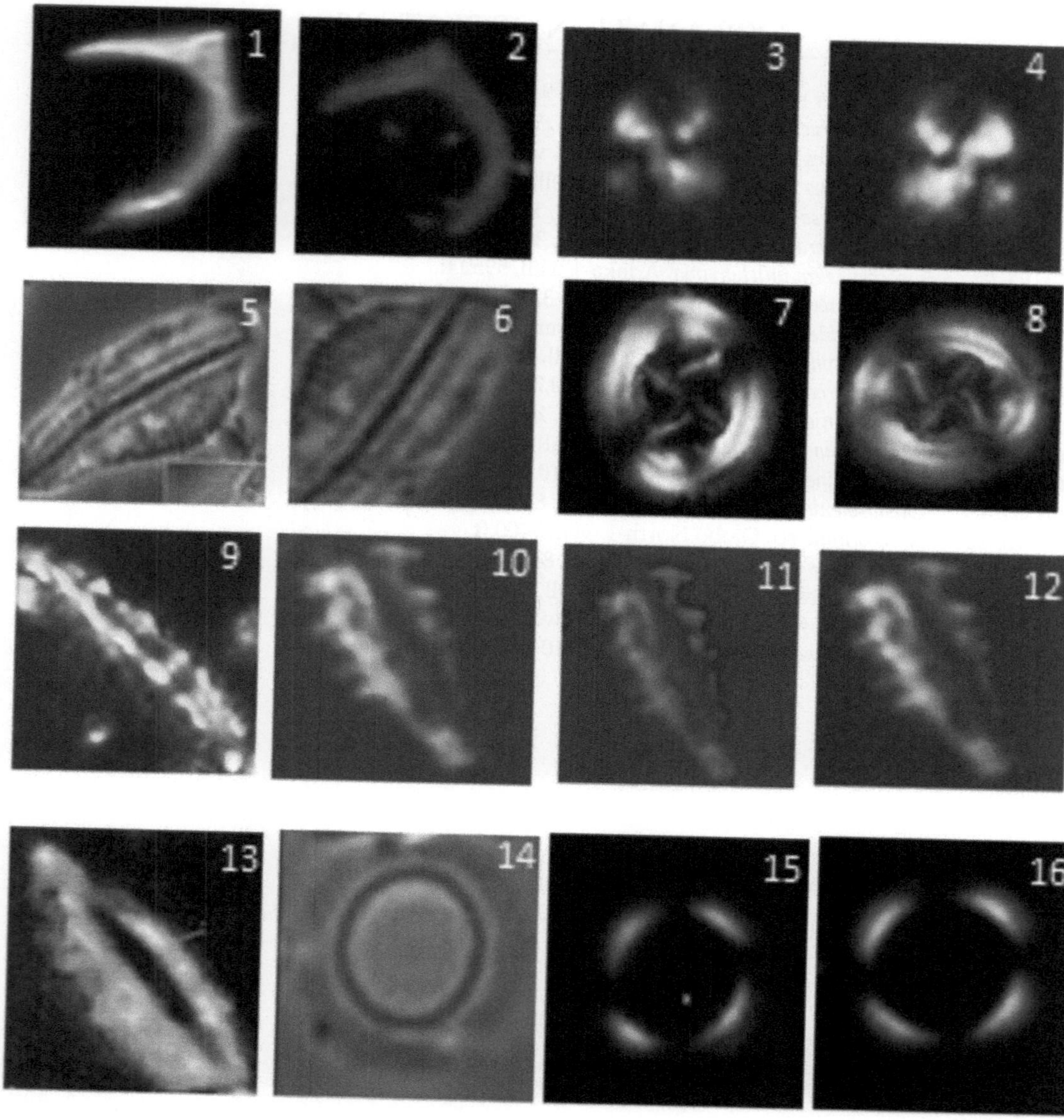
1
2
3
4
5
6
7
8
9
10
11
12
13
14
15
16

EXPLANATION OF PLATE 21

FIGURE 1	*Ceratolithus acutus*. Otobo N-1 well, 8650 ft.
FIGURE 2	*Ceratolithus acutus*. Otobo N-1 well, 8650 ft.
FIGURE 3	*Gephyrocapsa caribbeanica*. Anyala-8 well, 3460 ft.
FIGURE 4	*Gephyrocapsa caribbeanica*. Anyala-8 well, 3460 ft.
FIGURE 5	*Triquetrorhabdulus rugosus*. Fragbene-1 well, 5170 ft.
FIGURE 6	*Triquetrorhabdulus rugosus*. Fragbene-1 well, 5170 ft.
FIGURE 7	*Chiasmolithus bidens*. Otobo N-1 well, 8650 ft.
FIGURE 8	*Chiasmolithus bidens*. Otobo N-1 well, 8650 ft.
FIGURE 9	*Triquetrorhabdulus rugosus*. Fragbene-1 well, 7860 ft.
FIGURE 10	*Triquetrorhabdulus rugosus*. Fragbene-1 well, 7860 ft.
FIGURE 11	*Triquetrorhabdulus rugosus*. Fragbene-1 well, 7860 ft.
FIGURE 12	*Triquetrorhabdulus rugosus*. Fragbene-1 well, 7860 ft.
FIGURE 13	*Triquetrorhabdulus millowii*. Fragbene-1 well, 7050 ft.
FIGURE 14	*Reticulofenestra rotaria*. Inanga well, 9530 ft.
FIGURE 15	*Reticulofenestra rotaria*. Inanga well, 9530 ft.
FIGURE 16	*Reticulofenestra rotaria*. Inanga well, 9530 ft.

References

Achuthan, M.V., Stradner, H., 1969. Calcareous nannoplankton from the Wemmelian stratotype. In: Bronnimann, P., Renz, H.H. (Eds.), Proceedings of the First International Conference on Planktonic Microfossils, Geneva, vol. 1. E.J. Brill, Leiden, pp. 1–13.

Acosta, J.T., 1940. Algunos foraminifeos nuevos de la costas Cubanas, Torreia, Havana, Cuba, No. 5, pp. 1–6.

Adedokun, O.A., 1981. Petrology, provenance and depositional environments of the reservoir sandstones of Ossu Izombe oilfield, Imo State, Nigeria. J. Petr. Geol. 4, 35–56.

Adegoke, O.S., 1969. Eocene stratigraphy of Southern Nigeria. Colloque sur l'eocene, v. 111, Memoires, BRGM, No. 69, pp. 23–49, 6 text-figs. 1 pl.

Adegoke, O.S., 1975. Foraminiferal fauna of the polyhaline lagoons of the Gulf of Guinea. J. Min. Geol. 12 (1 and 2), 1–8.

Adegoke, O.S., 1977. Stratigraphy and paleontology of the Ewekoro Formation (Paleocene) of southwestern Nigeria. Bull. Am. Paleontol. 71, 295.

Adegoke, O.S., 2000. High resolution biostratigraphy, sequence stratigraphy and 3-D modeling: indispensable tools for E and P activities in the next Millennium. Keynote Paper, 2000, NAPE International Conference, NAPE Bulletin, vol. 16, No. 1, pp. 46–65.

Adegoke, O.S., Abibo, P.T., Mochu, I.N., Odebode, M.O., Nwachukwu, J.I., 1980a. Sediment distribution in the Lagos Lagoon and Harbour. Ann. Mines Geol., Tunis, No. 28, Tome III, pp. 413–425.

Adegoke, O.S., Arua, I., Oyegoke, O., 1980b. Two new nautiloids from Imo Shale (Paleocene) and Ameki Formation (middle Eocene) Anambra State, Nigeria. J. Min. Geol. 17 (1), 85–89.

Adegoke, O.S., Dessauvagie, T.F.J., Kogbe, C.A., 1971. Planktonic Foraminifera in Gulf of Guinea sediments. Micropaleontology 17 (2), 197–213, 4 pls.

Adegoke, O.S., Jeje, L.K., Durotoye, A.B., Adeleye, D.R., Ebukanson, E.J., 1980c. The geomorphology and aspects of the sedimentology of the coastal region of western Nigeria. J. Min. Geol. 17, 217–223.

Adegoke, O.S., Omatsola, M.E., Ramanathan, R., 1987. Deep sea foraminiferal biofacies off West Africa. In: Matheis, G., Schandelmeier, H. (Eds.), Current Research in African Earth Sciences. Balkema Publishers, Rotterdam, pp. 147–149.

Adegoke, O.S., Omatsola, M.E., Salami, M.B., 1976. Benthonic foraminiferal biofacies of the Niger Delta. Proceedings of the First International Symposium on Benthonic Foraminifera of Continental Margins, Part A: Geology and Biology. Marine Sediments, Special Publication No. 1, pp. 279–292.

Adegoke, O.S., Stanley, D.J., 1972. Mica and shell as indicators of energy level and depositional regime on the Nigerian Shelf. Mar. Geol. 13, 61–66.

Adegoke, O.S., Tevesz, M.J.S., 1974. Gastropod predation patterns in the Eocene of Nigeria. Lethaia 7, 17–24.

Adegoke, O.S., Ulu, O.K., Jan du Chene, R., Obilaja, O.T., 2011. Sequence stratigraphic correlation of the Uppermost Miocene to Pleistocene interval of the wells A1, A2 and A3, Niger Delta, Nigeria. NAPE Bull. 23 (2), 10–14.

Adeniran, B.V., 1997. Quantitiative Neogene planktic foraminiferal biostratigraphy of Western Niger Delta, Nigeria. NAPE Bull. 12 (1), 54–69.

AGIP S.P.A., 1982. Foraminifera Padani (Terziario e Quaternario) Atlanteiconograficoe ditribuzione stratigrafica. Milan.

AGIP S.P.A., 1988. Foraminifera of the Niger Delta. (S. Donato Milanese).

AGIP MINERARIA, 1957. Foraminiferi Padani (Terziarioe Quaternario) Atlante Iconografico und Distribuzione Stratigrafica. Agip Mineraria Milan.

Aiyuk, S.E., 1993. Stratigraphy, sedimentology and groundwater potentials of alluvial deposits east of the Niger River and north of Onitsha, Anambra State, Nigeria. MSc Thesis, Geol. Dept., Univ. of Nigeria, Nsukka, 214 p.

Ajakaiye, D.E., Bally, A.W., 2002. Course Manual and Atlas of Structural Styles on Reflection Profiles From the Niger Delta. AAPG Continuing Education Course Note Series No. 41, 107 p.

Akaegbobi, I.M., Anyaeji, U.O., Ogungbesan, G.O., 2009. Integrated chemostratigraphy and biostratigraphy of EZ-1 and BZ-1 wells, Western Niger Delta, Nigeria. NAPE Bull. 21 (1), 53–60.

Akers, W.H., 1955. Some planktonic Foraminifera of the American Gulf Coast and suggested correlations with the Caribbean Tertiary. J. Paleontol. 29, 647–664.

Akers, W.H., Dorman, J.H., 1964. Pleistocene Foraminifera of the Gulf Coast. Tulane Stud. Geol. 3 (1), 1–93.

Ako, B.D., Ojo, S.B., Okereke, C.S., Fieberg, F.C., Ajayi, T.R., Adepelumi, A.A., Afolayan, J.F., Afolabi, O., Ogunnusi, H.O., 2004. Some observations from gravity/magnetic data interpretation of the Niger Delta. NAPE Bull. 17 (1), 1–21.

Allen, J.R.L., 1964. The Nigerian continental margin: bottom sediments, submarine morphology, and geological evolution. Mar. Geol. 1, 289–332.

Allen, J.R.L., 1965a. Late Quaternary Niger Delta and adjacent areas: sedimentary environments and lithofacies. Am. Assoc. Petr. Geol. Bull. 49 (5), 547–600.

Allen, J.R.L., 1965b. A review of the origin and characteristics of recent alluvial sediments. Sedimentology 5, 89–191.

Allen, J.R.L., 1970. Sediments of the modern Niger Delta: summary and review. In: Morgan, J.P., (Ed.), Deltaic Sedimentation, Modern and Ancient. SEPM Special Publication No. 15, pp. 138–151.

Cenozoic Foraminifera and Calcareous Nannofossil Biostratigraphy of the Niger Delta. http://dx.doi.org/10.1016/B978-0-12-812161-0.00011-9

Andersen, H.V., 1961. Genesis and paleontology of the Mississippi River Mudlumps, Lower Mississippi River Delta, Louisiana. Louisiana Geological Society, Bulletin 35, pp. 1–208, 29 pls.

Anyanwu, N.P.C., Arua, I., 1990. Ichnofossils from the Imo Formation, and their paleoenvironmental significance. J. Min. Geol. 26, 1–4.

Applin, E.S., 1925. In: Applin, E.S., Ellisor, A.C., Kniker, H.T. (Eds.), Sub-surface stratigraphy of the coastal plain of Texas and Louisiana. American Association of Petroleum Geologists Bulletin, vol. 9, No. 1, pp. 79–122.

Armentrout, J.M., 1991. Paleontologic constraints on depositional modelling: examples of integration of biostratigraphy and seismic stratigraphy, Pliocene–Pleistocene, Gulf of Mexico. Seismic Facies and Sedimentary Processes of Submarine Fans and Turbidites SystemsSpringer-Verlag, New York, NY, (Chapter 7).

Armentrout, J.M., 1995. High resolution sequence biostratigraphy: examples from the Gulf of Mexico Plio-Pleistocene. In: Howell, J.A., Aitken, J.F. (Eds.), High Resolution Sequence Stratigraphy: Innovations and Applications. Geological Society, Special Publication No. 104, pp. 65–86.

Arua, I., 1980. Paleocene macrofossils from the Imo Shale in Anambra State. J. Min. Geol. 17, 81–84.

Asano, K., 1944. *Hanzawaia* a new genus of Foraminifera from the Pliocene of Japan. Geol. Soc. Japan J. 51 (606), 97–98.

Asano, K., 1958. Some Paleogene smaller Foraminifera from Japan. Tohoku Univ. Sci. Rep. Japan 29, 51, Ser. 2.

Asseez, L.O., Fayose, E.A., Omatsola, M.E., 1974. Ecology of the Ogun River estuary. Palaeogeogr. Palaeoclimatol. Palaeocol. 16, 243–260.

Aubert, T.J., Berggren, W.A., 1976. Paleocene benthic foraminiferal biostratigraphy and paleoecology of Tunisia. Bull. Centres Rech. Explor. Prod. Elf Aquitaine 10 (2), 379–469.

Aubry, M.P., 1992. In: Prothero, D.R., Berggren, W.A. (Eds.), Eocene–Oligocene Climate and Biotic Evolution. Princeton University Press, U.S.A., pp. 271–309.

Avbovbo, A.A., Ayoola, E.O., 1981. Petroleum prospects of southern Nigeria's Anambra Basin. Oil Gas J. 1981, 334–347.

Backman, J., 1978. Late Miocene–Early Pliocene nannofossil biochronology and biogeography in the Vera Basin, SE Spain. Stockholm Contrib. Geol. 32, 93–114.

Backman, J., 1980. Miocene–Pliocene nannofossil and sedimentation rates in the Hatton-Rockall basin, Northeastern Atlantic Ocean. Acta Univ. Stockholm Contrib. Geol. 36 (1), 1–91.

Backman, J., 1984. Cenozoic Calcareous Nannofossil Biostratigraphy From the Northeastern Atlantic Ocean. Initial Reports of the Deep Sea Drilling Project, Washington (US Government Printing Office), vol. 81, pp. 403–428.

Backman, J., Schneider, D.A., Rio, D., Okada, H., 1990. Neogene low-latitude magnetostratigraphy from Site 170 and revised age estimates of Miocene nannofosssil datum events. Proc. Ocean Drill. Prog. Sci. Results 115, 271–276.

Backman, J., Shackleton, N.J., 1983. Quantitative biochronology of Pliocene and Early Pleistocene calcareous nannofossils from the Atlantic, Indian and Pacific Oceans. Mar. Micropaleontol. 8, 141–170.

Bandy, O.L., 1949. Eocene and Oligocene forminifera from Little Stave Creek, Clarke County, Alabama. Bull. Am. Paleontol. 32, 35–240.

Bandy, O.L., 1954. Distribution of some shallow-water Foraminifera in the Gulf of Mexico. United States Geological Survey, Professional Paper No. 254-F, pp. 25–140.

Bandy, O.L., 1967. Relationships of Neogene planktonic foraminifera to paleoceanography and correlation. In: Proceedings of the First International Conference on Planktonic Microfossils. E.J. Brill, Leiden, pp. 46–57.

Bandy, O.L., Arnal, R.E., 1957. Distribution of recent foraminifera off the west coast of Central America. AAPG Bull. 41, 2037–2053.

Bandy, O.L., Casey, R.E., Wright, R.C., 1971. Late Neogene planktonic zonation magnetic reversals and radiometric dates. Antarctic to the tropics. Antartic Res. Serv. 15, 1–26.

Bandy, O.L., Frerichs, W.E., Vincent, E., 1967. Origin, development and geologic significance of *Neogloboquadrina* Bandy, Frerichs and Vincent, gen. nov. Contrib. Cushman Found. Foramin. Res. 18, 152–157.

Banner, F.T., Blow, W.H., 1959. The classification and stratigraphical distribution of the Globigerinacea. Palaeontology 21, 1–27.

Banner, F.T., Blow, W.H., 1960. The taxonomy, morphology and affinities of the genera included in the subfamily Hastigerininae. Micropaleontology 6, 19–31.

Banner, F.T., Blow, W.H., 1967. The origin, evolution and taxonomy of the foraminiferal genus *Pulleniatina* Cushman, 1927. Micropaleontology 13 (2), 133–162.

Barker, R.W., 1960. Taxonomic notes on the species figured by H. B. Brady in his report on the foraminifera dredged by H.M.S Challenger during the years 1873–1876. Society of Economic Paleontologist and Mineralogist, Special Publication No. 9, 238 p.

Batsch, A.I.G.C., 1791. Testaceorum arenulae marinae tabulae sex—(Sechs Kupfertafeln mit Conchylien des Seesandes, gezeichnet undgestochen). Jena, 6.

Baumann, K., 1993. Morphology of Quaternary *Coccolithus pelagicus*. Coccoliths from Northern North Atlantic and its Paleoceanographic significance. Proceedings of the Fifth International Nannoplankton Association Conference, Salamanca, 1993, pp. 11–21.

Be, A.W.H., 1977. An ecological, zoogeographic and taxonomic review of recent planktonic Foraminifera. Ramsay, A.T.S. (Ed.), Oceanic Micropaleontology, vol. 1, Academic Press, London, pp. 1–100.

Be, A.W.H., Hemleben, C., Anderson, O.R., Spindler, M., 1980. Pore structures in planktonic Foraminifera. J. Foramin. Res. 10, 2117–2128.

Beard, D.C., Weyl, P.K., 1973. Influence of texture on porosity and permeability of unconsolidated sand. AAPG Bull. 57, 349–369.

Becker, L.J., Dusenbury Jr., A.N., 1955. Mio-Oligocene Aquitanian Foraminifera from the Goajira Peninsula Columbia. Contrib. Cushman Foramin. Res., Special Publication No. 4, pp. 1–48.

Beckmann, J.P., 1957. *Chloguembelina* Leoblich and Tappan and related foraminifera from the Lower Tertiary of Trinidad, B.W.I. US Natl. Museum Bull. 215, 83–95.
Bender, H., 1989. Gehauseaufbau, Gehausegenese und Biologie agglutinierter Foraminiferen (Sarcodina:Textulariina). Jahrbuch Geologische Bundesanstalt 132 (2), 259–347.
Benkhelil, J., 1989. The origin and evolution of the Cretaceous Benue Trough, (Nigeria). J. Afr. Earth Sci. 8, 251–282.
Berggren, W.A., 1969. Paleogene biostratigraphy and planktonic Foraminifera of northern Europe. In: P., Bronnimann, Renz, H.H. (Eds.), Proceedings of the First International Conference on Planktonic Microfossils, vol. 1. E.J. Brill, Geneva, pp. 121–160.
Berggren, W.A., 1972. A Cenozoic time scale—some implications for regional geology and paleobiogeography. Lethaia 5, 195–215.
Berggren, W.A., 1979. Recent advances in Cenozoic in marine and continental planktonic Foraminiferal biostratigraphy geochronology and paleoclimatology of the last 15 million years in marine and continental stratigraphy. Paleogeogr. Paleoclimatol. 16 (1–2), 1–216.
Berggren, W.A., Amdurer, M., 1973. Late Paleogene (Oligocene) and Neogene planktonic foraminiferal biostratigraphy of the Atlantic Ocean (Lat. 30°N to Lat. 30°S). Rev. Ital. Paleontol. 79, 337–392.
Berggren, W.A., Aubert, J., 1975. Paleocene benthonic Foraminiferal biostratigraphy, paleobiogeography and paleoecology of Atlantic Tethyan regions: midway-type fauna. Palaeogeogr. Palaeoclimatol. Palaeobot. 18, 73–192.
Berggren, W.A., Kent, D.V., Flynn, J.J., van Couvering, J.J., 1985. Cenozoic geochronology. Am. Assoc. Petr Geol. Bull. 96, 1406–1418.
Berggren, W.A., Kent, D.V., Swisher, C.C., Aubry, M., 1998a. A Revised Cenozoic Geochronology and Chronostratigraphy. SEPM Special Publication No. 54, pp. 128–212.
Berggren, W.A., Lucas, S., Aubry, M.-P., 1998b. Late Paleocene–Early Eocene climatic and biotic evolution: an overview. In: Aubry, M.-P., Lucas, S., Berggren, W.A. (Eds.), Late Paleocene–Early Eocene Climatic and Biotic Events in the Marine and Terrestrial Records. Columbia University Press, New York, NY, pp. 1–17.
Berggren, W.A., Olsson, R.K., Reyment, R.A., 1967. Origin and evolutionary development of the foraminiferal genus *Pseudohastigerina* Banner and Blow, 1959. Micropaleontology 13, 265–288.
Berggren, W.A., van Couvering, J.A., 1974. The Late Neogene—Biostratigraphy, geochronology and paleoclimatology of the last 15 million years in marine and continental sequences. Paleogeogr. Paleoclimatol. Paleoecol 1 (1/2), 9–35, (special issue).
Berggren, W.A., Burckle, L.H., Cita, M.B., Cooke, H.B.S., Funnell, B.M., Gartner, S., Hays, J.D., Kennett, J.P., Opdyke, N.D., Pastouret, L., Shackleton, N.J., Takayanagi, Y., 1980. Towards a quaternary time scale. Q. Res. 13, 277–302.
Bermucez, P.J., 1937. Nuevas especies de foraminiferos del Eocene de Cuba. Memorias de la Sociedad Cubana de Historia Natural "Felipe Poey" 11 (2), 137–150.
Bermucez, P.J., 1949. Tertiary smaller Foraminifer of the Dominican Republic. Cushman Laboratory for Foraminiferal Research, Special Publication No. 28, p. 322.
Bermucez, P.J., 1961. Contribucion al estudio de las Globigerinidea de la region Caribe-Antillana (Paleacene–Reciente). Memoria de III Congreso Geologies Venezolano, Caracas, vol. 3. Bolatin de Geologia, Publicacion especial No. 3, 1960, 1.119–1.393.
Bermucez, P.J., Bolli, H.M., 1969. Consideraciones sobre los sedimentos del Mioceno Medio al Reciente de las costas central oriental de Venezuela. Boletin de Geologia, Venezuela 10, 137–223.
Bermucez, P.J., Seiglie, G.A., 1963. Estudio sistematico de los Foraminiferos del Golfo de Cariaco. Oriente Univ. Inst. Oceanogr. Bol. Cumana Venezuela 2 (2), 37–38.
Biolzi, M., Perch-Nielsen, K., 1982. *Helicosphaera truempyi*, a new Early Miocene calcareous nannofossil. Eclogae Geol. Helvet. 75, 171–175.
Black, M., 1967. New names for some coccolith taxa. Proceedings of the Geological Society, London, 1640, pp. 139–145.
Black, M., 1971. The systematics of coccoliths in relation to the paleontological record. In: Funnel, B.M., Riedel, W.R. (Eds.), The Micropaleontology of Oceans. Cambridge University Press, U.K., pp. 611–624.
Blow, W.H., 1956. Origin and evolution of the foraminiferal genus *Orbulina* d'Orbigny. Micropaleontology 2, 57–70.
Blow, W.H., 1959. Age, correlation, and biostratigraphy of the upper Tocuyo (San Lorenzo) and Poron Formations, Eastern Falcon, Venezuela. Bull. Am. Paleontol. 39, 67–251.
Blow, W.H., 1969a. Late Middle Eocene to recent planktonic foraminiferal biostigraphy. In: Bronnimann, P., Renz, H.H. (Eds.), Proceedings of the First International Conference on Planktonic Microfossils, vol. 1. E.J. Brill, Geneva, pp. 199–422.
Blow, W.H., 1969b. The Cenozoic Globigerinidavol. 1–3E.J. Brill, Leiden, 413 p.
Blow, W.H., 1979. Danian to Oligocene planktonic foraminiferal biostratigraphy. The Cainozoic Globigerinida. A Study of the Morphology, Taxonomy, Evolutionary Relationships and the Stratigraphical Distribution of Some Globigerinida (Mainly Globigerinacea). E.J. Brill, Leiden, pp. 753-1413.
Blow, W.H., Banner, F.T., 1962. The Mid-Tertiary (Upper Eocene to Aquitanian) Globigerinacea. In: Eames, F.E. et al., (Ed.), Fundamentals of Mid-Tertiary Stratigraphical Correlation. Cambridge University Press, New York, NY, pp. 61–151.
Blow, W.H., Banner, F.T., 1966. The morphology, taxonomy and biostratigraphy of *Globorotalia barisanensis* Le Roy, *Globorotalia fohsi* Cushman and Ellisor, and related taxa. Micropaleontology 12, 286–302.
Boboye, O.A., Adeleye, A.M., 2009. High resolution biostratigraphy of Early Pliocene–Late Miocene calcareous nannoplankton and Foraminifera, deep offshore, Niger Delta, Nigeria. Eur. J. Sci. Res. 34 (3), 308–325.
Boboye, O.A., Fowora, O., 2007. Biostratigraphy of calcareous nannofossils of well XH-1, deep offshore, Niger Delta, Nigeria. J. Min. Geol. 43 (1), 175–186.
Bolli, H.M., 1957a. The genera *Globigerina* and *Globorotalia* in the Paleocene—Lower Eocene Lizard Springs Formation of Trinidad, B.W.I. US Natl. Museum Bull. 215, 61–81.

Bolli, H.M., 1957b. Planktonic Foraminifera from the Oligocene—Miocene Cipero and Lengua Formation of Trinidad, etc. (mansion). In: Bolli, H.M., Saunders, J.B., Perch-Nielsen, K. (Eds.), Plankton Stratigraphy, vol. 1, Planktonic Foraminifera, Calcareous Nannofossils and Calpionellids. Cambridge University Press, Cambridge, U.K., pp. 155–262.
Bolli, H.M., 1957c. Planktonic Foraminifera from the Eocene Navet and San Fernando formations of Trinidad, B.W.I. US Natl. Museum Bull. 215, 155–172.
Bolli, H.M., 1966. Zonation of Cretaceous to Pliocene marine sediments based on planktonic Formainifera. Assoc. Venezolana Geol. Min. Petr. Bol. Informativo, 3–32.
Bolli, H.M., Beckmann, J.P., Saunders, J.B., 1994. Benthic Foraminifera Biostratigraphy of the South Caribbean Region. Cambridge University press, Cambridge, p. 258, fig. 64, 4–7.
Bolli, H.M., Bermudez, P.J., 1965. Zonation based on planktonic Foraminifera of Middle Miocene to Pliocene warm-water sediments. Assoc. Venezolana Geol. Min. Petr. Bol. Informativo 8 (5), 121–149.
Bolli, H.M., Cita, M.B., 1960. Globigerine e Globorotalie del Paleocene di Paderno d'Adda (Italia). Rivista Italiana di Paleontologiae Stratigrafia 66, 361–408.
Bolli, H.M., Loeblich, Jr., A.R., Tappan, H., 1957. Planktonic foraminiferal families Hantkeninidae, Orbulinidae, Globorotaliidae and Globotruncanidae. US Natl. Museum Bull. 215, 3–50.
Bolli, H.M., Saunders, J.B., 1985. Oligocene to Holocene low latitutude plankton stratigraphy. In: Bolli, H.M. et al., (Ed.), Plankton Stratigraphy. Cambridge University Press, Cambridge, pp. 155–262.
Boltovskoy, E., Guissani, G., Watanabe, S., Wright, R., 1980. Atlas of Benthic Foraminifera of the Southwest Atlantic. Dr. W. Junk b.v. Publishers, The Hague.
Boudreaux, J.E., Hay, W.W., 1969. Calcareous nannoplankton and biostratigraphy of the Late Pliocene–Pleistocene—recent sediments in the Submarex Cores. Rev. Espanola Micropaleontol. 1, 249–292.
Bowen, B.E., Hall, D.J., Rossen, R.N., Shaffer, B.L., 1994. Sequence stratigraphic and structural framework, southeast Niger Delta shelf. NAPE Bull. 9, 51–58.
Bown, P.R., 1999. Calcareous Nannofossil Biostratigraphy. British Micropalaeontological Society Publications SeriesKluwer Academic Publishers, London, U.K., 314 p.
Brady, H.B., 1871. On the rhizopodal fauna of the deep sea. Proceedings of the Royal Society of London, vol. 18, 1868, No. 114, pp. 59–62.
Brady, H.B., 1877. Supplementary note on the Foraminifera of the chalk (?) of the New Britain group. Geol. Mag. 4, 534–536.
Brady, H.B., 1879. Notes on some of the reticularian Rhizopoda of the 'Challenger' Expedition: Part I, on new or little-known arenaceous types. Q. J. Microsc. Sci. 19, 20–63.
Brady, H.B., 1882. Report on the Foraminifera. Proc. R. Soc. Edinb. 11, 708–717.
Brady, H.B., 1884. Report on the foraminifera dredged by HMS Challenger during the years 1873—1876. Report of Scientific Results of the Exploration Voyage of HMS Challenger, Zoology, vol. 9, pp. 1–814.
Brady, H.B., 1937. Cushman Laboratory Foraminifera Research, Special publication No. 8, p. 53.
Brady, H.B., 1954. Bull. Mus. Comp. Zool., Harvard, vol. 111, No. 10, p. 487.
Bramlette, M.N., Martini, E., 1964. The great change in calcareous nannoplankton fossils between the Maestrichtian and Danian. Micropalaeontology 10, 291–322.
Bramlette, M.N., Riedel, W.R., 1954. Stratigraphic value of discoasters and some other microfossils related to recent Coccolithophores. J. Paleontol. 28, 385–403.
Bramlette, M.N., Sullivan, F.R., 1961. Coccolithophorids and related nannoplankton of Early Tertiary in California. Micropaleontology 7, 129–188.
Bramlette, M.N., Wilcoxon, J.A., 1967a. Discoaster druggi nom. nov. pro Discoaster extensus Bramlette and Wilcoxon, 1967, non Hay, 1967. Tulane Stud. Geol. Paleontol. 5, 220.
Bramlette, M.N., Wilcoxon, J.A., 1967b. Middle Tertiary calcareous nannoplankton of the Cipero section, Trinidad, W.I. Tulane Stud. Geol. Paleontol. 5, 93–131.
Bronnimann, P., 1951. *Globigerinita naparimaensis* n. gen, n. sp. from the Miocene of Trinidad, B.W.I. Contrib. Cushman Found. Foramin. Res. 2, 16–18.
Bronnimann, P., 1953. Arenaceous Foraminifera from the Oligocene–Miocene of Trinidad. Contrib. Cushman Found. Foramin. Res. 4, 96.
Bronnimann, P., 1954. Appendix: Descriptions of new species. In: Todd, R., Cloud, P.E., Low, D., Schmidt, R.G. (Eds.), Probable Occurrence of Oligocene in Japan. Am. J. Sci. 252, 680.
Bronnimann, P., Bermudez, P.J., 1953. *Truncorotaloides*, a new foraminiferal genus from the Eocene of Trinidad, B.W.I. J. Paleontol. 27, 817–820.
Bronnimann, P., Resig, J., 1971. A Neogene Globigerinacean Biochronological Time-Scale of the Southwestern Pacific. Initial Reports of the Deep Sea Drilling Project, vol. 7, pp. 1235–1469.
Bronnimann, P., Stradner, H., 1960. Die Foraminiferen-und Discoasteriden-Zonen von kuba und ihre interkontinentale Korrelation. Erdoel-Zeitschrift fur Bohr-un Foerdertechnik 76, 364–369.
Brotzen, F., 1936. Foraminifera aus dem schwedischem untersen senon von Eriksdal in Schonen. Sver. Geol. Undersok 30 (3), 1–206, Ser. C (396).
Brotzen, F., 1940. Flintrannans och trindelrannans Geologi. Sver. Geol. Undersok 34 (2), 1–33, Ser. C (435).
Brotzen, F., 1942. The Swedish Paleocene and its foraminiferal fauna. Sver. Geol. Undersok 36 (8), 1–60, Ser. C (451).
Brotzen, F., 1948. The Swedish Paleocene and its foraminiferal fauna. Sver. Geol. Undersok. 42 (2), 1–140, C (493).
Brotzen, F., Pozaryska, 1961. Foraminiferes du Paleocene et de l'Eocene inferieur en Pologne septentrionale, remarques paleogeographiques. Rev. Micropaleontol. 4, 155–166.

Brown, L.F., Fisher, W.L., 1977. Seismic stratigraphic interpretation of depositional systems: examples from Brazilian rift and pull-apart basins. In: Payton, C.E (Ed.), Seismic Stratigraphy—Applications to Hydrocarbon Exploration, AAPG Memoir, vol. 26, pp. 213–248.
Brun, L., Chierici, M.A., Meijer, M., 1984. Evolution and morphological variations of the principal species of Bolivintidae in the Tertiary of the Gulf of Guinea. Geologie Méditerranée Tome XI, No. 1, pp. 13–57.
Brunnich, M.T., 1772. Brunnich Zoologie Fundamentals. Grunde 1, Dyrelorren, Hafniae at Lipsiae.
Bubik, M., 1995. Cretaceous to Paleogene agglutinated foraminifera of the Bile Karpaty unit (West Carpathian, Czech Republic). In: Kaminski, M.A., Geroch, S., Gasinski, M.A. (Eds.), Proceedings of the Fourth International Workshop on Agglutinated Foraminifera, pp. 71–116.
Bukry, D., 1971. Cenozoic calcareous nannofossils from the Pacific Ocean. San Diego Soc. Nat. Hist. Trans. 16, 303–327.
Bukry, D., 1973a. Low-Latitude Coccolith Biostratigraphic Zonation. Initial Reports of the Deep Sea Drilling Project, vol. 15, pp. 685–703.
Bukry, D., 1973b. Coccolith Stratigraphy, Eastern Equatorial Pacific, Leg 16, Deep Sea Drilling Project. Initial Reports of the Deep Sea Drilling Project, vol. 20, pp. 307–317.
Bukry, D., 1979. Neogene Coccolith Stratigraphy, Mid-Atlantic Ridge. Initial Reports of the Deep Sea Drilling Project, vol. 45, pp. 307–317.
Bukry, D., Bramlette, M.N., 1968. Stratigraphic significance of two genera of Tertiary calcareous nannofossils. Tulane Stud. Geol. Paleontol. 6, 149–155.
Bukry, D., Bramlette, M.N., 1969. Some new and stratigraphically useful calcareous nannofossils of the Cenozoic. Tulane Stud. Geol. Paleontol. 7, 131–142.
Bukry, D., Percival, S.F., 1971. New Tertiary calcareous nannofossils. Tulane Stud. Geol. Paleontol. 8, 123–146.
Burke, K., 1972. Longshore drift, submarine canyons and submarine fans in development of Niger Delta. Am. Assoc. Petr. Geol. Bull. 56, 1975–1983.
Burke, K., 1996. The African Plate. South Afr. J. Geol. 99, 341–409.
Burke, K., Dessauvagie, T.F.G., Whiteman, A.J., 1972. Geological history of the Benue valley and adjacent areas. In: Dessauvagie, T.F.G., Whiteman, A.J. (Eds.), African Geology. Ibadan University Press, Ibadan, pp. 187–205.
Bustin, R.M., 1988. Sedimntology and characteristics of dispersed organic matter in Tertiary Niger Delta: origin of source rocks in a deltaic environment. AAPG Bull. 72, 277–298.
Bykova, N.K., et al., 1959. Otryad Rotaliida. In: Rauzer-chernousova, D.M., Fursenko, A.V. (Eds.), Osnovy paleontologii. Oblichaya chast, Prosteishie. Akad. Nauk USSR, Moscow, pp. 265–307.
Carbonnel, G., de Klasz, I., de Klasz, S., Hovath, M., Lang, J., Oyede, M., 1996. Microfaunes et milieux de depots des niveaux oligocenes surmontant la 'discordance oligocene' sur la partite terrestre du bassin sedimentaire cotier du Benin. Actes Colloques Angers 1994, pp. 235–273.
Carpenter, W.B., Parker, W.K., Jones, T.R., 1862. Introduction to the study of the Foraminifera. Ray Society, London.
Chapman, F., Parr, W.K., 1937. Aunstralasian Antarctic Exped. 1911–1914, Sci. Repts. vol. 1, p. 153.
Chapman, F., Parr, W.J., Collins, A.C., 1934. Tertiary Foraminifera of Victoria, Australia—The Balcombian deposits of Port Phillip, Part 3. Zool. J. Linnean Soc. 38, 553–577.
Christopher, H., Kaminski, M.A., Wolfgang, K., Scott, D.B., 2012. Paleoecology, Biostratigraphy, Paleoceanography and Taxonomy of Agglutinated Foraminifera. Kluwer Academic Publishers, Springer Netherlands, p. 335.
Cifelli, R., Scott, G., 1986. Stratigraphic record of the Neogene globorotaliid radiation (planktonic Foraminifera). Smithsonian Contrib. Paleobiol. 58, 101.
Cole, W.J., 1927. A foraminiferal fauna from the Guayabal Formation in Mexico. Bull. Am. Paleontol. 14, 1–46.
Cordey, W.G., Berggren, W.A., Olsson, R.K., 1970. Phylogenetic trends in the planktonic foraminiferal genus *Pseudohastigerina* Banner and Blow. Micropaleontology 16, 235–242.
Core Laboratories, 1996. Reservoir properties of Tertiary Niger Delta sediments. Non-proprietary Report, Aberdeen, 7 volumes.
Costa, O.G., 1856. Paleontologia del regno di Napoli, Parte II—Academia Pontaniana, Atti, Napoli, 7, 2, 113-378.
Crescenti, U., 1966. Sulla biostratigrafia del Miocene affiorante al confine marchigiano 'abruzzese. Geol. Romana 5, 1–54.
Cushman, J.A., 1913. A monograph of foraminifera of the North Pacific Ocean: Part 3, Lagenidae. US Natl. Museum Bull. 71 (3), 125.
Cushman, J.A., 1918a. Some Pliocene and Miocene Foraminifera of the Coastal Plain of the U.S. US Geol. Surv. Bull. 676, 100.
Cushman, J.A., 1918b. The smaller fossil Foraminifera of the Panama Canal zone. US Natl. Museum Bull., Washington, DC, 103, 53.
Cushman, J.A., 1920a. Lower Miocene foraminifera of Florida. US Geol. Surv. Prof. Paper, Washington, United States No. 128B.
Cushman, J.A., 1920b. The Foraminifera of the Atlantic Ocean, Part 2. Lituolidae. US Natl. Museum Bull. 104 (2), 1–111.
Cushman, J.A., 1922. The Foraminifera of the Atlantic Ocean, Part 3. Textulariidae. US Natl. Museum Bull. 104, 1–149.
Cushman, J.A., 1923. The Foraminifera of the Atlantic Ocean. US Natl. Museum Bull. 104 (Pt. 4), 1–228.
Cushman, J.A., 1924. The Foraminifera of the Atlantic Ocean, Part 5—Chilostomellidae and Globigerinidae. US Natl. Museum Bull. 104 (Pt. 5), 1–55.
Cushman, J.A., 1925a. Some new Foraminifera from the Velasco Shale of Mexico. Contrib. Cushman Lab. Foramin. Res. 1, 18–23.
Cushman, J.A., 1925b. Some Textulariidae from the Miocene of California. Contrib. Cushman Lab. Foramin. Res. 1 (pt. 3).
Cushman, J.A., 1926a. The genus *Chilostomella* and related genera. Contrib. Cushman Lab. Foramin. Res. 1 (Pt. 4), 73–80.
Cushman, J.A., 1926b. Some fossil Bolivinas from Mexico. Contrib. Cushman Lab. Foramin. Res. 1 (Pt. 4), 81.

Cushman, J.A., 1926c. Some Pliocene Bolivinas from California. Contrib. Cushman Lab. Foramin. Res. 2 (29), 41.

Cushman, J.A., 1926d. Texas Jackson Foraminifera. Am. Assoc. Petr. Geol. Bull. 10 (2), 154–189.

Cushman, J.A., 1927. An outline of a reclassification of the Foraminifera. Contrib. Cushman Lab. Foramin. Res. 3 (Pt. 1), 1–105.

Cushman, J.A., 1929. A Late Tertiary fauna of Venezuela and other related regions. Contrib. Cushman Lab. Foramin. Res. 5 (Pt. 4), 77–101.

Cushman, J.A., 1930a. The Foraminifera of the Choctawhatchee Formation of Florida. Florida State Geol. Surv. Bull. 4, 5–63.

Cushman, J.A., 1930b. Note on some Foraminifera from Venezuela and Colombia. Contrib. Cushman Lab. Foramin. Res. 6 (Pt. 3), 64–69.

Cushman, J.A., 1931. New late Tertiary Foraminifera from Vitilevu, Fiji. Contrib. Cushman Lab. Foramin. Res. 7 (Pt. 2), 25–32.

Cushman, J.A., 1932. An Eocene foraminiferal fauna of Wilcox age from Alabama. Contrib. Cushman Lab. Foramin. Res. 8, 51–72.

Cushman, J.A., 1933a. New Foraminifera from the Upper Jackson Eocene of the Southeastern Coastal Plain region of the United States. Contrib. Cushman Lab. Foramin. Res. 9 (Pt. 1), 1–24.

Cushman, J.A., 1933b. An illustrated key to the genera of the Foraminifera. Cushman Laboratory for Foraminiferal Research, Special Publication No. 5, pp. 1–26.

Cushman, J.A., 1935. A new genus of Foraminifera from the Miocene of Venezuela. Contrib. Cushman Lab. Foramin. Res. 6 (Pt. 3), 64–69.

Cushman, J.A., 1936a. Some new species of *Nonion*. Contrib. Cushman Lab. Foramin. Res. 12, 66.

Cushman, J.A., 1936b. New genera and species of the families Verneuilinidae and Valvulinidae and of the subfamily Virgulininae. Cushman Laboratory for Foramininferal Research, Special Publication No. 6, pp. 1–71.

Cushman, J.A., 1936c. Middle Eocene Foraminifera from the Llajas Formation, Ventura County, California. J. Paleontol. 10 (6), 497–517.

Cushman, J.A., 1937a. Further new species of Foraminifera from the Eocene of Cuba. Contrib. Cushman Lab. Foramin. Res. 18, 1–20.

Cushman, J.A., 1937b. A monograph of the foraminiferal family Valvulinidae. Cushman Laboratory for Foraminiferal Research, Special Publication No. 8, pp. 1–210.

Cushman, J.A., 1937c. A monograph of the foraminiferal family Valvulinidae. Cushman Laboratory for Foraminiferal Research, Special Publication No. 8, p. 98.

Cushman, J.A., 1939. A Monograph of the foraminiferal family Nonionidae. United States Geological Survey, Professional Paper, No. 191, pp. 1–100.

Cushman, J.A., 1940. Midway Foraminifera from Alabama. Contrib. Cushman Lab. Foramin. Res. 16, 51–73.

Cushman, J.A., 1947a. The foraminiferal fauna of the Oligocene Ste. Croix formation of Trinidad, B.W.I. Cushman Laboratory for Foraminiferal Research, Special Publication No. 22, p. 46.

Cushman, J.A., 1947b. New species and varieties of Foraminifera from the southeastern Coast of the United States. Contrib. Cushman Lab. Foramin. Res. 23, 1–90.

Cushman, J.A., 1947c. Some new Foraminifera from the Paleocene of the Southern United States. Contrib. Cushman Lab. Foramin. Res. 23 (4), 81–86.

Cushman, J.A., 1951. Paleocene Foraminifera of the Gulf Coastal Region of the United States and adjacent area. US Geol. Surv. Prof. Paper 232, 1–75.

Cushman, J.A., Applin, E.R., 1946. Some Foraminifera of Woodbine age from Texas Mississippi Alabama and Georgia. Contr. Cushman Lab. Foramin. Res. 22 (2), 27–38.

Cushman, J.A., Bermudez, P.J., 1936. Additional new species of Foraminifera and a new genus from the Eocene of Cuba. Contrib. Cushman Lab. Foramin. Res. 12 (3), 55–63.

Cushman, J.A., Bermudez, P.J., 1948. Some Paleocene Foraminifera from the Madruga Formation of Cuba. Cushman Lab. Foramin. Res. Contrib. Sharon Mass. 24, 63–75, pt. 3, pls. 11–12.

Cushman, J.A., Bermudez, P.J., 1949. Some Cuban species of *Globorotalia*. Contrib. Cushman Lab. Foramin. Res. 3 (Pt. 1), 1–105.

Cushman, J.A., Bronnimann, P.J., 1948. Some new genera and species of Foraminifera from brackish water of Trinidad. Contrib. Cushman Lab. Foramin. Res. 24 (1), 15–21.

Cushman, J.A., Deaderisk, W.H., 1942. Cretaceous Foraminifera from the Brownstown Marl of Arkanas. Contrib. Cushman Lab. Foramin. Res. 18 (3), 50–66.

Cushman, J.A., Edwards, P.G., 1938. Notes on the Oligocene species of *Uvigerina* and *Angulogerina*. Contrib. Cushman Lab. Foramin. Res. 14 (Pt. 4), 74–89.

Cushman, J.A., Ellisor, A., 1932. Additional new Eocene Foraminifera. Contrib. Cushman Lab. Foramin. Res. 8, 40–43.

Cushman, J.A., Ellisor, A.C., 1939. New species of Foraminifera from the Oligocene and Miocene. Contrib. Cushman Lab. Foramin. Res. 15, 1–14.

Cushman, J.A., Ellisor, A., 1945. The Foraminifera fauna of the Anahuac Formation. J. Paleontol. 19 (6), 545–572.

Cushman, J.A., Harris, R.W., 1927. Some notes on the genus *Ceratobulimina*. Contrib. Cushman Lab. Foramin. Res. 3 (4), 171–179.

Cushman, J.A., Hedberg, H.D., 1941. Upper Cretaceous foraminifera from Santander del Norte Colombia, S.A. Contrib. Cushman Lab. Foramin. Res. 17 (4), 79–100.

Cushman, J.A., Jarvis, P.W., 1928. Cretaceous Foraminifera from Trinidad. Contrib. Cushman Lab. Foramin. Res. 4, 85–103.

Cushman, J.A., Jarvis, P.W., 1929a. A new foraminifera from Trinidad. Contrib. Cushman Lab. Foramin. Res. 5 (Pt. 1), 6–7.

Cushman, J.A., Jarvis, P.W., 1929b. The new Foraminifera from the Miocene, Bowden Marl of Jamaica. Contrib. Cushman Lab. Foramin. Res. 12, 3–5.

Cushman, J.A., Jarvis, P.W., 1930. Miocene Foraminifera from Buff Bay, Jamaica. J. Paleontol. 4, 353–368.

Cushman, J.A., Jarvis, P.W., 1932. Upper Cretaceous Foraminifera from Trinidad. US Natl. Museum Proc. 80 (14), 1–66.
Cushman, J.A., Jarvis, P.W., 1936. Three new Foraminifera from the Miocene Bowden Marl of Jamaica. Contrib. Cushman Lab. Foramin. Res. 12, 3–5.
Cushman, J.A., Kleinpell, R.M., 1934. New and unrecorded foraminifera from the California Miocene. Contrib. Cushman Lab. Foramin. Res. 10 (1), 1–23.
Cushman, J.A., Laiming, B., 1931. Miocene Foraminifera from Los Sauses Creek, Ventura County, California. J. Paleontol. 5, 79–120.
Cushman, J.A., LeRoy, L.W., 1938. A microfauna from the Vagueros Formation Lower Miocene, Simi Valley, Ventura County, California. J. Paleontol. 2, 117–126.
Cushman, J.A., Mcculloch, I., 1948. The species of *Bulimina* and related genera in the collections of the Allan Hancock Foundation. Allan Hancock Pacific Exped. 6 (5), 231–294.
Cushman, J.A., Parker, F.L., 1931. Recent Foraminifera from the Atlantic Coast of South America. Proc. US Natl. Museum 80, 1–24.
Cushman, J.A., Parker, F.L., 1936. Notes on some Cretaceous species of *Bulimina* and *Neobulimina*. Contrib. Cushman Lab. Foramin. Res. 12 (1), 1–10.
Cushman, J.A., Parker, F.L., 1937. Notes on some Oligocene species of *Bulimina* and *Buliminella*. Contrib. Cushman Lab. Foramin. Res. 13 (1), 36–40.
Cushman, J.A., Parker, F.L., 1938. *Bulimina* and related Foraminifera genera. Contrib. Cushman Lab. Foramin. Res. 13 (1), 61.
Cushman, J.A., Parker, F.L., 1947. *Bulimina* and related foraminiferal genera. US Geol. Surv. Prof. Paper 189-D, 101–119.
Cushman, J.A., Ponton, G.M., 1932. The Foraminifera of the Upper, Middle and part of the Lower Miocene of Florida. Florida State Geol. Surv. Bull. 8, 1–147.
Cushman, J.A., Renz, H.H., 1941. New Oligocene–Miocene Foraminifera from Venezuela. Contrib. Cushman Lab. Foramin. Res. 17 (Pt. 1), 1–27.
Cushman, J.A., Renz, H.H., 1942. Eocene Midway Foraminifera from Soldado Rock, Trinidad. Contrib. Cushman Lab. Foramin. Res. 18, 1–20.
Cushman, J.A., Renz, H.H., 1948. Eocene foraminifera of the Navet and Hospital Hill formations of Trinidad, B.W.I. Cushman Laboratory for Foraminiferal Research, Special Publication No. 24, pp. 1–42.
Cushman, J.A., Stainforth, R.M., 1945. The Foraminifera of the Cipero Marl Formation of Trinidad, British West Indies. Cushman Laboratory for Foraminiferal Research, Special Publication No. 14, pp. 1–75.
Cushman, J.A., Stainforth, R.M., 1947. A new genus and some new species of foraminifera from the Upper Eocene of Ecuador. Contrib. Cushman Lab. Foramin. Res. 23 (4), 77–80.
Cushman, J.A., Stainforth, R.M., 1951. Tertiary Foraminifera of coastal Ecuador: Part 1—Eocene. Contrib. Cushman Lab. Foramin. Res. 25 (2), 33.
Cushman, J.A., Stewart, R.E., Stewart, K.C., 1947. Upper Coaledo (Upper Eocene) Foraminifera from Yokan Point, Coos County, Oregon. Oregon Dep. Geol. Min. Ind. Bull. 36, 59–68.
Cushman, J.A., Todd, R., 1945. Miocene Foraminifera from Buff Bay, Jamaica. Cushman Laboratory for Foraminiferal Research, Special Publication No. 15, pp. 1–73.
Czjzek, J., 1848. Bettrag zur Kenntniss der fossili foraminiferen des Wiener. Natur Abh Wien Osteireich Bd. 2, Abth. 1.
d'Orbigny, A.D., 1826. Tableau methodique de la classe des cephalopodes. Ann. Sci. Nat. 7, 245–314.
d'Orbigny, A.D., 1839. Foraminiferes des Iles Canaries, In: Barker-Webb, P., Berthelot, S. (Eds.), Historie Naturelle des Canaries. Paris, vol. 2, Part 2, pp. 119–146.
d'Orbigny, A.D., 1840. Memoire sur les foraminifers de la craie blanche du basin de Paris. Soc. Geol. France Mem. 4 (Pt. 1), 1–51.
d'Orbigny, A.D., 1846. Foraminiferes fossiles du Bassin Tertiare de Vienne (Autriche). Gide et Comp. Paris, 1–312 p.
d'Orbigny, A.D., 1852. Prodrome de Paleontologie stratigraphique universelle des animaux mollusques et rayonnes Paris, France, v. Masson. Ann. Sci. Nat. 3, 1–194.
Daniel, A.R., Jose, E.A., Carlos, H.C., Andres, L.C., Carlos, E.M., Claudia, C., Carlos, J., 2007. Eocene–Pliocene planktonic foraminifera biostratigraphy from the continental margin of the southwest Caribbean. Stratigraphy 4 (4), 261–311.
Davidzon, R.M., 1976. Novyy paleogenovy rod planktonnykh foraminifer, VN6NI. Tadzhikskoe Otdelenie 183, 197–198.
de Blainville, H.M., 1825. Dictionnaire des Sciences naturelles par plusieurs professeurs du jardin du Roi et des principales Ecoles de Paris.
de Klasz, I., Kroon, D., Van Hinte, J., 1989. Notes on the foraminiferal genera *Laterostomella* De Klasz and Rerot and *Streptochilus* Bronnimann and Resig. Micropaleontology 8, 215–226.
de Klasz, I., Le Calvez, Y., Rerat, D., 1964a. Deux nouuveaux Buliminidae (foraminifers) du Paleogene de l'Afrique occidentale. C.R. Somm. Soc. Geol. France Fasc. 5, 208.
de Klasz, I., Le Calvez, Y., Rerat, D., 1964b. Deux importantes especes de foraminifers du Miocene inferieur de l'Afrique occidentale. C.R. Somm. Soc. Geol. France Fasc. 5, 194.
de Klasz, I., Le Calvez, Y., Rerat, D., 1964c. Deux Nouveaux genres de Foraminiferes du Gabon (Afrique equatoriale). C.R. Somm. Soc. Geol. France Fasc. 6, 236–237.
de Klasz, I., Le Calvez, Y., Rerat, D., 1970. Nouveaux foraminifers du basin sedimentarire du Gabon (Afrique equatoriale). Proceedings of Third African Micropaleontology Colloquium, Cairo, pp. 269–286.
de Klasz, I., Magne, J., Rerat, D., 1963. Quelques forme nouvelles de Buliminidae caracteristiques du Cretace superleur du Gabon (Afrique equatoriale). Rev. Micropaleontol. 6 (3), 145–152.

de Klasz, I., Rerat, D., 1962. Quelques nouveaux foraminiferes du Cretace et du Tertiare du Gabon (Afrique Equatoriale). Rev. Micropaleontol. 4, 175–189.

de Klasz, J., Rerat, D., 1963. *Mandjina* Nov. Gen. Nouveau Foraminifer de l'Eocene superieur du Gabon (Afrique equatoriale). C.R. Som. Séance Soc. Geol. France Fasc. 4, 115–116.

de Stefani, T., 1952. Sualcune manifestazioni di idrocarburi in provincia de Palermo e descrizione di formaminiferi nuovi. Plinia 3, 1–12.

Debenay, J.-P., 1990. Recent foraminiferal assemblages and their distribution relative to environmental stress in the paralic environments of West Africa (Cape Timiris to Ebrie Lagoon). J. Foramin. Res. 20, 267–282.

Debenay, J.-P., Basov, I., 1993. Distribution of recent benthic foraminifera on the West African shelf and slope. A synthesis. Rev. Paleobiol. 12, 265–300.

Debenay, J.-P., Beck-Eichler, B., Fernandez-Gonzalez, M., Mathieu, P., Bonetti, C., Duleba, W., 1996. Les foraminiferes paraliques des cotes d' Afrique et d' Amerique du sud, de part et d' autre de l'ocean l'Atlantique: comparison—discussion. Actes Colloques Angers 1994, 463–471.

Debenay, J.-P., Redois, F., 1997. Recent Foraminifera of the northern continental shelf of Senegal. Rev. Micropaleontol. 40, 15–38.

Deflandre, G., 1942. Coccolithophorides fossils d'Oranie. Genres *Scyphosphaera* Lohmann et *Thoracosphaera* Ostenfeld. Societe d'Histoire Naturelle de Toulouse Bulletin 77, 125–137.

Deflandre, G., 1947. *Calciodinellum* nov. gen. Premier représentant d'une famille nouvelle de Dinoflagellés fossiles á theque calcaire. Comptes Rendus (Hebdomadaires des Séances) de l'Académié Sciences 224, 1781–1782.

Deflandre, G., 1952. Class des Coccolithophoridés (Coccolithophoridea lohmann, 1902). In: Grassé, P.P. (Ed.), Traité de Zoologie. Anatomie, Systematique, Biologie, 1, Part 1, Phylogenie. Protozaires: Generalites, Flagellés. Masson and Cie, Paris, pp. 439–470.

Deflandre, G., 1959. Sur les nannofossils calcaires et leur systematique. Rev. Micropaleontol. 2, 127–152.

Deflandre, G., Fert, C., 1954. Observations surles Coccolithophoridés actuels et fossils en microscopie ordinaire et électronique. Ann. Paleontol. 40, 115–176.

DeFrance, M.J.L., 1824. In: de Blainville, H.M.D. (Ed.), Dictionnaire des Sciences Naturelles, vol. 32, Mollus-Morf. F.G. Levrault, Paris.

Dessauvagie, T.F.G., 1972. Biostratigraphy of the Odukpani Formation (Cretaceous) Type Section, Nigeria. In: Dessauvagie, T.F.G., Whiteman, A.J. (Eds.), African Geology. Ibadan University Press, Ibadan, pp. 207–218.

Dessauvagie, T.F.G., 1975. Explanatory note on the 1:1,000,000 Geological Map of Nigeria. J. Min. Geol. 9, 128.

Dmitrenko, O., 1993. Evolution of calcareous nannoflora in Cenozoic Atlantic and Indian Oceans. In: Young, J.R., Bown, P.R. (Eds.), Fifth International Nannoplankton Association Conference, Salamanca.

Doust, H., Omatsola, M.E., 1989. Niger Delta. In: Edwards, J.D., Santogrossi, P.A. (Eds.), Divergent/Passive Margin Basins. AAPG Memoir, vol. 48, pp. 201–238.

Du Preez, J.W., 1947. Further observations on the lignites of Onitsha Province. Ann. Rep. Geol. Surv. Nigeria 1946, 25–26.

Dublin-Green, C.O., 1996. Seasonal distribution of benthic foraminifera in the Bonny Estuary, Niger Delta. Actes Colloques Angers 1994, 175–189.

Durand, J., 1995. High resolution sequence stratigraphy (genetic stratigraphy) in reservoir sedimentology: examples from the Niger Delta. NAPE Bull. 10, 65–73.

Eames, F.E., et al., 1962. Fundamentals of Mid-Tertiary Stratigraphical Correlation. Cambridge University Press, Cambridge, 163 p.

Edwards, A.R., 1973. Key species of New Zealand calcareous nannofossils. NZ J. Geol. Geophys. 16, 68–89.

Ehinola, O.A., Ojo, E.A., Fadiya, S.L., 2009. Biozonation and correlation of BDX-1 and BDX-2 wells of deep offshore Niger Delta using calcareous nannofossils. NAPE Bull. 21 (1), 38–52.

Ehrenberg, G.C., 1838. Uber dem blossen Auge unsichtbare Kalkthierchen und Kieselthierchen als Hauptbestandheile der Kreidegebirge. Bericht uber die Zur Bekanntmachung geeigneten Verhanlungen del koniglichen Preussischen Akademie der Wissenschaften zu Berlin, pp. 192–200.

Ehrenberg, G.C., 1839. Uber die Bildung der Kreidefelsen und des kreideinergels durch unsichtbare organismen. Physikalische Abhandlungen der koniglichen A kademie der Wissenschaften zu Berlin, 1838 (1840: separate 1839), pp. 59–147.

Eimer, G.H.T., Ficckert, C., 1899. Die Artibildung und Verwandtschaft bei den Foraminiferen. Entwurf einer naturlichen Einthilung derselben. Z. Wiss. Zool. 65, 599–708.

Evamy, B.D., Haremboure, J., Kamerling, P., Knaap, W.A., Molloy, F.A., Rowlands, P.H., 1978. Hydrocarbon habitat of Tertiary Niger Delta. AAPG Bull. 62, 1–39.

Fadiya, S.L., 2012. Two new short-ranged *Calcidiscus* species from the offshore marine Neogene Niger Delta sequences. Micropaleontol. J. 58 (6), 539–542.

Fadiya, S.L., 2014. Impact of wellsite biostratigraphy on exploration drilling in the deepwater offshore Nigeria. J. Afr. Earth Sci. 100, 60–69.

Fadiya, S.L., Salami, M.B., 2012. Middle Miocene carbonate crash in the Niger Delta: evidence from calcareous nannofossils. J. Nannoplankton Res. 32 (2), 59–70.

Fahmy, S., 1975. Contribution to Eocene stratigraphy and Micropaleontology in the Nile Valley, Egypt. Proc. Vth Coll. Afr. Microp. Coll., Addis Ababa, 1972. Rev. Espan. Microp. Series 7, No. 3, pp. 293–334.

Fairchild, W.W., Wesendunk, P.R., Weaver, D.W., 1969. Eocene and Oligocene Foraminifera from the Santa Cruz Mountains, California. Geol. Sci. 81, 1–144.

Faulkner, J.S., de Klasz, I., Rerat, D., 1963. *Megastomella* nov. gen. Nouveau foraminifere de l'Afrique occidentale. Rev. Micropaleontol. 6 (1), 19–22.

Fayose, E.A., 1970. Stratigraphical paleontology of Afowo-1 well, Southwestern Nigeria. J. Min. Geol. 5 (1 and 2), 1–98.
Fayose, E.A., 1976. Some species of smaller benthonic foraminifera off the Ivory Coast, West Africa. Nigerian J. Sci. 10 (1 and 2), 217–226.
Fichtel, L.V., Moll, J.P.C.V., 1803. Testacea microscopica aliaque minuta ex generibus, Argonauta et Nautilus Adnaturam delineata et descripta; Milroskopische und Andere Kleine Schalthiere aus den Geschlectern Argonaute and Schiffer, nach der Natur gezeichnet und Beschrieben, Wien, Camesina, pp. 1–124.
Finlay, H.J., 1939. New Zealand Foraminifera: key species in stratigraphy—No. 2. Trans. R. Soc. NZ 69, 89–128.
Finlay, H.J., 1940. New Zealand Foraminifera: key species in stratigraphy—No. 4. Trans. R. Soc. NZ 69, 448–472.
Finlay, H.J., 1947. New Zealand Foraminifera: key species in stratigraphy—No. 5. NZ J. Sci. Technol. 28, 259–292.
Fleisher, R.L., 1974. Cenozoic Planktonic Foraminifera and Biostratigraphy, Arabian Sea Deep, Sea Drilling Project, Leg. 23A. Initial Reports of the Deep Sea Drilling Project, vol. 23, pp. 1001–1072.
Fornaciari, E., di Sterfano, A., Rio, D., Negri, A., 1996. Middle Miocene quantitative calcareous nannofossil biostratigraphy in the Mediterranean region. Micropaleontology 42, 38–64.
Fornasini, C., 1883. Note preliminare sui foraminiferi della marna Pliocenica del pouticallo di savana nel Holoehasa. Soc. Geol. Ital. 2, 1–183.
Fornasini, C., 1888. Tavola Paleo-protis tografica. Soc. Geol. Ital. 7, 1–48.
Fornasini, C., 1901. Le Bulimine e le Cassiduline fossili d'Italia. Soc. Geol. Ital. 20, 159–214.
Frankl, E.J., Cordry, E.A., 1967. The Niger Delta oil province—recent developments onshore and offshore. Seventh World Petroleum Congress, Mexico City, vol. 2, pp. 135–209.
Franzenau, A., 1884. Heterolepa egy uj genus a Foraminiferak rendjeben: Termeszetrajzi Furzetek Budapest, v. 8, pp. 181–184, pp. 214–217.
French, P., 2007. Deltaic and estuarine environments. In: Perry, C., Taylor, K. (Eds.), Environmental Sedimentology. Blackwell Publishing, Oxford, pp. 223–262.
Frenzel, P., 2000. The benthic Foraminifera of the Rugen Chalk Lower Maastrichtian, Northeastern Germany, Neve Palaontolische Abhandlungen, 3: Dresden 361 p, plate 14, fig. 1.
Gallagher, L.T., 1990. Calcareous nannofossil biostratigraphy of the Tertiary of the North Sea Basin. Newslett. Stratigr. 22, 21–41.
Gallitelli, E.M., 1957. A revision of the foraminiferal family Heterohelicidae. US Natl. Museum Bull. 215, 133–154.
Galloway, J.J., 1933. A manual of Foraminifera. Principia Press, Bloomington.
Galloway, G.E., 1975. Process framework for describing the morphologic evolution of deltaic depositional systems. In: Broussard, M.L. (Ed.), Deltas: Models for Exploration. Houston Geol. Soc., Houston, p. 8798.
Galloway, J.J., Hemingway, C.E., 1941. Tertiary Foraminifera of Porto Rico New York Academy Sciences. Sci. Surv. Porto Rico Virgin-Islands 3 (Pt. 4), 275–491.
Galloway, J.J., Morrey, M., 1929. A lower Tertiary foraminiferal fauna from Manta, Ecuador. Bull. Am. Paleontol. 15 (55), 7–56.
Galloway, J.J., Wissler, S.G., 1927. Pleistocene Foraminifera from the Lomita Quarry, Palos Verdes Hills, California. J. Paleontol. I, 35–87.
Gartner, Jr., S., 1969. Correlation of Neogene planktonic foraminifera and calcareous nannofossil zones. Trans. Gulf Coast Assoc. Geol. Soc. 19, 585–599.
Gartner, Jr., S., 1977. Nannofossils and biostratigraphy: an overview. Earth Sci. Rev. 13, 227–250.
Gartner, S., Bukry, D., 1967. Nannofossil species related to *Cycloccolithus leptoporus* (Murray and Backman). The University of Kansas. Paleontological Contributions. Paper 28, pp. 1–4.
Gartner, S., Bukry, D., 1975a. Morphology and phylogeny of the Coccolithophycean family Ceratolithaceae. US Geol. Surv. J. Res. 3, 451–465.
Gartner, S., Bukry, D., 1975b. *Ceratolithus acutus* Gartner and Bukry n. sp. and *Ceratolithus amplificus* Bukry and Perciral-Nomenclatural Clarification. Tulane Stud. Geol. Paleontol. 11, 115–118.
Gebhardt, H., 1997. Cenomanian to Turonian Foraminifera from Ashaka (northeastern Nigeria): quantitative analysis and paleoenvironmental interpretation. Cret. Res. 18, 17–36.
Geitzenauer, K.R., 1972. The Pleistocene Calcareous Nannoplankton of the Subantarctic Pacific Ocean. Initial Reports of the Deep Sea Drilling Project, vol. 19, pp. 45–61.
Genik, G.J., 1992. Regional framework, structural and petroleum aspects of rift basins in Niger, Chad and the Central African Republic (CAR). Tectonophysics 213, 169–185.
Genik, G.J., 1993. Petroleum geology of Cretaceous-Tertiary rift basins in Niger, Chad and Central African Republic. AAPG Bull. 77, 1405–1434.
Gianotti, A., 1953. Microfaune della serie Tortoniana del Rio Mazzapiedi Castellania (Tortona—Alessandria). Riv. Ital. Paleontol. Stratigr. 6, 167–301.
Glaessner, M.F., 1937. Studien uber Foraminiferen aus der Kreide und dem Tertidr. Des Kaukasus. I. Die Foraminiferen der altesten Tertiarschichten des Nordwest-Kaukasus. Problemy Paleontologii. Paleontologicheskaya Laboratoriya Moskovskogo Gosudarstvennogo Universiteta 2–3, 349–410.
Gradstein, F.M., Ogg, J.G., 2012. The Geologic Time Scale 2012, vol. 1–2, Elsevier, Oxford, U.K., p. 1129.
Gradstein, F.M., Ogg, J.G., Smith, A.G. (Eds.), 2004a. A Geologic Time Scale 2004. Cambridge University Press, Cambridge, p. 589.
Gradstein, F.M., Ogg, J.G., Smith, A.G., et al., 2004b. Cenozoic Time Scale (Mesozoic and Cenozoic) Sequence Stratigraphy of European Basins. SEPM Special Publication No. 60, Cambridge University Press.
Gradstein, F.M., Ogg, J.G., Smith, A.G., et al., 2010. Cenozoic Time Scale (Mesozoic and Cenozoic Sequence Stratigraphy of European Basins. SEPM Special Publication, No. 60, Cambridge University Press.

Gradstein, F.M., Ogg, J.G., Smith, A.G., Ogg, G.M. (Eds.), 2012. The Geologic Time Scale 2012. Elsevier Publications, Oxford, United Kingdom, p. 1129.
Graham, I., de Klasz, I., Rerat, D., 1965. Quelques importants Foraminiferes du Tertiaire du Gabon (Afrique equatoriale). Rev. Micropaleontol. 4, 175–189.
Gran, H.H., 1912. In: Murray, J., Hjort, J. (Eds.), Pelagic Plant Life, in the Depths of the Ocean. Macmillan, London, pp. 307–386.
Grassé, P.P., 1952. Traité de zoologie, Anatomie, systematique, biologie. Tome 1, Fasc. 1: Phylogene. Protozoaires: généralités, Flagellés. Mason and Cie, Paris.
Grove, A.T., 1951. Land Use and Soil Conservation in Parts of Onitsha and Owerri Provinces. Geol. Survey of Nigeria Bull. No. 21.
Haack, R.C., Sunderarraman, P., Diedjomahor, J.O., Xiao, H., Gant, N.J., May, E.D., Kelsch, K., 2000. Niger Delta Petroleum Systems, Nigeria. In: Mello, M.R., Katz, B.J. (Eds.), Petroleum Systems of South Atlantic Margins. AAPG Memoir No. 73, pp. 213–231.
Hadley, Jr., W.H., 1934. Some Tertiary Foraminifera from the north coast of Cuba. Bull. Am. Paleontol. 20, 1–40, 70A.
Haeckel, D., 1894a. Systematische Phylogenie. Entwurf eines Naturlichen systems der organismen auf Grund ihrer stammesgeschichte. Systematische Phylogenie der Protisten und Pflanzen, 1. Georg Reimer, Berlin.
Haeckel, E., 1894b. Systematicische Phylogenie der Protisten und Pflanzen. Reimer, Berlin.
Hamaoui, M., Chierici, M., Wenter, M., 1986. Biogeography and biostratigraphy of selected Upper Cretaceous-Lower Tertiary Agglutinated Foraminifera from West African Basins. EP/S/EXP/RAG. Lab. Pau. No. 86/63 RS.
Hansen, H.J., Lykke-Andersen, A., 1976. Wall structure and classification of fossil and recent elphidiid and nonionid Formainifera. Fossil and Strata, Oslo, No. 10, pp. 1–37.
Haq, B.U., 1966. Electron microscope studies on some upper Eocene calcareous nannoplankton from Syria. Stockholm Contrib. Geol. 15, 23–37.
Haq, B.U., Berggren, W.A., 1978. Late Neogene calcareous plankton biochronology of the Rio Grande Rise (South Atlantic Ocean). J. Paleontol. 52, 1167–1194.
Haq, B.U., Hardenbol, J., Vail, P.R., 1987. Chronology of fluctuating sea levels since the Triassic. Science 235, 1156–1167.
Haq, B.U., Hardenbol, J., Vail, P.R., 1988. Mesozoic and Cenozoic chronostratigraphy and cycles of sea level change. In: Wilgus, C.K., Kendall, C.G. St. C., Posamentier, H.W., Ross, C.A., Van Wagoner, J.C. (Eds.), Sea Level Changes—An Integrated Approach. Society of Economic Paleontologists and Mineralogists, Tulsa, OK, Special Publication No. 42, pp. 71–108.
Haque, A.F.M., 1956. The Foraminifera of the Ranikot and the Laki of the Nammal Gorge, Salt Range. Palaeontol. Pak. 1, 1–300.
Hardenbol, J., Thierry, J., Farley, M.B., Jacquin, T., de Graciansky, P.-C., Vail, P.R., 1998. Mesozoic and Cenozoic Sequence Chronostratigraphy of European Basin. SEPM Special Publication No. 60, pp. 3–13.
Harris, D.K., 1981. Paleoenvironmental interpretation and foraminiferal biostratigraphy of cores offshore from the Niger Delta, Unpublished Dissertation, University College, Swansea.
Hay, W.W., 1964. The use of electron microscope in the study of fossils. Annu. Rep. Smithsonian Inst. 1963, 409–415.
Hay, W.W., 1970. Calcareous Nannofossils From Cores Recovered on Leg 4. Initial Reports of the Deep Sea Drilling Project, vol. 4, pp. 455–501.
Hay, W.W., Mohler, H.P., 1967. Calcareous nannoplankton from early Tertiary rocks at Pont Labau, France, and Paleocene–Eocene correlation. J. Palaeontol. 41, 1505–1541.
Hay, W.W., Mohler, H.P., Roth, P.H., Schmidt, R.R., Boudreaux, J.E., 1967. Calcareous nannoplankton zonation of the Cenozoic of the Gulf Coast and Carribean-Antillean area and transoceanic correlation. Trans. Gulf Coast Assoc. Geol. Soc. 17, 428–480.
Hay, W.W., Mohler, H.P., Wade, M.E., 1966. Calcareous nannofossils from Nal'chik (northwest Caucasus). Eclogae Geol. Helvet. 59, 379–399.
Haynes, J., 1954. Taxonomic position of some British Paleocene Buliminidae. Contrib. Cushman Found. Foramin. Res. 5 (4), 185–191.
Heckel, H., 1968. Nannoplanktonhorizonte und, tekonishe Strukturen in der Flyschzone nördlich von Wien (Bisambergzug). Ib. Geol. Bundes Anstalt 111, 293–337.
Hedberg, H.D., 1937. Foraminifera of the middle Tertiary Carapita Formation of northeastern Venezuela. J. Paleontol. 11 (8), 661–697.
Heinio, P., Davies, R.J., 2006. Degradation of compressional fold belts: deep-water Niger Delta. AAPG Bull. 90, 753–770.
Heron-Allen, E., Earland, A., 1932. Foraminifera. Part. 1. The ice free area of the Falkland Islands and adjacent seas. Discov. Rep. 4, 291–460.
Hine, N., 1990. Late Cenozoic calcareous nannoplankton from the Northeast Atlantic. Unpublished PhD Thesis, University of East Anglia.
Hofker, J., 1951. The Foraminifera of the "Siboga" Expedition: Part III. Siboga Expedite, 4b (vii–xii), pp. 1–51.
Hornibrook, N. de B., 1961. Tertiary foraminifera from the Oamuru District (N.Z.); Pt. 1-Systematics and distribution. NZ Geol. Surv. Paleontol. Bull. 34, 1–192.
Hosni, F., et al., 1975. On the occurrence of *Altistoma* species in the Middle Eocene of Fayoum Province, Egypt. Proc. Vth Afr. Coll. Microp., Addis Ababa, 1972. Rev. Espana Microp. Series 8, No. 3.
Howe, H.V., 1939. Louisiana Cook Mountain Eocene Foraminifera. Bull. Geol. Surv. Louisiana 14, 1–122.
Howe, H.V., Wallace, W.E., 1935. Foraminifera of the Jackson Eocene at Danville on the Ouachita, Catahoula Parish, Louisiana. Louisiana Dep. Conserv. Geol. Bull. 2, 1–118.
Husezima, R., Maruhasi, M., 1944. A new genus and thirteen new species of foraminifera from the core sample of Kasiwazaki oil field, Niigata-ken. J. Sigenkagaku Kenkyusyo 1 (3), 391–400.
Hussey, K.M., 1949. Louisiana Cane River Eocene Foraminifera. J. Paleontol. 23, 109–144.

Ingle, J.C., 1967. Foraminiferal biofacies variation and the Miocene–Pliocene boundary in California. Bull. Am. Paleontol. 52, 210–394.
Ingle, J.C., 1973. Summary Comments on Neogene Biostratigraphy and Paleoceanography in the Marginal Northeastern Pacific Ocean. Initial Reports of the Deep Sea Drilling Project, vol. 18, pp. 949–960.
Ishizaki, K., 1939. On the Foraminifera of the so-called Upper Marine Fossil Beds in the vicinity of Kwansai, Sintiku Prefecture, Taiwan. Geol. Notes Formosa 10 (4), 110.
Jafar, S.A., Martini, E., 1975. On the validity of the calcareous nannoplankton genus *Helicosphaera*. Senckenbergiana Lethaea 56, 381–397.
Jenkins, D.G., 1960. Planktonic foraminifera from the Lakes Entrance oil shaft, Victoria, Australia. Micropaleontology 6, 345–371.
Jenkins, D.G., 1966. Planktonic foraminiferal zones and new taxa from the Danian to Lower Miocene of New Zealand. NZ J. Geol. Geophys. 8, 1088–1126.
Jenkins, D.G., 1971. New Zealand Cenozoic planktonic Foraminifera. Paleontol. Bull. Wellington 42, 1–278.
Jenkins, D.G., Murray, J.W., 1989. Stratigraphical Atlas of fossil Foraminifera. Ellis Horwood Publishers; John Wiley and Sons, New York, NY.
Jenkins, D.G., Orr, W.N., 1972. Planktonic Formainiferal Biostratigraphy of the East Equatorial Pacific—DSDP Leg 9. Initial Reports of the Deep Sea Drilling Project, vol. 9, pp. 1059–1193.
Jenkins, D.G., Orr, W.N., 1973. J. Foramin. Res. 3, 133–135.
Jerkovic, L., 1970. Noelaerhabolus nov. gen. Type d'une nouvelle famille de Coccolithophoridés fossiles. Noelaerhabdoceae du Miocéne supérieur de Yougoslavie. Comptes Rendus (Hebdomadaires des Séances) de l'Académie des Sciences, Paris, v. 270, pp. 468–470.
Jiang, M.J., Gartner, S., 1984. Neogene and Quaternary Calcareous Nannofossil Biostratigraphy of the Walvis Ridge (Southern Atlantic). In: Moore Jr., T.C., Rabinowitz, P.D., et al. (Eds.), Initial Reports of the Deep Sea Drilling Project, vol. 74, pp. 561–595.
Jones, T.R., 1875. In: Griffith, J.W., Henfrey, A. (Eds.), The Micrographic Dictionary. van Voorst, London.
Jones, B., 1948. The sedimentary rocks of Sokoto Province. Bull. Geol. Surv. Nigeria 18, 75.
Jones, T.R., Parker, W.K., 1876. Notice Sur les foraminiferes vivants et fossils de jamaique: Suivie de la description dune espece nouvelle [Tinoporus pillaris] des Couches Miocenes dela jamaique, par H.B. Brady: Soc Malacol. Belg., Ann., v. II, Mein., pp. 91–103.
Jordan, R.W., Kleijne, A., Heimdal, B.R., 1993. Proposed changes to classification system of living coccolithophorids, III. INA Newsl. 15, 18–22.
Kameo, K., Sato, T., 1993. Late Pliocene nannofossil biostratigraphy at Site 607 and its oceanographic significance. Initial Reports of the Deep Sea Drilling Project Leg 94 (North Atlantic Ocean). In: Young, J.R., Bown, P.R. (Eds.), Fifth International Nannoplankton Association Conference, Salamanca.
Kamptner, E., 1937. Neue und bermerkenswerte Coccolithineen aus dem Mittelmeer, Archiv fur Protistenkunde, v. 89, pp. 279–316.
Kamptner, E., 1943. Zur Revision der Coccolithinen-Spezies Pontosphaera hyxleyi Lohm. Akademie der Wissenschaften in Wien, Mathematisch-Maturwissenchaftliche klasse, 80, pp. 43–49.
Kamptner, E., 1948. Coccolithen aus dem Torton des Inneralpinen Wiener Beckens. Sitzungberichte der Osterreichischen Akademie der Wissenschaften. Mathematisch-Naturwissenschaftliche Klasse, Abt. 1, 157, pp. 1–16.
Kamptner, E., 1950. ber den submikroskopischen Aufbau der Coccolithen. Akademie der Wissenschuften in Wien, Mathematisch-Naturnissenschaftliche Klasse, 87, pp. 152–158.
Kamptner, E., 1954. Untersuchungen ber den Feinbau der Coccolithen. Arch. Protistenk 100, 1–90.
Kamptner, E., 1963. Coccolithineen-Skelettreste aus Tiefseeablagerungen des Pazifischen Ozeans. Naturhistorisches Museum in Wien, Annalen, v. 66, pp. 139–204.
Kellough, G.R., 1962. Paleoecology of the Foraminiferida of the Wills Point Formation (Midway Group) in Northeast Texas. Trans. Gulf Coast Assoc. Geol. Soc. 15, 73–153.
Kennett, J.P., 1960. Recognition and correlation of the Kapitean Stage (Upper Miocene, New Zealand). New Engl. J. Geol. Geophys. 10, 1051–1063.
Kiesel, Y., Lotsch, D., Strumper, E., 1969. Planktonic foraminifera from the Oligo/Miocene of the German Democratic Republic (GDR). In: Bronnimann, P., Renz, H.H. (Eds.), Proceedings of the First International Conference on Planktonic Microfossils, vol. II. E.J. Brill, Geneva, pp. 339–342.
Kleinpell, R.M., 1938. Miocene Stratigraphy of California. American Association of Petroleum Geologists, Tulsa, OK, pp. 1–450.
Knox, G.J., Omatsola, E.M., 1987. Development of the Cenozoic Niger Delta in terms of the "Escalator Regression" model and impact on hydrocarbon distribution. Proceedings of the KNGMG Symposium: Coastal Lowlands, Geology and Geotechnology. Kluwer Academic Publishers, The Netherlands, pp. 181–202.
Koch, R.E., 1923. Die jung-tertiare Foraminifere fauna von Kabu (Res. Surabaja, Java). Eclogae Geol. Helvet. 18, 342–357.
Kogbe, C.A., Mehes, K., 1986. Micropaleontology and biostratigraphy of the coastal basins of West Africa. J. Afr. Earth Sci. 1, 9.
Kogbe, C.A., Le Calvez, Y., et al., 1975. Microbiostratigraphy of Upper Cretaceous and Tertiary sediments penetrated by Gbekedo "B" well (Niger Delta, Nigeria). Proc. 5th. Coll. Afr., Addis Ababa, 1972, Rev. Espan. Microp., Series 7, No. 3, pp. 493–504.
Kohl, B., 1985. Early Pliocene benthic forminifers from the Salina Basin, Southeastern Mexico. Bull. Am. Paleontol. 88 (322), 1–173.
Kreisa, R.D., Unomah, G.I., Joiner, S.D., Oluwale, Y.R., Leninger, S.C., 1999. Improved reservoir description and development through quantitative integration of core data. NAPE Bull. 14, 1–17.
Lacroix, E., 1931. Microtexture du test des Textularidae. Bulletin de I'Institut Oceanographique Monaco 588, 1–18.

Lacroix, E., 1932. Textularidae du plateau continental Mediterraneen entre Saint-Raphael et Monaco. Bulletin de I'Institut Oceanographique Monaco 591, 1–28.
Lalicker, C.G., McCulloch, I., 1940. Some Textulariidae of the Pacific Ocean. Allan Hancock Pacific Exped. 6 (2), 1–143.
Lamarck, J.B., 1804. Suite des memoires sur les fossiles des environs de Paris. Ann. Museum Natl. Historie Nat. 5, 179–188.
Lamb, J.L., 1969. Planktonic foraminiferal datums and Late Neogene epoch boundaries in the Mediterranean Caribbean, and Gulf of Mexico. Trans. Gulf Coast Assoc. Geol. Soc. 19, 559–578.
Lamb, J.L., Beard, J.H., 1972. Late Neogene planktonic foraminifers in the Caribbean, Gulf of Mexico and Italian stratotypes. Kansas University Paleontological Contributions, Article 57 (Protozoa 8), pp. 1–67.
Lamb, J.L., Miller, T.H., 1984, Stratigraphic significance of Uvigerinid Foraminifers in the Western Hemisphere. The University of Kansas Paleontological Contributions, Article 66, p. 8, pl. 14: 1–3; fig. 1.
Larsen, A.R., 1982. Foraminifera From off the West Coast of Africa, Atlantide Report, No. 13, Copenhagen, pp. 1–147.
Le Calvez, Y., de Klasz, I., Brun, L., 1971. Quelques Foraminiferes des l'Afrique occidentale. Rev. Espan. Micropaleontol. 3 (3), 310.
Lemmermann, E., 1908. Flagellatae, Chlorophyceae, Coccosphaerales und Silicoflagellatae. In: Brandt, K., Apstein, C. (Eds.), Nordisches Plankton. Botanischer Teil. Lipsius and Tischer, Kiel, Leipzig, pp. 1–40.
LeRoy, L.W., 1939. Some small Foraminifera, ostracoda and otoliths from the Neogene ("Miocene") of the Rokan-Tapanoeli area, Central Sumatra. Natuurkwetenschappelijk Tijdschrift voor Nederlandsch Indie 99 (6), 215–296.
LeRoy, L.W., 1941. Small Foraminifera from LateTertiary of the Netherlands East Idies. Colorado Sch. Mines Q. Golden 36 (1), 13–62.
LeRoy, L.W., 1944. Miocene foraminifera from Sumatra and Java, Netherlands East Indies, Part. 1. Miocene foraminifera of Central Sumatra, Netherlands East Indies. Colorado Sch. Mines Q. 39, 1–69.
LeRoy, L.W., 1953. Biostratigraphy of the Magfi section, Egypt. Mem. Geol. Soc. Am. 54, 43.
LeRoy, D.O., Hodgkinson, K.A., 1975. Benthonic foraminifera and some Pteropoda from a deep-water dredge sample, northern Gulf of Mexico. Micropaleontology 21 (4), 420–447.
LeRoy, D.O., Levinson, S.A., 1974. A deep-water Pleistocene microfossil assemblage from a well in the northern Gulf of Mexico. Micropaleontology 20 (1), 1–37.
Li, Q., Radford, S., 1994. Notes on the topotype and deep-sea specimens of the planktonic foraminifer *Cassigerinelloita amekiensis* Stolk. J. Micropaleontol. 13, 157–159.
Linnaeus, C., 1758. Systema naturae per regna tria naturae, secundum classes, ordines, genera, species, cum characteribus, differentiis, synonymis, locis. In: Engleman G., tenth ed., vol. 1, pp. 1–824.
Linnaeus, C., 1767. Systema naturae, Tom.1. Pars II Editio Duodecima Reformata, 533-132, pp. 1–37. Holmiae salvius.
Lipps, J.H., 1969. *Triquetrorhabdulus* and similar calcareous nannoplankton. J. Paleontol. 43, 1029–1032.
Loeblich, Jr., A.R., Tappan, H., 1956. *Chiloguembelina* a new Tertiary genus of the Heterohelicidae (Foraminifera). J. Washington Acad. Sci. 46, 340.
Loeblich, Jr., A.R., Tappan, H., 1957. Planktonic foraminifera of Paleocene and Early Eocene age from the Gulf and Atlantic Coastal Plains. US Natl. Museum Bull. 215, 223–232.
Loeblich, Jr., A.R., Tappan, H., 1961. Cretaceous planktonic foraminifera. Part I: Cenomanian. Micropaleontology 7, 257–304.
Loeblich, Jr., A.R., Tappan, H., 1964. Treatise on Invertebrate Palaeontology, Part C. Protista 2. Sarcodina-Chiefly "Thecamoebians" and Foraminiferida, vols. 1 and 2, Geological Society of America; University of Kansas Press, Lawrence, KS.
Loeblich, Jr., A.R., Tappan, H., 1987. Foraminiferal Genera and Their Classification. Van Nostrand Reinhold Co., New York, NY.
Lohmann, H., 1902. Die Coccolithophoridae, eine Minographie der Coccolithen bildenden Flagellaten, zugleich ein Beitrat zur kenntnis des Mittelmeerauftriebs. Archiv fur Protistenkunde 1, 89–165.
Lys, M., 1961. Etudes paleontologiques et geologiques sur les falaises de Fresco (Cote d'Ivoire) 3. Foraminiferes. Ann. Fac. Sci. Univ. Dakar., Tome 6.
Macfayden, W.A., 1930. Miocene Foraminifera From the Clysmic Area of Egypt and Sinai. Egyptian Geological Survey, Egypt, pp. 1–149.
Marks, P., 1951. A revision of the smaller Foraminifera from the Miocene of the Vienna Basin. Contrib. Cushman Found. Foramin. Res. 2, 33–73.
Maron, P., 1969. Stratigraphical aspects of the Niger Delta. J. Min. Geol. 4 (1 and 2), 3–12.
Martini, E., 1959. Discoasteriden und verwandte Formen in NW-deutschen Eozän (Coccolithophorida). Z. Stratigraphische Auswertung. Senchenbergiana Lethaea 40, 137–157.
Martini, E., 1965. Mid-Tertiary calcareous nannoplankton from Pacific deep-sea cores. Calston Papers 17, 393–411.
Martini, E., 1969. Calcareous nannoplankton from the Kallo well. Mem. Explication Cartes Geol. Min. Belg. 11, 39–41.
Martini, E., 1970. Standard Palaeogene calcareous nannoplankton zonation. Nature 226, 560–561.
Martini, E., 1971. Standard Tertiary and Quaternary calcareous nannoplankton zonation. In: Farinacci, A. (Ed.), Proceedings of the Second Planktonic Conference, Roma, 1970, vol. 2, pp. 739–785.
Martini, E., 1976. Cretaceous to Recent calcareous nannoplankton from the Central Pacific Ocean (DSDP Leg 33). Initial Rep. DSDP 49, 533–549.
Martini, E., 1981a. Nannoplankton in der Ober-Kreide, im Atterbiar und im tieferen Juntertiar von suddeutschland und dem angren zenden Osterreich. Geol. Bavarica 82, 345–356.
Martini, E., Bramlette, M.N., 1963. Calcareous nannoplankton from the experimental Mohole drilling. J. Paleontol. 37, 845–855.
Martini, E., Worsley, T., 1970. Standard Neogene calcareous nannoplankton zonation. Nature 225, 289–290.

Matoba, Y., 1970. Distribution of recent shallow-water Foraminifera of Matsushima Bay. Miyagi Prefecture, northwest Japan. Tohuku Univ. Sci. Rep. 42 (1), 1–85, 2nd series (Geology).
Matthews, R.D., 1945. *Rectuvigerina*, a new genus of foraminifera from a restudy of *Siphogenerina*. J. Paleontol. 19 (6), 588–606.
McGowran, B., 1968. Reclassification of Early Tertiary *Globorotalia*. Micropaleontology 14, 179–198.
McHargue, T., Diedjomahor, J., Arowolo, I., Hobert, R., Onyia, V., 1993. Application of sequence stratigraphy to neritic sediments of the Niger Delta. Am. Asso. Petr. Geol. Bull. 77, 1389–1404.
Meijer, M.J., 1975. A brief outline of the Oligocene–Miocene planktonic biostratigraphy of the Cuanza Basin (Angola, Western Africa). Proc. Vth Afr. Coll, Microp., Addis Ababa, April 1972. Rev. Espana. Microp. Series 7, No. 3, pp. 635–658.
Merki, P., 1972. Structural geology of the Cenozoic Niger Delta. In: Dessauvagie, T.F.J., Whiteman, A.J. (Eds.), African Geology. Department of Geology, Ibadan University Press, Ibadan, pp. 635–646.
Mitchum, J.M., Vail, P.R., Thompson, S., 1977. The depositional sequence as a basic unit for stratigraphic analysis. In: Payton, C.E. (Ed.), Seismic Stratigraphy—Application to Hydrocarbon Exploration. AAPG Memoir, vol. 26, pp. 53–62.
Montfort, P., Denys, D.E., 1808. Conchyliologie Systematique et Classification Methodique des Conquilles, Paris, v. 1.
Morkhoven, F.P.C.M., Berggren, W.A., Edwards, A.S., 1986. Cenozoic cosmopolitan deep-water benthic foraminifera. Bull. Centres Rech. Explor. Prod. Elf Aquitaine, Mem. II, pp. 1–421.
Morozova, V.G., 1939. K stratigrafii verkhnego mela i paleogena Embenskoi oblasti po faune foraminifer. Byulletin Moskovskogo Obshchestva Ispytateley Prirody, Otdel Geologicheskii 23, 22–43.
Morris, R.E., 1965. Living cells of Ceratolithus cristatus (Coccolithophorineaceae). Archiv fůr Protistenkunde 108, 19–24.
Mosunmolu Limited, 1991. The Refinement of the Middle Miocene F9500 Zone Niger Delta (25 Well Study). Unpublished Report for SPDC.
Mosunmolu Limited, 1992a. The Refinement of the Lower Miocene F9300 Zone (10 Well Study). Unpublished Report for SPDC.
Mosunmolu Limited, 1992b. Sequence Stratigraphy Miocene Niger Delta Pilot Project with MICRO-STRAT Inc. Houston, TX and Prof. Peter R. Vail. Unpublished Report for SPDC.
Mosunmolu Limited, 1995. Establishment of the MPN Faunal Zones IV and V. Unpublished Report for Mobil Producing Nigeria Unltd.
Mosunmolu Limited, 1996. Refinement of Late Miocene F9600 and F9700 Microfaunal Zones, Niger Delta. Unpublished Report for SPDC.
Mosunmolu Limited, 2013. Chronostratigraphic Refinement of the NN5-NN4 Calcareous Nannofossil Zones in the Niger Delta, Nigeria. Unpublished Report for Shell Petroleum Development Company of Nigeria Ltd.
Mosunmolu Limited, 2014. Petroleum Systems Modeling of the Egbaoma Field, OML 38. Unpublished Report for Platform Petroleum Ltd.
Muller, C., 1974. Nannoplankton aus dem Mittel-Miozan von Walbersdorf (Burgenland). Senckenbergiana Lethaea 55, 389–405.
Muller, C., 1981. Beschreibung neuer *Helicosphaera*-Arten aus dem Miozan und Revision biostratigraphischer Reichweiten einiger neogener Nannoplankton-Arten. Senckenbergiana Lethaea 61, 427–435.
Mǔller, C., 1970. Nannoplankton-Zonen der Unteren-Meteeresmolasse Bayerns. Geol. Bav. 63, 107–118.
Murat, R.C., 1972. Stratigraphy and paleogeography of the Cretaceous and Lower Tertiary in southern Nigeria. In: Dessauvagie, T.F.J., Whiteman, A.J. (Eds.), African Geology. Ibadan University Press, Ibadan, pp. 251–266.
Murray, J.W., 1971. An Atlas of British Recent Foraminiferids. Elsevier Publishing Company, Amsterdam.
Murray, J.W., 1991. Ecology and Palaeoecology of Benthic Foraminifera. John Wiley and Sons Inc., New York, NY, 397 p.
Murray, G., Blackman, V.H., 1898. On the nature of the coccospheres and rhabdospheres. Philos. Trans. R. Soc. London 190B, 427–441.
Natland, M.L., 1938. New Species of Foraminifera From off the West Coast of North America and From the Later Tertiary of the Los Angeles Basin. California, Technical Series, vol. 4, pp. 1–144.
NEDECO, 1954. Western Niger Delta Report on Investigation, Consultant Report. Netherlands Engineering Consultants, The Hague, 143 p.
NEDECO, 1959. River Studies and Recommendations on Improvement of Niger and Benue. North Holland Publ. Co., Amsterdam, 1000 p.
NEDECO, 1961. The Waters of the Niger Delta. Netherlands Engineering Consultants, The Hague, 317 p.
Norris, R.E., 1965. Living cells of *Ceratolithus cristatus* (Coccolithophorinae). Archiv fur Protistenkunde 108, 19–24.
Nuttall, W.L.F., 1928. Tertiary Foraminifera from the Naparima region of Trinidad (British West Indies). Q. J. Geol. Soc. London 84 (33), 57–115.
Nuttall, W.L.F., 1930. Eocene Foraminifera from Mexico. J. Paleontol. 4, 271–293.
Nwachukwu, J.I., Odjegba, E.T., 2001. Compaction in reservoir sandstones of the Niger Delta Nigeria. J. Min. Geol. 37, 113–120.
Nwajide, C.S., 1979. A lithostratigraphic analysis of the Nanka Sand, southeastern Nigeria. Nig. J. Min. Geol. 16, 103–109.
Nwajide, C.S., 1980. Eocene tidal sedimentation in the Anambra Basin, southern Nigeria. Sediment. Geol. 25, 189–207.
Nwajide, C.S., 2013. Geology of Nigeria's Sedimentary Basins. CSS Books Ltd., Lagos, Nigeria, 565 p.
Nwajide, C.S., Reijers, T.J.A., 1996. Sequence architecture in outcrops: examples from the Anambra Basin. NAPE Bull. 11, 23–32.
Ogunjobi, O., 1994. Calcareous nannofossil biostratigraphy of the 'western' Niger Delta. Unpublished MSc Thesis, University College, London.
Ojo, O.A., 1996. Pre-drill prospect evaluation in deepwater Nigeria. NAPE Bull. 11, 15–22.
Okada, H., Bukry, D., 1980. Supplementary modification and introduction of code numbers to the low-latitude coccolith biostratigraphic zonation (Bukry, 1973; 1975). Mar. Micropaleontol. 5 (2), 321–325.
Okezie, C.N., Onuogu, S.A., 1985. The lignites of southeastern Nigeria. Geol. Surv. Nigeria Occasional Paper 10, 28.

Okosun, E.A., Akande, W.G., Onoduku, U.S., Alkali, Y.B., Obaje, S.O., 2015. Taxonomic notes on selected Late Paleocene to Early Eocene Foraminifera from Eastern Benin Basin, Southwestern Nigeria. J. Basic Appl. Res. Int. 8 (1), 41–55.
Okosun, E.A., Liebau, A., 1999. Foraminiferal biostratigraphy of eastern Niger Delta, Nigeria. NAPE Bull. 14, 136–156.
Olsson, R.K., 1964. *Praeorbulina* Olsson, a new foraminiferal genus. J. Paleontol. 38, 770–771.
Olsson, R.K., Hemleben, C., Berggren, W.A., Huber, B.T., 1999. Atlas of Paleocene planktonic Foraminifera. Smithsonian Contrib. Paleobiol. 85, 1–252.
Omatsola, M.E., Adegoke, O.S., 1982. Tectonic evolution and Cretaceous Stratigraphy of the Dahomey Basin. J. Mining Geol. 18 (1), 130–137.
Onuoha, K.M., 1981. Sediment loading and subsidence in the Niger Delta sedimentary basin. J. Min. Geol. 18 (1), 138–140.
Onyia, V., Adejobi, A., Ibie, E., Nkeme, U., Haack, R., 2002. Regional chronostratigraphic interpretation in northwestern Niger Delta. NAPE Bull. 16, 81–92.
Opara, G.E., 1981. A review of the Tertiary Unam "B" Field in OML 67, Niger Delta. J. Min. Geol. 18, 86–95.
Osterloff, P., Adegoke, O.S., Starkie, S., Adebiyi, A., Oyeyinka, G., Ndulue, P., Tiamiyu, A., Lawal, K., 2013. A Practical Application of Chronostratigraphic Refinement of the NN5-NN4 Calcareous Nannofossil Zones in the Niger Delta. NAPE Bull., Abstract Volume.
Osterloff, P., Ndulue, P., Adebiyi, A., Adegoke, O.S., Oyeyinka, G., Tiamiyu, A., Lawal, K., Starkie, S., 2015. Chronostratigraphic Refinement of the NN5-NN4 Calcareous Nannofossil Zones in the Niger Delta, Nigeria.
Oti, M.N., Postma, G. (Eds.), 1995. Geology of Deltas. A.A. Balkema Publishers, Rotterdam, p. 315.
Oyebamiji, A.S., 1997. Calcareous nannofossil biostratigraphy of a well in the Niger Delta, Nigeria. Unpublished MSc Thesis, University College London.
Ozumba, M.B., 1999. Middle to Late Miocene sequence stratigraphy of western Niger Delta. NAPE Bull. 14, 176–192.
Ozumba, M.B., 2011. Planktonic foraminiferal zonation of the Niger Delta. NAPE Bull. 23 (1), 44–52.
Ozumba, M.B., Amajor, L.C., 1999a. Evolutionary relationships in some benthonic foraminifera of the Middle to the Late Miocene, Niger Delta. NAPE Bull. 14 (2), 157–167.
Ozumba, M.B., Amajor, L.C., 1999b. Middle to Late Miocene Biozonation of the Western Niger Delta. NAPE Bull. 14 (2), 168–175.
Palmer, D.K., 1934. Some large fossil foraminifera from Cuba. Memorias de la Sociedad Cubana de Historia Natural 8, 235–264.
Parker, F.L., 1954. Distribution of the Foraminifera in the northeast Gulf of Mexico. Museum Comp. Zool. Bull. 111 (10), 453–588.
Parker, F.L., 1967. Late Tertiary biostratigraphy (planktonic foraminiferal) of tropical Indo-Pacific deep-sea cores. Bull. Am. Paleontol. 52, 115–208.
Parker, W.K., Jones, T.R., 1865. On the foraminifera from North Atlantic and Artic Oceans, including Davis straits and Baffin's Bay. Royal Society of London. Philosophical Transactions 155, 325–441.
Parker, W.K., Jones, T.R., Brady, H.B., 1871. On the nomenclature of the Foraminifera. Part XIV—The species enumerated by d'Orbigny in the 'Annales des Sciences, Naturelles' 1826, v. 7. Annals and Magazine of Natural History, London, Ser. 4, v. 8, 145, 179, pp. 283–266.
Parker, F.I., Phleger, F.B., Pierson, J.F., 1953. Ecology of Foraminifera from San Antonio Bay and environs, Southwest Texas. Cushman Foundation for Foraminiferal Research, Special Publication No. 2, pp. 1–75.
Parkinson, J., 1907. The post-Cretaceous stratigraphy of Southern Nigeria. Q. J. Geol. Soc. London 63, 311–320.
Parr, W.J., 1938. Upper Eocene Foraminifera from the deep borings in King's Park, Perth, Western Australia. J. R. Soc. West. Aust. 24, 69–101.
Parr, W.J., 1947. The Lagenid Foraminifera and their relationships. Proc. R. Soc. Victoria 58, 116–130.
Perch-Nielsen, K., 1971. Elektronenmikroscopische Untersuchungen an Coccolithen und verwandten formen aus dem EozSn von DSnemark. K. Danske Videnskabernes Selskab. Biol. Skrifter 18 (3), 1–75.
Perch-Nielsen, K., 1977. Albian to Pleistocene Calcareous Nannofossils From the Western South Atlantic, DSDP Leg 39. Initial Reports of the Deep Sea Drilling Project, vol. 39, pp. 699–823.
Perch-Nielsen, K., 1981. Nouvelles observations sur les nannofossils calcaires a la limite Cretace-Tertiaire pres de El Kef (Tunisie). Cahiers de Micropaleontologie 3, 25–36.
Perch-Nielsen, K., 1985a. Cenozoic calcareous nannofossil. In: Bolli, H.M., Saunders, J.B., Perch-Nielsen, K. (Eds.), Plankton Stratigraphy. Cambridge University Press, Cambridge, pp. 427–554.
Perch-Nielsen, K., 1985b. Mesozoic calcareous nannofossils. In: Bolli, H.M., Saunders, J.B., Perch-Nielsen, K. (Eds.), Plankton Stratigraphy. Cambridge University Press, Cambridge, pp. 329–426.
Petters, V., 1954. Tertiary and Upper Cretaceous Foraminifera from Colombia, S.A. Contrib. Cushman Lab. Foramin. Res. 5, 37–41.
Petters, S.W., 1982. Central West Africa Cretaceous-Tertiary benthic foraminifera and stratigraphy. Paleontographica. Abs. Part A 179, 1–104.
Petters, S.W., 1983. Gulf of Guinea planktonic foraminiferal biochronology and geological history of the South Atlantic. J. Foramin. Res. 13, 32–59.
Petters, S.W., 1984. An ancient submarine canyon in the Oligocene–Miocene of the Western Niger Delta. Sedimentology 31, 805–810.
Petters, S.W., 1985. Gulf of Guinea planktonic foraminiferal biochronology and geological history of the South Atlantic. J. Foramin. Res. 13, 32–59.
Petters, S.W., 1995. Foraminiferal biofacies in the Nigerian rift and continental margin deltas. In: Oti, M.N., Postma, G. (Eds.), Geology of Deltas. A.A. Balkema Publishers, Rotterdam, pp. 219–235.
Phleger, F.B., 1954. Ecology of Foraminifera and associated micro-organisms from Mississippi sound and environs. Am. Assoc. Petr. Geol. Bull. 38, 584–647.
Phleger, F.B., 1960. Ecology and Distribution of Recent Foraminifera. The Johns Hopkins Press, Baltimore, MD, 279 p.
Phleger, F.B., Parker, F.L., 1951. Ecology of foraminifera of northwest Gulf of Mexico. Geological Society of America, Memoir 46, Part 2, pp. 1–64.

Pijpeers, P.J., 1933. Geology and paleontology of Bonaire (Dutch West Indies). Geographische en Geologische Modedeelingen, Physiographische - Geologische Series, Utrecht 8, 1–103.
Plummer, H.J., 1927. Foraminifera of the Midway Formation in Texas. Bulletin Univ. Texas Bureau Econ. Geol. Technol. 2644, 1–206.
Poag, C.W., 1981. Ecologic Atlas of Benthic Foraminifera of the Gulf of Mexico. Marine Science International, Woods Hole, MA, pp. 1–153.
Poche, F., 1913. Das System der Protozoa. Archiv fur Protistenkunde 30, 125–321.
Pokorny, V., 1955. *Cassigerinella boudecensis* n. gen., n. sp. (Foraminifera, Protozoa) zolgocenu Zdanickeho flyse. *Vestnik*. Ustredniho Ustara Geologickeho 30, 136–140.
Poore, P.Z., 1979. Oligocene Through Quaternary Planktonic Foraminiferal Biostratigraphy of the North Atlantic: DSDP Leg 49. Initial Reports of the Deep Sea Drilling Project, vol. 49, pp. 447–517.
Poore, R.Z., Gosnell, L.B., 1985. Apertural features and surface texture of Upper Paleogene biserial planktonic foraminifers: links between *Chiloguembelina* and *Streptochilus*. J. Foramin. Res. 15, 1–5.
Postuma, J.A., 1971. Manual of Planktonic Foraminifera. Elsevier Publishing Co., Amsterdam.
Premoli Silva, I., Bolli, H.M., 1973. Late Cretaceous to Eoceneplanktonic Foraminifera and Stratigraphy of Leg 15 Sites in the Caribbean Sea. Initial Reports Deep Sea Drilling Project, vol. 15, pp. 449–547.
Puri, H.S., 1953. Contribution to the study of the Miocene of the Florida Panhandle. Florida State Geol. Surv. Bull. 36, 71–213, Part II, Foraminifera.
Raffi, I., Rio, D., d'Atri, A., Fornaciari, E., Rochetti, S., 1995. Quantitative distribution patterns and biomagnetostratigraphy of middle and late Miocene calcareous nannofossils from western equatorial India and Pacific Oceans. (Legs 115, 130 and 138). Proc. ODP Sci. Results 138, 479–502.
Reijers, T.J.A., 1995. Reservoir geological core description using standardised lithofacies and their associations in the Tertiary Niger Delta. NAPE Bull. 10, 27–39.
Reijers, T.J.A., 1996. Selected Chapters on Geology: Sedimentary Geology and Sequence Stratigraphy in Nigeria and Three Case Studies and a Field Guide. SPDC Corporate Reprographic Services, Warri, p. 197.
Reijers, T.J.A., 2011. Stratigraphy and sedimentology of the Niger Delta. Geologos 17 (3), 133–162.
Reijers, T.J.A., Petters, S.W., Nwajide, C.S., 1997. The Niger Delta basin. Selley, R.C. (Ed.), African Basins: Sedimentary Basins of the World, vol. 3, Elsevier Sc., B.V., Amsterdam, pp. 151–172.
Reiss, Z., 1963. Reclassification of perforate Foraminifera. Bull. Geol. Surv. Israel 35, 1–111.
Renz, H.H., 1984. Stratigraphy and fauna of the Agua Salada Group, State of Falcon, Venezuela. Geological Society of America, Memoir No. 32, pp. 1–219.
Reuss, A.E., 1850. Neues Foraminiferen aus den Schichten des Osterreichischen Tertiarbeckens. Denkschriften der Kaiserlichen Akademie der Wissenschaften, Mathematisch. Naturwissenschaftliche Classe 1, 365–390.
Reuss, A.E., 1851. Ueber die fossilen Foraminiferen un Entomostraceen der Septarienthone der Umgegend von Berlin. Deutsche geologische Gessellschaft, Zeitschrift 3, 49–91.
Reuss, A.E., 1860. Die foraminiferen der Westphalischen Kreideformation-*K*. Ahad wiss wien math Naturwiss CL Sitzungsber 40, 147–238.
Reuss, A.E., 1862. Entwurf evier systematischen Zusammenstellung der Foraminiferen-K. Adak wiss wien math Natur wiss. CL Sitzungber 44, 355–396.
Reyment, R.A., 1959. The foraminiferal genus *Afrobolivina* gen. nov. and *Bolivina* in the Upper Cretaceous and Lower Tertiary of West Africa. Stockh. Contrib. Geol. 3, 1–57.
Reyment, R.A., 1965. Aspects of the Geology of Nigeria: The Stratigraphy of the Cretaceous and Cenozoic Deposits. Ibadan University Press, Ibadan, 145 p.
Reyment, R.A., 1966. Studies on the upper cretaceous and Lower Tertiary Ostracoda. Part 3, stratigraphical, paleoecological and biometrical conclusions. Stockhom Contrib. Geol. 14, 151.
Reyment, R.A., Barber, W.M., 1956. Nigeria and Cameroons. Lexique Stratigraphique International 4, 35–39.
Rio, D., Fornaciari, E., Raffi, I., 1990. Late Oligocene through early Pleistocene calcareous nannofossils from western equatorial India Ocean (Leg 115). Proc. ODP Sci. Results 115, 172–235.
Robertson, D., 1891. *Trochammina bradyi n.n.* Ann. Mag. Nat. Hist. 6, 388.
Rogl, F., Bolli, H.M., 1970. Initial Reports of the Deep Sea Drilling Project, vol. IV, p. 564.
Rögl, F., Bolli, H.M., 1973. Holocene to Pleistocene Planktonic Foraminifera of Leg 15, Site 147 (Cariaco Basin (Trench), Caribbean Sea) and Their Climatic Interpretation. Initial Reports of the Deep Sea Drilling Project, vol. XV, pp. 553–615.
Romein, A.J.T., 1979. Lineages in early Paleogene calcareous nannoplankton. Utrecht Micropaleontol. Bull. 22, 1–230.
Roth, P.H., 1970. Oligocene calcareous nannoplankton biostratigraphy. Eclogae Geol. Helvet. 63, 799–881.
Roth, P.R., Thierstein, H.R., 1972. Calcareous Nannoplankton: Leg 14 of the Deep Sea Drilling Project. Initial Reports of the Deep Sea Drilling Project, vol. 14, pp. 421–485.
Rzehak, 1886. Die Foraminiferenfauna der Neogenformation der Umgebung von Mahr. Ostrau. Naturf. Ver. Brunn, verh. Brunn [Brno, Czechoslovakia], Bd. 24 (1885), p. 87, pl. 1, fig. 5a–c.
Samtleben, C., 1980. Die Evolution der Coccolithophoriden-Gattung *Gephyrocapsa* nach Befunden im Atlantik. Paleontology 54, 91–127.
Sansores, J.C., Flores-Covarrubias, C., 1972. Descripcion de seis nuevas especies de foraminiferos bentonicos de la Cuenca Salina del Istmo de Tehuantepec, Mexico. Revista del Instituto Mexicano de Petroleo 4 (4), 77–83, 1 pl.

Sars, M., 1869. Fortsatte bemaerkninger over det dyriske livs unbrending/havets dybber, For handlinger 1 videnskasseiskabet/kristiania, 1868, pp. 246–275.
Sars, M., 1910. US Natl. Museum Bull. 71, 105.
Scheltema, R.S., 1977. Dispersal of marine invertebrate organisms: paleobiogeographic and biostratigraphic implications. In: Kaufma, E.G., Hazel, J.E. (Eds.), Concepts and methods of Biostratigraphy. Dowden, Jutchinson and Ross, Stroudsburg, PA, pp. 73–100.
Schiller, J., 1925. Die planktonischen Vegetationen des adriatischen Meeres. A. Die Coccolithophoriden-Vegetation in den Jahren 1911–14. Archiv fur Protistenkunde 51, 1–130.
Schiller, J., 1930. Coccolithineae, In: Rabenhorst, L. (Ed.), Kryptogamen-Flora van Deutschland, Osterreich und der Schweiz. 10. Band, 2 Abt. Akademische Verlagsgesellschaft, Leipzig, pp. 89–267.
Schlumberger, C., 1881. Note sur les Foraminiferes: introduction La Feuille des jeunes Naturalistes, 12 (133), 2–6
Schlumberger, C., 1890. Note aur l'Adelosina polygonia. Bulletin de la Societe Zoologique de France 15, 1–139.
Schope, T.J.M., 1980. Paleocenography. Harvard University, Cambridge, MA, 341 p.
Schroeteer, J.S., 1803. Archiv. Zool. Zoot., Braunschweig Part 2, 63–66.
Schroeter, J.S., 1783. Einleitung in die Couchylien Kenntniss nach Linne, Bd. 1.-J. J. Gebauer, Halle, 860 p.
Schultze, M.S., 1854. Uber den Organismus de Polythalamien (Foraminiferen), nebst Bemerkungen uber die Rhizopoden im Allgemeinen. Whilhelm Engelmann, Leipzig, 68 p.
Schwager, C., 1866. Fassile Foraminiferen von Kar-Nicobar. Novara Exped., Geologisch Theil 2, 187–268.
Schwager, C., 1868. *Haplophragmium coprolithforme*. Beneck's Georg. Paleont. Beitrage, 1, 654.
Schwager, C., 1876. Saggio di una classificatione dei foraminiferi avuto riguardo alle lorofamiglie naturali. R. Comitato Geol. Ital. Bull. 11–12, 475–485.
Schwager, C., 1877. Quadro del proposto sistema de classificazione dei foraminiferi conguscio. R. Comitato Geol. Ital. Bull. 8 (1–2), 18.
Schwager, C., 1883. Die Foraminiferen aus den Eocaenablagerungen der Libyschen Wuste und Aegyptens. Palaeontographica 30, 79–154.
Schwarz, E.H.L., 1894. Coccoliths. Ann. Mag. Nat. History 14, 341–346, Ser. 6.
Seguenza, G., 1862. Die terreni terziarii del distretto di Messina, Parte II. Descrizione dei foraminiferi monotalamici delle marne mioceniche distretto di Messina 84, 2.
Selli, R., 1971. Messinien. Giornale di Geologia 37, 121–133, Ser. 2.
Shaffer, B.L., 1987. The potential of calcareous nannofossils for recognizing Pliocene—Pleistocene cycles and sequence boundaries on the Shelf. In: Proceedings of the Eighth Annual Research Conference of the Society of Economic Paleontologists and Mineralogists, pp. 142–145.
Short, K.C., Stauble, A.J., 1967. Outline of Geology of Niger Delta. Am. Assoc. Petr. Geol. Bull. 51 (5), 761–779.
Silvestri, A., 1903. Dimorfis oe nomenclatra d'una Spiroplecta. Altre notizie sulla struttura della Siphogenerina columcllaris. Atti della Pontificia Acaddemia Romana dei Nuovi Lincei 56, 59–66.
Silvestri, A., 1904. Ricerche strutturali su alcune forme dei Trubi di Bonfornello (Palermo). Accad. Pont. Nuovi Lincei Mem. 22, 235–270.
Simpson, A., 1955. The Nigerian coalfield: the geology of parts of Owerri and Benue Provinces. Bull. Geol. Surv. Nigeria 24, 1–85.
Slansky, M., 1962. Contribution a l'etude geologique du basin sedimentaire cotier du Dahomey et du Togo. Mem. Bur. Rech. Geol. Min., 1–270.
Sleep, E.H., 1971. Thermal effects of the formation of Atlantic continental margins by continental break-up. Geophys. J. R. Astron. Soc. 24, 325–350.
Slobodkin, L.B., Sanders, H.L., 1969. On the contribution of environmental predictability to species diversity. Brookhaven Symposia Biology 2, 82–92.
Souaya, F.J., 1965. Miocene Foraminifera of the Gulf of Suez region U.A.R. Part I, Systematics (Astorhizidae-Buliminoidea). Micropaleontology 11 (3), 31–334.
Souaya, F.J., 1966. Miocene Foraminifera of the Gulf of Suez region U.A.R. Part II, Systematics (Rotaloidea). Micropaleontology 12, 43–64.
SPDC Catalogue, 1980. Stereoscan Photos, Paleontological Type Collection, Nigeria. Benthonic Foraminifera, 33 pls.
Stacher, P., 1994. The Niger Delta hydrocarbon habitat. NAPE Bull. 9, 67–75.
Stacher, P., 1995. Present understanding of the Niger Delta hydrocarbon habitat. In: Oti, M.N., Postma, G. (Eds.), Geology of Deltas. A.A. Balkema Publishers, Rotterdam, pp. 257–267.
Stainforth, R.M., Lamb, J.L., Luterbacher, H.P., Beard, J.H., Jeffords, R.M., 1975. Cenozoic planktonic foraminiferal zonation and characteristics of index forms. Paleontological Contribution, University of Kansas, Article 62, pp. 163–425.
Steckler, M.S., Watts, A.B., 1978. Subsidence of the Atlantic-type continental margin off New York. Earth Planet. Sci. Letts. 41, 1–13.
Stewart, R.E., Stewart, K.C., 1930. Post-Miocene foraminifera from the Ventura Quadrangle, Ventura Co., California. J. Paleontol. 4 (1), 60–72, 2 pls.
Stolk, J., 1965. Contribution a l'etude des correlations microfauniques du Tertiaire inferieur de la Nigeria meridionale. Memoires du Bureau de Recherches Geologiques et Minieres 32, 247–267.
Stradner, H., 1959. Die Fossilen Discoasteriden Osterreichs. II. Teil. Erdoel-Zeitschrift 75, 472–488.
Stradner, H., 1960. Uber nannoplankton-Invasionen im sarmat des Wiener Beckens. Erdoel-Zeitschrift 76, 430–432.
Stradner, H., 1961. Vorkommen von Nannossilien im Mesozoikum und Alttertiar. Erdorl-Zeitschrift 77, 77–78.

Su, X., 1993. Quantitative studies of calcareous nannofosssils assemblage from Pliocene–Pleistocene sediments of the North Atlantic. In: Young, J.R., Bown, P.R. (Eds.), Fifth International Nannoplankton Association Conference, Salamanca, vol. 15(2), p. 96.
Subbotina, N.N., 1947. Foraminifery Datskikh i Paleogenovykh -otlozheniy severnogo Kavkaza (Foraminifera of Danian and Paleogene deposits of the northern Caucasus) in Mikrofauna Neftyanykh Mastorozhdenig Kavkaza, Embyi Sredny Azii. Leningrad: VNGRI, pp. 39–160.
Subbtoma, N.N., 1953. Iskopaemye Foraminifery SSR. Globigerinidy, Khantkeninidy i Globorotaliidy (Fossil Foraminifera of the USSR. Globigerinidae, Hantkeninidae and Globorotaliidae). Trudy Vsesoyuznogo Neftyanogo Nauchnoissledovatel skogo Geologorazvedochnogo Instituta (VNIGRI), Ser. 76, pp. 1–296.
Sullivan, F.R., 1962. Foraminifera from the type section of the San Lorenzo Formation, Santa Cruz County, California. University of California Publications in the Geological Sciences 37, 233–353.
Sullivan, F.R., 1964a. Lower Tertiary nannoplankton from the California Coast Ranges II. Eocene. University of California Publications Geological Sciences, vol. 53, pp. 1–74.
Sullivan, F.R., 1964b. Lower Tertiary nannoplankton from the California Coast Ranges. I. Palaocene. University of California Publications in Geological Sciences, vol. 44, pp. 163–227.
Swinton, W.E., 1930. On fossil Reptilia from Sokoto Province. Bull. Geol. Surv. Nigeria 13, 1–61.
Takayama, T., Sato, T., 1987. Coccolith Biostratigraphy of the North Atlantic Ocean. Initial Reports of the Deep Sea Drilling Project, vol. 94, pp. 651–702.
Takayanagi, Y., Saito, T., 1962. Planktonic Foraminifera from the Mobori Formation, Shikoku, Japan. Tohoku Univ. Sci. Rep. 5, 67–106, Ser. 2 (Geol.).
Tan Sin Hok, 1927. Discoasteridae *incertae sedis*. Proceedings of the Koninklijke Nederlandse Akkademie Van Wetenschappen, Sect. Sci, 30, pp. 411–419.
Tappan, H., 1940. Foraminifera from the Grayson Formation of northern Texas. J. Paleontol. 14, 93–126.
Tappan, H., 1943. Foraminifera from the Duck Creek Formation of Oklahoma and Texas. J. Paleontol. 17, 476–517.
Tattam, C.M., 1944. A review of Nigerian Stratigraphy. Rep. Geol. Surv. Nigeria 1943, 27–46.
Ten Dam, A., 1944. Die stratigraphische Gliederung des niederlandischen Palaozan und Eozans nach foraminiferen. Med. Geol. Stiching 3, 102.
Terquem, O., 1876. Essai sur le classement des animaux qui vivent sur la plage et dans les environs de Dunherque. L. Flault, Paris, pp. 55–100.
Terquem, O., 1882. Les foraminifers de l'Eocene des environs de Paris. Soc. Geol. France Mem. 3, 1–193, 3, 2.
Terquem, O., 1883. Sur un nouveau genre de Foraminiferes du Fuller's-earth de la Moselle. Soc. Geol. France Bull. 3, 37–39, 11, 1.
Thalmann, H.E., 1939. Mitteilungen ubder Foraminiferen, Part 4. Eclogae Geol. Helv. 31 (2), 327–333.
Theodoris, S., 1984. Calcareous nannofossil biostratigraphy of the Miocene and revision of the helicoliths and discoasters. Utrecht Micropaleontol. Bull. 32, 271.
Thiede, J., van Andel, T.H., 1977. The paleoenvironment of anaerobic sediments in the Late Mesozoic South Atlantic Ocean. Earth Planet Sci. Lett. 33, 301–309.
Thierstein, H.R., Geitzenauer, K.R., Molfino, B., 1977. Global synchroneity of Late Quaternary Coccolith datum levels: Validation by oxygen isotopes. Geology 5, 400–404.
Thomas, E., 1980. Details of *Uvigerina* development in the Cretan Mio-Pliocene. Utrecht Micropaleontol. Bulls. 23, 1–167.
Thomson, W., 1876. Comments; In: Murray, John: Preliminary reports to Professor Wyville Thomson, F.R.S., Director of the Civilian Scientific Staff, on work done abroad the 'Challenger'. Royal Society of London Proceedings, v. 24, pp. 1–534.
Todd, R., 1952. Vicksburg (Oligocene) smaller foraminifera from Mississippi. US Geol. Surv. Prof. Paper 241, 1–53.
Todd, R., Cloud Jr., P.E., Low, D. Schmidt, R.G., 1954. Appendix: probable occurrence of Oligocene on Saipan. Am. J. Sci., 252, 673–682.
Todd, R., 1957. Geology of Saipan, Mariana Island, Part 3, Paleontology. Chapter H, Smaller Foraminifera, US Geol. Surv. Prof. Paper, 280H, pp. 265–320.
Todd, R., 1958. Foraminifera from western Mediterranean deep-sea cores. Reports Swedish Deep-Sea Expedition 1947–1948, v. 8, Sediment Cores from the Mediterranean Sea and the Red Sea, Fasc. 2, (3), pp. 167–215.
Todd, R., Bronnimann, P., 1957. Recent Foraminifera and Thecamoebina from the eastern Gulf of Paria, Trinidad. Cushman Found Foramin. Res. 3, 1–43.
Todd, R., Low, D., 1976. Smaller Foraminifera from deep wells on Puerto Rico and St. Croix. United States Geological Survey, Professional Paper No. 863, pp. 1–32.
Turcotte, D.L., 1977. On the thermal and subsidence history of sedimentary basins. J. Geophys. Res. 82, 3762–3766.
Toulmin, L.D., 1941. Eocene smaller Foraminifera from the Salt Mountain Limestone of Alabama. J. Paleontol. 15, 567–611.
Toumarkine, M., Bolli, H.M., 1970. Evolution de *Globorotalia cerroazulensis* (Cole) dans l'Eocene moyen et superieur de Possagno (Italie). Rev. Micropaleontol. 13, 131–145.
Toumarkine, M., Luterbacher, H., 1985. Paleocene and Eocene planktic Foraminifera. In: Bolli, H.M., Saunders, J.B., Perch-Nielsen, K. (Eds.), Plankton Stratigraphy. Cambridge University Press, Cambridge, pp. 87–154.
Udo, R.K., 1970. Geographical Regions of Nigeria. Heinemann, London, 212 p.
Umeji, O.P., 2003. Palynological data from the road section at the Ogbunike toll gate, Onitsha, southeastern Nigeria. J. Min. Geol. 39, 95–102.
Unomah, G.I., Akinnagbe, T.A., Edet, J.T., 1999. Lithostratigraphy of the Qua Iboe collapse system (Abstr.). Niger Delta Stratigraphic Committee Workshop, Ijebu Ode, October, 1999.

Vail, P.R., Mitchum Jr., R.M., Thompson III, S., 1977. Seismic stratigraphy and global changes in sea level, Part 3: relative changes of sea level from Coastal onlap. In: Payton, C.E. (Ed.), Seismic Stratigraphy—Applications to Hydrocarbon Exploration. AAPG Memoir, vol. 26, pp. 63–81.

Van Wagoner, J.C., Mitchum Jr., J.C., Campion, K.M., Rahmanian, V.D., 1990. Siliciclastic sequence stratigraphy in well logs, cores and outcrops. AAPG Methods in Exploration Series No. 7.

Van Wagoner, J.C., Posamentier, H.W., Mitchum, R.M., et al., 1988. An overview of the fundamentals of sequence stratigraphy and key definitions. In: Wilgus, C.K., Hastigns, B.S., Kendall, C.G., Posamentier St., H.W., et al. (Eds.), Sea-level changes—An Integrated Approach. Special Publication of the Society of Economic Paleontologists and Mineralogists, Tulsa, OK, No. 42, pp. 39–45.

Vannier, F., Durand, J., 1994. High resolution stratigraphy in a Miocene Reservoir of the Niger Delta. Bull. Centres Rech. Explor. Prod. Elf Aquitaine 18, 1–78.

Varol, O., 1982. Calcareous nannofossils from the Antalya Basin, Turkey, Neues Jahrbuch fur Geologie und Palaontologie. Monatshefte 4, 244–256.

Varol, O., 1998. Palaeogene. In: Bown, P.R (Ed.), Calcareous Nannofossil Biostratigraphy. British Micropalaeontological Society Publication Series, pp. 267–310.

Wade, B.S., Pearson, P.N., Berggren, W.A., Palike, H., 2011. Review and revision of Cenozoic tropical planktonic foraminiferal biostratigraphy and calibration to the geomagnetic polarity and astronomical time scale. Earth Sci. Rev. 104, 111–142.

Walker, G., Jacob, E., 1798. In: Kanmacher, F. (Ed.), Adams—Essays on the microscope. second ed., with considerable additions and improvements. Dillon and Keating, London, pp. 1–712.

Wallich, G.C., 1877. Observations on the coccosphere. Ann. Mag. Nat. History 19, 342–350.

Weber, K.J., 1971. Sedimentological aspects of oil fields in the Niger Delta. Geologie en Mijnbouw 50, 59–67.

Weber, K.J., Daukoru, E.M., 1975. Petroleum geology of the Niger Delta. Proc. Nineth World Petrol. Congr., Tokyo (Geol.) 2, pp. 209–221.

Weber, K.J., Mandl, G., Pillaar, W.F., Lehner, B.V.F., Precious, R.G., 1978. The role of faults in hydrocarbon migration and trapping in Nigerian growth fault structures. Proc. Tenth Offshore Technology. Conf., Houston, OTC. 3356, pp. 2643–2653.

Wei, W., Peleo-Alampay, A., 1993. Updated Cenozoic nannofossil magneto-biochrono stratigraphy. Fifth International Nannoplankton Association Conference, Salamanca, vol. 15(1), pp. 15–17.

Weinzierl, L.L., Applin, E.R., 1929. The Clairborne formation on the coastal domes. J. Paleontol. 3, 42–44.

White, E.I., 1926. Eocene fishes of Nigeria. Bull. Geol. Surv. Nigeria 10, 78.

White, M.P., 1928. Some index foraminifera of the Tampico Embayment area of Mexico. J. Paleontol. 2, 177–215.

Whiteman, A.J., 1982. Nigeria: Its Petroleum Geology, Resources and Potentialvols. I and IIGraham and Trotman Ltd., London, 394 p.

Wiesner, H., 1931. Die Foraminiferen der deutschen Sudpoplar Expedition 1901-1993 Deutsche Sudpolar Expedition 1991–1993, Herausgegeben von Erich von Drygalski, v. 20, Zool., v. 12, pp. 53–165.

Williamson, W.C., 1858. On the recent Foraminifera of Great Britain. Ray Society, London, pp. 1–107.

Wilson, R.C., 1925. The geology of the Eastern Railway: section 1. Port Harcourt to Enugu. With notes by A.D.N. Bain. Bull. Geol. Surv. Nigeria 8, 95.

Wilson, R.C., Bain, A.D.N., 1928. The Nigerian Coal field, section II, Parts of Onitsha and Owerri Provinces. With an Appendix by Spath, L.F. on the Albian Ammonoidea of Nigeria. Bull. Geol. Surv. Nigeria (12), 54.

Winker, C.D., 1982. Cenozoic shelf margins northwestern Gulf of Mexico Basin. Trans. Gulf Coast Assoc. Geol. Soc. 32, 427–448.

Wise, S.W., 1973. Calcareous Nannofossils From Cores Recovered During Leg 18, Deep Sea Drilling Project: Biostratigraphy and Observations on Diagenesis. Initial Reports of the Deep Sea Drilling Project, vol. 18, pp. 569–615.

Wornardt, W.W., Vail, P.R., 1990. Revision of Plio-Pleistocene cycles and their application to sequence stratigraphy of shelf and slope sediments in the Gulf of Mexico. Gulf Coast Section Society of Economic Paleontologists and Mineralogists, 12th Annual Research Conference.

Young, J.R., 1990. Size variation of Neogene *Reticulofenestra* Coccoliths from the Indian Ocean. Deep Sea Drilling Project Cores. J. Micropaleontol. 9 (1), 71–86.

Young, J.R., 1998. Neogene calcareous nannofossil biostratigraphy. In: Bown, P.R. (Ed.), Calcareous Nannofossil Biostratigraphy, British Micropaleontological Society Series. Kluwer Academic Publishers, London, U.K., pp. 225–283.

Young, J.R., Bown, P.R., 1997. Cenozoic calcareous nannoplankton classification. J. Nannoplankton Res. 19, 36–47.

Young, J.R., Flores, J.A., Wei, W., 1994. A summary chart of Neogene nannofossil magnetobiostratigraphy. J. Nannoplankton Res. 16 (1), 21–27.

Zachariasse, W.J., 1990. Short course in Cretaceous-Recent planktonic foraminiferal biostratigraphy. University of Utrecht Department of Geoscience, 200 p.

Taxonomic Index

C

D

H

N

O

T

Author Index

Subject Index

D

E

G

U

Kolibri 360 GmbH
Ferdinand-Jühlke-Straße 7
99095 Erfurt, Deutschland
produktsicherheit@kolibri360.de